Ecohydraulics

By Ian Maddock:
For Katherine, Ben, Joe and Alice.

By Atle Harby:
Dedicated to Cathrine, Sigurd and Brage.

By Paul Kemp:
Dedicated to Clare, Millie, Noah and Florence.

By Paul Wood:
For Maureen, Connor and Ryan.

Ecohydraulics
An Integrated Approach

EDITED BY

Ian Maddock
Institute of Science and the Environment, University of Worcester, UK

Atle Harby
SINTEF Energy Research, Trondheim, Norway

Paul Kemp
International Centre for Ecohydraulics Research, University of Southampton, UK

Paul Wood
Department of Geography, Loughborough University, Leicestershire, UK

WILEY Blackwell

Library of Congress Cataloging-in-Publication Data

Maddock, Ian (Ian Philip)
  Ecohydraulics : an integrated approach / Ian Maddock, Atle Harby, Paul Kemp, Paul Wood.
    pages cm
  Includes bibliographical references and index.
  ISBN 978-0-470-97600-5 (cloth)
  1. Ecohydrology.   2. Aquatic ecology.   3. Wetland ecology.   4. Fish habitat improvement.   5. Stream conservation.   I. Harby, Atle, 1965–   II. Kemp, Paul, 1972–
III. Wood, Paul J.   IV. Title.
  QH541.15.E19M33 2013
  577.6–dc23
                                                                      2013008534

A catalogue record for this book is available from the British Library.

Wiley also publishes its books in a variety of electronic formats. Some content that appears in print may not be available in electronic books.

Cover image: Images supplied by Author
Cover design by Dan Jubb

Set in 9.25/11.5pt Minion by Aptara® Inc., New Delhi, India
Printed and bound in Singapore by Markono Print Media Pte Ltd

1   2013

Contents

List of Contributors

Michael C. Acreman
Centre for Ecology and Hydrology
Maclean Building
Benson Lane
Wallingford
Oxfordshire
OX10 8BB
UK

Donald J. Baird
Environment Canada
Canadian Rivers Institute
Department of Biology
10 Bailey Drive
P.O. Box 4400
University of New Brunswick
Fredericton
New Brunswick
E3B 5A3
Canada

Rohan Benjankar
Center for Ecohydraulics Research
University of Idaho
322 E. Front Street
Boise
ID 83702
USA

Melanie Bickerton
Geography, Earth and Environmental Sciences
University of Birmingham
Edgbaston
Birmingham
B15 2TT
UK

Bernadette Blamauer
Christian Doppler Laboratory for Advanced Methods in
River Monitoring, Modelling and Engineering
Institute of Water Management, Hydrology and
Hydraulic Engineering
University of Natural Resources and Life Sciences Vienna
Muthgasse 107
1190 Vienna
Austria

Gudrun Bornette
Université Lyon 1
UMR 5023 Ecologie des hydrosystèmes naturels et
anthropisés (Université Lyon 1; CNRS; ENTPE)
43 boulevard du 11 novembre 1918
69622 Villeurbanne Cedex
France

Rocko A. Brown
Department of Land, Air, and Water Resources
University of California
One Shields Avenue
Davis
CA 95616
USA

Anthonie D. Buijse
Deltares
P.O. Box 177
2600 MH Delft
The Netherlands

Olle Calles
Department of Biology
Karlstad University
S-651 88 Karlstad
Sweden

Sung-Uk Choi
Department of Civil and Environmental Engineering
Yonsei University
134 Shinchon-dong
Seodaemun-gu
Seoul
Korea

Claudio Comoglio
Turin Polytechnic
Corso Duca degli Abruzzi 24
c/o DITAG
10129 Torino
Italy

Bernard De Baets
Department of Mathematical Modelling
Statistics and Bioinformatics
Ghent University
Coupure links 653
9000 Gent
Belgium

Harm Duel
Deltares
P.O. Box 177
2600 MH Delft
The Netherlands

Lynda R. Eakins
International Centre for Ecohydraulics Research
University of Southampton
Highfield
Southampton
SO17 1BJ
UK

Gregory Egger
Environmental Consulting Ltd
Bahnhofstrasse 39
9020 Klagenfurt
Austria

Teresa Ferreira
Forest Research Centre
Instituto Superior de Agronomia
Technical University of Lisbon
Tapada da Ajuda 1349-017
Lisbon
Portugal

Virginia Garófano-Gómez
Institut d'Investigació per a la Gestió Integrada de Zones Costaneres (IGIC)
Universitat Politècnica de València
C/ Paranimf, 1
46730 Grau de Gandia (València)
Spain

Gertjan W. Geerling
Deltares
P.O. Box 177
2600 MH Delft
The Netherlands

Peter Goethals
Aquatic Ecology Research Unit
Department of Applied Ecology and Environmental Biology
Ghent University
J. Plateaustraat 22
B-9000 Gent
Belgium

Javier Gortázar
Ecohidráulica S.L.
Calle Rodríguez San Pedro 13
4°7
28015 Madrid
Spain

Larry Greenberg
Department of Biology
Karlstad University
S-651 88 Karlstad
Sweden

Helmut Habersack
Christian Doppler Laboratory for Advanced Methods in River Monitoring, Modelling and Engineering
Institute of Water Management, Hydrology and Hydraulic Engineering
University of Natural Resources and Life Sciences Vienna
Muthgasse 107
1190 Vienna
Austria

Atle Harby
SINTEF Energy Research
P.O. Box 4761 Sluppen
7465 Trondheim
Norway

Thomas B. Hardy
The Meadows Center for Water and Environment
Texas State University
601 University Drive
San Marcos
Texas 78666
USA

Jan Heggenes
Telemark University College
Department of Environmental Sciences
Hallvard Eikas Plass
N-3800 Bø i Telemark
Norway

Graham Hill
Institute of Science and the Environment
University of Worcester
Henwick Grove
Worcester
WR2 6AJ
UK

Alice Howe
School of Engineering
The University of Newcastle
Callaghan
NSW 2308
Australia

Ari Huusko
Finnish Game and Fisheries Research Institute
Manamansalontie 90
88300 Paltamo
Finland

Georg A. Janauer
Department of Limnology
University of Vienna
Althanstrasse 14
A-1090 Vienna
Austria

Klaus Jorde
KJ Consulting/SJE Schneider & Jorde Ecological Engineering
Gnesau 11
A-9563 Gnesau
Austria

Paul Kemp
International Centre for Ecohydraulics Research
University of Southampton
Highfield
Southampton
SO17 1BJ
UK

James R. Kerr
International Centre for Ecohydraulics Research
University of Southampton
Highfield
Southampton
SO17 1BJ
UK

Aleksandra Krivograd Klemenčič
University of Ljubljana
Faculty of Health Sciences
Department of Sanitary Engineering
SI-1000 Ljubljana
Slovenia

Ian Maddock
Institute of Science and the Environment
University of Worcester
Henwick Grove
Worcester
WR2 6AJ
UK

Wendy A. Monk
Environment Canada
Canadian Rivers Institute
Department of Biology
10 Bailey Drive
P.O. Box 4400
University of New Brunswick
Fredericton
New Brunswick
E3B 5A3
Canada

Ans Mouton
Research Institute for Nature and Forest
Department of Management and Sustainable Use
Ghent University
Kliniekstraat 25
B-1070 Brussels
Belgium

Markus Noack
Federal Institute of Hydrology
Department M3 – Groundwater, Geology, River
Morphology
Mainzer Tor 1
D-56068 Koblenz
Germany

Jessica M. Orlofske
Canadian Rivers Institute
Department of Biology
10 Bailey Drive
P.O. Box 4400
University of New Brunswick
Fredericton
New Brunswick
E3B 5A3
Canada

Piotr Parasiewicz
Rushing Rivers Institute
592 Main Street
Amherst
MA 01002
USA;
The Stanisław Sakowicz Inland Fisheries Institute
ul. Oczapowskiego 10
10-719 Olsztyn 4
Poland

Gregory B. Pasternack
Department of Land, Air, and Water Resources
University of California
One Shields Avenue
Davis
CA 95616
USA

Mark Pegg
University of Nebraska
402 Hardin Hall
Lincoln
NE 68583-0974
USA

Adam T. Piper
International Centre for Ecohydraulics Research
University of Southampton
Highfield
Southampton
SO17 1BJ
UK

Emilio Politti
Environmental Consulting Ltd
Bahnhofstrasse 39
9020 Klagenfurt
Austria

Sara Puijalon
Université Lyon 1
UMR 5023 Ecologie des hydrosystèmes naturels et
anthropisés (Université Lyon 1; CNRS; ENTPE)
43 boulevard du 11 novembre 1918
69622 Villeurbanne Cedex
France

Walter Reckendorfer
WasserCluster Lunz – Biologische Station GmbH
Dr Carl Kupelwieser Promenade 5
A-3293 Lunz am See
Austria

Rui Rivaes
Forest Research Centre
Instituto Superior de Agronomia
Technical University of Lisbon
Tapada da Ajuda 1349-017
Lisbon
Portugal

Peter Rivinoja
Department of Wildlife, Fish and Environmental Studies
SLU (Swedish University of Agricultural Sciences)
Umeå 901 83
Sweden

José F. Rodríguez
School of Engineering
The University of Newcastle
Callaghan
NSW 2308
Australia

Joseph N. Rogers
Rushing Rivers Institute
592 Main Street
Amherst
MA 01002
USA

Udo Schmidt-Mumm
Department of Limnology
University of Vienna
Althanstrasse 14
A-1090 Vienna
Austria

Matthias Schneider
Schneider & Jorde Ecological Engineering GmbH
Viereichenweg 12
D-70569 Stuttgart
Germany

Thomas Seager
Rushing Rivers Institute
592 Main Street
Amherst
MA 01002
USA

Thomas A. Shaw
U.S. Fish and Wildlife Service
Arcata Fish and Wildlife Office
1655 Heindon Road
Arcata
California 95521
USA

Antonius J.M. Smits
DSMR
Radboud University
P.O. Box 9010
6500 GL Nijmegen
The Netherlands

Nataša Smolar-Žvanut
Institute for Water of the Republic of Slovenia
Hajdrihova 28c
SI-1000 Ljubljana
Slovenia

Michael Stewardson
Department of Infrastructure Engineering
Melbourne School of Engineering
The University of Melbourne
Melbourne 3010
Australia

Morten Stickler
Statkraft AS
Lilleakerveien 6
0216 Oslo
Norway

Daniele Tonina
Center for Ecohydraulics Research
University of Idaho
322 E Front Street
Suite 340
Boise
ID 83702
USA

Teppo Vehanen
Finnish Game and Fisheries Research Institute
Paavo Havaksen tie 3
90014 Oulun yliopisto
Finland

Paolo Vezza
Turin Polytechnic
Corso Duca degli Abruzzi 24
c/o DITAG
10129 Torino
Italy

Fleur Visser
Institute of Science and the Environment
University of Worcester
Henwick Grove
Worcester
WR2 6AJ
UK

Andrew S. Vowles
International Centre for Ecohydraulics Research
University of Southampton
Highfield
Southampton
SO17 1BJ
UK

Silke Wieprecht
Institute for Modelling Hydraulic and Environmental
Systems
Department of Hydraulic Engineering and Water
Resources Management
University of Stuttgart
Pfaffenwaldring 61
D-70569 Stuttgart
Germany

Martin A. Wilkes
Institute of Science and the Environment
University of Worcester
Henwick Grove
Worcester
WR2 6AJ
UK

Wiesław Wiśniewolski
The Stanisław Sakowicz Inland Fisheries Institute
ul. Oczapowskiego 10
10-719 Olsztyn 4
Poland

Jens Wollebæk
The Norwegian School of Veterinary Science
Department of Basic Sciences and Aquatic Medicine
Box 8146
Dep. 0033 Oslo
Norway

Hyoseop Woo
Korea Institute of Construction Technology
2311 Daehwa-dong
Illsanseo-gu
Goyang-si
Gyeonggi-do
Korea

Paul Wood
Department of Geography
Loughborough University
Leicestershire
LE11 3TU
UK

Sarah Yarnell
Center for Watershed Sciences
University of California, Davis
One Shields Avenue
Davis
CA 95616
USA

Elisa Zavadil
Alluvium Consulting Australia
21–23 Stewart Street
Richmond
Victoria 3121
Australia

1

Ecohydraulics: An Introduction

Ian Maddock[1], Atle Harby[2], Paul Kemp[3] and Paul Wood[4]

[1] Institute of Science and the Environment, University of Worcester, Henwick Grove, Worcester, WR2 6AJ, UK
[2] SINTEF Energy Research, P.O. Box 4761 Sluppen, 7465 Trondheim, Norway
[3] International Centre for Ecohydraulics Research, University of Southampton, Highfield, Southampton, SO17 1BJ, UK
[4] Department of Geography, Loughborough University, Leicestershire, LE11 3TU, UK

1.1 Introduction

It is well established that aquatic ecosystems (streams, rivers, estuaries, lakes, wetlands and marine environments) are structured by the interaction of physical, biological and chemical processes at multiple spatial and temporal scales (Frothingham *et al.*, 2002; Thoms and Parsons, 2002; Dauwalter *et al.*, 2007). The need for interdisciplinary research and collaborative teams to address research questions that span traditional subject boundaries to address these issues has been increasingly recognised (Dollar *et al.*, 2007) and has resulted in the emergence of new 'sub-disciplines' to tackle these questions (Hannah *et al.*, 2007). Ecohydraulics is one of these emerging fields of research that has drawn together biologists, ecologists, fluvial geomorphologists, sedimentologists, hydrologists, hydraulic and river engineers and water resource managers to address fundamental research questions that will advance science and key management issues to sustain both natural ecosystems and the demands placed on them by contemporary society.

Lotic environments are naturally dynamic, characterised by variable discharge, hydraulic patterns, sediment and nutrient loads and thermal regimes that may change temporally (from seconds to yearly variations) and spatially (from sub-cm within habitat patches to hundreds of km^2 at the drainage basin scale). This complexity produces a variety of geomorphological features and habitats that sustain the diverse ecological communities recorded in fresh, saline and marine waters. Aquatic organisms, ranging from micro-algae and macro-

phytes to macroinvertebrates, fish, amphibians, birds and mammals, have evolved adaptations to persist and thrive in hydraulically dynamic environments (Lytle and Poff, 2004; Townsend, 2006; Folkard and Gascoigne, 2009; Nikora, 2010). However, anthropogenic impacts on aquatic systems have been widespread and probably most marked on riverine systems. A report by the World Commission on Dams (2000) and a recent review by Kingsford (2011) suggested that modification of the river flow regime as a result of regulation by creating barriers, impoundment and overabstraction, the spread of invasive species, overharvesting and the effects of water pollution were the main threats to the world's rivers and wetlands and these effects could be compounded by future climate change.

The impacts of dam construction, river regulation and channelisation have significantly reduced the natural variability of the flow regime and channel morphology. This results in degradation, fragmentation and loss of habitat structure and availability, with subsequent reductions in aquatic biodiversity (Vörösmarty *et al.*, 2010). Recognition of the long history, widespread and varied extent of human impacts on river systems, coupled with an increase in environmental awareness has led to the development of a range of approaches to minimise and mitigate their impacts. These include river restoration and rehabilitation techniques to restore a more natural channel morphology (e.g. Brookes and Shields Jr, 1996; de Waal *et al.*, 1998; Darby and Sear, 2008), methods to define ways to reduce or mitigate the impact of abstractions and river regulation through the definition and application of instream

---

Ecohydraulics: An Integrated Approach, First Edition. Edited by Ian Maddock, Atle Harby, Paul Kemp and Paul Wood.
© 2013 John Wiley & Sons, Ltd. Published 2013 by John Wiley & Sons, Ltd.

or environmental flows (Dyson *et al.*, 2003; Acreman and Dunbar, 2004; Annear *et al.*, 2004; Acreman *et al.*, 2008), and the design of screens and fish passes to divert aquatic biota from hazardous areas (e.g. abstraction points) and to enable them to migrate past physical barriers, especially, but not solely associated with dams (Kemp, 2012).

Key legislative drivers have been introduced to compel regulatory authorities and agencies to manage and mitigate historic and contemporary anthropogenic impacts and, where appropriate, undertake restoration measures. The EU Water Framework Directive (Council of the European Communities, 2000) requires the achievement of 'good ecological status' in all water bodies across EU member states by 2015 (European Commission, 2012). This, in turn, has required the development of methods and techniques to assess the current status of chemical and biological water quality (Achleitner *et al.*, 2005), hydromorphology and flow regime variability, and identify ways of mitigating impacts and restoring river channels and flow regimes where they are an impediment to the improvement of river health (Acreman and Ferguson, 2010). Similar developments have occurred in North America with the release of the United States Environmental Protection Agency guidelines (US EPA, 2006). In Australia, provision of water for environmental flows has been driven by a combination of national policy agreements including the National Water Initiative in 2004, national and state level legislation and government-funded initiatives to buy back water entitlements from water users including the 'Water for the Future' programme (Le Quesne *et al.*, 2010). Important lessons can be learned from South Africa, where implementation of the National Water Act of 1998 is recognised as one of the most ambitious pieces of water legislation to protect domestic human needs and environmental flows on an equal footing ahead of economic uses. However, Pollard and du Toit (2008) suggest that overly complicated environmental flow recommendations have inhibited their implementation. This provides a key message for ecohydraulic studies aimed at providing environmental flow or indeed other types of river management recommendations (e.g., river restoration) worldwide.

1.2 The emergence of ecohydraulics

During the 1970s and 1980s it was common for multidisciplinary teams of researchers and consultants to undertake pure and/or applied river science projects and to present results collected as part of the same study independently to stakeholders and regulatory/management authorities, each from the perspective of their own disciplinary background. More recently, there has been a shift towards greater interdisciplinarity, with teams of scientists, engineers, water resource and river managers and social scientists working together in collaborative teams towards clearly defined common goals (Porter and Rafols, 2009). Developments in river science reflect this overall pattern, with the emergence of ecohydrology at the interface of hydrology and ecology (Dunbar and Acreman, 2001; Hannah *et al.*, 2004; Wood *et al.*, 2007) and hydromorphology, which reflects the interaction of the channel morphology and flow regime (hydrology and hydraulics) in creating 'physical habitat' (Maddock, 1999; Orr *et al.*, 2008; Vaughan *et al.*, 2009).

Like 'ecohydrology', 'ecohydraulics' has also developed at the permeable interface of traditional disciplines, combining the study of the hydraulic properties and processes associated with moving water typical of hydraulic engineering and geomorphology and their influence on aquatic ecology and biology (Vogel, 1996; Nestler *et al.*, 2007). Ecohydraulics has been described as a subdiscipline of ecohydrology (Wood *et al.*, 2007) although it has become increasingly distinct in recent years (Rice *et al.*, 2010). Hydraulic engineers have been engaged with design criteria for fish passage and screening facilities at dams for many years. Recognition of the need to solve river management problems like these by adopting an interdisciplinary approach has been the driver for the development of ecohydraulics. Interdisciplinary research that incorporates the expertise of hydrologists, fluvial geomorphologists, engineers, biologists and ecologists has begun to facilitate the integration of the collective expertise to provide holistic management solutions. Ecohydraulics has played a critical role in the development of methods to assess and define environmental flows (Statzner *et al.*, 1988). Although pre-dating the use of the term 'ecohydraulics', early approaches, such as the Physical Habitat Simulation System (PHABSIM) in the 1980s and 1990s, were widely applied (Gore *et al.*, 2001) but often criticised due to an over-reliance on simple hydraulic models and a lack of ecological relevance because of the way that habitat suitability was defined and calculated (Lancaster and Downes, 2010; Shenton *et al.*, 2012). State-of-the-art developments associated with ecohydraulics are attempting to address these specific gaps between physical scientists (hydraulic engineers, hydrologists and fluvial geomorphologists) and biological scientists (e.g. aquatic biologists and ecologists) by integrating hydraulic and biological tools to analyse and predict ecological responses to hydrological and hydraulic variability and change (Lamouroux *et al.* in press). These developments intend to support water resource management and the decision-making process by providing ecologically relevant and environmentally sustainable solutions to issues associated with hydropower operations, river restoration and the delineation of environmental flows (Acreman and Ferguson, 2010).

The growing worldwide interest in ecohydraulics can be demonstrated by increasing participation in the international symposia on the subject. The first symposium (then titled the 1st International Symposium on Habitat Hydraulics) was organised in 1994 in Trondheim, Norway by the Foundation for Scientific and Industrial Research (SINTEF), the Norwegian University of Science and Technology (NTNU) and the Norwegian Institute of Nature Research (NINA) with about 50 speakers and 70 delegates. Subsequent symposia in Quebec City (Canada, 1996), Salt Lake City (USA, 1999), Cape Town (South Africa, 2002), Madrid (Spain, 2004), Christchurch (New Zealand, 2007), Concepción (Chile, 2009), Seoul (South Korea, 2010) and most recently in Vienna (Austria, 2012) have taken the scientific community across the globe, typically leading to more than 200 speakers and approximately 300 delegates at each meeting.

A recent bibliographic survey by Rice *et al.* (2010) indicated that between 1997 and the end of 2009 a total of 146 publications had used the term 'ecohydraulic' or a close variant (eco hydraulic, ecohydraulics or eco-hydraulics) in the title, abstract or keywords (ISI Web of Knowledge, http://wok.mimas.ac.uk/). This meta-analysis indicated greater use of the term 'ecohydraulics' amongst water resources and engineering journals (48%) and geoscience journals (31%) compared to a more limited use in (21%) biological or ecological journals. By the end of 2011 this figure had risen to 211 publications, with 65 papers being published between 2010 and the end of 2011 (Figure 1.1). This suggests a significant increase in the use of the terms more recently, and strongly mirrors the rapid rise in the use of the term 'ecohydrology', which has been used in the title, abstract or as a keyword 635 times since 1997 (186 between 2010 and 2011). However, bibliographic analysis of this nature only identifies those publications that have specifically used one of the terms and there is an extensive unquantified literature centred on ecohydraulics and ecohydrology that has not specifically used these terms.

Porter and Rafols (2009) suggested that interdisciplinary developments in science have been greatest between closely allied disciplines and less well developed and slower for fields with a greater distance between them.

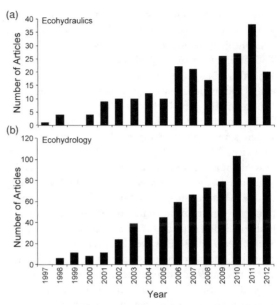

Figure 1.1 Number of peer-reviewed articles using the terms (a) ecohydraulic(s), eco-hydraulic(s) or eco hydraulic(s) and (b) ecohydrology, eco-hydrology or eco hydrology 1997–2012 as listed on Thomson Reuters ISI Web of Knowledge (http://wok.mimas.ac.uk/). Note: WoK data for 2012 compiled on 22/11/2012.

This appears to be the case when comparing developments in ecohydrology and ecohydraulics. Ecohydrology has increasingly been embraced by an interdisciplinary audience and even witnessed the launch of a dedicated journal, *Ecohydrology*, in 2008 (Smettem, 2008), drawing contributions from across physical, biological and social sciences as well as engineering and water resources management. In contrast, publications explicitly referring to 'ecohydraulics' predominately appeared in water resources, geosciences and engineering journals and the affiliation of the primary authors remains firmly within engineering and geosciences departments and research institutes. However, the greatest number of papers has appeared in the interdisciplinary journal *River Research and Applications* (17 papers since 2003). This figure includes five out of ten papers within a special issue devoted to ecohydraulics in 2010 (Rice *et al.*, 2010) and two out of nine papers within a special issue devoted to 'Fish passage: an ecohydraulics approach' in 2012 (Kemp, 2012), and clearly demonstrates that many authors do not routinely use the term 'ecohydraulics'. Biologists have been investigating organism responses to their abiotic environments, including the role of fluid dynamics on aquatic communities, for decades and well before the term 'ecohydraulics' was coined. For example, from an environmental flow perspective, biological scientists have been involved with determining the relationship between fish (and other biota) and hydraulics since at least the 1970s (e.g. Bovee and Cochnauer, 1978). What this bibliographic analysis highlights is that geoscientists and engineers have more readily adopted the terms than colleagues in biology and ecology.

The dominance of physical scientists and engineers within some studies, many of them using modelling approaches, has been highlighted as a potential weakness of some research. It is argued they rely on faulty assumptions and lack any ecological or biological reality due to inadequate consideration of biological interactions between organisms (inter- or intra-specific), or natural population dynamics (Lancaster and Downes, 2010; Shenton *et al.*, 2012). However, these criticisms have been contested and there is growing evidence that interdisciplinarity is being embraced more widely (Lamouroux *et al.*, 2010; Lamouroux *et al.*, Lamouroux *et al.*, in press). This issue is discussed further in the concluding chapter of this volume.

1.3 Scope and organisation of this book

The aim of this research-level edited volume is to provide the first major text to focus on ecohydraulics. It is comprised of chapters reflecting the range and scope of research being undertaken in this arena (spanning engineering, geosciences, water resources, biology, ecology and interdisciplinary collaborations). Individual chapter authors have provided overviews of cutting-edge research and reviews of the current state of the art in ecohydraulics. In particular, authors have been encouraged to demonstrate how their work has been informed by and is influencing the on-going development of ecohydraulics research. The contributions use case study examples from across the globe, highlighting key methodological developments and demonstrating the real-world application of ecohydraulic theory and practice in relation to a variety of organisms ranging from riparian vegetation and instream algae, macrophytes, macroinvertebrates and fish to birds and amphibians. The chapters reflect a spectrum of research being undertaken within this rapidly developing field and examine the interactions between hydraulics, hydrology, fluvial geomorphology and aquatic ecology on a range of spatial (individual organism in a habitat patch to catchment) and temporal scales.

The book is structured into four parts: Part One considers the range and type of methods and approaches

used in ecohydraulics research, with a particular focus on aquatic habitat modelling; Part Two considers a range of species–habitat relationships in riverine and riparian habitats; Part Three consists of detailed ecohydraulics case studies that have a clear management application, mostly, but not exclusively, relating to environmental flow determination, fish passage design, river channel and habitat restoration and ecosystem assessment. The final chapter (Part Four) aims to draw together the work contained in the book to outline key research themes and challenges in ecohydraulics and discuss future goals and directions. A number of chapters involve methods, species–habitat relationships and case studies and therefore could have been located in more than one part of the book. The final decision regarding which part to place them in was in some cases clear-cut and in others fairly arbitrary.

We realise that the coverage provided in this volume is not complete and are conscious that the chapters are almost exclusively centred on freshwater, riverine ecosystems. Indeed there has been a considerable volume of research centred on marine (e.g. Volkenborn *et al.*, 2010), estuarine (e.g. Yang *et al.*, 2012) and lentic (lake) ecosystems (e.g. Righetti and Lucarelli, 2010), where equally challenging and exciting ecohydraulic research questions are being addressed. Their exclusion is driven by a desire to keep this book within a manageable size and scope rather than a view that these other parts of the natural environment are somehow less important than riverine ecosystems.

Research currently being undertaken in the arena of ecohydraulics is developing rapidly and is becoming increasingly interdisciplinary, drawing on a range of academic and practitioner traditions and addressing real-world problems. As this interdisciplinary science matures there is a growing demand from river managers and end users to be involved not just at the inception and conclusion, but throughout the studies to enhance the possibility that any management recommendations can be implemented successfully. The occurrence of this would signal a move from interdisciplinarity (between traditional disciplines) to 'transdisciplinarity' (that also engages with managers and end users during the research). The editors hope that the realisation of this development will be one mark of this book's success.

References

Achleitner, S., de Toffol, S., Engelhard, C. and Rauch, W. (2005) The European Water Framework Directive: water quality classification and implications to engineering planning. *Environmental Management*, 35: 517–525.

Acreman, M. and Dunbar, M.J. (2004) Defining environmental flow requirements – a review. *Hydrology and Earth System Sciences*, 8: 861–876.

Acreman, M., Dunbar, M., Hannaford, J., Mountfield, O., Wood, P., Holmes, N., Cowx, I., Noble, R., King, J., Black, A., Extence, C., Aldrick, J., Kink, J., Black, A. and Crookall, D. (2008) Developing environmental standards for abstractions from UK rivers to implement the EU Water Framework Directive. *Hydrological Sciences Journal*, 53: 1105–1120.

Acreman, M. and Ferguson, A.J.D. (2010) Environmental flows and the European Water Framework Directive. *Freshwater Biology*, 55: 32–48.

Annear, T., Chisholm, I., Beecher, H., Locke, A. *et al.* (2004) *Instream Flows for Riverine Resource Stewardship*, (revised edition). Instream Flow Council, Cheyenne, WY.

Bovee, K.D. and Cochnauer, T. (1978) *Development and evaluation of weighted criteria, probability-of-use curves for instream flow assessment: fisheries.* Instream Flow Information Paper No. 3. Cooperative Instream Flow Service Group, Western Energy and Land Use Team, Office of Biological Services, Fish and Wildlife Service, U.S. Dept. of the Interior.

Brookes, A. and Shields Jr., F.D. (eds) (1996) *River Channel Restoration: Guiding Principles for Sustainable Projects*, John Wiley & Sons, Ltd, Chichester, UK.

Council of the European Communities (2000) Directive 2000/60/EC of the European Parliament and of the Council of 23 October 2000 establishing a framework for Community action in the field of water policy. *Official Journal of the European Communities*, L327: 1–73.

Darby, S. and Sear, D. (eds) (2008) *River Restoration: Managing the Uncertainty in Restoring Physical Habitat*, John Wiley & Sons, Ltd, Chichester, UK.

Dauwalter, D.C., Splinter, D.K., Fisher, W.L. and Marston, R.A. (2007) Geomorphology and stream habitat relationships with smallmouth bass (*Micropterus dolomieu*) abundance at multiple spatial scales in eastern Oklahoma. *Canadian Journal of Fisheries and Aquatic Sciences*, 64: 1116–1129.

de Waal, L.C., Large, A.R.G. and Wade, P.M. (eds) (1998) *Rehabilitation of Rivers: Principles and Implementation*, John Wiley & Sons, Ltd, Chichester, UK.

Dollar, E.S.J., James, C.S., Rogers, K.H. and Thoms, M.C. (2007) A framework for interdisciplinary understanding of rivers as ecosystems. *Geomorphology*, 89: 147–162.

Dunbar, M.J. and Acreman, M. (2001) Applied hydro-ecological science for the twenty-first century. In Acreman, M. (ed.) *Hydro-Ecology: Linking Hydrology and Aquatic Ecology*. IAHS Publication no. 288. pp. 1–17.

Dyson, M., Bergkamp, G. and Scanlon, J. (eds) (2003) *Flow: The Essentials of Environmental Flows*. IUCN, Gland, Switzerland and Cambridge, UK.

European Commission (2012) The EU Water Framework Directive: integrated river basin management for Europe. Available at: http://ec.europa.eu/environment/water/water-framework/index_en.html [Date accessed: 20/7/12].

Folkard, A.M. and Gascoigne, J.C. (2009) Hydrodynamics of discontinuous mussel beds: Laboratory flume simulations. *Journal of Sea Research*, 62: 250–257.

Frothingham, K.M., Rhoads, B.L. and Herricks, E.E. (2002) A multiple conceptual framework for integrated ecogeomorphological research to support stream naturalisation in the agricultural Midwest. *Environmental Management*, 29: 16–33.

Gore, J.A., Layzer, J.B. and Mead, J. (2001) Macroinvertebrate instream flow studies after 20 years: A role in stream management and restoration. *Regulated Rivers: Research and Management*, 17: 527–542.

Hannah, D.M., Wood, P.J. and Sadler, J.P. (2004) Ecohydrology and hydroecology: a new paradigm. *Hydrological Processes*, 18: 3439–3445.

Hannah, D.M., Sadler, J.P. and Wood, P.J. (2007) Hydroecology and ecohydrology: a potential route forward? *Hydrological Processes*, 21: 3385–3390.

Kemp, P. (2012) Bridging the gap between fish behaviour, performance and hydrodynamics: an ecohydraulics approach to fish passage research. *River Research and Applications*, 28: 403–406.

Kingsford, R.T. (2011) Conservation management of rivers and wetlands under climate change – a synthesis. *Marine and Freshwater Research*, 62: 217–222.

Lamouroux, N., Merigoux, S., Capra, H., Doledec, S., Jowette, I.G. and Statzner, B. (2010) The generality of abundance–environment relationships in micro-habitats: a comment on Lancaster and Downes (2009). *River Research and Applications*, 26: 915–920.

Lamouroux, N., Merigoux, S., Doledec, S. and Snelder, T.H. (in press) Transferability of hydraulic preference models for aquatic macroinvertebrates. *River Research and Applications*, DOI: 10.1002/rra.2578.

Lancaster, J. and Downes, B.J. (2010) Linking the hydraulic world of individual organisms to ecological processes: putting ecology into ecohydraulics. *River Research and Applications*, 26: 385–403.

Le Quesne, T., Kendy, E. and Weston, D. (2010) *The Implementation Challenge: taking stock of government policies to protect and restore environmental flows*. The Nature Conservancy, World Wide fund for Nature Report, 2010. Available at: http://19assets.dev.wwf.org.uk/downloads/global_flows.pdf [Date accessed: 19/10/12].

Lytle, D.A. and Poff, N.L. (2004) Adaptation to natural flow regimes. *Trends in Ecology and Evolution*, 19: 94–100.

Maddock, I. (1999) The importance of physical habitat assessment for evaluating river health. *Freshwater Biology*, 41: 373–391.

Nestler, J.M., Goodwin, R.A., Smith, D.L. and Anderson, J.J. (2007) A mathematical and conceptual framework for ecohydraulics. In Wood, P.J., Hannah, D.M. and Sadler, J.P. (eds) *Hydroecology and Ecohydrology: Past, Present and Future*, John Wiley & Sons, Ltd, Chichester, UK, pp. 205–224.

Nikora, V. (2010) Hydrodynamics of aquatic ecosystems: An interface between ecology, biomechanics and environmental fluid mechanics. *River Research and Applications*, 26: 367–384.

Orr, H.G., Large, A.R.G., Newson, M.D. and Walsh, C.L. (2008) A predictive typology for characterising hydromorphology. *Geomorphology*, 100: 32–40.

Pollard, S. and du Toit, D. (2008) Integrated water resource management in complex systems: how the catchment management strategies seek to achieve sustainability and equity in water resources in South Africa. *Water SA 34 (IWRM Special Edition)*: 671–679. Available at: http://www.scielo.org.za/pdf/wsa/v34n6/a03v34n6.pdf [Date accessed: 19/10/12].

Porter, A.L. and Rafols, I. (2009) Is science becoming more interdisciplinary? Measuring and mapping six research fields over time. *Scientometrics*, 81: 719–745.

Rice, S.P., Little, S., Wood, P.J., Moir, H.J. and Vericat, D. (2010) The relative contributions of ecology and hydraulics to ecohydraulics. *River Research and Applications*, 26: 1–4.

Righetti, M. and Lucarelli, C. (2010) Resuspension phenomena of benthic sediments: the role of cohesion and biological adhesion. *River Research and Applications*, 26: 404–413.

Shenton, W., Bond, N.R., Yen, J.D.L. and MacNally, R. (2012) Putting the "Ecology" into environmental flows: ecological dynamics and demographic modelling. *Environmental Management*, 50: 1–10.

Smettem, K.R.J. (2008) Editorial: Welcome address for the new 'Ecohydrology' Journal. *Ecohydrology*, 1: 1–2.

Statzner, B., Gore, J.A. and Resh, J.V. (1988) Hydraulic stream ecology – observed patterns and potential applications. *Journal of the North American Benthological Society*, 7: 307–360.

Thoms, M.C. and Parsons, M. (2002) Eco-geomorphology: an interdisciplinary approach to river science. In Dyer, F.J., Thoms, M.C. and Olley, J.M. (eds) The Structure, Function and Management Implications of Fluvial Sedimentary Systems (Proceedings of an international symposium held at Alice Springs, Australia, September 2002) *International Association of Hydrological Sciences*, 276: 113–119.

Townsend, S.A. (2006) Hydraulic phases, persistent stratification, and phytoplankton in a tropical floodplain lake (Mary River, northern Australia). *Hydrobiologia*, 556: 163–179.

USEPA (2006) *Guidance for 2006 Assessment, Listing and Reporting Requirements Pursuant to Sections 303(d), 305(b) and 314 of the Clean Water Act.* http://water.epa.gov/lawsregs/lawsguidance/cwa/tmdl/upload/2006irg-report.pdf

Vaughan, I.P., Diamond, M., Gurnell, A.M., Hall, K.A., Jenkins, A., Milner, N.J., Naylor, L.A., Sear, D.A., Woodward, G. and Ormerod, S.J. (2009) Integrating ecology with hydromorphology: a priority for river science and management. *Aquatic Conservation: Marine and Freshwater Ecosystems*, 19: 113–125.

Vogel, S. (1996) *Life in moving fluids: the physical biology of flow.* Princeton University Press, Princeton.

Volkenborn, N., Polerecky, L., Wethey, D.S. and Woodin. S.A. (2010) Oscillatory porewater bioadvection in marine sediments induced by hydraulic activities of *Arenicola marina. Limnology and Oceanography*, 55: 1231–1247.

Vörösmarty, C.J., McIntyre, P.B., Gessner, M.O., Dudgeon, D., Prusevich, A., Green, P., Glidden, S., Bunn, S.E., Sullivan, C.A., Reidy Liermann, C. and Davies, P.M. (2010) Global threats to human water security and river biodiversity. *Nature*, 467: 555–561.

Wood, P.J., Hannah, D.M. and Sadler, J.P. (eds) (2007) *Hydroecology and Ecohydrology: An Introduction.* In Wood, P.J., Hannah, D.M. and Sadler, J.P. (eds) *Hydroecology and Ecohydrology: Past, Present and Future*, John Wiley & Sons, Ltd, Chichester, UK, pp. 1–6.

World Commission on Dams (2000) *Dams and Development: a new framework for decision-making.* The report of the World Commission on Dams. Earthscan.

Yang, Z., Wang, T., Khangaonkar, T. and Breithaupt, S. (2012) Integrated modelling of flood flows and tidal hydrodynamics over a coastal floodplain. *Environmental Fluid Mechanics*, 12: 63–80.

Methods and Approaches

2

Incorporating Hydrodynamics into Ecohydraulics: The Role of Turbulence in the Swimming Performance and Habitat Selection of Stream-Dwelling Fish

Martin A. Wilkes[1], Ian Maddock[1], Fleur Visser[1] and Michael C. Acreman[2]

[1] Institute of Science and the Environment, University of Worcester, Henwick Grove, Worcester, WR2 6AJ, UK
[2] Centre for Ecology and Hydrology, Maclean Building, Benson Lane, Wallingford, Oxfordshire, OX10 8BB, UK

2.1 Introduction

The complexity and dynamism of river systems, the strength of their biophysical linkages and the need to respond to adverse anthropogenic impacts has led to the emergence of *hydroecology* as a key area of interdisciplinary research (Hannah *et al.*, 2007). Wood *et al.* (2007) provide an outline of the target elements of hydroecology in which they emphasise the bi-directional nature of physical–ecological interactions and the need to identify causal mechanisms rather than merely establishing statistical links between biota, ecosystems and environments. Such causal mechanisms operate in the realm of the physical habitat (Harper and Everard, 1998). A sub-discipline of hydroecology known as *ecohydraulics* has emerged from the scientific literature in recent decades (Leclerc *et al.*, 1996) and, as a contemporary science, has its roots in the hydraulic stream ecology paradigm (Statzner *et al.*, 1988). Ecohydraulics relies on the assumption that flow

forces are ecologically relevant (i.e. that they influence the fitness of individual organisms and, therefore, the structure and function of aquatic communities). It lies at the interface of hydraulics and ecology where new approaches to research are required to reconcile the contrasting conceptual frameworks underpinning these sciences, which can be seen respectively as Newtonian (reductionist) and Darwinian (holistic) (Hannah *et al.*, 2007). Harte (2002) has identified elements of synthesis for integrating these disparate traditions which include the use of simple, falsifiable models and the search for patterns and laws. Newman *et al.* (2006) suggested that hierarchical scaling theory, whereby reductionist explanations are considered at different levels of organisation, could be used to integrate these two approaches. River habitat is structured at a number of scales (Frissell *et al.*, 1986) but it is at the microscale ($<10^{-1}$ m) of the hydraulic environment where reductionist explanations for ecological phenomena are most often sought (e.g. Enders *et al.*, 2003; Liao *et al.*, 2003a).

Ecohydraulics: An Integrated Approach, First Edition. Edited by Ian Maddock, Atle Harby, Paul Kemp and Paul Wood.
© 2013 John Wiley & Sons, Ltd. Published 2013 by John Wiley & Sons, Ltd.

Table 2.1 Common terms used to describe the flow environment.

Term	Description	Notes
h	Flow depth	
y	Height above bed datum	
A	Cross-sectional area of flow	
P	Wetted perimeter	
R	Hydraulic radius	$= A/P$
S	Longitudinal bed slope	
ρ	Fluid density of water	Taken as 1000 kg m^{-3}
g	Acceleration due to gravity	9.81 m s^{-2}
k	Height of surface roughness elements	Various methods to quantify k provided by Statzner *et al.* (1988). Typically based on particle size (D) distributions for gravel-bed rivers (e.g. $3.5D_{84}$) (Clifford *et al.*, 1992)
v	Kinematic viscosity	1.004×10^{-6} m^2 s^{-1} at 20°C
U	Mean streamwise column velocity	Measured at $y/h = 0.4$ or depth-averaged
Fr	Froude number $= U/\sqrt{gh}$	$Fr < 1 \rightarrow$ sub-critical flow $Fr = 1 \rightarrow$ critical flow $Fr > 1 \rightarrow$ super-critical flow
Re	Bulk flow Reynolds number $= Uh/v$	$Re < 500 \rightarrow$ laminar flow $500 < Re < 10^3$–$10^4 \rightarrow$ transitional flow $Re > 10^3$–$10^4 \rightarrow$ turbulent flow
τ	Shear stress (section- or reach-averaged) $= PgRS$	Point measurements can be made using fliesswasserstammtisch (FST) hemispheres
U_*	Shear velocity or friction velocity $= \sqrt{\tau/\rho}$	Calculated from point measurements of shear stress or estimated from near-bed velocity profile
Re^*	Roughness Reynolds number $= U_*k/v$	$Re^* < 5 \rightarrow$ hydraulically smooth flow $5 < Re^* < 70 \rightarrow$ transitional flow $Re^* > 70 \rightarrow$ hydraulically rough flow
δ	Thickness of laminar sublayer $= 11.5v/U_*$	$\delta/k < 1 \rightarrow$ hydraulically smooth flow $\delta/k > 1 \rightarrow$ hydraulically rough flow

2.1.1 'Standard' ecohydraulic variables

Much research has focused on the relationship between instream biota and the 'standard' ecohydraulic variables of flow depth (h), mean streamwise velocity (U) and combinations of these. These simple hydraulic quantities, and indices derived from them (e.g. Froude number, $U{:}h$), have traditionally been used to classify a range of mesoscale (10^{-1}–10^1 m) units of instream habitat (e.g. channel geomorphic units, hydraulic biotopes, functional habitats) for habitat assessment and design purposes (Jowett, 1993; Padmore, 1997; Wadeson and Rowntree, 1998; Kemp *et al.*, 2000). U is typically measured at 'point six' depth ($y/h = 0.4$, where y is height above the bed) and (ensemble) averaged over 10–60 s. Other commonly used variables describing the bulk flow are the Froude number (Fr, ratio of inertial to gravitational forces) and the Reynolds number (Re, ratio of inertial to viscous forces) (Table 2.1). These are dimensionless variables representing gradients from tranquil (sub-critical)

to shooting (super-critical) and laminar to fully developed (turbulent) flow respectively. Because the flow environment experienced by benthic organisms living very close to the bed differs markedly to that farther up in the water column (Statzner *et al.*, 1988), the inner region (see Figure 2.1) has often been characterised by

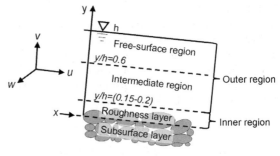

Figure 2.1 Co-ordinate system for three-dimensional flows and structure of flow over rough, permeable boundaries.

a different set of variables. They include bed shear stress (τ), shear velocity (U_*), roughness Reynolds number (Re^*) and the thickness of the laminar sublayer (δ). U_* is related to τ (Table 2.1) which, in turn, is responsible for the appearance of a mean gradient in the vertical velocity profile. U_* can be interpreted as a velocity scale for flow statistics in the inner region. Re^* describes the 'roughness' of the near-bed flow environment. Finally, δ approximates the thickness of the laminar sublayer where viscous forces predominate over inertial forces. In rivers with coarse bed material (i.e. gravel-bed rivers) which are characterised by hydraulically rough flow ($Re^* > 70$), however, δ is typically very small in comparison to roughness size (k) (Davis and Barmuta, 1989; Kirkbride and Ferguson, 1995), rendering it irrelevant to the study of all but the smallest organisms (Allan, 1995).

Flow forces are reported to be the dominant factors influencing the processes of dispersal, reproduction, habitat use, resource acquisition, competition and predation in river ecosystems (Table 2.2). The passive dispersal of benthic organisms is controlled by the same mechanisms as sediment transport (Nelson et al., 1995; McNair et al., 1997), although many invertebrates actively enter the water column and are able to swim back to the substrate (Waters, 1972; Mackay, 1992). Hydraulic limitations to fish migration are related to body depth and maximum sustained and burst swimming speeds $V_{\max}$, which vary considerably between species and with water temperature (Beamish, 1978). h and U are key factors in the segregation of rheophilic species (e.g. Bisson et al., 1988), whilst the distribution of benthic organisms has been related to δ, Fr, τ and Re^* (e.g. Statzner, 1981a, 1981b; Scarsbrook and Townsend, 1993; Brooks et al., 2005). Most instream biota exhibit a subsidy-stress response to flow as resources (e.g. food, nutrients, oxygen) may be limiting at low U, whilst at high U drag disturbance and mass transfer may be the limiting factors (Hart and Finelli, 1999; Nikora, 2010). Thus, for example, the energetic cost of swimming for juvenile Atlantic salmon (*Salmo salar*) is negatively related to U, whilst prey delivery is positively related to U (Godin and Rangeley, 1989). Some of these examples offer mechanistic explanations for flow–biota interactions on which predictive models may be built (e.g. Hughes and Dill, 1990) but ecohydraulic research more often relies on correlative techniques to describe abundance–environment relationships. Whilst correlative approaches may represent a pragmatic compromise in the absence of detailed mechanistic knowledge (Lamouroux et al., 2010), ecohydraulics should strive to establish a more ecologically realistic foundation for modelling the response of populations

to environmental change and management interventions (Lancaster and Downes, 2010; Frank et al., 2011).

In this chapter we argue that the inclusion of higher order (turbulent) properties of the flow constitutes a more complete and ecologically relevant characterisation of the hydraulic environment that biota are exposed to than standard ecohydraulic variables alone. The use of turbulent flow properties in ecohydraulics, therefore, has the potential to contribute towards achieving river research and management goals (e.g. river habitat assessment, modelling, rehabilitation) but more information on the mechanisms by which turbulence affects biota is required before this potential can be realised. After outlining the theory, structure and measurement of turbulent flow in open channels we focus on the swimming performance and habitat selection of stream-dwelling fish as an example of how the hydrodynamics of river ecosystems may affect resident biota. The discussion is biased towards salmonids (*S. salar*, *S. trutta*, *Oncorhynchus mykiss*) as most research has focused on these species due to their ecological (Wilson and Halupka, 1995; Jonsson and Jonsson, 2003) and socio-economic (e.g. Murray and Simcox, 2003) importance and our ability to measure turbulence at the focal point of these organisms, although the turbulent flow properties discussed are likely to be relevant to a range of other aquatic biota. Our scope is generally confined to small to medium (second–fourth order) lowland gravel-bed rivers, although there may well be wider applicability both in terms of river size and type. We acknowledge that many factors (e.g. physico-chemical, biological) make up the multidimensional niche of biota (e.g. Kohler, 1992; Sweeting, 1994; Lancaster and Downes, 2010) but ecohydraulics serves to emphasise the physical environment, which many have cited as the dominant factor in the ecology of lotic communities (e.g. Statzner et al., 1988; Hart and Finelli, 1999; Thompson and Lake, 2010). The discussion, therefore, is restricted to the hydraulics of river habitats.

2.2 Turbulence: theory, structure and measurement

Turbulence in fluid flows was recognised by Leonardo Da Vinci as early as 1513 and is a ubiquitous phenomenon in river ecosystems, where $Re \gg 500$ (Davidson, 2004). Despite this, however, there is still no formal definition of turbulence, although a number of key qualities have been identified. Turbulent flow exhibits seemingly random

Table 2.2 Some examples of flow-biota links identified in the ecohydraulics literature.

Reference	Variable(s)	Species/community/process influenced by variable
Dispersal and reproduction		
Silvester and Sleigh (1985); Reiter and Carlson (1986); Biggs and Thomsen (1995)	τ, U_*	Positively correlated with loss of biomass of filamentous and matt-forming algal communities
Stevenson (1983); Peterson and Stevenson (1989)	U	Negatively correlated with diatom colonisation rates on clean ceramic tiles
Deutsch (1984); Becker (1987) cited in Statzner *et al.* (1988)	*Re*, *Fr*	Oviposition sites of certain caddis fly (Trichoptera) genera correlated with *Re* and *Fr*
McNair *et al.* (1997)	U_*	Transport distance positively related to Rouse number ($= V_s/U_*$, where V_s is settling velocity)
Beamish (1978); Crisp (1993); Hinch and Rand (2000)	h, U	Fish migration inhibited when $h \ll$ body depth and/or when $U > V_{\max}$
Habitat use		
Biggs (1996)	U	Growth rate and organic matter accrual of periphyton and macrophytes enhanced at intermediate U
Scarsbrook and Townsend (1993); Lancaster and Hildrew (1993)	τ	Macroinvertebrate community structure related to spatial and temporal variation in τ
Statzner (1981a)	δ	Body length of freshwater snails (Gastropoda) and shrimps (Gammarus) positively correlated with δ
Statzner (1981b)	δ, *Fr*	Abundance of *Odagmia ornata* (Diptera:Simuliidae) negatively correlated with δ and positively correlated with *Fr*
Statzner *et al.* (1988)	$Re > U > \delta > Re_* > Fr$	Order of best explanatory variables to predict distribution of water bug *Aphelocheirus aestivalis*
Brooks *et al.* (2005)	Re_*	Strongest (negative) correlation with macroinvertebrate abundance and species richness
Bisson *et al.* (1988); Lamouroux *et al.* (2002); Moir *et al.* (1998, 2002); Sagnes and Statzner (2009)	h, U, *Fr*	Fish species and life stages segregated by hydraulic variables due to morphological and ecological traits
Resource acquisition, competition and predation		
Wiley and Kohler (1980); Eriksen *et al.* (1996); Stevenson (1996)	U, δ	U controls the delivery of limiting resources. Laminar sublayer (δ) limits rate of molecular diffusion.
Godin and Rangeley (1989); Hayes and Jowett (1994); Heggenes (1996)	U, h	U positively correlated with prey delivery and negatively correlated with capture rates for salmonids; velocity gradients determine energetic costs of drift-feeding by insectivorous fish; high h provides refuge from predators and competition
Peckarsky *et al.* (1990); Malmqvist and Sackman (1996); Hart and Merz (1998)	U	High U serves as a refuge from predators for blackflies (Simuliidae) and stoneflies (Plecoptera)
Poff and Ward (1992, 1995); DeNicola and McIntire (1991)	U	Negatively correlated with rates of algal consumption by snails and certain caddis flies (Trichoptera)
Matczak and Mackay (1990); Hart and Finelli (1999)	U	Higher U reduces competition and increases carrying capacity of filter-feeding macroinvertebrates

behaviour, has three-dimensionality and rotationality and is intermittent in time and space over a range of scales (Nikora, 2010). Turbulent fluctuations in flow velocities have been implicated in suspended sediment transport (e.g. Bagnold, 1966), bedload transport and the development of bed morphology (e.g. Best, 1993), mixing of dissolved and particulate substances (e.g. Zhen-Gang, 2008), primary productivity and the growth and destruction of algae (e.g. Stoecker *et al.*, 2006; Labiod *et al.*, 2007), biomechanics and bioenergetics (e.g. Enders *et al.*, 2003; Liao *et al.*, 2003a) and the distribution of aquatic organisms (e.g. Cotel *et al.*, 2006; Smith *et al.*, 2006). Because the hydraulic variables typically used in ecohydraulics are based on time-averaged velocity and relate to bulk characteristics of the flow, they do not fully describe all ecologically relevant aspects of the flow environment. Recent advances in field instrumentation now mean that the widespread measurement of turbulence is feasible (Kraus *et al.*, 1994; Voulgaris and Trowbridge, 1998). Furthermore, some have reported that turbulent flow properties are poorly correlated with standard ecohydraulic variables (e.g. h, U), suggesting that turbulence may be considered a distinct parameter in habitat assessment and modelling applications (Smith and Brannon, 2007; Roy *et al.*, 2010). For these reasons, ecohydraulic research has increasingly focused on the hydrodynamics of aquatic ecosystems (Nikora, 2010). This requires a firm knowledge of turbulence in open channel flows and necessitates the use of a consistent coordinate system (Figure 2.1).

Research in the past century has focused on two complementary frameworks within which to study turbulence in open channel flows. The statistical framework treats turbulence as a random phenomenon and focuses on descriptions of the bulk statistical properties of the flow (Richardson, 1922; Kolmogorov, 1941), whereas the deterministic framework emphasises the structural coherency of turbulent flows at a number of spatiotemporal scales (Robinson, 1991).

2.2.1 Statistical descriptions of turbulence

Water behaves as an uncompressible, homogenous, Newtonian fluid in rivers and its flow is governed by equations describing the conservation of mass, momentum and energy. These mass–momentum (Navier–Stokes) and energy equations are set out by Tonina and Jorde (see Chapter 3). The basic principles underlying fluid mechanics are described in any introductory-level text on hydraulics (e.g. Kay, 2008). The full set of equations

describing turbulent flow is provided by Nezu and Nakagawa (1993) and several other research-level texts. The turbulence intensity $\overline{u_i' u_i'}$ is a vector quantity, with each component ($u_i = u, v, w$) derived from the three normal Reynolds stress terms ($\rho \overline{u'u'}$, $\rho \overline{v'v'}$, $\rho \overline{w'w'}$) in the Reynolds-averaged Navier–Stokes (RANS) equation:

$$\rho \frac{\partial \bar{u}_i \bar{u}_j}{\partial x_j} = \rho g_i - \bar{f}_i + \frac{\partial}{\partial x_j}$$
$$\times \left[-p\delta + \mu \left(\frac{\partial \bar{u}_i}{\partial x_j} + \frac{\partial \bar{u}_j}{\partial x_i} \right) - \rho \overline{u_i' u_j'} \right]$$

(2.1)

where $\bar{f}$ is body force per unit volume of fluid (N m^{-3}) and p is isotropic hydrostatic pressure force (N m^{-3}). According to Reynolds decomposition, the instantaneous velocity (time series) at a point can be separated into mean and fluctuating components in the streamwise (u), vertical (v) and spanwise (w) directions:

$$u = U + u', v = V + v', w = W + w'. \quad (2.2)$$

where U, V and W are time-averaged velocities and primes denote turbulent fluctuations. Reynolds decomposition requires strict stationarity of the mean so that the fluctuating components only describe turbulence and do not include variation of the mean flow. Turbulence intensity may be characterised in a number of ways, including standard deviation ($\sigma_{u,v,w}$), relative turbulence intensity ($TI_{u,v,w}$):

$$TI_u = \sigma_u/U, \ TI_v = \sigma_v/V, \ TI_w = \sigma_w/W. \quad (2.3)$$

and root-mean-squared ($RMS_{u,v,w}$) values:

$$RMS_u = \sqrt{\frac{1}{n}(u_1'^2 + u_2'^2 + \cdots + u_n'^2)},$$
$$RMS_v = \sqrt{\frac{1}{n}(v_1'^2 + v_2'^2 + \cdots + v_n'^2)}, \quad (2.4)$$
$$RMS_w = \sqrt{\frac{1}{n}(w_1'^2 + w_2'^2 + \cdots + w_n'^2)}$$

where n is the number of individual observations within a velocity time series. *RMS* values reflect the normal Reynolds stresses included in the final term of Equation (2.1), whilst the diagonal Reynolds shear stresses (τ_{ij}) are given by:

$$\tau_{uv} = \rho \overline{u'v'}, \ \tau_{uw} = \rho \overline{u'w'}, \ \tau_{vw} = \rho \overline{v'w'}. \quad (2.5)$$

These represent the turbulent flux of momentum within a fluid which is related to force by Newton's second law.

A summary of overall turbulence is given by Turbulent Kinetic Energy (*TKE*):

$$TKE = 0.5(RMS_u^2 + RMS_v^2 + RMS_w^2) \qquad (2.6)$$

which, as a scalar quantity, is a useful descriptor of turbulence in complex three-dimensional flows. The order $RMS_u > RMS_w > RMS_v$ has been found to hold throughout the water column with the following ratios (Nezu and Nakagawa, 1993; Song and Chiew, 2001):

$$\frac{RMS_w}{RMS_u} = (0.71 - 0.75), \quad \frac{RMS_v}{RMS_u} = (0.5 - 0.55). \quad (2.7)$$

The above quantities used to describe turbulence intensity are often non-dimensionalised by dividing through U or U_*. Nezu and Nakagawa (1993) derived semi-empirical equations to describe the distribution of turbulence intensities and *TKE* throughout the flow depth:

$$\begin{aligned}
\sigma_u/U_* &= 2.30 \exp(-y/h), \\
\sigma_v/U_* &= 1.27 \exp(-y/h), \\
\sigma_w/U_* &= 1.63 \exp(-y/h), \\
TKE/U_*^2 &= 4.78 \exp(-2y/h).
\end{aligned} \qquad (2.8)$$

These semi-theoretical curves (Equation (2.8)) are based on flows at a range of Re and Fr and provide a good fit when limited to the intermediate flow region ($0.1 < y/h < 0.6$) of fully developed flows.

An essential feature of turbulent flows is that they are rotational or, in other words, they are characterised by non-zero vorticity. Vorticity (ω) describes the curl (curve) of the velocity vector and is equal to twice the angular velocity (rate of rotation of the fluid at a point). An eddy can be defined as a region of flow with finite vorticity (Webb and Cotel, 2010). The fundamental concept underpinning the statistical description of turbulence is the eddy or energy cascade (EC). The EC states that turbulence is initiated in the production range at an external scale of the flow (i.e. h). The depth of the largest eddies (L_y) in open channel flows, therefore, is comparable to h. The largest eddies are anisotropic and, when point sampled velocity time series data are available, their integral length scale (L_x) must be determined by integrating the autocovariance function, to give the integral time scale (ITS), and applying Taylor's (1935) frozen turbulence approximation (Clifford and French, 1993a), which states that:

$$L = Ut \qquad (2.9)$$

where L is length and t is time scale. The large eddies are unstable and transfer their energy to successively smaller eddies in the inertial subrange until eddies become so

small that viscous forces in the dissipation range finally cause kinetic energy to be dissipated to heat at Kolmogorov's micro-scale (η):

$$\eta = (v^3 \varepsilon)^{1/4} \qquad (2.10)$$

where v is kinematic viscosity and ε is the rate of turbulent energy dissipation, which should ideally be estimated from the scaling of velocity spectra in the inertial subrange (Pope, 2000) but is more often estimated by assuming isotropic tendency:

$$\varepsilon = 15 v \overline{\left(\frac{\partial u}{\partial x}\right)^2} = 15 v \left(\frac{RMS_u}{\lambda}\right)^2 \qquad (2.11)$$

where λ is the Taylor microscale denoting the boundary between the inertial and dissipation ranges.

The extent of the inertial subrange can be defined by application of the Kolmogorov law describing the one-dimensional energy spectrum, which states that the frequency spectrum of eddies decays according to a power law of $-5/3$ in the inertial subrange (Figure 2.2). This subrange loosely corresponds to the intermediate region where energy generation (G) and ε are in quasi-equilibrium. $G > \varepsilon$ in the inner region whereas $G < \varepsilon$ in the free-surface region. Turbulence is therefore said to be exported from near the bed towards the surface (Nezu and Nakagawa, 1993).

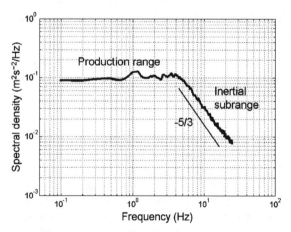

Figure 2.2 Power spectrum for the vertical velocity component in the wake of a submerged boulder showing production range and inertial subrange as defined by Kolmogorov's $-5/3$ power law. F. Breton (unpublished data).

Average eddy frequency ($f_{u,v,w}$) can be determined from a time series by fitting a second order autoregressive model of the form:

$$u_t = a_1 u_{t-1} + a_2 u_{t-2} + e_t \qquad (2.12)$$

where a_1 and a_2 are coefficients of the velocity at a given time lag and e_t is a random component (Clifford and French, 1993a). Alternatively, dominant eddy frequencies can be identified through examination of peaks in velocity power spectra (Figure 2.2) or from the results wavelet analysis (e.g. Torrence and Compo, 1998; Hardy *et al.*, 2009). These frequencies may be converted to dominant or average eddy dimensions ($L_{u,v,w}$) by applying Equation (2.9).

2.2.2 Coherent flow structures

Another description of turbulence based on coherent flow structures (CFSs) has emerged due to the fact that most statistical descriptions ignore the presence of quasi-periodic patterns of coherent motion in the flow (Robinson, 1991). Nikora (2010, p. 373) broadly defines a CFS as 'a three-dimensional flow region over which at least one fundamental flow variable exhibits significant correlation with itself or with another variable over a range of space and/or time'. Research into CFSs has progressed through flow visualisations (e.g. Kline *et al.*, 1967; Shvidenko and Pender, 2001), direct numerical simulations (e.g. Hardy *et al.*, 2007) and analysis of turbulent flow time series in the space and/or time (frequency) domains (e.g. Buffin-Bélanger and Roy, 1998; Lacey and Roy, 2007). CFSs contain most of the turbulent energy and are generally found in the productive subrange (Nezu and Nakagawa, 1993). They can be categorised into two broad scales. At a relatively small scale, CFSs are generated by vortex shedding from protuberant roughness elements (e.g. pebble clusters) and the separation zones in lee of them. The basic forms of such CFSs are horseshoe and hairpin vortices as well as the Kármán vortex street, a region with alternating passages of clockwise and anti-clockwise eddies rotating on a vertical axis (Figure 2.3). At a larger scale, turbulent fluctuations are manifested in high- and low-speed wedges occupying the full depth of the flow.

Clifford and French (1993b) provided evidence that dominant eddy frequencies in gravel-bed rivers could be linked to bed particle sizes by means of the Strouhal relationship, which states that:

$$S_l = \frac{SU}{f} \qquad (2.13)$$

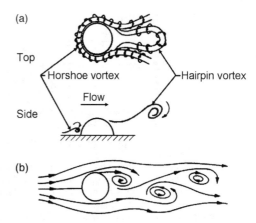

Figure 2.3 (a) Illustration of horseshoe and hairpin vortices over a hemispherical body. Reproduced from Acarlar and Smith (1987) by permission of Cambridge University Press. (b) Top view of streamlines associated with the Kármán vortex street. Reproduced from Davidson (2004) by permission of Oxford University Press.

where S_l is the diameter of a theoretical body responsible for vortex shedding, S is the Strouhal number and f is the frequency of interest. Assuming $S = 0.2$ (Schlichting, 1979), it was found that values of S_l associated with peaks in the power spectrum were of the same order of magnitude as roughness characteristics derived from bed particle size (D) distributions, including $3.5D_{84}$ which reflects typical pebble cluster dimensions. Harvey and Clifford (2009) provided support for this relationship, this time relating average eddy frequencies to particle size distributions in two reaches of a mixed-bed river. Lacey and Roy (2008) used $S = 0.18$ (Achenbach, 1974) and found that the predicted eddy shedding frequency was in good agreement with the frequency (1 Hz) of small-scale vortices observed using flow visualisation in the wake of a submerged pebble cluster. In addition to this high frequency mode, lower frequency fluctuations caused by the intermittent interaction and amalgamation of small-scale vortices were identified, a phenomenon also reported from wavelet analysis of flow over a naturalised gravel bed in the laboratory (Hardy *et al.*, 2009). Tritico and Hotchkiss (2005), on the other hand, found that $S = 0.2$ gave estimates of f which were an order of magnitude lower than the frequency of vortices observed to shed from emergent boulders. The Strouhal relationship, however, may only apply to submerged roughness elements (Franca and Lemmin, 2007). Even in these cases, there is much doubt as to the universality of the scaling in natural settings or naturalised flows in the laboratory, with reported values of S ranging from 0.1 to 0.25 (Venditti and Bauer, 2005).

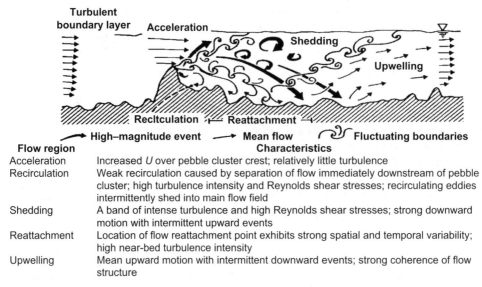

Flow region	Characteristics
Acceleration	Increased U over pebble cluster crest; relatively little turbulence
Recirculation	Weak recirculation caused by separation of flow immediately downstream of pebble cluster; high turbulence intensity and Reynolds shear stresses; recirculating eddies intermittently shed into main flow field
Shedding	A band of intense turbulence and high Reynolds shear stresses; strong downward motion with intermittent upward events
Reattachment	Location of flow reattachment point exhibits strong spatial and temporal variability; high near-bed turbulence intensity
Upwelling	Mean upward motion with intermittent downward events; strong coherence of flow structure

Figure 2.4 Flow regions associated with the presence of a pebble cluster. Reprinted from Buffin-Bélanger and Roy (1998). Copyright 1998, with permission from Elsevier.

A number of studies in gravel-bed rivers have shown that roughness elements such as pebble clusters are associated with distinct zones of turbulent flow conditions (e.g. Buffin-Bélanger and Roy, 1998; Lawless and Robert, 2001a; Lacey and Roy, 2007) (Figure 2.4), which do not closely correspond with the structures illustrated in Figure 2.3 due to the depth-limited nature and high Re of flow over rough gravel beds. These zones may be considered CFSs. In addition to the streamwise and vertical patterns of flow over roughness elements identified by Buffin-Bélanger and Roy (1998), Lawless and Robert (2001b) found that flow around pebble clusters recreated in a laboratory flume was also associated with spanwise flow perturbations, resulting in patterns of flow divergence and convergence. Given that pebble clusters may comprise as much as 10% of the area of the bed (Naden and Brayshaw, 1987), one would expect them to have a substantial effect on reach-scale turbulence characteristics. Other microbedforms (e.g. transverse ribs, stone cells) typical of gravel-bed rivers (Hassan and Reid, 1990; Tribe and Church, 1999) may also be expected to influence turbulence at the reach scale. Lamarre and Roy (2005) and Legleiter et al. (2007), however, found that the effects of such bedforms on distributions of turbulent flow statistics were only localised ($<25D_{84}$ downstream), with reach scale turbulence largely influenced by gross channel morphology (e.g. pool-riffle sequences, meander bends). Despite this, pebble clusters have a consider-

able localised effect on the magnitude of turbulent flow properties (Lacey et al., 2007). Working in a gravel-bed river, for example, Buffin-Bélanger et al. (2006) found that TKE, RMS_v and RMS_w were 100%, 80% and 30% greater respectively with a pebble cluster than without it. Lacey and Roy (2008) reported over a fourfold increase in TKE and τ_{uv} values in the wake region of a pebble cluster compared to background levels. The effects of pebble clusters may persist for a downstream distance of $8.5h_s$ (Buffin-Bélanger et al., 2006) to $15h_s$ (Buffin-Bélanger and Roy, 1998), where h_s is bedform height, in lee of the topographic high.

A parallel strand of research into CFSs has focused on larger-scale structures which take the form of alternating high- and low-speed fluid wedges inclined at an angle to the bed. Though they occur over both rough and smooth beds, the mechanism by which they are formed may be different in each of these environments. Over smooth beds they are thought to originate from the bursting of streamwise streaks of low-speed fluid in the viscous sublayer into the outer region, which triggers a subsequent high-speed sweep towards the bed (Nezu and Nakagawa, 1993). Flow visualisations and direct numerical simulations have shown them to be formed by the concatenation of hairpin vortices which induce regions of coherent flow velocities over the whole flow depth (Adrian, 2007). Since a viscous sublayer does not form over rough (e.g. gravel) beds, their origin in such cases is uncertain. Nevertheless,

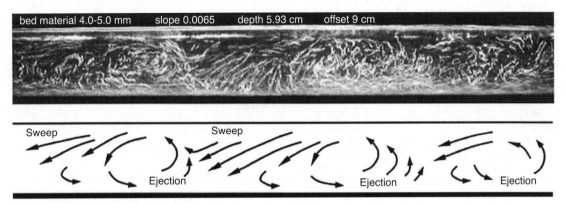

bed material 4.0–5.0 mm   slope 0.0065   depth 5.93 cm   offset 9 cm

Sweep   Sweep

Ejection   Ejection   Ejection

Figure 2.5 Side view of macroturbulent eddies over uniform gravel detected by particle image analysis in a flume. Flow is from right to left. Reprinted from Schvidenko and Pender (2001), with permission from American Geophysical Union.

numerous studies have shown that they do exist in flows over rough beds (e.g. Shvidenko and Pender, 2001; Roy *et al.*, 2004), where they are known as ejections (low-speed, upward motion) and sweeps (high-speed, downward motion). Flow visualisations over a uniform gravel bed by Shvidenko and Pender (2001) suggest that they are caused by the passage of large, macroturbulent eddies rotating on a spanwise axis (Figure 2.5). These structures interact with CFSs in the wake of submerged roughness elements, causing the vertical expansion (ejections) or contraction (sweeps) of the recirculation zone (Buffin-Bélanger and Roy, 1998; Buffin-Bélanger *et al.*, 2001; Lacey and Roy, 2008). Ejections, sweeps and other flow events are traditionally detected using conditional sampling techniques (e.g. Lu and Willmarth, 1973; Blackwelder and Kaplan, 1976; Keylock, 2007). One commonly applied method technique is that of Lu and Willmarth (1973), known as quadrant analysis. This involves attributing events to one of four quadrants (e.g. Q_2) depending on the joint variation of u' and v' around the mean (Figure 2.6), usually

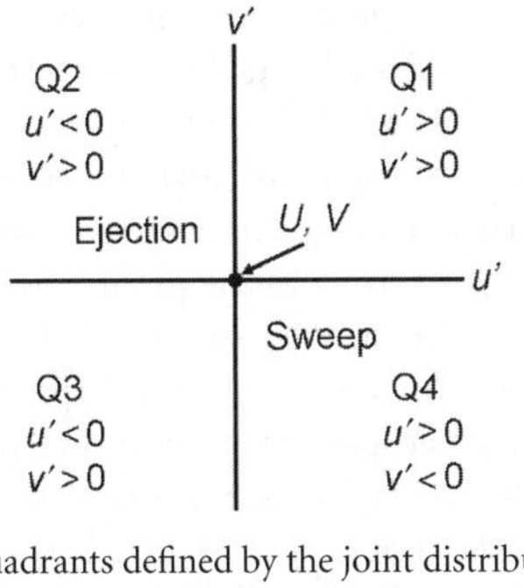

with an amplitude threshold value or 'hole size' (e.g. $2\sigma_{uv}$; Harvey and Clifford, 2009) so that only the stronger events are detected. Several variables may be derived from such analyses, including time in each quadrant for a given hole (H) size (e.g. $T_{Q2}T_{H:2}$), event frequency (e.g. $f_{Q2}T_{H:2}$) and fractional contribution to Reynolds shear stress (e.g. $\tau_{uvQ2}T_{H:2}$) (Lacey *et al.*, 2007; Harvey and Clifford, 2009; Roy *et al.*, 2010).

The importance of macroturbulent structures lies in the fact that they dominate energy production, with ejections and sweeps contributing most to Reynolds shear stress (Williams *et al.*, 1989; Clifford and French, 1993b; Roy *et al.*, 2004). Ejections have been found to dominate fractional contributions to τ_{uv} except in the roughness sublayer (Dancey *et al.*, 2000), the near-wake region immediately in lee of roughness elements (Lacey and Roy, 2008) and in relatively shallow flow, whereas the strength of both ejections and sweeps may be reduced in relatively deep areas (Hardy *et al.*, 2007). The spatial and temporal organisation of high-magnitude events leads to patterns of scour due to bedload transport (Best, 1992; Shvidenko and Pender, 2001). An understanding of the dynamics and dimensions of macroturbulent structures, therefore, is crucial to our understanding of bedform development and possible implications for biota. Studies in a wide range of flow conditions have found that these structures scale with h (Table 2.3). Working in a gravel-bed river, for instance, Roy *et al.* (2004) found that scalings were very similar to those reported from laboratory studies over smooth beds. Furthermore, they calculated that the spatial persistence of high- and low-speed wedges was over $5.6h$ and that they were advected downstream at a velocity close to U, with the convective velocity (U_c) of high-speed wedges approximately 10% higher than low-speed regions.

v'

Q2	Q1
$u' < 0$	$u' > 0$
$v' > 0$	$v' > 0$
Ejection	U, V

u'

Sweep	
Q3	Q4
$u' < 0$	$u' > 0$
$v' > 0$	$v' < 0$

Figure 2.6 Quadrants defined by the joint distribution of the velocity fluctuations from the mean for the streamwise (u') and (v') vertical components.

Table 2.3 Dimensions of macroturbulent structures from laboratory experiments (except *). Dimensions scaled by flow depth (h).

Reference	Re	Length	Width	Height
Nakagawa and Nezu (1981)	4200–12 000	1.5	—	0.5–1
Inamoto and Ishigaki (1987)	6100–7800	2	1	1
Komori et al. (1989)	11 000	2	1	0.5
Yalin (1992)	Not given	6	2	1
Shvidenko and Pender (2001)	12 000–98 000	4–5	2	1
Liu et al. (2001)	10 756–59 870	1–2	—	0.25
Roy et al. (2004)*	150 000–200 000	3–5	0.5–1	1

2.2.3 Measuring turbulence in the field

At least four key aspects of the measuring device and sampling protocol are fundamental to the accuracy and completeness of turbulence measurements which are to be studied within both of the complementary frameworks outlined above: the degree of disturbance introduced into the flow; the digitisation rate; the size of the sampling volume; and the record length. Any device which measures flow around or in close proximity to a physical sensor will interfere with the flow (e.g. Lane et al., 1993). Devices which are able to record flow velocities in a remote volume of fluid, therefore, are preferable.

The digitisation rate (f_D) determines the highest frequency of velocity fluctuation that can be resolved, which is equal to the Nyquist frequency (f_N):

$$f_N = 0.5 f_D \qquad (2.14)$$

in order to avoid aliasing effects (Bendat and Piersol, 2000). Nezu and Nakagawa (1993) provide an estimate of minimum f_D based on turbulence theory:

$$f_D > \frac{(50/\pi)}{(U/h)} \qquad (2.15)$$

An approximation of maximum useful digitisation rate beyond which additional data will be redundant is given by C. M. Garcia (personal communication):

$$f_D < \frac{U_C}{D_S} \qquad (2.16)$$

where D_S is the characteristic length (maximum dimension of the sampling volume). 20 Hz is often used as a minimum f_D for in situ measurements of turbulence (Buffin-Bélanger and Roy, 2005), as recommended by Clifford and French (1993a). According to Equations (2.15–2.16), however, this may not always be sufficient to resolve higher frequency fluctuations in rivers and there is scope for the

use of much higher f_D, depending on D_s, which limits the maximum useful f_D due to spatial averaging effects.

If $D_S > \eta$ then the device will fail to resolve turbulence down to the dissipation range. η may be estimated according to (Nezu and Nakagawa, 1993):

$$\eta \approx \frac{h}{Re^{*0.75}} \qquad (2.17)$$

Nikora (2010) asserts that η in a typical river may be as large as 3 mm. Devices which have sampling volumes with maximum dimensions greater than 3 mm, therefore, are unlikely ever to resolve the finest turbulent structures in rivers. As larger scales of the flow contain most of the turbulent energy (Davidson, 2004), however, resolution of the smallest scales may not be necessary to obtain accurate measurements of certain turbulence quantities (e.g. TKE, τ_{uv}).

Whilst f_D and D_s limit the finest detail that can be resolved from turbulence measurements, record length (RL), a function of f_D and time series duration (t):

$$RL = f_D t \qquad (2.18)$$

determines the largest flow structures that can be detected and influences the precision of the resulting turbulent flow statistics. Buffin-Bélanger and Roy (2005) provided an empirical assessment of optimum RL by performing a bootstrapping technique on 19 long time series (24 000 time steps) to derive sample time series of various lengths. They defined the optima as the point at which the standard error of turbulence statistics levelled off. The overall mean optimum RL was 1300 time steps, whereas 3500 was sufficient to encapsulate optima for all turbulent flow properties (Figure 2.7). Given a typical f_D of 20–25 Hz the optimal time series duration to achieve low standard errors with minimum sampling effort was recommended as 60–90 s.

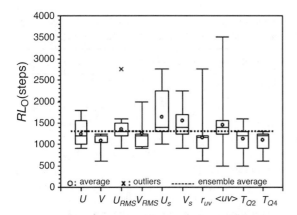

Figure 2.7 Distributions of optimum record length (RL_o) derived for 19 time series for turbulent flow properties, including skewness coefficients (U_s, V_s), Pearson correlation coefficient between u and v (r_{uv}), Reynolds shear stress ($<uv>$) and proportion of time spent in ejections (T_{Q2}) and sweeps (T_{Q4}). The dashed line represents mean overall optimum record length and vertical bars represent ranges excluding outliers. Reprinted from Buffin-Bélanger and Roy (2005). Copyright 2005, with permission from Elsevier.

A range of apparatus has been developed for point measurements of turbulence in the laboratory. These include total head or pitot-static tubes (e.g. Ippen and Raichlen, 1957), hot-film anemometers (e.g. Nakagawa et al., 1975), laser-Doppler velocimeters (e.g. Nezu and Rodi, 1985) and acoustic Doppler velocimeters (ADVs). More recently, particle imaging velocimeters (PIVs) have emerged as a useful tool in laboratory studies. PIVs provide information on the flow field by recording the displacement of particles suspended in a region of fluid (Raffel et al., 2007), thus avoiding the need to rely on Taylor's frozen turbulence approximation (Equation (2.9)) and allowing direct measurement of eddy dimensions. The aforementioned devices, however, are difficult to deploy in the field due to their high sensitivity to environmental variation or the requirement for careful positioning and orientation relative to the boundary (Nezu and Nakagawa, 1993; Nezu, 2005), although submersible miniature PIVs have been developed and tested in a limited range of environmental conditions (e.g. Tritico et al., 2007; Liao et al., 2009). Instead, field investigations have often relied on point measurements using electromagnetic current meters (ECMs) (e.g. Clifford and French, 1993b; Harvey and Clifford, 2009) due to their physical robustness, yet these devices are intrusive and modify flow patterns in the vicinity of the

probe. Furthermore, they are not capable of simultaneous measurement of three-dimensional velocity components and do not satisfy the criteria for f_D and D_S outlined in the above section.

Originally developed for use in the laboratory, ADVs have become an appealing alternative for turbulence measurement in natural river settings since the 1990s (Lane et al., 1998) as data on all three velocity components are recorded in a small sampling volume which is remote (50–100 mm) from the sensing probe, thus minimising the effects of flow intrusion (Kraus et al., 1994). Commercially available second generation ADVs are capable of digitisation rates of up to 200 Hz, have maximum sensor dimensions of 6 mm (Rusello et al., 2006) and can provide reliable estimates of turbulence quantities at distances less than 10 mm from a solid boundary (P. Rusello, personal communication). Despite these obvious advantages, ADV measurements are subject to a number of errors that are controlled by probe positioning, instrument settings and local flow properties (McLelland and Nicholas, 2000). Close attention to instrument settings and carefully designed data collection and processing procedures, therefore, are critical to obtaining reliable results with ADVs. Probe positioning and orientation in relation to local site coordinates may be particularly important if field data are collected for certain purposes (e.g. model validation), in which case an appropriate surveying method should be incorporated into the data collection process (e.g. Lane et al., 1998). For ecohydraulic studies it may be sufficient to rotate the data during post-processing so that $V = W = 0$. As with any measurement of turbulence in potentially unsteady flows, the stationarity of the mean must be tested using an appropriate method, such as a reverse arrangement test (Bendat and Piersol, 2000), and non-stationary time series detrended using linear or low order polynomial regressions before residuals are calculated (Clifford and French, 1993b).

Four further sources of error can contaminate the signal and introduce bias into the resulting turbulent flow statistics (Voulgaris and Trowbridge, 1998). First, Doppler noise caused by random scattering motions in the sampling volume is intrinsic to ADVs. As this noise is normally distributed, it has no effect on mean velocities. Vertical stress components are also relatively unaffected due to the sensor's geometrical characteristics but horizontal components and *TKE* will be biased high (Lane et al., 1998; Nikora and Goring, 1998). The frequency at which the signal is dominated by Doppler noise, termed the noise floor, can be seen as a flattening in the power spectra at high frequencies and may be as low as 4–10 Hz

(Nikora and Goring, 1998). Several methods have been proposed to detect and filter out Doppler noise (e.g. Lane *et al.*, 1998; Nikora and Goring, 1998; Voulgaris and Trowbridge, 1998; McLelland and Nicholas, 2000). Second, due to internal spatial and temporal averaging, ADVs produce a reduction in all of the even moments in the velocity signal (Garcia *et al.*, 2005). Despite these potential sources of error, Garcia *et al.* (2005) have shown that ADVs yield a good description of turbulence when:

$$\frac{f_D h}{U_C} => 20 \qquad (2.19)$$

Third, errors due to phase shift uncertainties, when the phase shift between outgoing and incoming pulses lies outside the range $-180°$ to $+180°$, results in intermittent spikes in the time series when flow velocities approach or exceed the velocity range set by the user. This type of error, commonly known as phase wraparound, can bias estimates of mean and turbulent flow statistics and methods to detect, filter and replace spikes (e.g. Goring and Nikora, 2002; Parsheh *et al.*, 2010) are required to minimise its effects. Finally, velocity shear in the sampling volume may contribute a significant proportion of the overall measurement error close to the boundary (McLelland and Nicholas, 2000). As an indication of the overall quality of data at the time of collection, ADV user interfaces report the average and instantaneous velocity correlation (R^2) between successive radial velocities for each receiver as well as the signal-to-noise ratio (SNR), which is related to the concentration and quality of seeding particles in the flow. Commonly applied quality control thresholds for the estimation of turbulent flow statistics in ecohydraulic studies are average $R^2 > 0.7$ and SNR > 20 (e.g. Smith *et al.*, 2006; Enders *et al.*, 2009).

2.3 The role of turbulence in the swimming performance and habitat selection of river-dwelling fish

Research into the link between fish and turbulence has focused on swimming performance and habitat selection. Swimming stability and kinematics have been used as surrogates for the energetic costs of swimming in turbulent flow in order to supplement the few studies that have measured energetics directly. Field studies evaluating the role of turbulence in the habitat selection of fish are extremely rare and currently limited to brown trout and Atlantic salmon, although several large-scale flume experiments have the potential to contribute towards a greater under-

Table 2.4 The IPOS framework for studying fish-turbulence links. Modified from Lacey *et al.* (2012).

Relevant turbulent flow properties	
Intensity	$\sigma_{u,v,w}$
	TI
	$RMS_{u,v,w}$
	τ_{uv}, τ_{uw}
	Vorticity (ω)
	Eddy maximum angular momentum (Π_e)
Periodicity	$f_{u,v,w}$
	ITS
	Spectral peaks and flatness
Orientation	Axis of eddy rotation (x–y, x–z, y–z)
	Vector of dominant flow fluctuation (x, y, z)
Scale	Average eddy dimensions ($L_{u,v,w}$)
	Integral scales ($L_{x,y,z}$)
	Re

standing in this area. Lacey *et al.* (2012) have emphasised the need to consider four aspects of turbulent flow (intensity, periodicity, orientation and scale) when examining the links between fish and turbulence (Table 2.4).

2.3.1 Swimming performance

Some early experimental research into fish-turbulence links was reviewed by Pavlov *et al.* (2000). They reported that critical U thresholds at which fish were displaced downstream for gudgeon (*Gobio gobio*), roach (*Rutilus rutilus*) and perch (*Perca fluviatilis*) were negatively related to TI_u. Furthermore, larger fish of a given species had higher critical U thresholds for a given TI_u. Linear regression revealed a significant ($p < 0.002$) relationship between eddy length (L_u) and critical U threshold. The critical L_u was equal to $0.66bl$, where bl is fish body length. The mechanism behind this relationship was cited as the distribution of hydrodynamic forces acting on the body of a fish. When $L_u \ll bl$, the moments of the forces were evenly distributed along the body of the fish. When $L_u > 0.66bl$, fish were destabilised and actively moved their pectoral fins to correct their position, thus creating greater hydrodynamic resistance and presumably increasing energy expenditure. This result was confirmed by Lupandin (2005) in the case of perch. Pectoral fins are also known to be important in the swimming stability of salmonids (McLaughlin and Noakes, 1998; Drucker and Lauder, 2003) and are a particularly distinctive feature of the station-holding behaviour of Atlantic salmon parr

(Arnold *et al.*, 1991), which have larger pectoral fins than other salmonids allowing them to maintain position in lower velocities on the substrate rather than in the water column.

Two essential features of an eddy, its orientation and intensity, are ignored by the Pavlov *et al.* (2000) model. The orientation of perturbations (e.g. eddies) is a critical factor in the swimming stability of fish (Webb, 2004; Liao, 2007), yet Pavlov *et al.* (2000) only considered eddies rotating on a horizontal axis. Tritico and Cotel (2010) found that the swimming stability of creek chub (*Semotilus atromaculatus*) in a flume was related to both eddy size and orientation. Instances where fish lost postural control (spills) were not observed until the 95th percentile of eddy diameter, determined using PIV, reached $0.76bl$. Spills were more than twice as frequent and lasted 24% longer in flow fields dominated by eddies rotating on a horizontal axis. The resumption of steady swimming after disturbance from horizontal eddies required additional rolling movements in comparison to recovery from vertical eddies. It has been suggested that susceptibility to destabilisation from eddies of different orientation is related to body morphology, with laterally and dorso-ventrally compressed fish more susceptible to horizontal and vertical eddies respectively (Lacey *et al.*, 2012). In addition to L_u and eddy orientation, several commentators have suggested that the potential for an eddy to destabilise a fish is also a function of the ratio of eddy momentum to fish momentum (Webb *et al.*, 2010; Webb and Cotel, 2010; Lacey *et al.*, 2012). Tritico and Cotel (2010) quantified the maximum angular momentum of eddies (Π_e), given by:

$$\Pi_e = \frac{m_e \Gamma_e}{4\pi} \qquad (2.20)$$

(where m_e is eddy mass), and found that the occurrence of spills increased as a function of Π_e above a threshold of 30 000 g cm^2 s^{-1}.

Though they have been cited with respect to juvenile salmonids (e.g. Enders *et al.*, 2005a) the findings of Pavlov *et al.* (2000) and Lupandin (2005) may have limited relevance for this species and life stage. The reason for this relates to the eddy sizes and orientation covered in the studies. Given a typical bl of Atlantic salmon parr of 70–150 mm (Gibson and Cutting, 1993), for instance, and the fact that their habitat preference typically means that $h \gg bl$ (Crisp, 1993; Armstrong *et al.*, 2003), the dominant energy-containing (macroturbulent) eddies rotating on a horizontal axis (Figure 2.5) are likely to be much larger than bl if the scalings provided in Table 2.3 are correct. Webb and Cotel (2010) postulated that when $L_u \gg bl$ fish perceive the flow as rectilinear, responding only to a variation of the mean flow vector. Enders *et al.* (2005b) found that the feeding behaviour of Atlantic salmon parr was not related to the passage of macroturbulent structures, suggesting that they do not respond to the largest horizontal eddies. Furthermore, the station-holding microhabitat of juvenile salmonids is typically in lee of a home rock (Cunjak, 1988; Guay *et al.*, 2000) where eddies are shed on a vertical axis in the Kármán vortex street (Figure 2.3b). By simplifying this microhabitat in a smooth-walled laboratory flume, Liao *et al.* (2003b) showed that rainbow trout could attune their swimming kinematics (body amplitude, tail-beat frequency) to the frequency of vertical eddies ($0.25 < L_u/bl < 0.5$) shed from a cylinder. They termed this swimming behaviour 'Kármán gaiting' and Liao (2006) found that four fish spent the majority of time in locations of the flume where this gait was possible (Figure 2.8).

Using electromyography, Liao *et al.* (2003a) and Liao (2004) revealed that the Kármán gait was associated with lower muscle activity in rainbow trout swimming in the Kármán vortex street than those swimming in the free stream with no cylinder (Figure 2.9). The Re of flows used in these experiments was not reported but calculations based on reported flume dimensions and discharges

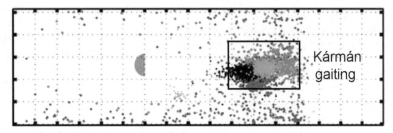

Figure 2.8 Head locations of four rainbow trout (*Oncorhynchus mykiss*) in a flume in relation to the area with suitable conditions for Kármán gaiting. Individual fish illustrated by four different shades. Locations tracked every 5 s for 1 h. Modified from Liao (2006). Reproduced by permission of *The Journal of Experimental Biology*.

(a)   (b)

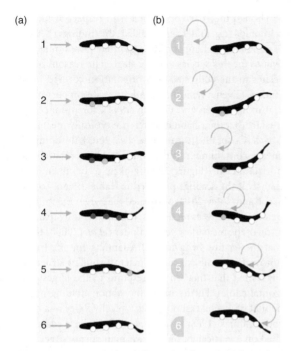

Figure 2.9 Time series (1–6) illustrating that red axial muscle activity measured in a flume using electromyography differed between rainbow trout (*Oncorhynchus mykiss*) swimming in the free stream (a) and behind a cylinder (b). Circles denote electrode positions with no (open), intermediate (grey) or high (closed) muscle activity. From Liao *et al.* (2003a). Reprinted with permission from American Association for the Advancement of Science.

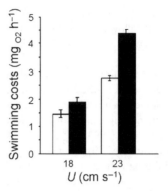

Figure 2.10 Swimming costs of Atlantic salmon parr for four experimental treatments. Low turbulence conditions ($\sigma_u = 5$ cm s^{-1}) are represented by open bars and high turbulent conditions ($\sigma_u = 8$ cm s^{-1}) are represented by solid bars. Vertical lines represent 95% confidence intervals. From Enders *et al.* (2003). Reproduced by permission Canadian Science Publishing.

(Liao *et al.*, 2003b) suggest that the work was carried out at approximately $7500 < Re < 14\,000$. In these conditions ($Re > 400$) the organised structure of the Kármán vortex street is disrupted by three-dimensional instabilities, although the periodicity and rotationality of the vortices may remain detectable (Davidson, 2004). In gravel-bed rivers, regions of flow in lee of submerged roughness elements (e.g. pebble clusters) are characterised by highly dynamic zones of eddy shedding and complex three-dimensional flow patterns (Buffin-Bélanger and Roy, 1998; Lawless and Robert, 2001b). Despite the occurrence of vertical eddies, the turbulent energy in these flow regions is dominated by spanwise macroturbulent structures (Lacey and Roy, 2008), raising questions over the prevalence of suitable conditions for Kármán gaiting. Kármán gaiting nevertheless remains a plausible explanation for the highly efficient upstream migration of adult sockeye salmon (*Oncorhynchus nerka*), as suggested by Hinch and Rand (2000) and Standen *et al.* (2004).

Evidence for the opposite relationship between turbulence and swimming energetics to that suggested by Liao *et al.* (2003a; 2003b) has emerged from experiments in respirometers. Using four combinations of U and σ_u, Enders *et al.* (2003) found that swimming costs (rate of oxygen consumption) for Atlantic salmon parr increased significantly ($p < 0.05$) with σ_u for a given U (Figure 2.10). This relationship was also reported by Enders *et al.* (2005a), who found that σ_u contributed 14% of the explained variation in swimming costs in a model which included temperature (2% of variation), fish body mass (31%) and U (46%). They also reported that existing bioenergetic models based on forced swimming (e.g. Boisclair and Tang, 1993) underestimated swimming costs under highly turbulent conditions ($\sigma_u = 10$ cm s^{-1}) by a factor of 14. Underlining the equivocal nature of the evidence further, Nikora *et al.* (2003) found that turbulence had no effect on the swimming performance (time-to-fatigue) of inanga (*Galaxius maculatus*). Turbulence in these studies, however, was produced using pumps or artificial structures and may not be comparable to the conditions created by Liao *et al.* (2003a; 2003b) in terms of intensity, orientation and scale, having no effect on fish or possibly impeding rather than enhancing swimming performance (Lacey *et al.*, 2012).

2.3.2 Habitat selection

Very few studies have examined fish habitat selection with respect to turbulence. These have most often been

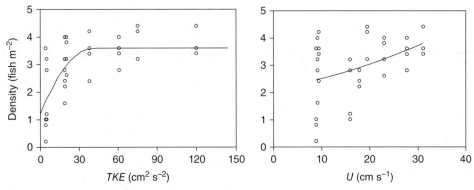

Figure 2.11 Comparison of models to predict volitional rainbow trout density in response to experimental treatments in a flume. Reproduced from Smith *et al.* (2006) by permission of The American Fisheries Society.

undertaken in artificial settings in the laboratory. Smith *et al.* (2005), for instance, recorded the microhabitat positions of juvenile rainbow trout in a flume with cover provided in the form of bricks. By measuring mean (U, V, W) and turbulent ($RMS_{u,v,w}$, TKE, $\tau_{uv,uw}$, L_x, L_y) flow properties at focal positions taken up by fish and those available throughout the flume, they found that fish selected positions with significantly lower V ($p = 0.01$) and L_x ($p < 0.01$) than available during a low-discharge treatment and lower U ($p < 0.01$), τ_{uv} ($p < 0.01$) and L_x ($p = 0.03$) during a separate high-discharge treatment. This study was performed with individual fish in each trial, thus ignoring the effects of competition. In the same flume, this time with three cover treatments to create varying levels of turbulence and three discharge treatments (0.03–0.11 m^3 s^{-1}), Smith *et al.* (2006) placed 30 fish in the test section for each of four replicate trials. The number of individuals that chose to remain in the test section after 24 hours was better predicted by TKE than U (Figure 2.11). The reason cited for this was that TKE was more sensitive to the experimental treatments. Smith and Brannon (2007) subsequently found that turbulent flow properties ($RMS_{u,w}$, TKE) were better able to detect the presence of cover types used by fish (boulders, woody debris, scour holes) than U in four gravel/cobble-bed rivers. The intensity of turbulence in these flume studies, however, was not comparable to that typically found in gravel-bed rivers. Maximum values of TKE and τ_{uv}, for instance, were an order of magnitude lower than those found by Tritico and Hotchkiss (2005) in lee of a boulder.

Flume experiments examining the routes taken by fish during migration through artificial structures have begun to reveal a highly species-specific relationship between turbulence and habitat selection. Russon *et al.*

(2010) found that most (63.3%) approaches by downstream migrating adult European eels (*Anguilla anguilla*) towards bar racks designed to screen fish from hydroelectric power turbines were associated with zones of highest TI_u. Conversely, Kemp and Williams (2008) reported that downstream migrating Chinook salmon (*Oncorhynchus tshawytscha*) smolts preferred a smooth culvert with significantly lower TI_u ($p < 0.001$) to treatment culverts augmented with corrugated sheet and cobbles. Silva *et al.* (2011) similarly found that migrants avoided highly turbulent conditions, in this case by observing Iberian barbel (*Luciobarbus bocagei*) moving upstream through orifices associated with an experimental pool-type fishway. Most barbel used areas of lowest U and TKE and there were negative correlations between turbulent flow properties and fish transit time, with the most significant being τ_{uv} ($p < 0.001$) and TKE ($p < 0.01$). Fish were observed to use pectoral fins for postural control more frequently under the most turbulent conditions. For lamprey, which lack paired fins to facilitate control, turbulence is likely to present a greater challenge to swimming stability. Kemp *et al.* (2011) presented indirect evidence suggesting that upstream migrating river lamprey (*Lampetra fluviatilis*) were able to pass experimental weirs under highly turbulent conditions by altering swimming trajectories to closely follow the substrate and channel walls. This strategy allowed lamprey to hold station by using the oral disk to attach to the structure, possibly as part of a burst-and-attach swimming strategy.

Field studies of fish habitat selection with respect to turbulence are limited to two examples focusing on salmonids. At the microscale in a third-order sand-bed stream, Cotel *et al.* (2006) observed the locations of brown trout by snorkel surveying at summer low flow over three

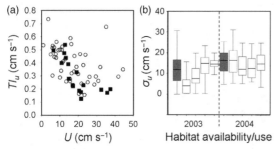

Figure 2.12 Relationships between turbulence and salmonids in rivers. (a) Turbulence intensity (TI_u) and nose velocity (U) for various locations with brown trout (closed symbols) and similar locations with no fish (open symbols). Reproduced from Cotel *et al.* (2006), by permission of The American Fisheries Society; (b) reach-scale habitat availability (grey boxes) and locations of four tagged Atlantic salmon parr (open boxes) in relation to σ_u over two years. Vertical bars represent upper and lower 5th percentiles. Reproduced from Enders *et al.* (2009), by permission of The American Fisheries Society.

years. They also identified areas of the channel with similar habitat conditions (h, U, cover) and took hydraulic measurements in these representative locations as well as those used by brown trout. Though results were variable, in general fish occupied locations with reduced TI_u compared with otherwise similar locations without fish (Figure 2.12(a)). TI_u, however, was the only turbulent flow property reported and there is debate as to whether this relative measure of turbulence, which assumes that the swimming stability of fish increases with mean velocity, is the most relevant (Smith *et al.*, 2005; Webb, 2006). At the reach scale of a gravel-bed river in Canada, Enders *et al.* (2009) used radiotelemetry to track the locations of four Atlantic salmon parr over two summer low flow periods. They characterised habitat available in an 80 m reach by taking hydraulic measurements at the nodes of a 2 m × 2 m grid. Though statistically significant differences between available and used distributions of σ_u and *TKE* were found for some fish in one or more of the study years, the sign of the relationship was not consistent (Figure 2.12(b)), leading the authors to conclude that there was no link between Atlantic salmon parr habitat selection and turbulence at the reach scale.

2.4 Conclusions

Ecohydraulics has suffered from an overreliance on correlative approaches based on relatively simple, mean characteristics of the flow, yet turbulence is a ubiquitous phenomenon in rivers. The inclusion of turbulent

flow properties in ecohydraulic research, therefore, should enhance our mechanistic knowledge of physical–ecological interactions. The hydrodynamic environment is composed of a range of turbulent structures of varying intensity, periodicity, orientation and scale from across the EC. The development of ADVs as field tools has helped to advance our knowledge of these structures in natural settings (e.g. Lacey and Roy, 2008; Roy *et al.*, 2010) but great care must be taken to ensure that ADVs provide a reliable description of the turbulence (Garcia *et al.*, 2005). Information on the flow field using PIVs (Tritico *et al.*, 2007; Tritico and Cotel, 2010) and two- or three-dimensional hydrodynamic models (e.g. Crowder and Diplas, 2002, 2006; see also Chapter 3 in this volume), rather than point measurements of turbulence, could represent the next major step forward in our understanding of how turbulence affects instream biota. Work in the laboratory as well as the field is likely to be useful in this regard, but flume studies must ensure that the relevant characteristics of turbulent flow are recreated to closely mimic the hydraulic habitat of target biota (Lacey *et al.*, 2012).

The mechanisms by which turbulence may affect the fitness of individual fish have been quantified in several ways, including critical eddy length (e.g. Pavlov *et al.*, 2000) and the energetic costs of swimming in turbulent flow (e.g. Enders *et al.*, 2003). These mechanisms may, in turn, influence the structure of fish communities by determining habitat preference with respect to turbulence (e.g. Cotel *et al.*, 2006; Smith *et al.*, 2006). Hydraulic research has highlighted two scales of turbulence – scaling with h (Roy *et al.*, 2004) and the size of microbedforms (Clifford and French, 1993b; Lacey and Roy, 2008) respectively – which could be particularly relevant to stream-dwelling fish. Laboratory work with a limited number of species suggests that high-magnitude spanwise (macro-turbulent) eddies may, depending on the ratio of eddy size and momentum to fish body length and momentum, destabilise fish and result in increased energetic costs (Pavlov *et al.*, 2000; Lupandin, 2005; Tritico and Cotel, 2010), whereas smaller, vertically oriented eddies could enhance swimming efficiency (Liao *et al.*, 2003a; Liao, 2006). Research with Atlantic salmon parr has revealed a negative relationship between swimming costs and turbulence in a highly artificial setting (Enders *et al.*, 2003) and field observations on the summer habitat preference of brown trout in a sand-bed stream suggest that this could influence position choice at the microscale (Cotel *et al.*, 2006). A tagging study found no relationship between turbulence and the summer habitat preference of Atlantic salmon parr at the reach scale of a gravel-bed river (Enders *et al.*, 2009). Any link between fish and turbulence, therefore, may be scale-dependent.

Evidence on the effects of turbulence on the microhabitat selection of non-salmonid species is lacking and no attention has been paid to fish-turbulence links at the mesoscale ($10^{-1} - 10^{1}$ m), where the results of ecohydraulic research are often applied (Harper and Everard, 1998; Newson and Newson, 2000). Even for salmonid species the accumulated knowledge is sparse. What knowledge we do have from the field is limited to summer low flow periods, yet the feeding behaviour of juvenile salmonids is known to vary seasonally (e.g. Heggenes, 1996) and critical population bottlenecks may occur at other times of the year (e.g. Armstrong *et al.*, 2003). Understanding gained from the laboratory on the mechanisms affecting fish swimming in turbulent flow (e.g. Pavlov *et al.*, 2000; Liao *et al.*, 2003a; Tritico and Cotel, 2010) must be validated in the field. Revealing how fish respond to a range of CFSs in natural settings is another key research priority which has received little attention (but see Enders *et al.*, 2005b). More detailed information is needed before turbulence can be integrated as a distinct ecological variable in river research and management activities (e.g. habitat modelling, assessment and rehabilitation) targeted at fish and other biota. *TKE*, dominant axes of eddy rotation and metrics describing eddy diameter and momentum are likely to be among the most important turbulent flow properties to focus on.

Acknowledgements

The authors would like to thank Peter J. Rusello (Research Scientist, Nortek USA) and Carlos M. Garcia (Centro de Estudios y Tecnologia del Agua, Universidad Nacional de Cordoba, Argentina) for their helpful advice on ADV performance, Felipe Breton (Universidad de Concepcion, Chile), Eva Enders (University of Alberta, Canada) and David L. Smith (US Army Corps of Engineers) for help with reproducing figures, and two anonymous reviewers for their helpful comments and suggestions.

References

Acarlar, M.S. and Smith, C.R. (1987) A study of hairpin vortices in a laminar boundary layer. Part 1. Hairpin vortices generated by a hemisphere protuberance. *Journal of Fluid Mechanics*, **175**: 1–41.

Achenbach, E. (1974) Vortex shedding from spheres. *Journal of Fluid Mechanics*, **62**: 209–221.

Adrian, R.J. (2007) Hairpin vortex organization in wall turbulence. *Physics of Fluids*, **19**(4): 041301.

Allan, J.D. (1995) *Stream Ecology: Structure and function of running waters*, Chapman & Hall, London.

Armstrong, J.D., Kemp, P.S., Kennedy, G.J.A., Ladle, M. and Milner, N.J. (2003) Habitat requirements of Atlantic salmon and brown trout in rivers and streams. *Fisheries Research*, **62**: 143–170.

Arnold, G.P., Webb, P.W. and Holford, B.H. (1991) The role of pectoral fins in station-holding behaviour of Atlantic salmon parr (*Salmo Salar L.*). *Journal of Experimental Biology*, **156**: 625–629.

Bagnold, R.A. (1966) An approach to the sediment transport problem from general physics. *US Geological Survey Professional Paper 422-I*, Washington.

Beamish, F.W.H. (1978) Swimming capacity. In Hoar, W.S. and Randall, D.J. (eds) *Fish Physiology*, volume VII, Academic Press, New York.

Bendat, J. and Piersol, A. (2000) *Random Data*, 2nd edition, John Wiley & Sons, Inc., Hoboken, NJ.

Best, J. (1992) On the entrainment of sediment and initiation of bed defects: Insights from recent developments within turbulent boundary layer research. *Sedimentology*, **39**: 797–811.

Best, J.L. (1993) On the interactions between turbulent flow structure, sediment transport and bedform development: some considerations from recent experimental research. In Clifford, N.J., French, J.R. and Hardisty, J. (eds) *Turbulence: Perspectives on Flow and Sediment Transport*, John Wiley & Sons, Ltd, Chichester, UK.

Biggs, B.J.F. (1996) Patterns in benthic algae of streams. In Stevenson, R.J., Bothwell, M.L. and Lowe, R.L. (eds) *Algal Ecology: Freshwater Benthic Ecosystems*, Academic Press, New York.

Biggs, B.J.F. and Thomsen, H. (1995) Disturbance of stream periphyton by perturbations in shear stress: time to structural failure and differences in community composition. *Journal of Phycology*, **31**: 233–241.

Bisson, P.A., Sullivan, K. and Nielsen, J.L. (1988) Channel hydraulics, habitat use, and body form of juvenile coho salmon, steelhead, and cutthroat trout in streams. *Transactions of the American Fisheries Society*, **117**: 262–273.

Blackwelder, R.F. and Kaplan, R.E. (1976) On the bursting phenomenon near the wall in bounded turbulent shear flows. *Journal of Fluid Mechanics*, **76**: 89–112.

Boisclair, C. and Tang, M. (1993) Empirical analysis of the influence of swimming pattern on the net energetic cost of swimming in fishes. *Journal of Fish Biology*, **42**: 169–183.

Brooks, A.J., Haeusler, T., Reinfelds, I.A. and Williams, S. (2005) Hydraulic microhabitats and the distribution of macroinvertebrate assemblages in riffles. *Freshwater Biology*, **50**: 331–344.

Buffin-Bélanger, T. and Roy, A.G. (1998) Effects of a pebble cluster on the turbulent structure of a depth-limited flow in a gravel-bed river. *Geomorphology*, **25**: 249–267.

Buffin-Bélanger, T. and Roy, A.G. (2005) 1 min in the life of a river: selecting the optimal record length for the measurement of turbulence in fluvial boundary layers. *Geomorphology*, **68**: 77–94.

Buffin-Bélanger, T., Roy, A.G. and Levasseur, M. (2001) Interactions entre les structures d'échappement et les structures à grande échelle dans l'écoulement des rivières à lit de graviers. *Revue des Sciences de l'eau*, **14**: 381–407.

Buffin-Bélanger, T., Rice, S., Reid, I. and Lancaster, J. (2006) Spatial heterogeneity of near-bed hydraulics above a patch of river gravel. *Water Resources Research*, **42**: W04413.

Clifford, N.J. and French, J.R. (1993a) Monitoring and modelling turbulent flow: historical and contemporary perspectives. In Clifford, N.J., French, J.R. and Hardisty, J. (eds) *Turbulence: Perspectives on Flow and Sediment Transport*, John Wiley & Sons, Ltd, Chichester, UK.

Clifford, N.J. and French, J.R. (1993b) Monitoring and analysis of turbulence in geophysical boundaries: some analytical and conceptual issues. In Clifford, N.J., French, J.R. and Hardisty, J. (eds) *Turbulence: Perspectives on Flow and Sediment Transport*, John Wiley & Sons, Ltd, Chichester, UK.

Clifford, N.J., Robert, A. and Richards, K.S. (1992) Estimation of flow resistance in gravel-bedded rivers: a physical explanation of the multiplier of roughness length. *Earth Surface Processes and Landforms*, **17**: 111–126.

Cotel, A.J., Webb, P.W. and Tritico, H. (2006) Do Brown Trout Choose Locations with Reduced Turbulence? *Transactions of the American Fisheries Society*, **135**: 610–619.

Crisp, D.T. (1993) The environmental requirements of salmon and trout in fresh water. *Freshwater Forum*, **3**(3): 176–202.

Crowder, D.W. and Diplas, P. (2002) Vorticity and circulation: spatial metrics for evaluating flow complexity in stream habitats. *Canadian Journal of Fisheries and Aquatic Sciences*, **59**: 633–645.

Crowder, D.W. and Diplas, P. (2006) Applying spatial hydraulic principles to quantify stream habitat. *River Research and Applications*, **22**: 79–89.

Cunjak, R.A. (1988) Behaviour and microhabitat of young Atlantic salmon (*Salmo salar*) during winter. *Canadian Journal of Fisheries and Aquatic Sciences*, **45**: 2156–2160.

Dancey, C.L., Balakrishnan, M., Diplas, P. and Papanicolaou, A.N. (2000) The spatial inhomogeneity of turbulence above a fully rough, packed bed in open channel flow. *Experiments in Fluids*, **29**: 402–410.

Davidson, P.A. (2004) *Turbulence: An Introduction for Scientists and Engineers*, University Press, Oxford.

Davis, J.A. and Barmuta, L.A. (1989) An ecologically useful classification of mean and near-bed flows in streams and rivers. *Freshwater Biology*, **21**: 271–282.

DeNicola, D.M. and McIntire, C.D. (1991) Effects of hydraulic refuge and irradiance on grazer–periphyton interactions in laboratory streams. *Journal of the North Amercan Benthological Society*, **10**: 251–262.

Deutsch, W.G. (1984) Oviposition of Hydropsychidae (Trichoptera) in a large river. *Canadian Journal of Zoology*, **62**: 1988–1994.

Drucker, E.G. and Lauder, G.V. (2003) Function of pectoral fins in rainbow trout: behavioural repertoire and hydrodynamic forces. *The Journal of Experimental Biology*, **206**: 813–826.

Enders, E.C., Boisclair, D. and Roy, A.G. (2003) The effect of turbulence on the cost of swimming for juvenile Atlantic salmon (*Salmo salar*). *Canadian Journal of Fisheries and Aquatic Sciences*, **60**: 1149–1160.

Enders, E.C., Boisclair, D. and Roy, A.G. (2005a) A model of total swimming costs in turbulent flow for juvenile Atlantic salmon (*Salmo salar*). *Canadian Journal of Fisheries and Aquatic Sciences*, **62**: 1079–1089.

Enders, E.C., Buffin-Bélanger, T., Boisclair, D. and Roy, A.G. (2005b) The feeding behaviour of juvenile Atlantic salmon in relation to turbulent flow. *Journal of Fish Biology*, **66**: 242–253.

Enders, E.C., Roy, M.L., Ovidio, M., Hallot, E.J., Boyer, C., Petit, F. and Roy, A.G. (2009) Habitat choice by Atlantic salmon parr in relation to turbulence at a reach scale. *North American Journal of Fisheries Management*, **30**: 1819–1830.

Eriksen, C.H., Lamberti, G.A. and Resh, V.H. (1996) Aquatic insect respiration. In Merritt, R.W. and Cummins, K.W. (eds) *An Introduction to the Aquatic Insects of North America*, Kendall/Hunt, Dubuque, Iowa.

Franca, M.J. and Lemmin, U. (2007) Discussion of "Unobstructed and Obstructed Turbulent Flows in Gravel Bed Rivers" by Hans M. Tritico and Rollin H. Hotchkiss. *Journal of Hydraulic Engineering*, **133** (1): 116–117.

Frank, B.M., Piccolo, J.J. and Baret, P.V. (2011) A review of ecological models for brown trout: towards a new demogenetic model. *Ecology of Freshwater Fish*, **20**: 167–198.

Frissell, C.A., Liss, W.J., Warren, C.E. and Hurley, M.D. (1986) A hierarchical framework for stream habitat classification: viewing streams in a watershed context. *Environmental Management*, **10**(2): 199–214.

Garcia, C.M., Cantero, M.I., Niño, Y. and Garcia, M.H. (2005) Turbulence measurements with acoustic Doppler velocimeters. *Journal of Hydraulic Engineering*, **131**(12): 1062–1073.

Gibson, R.J. and Cutting, R.E. (1993) *Production of Juvenile Atlantic Salmon, Salmo salar, in Natural Waters*, NRC Research Press, Ontario.

Godin, J.J. and Rangeley, R.W. (1989) Living in the fast lane: effects of cost of locomotion on foraging behaviour in juvenile Atlantic salmon. *Animal Behaviour*, **37**: 943–954.

Goring, D.G. and Nikora, V.I. (2002) Despiking acoustic Doppler velocimeter data. *Journal of Hydraulic Engineering*, **128**(1): 117–125.

Guay, J.C., Boisclair, D., Rioux, D., Leclerc, M., Lapointe, M. and Legendre, P. (2000) Development and validation of numerical habitat models for juveniles of Atlantic salmon (*Salmo salar*). *Canadian Journal of Fisheries and Aquatic Sciences*, **57**: 2065–2075.

Hannah, D.M., Sadler, J.P. and Wood, P.J. (2007) Hydroecology and Ecohydrology: Challenges and Future Prospects. In Wood, P.J., Hannah, D.M. and Sadler, J.P. (eds) *Hydroecology and Ecohydrology: Past, Present and Future*, John Wiley & Sons, Ltd, Chichester, UK.

Hardy, R.J., Best, J.L., Lane, S.N. and Carbonneau, P.E. (2009) Coherent flow structures in a depth-limited flow over a gravel surface: the role of near-bed turbulence and influence of Reynolds number. *Journal of Geophysical Research*, **114**: F01003.

Hardy, R.J., Lane, S.N., Ferguson, R.I. and Parson, D.R. (2007) Emergence of coherent flow structures over a gravel surface: A numerical experiment. *Water Resources Research*, **43**: W03422.

Harper, D.M. and Everard, M. (1998) Why should the habitat-level approach underpin holistic river survey and management? *Aquatic Conservation-Marine and Freshwater Ecosystems*, **8**(4): 395–413.

Hart, D.D. and Finelli, C.M. (1999) Physical–biological coupling in streams: The pervasive effects of flow on benthic organisms. *Annual Review of Ecology and Systematics*, **30**: 363–395.

Hart, D.D. and Merz, R.A. (1998) Predator–prey interactions in a benthic stream community: field test of flow-mediated refuges. *Oecologia*, **114**: 263–273.

Harte, J. (2002) Towards a synthesis of Newtonian and Darwinian worldviews. *Physics Today*, **55**(10): 29–43.

Harvey, G.L. and Clifford, N.J. (2009) Microscale hydrodynamics and coherent flow structures in rivers: Implications for the characterization of physical habitat. *River Research and Applications*, **25**(2): 160–180.

Hassan, M. and Reid, I. (1990) The influence of microform bed roughness elements on flow and sediment transport in gravel bed rivers. *Earth Surface Processes and Landforms*, **15**: 739–750.

Hayes, J.W. and Jowett, I.G. (1994) Microhabitat models of large drift-feeding brown trout in three New Zealand rivers. *North American Journal of Fisheries Management*, **14**: 710–725.

Heggenes, J. (1996) Habitat selection by brown trout (*Salmo trutta*) and young Atlantic salmon (*S. salar*) in streams: static and dynamic hydraulic modelling. *Regulated Rivers: Research and Management*, **12**: 155–169.

Hinch, S.G. and Rand, P.S. (2000) Optimal swimming speeds and forward-assisted propulsion: energy-conserving behaviours of upriver-migrating adult salmon. *Canadian Journal of Fisheries and Aquatic Sciences*, **57**: 2470–2478.

Hughes, N.F. and Dill, L.M. (1990) Position choice by drift-feeding salmonids: model and test for Arctic grayling (*Thymallus arcticus*) in subarctic mountain streams, interior Alaska. *Canadian Journal of Fisheries and Aquatic Sciences*, **47**: 2039–2048.

Inamoto, H. and Ishigaki, T. (1987) Visualization of longitudinal eddies in an open channel flow. In Véret, C. (ed.) *Flow Visualisation IV*, Hemisphere, New York.

Ippen, A.T. and Raichlen, F. (1957) Turbulence in civil engineering: measurements in free surface streams. *Journal of the Hydraulics Division*, **83**(5): 1–27.

Jonsson, B. and Jonsson, N. (2003) Migratory Atlantic salmon as vectors for the transfer of energy and nutrients between freshwater and marine environments. *Freshwater Biology*, **48**: 21–27.

Jowett, I.G. (1993) A method for objectively identifying pool, run and riffle habitats. *New Zealand Journal of Marine and Freshwater Research*, **27**: 241–248.

Kay, M. (2008) *Practical Hydraulics*, 2nd edition, Taylor & Francis, Oxon.

Kemp, J.L., Harper, D.M. and Crosa, G.A. (2000) The habitat-scale ecohydraulics of rivers. *Ecological Engineering*, **16**: 17–29.

Kemp, P.S. and Williams, J.G. (2008) Response of migrating Chinook salmon (*Oncorhynchus tshawytscha*) smolts to in-stream structure associated with culverts. *River Research and Applications*, **24**: 571–579.

Kemp, P.S., Russon, I.J., Vowles, A.S. and Lucas, M.C. (2011) The influence of discharge and temperature on the ability of upstream migrant adult river lamprey (*Lampetra fluviatilis*) to pass experimental overshot and undershot weirs. *River Research and Applications*, **27**: 488–498.

Keylock, C.J. (2007) The visualization of turbulence data using a wavelet-based method. *Earth Surface Processes and Landforms*, **32**: 637–647.

Kirkbride, A.D. and Ferguson, R. (1995) Turbulent flow structure in a gravel-bed river: Markov chain analysis of the fluctuating velocity profile. *Earth Surface Processes and Landforms*, **20**: 721–733.

Kline, S.J., Reynolds, W.C., Schraub, F.A. and Runstadler, P.W. (1967) The structure of turbulent boundary layers. *Journal of Fluid Mechanics*, **30**: 741–773.

Kohler, S.L. (1992) Competition and the structure of a benthic stream community. *Ecological Monographs*, **62**: 165–188.

Kolmogorov, A.N. (1941) The local structure of turbulence in incompressible viscous fluid for very large Reynolds numbers. *Dolkady Akademii Nauk SSSR*, **30**: 301–305.

Komori, S., Murakami, Y. and Ueda, H. (1989) The relationship between surface-renewal and bursting motions in an open-channel flow. *Journal of Fluid Mechanics*, **203**: 103–123.

Kraus, N.C., Lohrmann, A. and Cabrera, R. (1994) New acoustic meter for measuring 3D laboratory flows. *Journal of Hydraulic Engineering*, **120**: 406–412.

Labiod, C., Godillot, R. and Caussade, B. (2007) The relationship between stream periphyton dynamics and near-bed turbulence in rough open-channel flow. *Ecological Modelling*, **209**(2–4): 78–96.

Lacey, R.W.J. and Roy, A.G. (2007) A comparative study of the turbulent flow field with and without a pebble cluster in a gravel bed river. *Water Resources Research*, **43**: W05502.

Lacey, R.W.J. and Roy, A.G. (2008) Fine-scale characterization of the turbulent shear layer of an instream pebble cluster. *Journal of Hydraulic Engineering*, **134**(7): 925–936.

Lacey, R.W.J., Legendre, P. and Roy, A.G. (2007) Spatial-scale partitioning of in situ turbulent flow data over a pebble cluster in a gravel-bed river. *Water Resources Research*, **43**: W03416.

Lacey, R.W.J., Neary, V.S., Liao, J.C., Enders, E.C. and Tritico, H.M. (2012) The IPOS framework: linking fish swimming performance in altered flows from laboratory experiments to rivers. *River Research and Applications*, **28**(4): 429–443.

Lamarre, H. and Roy, A.G. (2005) Reach scale variability of turbulent flow characteristics in a gravel-bed river. *Geomorphology*, **68**: 95–113.

Lamouroux, N., Poff, N.L. and Angermeier, P.L. (2002) Intercontinental convergence of stream fish community traits along geomorphic and hydraulic gradients. *Ecology*, **83**(7): 1792–1807.

Lamouroux, N., Mérigoux, S., Capra, H., Dolédec, S., Jowett, I.G. and Statzner, B. (2010) The generality of abundance–environment relationships in microhabitats: a comment on Lancaster and Downes (2009) [sic]. *River Research and Applications*, **26**: 915–920.

Lancaster, J. and Downes, B.J. (2010) Linking the hydraulic world of individual organisms to ecological processes: putting ecology into ecohydraulics. *River Research and Applications*, **26**: 385–403.

Lancaster, J. and Hildrew, A.G. (1993) Flow refugia and microdistribution of lotic macroinvertebrates. *Journal of the North American Benthological Society*, **12**: 385–393.

Lane, S.N., Richards, K.S. and Warburton, J. (1993) Comparison of high-frequency velocity records obtained with spherical and discoidal electromagnetic current meters. In Clifford, N.J., French, J.R. and Hardisty, J. (eds) *Turbulence: Perspectives on Flow and Sediment Transport*, John Wiley & Sons, Ltd, Chichester, UK.

Lane, S.N., Biron, P.M., Bradbrook, K.F., Butler, J.B., Chandler, J.H., Crowell, M.D., McLelland, S.J., Richards, K.S. and Roy, A.G. (1998) Three-dimensional measurement of river channel flow processes using acoustic Doppler velocimetry. *Earth Surface Processes and Landforms*, **23**: 1247–1267.

Lawless, M. and Robert, A. (2001a) Scales of boundary resistance in coarse-grained channels: turbulent velocity profiles and implications. *Geomorphology*, **39**: 221–238.

Lawless, M. and Robert, A. (2001b) Three-dimensional flow structure around small-scale bedforms in a simulated gravel-bed environment. *Earth Surface Processes and Landforms*, **26**: 507–522.

Leclerc, M., Capra, H., Valentin, S., Boudreault, A. and Cote, Y. (eds) (1996) *Ecohydraulics 2000: proceedings of the 2nd IAHR symposium on habitat hydraulics, Quebec, Canada.*

Legleiter, C.J., Phelps, T.L. and Wohl, E.E. (2007) Geostatistical analysis of the effects of stage and roughness on reach-scale spatial patterns of velocity and turbulence intensity. *Geomorphology*, **83**: 322–345.

Liao, J.C. (2004) Neuromuscular control of trout swimming on a vortex street: implications for energy economy during the Kármán gait. *Journal of Experimental Biology*, **207**: 3495–3506.

Liao, J.C. (2006) The role of lateral line and vision on body kinematics and hydrodynamics preference of rainbow trout in turbulent flow. *Journal of Experimental Biology*, **209**: 4077–4090.

Liao, J.C. (2007) A review of fish swimming mechanics and behaviour in altered flows. *Philosophical Transactions of the Royal Society B*, **362**(1487): 1973–1993.

Liao, J.C., Beal, D.N., Lauder, G.V. and Triantafyllou, M.S. (2003a) Fish exploiting vortices decrease muscle activity. *Science*, **302**: 1566–1569.

Liao, J.C., Beal, D.N., Lauder, G.V. and Triantafyllou, M.S. (2003b) The Kármán gait: novel body kinematics of rainbow trout swimming in a vortex street. *Journal of Experimental Biology*, **206**: 1059–1073.

Liao, Q., Bootsma, H.A., Xiao, J., Klump, J.V., Hume, A., Long, M.H. and Berg, P. (2009) Development of an in situ underwater particle image velocimetry (UWPIV) system. *Limnology and Oceanography: Methods*, **7**: 169–184.

Liu, Z., Adrian, R.J. and Hanratty, T.J. (2001) Large-scale mode of turbulent channel flow: transport and structure. *Journal of Fluid Mechanics*, **448**: 53–80.

Lu, S.S. and Willmarth, W.W. (1973) Measurements of the structure of Reynolds stress in a turbulent boundary layer. *Journal of Fluid Mechanics*, **60**: 481–511.

Lupandin, A.I. (2005) Effect of flow turbulence on swimming speed of fish. *Biology Bulletin*, **32**(5): 461–466.

Mackay, R.J. (1992) Colonization by lotic macroinvertebrates: a review of processes and patterns. *Canadian Journal of Fisheries and Aquatic Sciences*, **49**: 617–628.

Malmqvist, B. and Sackman, G. (1996) Changing risk of predation for a filter-feeding insect along a current velocity gradient. *Oecologia*, **108**: 450–458.

Matczak, T.Z. and Mackay, R.J. (1990) Territoriality in filter-feeding caddisfly larvae: laboratory experiments. *Journal of the North American Benthological Society*, **9**: 26–34.

McLaughlin, R.L. and Noakes, D.L.G. (1998) Going against the flow: an examination of the propulsive movements made by young brook trout in streams. *Canadian Journal of Fisheries and Aquatic Sciences*, **55**: 853–860.

McLelland, S.J. and Nicholas, A.P. (2000) A new method for evaluating errors in high-frequency ADV measurements. *Hydrological Processes*, **14**: 351–366.

McNair, J.N., Newbold, J.D. and Hart, D.D. (1997) Turbulent transport of suspended particles and dispersing benthic organisms: How long to hit bottom? *Journal of Theoretical Biology*, **188**: 29–52.

Moir, H.J., Soulsby, C. and Youngson, A. (1998) Hydraulic and sedimentary characteristics of habitat utilized by Atlantic salmon for spawning in the Girnock Burn, Scotland. *Fisheries Management and Ecology*, **5**: 241–254.

Moir, H.J., Soulsby, C. and Youngson, A. (2002) Hydraulic and sedimentary control on the availability and use of Atlantic salmon (*Salmo salar*) spawning habitat in the River Dee system, north-east Scotland. *Geomorphology*, **100**: 527–548.

Murray, M. and Simcox, H. (2003) *Use of wild living resources in the United Kingdom – a Review*. UK Committee for IUCN, MGM Environmental Solutions, Edinburgh.

Naden, P.M. and Brayshaw, A.C. (1987) Small and medium scale bedforms in gravel-bed rivers. In Richards, K.S. (ed.) *Institute of British Geographers Special Publication*, Basil Blackwell, Oxford.

Nakagawa, H. and Nezu, I. (1981) Structure of space–time correlations of bursting phenomena in an open-channel flow. *Journal of Fluid Mechanics*, **104**: 1–43.

Nakagawa, H., Nezu, I. and Ueda, H. (1975) Turbulence of open channel flow over smooth and rough beds. *Proceedings of the Japan Society of Civil Engineers*, **241**: 155–168.

Nelson, J.M., Shreve, R.L., McLean, S.R. and Drake, T.G. (1995) Role of near-bed turbulence structure in bed load transport and bed form mechanics. *Water Resources Research*, **31**: 2071–2086.

Newman, B.D., Wilcox, B.P., Archer, A.R., Breshears, D.D., Dahm, C.N., Duffy, C.J., McDowell, N.G., Phillips, F.M., Scanlon, B.R. and Vivoni, E.R. (2006) Ecohydrology of water-limited environments: A scientific vision. *Water Resources Research*, **42**: W06302.

Newson, M.D. and Newson, C.L. (2000) Geomorphology, ecology and river channel habitat: mesoscale approaches to basin-scale challenges. *Progress in Physical Geography*, **24**(2): 195–217.

Nezu, I. (2005) Open-channel flow turbulence and its research prospects in the 21st century. *Journal of Hydraulic Engineering*, **131**: 229–246.

Nezu, I. and Nakagawa, H. (1993) *Turbulence in Open-Channel Flows*, Balkema, The Netherlands.

Nezu, I. and Rodi, W. (1985) Experimental study on secondary currents in open-channel flow. *Proceedings of the 21st International Association for Hydraulic Research Congress*, Volume 2, Delft, The Netherlands.

Nikora, V. (2010) Hydrodynamics of aquatic ecosystems: an interface between ecology, biomechanics and environmental fluid mechanics. *River Research and Applications*, **26**: 367–384.

Nikora, V.I. and Goring, D.G. (1998) ADV measurements of turbulence: can we improve their interpretation? *Journal of Hydraulic Engineering*, **124**(6): 630–634.

Nikora, V.I., Aberle, J., Biggs, J.F., Jowett, I.G. and Sykes, J.R.E. (2003) Effects of fish size, time-to-fatigue and turbulence on swimming performance: a case study of *Galaxias maculates*. *Journal of Fish Biology*, **63**: 1365–1382.

Padmore, C.L. (1997) Biotopes and their hydraulics: a method for defining the physical component of freshwater quality. In Boon, P.J. and Howell, D.L. (eds) *Freshwater Quality: Defining the Indefinable?* The Stationery Office, Edinburgh.

Parsheh, M., Sotiropoulos, F. and Porté-Agel, F. (2010) Estimation of Power Spectra of Acoustic-Doppler Velocimetry Data Contaminated with Intermittent Spikes. *Journal of Hydraulic Engineering*, **136**(6): 368–378.

Pavlov, D.S., Lupandin, A.I. and Skorobogatov, M.A. (2000) The effects of flow turbulence on the behaviour and distribution of fish. *Journal of Icthyology*, **40**(S2): S232–S261.

Peckarsky, B.L., Horn, S.C. and Statzner, B. (1990) Stonefly predation along a hydraulic gradient: a field test of the harsh-benign hypothesis. *Freshwater Biology*, **24**: 181–191.

Peterson, C. and Stevenson, R.J. (1989) Substratum conditioning and diatom colonization in different current regimes. *Journal of Phycology*, **25**: 790–793.

Poff, N.L. and Ward, J.V. (1992) Heterogeneous currents and algal resources mediate in-situ foraging activity of a mobile stream grazer. *Oikos*, **65**: 465–478.

Poff, N.L. and Ward, J.V. (1995) Herbivory under different flow regimes: a field experiment and test of a model with a benthic stream insect. *Oikos*, **71**: 179–188.

Pope, S.B. (2000) *Turbulent Flows*, University Press, Cambridge.

Raffel, M., Willert, C., Wereley, S. and Kompenhans, J. (2007) *Particle Image Velocimetry: A Practical Guide*, 2nd edition, Springer, Berlin.

Reiter, M.A. and Carlson, R.E. (1986) Current velocity in streams and the composition of benthic algal mats. *Canadian Journal of Fisheries and Aquatic Sciences*, **43**: 1156–1162.

Richardson, L.F. (1922) *Weather Prediction by Numerical Process*, University Press, Cambridge.

Robinson, S.K. (1991) Coherent motion in the turbulent boundary layer. *Annual Review of Fluid Mechanics*, **23**: 601–639.

Roy, A.G., Buffin-Bélanger, T., Lamarre, H. and Kirkbride, A.D. (2004) Size, shape and dynamics of large-scale turbulent flow structures in a gravel-bed river. *Journal of Fluid Mechanics*, **500**: 1–27.

Roy, M.L., Roy, A.G. and Legendre, P. (2010) The relations between 'standard' fluvial habitat variables and turbulent flow at multiple scales in morphological units of a gravel-bed river. *River Research and Applications*, **26**: 439–455.

Rusello, P.J., Lohrmann, A., Siegel, E. and Maddux, T. (2006) Improvements in acoustic Doppler velocimetry. *Proceedings of the 7th International Conference on Hydroscience and Engineering*, September 2006, Philadephia, USA.

Russon, I.J., Kemp, P.S. and Calles, O. (2010) Response of downstream migrating adult European eels (*Anguilla anguilla*) to bar racks under experimental conditions. *Ecology of Freshwater Fish*, **19**: 197–205.

Sagnes, P. and Statzner, B. (2009) Hydrodynamic abilities of riverine fish: a functional link between morphology and habitat use. *Aquatic Living Resources*, **22**: 79–91.

Scarsbrook, M.R. and Townsend, C.R. (1993) Stream community structure in relation to spatial and temporal variation: a habitat templet study of two contrasting New Zealand streams. *Freshwater Biology*, **28**: 395–410.

Schlichting, H. (1979) *Boundary-Layer Theory*, 7th edition, McGraw-Hill, New York.

Shvidenko, A.B. and Pender, G. (2001) Macroturbulent structure of open-channel flow over gravel beds. *Water Resources Research*, **37**(3): 709–719.

Silva, A.T., Santos, J.M., Ferreira, M.T., Pinheiro, A.N. and Katopodis, C. (2011) Effects of water velocity and turbulence on the behaviour of Iberian barbel (*Luciobarbus bocagei*, Steindachner 1864) in an experimental pool-type fishway. *River Research and Applications*, **27**: 360–373.

Silvester, N.R. and Sleigh, M.A. (1985) The forces on microorganisms at surfaces in flowing water. *Freshwater Biology*, **15**: 433–448.

Smith, D.L. and Brannon, E.L. (2007) Influence of cover on mean column hydraulic characteristics in small pool riffle morphology streams. *River Research and Applications*, **23**: 125–139.

Smith, D.L., Brannon, E.L. and Odeh, M. (2005) Response of juvenile rainbow trout to turbulence produced by prismatoidal shapes. *Transactions of the American Fisheries Society*, **134**: 741–753.

Smith, D.L., Brannon, E.L., Shafii, B. and Odeh, M. (2006) Use of the average and fluctuating velocity components for estimation of volitional rainbow trout density. *Transactions of the American Fisheries Society*, **135**: 431–441.

Song, T. and Chiew, Y.M. (2001) Turbulence measurement in nonuniform open-channel flow using acoustic Doppler velocimeter (ADV). *Journal of Engineering Mechanics*, **127**(3): 219–233.

Standen, E.M., Hinch, S.G. and Rand, P.S. (2004) Influence of river speed on path selection by migrating adult sockeye salmon (*Oncorhynchus nerka*). *Canadian Journal of Fisheries and Aquatic Sciences*, **61**: 905–912.

Statzner, B. (1981a) The relation between "hydraulic stress" and microdistribution of benthic macroinvertebrates in a lowland running water system, the Schierenseebrooks (north Germany). *Arhiv für Hydrobiologie*, **91**: 192–218.

Statzner, B. (1981b) A method to estimate the population size of benthic macroinvertebrates in streams. *Oecologia*, **51**: 157–161.

Statzner, B., Gore, J.A. and Resh, V.H. (1988) Hydraulic stream ecology: observed patterns and potential applications. *Journal of the North American Benthological Society*, **7**: 307–360.

Stevenson, R.J. (1983) Effects of current and conditions simulating autogenically changing microhabitats on benthic diatom immigration. *Ecology*, **64**: 1514–1524.

Stevenson, R.J. (1996) The stimulation and drag of current. In Stevenson, R.J., Bothwell, M.L. and Lowe, R.L. (eds) *Algal Ecology: Freshwater Benthic Ecosystems*, Academic Press, New York.

Stoecker, D.K., Long, A., Suttles, S.E. and Sanford, L.P. (2006) Effect of small-scale shear on grazing and growth of the dinoflaggelate *Pfiesteria piscicida*. *Harmful Algae*, **5**: 407–418.

Sweeting, R.A. (1994) River pollution. In Calow, P. and Petts, G.E. (eds) *The Rivers Handbook: Hydrological and Ecological Principles*, Volume Two, Blackwell, Oxford, UK.

Taylor, G.I. (1935) Statistical theory of turbulence. *Proceedings of the Royal Society of London, Series A*, **151**: 421–444.

Thompson, R.M. and Lake, P.S. (2010) Reconciling theory and practice: the role of stream ecology. *River Research and Applications*, **26**: 5–14.

Torrence, C. and Compo, G.P. (1998) A practical guide to wavelet analysis. *Bulletin of the American Meteorological Society*, **79**: 61–78.

Tribe, S. and Church, M. (1999) Simulations of cobble structure on a gravel streambed. *Water Resources Research*, **35**(1): 311–318.

Tritico, H.M. and Cotel, A.J. (2010) The effects of turbulent eddies on the stability and critical swimming speed of creek chub (*Semotilus atromaculatus*). *Journal of Experimental Biology*, **213**: 2284–2293.

Tritico, H.M. and Hotchkiss, R.H. (2005) Unobstructed and obstructed turbulent flow in gravel bed rivers. *Journal of Hydraulic Engineering*, **131**(8): 635–645.

Tritico, H.M., Cotel, A.J. and Clarke, J.N. (2007) Development, testing and demonstration of a portable submersible particle imaging velocimetry device. *Measurement Science and Technology*, **18**: 2555–2562.

Venditti, J.G. and Bauer, B.O. (2005) Turbulent flow over a dune: Green River, Colorado. *Earth Surface Processes and Landforms*, **30**(3): 289–304.

Voulgaris, G. and Trowbridge, J.H. (1998) Evaluation of the acoustic Doppler velocimeter (ADV) for turbulence measurements. *Journal of Atmospheric and Oceanic Technology*, **15**: 272–289.

Wadeson, R.A. and Rowntree, K.M. (1998) Application of the hydraulic biotope concept to the classification of instream habitats. *Aquatic Ecosystem Health and Management*, **1**: 143–157.

Waters, T.F. (1972) The drift of stream insects. *Annual Review of Entomology*, **17**: 253–272.

Webb, P.W. (2004) Response latencies to postural differences in three species of teleostean fishes. *Journal of Experimental Biology*, **207**: 955–961.

Webb, P.W. (2006) Stability and maneuverability. In Shadwick, R.E. and Lauder, G.V. (eds) *Fish Physiology*, Elsevier, California.

Webb, P.W. and Cotel, A.J. (2010) Turbulence: does vorticity affect the structure and shape of body and fin propulsors? *Integrative and Comparative Biology*, **50**(6): 1155–1166.

Webb, P.W., Cotel, A. and Meadows, L.A. (2010) Waves and eddies: effects on fish behaviour and habitat distribution. In Domenici, P. and Kapoor, B.G. (eds) *Fish Locomotion: An Eco-Ethological Perspective*, Science Publishers, Enfield, New Hampshire.

Wiley, M.J. and Kohler, S.L. (1980) Positioning changes of mayfly nymphs due to behavioural regulation of oxygen consumption. *Canadian Journal of Zoology*, **58**: 618–622.

Williams, J.J., Thorne, P.D. and Heathershaw, A.D. (1989) Measurements of turbulence in the benthic boundary-layer over a gravel bed. *Sedimentology*, **36**(6): 959–971.

Wilson, M.F. and Halupka, K.C. (1995) Anadromous fish as keystone species in vertebrate communities. *Conservation Biology*, **9**(3): 489–497.

Wood, P.J., Hannah, D.M. and Sadler, J.P. (2007) Ecohydrology and hydroecology: an introduction. In Wood, P.J., Hannah, D.M. and Sadler, J.P. (eds) *Hydroecology and Ecohydrology: Past, Present and Future*, John Wiley & Sons Ltd, Chichester, UK.

Yalin, M.S. (1992) *River Mechanics*, Pergamon, New York.

Zhen-Gang, J. (2008) *Hydrodynamics and Water Quality: Modelling Rivers, Lakes and Estuaries*, Wiley-Interscience, Hoboken, NJ.

3

Hydraulic Modelling Approaches for Ecohydraulic Studies: 3D, 2D, 1D and Non-Numerical Models

Daniele Tonina[1] and Klaus Jorde[2]

[1]Center for Ecohydraulics Research, University of Idaho, 322 E Front Street, Suite 340, Boise, ID 83702, USA
[2]KJ Consulting/SJE Schneider & Jorde Ecological Engineering, Gnesau 11, A-9563 Gnesau, Austria

3.1 Introduction

Ecohydraulics (Leclerc et al., 1996) as well as ecohydrology (Richter et al., 1996), hydromorphology (Orr et al., 2008) and hydrodynamics of aquatic ecosystems (Nikora, 2010) is rooted in the river continuum concept, which establishes a connection between abiotic processes and the biotic environment (Vannote et al., 1980). They stem from the principle that the structure and function of biological communities, which define aquatic ecosystems, depend on the interplay between biological, physical and chemical processes in aquatic environments, such as rivers, lakes, estuaries and seas (Frissell et al., 1986; Leclerc et al., 1996; Osmundson et al., 2002; Statzner, 2008). Ecohydraulics is a bottom up method and predicts the status of the ecosystem based on knowledge of the physical environment coupled with biological requirements (Newson and Newson, 2000). It investigates the effects of physical properties (such as flow velocity, depth, shear stress, turbulence and temperature) on ecosystems (Bovee, 1982; Capra et al., 1995; Roy et al., 2009; Lancaster and Downes, 2010;) as well as the influence of organism community structure on the physical environment (Statzner et al., 1996, 1999, 2000, 2003; Hassan et al., 2008; Tonina and Buffington, 2009). Consequently, several approaches for characterizing the physical habitat have been pro-

posed. They include hydrologic (Richter et al., 1996, 1997; Poff et al., 1997; Poff, 2002), geomorphologic (Kondolf, 1995; Payne and Lapointe, 1997; Baxter and Hauer, 2000; Buffington et al., 2004; Orr et al., 2008) and hydraulic (Benjankar et al., 2013; Bovee, 1982; Kondolf et al., 2000; Pasternack et al., 2004; Tonina et al., 2011) methods. For example, aquatic habitat modelling has, for a number of years, been used to provide criteria for riverine habitat enhancement, rehabilitation and restoration. It couples flow properties, such as flow velocity, depth and shear stress, with biological requirements, typically expressed as univariate curves (Bjornn and Reiser, 1991; Payne and Allen, 2009), to define habitat availability or to quantify flow-related ecological functions (Escobar-Arias and Pasternack, 2010). Stream flow properties, in one, two or three dimensions, can be obtained with field measurements, statistical methods (Lamouroux et al., 1995; 1998), analytical solutions (Brown and Pasternack, 2009) and numerical hydrodynamic models (Daraio et al., 2010; Tonina and McKean, 2010; Pasternack and Senter, 2011). The choice of a method for characterizing flow properties depends on several factors, which include the processes to be modelled, the questions to be answered and the temporal and spatial scales of the process and its resolution. For instance, traditional application of the instream flow incremental methodology (IFIM) (Bovee, 1978, 1982; Bovee et al., 1998) only uses the statistical

Ecohydraulics: An Integrated Approach, First Edition. Edited by Ian Maddock, Atle Harby, Paul Kemp and Paul Wood.
© 2013 John Wiley & Sons, Ltd. Published 2013 by John Wiley & Sons, Ltd.

distribution of the physical properties (e.g. the fraction of the aquatic habitat with flow properties such as depth and velocity within certain ranges adequate for a given species and a given life stage (Payne and Allen, 2009)) and does not use their spatial distribution to generate flow–habitat relationships such as the weighted usable area (WUA) (Bovee, 1978; Bovee et al., 1998; Payne, 2003). Consequently, spatially explicit flow models, such as two- or three-dimensional hydraulic models, may not be necessary if field measurements at several sampling points in a river or analytical or statistical flow models can provide the flow field statistics (Lamouroux et al., 1998). Conversely, new advances on the riverscape concept emphasize the use of spatially varying flow characteristics (Fausch et al., 2002). New stream fish management methods require explicit habitat mapping (Le Pichon et al., 2006; 2009). Fish behaviour modelling around hydraulic structures (Nestler et al., 2008), organism dispersal and food drifting may require spatial knowledge of the flow field (Hayes et al., 2007; Daraio et al., 2010). Additionally, spatially explicit models have been shown to provide a more accurate statistic of habitat quality, such as WUA values, than 1D or non-numerical modelling for aquatic habitat modelling (e.g. Brown and Pasternack, 2009).

Here, we confine our scope to hydraulic modelling for predicting flow properties used in ecohydraulics, in its broader sense, which encompasses aquatic habitat modelling, ecological assessment and the interaction between flow and organisms. The aim of this chapter is to provide the reader with the basic knowledge and terminology on hydraulic modelling. In particular, we focus on computational fluid dynamics (CFD) modelling and its application in ecohydraulics. This chapter describes the fundamental equations of CFD, defines their salient points and refers the reader to appropriate literature for detailed information on discretization (e.g. Peiró and Sherwin, 2005; Hirsch, 2007), numerical techniques (e.g. Ferziger and Peric, 2002), and turbulence closure (e.g. Rodi, 2000).

3.2 Types of hydraulic modelling

Although the basic mechanical principles of river hydraulics are well established, the complexity of flows in natural streams, whose beds are often permeable, mobile and display irregularity at different scales ranging from grain size (few millimetres) to the distance between meanders (which could be several hundred metres), has precluded the development of analytical solutions, except under idealized and simplified conditions (Chaudhry, 1993). Consequently, computational fluid

dynamics (CFD) has emerged as an alternative technique to provide quantitative spatial and temporal predictions of flow properties in natural streams (Leclerc et al., 1995; Hardy, 1998; Lane and Bates, 1998; Crowder and Diplas, 2000b; Lacey and Millar, 2004; Rice et al., 2010; Tonina and McKean, 2010; Tonina et al., 2011).

The application of CFD in ecohydraulics has been increasingly supported by advances in data-acquisition methods (e.g. McKean et al., 2009), the development of efficient and robust numerical codes (e.g. FaSTMECH, SRH-2D and Delft3D) and the availability of fast and affordable computers (Hardy, 1998). Here, we divide the methods for predicting flow properties into four categories: (1) three-dimensional (3D) hydraulic models, (2) two-dimensional (2D) hydraulic models, (3) one-dimensional (1D) hydraulic models and (4) non-numerical hydraulic modelling. The first three of these categories are based on spatial information of the hydraulic modelling in three, two and one dimensions respectively. 3D hydraulic models represent flow properties in the three-directions, longitudinal (downstream direction along x) transversal (along y) and vertical (along z). 2D models may represent flow properties in two dimensions: along the longitudinal and the transversal or the longitudinal and the vertical directions. 1D hydraulic models provide flow properties only in the downstream direction. The fourth category is comprised of non-numerical hydraulic modelling, which includes field measurements, analytical solutions and statistical analysis of the flow field.

1D through 3D models solve momentum (or energy) and conservation of mass equations. 3D modelling is still rare in ecohydraulic studies and few examples are available (Weber et al., 2006; Shen and Diplas, 2008; Daraio et al., 2010). This is because of the lack of commensurate ecological and geomorphic understanding of processes at the ecological scale, to make use of the three dimensions in model results, and the high computational requirements still needed. However, new methods for interpreting 3D model results, such as the weighted usable volume, are emerging (Mouton et al., 2007). 3D models may provide important information on spatial flow property variations, especially at small scales (e.g. grain diameter or particle clusters) and when dealing with vegetation (Wilson et al., 2006; Rice et al., 2009; Nikora, 2010). Furthermore, reducing 3D model results down to 2D yields better predictions than using a 2D model directly (Lane et al., 1999; MacWilliams et al., 2006).

In ecohydraulics, 2D hydraulic modelling is currently a more affordable approach compared to 3D modelling because of its computational running time requirements

(Leclerc *et al.*, 1995; Ghanem *et al.*, 1996; Crowder and Diplas, 2000a, 2000b, 2002; Shen and Diplas, 2008). It provides spatial and temporal variations of flow properties along the transversal and the longitudinal direction at different ecological scales (e.g. 10^{-1} to 10 m^2). It has also been suggested that 2D modelling can be sufficient for most ecohydraulics applications (Rodi *et al.*, 1981). For instance, 2D modelling could be applied at the 1 m^2 scale, which is comparable with the fish microhabitat scale (Grant and Kramer, 1990; Leclerc *et al.*, 1995; Pasternack and Senter, 2011).

1D hydraulic models provide cross-sectional averaged velocity distribution and water surface elevation at selected longitudinal locations along a river. They have also been used to provide flow information for aquatic habitat models such as the Physical Habitat Simulation Model (PHABSIM) (USGS, 2001) and the Computer Aided Simulation Model for Instream Flow Requirements (CASiMiR) (Schneider *et al.*, 2008). 1D models do not provide the transversal structure of downstream-directed flow properties, let alone cross-stream or vertically directed components of flow properties (Brown and Pasternack, 2009). The first of these properties can be approximated from 1D model solutions using streambed bathymetry and simple engineering rules such as local application of the uniform flow law (e.g. Manning's equation) along the transversal direction (e.g. Ghanem *et al.*, 1996; Schneider *et al.*, 2008), but the others cannot be obtained without explicit 2D or 3D modelling. Compared to 2D and 3D modelling, which use continuous bathymetry to describe the streambed topography, 1D modelling uses discrete river cross-sections. Transects sample the bathymetry at discrete locations, typically several channel widths apart (Pasternack and Senter, 2011). Cross-section location and spacing should depend on the ecohydraulics questions under investigation because model results and accuracy depend on transect position (Samuels, 1990; Payne *et al.*, 2004; Williams, 2010; Ayllón *et al.*, 2011). Investigations show that 1D and 2D models may provide comparable cross-sectional averaged properties when the topography mostly varies longitudinally but they diverge when transversal topographical features are present, causing large flow gradients (Brown and Pasternack, 2009).

Non-numerical hydraulic modelling relies on field measurements (Stone, 2005; Moir and Pasternack, 2009; Whited *et al.*, 2011), which includes the so-called zero models (Parasiewicz and Dunbar, 2001), analytical solutions (e.g. uniform flow equation) (Brown and Pasternack, 2009) and statistical descriptions (Lamouroux *et al.*, 1998) of the flow field. These methods

are traditionally based on cross-sectional measurements. Field measurements of flow quantities are taken at discrete representative locations and at different discharges. Hydraulic parameters are interpolated spatially and between discharges. This limits their utility in complex streams (e.g. braided, pool-riffle and meandering reaches) within the range of measured discharges and within the spatial limits of their surveys. Similar limitations apply to statistical models, which require field data to evaluate their parameters (Lamouroux *et al.*, 1995; 1998). Conversely, analytical solutions can be applied in different streams once the appropriate parameters are selected (Brown and Pasternack, 2009). Non-numerical methods, especially empirical models, have been used extensively with PHABSIM (USGS, 2001).

Although applications of CFD modelling are based on well-defined procedures (Anderson and Woessner, 1992), the choice of a CFD model depends upon the project definition, the modeller's knowledge of the study site, the modeller's understanding of flow hydraulics and of the mathematical and numerical behaviour of the model. Figure 3.1 modifies Anderson and Woessner's (1992) procedure and shows the major choices required in setting up a CFD model. In the following sections, we will introduce the fundamental topics and concepts of numerical modelling (Section 3.3), then we will describe the features of and differences among hydraulic model approaches (Sections 3.4, 3.5, 3.6, 3.7 and 3.8) and the major elements needed to set up numerical models (Sections 3.4.1, 3.5.1 and 3.6.1). We will then present four case studies (Section 3.9) followed by a summary highlighting the key points (Section 3.10).

3.3 Elements of numerical hydrodynamic modelling

Stream flows may be categorized depending on spatial and temporal variations of flow properties (Table 3.1). Non-uniform, steady flows are the most common flow type in ecohydraulic applications. Unsteady flows are typically modelled as a succession of local steady-state flows (Nelson and Smith, 1989), because hydrograph time scales are typically large enough that local accelerations may be neglected.

3.3.1 Mathematical model
Mass–momentum equations, known as Navier–Stokes (NS) equations, and energy equations for uncompressible, homogenous and Newtonian fluids (linear relationship between shear stress and strain rate, i.e. the velocity gradient perpendicular to the direction of shear) with negligible

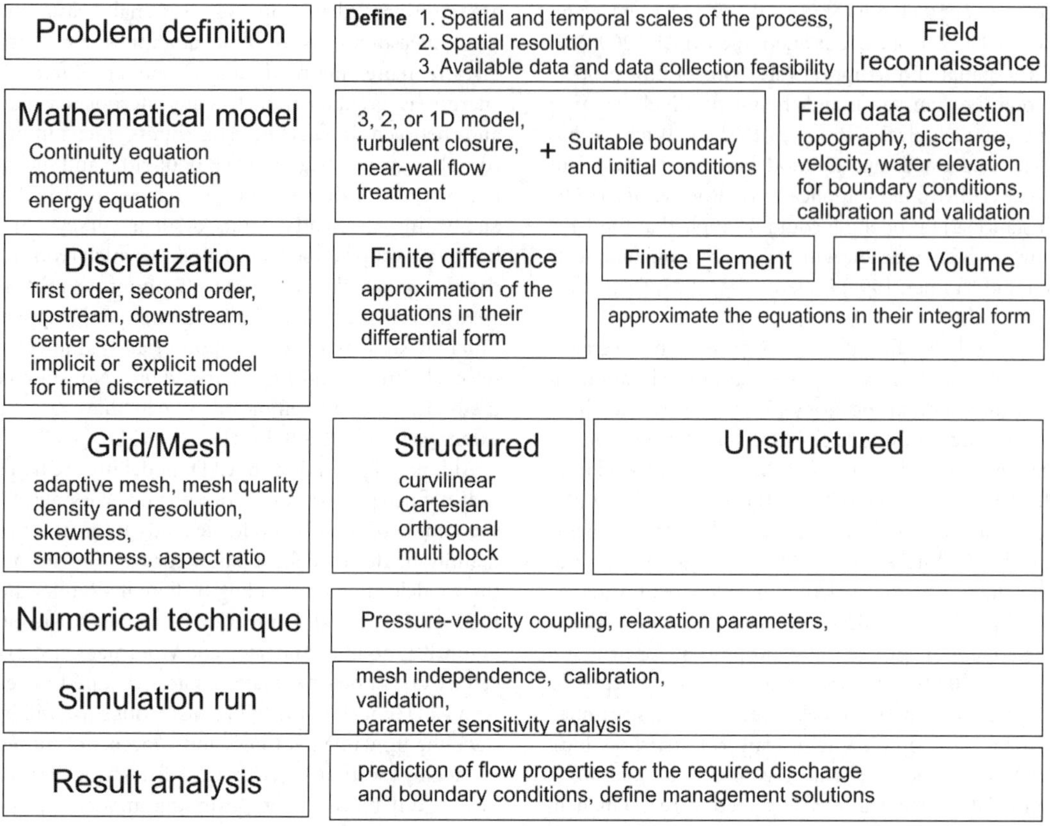

Problem definition	Define 1. Spatial and temporal scales of the process, 2. Spatial resolution 3. Available data and data collection feasibility	Field reconnaissance

Mathematical model Continuity equation momentum equation energy equation	3, 2, or 1D model, turbulent closure, + Suitable boundary near-wall flow and initial conditions treatment	Field data collection topography, discharge, velocity, water elevation for boundary conditions, calibration and validation

Discretization first order, second order, upstream, downstream, center scheme implicit or explicit model for time discretization	Finite difference approximation of the equations in their differential form	Finite Element	Finite Volume
		approximate the equations in their integral form	

Grid/Mesh adaptive mesh, mesh quality density and resolution, skewness, smoothness, aspect ratio	Structured curvilinear Cartesian orthogonal multi block	Unstructured

Numerical technique	Pressure-velocity coupling, relaxation parameters,

Simulation run	mesh independence, calibration, validation, parameter sensitivity analysis

Result analysis	prediction of flow properties for the required discharge and boundary conditions, define management solutions

Figure 3.1 Aspects to be considered when selecting a numerical model for predicting open channel flow hydraulics.

Table 3.1 Flow type classifications and examples.

Variation in time	Variation in space	Mode	Examples
1. **Steady** Flow parameters do not change in time $\frac{\partial}{\partial t} = 0$	a. **Uniform** Flow parameters do not change in downstream direction $\frac{\partial}{\partial x} = 0$		**1.a.**: flow in a smooth trapezoidal channel with constant area and discharge **1.b.**: constant discharge in natural channels with bed form and width variations **1.b.I**: localized process, hydraulic jumps, bridge piers **1.b.II**: backwater
2. **Unsteady** Flow parameters change in time $\frac{\partial}{\partial t} \neq 0$	b. **Non-uniform** Flow parameters change in downstream direction $\frac{\partial}{\partial x} \neq 0$	I. **Rapidly** $\frac{\partial}{\partial x}$ or $\frac{\partial}{\partial t}$ have high values II. **Gradually** $\frac{\partial}{\partial x}$ or $\frac{\partial}{\partial t}$ have low values	**2.a.**: not occurring in natural rivers, rare and only in full pipes with sudden flow variations **2.b.**: flows with varying hydrographs in natural streams **2.b.I.**: flash floods, hydro-peaking **2.b.II.**: snow melt hydrograph

Coriolis forces (apparent forces due to Earth's rotation) are the mathematical basis for open channel CFD (Johnson, 1998). Water behaves as a homogeneous and Newtonian fluid in most riverine applications without strong stratifications (Hodskinson, 1996). The uncompressible assumption is appropriate for flows with Mach numbers (the ratio between flow velocity and the speed of sound) less than 0.3 (Anderson, 1995). With these assumptions, the continuity and NS equations in the Cartesian coordinate system have the following form:

$$\frac{\partial u^*}{\partial x} + \frac{\partial v^*}{\partial y} + \frac{\partial w^*}{\partial z} = 0 \qquad a.$$

$$\frac{\partial u^*}{\partial t} + \frac{\partial u^{*2}}{\partial x} + \frac{\partial u^* v^*}{\partial y} + \frac{\partial u^* w^*}{\partial z} + \frac{1}{\rho}\frac{\partial P^*}{\partial x}$$
$$-g_x - \frac{\mu}{\rho}\left(\frac{\partial^2 u^*}{\partial x^2} + \frac{\partial^2 u^*}{\partial y^2} + \frac{\partial^2 u^*}{\partial z^2}\right) = 0 \qquad b.$$

$$\frac{\partial v^*}{\partial t} + \frac{\partial u^* v^*}{\partial x} + \frac{\partial v^{*2}}{\partial y} + \frac{\partial v^* w^*}{\partial z} + \frac{1}{\rho}\frac{\partial P^*}{\partial y} \qquad (3.1)$$
$$-g_y - \frac{\mu}{\rho}\left(\frac{\partial^2 v^*}{\partial x^2} + \frac{\partial^2 v^*}{\partial y^2} + \frac{\partial^2 v^*}{\partial z^2}\right) = 0 \qquad c.$$

$$\frac{\partial w^*}{\partial t} + \frac{\partial w^* u^*}{\partial x} + \frac{\partial w^* v^*}{\partial y} + \frac{\partial w^{*2}}{\partial z} + \frac{1}{\rho}\frac{\partial P^*}{\partial z}$$
$$-g_z - \frac{\mu}{\rho}\left(\frac{\partial^2 w^*}{\partial x^2} + \frac{\partial^2 w^*}{\partial y^2} + \frac{\partial^2 w^*}{\partial z^2}\right) = 0 \qquad d.$$

where u^*, v^* and w^* are the instantaneous velocity components along the x, y and z directions, ρ is the water density, P^* is the instantaneous pressure, g is the acceleration of gravity with its components g_x, g_y and g_z along the x, y and z directions and μ is the fluid dynamic viscosity. The values of g_x, g_y and g_z depend on the orientation of the gravitational acceleration with respect to the coordinate system used and of the streambed surface; g is typically negative (downward direction) along the vertical axis z. A more common term is the kinematic viscosity, v, which is the ratio between the dynamic viscosity and the density of the fluid, $v = \mu/p$. This system of equations is a well-posed problem of four unknowns (u^*, v^*, w^* and P^*) and four equations (conservation of mass or continuity equation (Equation (3.1a)) and the momentum equations along the x (Equation (3.1b)), y (Equation (3.1c)) and z (Equation (3.1d)) directions). However, its solution is extremely difficult in natural streams due to the limitations in resolving the temporal and spatial fluctuations of the turbulence (see Section 3.3.2). Resolution of temporal fluctuations requires unsteady flows with time steps comparable to the smallest turbulent fluctuations, which could be tens of fluctuations per second (as fast as 200 Hz) (Soulsby, 1980; Lane, 1998; Rodi, 2000; Buffin-Bélanger and Roy, 2005; Köse, 2011). Resolution of the spatial fluctuations requires a mesh size of the computational grid smaller than the smallest tur-

bulent eddy size, which could be as small as 10^{-2} mm (Hervouet and Van Haren, 1996). Turbulence causes irregular, chaotic and random velocity and pressure fluctuations at very high frequencies (Soulsby, 1980; Lane, 1998; Rodi, 2000; Buffin-Bélanger and Roy, 2005; Köse, 2011) and it is a property of any rotational, dissipative and diffusive flows with high Reynolds numbers. It is a diffusive (spreads momentum, mass and energy within the flow) and dissipative (consumes energy) process, which converts kinetic energy to heat by viscosity through a cascade of vortices. Direct numerical simulation (DNS) solves the equation system (3.1) at the smallest spatio-temporal scales of turbulence at the expense of extremely intensive computational and memory requirements available only on cluster computing (Moin and Mahesh, 1998). However, most engineering and ecohydraulic applications are not interested in the solution of the turbulent fluctuations, but rather in mean flow properties (Lane, 1998). Consequently, Equations (3.1) are averaged over the turbulence time scale (symbol < > means averaging over the turbulence time scale or turbulence ensemble) with the dependent variables (u^*, v^*, w^* and P^*) written as Reynolds decompositions. Reynolds decompositions express a variable as the sum of its turbulent ensemble mean (e.g. u) and its fluctuations denoted by primes (') (e.g. $u^* = u + u'$) (Tennekes and Lumley, 1972). After the application of the Reynolds decomposition and performing the ensemble averaging (e.g. $<u^*> = u + <u'> = u$ and $<u^* v^*> = uv + <u'v'>$), the NS equations result in a new set of equations known as the Reynolds averaged Navier–Stokes (RANS) equations (Ferziger and Peric, 2002):

$$\frac{\partial u}{\partial x} + \frac{\partial v}{\partial y} + \frac{\partial w}{\partial z} = 0 \qquad a.$$

$$\frac{\partial u}{\partial t} + \frac{\partial u^2}{\partial x} + \frac{\partial uv}{\partial y} + \frac{\partial uw}{\partial z} + \frac{1}{\rho}\frac{\partial P}{\partial x} - g_x$$
$$-\frac{\mu}{\rho}\left(\frac{\partial^2 u}{\partial x^2} + \frac{\partial^2 u}{\partial y^2} + \frac{\partial^2 u}{\partial z^2}\right) + \frac{\partial <u'u'>}{\partial x}$$
$$+\frac{<u'v'>}{\partial y} + \frac{<u'w'>}{\partial z} = 0 \qquad b.$$

$$\frac{\partial v}{\partial t} + \frac{\partial uv}{\partial x} + \frac{\partial v^2}{\partial y} + \frac{\partial vw}{\partial z} + \frac{1}{\rho}\frac{\partial P}{\partial y} - g_y$$
$$-\frac{\mu}{\rho}\left(\frac{\partial^2 v}{\partial x^2} + \frac{\partial^2 v}{\partial y^2} + \frac{\partial^2 v}{\partial z^2}\right) + \frac{\partial <v'u'>}{\partial x} \qquad (3.2)$$
$$+\frac{<v'v'>}{\partial y} + \frac{<v'w'>}{\partial z} = 0 \qquad c.$$

$$\frac{\partial w}{\partial t} + \frac{\partial wu}{\partial x} + \frac{\partial wv}{\partial y} + \frac{\partial w^2}{\partial z} + \frac{1}{\rho}\frac{\partial P}{\partial z} - g_z$$
$$-\frac{\mu}{\rho}\left(\frac{\partial^2 w}{\partial x^2} + \frac{\partial^2 w}{\partial y^2} + \frac{\partial^2 w}{\partial z^2}\right) + \frac{\partial <w'u'>}{\partial x}$$
$$+\frac{<w'v'>}{\partial y} + \frac{<w'w'>}{\partial z} = 0 \qquad d.$$

RANS equations present nine new terms, which symmetry reduces to six new unknowns called Reynolds stresses:

$$\begin{aligned}
\tau_{xx} &= - < u'u' > \\
\tau_{yy} &= - < v'v' > \\
\tau_{zz} &= - < w'w' > \\
\tau_{xy} &= \tau_{yx} = - < u'v' >= - < v'u' > \\
\tau_{xz} &= \tau_{zx} = - < u'w' >= - < w'u' > \\
\tau_{yz} &= \tau_{zy} = - < v'w' >= - < w'v' >
\end{aligned}$$

(3.3)

These six terms are the dominant resistance factors in turbulent flows and their study is the focus of the turbulence closure problem, which models them via flow ensemble quantities, which are u, v and w (Rodi, 2000).

Models for the turbulence closure need to be accurate, applicable on a wide range of flow conditions, simple and computationally economical (ASCE Task Committee on Turbulence Models in Hydraulics Computations, 1988). Several turbulence models are available, ranging from simple zero-equation models, based on turbulent mixing length and a representative velocity to the Large Eddy Simulations (LES) (Ingham and Ma, 2005) (Table 3.2). The latter resolve the turbulence fluctuations over a scale that ranges from the domain size (with the large eddy sizes comparable to water depth or channel width) to a cut-off filter size, which is typically larger than the computational grid size (Sagaut, 2006). Turbulence effects at scales smaller than the cut-off filter size are not modelled directly, but they are parameterized with an opportune sub-grid turbulence model.

Many CFD models for ecohydraulic applications adopt the Boussinesq hypothesis, which models turbulence as if it were a viscous process characterized by an eddy or turbulent viscosity, v_T:

$$\tau_{ij} = -\rho < u_i u_j >= \rho v_T \left(\frac{\partial u_i}{\partial x_j} + \frac{\partial u_j}{\partial x_i} \right)$$

$$-\frac{2}{3}\rho k\delta_{ij} = \rho v_T S_{ij} - \frac{2}{3}\rho k\delta_{ij} \quad \text{with } i, j = x, y, z$$

(3.4)

which is a property of flow field. Consequently, v_T is heterogeneous (varies spatially) and anisotropic (varies directionally). In Equation (3.4), k is the turbulence kinematic energy ($k = <u_i^2>/2$) and δ_{ij} is Kronecker's delta ($\delta_{ij} = 1$ if $i = j$ and 0 otherwise) (Sotiropoulos, 2005). Several methods (e.g. zero-equation, $k-\varepsilon$ and $k-\omega$) are available to quantify the value of v_T and their performance in predicting accurate flow fields depends on the type and complexity of the flow field itself (Miller and Cluer, 1998; Wilson et al., 2002; Vionnet et al., 2004; van Balen et al., 2010; Papanicolaou et al., 2011). The zero-equation

model assumes that turbulence dissipation occurs at the same location where it is generated. Thus, the definition of 'zero-equation' stems from the fact that no partial differential equations (PDEs) are solved for the transport of turbulence. It quantifies the eddy viscosity with a relationship based on local length and velocity scales:

$$v_T(x, y, z_*) = \kappa\zeta(z_*) u (x, y)_* h (x, y)$$

(3.5)

where κ is the von Karman constant (typically equal to 0.408), u_* is the shear velocity $u_* = \sqrt{\tau_b/\rho}$ ($\tau_b = \sqrt{\tau_{xb}^2 + \tau_{yb}^2}$ is the total shear stress at the channel bottom), h is the water depth and $\zeta(z_*)$ is the shape function of the vertical turbulence profile with z_* the dimensionless vertical coordinate ($z_* = z/h$). Commonly used shapes for the shape function are the parabolic vertical profile and a combined parabolic and constant value for the lower 20% and upper 80% of the depth, respectively (Smith and McLean, 1984). For 2D simulations, the eddy viscosity can be specified as a constant for the entire domain as a function of the reach-averaged u_* and h or spatially variable as a function of local $u(x,y)_*$ and $h(x,y)$ (Wilson et al., 2002; Papanicolaou et al., 2011). The turbulence viscosity can also be defined from the mixing length and a set of velocity gradients, which depends on model closures (such as Prandtl's mixing length and Smagorinsky's sub-grid eddy viscosity) (Smagorinsky, 1963; Absi, 2006). The mixing length, which is $l = \kappa z$ near the wall, is the distance that a turbulent eddy can, on average, influence the flow field from a wall.

Other models account for the transport (advection and diffusion) of turbulence properties, which results in more complexity and computational requirements. The standard $k-\varepsilon$ (Harlow and Nakayama, 1968) and its variations (RNG (Yakhot and Orszag, 1986) and Realizable (Shih et al., 1995)) and standard $k-\omega$ (Wilcox, 1988) and its variations (modified (Wilcox, 2006) and SST (Menter, 1994)) models are two-equation turbulence closures. They account for the transport of the turbulent kinetic energy k and the turbulence dissipation ε or the specific turbulence dissipation ω via two additional PDEs. The number of additional PDEs increases with the complexity of turbulence closure, as does the number of associated parameters that are required. The zero-equation model is probably the most commonly used turbulence closure for 2D models (Miller and Cluer, 1998; Wheaton et al., 2004; Pasternack et al., 2006; Tonina and McKean, 2010; Tonina et al., 2011), whereas, the two-equation closure is more common for 3D applications (Hodskinson, 1996; Hodskinson and Ferguson, 1998; Nicholas and Sambrook Smith, 1999; Nicholas, 2001; Daraio et al., 2010).

Table 3.2 Turbulence modelling (Lane, 1998; Lane *et al.*, 1999; Rodi, 2000; Fluent Inc., 2003; Ingham and Ma, 2005; Sotiropoulos, 2005; Wilcox, 2006).

		Model	Strength	Weakness
RANS		0-equation	Eddie viscosity ν_T	Economical, fast calculation, Good for flow dominated by simple boundary layer turbulence where turbulence length and velocity scales are known, such as in shallow rivers, where turbulence is primarily generated by the streambed	Do not account for turbulence history and transport. Provide good predictions of water surface elevation and vertically averaged velocity magnitude in rivers with large width:depth ratio (larger than 5) (Nuze *et al.*, 1993) but away from confluences. Inadequate where there are multiple turbulence length scales, flows with separations and circulations.
		1-equation Spalart-Allmaras, Prandtl, Baldwin-Barth	Solve 1 extra PDE (advection dispersion) for ν_T	Economical, flow with mild separation	Inadequate for flow with moderate and high separation
		2-equation 2 extra PDE, 1 for k and one for ε or ω	$k-\varepsilon$ standard	Simple, good for many simple flows	Poor for circulation, eddy, and for flow with strong separation and with high strain
			$k-\varepsilon$ RNG	Improved for moderate flow separation swirl and secondary flow. Applied in river modelling	Problems linked to isotropic turbulence and in estimating round jet flows
			$k-\varepsilon$ Realizable	Like $k-\varepsilon$ RNG but better for round jet flows	Problems linked to isotropic turbulence
			$k-\omega$ Standard	Simple, good for many simple flows	Poor for circulation, eddy, and for flow with strong separation and with high strain and free surface
			$SST\ k-\omega$	As $k-\omega$ but less sensitive to inlet free surface boundary, shows promising application in river modelling	Resolved the problem of free surface of the standard $k-\omega$ and improved description of circulation of eddy
	NS	RSM	7 additional equations	Anisotropic turbulence	High CPU effort
	NS		LES	Solve the turbulence at the large scale and parameterize it at the sub-grid level	Very high CPU effort
			DNS	Solve directly the NS and continuity equation	Extremely high CPU effort

Increasing Complexity (arrow, left margin)

Isotropic turbulence / *Anisotropic turbulence* (right margin)

The Reynolds stress model (RSM) is a seven-equation closure, which does not use the eddy viscosity approach, but instead directly solves the six Reynolds stresses plus an equation for the turbulent dissipation (Rodi, 2000). This model avoids the isotropic eddy viscosity assumption of the previous models (Shen and Diplas, 2008). It performs well in flow with high curvature and flow separation. It reduces to the $k-\varepsilon$ model for low-strain flows (Rodi, 2000).

The work of van Balen *et al.* (2010) shows evidence that turbulence is mostly isotropic in open channels with shallow water (width to depth ratio larger than 5 (Nuze *et al.*, 1993) but see Sukhodolov *et al.*, (1998)), which supports the use of zero- or two-equation closures that assume isotropic turbulence. Their findings for a flume experiment with complex topography also show that momentum transport is affected mainly by the complex topography and bend curvature and only secondarily by turbulence momentum transport. This result supports and explains the observations that zero-equation models perform similarly to two-equation models in 2D simulations of rivers (Section 3.5) (Tingsanchali and Maheswaran, 1990; Simões and Wang, 1997; Lane and Richards, 1998; Wilson et al., 2002; Papanicolaou et al., 2011). The zero-model with spatially variable eddy viscosity (e.g. dependence on local u_* and h) performs better than that with reach-scale constant ν_T. However, the latter has been reported to perform reasonably well in predicting mean water surface elevation and magnitude of the depth-averaged velocity in natural rivers (Lane et al., 1999; MacWilliams et al., 2006; Pasternack et al., 2006). These observations support the conclusion that turbulence closure uncertainties may be less important than those of stream topography, boundary condition information and mesh quality (USACE, 1996; Lane, 1998; Lane et al., 2004) to provide accurate modelling results.

Once the mathematical model is developed, its equations are transformed into a set of algebraic equations via a discretization technique. Numerical modelling requires three types of discretization: (1) equation discretization (e.g. finite element, finite differences and finite volume), (2) physical domain discretization (e.g. structured, unstructured and block structured meshes) and (3) time discretization (e.g. implicit and explicit methods), which are explored in the following sections.

3.3.2 Discretization methods

Discretization of a closed-form mathematical equation, such as the continuity equation, is the essence of CFD (Anderson, 1995; Peiró and Sherwin, 2005; Hirsch, 2007). Discretization is the mathematical process that transforms the integral or the differential form of a partial differential equation (PDE) into an equivalent but different set of algebraic equations, which provide results at a finite number of points within the numerical domain.

$$\text{a. } \underbrace{\frac{\partial \rho}{\partial t} + \nabla \cdot (\rho \mathbf{V}) = 0}_{\text{differential form}}, \quad \text{b. } \underbrace{\frac{\partial}{\partial t} \iiint_{\Psi} \rho d\Psi + \iint_{S} \rho \mathbf{V} \cdot dS = 0}_{\text{integral form}}$$

$$(3.6)$$

Equations (3.6a) and (3.6b) are the differential and integral forms of the conservation of mass equation respectively, where $\mathbf{V}$ is the velocity vector, Ψ is the control volume and S is the surface boundary of the control volume. The most popular CFD discretization approaches are the finite difference (FDM), finite element (FEM) and finite volume (FVM). The FVM discretization approach is

the most popular in CFD for 3D modelling and is used in almost 85% of models (Ferziger and Peric, 2002). Other methods, such as smooth-particle hydrodynamics (SPH) (Monaghan, 1988; Kristof et al., 2009), boundary element (Cheng and Cheng, 2005; Ang, 2007), spectral methods (Hussaini and Zang, 1987) and vorticity-based methods (Leonard, 1980; Qian and Vezza, 2001) are usually applied for specific applications and are rarely used in ecohydraulics. However, SPH has recently seen increased attention in fluid mechanics because of its benefits in intrinsically conserving mass, computing pressure and defining the water surface elevation. Among the CFD models, it has also the advantage of being mesh free. SPH models solve the flow field in a Langrangian framework by dividing the fluid into a large, but finite, number of particles and by tracking their paths within the domain. Fluid and flow properties are derived by the composite properties of the fluid particles within a certain control volume.

FDM is the oldest discretization method and approximates the partial differential terms via algebraic differences (Peiró and Sherwin, 2005). These approximations are typically derived from Taylor's series expansions, as in the following example where the value of u at point $(i+1, j)$ is predicted from its value at point (i, j)

$$u_{i+1,j} = \underbrace{u_{i,j} + \left.\frac{\partial u}{\partial x}\right|_{i,j} \Delta x}_{\text{linearization term}} + \underbrace{\left.\frac{\partial^2 u}{\partial x^2}\right|_{i,j} \frac{(\Delta x)^2}{2} + \left.\frac{\partial^3 u}{\partial x^3}\right|_{i,j} \frac{(\Delta x)^3}{6} + O(\Delta x)^3}_{\text{truncation error}}$$

$$\Downarrow$$

$$\left.\frac{\partial u}{\partial x}\right|_{i,j} = \frac{u_{i+1,j} - u_{i,j}}{\Delta x} + \varepsilon_T$$

$$(3.7)$$

where Δx is the spacing between the two points and ε_T is the truncation error. The first term on the right side is the finite difference form of the partial differential equation with first order accuracy, because the truncation occurs at the first power of Δx.

Second order accuracy would have Δx at the second power and so forth. Note that only information on the right side of the point (i, j) approximates the derivative and this is referred to as the forward or downwind difference (Figure 3.2). Discretization may involve only left-side information, which leads to rearward, backward or upwind differences, or both sides, leading to central differences (Figure 3.2). Approximations of the PDE at higher orders of accuracy can be obtained with information from additional neighbouring points.

Application of the forward and backward difference to approximate the time derivative leads to two different schemes: the explicit and implicit models. Explicit schemes are based on the forward difference in time

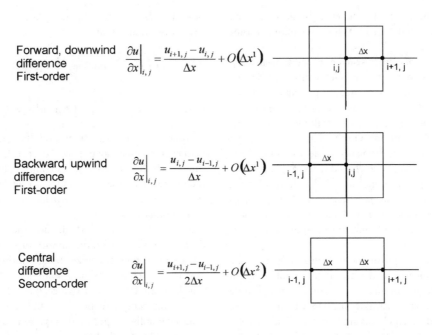

Forward, downwind difference First-order	$\dfrac{\partial u}{\partial x}\bigg	_{i,j} = \dfrac{u_{i+1,j} - u_{i,j}}{\Delta x} + O\left(\Delta x^1\right)$
Backward, upwind difference First-order	$\dfrac{\partial u}{\partial x}\bigg	_{i,j} = \dfrac{u_{i,j} - u_{i-1,j}}{\Delta x} + O\left(\Delta x^1\right)$
Central difference Second-order	$\dfrac{\partial u}{\partial x}\bigg	_{i,j} = \dfrac{u_{i+1,j} - u_{i-1,j}}{2\Delta x} + O\left(\Delta x^2\right)$

Figure 3.2 Finite difference expressions and scheme.

such that the solution for u at time $t = n + 1$ (the superscript indicates the time step and the subscript the x and y locations, respectively) is obtained directly only from its known values at time $t = n$

$$\frac{\partial u}{\partial t}\bigg|_{i,j}^{n} = \frac{u_{i,j}^{n+1} - u_{i,j}^{n}}{\Delta t} + \varepsilon_T \qquad (3.8)$$

whereas the backward difference leads to an implicit scheme

$$\frac{\partial u}{\partial t}\bigg|_{i,j}^{n+1} = \frac{u_{i,j}^{n+1} - u_{i,j}^{n}}{\Delta t} + \varepsilon_T \qquad (3.9)$$

which requires the solution of u within the entire domain also at the next time step $(n + 1)$. The first order wave equation with constant wave celerity c

$$\frac{\partial u}{\partial t} + c\frac{\partial u}{\partial x} = 0 \qquad (3.10)$$

can be approximated with a forward difference in time and central difference in space as

$$\frac{u_i^{n+1} - u_i^{n}}{\Delta t} + c\frac{u_{i+1}^{n} - u_{i-1}^{n}}{\Delta x^2} = 0 \Rightarrow$$
$$\qquad\qquad\qquad\qquad\qquad (3.11)$$
$$u_i^{n+1} = u_i^{n} - c\frac{\Delta t}{\Delta x^2}\left(u_{i+1}^{n} - u_{i-1}^{n}\right)$$

and with a backward difference in time and central second difference in space as

$$\frac{u_i^{n+1} - u_i^{n}}{\Delta t} + c\frac{u_{i+1}^{n+1} - u_{i-1}^{n+1}}{\Delta x^2} = 0 \qquad (3.12)$$

Whereas explicit schemes allow solutions for the next time step, u_i^{n+1}, from information only from the previous time step, Equation (3.12) presents two additional unknowns, u_{i-1}^{n+1} and u_{i+1}^{n+1}. Thus, solution at the next time step for one point requires a solution at the same time for all points within the domain. For a channel modelled with N computational nodes, this leads to the simultaneous solution of $2N$ (typically nonlinear) equations. A special implicit method widely used in CFD is the Crank–Nicolson scheme (Anderson, 1995).

The implicit technique is more robust and allows larger time steps than explicit schemes but it is less computationally economic. However, the Courant–Friedrichs–Lewy (CFL) criterion, which is a necessary but not sufficient condition for stability, limits the size of the largest time step for an explicit scheme (Peiró and Sherwin, 2005):

$$C_N = \frac{c_i}{\dfrac{\Delta x}{\Delta t}} < 1 \quad \Rightarrow \quad \text{largest time step} \Delta t < \frac{\Delta x}{c_i} \qquad (3.13)$$

where C_N is the Courant number, c_i is the local wave celerity at node i and Δx and Δt are the length and time steps, respectively.

The advantage of explicit schemes is that they are relatively simple to develop and run but the time step Δt for a given spatial step Δx needs to be shorter than a threshold value imposed by stability requirements. In some circumstances, Δt must be very small, resulting in very long computing (CPU) runtime. On the other hand, implicit schemes may be unconditionally stable and allow large time steps regardless of the selected space step. However, the time step cannot be increased indiscriminately due to truncation errors. This is especially true when transitional processes are important. However, they are more complex to program and require the manipulation of large matrices.

FEM and FVM approximate the integral form of the continuity, momentum and energy PDE (e.g. of Equation (3.2)). FEM divides the numerical domain into a finite number of elements (Zienkiewicz and Cheung, 1965). Continuity and NS equations are then solved in each element with the prescribed boundary conditions. Physical parameters are approximated within each element by interpolation functions, which are typically polynomials, and their nodal values such that the velocity component u within the element h (indicated in Equation (3.14) by the superscript (h)) can be expressed as:

$$u^{(h)}(\mathbf{x}, t) = \sum_{i=1}^{n_p} u_i(t) N_i(\mathbf{x}) \qquad (3.14)$$

where n_p is the number of nodes describing the element, u_i are the values at the nodes and N_i are the interpolation functions, which depend solely on the geometry of the element. The FEM forms of the original PDE are derived with two main methods: the variational and weighted residual methods (Finlavson, 1972). The former is typically adopted in mechanical engineering, whereas the latter approach is the most common in fluid mechanics. The FEM form of the conservation of mass equation for incompressible homogeneous fluids has the following form:

$$\iiint_{\Psi} \delta \cdot \mathbf{V} \chi(\mathbf{x}) d\Psi = 0 \qquad (3.15)$$

where χ represents various weighting functions. The most popular weighted residual method is Galerkin's approach. This approach assumes that the weighting func-

tions ($\chi(\mathbf{x})$) are equal to the shape functions ($N(\mathbf{x})$), $\chi(\mathbf{x}) = N(\mathbf{x})$, such that

$$\iiint_{\Psi} \sum_{i-1}^{n_p} \left(u_i(t) \frac{\partial N_i(\mathbf{x})}{\partial x} + v_i(t) \frac{\partial N_i(\mathbf{x})}{\partial x} \right. \\ \left. + w_i(t) \frac{\partial N_i(\mathbf{x})}{\partial x} \right) N_i(\mathbf{x}) d\Psi = 0 \qquad (3.16)$$

After the shape functions and their derivatives are defined, the model can be solved over each element. Continuity, which states that nodes common to adjacent elements must have the same values, is used to assemble the solutions of each element to provide the solution over the entire domain.

Similar to FEM, FVM divides the computational domain into an arbitrary number of control volumes. Equations are then discretized by accounting for the fluxes crossing the control volume boundaries (Versteeg and Malalasekera, 1995). For instance, the FVM integral form of the continuity equation over the control volume V assumes that the shape function is a constant of value 1 in Equation (3.15) such that

$$\iint_s u(\mathbf{x})|_i - u(\mathbf{x})|_{i+1} dydz + \iint_s v(\mathbf{x})|_i - v(\mathbf{x})|_{i+1} \\ dxdz + \iint_s w(\mathbf{x})|_i - w(\mathbf{x})|_{i+1} dxdy = 0 \qquad (3.17)$$

where the velocity components are evaluated at the boundary of the elements and i denotes the position along x. Several methods (e.g. first-order upwind, central differencing, quadratic interpolation QUICK scheme and power law) are available to approximate the dependent variables within the volumes. The node-centred and cell-centred approaches are the two main techniques used in defining the relationship between control volume shape and grid points where physical variables are calculated. Node-centred schemes place grid nodes at the centroids of each control volume such that control volumes coincide with grid cells. Conversely, cell-centred schemes define control volumes by connecting adjacent grid nodes.

Truncation errors, ε_T, round-off errors, ε_R, and numerical diffusion, ε_D, affect FDM, FEM and FVM solutions. Round-off error is due to rounding numbers to a fixed decimal position. Numerical diffusion is an artifact of discretization of the PDE and it is especially important in advection-dominated flows. Truncation error can be minimized by increasing the discretization order, by decreasing the cell size (known as h-type approximation) or, for FEM and FVM, by using polynomial functions of higher

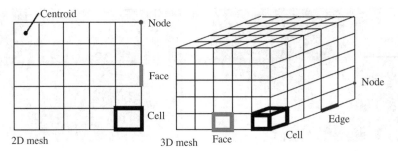

Figure 3.3 Components of a mesh.

order (e.g. power law or QUICK schemes) (known as p-type approximation) (Löhner, 2001; Peiró and Sherwin, 2005). Numerical diffusion can be reduced by using appropriate CFD techniques, high-order interpolation schemes and fine grids. In general, the term 'discretization errors' refers to those stemming from discrepancy between analytical and numerical solutions due to the discretization of the PDE. They do not include the round-off errors but include the truncation and numerical diffusion errors, which are errors that depend on mesh size and cell arrangement, truncations, polynomial functions and numerical scheme.

These errors affect convergence of the numerical results to the analytical solution, which is guaranteed for FDM by the Lax Equivalence Theorem, which states that a stable numerical model that solves a consistent set of algebraic equations with well-posed boundary conditions is convergent. Stability ensures that dependent variables computed at each node are bounded within reasonable numerical values during convergence, and consistency ensures that the algebraic equations tend to the analytical equations as mesh size and time step tend to zero (Hirsch, 2007).

The major advantage of FDM over FEM and FVM is its simple interpretation and implementation. However, FDM does not conserve momentum, mass and energy with coarse grids and its applicability is limited to problems where structured grids are applicable. Conversely, FEM performs well even with coarse grids and in suppressing numerical diffusion but it may not conserve mass locally (Oliveira *et al.*, 2000). It can use structured and unstructured grids (see the next section) and can accurately approximate the physical domain by refining the mesh where needed. However, it has the disadvantage of slow convergence for very large problems and poor performance in turbulent flows. FVM conserves mass, momentum and energy on coarse grids and it can be applied to virtually any control volume shape. It overcomes the

problem of slow convergence of FEM for large meshes, but it does not suppress numerical diffusion as efficiently as FEM.

3.3.3 Mesh

Whereas analytical solutions of a PDE provide continuous values of physical quantities (e.g. velocity, density and temperature), CFD provides numerical results only at a finite number of points within the numerical domain. A set of points, called nodes, whose arrangement forms a grid or mesh, defines the numerical domain, which is an approximation of the physical domain (Ferziger and Peric, 2002). Meshes or grids are at the heart of any CFD technique regardless of the dimensionality of the problem. Grids, which are grouped into structured and unstructured meshes (Figure 3.4), are composed of nodes, edges or nodal lines, faces or nodal planes, and cells or elements (Figure 3.3). Structured grids have nodes arranged in an orderly fashion following an indexing pattern. They have typically quadrilateral (e.g. square and rectangular) cells with Cartesian (Figure 3.4a) or curvilinear (Figure 3.4b) coordinate systems. Multi-block structured grids respond to the need to have fine meshes in zones where gradients are large and coarse meshes where the gradients are small. This provides some limited flexibility to adapt the mesh to the flow field. FDM only uses structured grids, whereas FEM and FVM can be applied also to unstructured grids (Figure 3.4c). These do not have an indexing pattern and do not have any constraints on cell layout. Potentially, they can accurately represent very complex physical domains.

Unstructured grids may use triangles and two-dimensional prisms in 2D grids and tetrahedrons, pyramids, hexahedrons and wedges or any arbitrary polyhedrons or a combination of them in 3D grids (Figure 3.5). CPU overhead for unstructured referencing schemes counterbalances their flexibility, favouring structured meshes in some applications. Both structured

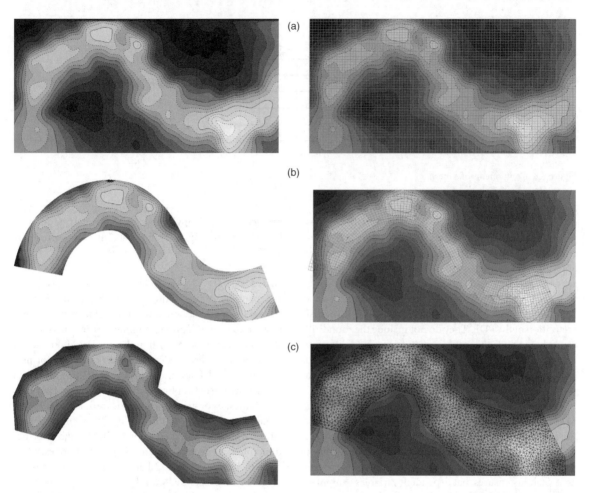

Figure 3.4 Structured (a) Cartesian, (b) curvilinear and (c) unstructured triangular meshes for a natural reach. The averaged cell area is 1 m² and the averaged topographic survey is resolution 0.25pt/m². The right column shows the topography and the mesh and the left column the numerical domain for the three different mesh types. Colour versions of the right-hand images for (a) and (b) appear in the colour plate section of the book.

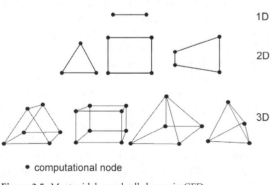

Figure 3.5 Most widely used cell shapes in CFD.

and unstructured meshes can be adaptive grids, which allow nodes to move to increase or decrease spatial mesh resolution in certain locations during computations. The streambed topography of most rivers has very complex geometric boundaries because of several scales of irregularities, from the macroscale (e.g. pool and riffles) to the microscale (e.g. grain roughness). Thus, grid generation to adhere to such topography can be difficult and an unstructured mesh may be preferred. To overcome this limitation for structured grids, the porosity mesh technique, which describes the topography by blocking the cells outside the boundary by using a porosity term, has been proposed and tested successfully for studying 3D gravel-bed rivers (Lane *et al.*, 2004).

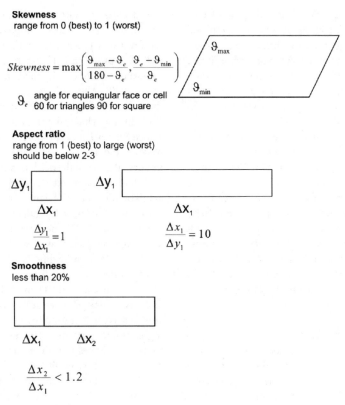

Skewness
range from 0 (best) to 1 (worst)

$$Skewness = \max\left(\frac{\vartheta_{max} - \vartheta_e}{180 - \vartheta_e}, \frac{\vartheta_e - \vartheta_{min}}{\vartheta_e}\right)$$

ϑ_e angle for equiangular face or cell
60 for triangles 90 for square

Aspect ratio
range from 1 (best) to large (worst)
should be below 2-3

Δy_1 Δy_1

Δx_1 Δx_1

$$\frac{\Delta y_1}{\Delta x_1} = 1$$ $$\frac{\Delta x_1}{\Delta y_1} = 10$$

Smoothness
less than 20%

Δx_1 Δx_2

$$\frac{\Delta x_2}{\Delta x_1} < 1.2$$

Figure 3.6 Criteria for mesh quality check.

3.3.4 Mesh quality

Because the shape and resolution of the mesh strongly affect convergence, stability and accuracy of any numerical result, its quality needs to be checked thoroughly before running any simulation, especially for 3D modelling (USACE, 1996). This usually includes checking for resolution (node density), skewness, aspect ratio and smoothness (Figure 3.6) (Bernard, 1992). Optimal grids have equilateral cells (equilateral triangles and squares) whose dimensions change smoothly through the domain. They may have fine meshes where gradients of physical parameters (e.g. velocity and water depth) are large within the domain and present coarse grids in zones where gradients are small. Cell skewness, which quantifies the divergence from equilateral shape for triangular cells and from orthogonal angles for quadrilaterals, should be limited to improved mesh performance. Cells' angles should be between 45 and 135 degrees (Bernard, 1992) and those less than 30 degrees should be avoided (Lane *et al.*, 1999). Similarly, the elongation of cells (the ratio between the longest and shortest edges) should also be limited. Aspect ratios should be close to 1 where flow is multidirectional. Ratios of 2–3 can be used with the longest edge along the primary

flow direction (Duan, 2004). Ratios of 10–50 would slow down extremely and may impair convergence. Smoothness refers to the change in length (expansion ratio), area or volume between adjacent cells. It has been shown that changes lower than 1.2 (20% increase/decrease) do not affect solutions whereas ratios larger than 10 cause physically unrealistic results.

Mesh resolution should be high enough to capture flow features, eddies and boundary conditions, which will be modelled by numerical simulations. For instance, large boulders or cobble clusters are important topographical features and their ecological function should be studied at a comparable mesh scale. Thus, both the topographical survey and the mesh size should be fine enough to capture their geometry. Burrows and Steffler (2005) suggest a mesh with 7.5 nodes per boulder diameter to capture the flow field generated by the boulders. The model of Waddle (2009) uses four times that density over boulders and twice in the boulder wake in order to minimize discretization errors. Node spacing (cell size) should be comparable to the topographic resolution of the survey (Lane and Richards, 1998). Lane and Richards (1998) suggest that flow predictions, with a grid much finer than the topographical survey, will be affected by the topographic interpolation, because topographical irregularities smaller (e.g. grain roughness, grain cluster and micro-topography) than the survey scale will not be represented in the model. Consequently, they use a mesh spacing of 10 cm for a topographic survey of 50 cm point spacing, even though they found that a mesh of 5 cm would be mesh-independent.

Results are typically considered mesh-independent if they are less than 3% different from those computed with a 30% finer mesh (Tonina and Buffington, 2007). Alternatively, a grid convergence index (GCI) analysis could be performed to investigate the effect of mesh size by comparing solution results for a coarser and finer (at least 10%) mesh (Roache, 1997). Conversely, flow predictions with a coarser grid than the topographical survey may miss important topographic information (Lane and Richards, 1998). Consequently, testing for mesh-independency in natural streams by comparing them with those computed with a finer mesh may be challenging.

3.3.5 Boundary conditions

All CFD models require information about the dependent variables (e.g. flow properties) at the boundary of the numerical domain. These values, called boundary conditions, are used to predict the dependent values within the domain (Anderson, 1995). Boundary conditions are typically of three types: Dirichlet conditions, which specify the values of the dependent variables at the boundary (e.g. water depth and velocity); Neumann conditions, which specify the gradient perpendicular to the boundary of the dependent variables (e.g. no flux for impermeable boundary); and mixed Neumann and Dirichlet conditions (Anderson, 1995). Boundary data in rivers include (1) terrestrial topography and aquatic bathymetry, which are usually treated as impervious surfaces (no flux through them) with zero tangential flow velocity (no-slip condition); (2) upstream flow conditions (flow entering the domain from all inlets), which may include discharge, cross-channel (for 2D and 3D) and vertical (for 3D) velocity profiles, and turbulence properties of the flow depending on turbulence modelling complexity (Fluent Inc., 2003); (3) downstream flow conditions (flow exiting the domain at all outlets), typically specified as a water surface elevation across the channel or as fully developed flows and (4) the water surface elevation, which is considered an impervious boundary with zero resistance in some cases (use of fix-lid method, see below) for 3D simulations.

Particular attention is required for the flow treatment near the streambed and bank boundaries, where veloc-ities decrease rapidly to zero due to the no-slip condition (Munson *et al.*, 2006). Large velocity gradients would require a very fine grid near the wall to predict the near-wall flow properties (Fluent Inc., 2003), which may be important for describing the benthic environment (Rice *et al.*, 2009). Alternatively, the processes occurring near the wall could be parameterized with the law-of-the-wall (logarithmic velocity distribution between the wall, where velocity is zero, and the outer flow (first node from the wall)) in the first grid cell, if the main body of the flow is important and not the near streambed conditions (Fluent Inc., 2003; Ingham and Ma, 2005). Roughness is specified at this boundary and its significance and the value adopted in the model strongly depend on the dimensionality of the model (1, 2 and 3D), mesh resolution and type of turbulence closure, such that physical and model roughness heights could be quite different (Morvan *et al.*, 2008).

The water surface boundary also requires attention in CFD, especially in 3D modelling. When water surface elevations are known and quasi-stationary, the fix-lid approach is used for simplifying convergence and decreasing computational time (Ingham and Ma, 2005). This technique defines the water surface as an impermeable boundary with typically zero shear stress and its elevation from field measurements. Two methods – the porosity (Spalding, 1985) and the mesh deformation (Olsen and Kjellesvig, 1998) models – are available to improve the accuracy of this technique to define the water surface elevation. An alternative approach is the volume of fluid (VOF) model, which computes the water surface elevation where the fluid pressure is equal to the atmospheric pressure in solving the momentum and continuity equations for both air and water (Hirt and Nichols, 1981). This approach is computationally expensive and may generate instability if the mesh is not designed carefully (Ingham and Ma, 2005). Different techniques are then implemented to smooth and render the water surface, as this needs to be interpolated within cells, which typically are partially filled with fluid and air.

The water surface elevation determines the extent of the submerged topography, or wetted area, in 2D modelling. The wetted area may vary during convergence for steady-state simulations and temporally for unsteady flow simulations. Thus, the computational domain may change with nodes becoming dry or wet. Numerical models should be able to account for this process, as water surface is typically an important factor to predict for ecological applications. Wetting and drying techniques can be divided into two groups: (1) moving and (2) fixed grids (Ghanem, 1995; Balzano, 1998; Lane, 1998). The moving grid technique adjusts the boundary cells to follow the edge of the water (Akanbi and Katapodes, 1988). Boundary dividing is a technique that divides the boundary cells into sub-elements to minimize the low-quality mesh resulting at the boundary as the mesh adapts to the edge of water (Lane, 1998). For fixed grids, a node is blocked out of the solution if its depth is below a prescribed minimum depth (drying criterion) and it is wet if it is above a wetting depth (wetting criterion). Typically, these values for riverine applications are approximately 0.01 m and 0.05 m for drying and wetting depths, respectively. A cell becomes partially wet if one of its nodes is dry. Several algorithms, which include marsh porosity, water surface elevation adjustment and element elimination, are available for dealing with these cells. The marsh porosity retains all the partially wet cells until all the nodes are dry and uses a porosity coefficient to adjust the flow to the amount of the cell carrying water. The element elimination blocks an entire cell when one of its nodes is dry and the elevation adjustment method modifies the water surface elevation within the partially wet cell (Horritt, 2002). The fixed grid approach, with the different treatments of partially wet cells, may introduce mass and momentum conservation errors, some unrealistic detached wetted areas and lobes (wiggles) at the boundary, which may lead to numerical instability. It is limited by the available knowledge on resistance in very shallow flows (Balzano, 1998; Lane, 1998).

3.3.6 Initial conditions

Whereas boundary conditions fix the values of the dependent variable at the boundary of the domain, initial conditions specify the initial values of any dependent variables for unsteady and steady state simulations at each node within the numerical domain at the beginning of a simulation. The most critical variable is the specification of the water surface elevation, which may be defined from values measured in the field, predicted from a 1D simulation or linearly interpolated between an upstream and downstream elevation (DHI, 2001; McDonald et al., 2005). The values of the other dependent variables, like flow velocity and depth, are then derived from the continuity or the uniform flow equation once the water surface is known. Some models have a 'hot-start' option, which defines the model variables within the entire domain from a previous simulation.

3.3.7 Model parameters

The previous sections show the use of several parameters in numerical modelling of river hydraulics. Two sets of model parameters, which are the boundary roughness and turbulent coefficients, require special consideration.

The significance of boundary roughness and its effects on model results strongly depend on model dimensionality (Morvan et al., 2008) and mesh resolution (Nicholas, 2001; Lane et al., 2004; Morvan et al., 2008). The effect of boundary roughness on results decreases with model dimensionality.

In 3D modelling, boundary roughness is part of the near-wall treatment, which could use a standard or non-equilibrium law-of-the-wall function (Lane et al., 2004), and accounts for all the topographical variations, which may include grain size and grain cluster formation within the size of a grid cell (sub-grid topography). Boundary roughness of natural rivers is typically quantified by upscaling the height from the wall at which the no-slip condition occurs, namely z_0, where the velocity is considered zero (see also Nikora et al., 1998; Nikora et al., 2001; Nikora et al., 2007) (Table 3.3). This height is usually associated with the equivalent sand roughness, ks, proposed by Nikuradse for a rough boundary, such as $z_0 = ks/30$ and with grain size, $z_0 = 0.1d_{84}$, (Whiting and Dietrich, 1991), $ks = 3.5d_{84}$ or $6.8d_{50}$. The diameters d_{84} and d_{50} are those for which 84% and 50% of the particles within a heterogeneous distribution are finer, respectively. The model roughness height (the value used in the model) should include all the surface irregularity not captured by the topographic survey or by the mesh discretization of the physical domain, such that it may be different from the height derived from a representative grain (physical height) (Lane et al., 2004). Consequently, the model z_0 could be a function of both mesh size and physical height. Clifford et al. (1992) report ks between $0.3d_{50}$ and $0.5d_{50}$ for only the grain roughness, which could be used with mesh of adequate resolution. Lumping into the z_0 parameter losses other than those associated with the grain height may result in inappropriate and unrealistic roughness height values (Lane et al., 2004; Morvan et al., 2008). This is particularly true in modelling flow over or around vegetation, which has been traditionally parameterized with fictitiously high resistance heights (Booker et al., 2001; Wilson et al., 2005, 2006). Following the concept of a turbulence closure, Wilson et al. (2005) call for a 'roughness closure' where process-based laws should replace friction parameters, which lump together several not-well-understood momentum loss mechanisms (Morvan et al., 2008). Separation of the momentum losses due to processes, which may include those due to vegetation, streambed irregularity, grain cluster and grain size, at scales smaller than

Table 3.3 Quantification of z_0, the height of the no-slip condition; u is the flow velocity.

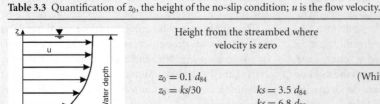

	Height from the streambed where velocity is zero	
$z_0 = 0.1\,d_{84}$		(Whiting and Dietrich, 1991)
$z_0 = ks/30$	$ks = 3.5\,d_{84}$	
	$ks = 6.8\,d_{50}$	
	$ks = 0.3\text{--}0.5d_{50}$	Only for grain roughness (Clifford *et al.*, 1992)

the cell size will improve the ability to define the benthic habitat (Jorde and Bratrich, 1998), food and particle drifting (Booker *et al.*, 2004) and sediment transport (Dietrich and Whiting, 1989). Consequently, the selection of a near-wall treatment and the partition of shear stress at the wall require careful consideration. The importance of these issues increases with lower dimensionality, because the effects of the boundary roughness extend through the entire water depth.

In 2D models, roughness expresses the resistance applied at the base of a vertical water column within a grid cell. It accounts for local losses due to grain size, surface texture and turbulence (Nelson and Smith, 1989; Morvan *et al.*, 2008) and it effectively behaves as a momentum sink. It is commonly estimated with a Manning's n, Gauckler–Strickler k_s ($k_s = 1/n$), Chezy C_z ($C_z = R_H^{1/6}/n$, where R_H is the hydraulic radius, which is the ratio between the cross-sectional area and the wetted perimeter), or Darcy–Weisbach f ($f = n^2\,8g/R_H^{1/3}$) resistance factor. Alternatively, if the vertical structure of the velocity is known, resistance could be quantified with a drag coefficient, C_d. For instance, if the standard law-of-the-wall flow structure, which assumes a logarithmic distribution of the velocity near the wall, is extended through the entire water depth, a drag coefficient can be estimated with the following equation (Nelson and Smith, 1989):

$$C_d = \left(\frac{1}{\kappa}\ln\left(\frac{h}{z_0}\right) - \frac{h - z_0}{\kappa\,h}\right)^{-2} \quad (3.18)$$

where κ is the von Karman constant (0.408) and h the local water depth. Several other equations are reported in the literature (e.g. Morvan *et al.*, 2008).

In 1D modelling, the roughness coefficient (n, k_s, C_z, or f) expresses the resistance along the entire wetted area of a reach. It lumps several processes together and it depends on channel geometry and flow characteristics (water surface elevation) (Morvan *et al.*, 2008). The model roughness coefficient represents such processes as (1) local losses due to streambed surface textural and grain roughness, (2) channel-width scale losses due to cross-sectional irregularities and bedforms, (3) reach-scale losses due to reach irregularities, contraction and expansion energy dissipations (unless specified separately in some models) and (4) topographical uncertainty due to spacing, location and resolution of cross-sections (Chow, 1959; Morvan *et al.*, 2008).

Whereas advances on characterizing the near bed roughness closure are just beginning, several turbulence closures have been proposed in the last 20 years. The number of turbulence parameters increases with closure complexity but typically only the zero-equation model requires the adjustment of a constant (Wilson *et al.*, 2002). The factor $\kappa\zeta(z_*)$ in Equation (3.5) is typically averaged vertically and expressed as a constant, whose values range between 0.07 (wide prismatic channels) and 0.8 (meandering with irregular banks), depending on stream type and irregularities (Table 3.4).

The cell dimensionless Peclet number, Pe_C, which is an index of advective (flow property transported by flow velocity along the flow direction) over diffusive processes (flow property transported in all directions proportionally to its gradient) at the scale of the grid cell, could be used for estimating a spatially variable eddy viscosity

$$Pe_C = \frac{uL_n}{v_T} \quad (3.19)$$

where u and v_T are the velocity magnitude and eddy viscosity at the cell and L_n is the maximum distance between nodes. Recommended Pe_C values range between 10 and 40 (Miller and Cluer, 1998; Crowder and Diplas, 2000b; Papanicolaou *et al.*, 2011) for natural streams.

The role of the lateral eddy viscosity is to act as an additional resistance factor (Miller and Cluer, 1998) between adjacent cells. Consequently, large v_T values tend to (1) smooth transversal velocity gradients, (2) increase

Table 3.4 Lateral eddy viscosity.

$\varphi = \int \kappa \, \zeta(z_*) \mathrm{d}z_*$	
0.03–0.12	For regular boundary from laboratory experiments (Jenkins, 1993; Simões and Wang, 1997)
0.15–0.45	For natural rivers with irregular boundaries (Jenkins, 1993; Simões and Wang, 1997)
0.08–0.11	From compound and shallow channels (Alavian and Chu, 1985)
0.1–0.2	From flume experiments (Fischer *et al.*, 1979 pp. 107–108)
0.11	Suggested for most rivers (Keefer, 1971; Olsen, 1999)
0.24–0.25	Irregular canal (Fischer *et al.*, 1979 pp. 107–108)
0.3–0.7	Flume experiment with irregular banks (Holley and Abraham, 1973)
0.4–0.8	From field studies reported by Fischer *et al.* (1979 pp. 111–112)
0.6	Suggested for natural channels (Froehlich, 1989)
0.01	With reach averaged velocity and depth (Barton *et al.*, 2005)

low velocities at the sides of the channel, (3) reduce high velocities within the channel (near the thalweg) and (4) suppress flow separation and recirculation zones. The high flow velocity reduction is compensated by an increase in low flow velocity and in water surface elevation. Doubling v_T values typically results in a change in water surface elevation of only a few centimetres or a small percentage when compared to the originally computed water surface (Ghanem *et al.*, 1996; Miller and Cluer, 1998). Conversely, very low v_T values enhance (1) transversal velocity gradients (as flow cells behave more independently from each other), (2) flow separation and (3) the formation of recirculation zones, which may not be realistic. The increased velocity variability may cause numerical instability, which may lead to solution divergence. Thus, large values of v_T are sometimes used to damp velocity fluctuations and reach solution convergence (Froehlich, 1989).

3.3.8 Model parameterization

Model parameterization is the process that specifies the model parameters, which are typically the turbulence parameters and the streambed roughness, for the specific study site. The roughness height is the most important parameter, especially for 1D and 2D modelling, as it affects the entire flow field. 3D model accuracy has been found to be more susceptible to boundary resolution than roughness parameterization (Lane and Richards, 1998; Lane *et al.*, 2005). Parameter values are estimated via field measurements, theoretical analysis, calibration or a combination of these methods.

Modellers of 2D and 3D simulations tend to use a combination of field measurements and theoretical analysis for estimating model parameters (Lane, 1998; Crowder and Diplas, 2000b; Pasternack *et al.*, 2004; Shen and Diplas, 2008) followed by an assessment of model performance (Lane *et al.*, 2005) rather than optimizing parameters via calibration. Field measurements of grain size, protrusion height distributions, vegetation density, aerial images and field observations may provide the possible range of variability of the roughness heights (e.g. Whiting and Dietrich, 1991) or the resistance coefficients (e.g. Cowan, 1956; Chow, 1959; Barnes, 1967; Arcement and Schneider, 1984). However, streambed roughness may vary greatly in natural streams, making its evaluation difficult (Chow, 1959; Barnes, 1967). Theoretical considerations and field observations may provide the range for the turbulence constant of the zero-equation model (Fischer *et al.*, 1979; Miller and Cluer, 1998; Vionnet *et al.*, 2004).

Calibration consists of varying the parameters (lateral eddy viscosity and streambed roughness) to minimize differences between predicted values and measured data, which are typically the water surface elevation and/or velocity distribution at selected locations (Waddle, 2009). Calibration is a legacy from hydrological models with parameter lumping processes at different scales and it is applied mostly in 1D modelling (Vidal *et al.*, 2007). The reason for not optimizing a set of parameters against field measurements for 2D and 3D models is the risk that optimized parameters may compensate for poor process representation, poor boundary information (e.g. topography resolution and flow values), poor mesh resolution and quality and measurement uncertainty (Lane *et al.*, 2005). Additionally, calibration may be affected by equifinality, where different parameter combinations may cause the model to converge to the same observed data (Beven and Freer, 2001). Consequently, optimizing parameters through calibration may reduce model predictive capability under different boundary conditions, such as a discharge that differs greatly from the discharge used to calibrate the model. A method to limit this problem is to use measurement information and theoretical analysis coupled with a comparison between predicted and observed values to facilitate the adjustment of the parameters within a realistic range.

3.3.9 Validation

Whereas model verification, which is benchmarked with analytical solutions, tests that the numerical results are accurate solutions of the original partial differential equations (AIAA, 1998; Grace and Taghipour, 2004), validation is the process of testing the accuracy of the model in representing the physical variables that it is designed to predict (AIAA, 1998). Traditionally, validation follows model calibration by comparing model results against a set of data defined as independent because it was not used during the calibration process. Comparisons can be reported as histograms of observed minus predicted values (Waddle, 2009), graphs of predicted versus observed data (Lane *et al.*, 1999) and visualization of predicted and observed data at their locations within the domain (Pasternack *et al.*, 2004). However, validation does not necessarily guarantee that a model is correct, because an invalid model may provide adequate results under certain conditions or for part of the domain (Lane *et al.*, 2005). Moreover, the definition of validation is more than just a single test for validity.

Firstly, validation is a process and not just a comparison between predicted and observed data. Secondly, it requires a benchmark for accuracy, which is typically done by comparing between 20 and 160 measured and predicted values of the variables of interest (Lane *et al.*, 1999; Stewart, 2000; Pasternack *et al.*, 2006; Tonina and McKean, 2010). For both 2 and 3D models of natural streams, the available literature reports that the explained variability (coefficient of determination R^2 of the regression line for the observed versus predicted velocities) between predicted and measured depth-averaged streamwise velocities may be approximately 80% (Lane and Richards, 1998; Nicholas, 2001; Lane *et al.*, 2005; Pasternack *et al.*, 2006; Waddle, 2009). Pasternack and Senter (2011) report R^2 of 0.5 and 0.7 of the regression line for observed versus predicted depth-averaged velocities and average errors of 20–30% for the magnitude of the depth-averaged velocity for a 12-km long highly complex bedrock-boulder stream. Errors could be smaller in less complex systems with gradually varying topography. The transversal velocity component typically has lower precision than the longitudinal component of the velocity magnitude. The vertical velocity component has been reported with explained variability even lower than 25% (Lane *et al.*, 1999). Thirdly, it needs to account for uncertainty of measured data. Consequently, less accurate observations would facilitate validation (Grace and Taghipour, 2004; Lane *et al.*, 2005). The magnitude of the vertically averaged velocities may present errors between 7% and 15% when measured in the field with current meters (Kondolf *et al.*, 2000). Flow

directions have been reported with standard deviations of 4.5 and 7.5 degrees (Lane *et al.*, 1999) and depth and width measurement errors of 2.5% and 0.5%, respectively (Kondolf *et al.*, 2000). Uncertainty on the exact position of the measurement also increases model uncertainty, especially for those variables like water depth, velocity and shear stress that vary rapidly in space. Thus, water surface elevations are typically used because they tend to vary gradually spatially. Fourthly, the variables of interest should be validated. For instance, if the model is used to estimate vertically averaged velocity and depth distributions for assessing aquatic habitats, those variables should be tested and not water surface elevation alone (Lane *et al.*, 2005). The same selection of variables should be used for assessing mesh independency, as some variables may converge with a coarser grid whereas other variables may converge only with a finer grid. For instance, Hardy *et al.* (2003) analyzed the effect of halving and doubling of the mesh size for a 3D open channel flow application on grid convergence index (GCI) values. They reported increased values of GCI from streamwise, transversal and vertical components of the velocity and a very large GCI value of the turbulence kinematic energy. They calculated the GCIs with a safety coefficient of 3 for the three velocity components of 0.53, 1.67 and 6.05% for halving, and 1.21, 13.91 and 28.11% for doubling the mesh size (Hardy *et al.*, 2003). The coarser mesh could be sufficient if velocity magnitudes are important, but the finer mesh should be used if transversal velocities are important, for instance, in studying drift or solute mixing. This shows the importance of checking the mesh independency for the variables of interest (Hardy *et al.*, 2003; Legleiter *et al.*, 2011).

Additionally, other factors may explain the uncertainty between measured and predicted values. Firstly, field measurements are typically point measurements or over a sampling volume, whereas numerical simulations provide values averaged over the cell size. Secondly, uncertainty on the boundary conditions for inflow and outflow discharges may affect results. Discharge measurement errors may range from 0.1–3% for fixed installations like flumes and weirs (Ackers *et al.*, 1978) to 5–10% for stage-gauge relations (USGS, 1992). These points are highlighted by USACE RMA2 user guide (USACE, 1996), which reports that appropriate quantification of streambed topography, boundary condition information and mesh quality explain 80% of the predictive ability of a 2D hydraulic model. Consequently, Lane *et al.* (2005) suggest that validation should be accompanied by '(i) conceptual model assessment; (ii) sensitivity analysis; and (iii) visualization' (page 185 in Lane *et al.*, 2005).

3.3.10 Scaling and averaging

Simplified models or averaged models could be considered depending on the problem definition and available topographic and hydraulic information. Scaling and averaging techniques are the two main methods to simplify equations. The former approach uses flow considerations and dimensional analysis to select and drop from the equations those processes that are negligible under certain conditions compared to others. Scaling analysis has been used to derive the shallow water equations, which are simplified 3D versions of the full 3D RANS equations (Section 3.4) and several simplified wave equations from the full 1D De Saint–Venant equations (Section 3.6). The latter approach averages dependent variables along one or two directions. Flow processes and structures occurring along the averaging directions are lost regardless of their importance. Parameterization of important processes lost due to averaging leads to the so-called pseudo-2D and quasi-3D models, often termed 1.5D and 2.5D models. The 2.5D models are vertically averaged, two-dimensional models with parameterized secondary flows (Johannesson and Parker, 1989; Bernard, 1992), which are characterized by zero net-mass movement within the cross-section, like helicoidal flows at bends (Dietrich and Smith, 1983). They are probably the most promising models in ecohydraulics applications because they capture many important processes while retaining simplicity and relative field-data economy (e.g. FAST-MECH (McDonald *et al.*, 2005)).

3.4 3D modelling

3D modelling solves the full three-dimensional Navier–Stokes (DNS) or the 3D RANS equations with appropriate turbulence closures (e.g. $k-\varepsilon$ or $k-\omega$ models) and near-wall (e.g. law-of-the-wall or double-layer) treatment. 3D models are still rare in ecohydraulics and are limited to steady-state simulations because of their high computational requirements (Lane, 1998; Lane *et al.*, 1999). They have been used to define flow fields for studying fish behaviour (Nestler *et al.*, 2008), food (Booker *et al.*, 2004) and organism drifts (Daraio *et al.*, 2010), fish aquatic habitat (Alfredsen *et al.*, 2004; Shen and Diplas, 2008), habitat rehabilitation (Mouton *et al.*, 2007) and biotic structures, such as fish redds (the salmonid egg nests) (Tonina and Buffington, 2009). They are mostly applied at a local scale of a few channel widths (e.g. Lane *et al.*, 1999) (Table 3.5).

3D modelling has the advantage of its ability (1) to predict the distribution of physical quantities in all directions, (2) to predict x, y and z vector components and (3) to account for non-hydrostatic pressure distribution

(Shen and Diplas, 2008) (Table 3.6). The first property preserves the flow's vertical, transversal and longitudinal structures and is fundamental for capturing secondary flows. Secondary flows strongly affect sediment (erosion and deposition and bar-pool formation) and solute transport (Figure 3.7) (Fischer *et al.*, 1979; Nelson *et al.*, 2003). 3D models also require that the velocity-pressure coupling technique be defined to solve for the pressure field. The most common techniques are fractional step methods, PISO, SIMPLE and SIMPLER models (Ferziger and Peric, 2002). However, most studies in ecohydraulics have natural streams with low streambed slopes (less than 3%), gradually varying morphology and low depth–width ratios (less than 0.1), which allows the simplification of the equations with the shallow water hypothesis.

The shallow water hypothesis assumes that streambed slope is small, such that $\sin(s)=\tan(s)=s$, where s is the local slope of the streambed, (less than 10% (Miller and Cluer, 1998)) and that horizontal length scales (width and longitudinal lengths of the channel, such as the width to depth ratio larger than 10 (Woodhead *et al.*, 2006)) are large compared to water depths. These conditions ensure that gravity acts primarily along z ($g_z = -g$) and flows can be regarded as parallel to the bottom, which results in negligible vertical accelerations with respect to the other terms in the momentum equation along the vertical direction and hydrostatic pressure variations (Lane, 1998). The latter hypothesis holds in most natural rivers over the reach scale because topography changes gradually; however, it may not hold locally, where abrupt variations may occur, for instance, at bridge piers, around boulders and other obstructions (e.g. Shen and Diplas, 2008). Under the shallow water assumption, the RANS equations simplify to the following form:

$$\frac{\partial u}{\partial x} + \frac{\partial v}{\partial y} + \frac{\partial w}{\partial z} = 0 \qquad\qquad a.$$

$$\frac{\partial u}{\partial t} + \frac{\partial u^2}{\partial x} + \frac{\partial uv}{\partial y} + \frac{\partial uw}{\partial z} + g\frac{\partial Y}{\partial x}$$
$$- \frac{1}{\rho}\left(\frac{\partial \tau_{xx}}{\partial x} + \frac{\partial \tau_{xy}}{\partial y} + \frac{\partial \tau_{xz}}{\partial z}\right) = 0 \quad b.$$

$$\frac{\partial v}{\partial t} + \frac{\partial uv}{\partial x} + \frac{\partial v^2}{\partial y} + \frac{\partial vw}{\partial z} + g\frac{\partial Y}{\partial y}$$
$$- \frac{1}{\rho}\left(\frac{\partial \tau_{yx}}{\partial x} + \frac{\partial \tau_{yy}}{\partial y} + \frac{\partial \tau_{yz}}{\partial z}\right) = 0 \quad c.$$

$$\frac{1}{\rho}\frac{\partial P}{\partial z} = -g \qquad\qquad\qquad d.$$

$$(3.20)$$

where Equation (3.20a) is the continuity equation and Equations (3.20b), (3.20c) and (3.20d) are the momentum

Table 3.5 Examples of hydraulic models and their main features.

	Software name	Discretization/ equation	Turbulence closure	Mesh	Application
3D	SSIIM (Sediment Simulation in Intakes with Multiblock Option)	Finite volume	$k-\varepsilon$ $k-\omega$	SSIIM 1 uses structured grid SSIIM 2 unstructured grid	Hydraulics, sediment transport convection–diffusion model, water quality module including temperature. (Olsen and Stokseth, 1995; Alfredsen et al., 2004; Booker et al., 2004; Clifford et al., 2005; Kilsby, 2008; Olsen, 2010)
3D, 2D	TELEMAC-3D and 2D	Finite element/3D shallow water equation with an option for dynamic pressure	3D Constant eddy viscosity, mixing length (vertical) with several variations, Smagorinsky (horizontal) $k-\varepsilon$ model 2D Constant, Elder, $k-\varepsilon$ model and Smagorinsky	Unstructured mesh	Hydraulics, sediment transport convection–diffusion model, water quality module including temperature. (Bourban et al., 2006) 2D (Wilson et al., 2002; Vionnet et al., 2004)
3D-2D	ANSYS Fluent	Finite volume	Several turbulence closures	Several mesh types	Hydraulics, sediment transport and water quality applications (Hodskinson and Ferguson, 1998; Nicholas and McLelland, 1999; Tonina and Buffington, 2009)
3D-2D	ANSYS CFX (previously Flow3D)	Finite element	Several turbulence closures	Several mesh types	Hydraulics, sediment transport and water quality applications (Shen and Diplas, 2008)
3D, 2D and 1D	MIKE31, MIKE21 MIKE11 MIKEFLOOD (1D channel 2D floodplain MIKE GIS	Finite difference	Constant eddy viscosity, Smagorinsky subgrid scale model, k model, $k-\varepsilon$ model, or a mixed Smagorinsky/$k-\varepsilon$ model	3D: rectilinear grid, a curvilinear grid, a triangular element mesh Structures 2D: Cartesian (MIKE21), Curvilinear (MIKE21c) flexible mesh (MIKE21fm)	Hydraulics, sediment transport and water quality applications (Burke et al., 2009; Caamaño et al., 2010)
3D, 2D and 1D	Delft 3D	Finite difference/ shallow water approximation	Zero-equation and the $k-\varepsilon$ model	Curvilinear and rectilinear grids, spherical grids, domain decomposition	Flow, sediment transport and morphology, waves, water quality and ecology (Geleynse et al., 2011)
3D-2D	COMSOL	Finite element	Several turbulence closures	Several mesh types	Hydraulics, sediment transport and water quality applications (Cardenas and Wilson, 2007)

Table 3.5 (*Continued*)

	Software name	Discretization/ equation	Turbulence closure	Mesh	Application
2D-2.5D	iRIC - FASTEMECH	Finite difference /steady flow	Zero-equation	Curvilinear	Flow, sediment transport and simple microhabitat computation (Tonina and McKean, 2010; Simões, 2011; Tonina *et al.*, 2011)
	iRIC MORPHO2D	Finite difference/ vegetation model, mixed-grain transport	Zero-equation	Structured	
	iRIC NAYS	Finite difference/ vegetation model, mixed-grain transport, bank erosion model	Horizontal LES and other turbulence closures	Structured	
	iRIC STORM	Finite volume fully unsteady	Zero-equation	Unstructured triangular grid	
	FESWMS-2DH	Finite element	Zero-equation	Flexible mesh	Hydraulics (Pasternack *et al.*, 2004, 2006; Brown and Pasternack, 2009; Papanicolaou *et al.*, 2011)
2D	RMA2	Finite element	Zero-equation	Flexible mesh	(Crowder and Diplas, 2000a; 2000b)
2D	SRH-2D	Finite volume/ no mesh generation	Zero-equation and $k-\varepsilon$ model	Unstructured mesh	Hydraulics (Tolossa *et al.*, 2009)
2D-1D	Sobek2D, 1D and Sobek1D/2D channel floodplain link	Finite difference	2D Zero-equation and the $k-\varepsilon$ model	Structured	Hydraulics (Carrivick, 2007)
2D-1D	TUFLOW	Finite difference	Smagorinsky	Orthogonal grid	Hydraulics (Baart *et al.* 2010)
2D	RIVER2D	Finite element	Zero-equation	Unstructured mesh	Hydraulics (Lacey and Millar, 2004; Hayes *et al.*, 2007; Clark *et al.*, 2008; Waddle, 2009)
1D	FEQ	Implicit finite-difference	Not applicable	Not applicable	Hydraulics, loops, structures (Wolterstorff *et al.*, 2003)
1D	SRH-1D	Explicit and implicit, finite difference	Not applicable	Not applicable	Hydraulics, sediment transport (USBR, 2011)
1D	HEC-RAS	Implicit finite differences	Not applicable	Not applicable	Hydraulics, loops, structures, sediment transport (García *et al.*, 2011)

equations along x, y and z, respectively. The variable Y is the local water surface elevation, and the pressure within the water column is expressed as $P = \rho g\,(Y - z)$, with the pressure at the water surface equal to 0 when $z = Y$ (gauge pressure) and the pressure at the stream bottom $P = \rho g h$ with h the local water depth. It is important to notice that the momentum equation along the vertical (z direc-

tion) simplifies to the hydrostatic pressure distribution of the water. These equations are the most commonly used 3D modelling in ecohydraulics because of their simplicity with respect to the full version (Lane, 1998).

3D public domain models, which include SSIIM (Olsen, 2011), TELEMAC-3D, OpenFOAM (OpenFOAM, 2011) and Delft3D (Deltares, 2011) have

Table 3.6 Typical CFD results for ecohydraulic applications. XS stands for cross-section, x, y and z stand for the longitudinal, transverse and vertical directions respectively.

			Outputs			
			Velocity		Shear stress	
	Water elevation	Depth	Magnitude	Components	Magnitude	Direction
3D	Local grid cell averaged	Local depth	Point averaged on the grid	3D x, y and z components	At the mesh scale; at the wall skin, grain and form drag of roughness elements	3D x, y and z components
2.5D	Local grid cell averaged	Local depth	Depth averaged + vertical velocity profile	2D horizontal x and y components	At the streambed; skin and form drag at the cell scale	2D horizontal x and y components
2D	As above	As above	Depth averaged	As above	As above	As above
1.5D	XS	Local along the XS interpolated between XS	Transversal profile derived applying uniform flow equations based on local depth	1D along x	Transversal profile derived applying uniform flow equations based on local depth	1D along x
1D	XS	XS mean depth	XS averaged	1D along x	XS averaged it combines: skin, form drag and bedform, reach scale losses and turbulence	XS averaged 1D along x
Non-numerical	Interpolated from measured values, predicted from analytical or empirical equations	Interpolated from measured values, predicted from analytical or empirical equations	Measured or predicted from analytical solutions	Measured, if predicted from analytical solutions perpendicular to the cross-section	Measured or derived from uniform flows	Measured, if predicted from analytical solutions perpendicular to the cross-section

not been commonly applied to date and have been developed for specific research questions. Commercial software packages include Fluent (Fluent Inc., 2003), CFX and MIKE31 (DHI, 2001).

3.4.1 Model setup

Field measurements include detailed bathymetry surveys at a resolution finer than the smallest topographical structure, whose effects on the flow are modelled (Table 3.7). For instance, the survey and the mesh resolutions should be fine enough to define boulder geometry if the ecological question requires the prediction of the flow field around boulders. Research suggests having at least 7.5

points per boulder (Crowder and Diplas, 2000b; Burrows and Steffler, 2005) for isolated boulders or 1–10 points per square metre in bedrock-boulder streams (Vallé and Pasternack, 2006). Other research reports one point every 5 m for mapping submerged topography, while the dry topography has point density as above (Pasternack and Senter, 2011 pp. 44–47).

In the emerging discipline of ecohydraulics, the goal of collecting topographic data is to define the aquatic and terrestrial physical environments, which are presented as bathymetric and terrestrial continuous surfaces. Terrestrial topography and aquatic bathymetry are collected with one or more of the following methods: traditional

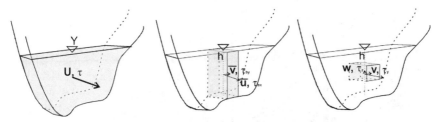

Figure 3.7 Velocity, shear stress and water depth results in 1D, 2D and 3D numerical simulation.

Table 3.7 Typical data necessary for simulating open channel CFD and set-up non-numerical methods; TC stands for turbulence closure.

			Inputs			Parameters and model assessment
				Boundary conditions		
	Topography	Roughness	Turbulence	Upstream	Downstream	
3D	Bathymetry	Local roughness function of cell size; depends on near-wall flow treatment	Yes, modelled depending on closure: DNS, RANS with RSM, $k-\varepsilon$, $k-\omega$	1. Velocity distribution at the inlets 2. turbulent quantities distribution depending on TC	Pressure distribution or water elevation or fully developed flow assumption	Water elevations along the reach and point velocity measurements at several locations especially close to points of interest
2.5D	Bathymetry + Model for vertical velocity profile	Bottom roughness at the cell scale + some turbulence n, C_z, k_s, f or C_d spatially variable or constant	Partially modelled, typically $k-\varepsilon$ or zero-equation models $v_T = \varphi u_* h$ with local or reach averaged u_* and h	1. Discharge or transversal depth-averaged velocity profile 2. Turbulent quantities depending on TC	Water elevation	Water elevation along the reach, depth-averaged velocity profiles at several locations, flow directions for one or two discharges
2D	Bathymetry	As above	As above	As above	As above	As above
1.5D	XS + Model for transversal velocity variation	Lump roughness at the reach scale spatially varying or constant in channel, different between channel and floodplain or vegetated areas n, C_z, k_s, or f	Lumped in the roughness	1. Discharge 2. Water elevation only if supercritical flows	Water elevation if subcritical or mixed flows	Water elevation along the reach for low, medium and high discharge[1]
1D	XS	As above	As above	As above	As above	As above
Non-numerical	XS or single points	As above for analytical solution	N/A			

[1]within the modelled flow range

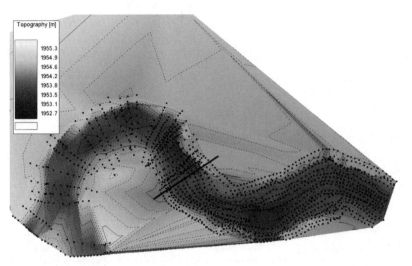

Figure 3.8 Cross-section versus morphological survey. Square markers are survey points overlaying a TIN generated topography. Flow is moving left to right. Upstream of the solid black line (left side of the figure) presents a cross-section based survey and downstream of the solid black line (right side of the figure) presents the survey with the suggested method.

ground-based surveying (total station, engineering level and/or RTK GPS) (e.g. Brasington *et al.*, 2000; Elkins *et al.*, 2007) sonar (e.g. Stewart, 2000; Conner, 2011), optical sensors (Fonstad and Marcus, 2005; Walther *et al.*, 2011) and recently with ground-based and airborne LiDAR (e.g. Hilldale and Raff, 2007; Heritage *et al.*, 2009; Benjankar *et al.*, 2011; Pasternack and Senter, 2011). Most recently, research is being conducted to obtain submerged topography using airborne EAARL (Experimental Advanced Airborne Research LiDAR) techniques (McKean *et al.*, 2009; Tonina and McKean, 2010; Tonina *et al.*, 2011), but this approach is still experimental and not readily commercialized or dependable (Kinzel III *et al.*, 2007). Sonar and LiDAR provide a point cloud of streambed elevations from which bathymetry is derived with different techniques (e.g. Triangular Irregular Network (TIN), Kriging and nearest neighbouring). Conversely, ground surveys traditionally collect point elevations along cross-sections (e.g. Brasington *et al.*, 2000), although recently points have been collected wherever one walks, which, in some cases, would be on cross-sections, from which bathymetry is interpolated (Pasternack and Senter, 2011).

Cross-section derived bathymetries have poor quality for 3 and 2D modelling, especially in streams with complex morphology, unless an adequate number of cross-sections is collected (Conner, 2011; Glenn, 2011). Recent investigations in single-thread channels show that transects should be placed at least 0.5 or 1 channel width apart depending on stream complexity (e.g. meandering

with pools, riffle and runs versus plane-bed morphology) and interpolation should be performed in a curvilinear coordinate system (Legleiter and Kyriakidis, 2008; Merwade *et al.*, 2008; Glenn, 2011). Other research shows that point elevation along the transversal, where topographical changes occur more rapidly than the longitudinal direction, should be dense enough to model the changes in slope (Horritt *et al.*, 2006). Cross-sectional derived bathymetry is typically improved by adding breaklines, points where the topographical contour changes abruptly, like the top of the banks or a riffle crest (Brasington *et al.*, 2000; French and Clifford, 2000; Vallé and Pasternack, 2006). The authors recommend an alternative and efficient survey method based on river morphology to avoid the transect-based survey limitations. In this method, the survey is conducted by walking along longitudinal paths following the natural stream shape, for instance, following major morphological breaks like bank top and bottom and bankfull edges, as shown on the right side of Figure 3.8. Point density along a path line and distance between path lines may vary spatially to capture the salient hydraulic and topographical features. Point density should be large where topographical gradients are large and at important features for the project. It is important to extend data collection one or two channel widths upstream and downstream of the study site. The upstream and downstream ends should be placed in simple and straight reaches without flow obstructions at the upstream end and with almost fully developed flow at the downstream end. Fully developed flows have flow properties that change slowly in the downstream direction and hence are uniform flows. They are typically present in straight, simple sections of a river. Extensions at the upstream or downstream ends can also be added with a synthetic reach to develop the incoming and exiting flows (Lane *et al.*, 1999).

The boundary roughness can be characterized from the grain size distributions, which can be measured with the Wolman pebble count for surface material (Wolman, 1954), bulk sieving techniques (Bunte and Abt, 2001) or photo-sieving (Ibbeken and Schleyer, 1986; Dugdale *et al.*, 2010). Sub-grid topographical roughness may also be modelled from the geometric characteristics of the bed forms, such as amplitude and wavelength for dune-like irregularities (Barton *et al.*, 2005). Roughness information may be spatially distributed or constant over the reach.

3D modelling requires detailed information of velocity and in some cases of turbulence quantity profiles at the upstream inlet and pressure distribution or (rarely) outflow velocity at the outlet. At the outlet, if field information is not available, the fully developed flow assumption could be used, namely the flow is uniform (Fluent Inc., 2003). Water surface elevations, velocity distributions and, in some cases, turbulent quantities through the domain should be collected for parameter specification and/or validation. Velocity measurements are usually collected along cross-sections and close to important hydraulic features and the flow elements of interest. They could be measured with a mechanical velocimeter such as Price AA or Price pygmy, acoustic Doppler velocimeter (ADV) (Rehmel, 2007) or acoustic Doppler current profiler (ADCP) (Nystrom *et al.*, 2002; Barton *et al.*, 2005). The last two provide velocity component and water depths and may provide turbulence information (Vermeulen *et al.*, 2011).

Model performance depends on mesh resolution, mesh type, solver scheme, turbulence-closure models and model roughness. 3D models do not perform well if mesh, topography and hydraulic boundary information are not of good quality regardless of parameter selection. Performance is typically assessed by comparing measured and predicted depths, water surface elevation and velocity magnitude and velocity components. Convergence of the model is evaluated by monitoring residuals of the variables of interest to check that they decrease below defined thresholds. Flow properties (e.g. shear stresses at the wall, pressure values and velocities) should also be monitored during convergence at important locations, as their stationarity between iterations may be a sign of convergence.

3.5 2D models

2D models are typically used at the geomorphic unit and reach scales (10–50 channel widths) because they require detailed topographical surveys with a resolution finer than the investigated topographic formation and have high computational requirements. However, the application of 2D modelling with high (e.g. 10^{-1} channel widths or 1–10 m^2) resolution on reaches longer than several kilometres are no longer exceptions due to advances in numerical modelling and survey techniques (Barton *et al.*, 2005; Tonina and McKean, 2010; Pasternack and Senter, 2011; Tonina *et al.*, 2011). 2D models are averaged on the transversal or on the vertical directions. The former models preserve the vertical but not the transversal flow structure. In this mode, they are usually developed to study water quality at the inlet of water bodies such as lakes and oceans, where lateral variations are negligible compared to those occurring along the vertical direction. These models show important information like flow stratifications and velocity reversal (Javan and Eghbalzadeh, 2011). Conversely, vertically averaged models are most common in studying stream hydraulics where transversal and longitudinal spatial distributions of water depths, bottom shear stresses and vertically averaged velocities are important. Whereas any vertical flow structure is lost in averaging, these models have three advantages as they provide (1) planimetric maps of physical quantities, like vertically averaged velocities, bottom shear stresses, water depths and other derived quantities, e.g. stream energy gradient, vorticity and circulation, which may be important ecohydraulic indexes (Nestler *et al.*, 2008; Shen and Diplas, 2008); (2) vector longitudinal and transversal components; and (3) transversal structure of flow features and water surface elevation (Table 3.6). Consequently, they are becoming the preferred tool in ecohydraulics including aquatic habitat quality modelling, river rehabilitation, organism dispersal and drift and organism behaviour studies (e.g. Leclerc *et al.*, 1995; Crowder and Diplas, 2000a; Kondolf *et al.*, 2000; Pasternack *et al.*, 2004; Hayes *et al.*, 2007; Tonina *et al.*, 2011). Results derived from a 2D model are not point measurements but are averaged values over the grid and hence they depend on grid size. Care should be devoted in analyzing and comparing observed (typically point measurements) with predicted values, especially velocities and shear stresses, which may change spatially more rapidly than quantities such as water surface elevation (MacWilliams *et al.*, 2006; Pasternack *et al.*, 2006).

The vertically averaged RANS are derived from the three-dimensional shallow water equations, which assume negligible vertical accelerations and consequently a hydrostatic pressure distribution. Thus, 2D models would provide inaccurate results where vertical accelerations are important (Shen and Diplas, 2008) and for channels with slopes exceeding 10% (Ghanem et al., 1996; Miller and Cluer, 1998). The general form of the continuity equation (3.21a), and momentum equations along the x-axis (3.21b) and y-axis (3.21c) in 2D is

$$\frac{\partial h}{\partial t} + \frac{\partial \bar{u} h}{\partial x} + \frac{\partial \bar{v} h}{\partial y} = 0 \qquad\qquad a.$$

$$\frac{\partial \bar{u} h}{\partial t} + \frac{\partial \beta_1 \bar{u}^2 h}{\partial x} + \frac{\partial \beta_2 \bar{u}\bar{v} h}{\partial y} + gh\frac{\partial Y}{\partial x} - \frac{1}{\rho}(\tau_{xs} + \tau_{xb})$$

$$- \frac{1}{\rho}\left(\frac{\partial \bar{\tau}_{xx}}{\partial x} + \frac{\partial \bar{\tau}_{xy}}{\partial y}\right) = 0 \qquad b. \quad (3.21)$$

$$\frac{\partial \bar{v} h}{\partial t} + \frac{\partial \beta_2 \bar{u}\bar{v} h}{\partial x} + \frac{\partial \beta_3 \bar{v}^2 h}{\partial y} + gh\frac{\partial Y}{\partial y} - \frac{1}{\rho}(\tau_{ys} + \tau_{yb})$$

$$- \frac{1}{\rho}\left(\frac{\partial \bar{\tau}_{yx}}{\partial x} + \frac{\partial \bar{\tau}_{yy}}{\partial y}\right) = 0 \qquad c.$$

where the symbol $^-$ identifies vertically averaged quantities, h is the local water depth, Y is the local water surface elevation and u and v are the vertically averaged x and y velocity components respectively. The β terms, called correlation or dispersion parameters, account for secondary momentum transfer due to vertical velocity profiles. The surface shear stresses, τ_{xs} and τ_{ys}, are typically due to wind resistance, which, for a wind of speed W, air density ρ_a, and drag coefficient C_w, may have the following form:

$$\tau_{is} = \frac{C_w \rho_a |W| W_i}{h} \quad \text{with } i = x, y \quad (3.22)$$

where the vertical lines | | denote the absolute value of the quantity between them.

Similarly, the bottom shear stresses, τ_{xb} and τ_{yb} with total shear stress $\tau_b = \sqrt{\tau_{xb}^2 + \tau_{yb}^2}$ can be modelled with a drag coefficient or with a roughness coefficient (e.g. Manning's n, Gauckler–Strickler k_s, Chezy C_z or Darcy–Weisbach). 2D models may also explicitly account for turbulence losses, which are represented by the last set of terms on the right side of Equations (3.21b) and (3.21c).

3.5.1 Model setup

Field measurements for 2D modelling are similar to those for 3D models: detailed bathymetry, water surface elevations along the reach and at the downstream end of the model, grain size distribution and streambed mate-

rial patches and vertically averaged velocity distributions (Table 3.7).

The upstream boundary should be in a simple reach with banks slightly diverging without any obstructions in the first (10–15) nodes. The downstream boundary should be in a reach with slightly converging banks to avoid recirculating eddies, which may create instability and/or low convergence. Mesh resolution should be comparable to the investigated topographical and hydraulic features, thus it should be equal to or finer than the topographical information. In narrow streams (width less than 10–15 m), it is recommended to have at least 10–13 nodes on the cross-section in order to capture lateral flow variations.

The upstream boundary conditions consist of a constant discharge or a hydrograph and/or the transversal profile of vertically averaged velocities and, depending on the turbulence model, boundary values of turbulence quantities. Water surface elevations as time series or a stage–discharge relationship is necessary at all downstream exits.

Calibration is primarily based on minimizing the error between predicted and observed water surface elevations along the study site by adjusting the roughness coefficient. It is advantageous to test the performance of a model first with a constant roughness coefficient for the entire domain and then successively use a spatially distributed roughness if the system warrants it and information is available (Pasternack et al., 2006; Legleiter et al., 2011; Logan et al., 2011). Sensitivity of Manning's n on numerical models could be tested with 0.005 increments around the suggested values because model results are typically insensitive to increments less than 0.001 but are sensitive to changes larger than 0.01 (Nicholas and Mitchell, 2003; Pasternack and Senter, 2011). Roughness coefficients should be increased if predicted water surface elevations are, on average, lower than those measured and decreased if the predicted water surface elevations are higher than those measured (Table 3.8). However, if predicted velocities are, on average, higher than the field measurements, then the roughness coefficient should be increased, and *vice versa* if the observed data are higher than the predicted values. The input discharge should be checked along with its accuracy when both water surface elevation and flow velocity are over or underestimated, because flow velocity and water surface elevation have inverse responses to roughness. When roughness changes with discharge, which could be due to wetted or drying areas with spatially varying roughness (Conner, 2011) or a change in bedform roughness with discharge (Barton et al., 2005), calibration or parameter selection may need to be performed at different discharges.

Table 3.8 Rules for adjusting roughness and eddy viscosity coefficient by comparing predicted and observed velocities and water surface elevations.

Roughness coefficient	Increased if predicted water surface elevations are, on average, lower than those measured
	Increased if predicted velocities are, on average, higher than the field measurements
	Decreased if predicted water surface elevations are, on average, higher than those measured
	Decreased if predicted velocities are, on average, lower than the field measurements
Eddy viscosity coefficient for zero-equation models	Increased if the model predicts formation of eddies where field data do not show them
	Increased if predicted low velocities are lower than those observed and predicted high velocities are higher than those observed
	Decreased if the model does not predict formation of eddies where field data observe them
	Decreased if predicted low velocities are higher than those observed and predicted high velocities are lower than those observed

Model performance could be evaluated instead of calibration, especially when information on the spatial distribution of roughness (e.g. from grain size patches) is available (Legleiter *et al.*, 2011). Selection of the constant for the zero-equation turbulence closure (Table 3.4) should be followed by inspection of the flow field for recirculating eddies' downstream obstructions, such as boulders, or eddies along sharp bends. Because large turbulence coefficients tend to suppress the formation of recirculating eddies, the coefficient should be reduced if the model does not predict their formation where field data locate them (Table 3.8). On the other hand, its value should be increased if the model predicts recirculating zones where they are not present (Miller and Cluer, 1998). Additionally, a comparison between observed and predicted velocities may provide other insights in selecting the constant for the lateral eddy viscosity. Its value should be increased if predicted low velocities are lower than observed and high velocities are higher than observed. However, it could be decreased if predicted low velocities are higher than measured and predicted high velocities are lower than measured (Pasternack, 2011).

Validation is done by comparing predicted and measured vertically averaged velocities, water depth and water surface elevations. Similarly to 3D models, 2D models are very sensitive to topography and mesh quality and their performance is low with poor mesh and bathymetry resolution and/or quality (Pasternack *et al.*, 2004, 2006; Horritt *et al.*, 2006; Conner, 2011).

2D models with parameterized secondary flows (Johannesson and Parker, 1989; Bernard and Schneider, 1992) are called quasi-3D, pseudo-3D or 2.5D models. This parameterization allows one to account for some vertical structure of the velocity field, which leads to better predictions of direction and magnitude of shear stresses and velocities than simple 2D models (Lane *et al.*, 1999; Nelson *et al.*, 2003). Several public domain 2D and 2.5D models are available and they include FAST-MECH (McDonald *et al.*, 2005), STORM (Simões, 2011), MORPHO2D and NAYS, which are now available under the graphical interface iRIC platform, FESWMS-2DH (Froehlich, 1989), RMA2, TELEMAC2D and SRH-2D (Lai, 2009). Commercial software packages include MIKE21, Sobek 2D, TUFLOW, and RIVER2D (Steffler and Blackburn, 2002) (Table 3.5).

3.6 1D models

Numerical simulations of steady and unsteady flows at large spatial (stream network) and time scales (from several days to years) are typically studied with 1D models when cross-sectional averaged water elevation, water depths, velocities and shear stresses at each mesh node are the important hydraulic variables. Averaging the shallow water equations over the cross-sectional area leads to the flow equations known as the De Saint–Venant equations (Cunge *et al.*, 1980):

$$\frac{\partial \Omega}{\partial t} + \frac{\partial Q}{\partial x} = 0 \quad a. \text{ continuity equation}$$

$$\frac{\partial Q}{\partial t} + \frac{\partial}{\partial x}\left(\beta_\Omega U^2 \Omega\right) + g\Omega \frac{\partial h}{\partial x} = \underbrace{g\Omega\left(s_o - s_f\right)}_{c.\ kinematic\ wave}$$

$$\underbrace{\hspace{6cm}}_{d.\ diffusive\ wave} \quad (3.23)$$

$$\underbrace{\hspace{7cm}}_{e.\ dynamic,\ quasi\text{–}steady\ wave}$$

$$\underbrace{\hspace{8cm}}_{f.\ simple\ wave}$$

$$b.\ dynamic\ wave\ equation$$

where Q is discharge, Ω is the cross-sectional area, U is the mean cross-sectional velocity, h is the water depth, which corresponds in 1D to the hydraulic depth $h_h = \Omega/B$ where B is the surface water width, and β_Ω is a parameter that is typically set equal to 1 and accounts for secondary momentum transfer within the flow. The resistance term s_f in the De Saint–Venant equations accounts for all the energy losses caused by turbulence, shear stress at the bottom, cross-section irregularities, longitudinal cross-sectional variations, contraction–expansion losses and surface resistance (wind) (Morvan et al., 2008). The resistance term is typically expressed in terms of a roughness coefficient like n, C_z, k_s or f:

$$s_f = n^2 \frac{Q\,|Q|\,P_W^{4/3}}{\Omega^{10/3}} = \frac{Q\,|Q|\,P_W}{C_z^2\Omega^3} = f\frac{Q\,|Q|\,P_W}{8g\Omega^3} \quad (3.24)$$

Scaling analysis has shown that some terms may be neglected in certain cases because they are small compared to others or stream topographical resolution is too coarse to capture their significant effects. This leads to the parameterization of certain processes, which is usually accomplished in 1D modelling by adjusting the resistance coefficient. Simplified solutions for steady-state flows at the reach scale lead to the back-water and direct-step methods for subcritical and supercritical flows for prismatic channels and to the Bakhmeteff classification of water profiles (steep, mild, horizontal, adverse slope profiles) for gradually varying flows (Chaudhry, 1993). Unsteady flows have been studied extensively and have several simplified forms. The kinematic wave (Equation (3.23c)), which approximates the (kinematic) wave speed (wave celerity) $c = 5/3$ U for wide streams, retains only the resistance and gravity terms (s_0 and s_f). The parabolic or diffusive wave (Equation (3.23d)), which provides a (dynamic) wave speed $c = \sqrt{gh}$ in large streams, also retains the pressure term ($\partial h/\partial x$). The importance of diffusion along a wide reach of length L and slope s_0 can be quantified with the equation $C_D = 0.3\ h/(s_0\ L)$ where h is the water depth. Low values of C_D, typical of shallow and steep streams, suggest diffusion is negligible, and thus a kinematic wave is a good approximation in steep streams; large values, typical of deep and low-gradient streams, suggest diffusion is important (dynamic wave) such that the wave will spread and the peak will decrease significantly over the reach L. The quasi-steady wave (Equation (3.23e)) is applied when local accelerations ($\partial Q/\partial t$) are small compared to other terms, typical of a long hydrograph and local advective accelerations ($\partial\left(U^2\Omega\right)/\partial x$) are strong due to complex stream geometry modelled with detailed enough topography. The simple wave equation (Equation (3.23f)) is used for fast hydrographs, like dam breaks, where inertial and local accelerations are large compared to other terms (Chaudhry, 1993).

The De Saint–Venant equations are solved with both implicit and explicit schemes. The implicit scheme has shown good performance for subcritical flows whereas explicit schemes have proven to perform well in supercritical and mix flow conditions, although limited by the CFL and other stability conditions specific for each explicit scheme.

Several 1D models are available in the public domain; these include HEC-RAS (USACE, 2002), Sedimentation and River Hydraulics 1D (SRH-1D) (Huang and Greimann, 2010), FEQ (Franz and Melching, 1997) and as commercial software MIKE-11 and Sobek 1D (Table 3.5).

3.6.1 1D model setup

All these models require boundary conditions, cross-section geometry, resistance factors (calibration parameter) and initial conditions, which are typically uniform flows or a constant water surface elevation. Boundary conditions are water surface elevations (time series or discharge–stage relationships) at the downstream end in subcritical or mixed flows, which is typically the case in natural streams, or at the upstream end for supercritical flows (rare case and usually localized) and discharge at the upstream end and at each tributary (Table 3.7).

Stream topography is collected at cross-sections with a total station, engineering level and/or GPS (global positioning system) in wadable streams, and in large streams with sonar or airborne LiDAR (McKean et al., 2009). Topographical measurements of cross-sections should be extended upstream and downstream of the study area in order to minimize the influence of boundary conditions within the study site. Because 1D models provide results (cross-sectional average water elevation, velocities, shear stresses and discharge) only at locations where cross-sections are available, cross-section selection should be planned carefully, especially if transects are several channel widths apart (Castellarin et al., 2009; Pasternack and Senter, 2011). It is important to note that 1D models may have additional computational points, which are generated by interpolation between cross-sections along the stream.

Guidelines for selecting cross-section locations for 1D hydraulic modelling include references to topographical and hydraulic criteria (Cunge et al., 1980; Samuels, 1990), which includes selecting features that affect hydraulic parameters such as flow velocity and depth. These features include structures, changes in channel cross-section geometry and floodplains. Guidelines based upon a combination of theoretical analysis and experience were presented by Samuels (1990) and tested in a more recent study by Castellarin *et al.* (2009). Samuels (1990) recommended placing cross-sections at the beginning and end of the project, at the crest of any rapids, riffles or glides, at the beginning and end of transitions and at the beginning, middle and end of any meander. The maximum distance between sections is given as $\Delta x = \alpha B_f$ where α is a constant (with a recommended range from 10 to 20) and B_f is the bankfull water surface width (Samuels, 1990). The recommendation is to have at least four cross-sections within the backwater length for a river flowing subcritically, resulting in

$$\Delta x < 0.2\frac{\left(1 - F^2\right)h}{s_0} \approx 0.2\frac{h}{s_0} \text{ when } F^2 \to 0 \quad (3.25)$$

where F is the Froude number, $F = U/\sqrt{gh}$. Besides the hydraulic requirements, transect locations should also reflect the ecological applications of the 1D model (Payne *et al.*, 2004). For instance, applications of 1D models with PHABSIM analysis would require that equal effort is spent in terms of the number of cross-sections per morphological unit (e.g. pools, riffles and runs or mesohabitats) as well as in distributing transects within each morphological unit to obtain a proper statistical sampling of hydraulics within each habitat type (Payne *et al.*, 2004). However, others suggest placing transect locations randomly such that statistical analysis such as bootstrapping could be used to estimate WUA values and its uncertainty due to transect locations within the domain (Williams, 2010).

Discharge–stage relationships or time series of water surface elevations are required for unsteady flows as boundary conditions, typically at the last downstream cross-section. For steady-state simulations, only the stage at the associated discharge is required. Water elevations along the stream network at specified cross-sections should be measured at low, medium and high discharges within the range of investigated flows for calibration, with an intermediate discharge reserved for validation.

Calibration of 1D models is typically accomplished by adjusting the resistance factor, the first estimate of which should be based on field observations (Cowan, 1956; Chow, 1959; Arcement and Schneider, 1984), to minimize the difference between predicted and measured water surface elevations at a given discharge. Calibration can be done using one resistance factor for the entire stream or by using several values after dividing the stream into reaches with homogenous morphodynamic characteristics, e.g. the same bedforms and discharge (Wasantha Lal, 1995). Extension of the model beyond the calibra-

tion range could provide misleading results due to the inadequacy of the resistance factor (Morvan *et al.*, 2008).

1D models can provide good predictions of water elevations even with poor resolution of the cross-section survey, because calibration of the roughness coefficient may partially mask missing topographic information. This is especially true for models with reach-averaged cross-sections (Wiele *et al.*, 2007). Mean depth, velocity and shear stresses are predicted at the cross-section scale, but they are actually at a scale larger than the cross-section because cross-sections are constrained to vary gradually.

1D models do not provide local information and thus may be less adequate than 2D models for studying microhabitat or local scour and deposition depths. Some postprocessing in large streams with changes in water elevations that are small compared to water depths can provide local water depths, velocities and shear stresses by coupling 1D water elevations with high-resolution bathymetry. This technique is usually based on uniform flow equations (Manning's, Chezy or Darcy–Weisbach) used at the local scale with local water depths and provides pseudo-2D or 1.5D models (Ghanem *et al.*, 1996).

3.7 River floodplain interaction

Investigation of river and floodplain interactions may extend several hundred square kilometres such that application of 2D modelling with high-resolution grid size (e.g. 1–16 m^2) may be limited by computational time. Conversely, 1D models are fast but are not able to predict the flood wave accurately with large lateral inflows and outflows, which typically occur for discharges larger than bankfull as water inundates floodplains. This limitation has driven the development of another type of pseudo-2D model, which has been applied in studying floodplain inundation areas and processes (e.g. riparian vegetation succession). Models of this type have a 1D model for in-channel hydraulics but use different approaches for schematizing floodplain processes. Three techniques have been used so far: geographic information system (GIS) models, raster or compartment models (e.g. Bates *et al.*, 2010) and quasi-two-dimensional hydrodynamic models (e.g. Benjankar *et al.*, 2011). The first extends in-channel water levels predicted by 1D models over the floodplain regardless of any mass or momentum conservation. The second divides the floodplain into cells with size equal to raster cells or larger compartments delineated by natural or man-made obstructions. Water levels within cells are calculated with the mass conservation equation.

Fluxes among cells and with the stream are quantified with appropriate equations, which could be the uniform flow, gradually varied flow or weir equations (Bates *et al.*, 2010). The third type dynamically couples a 1D model in the channel with a 2D model in the floodplain (Benjankar *et al.*, 2011).

1.5D models have the advantage of covering large areas while minimizing computational time with respect to fully 2D models. Examples of the first type of model described above are HEC-RAS and MIKE-GIS; examples of the second type are LISFLOOD-FP (which is a 2D raster model based on Manning's equation) (Bates *et al.*, 2010); quasi-2D hydrodynamic models include Sobek 1D-2D and MIKE-FLOOD (Table 3.5).

3.8 Non-numerical hydraulic modelling

A number of alternative methods to numerical hydrodynamic models for the description of river hydraulics have been developed.

Traditional field surveys, which are based on selecting morphologically and hydrologically representative reaches within the entire project site, are still the most common approach. They use measurements taken along transects, the location of which typically depends on stakeholder consensus and is constrained by the desire for proportionate sampling of mesohabitats (Bovee *et al.*, 1998; Payne *et al.*, 2004). Other literature suggests that measurements should be taken at random spots or random cross-sections in order to support statistical analysis (Williams, 2010). However, field measurement techniques have recently seen a dramatic advancement with the potential to provide continuous mapping of stream physical properties at high resolution over the entire domain (Carbonneau *et al.*, 2012). They include ADV and ADCP for measuring 3D flow structures, flow depth and flow turbulence properties at single points or continuous mapping of the flow field (Buffin-Bélanger and Roy, 2005). Recently, ADCP have been used to characterize streambed material (Shields, 2010) along with high-resolution imagery (Dugdale *et al.*, 2010) and LiDAR (Hodge *et al.*, 2009). GPS-based (Stockdale *et al.*, 2008) and remote sensing (Plant *et al.*, 2005) techniques have shown promising results to map surface water velocity, which can be related via a multiplier to depth-averaged flow velocity (Pasternack and Senter, 2011, pp. 59–60). Remote sensing has been used to detect water elevations and river bathymetry (McKean *et al.*, 2009; Walther *et al.*, 2011). The use of this multidimensional and integrative data collection equipment (Stone, 2005) will provide the

tools to delineate the riverscape continuously in space (Carbonneau *et al.*, 2012).

The results of these measurements are then related to site-specific habitat suitability curves to quantify habitat quality for aquatic and semi-aquatic organisms (White *et al.*, 2002; Moir and Pasternack, 2009; Whited *et al.*, 2011). Furthermore, a standard methodology has been developed from these criteria to obtain a set of benchmark river discharges, as suggested by Petts (1996): threshold ecological flow, acceptable ecological flow, desirable ecological flow, optimum ecological flow, channel maintenance and habitat maintenance flows. Each of these is defined and differentiated based on its impact on hydraulic habitats (Parasiewicz and Dunbar, 2001).

Usually, measurements are taken for a series of flows, and intermediate flows are evaluated by interpolations. The major issue of measurement-based models is their limitation for scenario-based or objective testing analyses, because the predictive capacity of these models is restricted within the range of measured flows and they do not allow for a change of topography, which is necessary for potential rehabilitation design actions likely after baseline ecohydraulic analysis (Elkins *et al.*, 2007).

Recently, Escobar-Arias and Pasternack (2010) proposed the functional flows model that assesses the time domain of the hydrological regime to achieve the local hydraulics necessary for specific ecological functions. They used analytical solutions based on uniform flow to characterize the flow hydraulics.

Lamouroux *et al.* (1995) used simple input data on a reach or mesohabitat scale to derive frequency distributions of various hydraulic variables, e.g. relative local point flow velocities $f(u/U)$, where u is the local flow velocity and U is the mean flow velocity of a river reach. Input parameters for the frequency distribution of the flow properties are reach-averaged Froude numbers and the roughness/depth ratios, mean roughness expressed in particle size. This means that discharge, depth, width and grain size have to be measured or known. Lamouroux *et al.* (1995) applied their model mostly in natural streams and it is not known how the model would perform in channelized rivers.

3.9 Case studies

This section describes four case studies where the underlying hydrodynamic processes drive certain ecological functions, such as the availability of habitats for certain fish or types of vegetation. The choice of the appropriate model consists firstly in the identification of the primary physical processes that affect a certain biotic system and thus must be modelled with sufficient accuracy and with adequate simulation time. Secondly, input data need to be adequate to the spatial and temporal range of the selected physical processes. Thirdly, software and trained personnel must be available and sufficient time and effort must be dedicated for the study.

The typical aim is to have the physical processes simulated very accurately, but this is not always achievable and it is not always necessary. It is not yet feasible to simulate local three-dimensional flow fields for large rivers over several hundred kilometres. Additionally, simulation run time should be within the expected project duration. For instance, the simulation time of real-time flood forecasting simulations must be significantly shorter than the flood itself, otherwise results cannot be used to identify flood management activities. Consequently, these limitations presently require simplifying the system, while important processes should be retained. This often leads to hybrid or mixed-model approaches where large-scale processes are simulated with one model, the results of which are then used as boundary conditions for smaller-scale, high-resolution models, which are applied at short river reaches. Meanwhile, several standard situations such as simple in-stream flow investigations of diversion type hydropower plants in single-channel rivers may not require such considerations. The following case studies explain the processes in selecting hydrodynamic numerical models for their ecohydraulic applications.

3.9.1 1D modelling of the Kootenai River, Idaho, USA

The Kootenai River is one of the largest tributaries of the Columbia River in the Pacific Northwest of Canada and the USA. Libby Dam (Montana, USA) entirely regulates the flow regime of the lower part of the Kootenai River for more than 200 km downstream of the dam. Numerous research activities have been investigating the effects of Libby Dam operation along the downstream reaches of the Kootenai River on multiple issues such as sturgeon reproduction, nutrient depletion, sediment transport (Barton et al., 2005), cottonwood (Populus spp) recruitment (Burke et al., 2009) and vegetation dynamics (Benjankar, 2009; Benjankar et al., 2011).

Cottonwood recruitment succeeds only in locations where specific hydraulic processes take place during a given year. These processes can be described in a very simplified manner by the presence of barren ground (sand or fine gravel, freshly cleaned by spring freshet), which must be wet when the seeds fall on it, and water levels

receding at a rate of approximately 21 cm per week to allow cottonwood seedlings to grow (Burke et al., 2009). The timing of these conditions needs to coincide with the period when cottonwoods shed their seeds, which is typically between May and June. Consequently, cottonwood recruitment modelling requires predicting water level fluctuations over long spatial (for several tens of kilometres of stream segments) and temporal (for several months) scales. A 1D unsteady model would be an adequate choice, because it provides water surface elevation fluctuations at the cross-sectional scale with a minimal amount of input data, which consists of measured cross-sections (bathymetry) and measured water surface elevations (for calibration). The model can run for an entire recruitment period in a reasonably short time to provide wetted areal extension and water surface elevation recession along each bank. The influence of the dam operation is represented through different inflow hydrographs at the upstream boundary of the model. Figure 3.9 shows flow stage fluctuations simulated with a 1D unsteady flow model (MIKE11) for one year at four locations along the river reach downstream of Libby Dam. The recession rates of the water surface elevation can be calculated at each cross-section during the period of interest, as shown in Figure 3.9.

3.9.2 Pseudo-2D modelling of the Biobío River, Chile

Several dams in the upstream catchment of the Biobío River (Chile) severely influence the stream. Dam operations change the natural hydrological regime and impose a hydro-peaking regime with very rapid discharge fluctuations onto the downstream reaches of the river. García et al. (2011) investigated the change of habitat quantity and quality for a native fish species (Trichomycterus areolatus) between natural and dam-regulated flow regimes. They used the fish aquatic habitat software CASiMiR to evaluate fish habitat along a 2-km-long braided reach, which was considered morphologically and hydraulically representative of the entire lower section of the river downstream of the dams.

The aerial pictures of the stream reach show a complex multi-thread reach with branches and vegetated islands where 1D models would not yield good results (Figure 3.10). However, the research team (García et al., 2011) did not have access to and had little experience of working with 2D models, but they had experience with 1D models. Thus, they envisioned an alternative methodology to study this system by splitting the braided sections into individual single-thread channels, shown in Figure 3.10. They used a series of flow measurements simultaneously

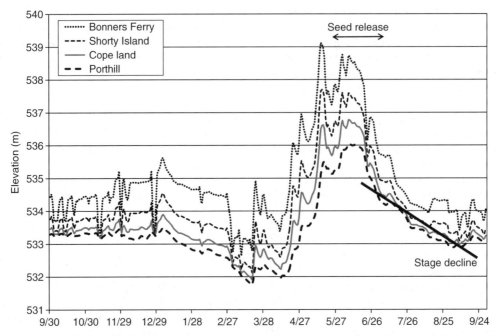

Figure 3.9 Annual stage hydrographs for water year 1997 for locations adjacent to Bonners Ferry, Shorty's Island, Copeland and Porthill, Kootenai River, Idaho, USA. Modified from Burke (2006).

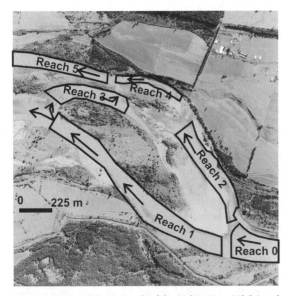

Figure 3.10 Braided river reach of the Biobío River (Chile) and six single-thread reaches used for hydrodynamic and fish habitat modelling; arrows indicate the flow direction. Modified from García, *et al.* (2011).

collected along the channel upstream and along the individual channels of the study reach to build a simple quantitative distribution model of the water flowing in each individual channel at any given discharge. Successively, they applied a steady-state 1D model (HEC-RAS) to determine water surface elevations and cross-section flow velocities in each individual channel. Results of the 1D simulations were used as input values for CASiMiR, which uses an integrated rule-based approach to calculate local 1D velocities based on cross-section geometry and local depth, hence providing pseudo-2D results. The pseudo-2D results for velocities and depths were coupled with a fuzzy logic biological approach to quantify the spatial distribution of the aquatic habitat status expressed with a cell suitability index (SI) at different discharges (Figure 3.11).

3.9.3 2D modelling of the Saane River, Switzerland

The Saane River (Switzerland) is an alpine river impacted by hydropower operations, which may cause fish stranding during rapid drawdown. A critical section of the river for stranding of juvenile bullhead was the confluence between the Saane and one of its tributaries, the Sense River, which enters from the right side of Figure 3.12 and is located 10 km downstream of a hydropower facility.

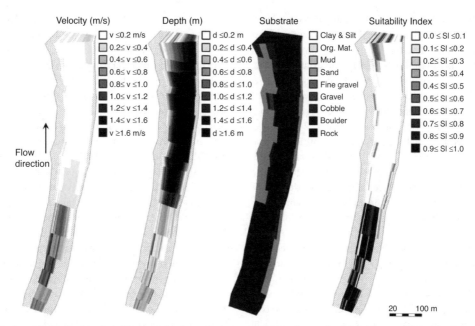

Figure 3.11 Local flow velocities in 1.5D, water depths and other parameters relevant for fish habitat modelling for the upper left segment. Reproduced from García *et al.* (2011), by permission of John Wiley & Sons.

Conceptually, stranding risk increases with quick shallowing of water depths during down-ramping operations. Stranding risk reaches a critical threshold at a location where both the water depth is shallow for a bullhead and water depth shoals faster than 2 cm in a ten-minute period. The main objective of this investigation was to identify streambed locations with high stranding risk for juvenile bullhead.

The confluence is a complex section, which shows braided flow patterns during low flows, and which becomes a single-thread channel during high flows. Figure 3.12 shows the local flow velocities and their directions based on the results of 2D hydrodynamic simulations conducted with Hydro-AS software. The complex flow pattern with eddies, lateral flow divergence and convergence (Figure 3.12) cannot be simulated with a 1D model. The

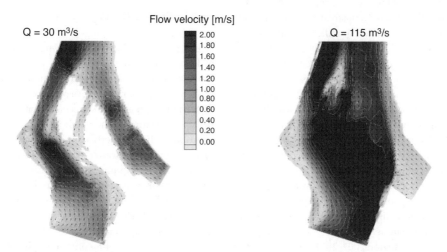

Figure 3.12 Local flow velocities in a test reach of the Saane, Switzerland, at two different flows. Contours of the same velocity (isovels) are at 0.2 m/s intervals. Modified from an unpublished report by SJE Schneider & Jorde Ecological Engineering Ltd.

approach adopted for the Biobío River analysis would not easily be applicable in this section due to the complex flow and unsteady flow behaviour during down-ramping.

As a result, this analysis requires a 2D hydraulic model. The size of study reach allowed a detailed survey of the channel bathymetry and flow measurements, while results of a large-scale 1D model provided the upstream and downstream boundary conditions. Figure 3.12 shows the depth-averaged flow distributions at the confluence for a constant discharge of 15 m^3/s in the Sense River and 30 and 115 m^3/s for the Saane River on the left and right panel, respectively. The model predicted water surface elevations within 3 cm accuracy as compared to measurements at the two different flows (SJE Schneider & Jorde Ecological Engineering, 2007).

The routed flood waves from the reservoir releases (hydrographs) during rapid up- and down-ramping of the hydropeaking were used for the investigations. The results of the hydrodynamic models were coupled with CASiMiR to identify the spatial distributions of the maximum decrease rates of water depths and areas that detach hydraulically from the main channel. These results are indicators for stranding risk levels. Water depth decreases reached maximum rates of 15 cm in 10 min in some locations, posing a very high risk of stranding. The results were used for mitigating down-ramping operations at the hydropower facility.

3.9.4 3D modelling of salmon redds

During construction of their egg nests (redds), female salmon dig a pit in which they lay their eggs and then cover them with the spoils of a second upstream pit (Bjornn and Reiser, 1991). Redd construction modifies channel topography, creating a pit and a downstream hump (tailspill), the dimensions of which can be comparable to macro-scale bedforms, like dunes. Additionally, spawning activity winnows fine grains from the streambed, resulting in relatively coarser and more porous sediment, with higher hydraulic conductivity than the undisturbed bed material (Hassan et al., 2008). Tonina and Buffington (2009) used two 3D CFD models to simulate channel hydraulics (surface water) and shallow hyporheic flow (groundwater) through a single pool-riffle sequence with and without a redd placed at the pool tail (a typical spawning location because of downwelling hyporheic flow induced by near-bed pressure variations caused by pool-riffle topography (Tonina and Buffington 2007, 2011; Marzadri et al., 2010)). Hyporheic flows are mostly stream water flowing within the sediment beneath and alongside streams (e.g. Tonina and Buffington, 2009; Tonina, 2012). Their

goal was to investigate the effects of the redd on surface and subsurface flows. Their study examined whether the presence of salmon nests significantly changes local river hydraulics and local patterns of upwelling and downwelling fluxes in and out of the sediment. The latter are important because they are the major mechanism that brings dissolved oxygen to the egg pockets. They decoupled the surface and subsurface flows because the mass and momentum exchange at the surface–subsurface boundary (Ruff and Gelhar, 1972; Vollmer et al., 2002; Packman et al., 2004) is typically small (e.g. Packman and Bencala, 1999). Thus, the streambed was treated as an immobile and impermeable wall. The stream flow was modelled with the RANS equations and the subsurface as a Darcian flow (Tonina and Buffington, 2009). The surface water model was simulated first because it provided the near-bed pressure distribution, which was used as a boundary condition for the groundwater model. Because previous investigations showed that flow separation may occur at the pit and crest of the redd, a full 3D model for the stream flow was warranted and not a 2D model, which would not have captured the flow separation and thus would have provided a different pressure distribution at the stream bottom. They used Fluent 6.1 with realizable $\kappa-\varepsilon$ model (Shih et al., 1995) and standard wall function for the near-wall treatment to simulate the surface flow. The volume of fluid option was selected to predict the water surface (Hirt and Nichols, 1981). Figure 3.13 shows the predicted near-bed pressure distributions with and without a redd located near the riffle crest.

3.10 Conclusions

There is no general rule on when, why and which hydraulic model should be applied for ecohydraulic studies. Advances in computational capability and in numerical hydrodynamic model development have made multi-dimensional models more accessible than in the past, and will continue to accelerate their efficiency and accessibility. Models that formerly could only run on computers operated by large research facilities are today running on any personal computer, and progress continues. However, users of hydrodynamic numerical models should have a good understanding of the reasons for and the implications of selecting a certain type of hydrodynamic model for their specific task.

The most important component in any numerical model is accurate topographical survey at the resolution of the scale of the processes investigated (Casas et al., 2010). Flow predictions will have low performance if mesh and

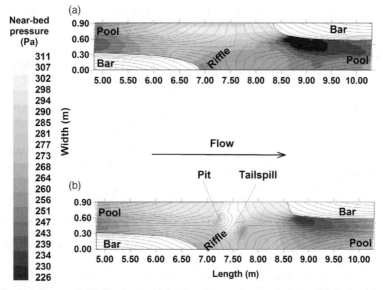

Figure 3.13 Predicted near-bed pressure distributions, which drive stream water into and out of the sediment, (a) without and (b) with a redd at the tail of the pool. Note that the redd generates a high-pressure zone along the tailspill, between the pit and the redd crest. This high-pressure zone drives oxygen-rich stream waters into the sediment toward the embryos. Whereas the low-pressure zone after the redd crest drives low-oxygen hyporheic waters from the sediment into the stream. Modified from Tonina and Buffington (2009). Reproduced by permission of Canadian Science Publishing.

topography have poor resolution (Horritt *et al.*, 2006). Mismatch between the scales of the ecological processes and those of the hydraulic model must be avoided or minimized (Railsback, 1999). Field data must be collected at spatial scales comparable to the resolution of the hydraulic model.

3D modelling is considered the approach with the most predictive capability, because it explicitly accounts for most of the flow processes, with the exception of turbulence, which may be parameterized. However, applications have shown that uncertainty due to the turbulence closure may have fewer effects on model results than those of stream topography, boundary condition information and mesh quality (USACE, 1996; Lane, 1998; Lane *et al.*, 2004; Pasternack and Senter, 2011). Their use is still limited to some special cases because little ecological and aquatic habitat knowledge is available to take advantage of the full dimensionality and the long computational time. On the other hand, 2D models provide flow properties which better match the present available ecological and biological knowledge. Consequently, they are becoming the mainstream tools for ecohydraulic applications both at the reach (10–50 channel width) and segment (50–1000 channel width) scales. They use simplified turbulence closures based on isotropic turbulence, which may be adequate for most applications.

1D modelling uses rather crude engineering techniques to calculate water surface elevations and then calculate individual cell velocities as described for the 1.5 or pseudo-2D approaches. They are cross-sectional based and thus selection of cross-section locations is very important (Pasternack and Senter, 2011). They require calibration because the roughness term lumps together several energy losses from the grain to the reach scales. Therefore, predictions outside the range of observed discharges are questionable. In the case of low flows, modelling becomes more inaccurate because of the increasing roughness of the streambed. 1D models are efficient tools when water surface elevations, wetted area extensions and flood durations are the major information required for the ecohydraulic applications (Benjankar *et al.*, 2011; Benjankar *et al.*, 2012). 1D models yield reasonable flow velocity results where flows are primarily unidirectional. Consequently, comparison between 1D and 2D approaches in aquatic habitat modelling shows rather small differences where the flow is primarily unidirectional (Schneider, 2001), which is often the case in regulated and channelized rivers, and substantial where it is not (Brown and Pasternack, 2009), typically in natural and pristine rivers. Differences of aquatic habitat suitability index results predicted with 1D and 2D hydraulic models may be smaller than the actual flow difference because flow velocities are only one among several attributes (e.g. depth, substrate and distance from cover). Therefore, careful selection of the model by evaluating the stream characteristics, questions and data available is recommended.

Acknowledgements

The authors thank Dr A. Garcia-Lancaster, Dr R. Benjankar and M. Burke for providing their data and Prof. G. Pasternack and Prof. S. Lane for their constructive suggestions and comments on an earlier version.

References

Absi, R. (2006) A roughness and time dependent mixing length equation. *Doboku Gakkai Ronbunshuu B (Japan Society of Civil Engineers)*, **62**(4): 437–446. doi:10.2208/jscejb.62.437.

Ackers, P., White, W.R., Perkins, J.A. and Harrison, A.J.M. (1978) *Weirs and Flumes for Flow Measurement*, John Wiley & Sons, Ltd, Chichester, UK.

AIAA (1998) *AIAA guide for the verification and validation of computational fluid dynamic simulations*, American Institute of Aeronautics and Astronautics.

Akanbi, A.A. and Katapodes, N.D. (1988) Model for flood propagation on initial dry land. *Journal of Hydraulic Engineering*, **114**(7): 689–706.

Alavian, V. and Chu, V. (1985) Turbulent exchange flow in shallow compound channel. *Proceedings of the 21st Congress of the IAHR*, IAHR, Melbourne, Australia.

Alfredsen, K., Borsanyi, P., Harvy, A., Fjeldstad, H.-P. and Wersland, S.-E. (2004) Application of habitat modelling in river rehabilitation and artificial habitat design. *Hydroécologie Appliquée*, **14**(2): 105–117. doi: 10.1051/hydro:2004007.

Anderson, J.D. (1995) *Computational Fluid Dynamics: The Basics with Applications*, McGraw Hill.

Anderson, M.P. and Woessner, W.W. (1992) *Applied Groundwater Modeling: Simulation of Flow and Advective Transport*, Academic Press, New York, NY.

Ang, W.T. (2007) *A Beginner's Course in Boundary Element Methods*, Universal Publishers, Boca Raton, FL.

Arcement, G.J. and Schneider, V.R. (1984) *Guide for selecting Manning's roughness coefficients for natural channels and flood plains*. Final Report FHWA-TS-84e204, Federal Highway Administration.

ASCE Task Committee on Turbulence Models in Hydraulics Computations (1988) Turbulence modeling of surface water flow and transport. *Journal of Hydraulic Engineering*, **114**(9): 970–1073.

Ayllón, D., Almodóvar, A., Nicola, G.G. and Elvira, B. (2011) The influence of variable habitat suitability criteria on PHAB-SIM habitat index results. *River Research and Applications*. doi: 10.1002/rra.1496.

Baart, I., Gschöpf, C., Blaschke, A.P., Preiner, S. and Hein, T. (2010) Prediction of potential macrophyte development in response to restoration measures in an urban riverine wetland. *Aquatic Botany*, **93**(3): 153–162. doi:10.1016/j.aquabot.2010.06.002.

Balzano, A. (1998) Evaluation of methods for numerical simulation of wetting and drying in shallow water flow model. *Coastal Engineering*, **34**: 83–107.

Barnes, H.H.J. (1967) *Roughness Characteristics of Natural Channels*, United States Government Printing Office, Washington, DC.

Barton, G.J., McDonald, R.R., Nelson, J.M. and Dinehart, R.L. (2005) *Simulation of flow and sediment mobility using a multidimensional flow model for the white sturgeon critical-habitat reach, Kootenai River near Bonners Ferry, Idaho*. Scientific Investigations Report 2005-5230, U.S. Geological Survey.

Bates, P.D., Horritt, M.S. and Fewtrell, T.J. (2010) A simple inertial formulation of the shallow water equations for efficient two-dimensional flood inundation modelling. *Journal of Hydrology*, **387**(1–2): 33–45. doi: 10.1016/j.jhydrol.2010.03.027.

Baxter, C.V. and Hauer, R.F. (2000) Geomorphology, hyporheic exchange, and selection of spawning habitat by bull trout (*Salvelinus confluentus*). *Canadian Journal of Fisheries and Aquatic Sciences*, **57**: 1470–1481.

Benjankar, R. (2009) *Quantification of reservoir operation-based losses to floodplain physical processes and impact on the floodplain vegetation at the Kootenai River, USA*. Unpublished PhD dissertation, University of Idaho, Moscow, Idaho.

Benjankar, R., Egger, G., Jorde, K., Goodwin, P. and Glenn, N.F. (2011) Dynamic floodplain vegetation model development for the Kootenai River, USA. *Journal of Environmental Management*, **92**(12): 3058–3070. doi: 10.1016/j.jenvman.2011.07.017.

Benjankar, R., Jorde, K., Yager, E.M., Egger, G. Goodwin, P. and Glenn, N.F. (2012) The impact of river modification and dam operation on floodplain vegetation succession trends in the Kootenai River, USA. *Ecological Engineering*, **46**:88–97. http://dx.doi.org/10.1016/j.ecoleng.2012.05.002.

Benjankar, R. Koenig, F. and Tonina, D. (2013) Comparison of hydromorphological assessment methods: application to the Boise River, USA. *Journal of Hydrology*. doi: 10.1016/j.jhydrol.2013.03.017.

Bernard, R.S. (1992) *STREMR: numerical model for depth-averaged incompressible flow*. US Army Corps Engineers, Waterways Experiment Research Station, Vicksburg, MS.

Bernard, R.S. and Schneider, M.L. (1992) *Depth-averaged numerical modelling for curved channels*. US Army Corps of Engineers, Waterways Experiment Research Station, Vicksburg, MS.

Beven, K. and Freer, J. (2001) Equifinality, data assimilation, and uncertainty estimation in mechanistic modelling of complex environmental systems using the GLUE methodology. *Journal of Hydrology*, **249**(1–4): 11–29. doi: 10.1016/S0022-1694(01)00421-8.

Bjornn, T.C. and Reiser, D.W. (1991) Habitat requirements of salmonids in streams. In Meehan, W.R. (ed.) Influence of Forest and Rangeland Management on Salmonid Fishes and Their Habitats, *American Fisheries Society Special Publication*, **19**: 83–138.

Booker, D.J., Dunbar, M. and Ibbotson, A. (2004) Predicting juvenile salmonid drift-feeding habitat quality using a three-dimensional hydraulic-bioenergetic model. *Ecological Modelling*, **177**: 157–177.

Booker, D.J., Sear, D.A. and Payne, A.J. (2001) Modelling three-dimensional flow structures and patterns of boundary shear stress in a natural pool-riffle sequence. *Earth Surface Processes and Landforms*, **26**(5): 553–576.

Bourban, S., Durand, N., Colameo, S. and Tchamen, G. (2006) 3D modelling as an alternative to comprehend the hydraulic behaviour of spillways. *Proceedings of the Annual Canadian Dam Association*, 30 September–5 October, 2006.

Bovee, K.D. (1978) The incremental method of assessing habitat potential for cool water species, with management implications. *American Fisheries Society Special Publication*, **11**: 340–346.

Bovee, K.D. (1982) *A guide to stream habitat analysis using the instream flow incremental methodology. Instream Flow Information*, US Fish and Wildlife Service, Fort Collins, Colorado.

Bovee, K.D., Lamb, B.L., Bartholow, J.M., Stalnaker, C.B., Taylor, J. and Henriksen, J. (1998) *Stream habitat analysis using the instream flow incremental methodology*, U.S. Geological Survey, Biological Resources Division Information and Technology.

Brasington, J., Rumsby, B.T. and McVey, R.A. (2000) Monitoring and modelling morphological change in a braided gravel-bed river using high resolution GPS-based survey. *Earth Surface Processes and Landforms*, **25**(9): 973–990.

Brown, R.A. and Pasternack, G.B. (2009) Comparison of methods for analyzing salmon habitat rehabilitation designs for regulated rivers. *River Research and Applications*, **25**: 745–772. doi: 10.1002/rra.1189.

Buffin-Bélanger, T. and Roy, A.G. (2005) 1 min in the life of a river: selecting the optimal record length for the measurement of turbulence in fluvial boundary layers. *Geomorphology*, **68**: 77–94.

Buffington, J.M., Montgomery, D.R. and Greenberg, H.M. (2004) Basin-scale availability of salmonid spawning gravel as influenced by channel type and hydraulic roughness in mountain catchments. *Canadian Journal of Fisheries and Aquatic Sciences*, **61**: 2085–2096. doi: 10.1139/F04-141.

Buffington, J.M. and Tonina, D. (2009) Hyporheic exchange in mountain rivers II: Effects of channel morphology on mechanics, scales, and rates of exchange. *Geography Compass*, **3**(3):1038–1062

Bunte, K. and Abt, S.R. (2001) *Sampling surface and subsurface particle-size distributions in wadable gravel- and cobble-bed streams for analyses in sediment transport, hydraulics, and streambed monitoring*, U.S. Department of Agriculture, Forest Service, Rocky Mountain Research Station, Fort Collins, CO.

Burke, M.P. (2006) *Linking hydropower operation to modified fluvial processes downstream of Libby Dam, Kootenai River, USA and Canada.* Unpublished MS thesis, University of Idaho, Moscow, ID.

Burke, M.P., Jorde, K. and Buffington, J.M. (2009) Application of a hierarchical framework for assessing environmental impacts of dam operation: Changes in streamflow, bed mobility and recruitment of riparian trees in a western North American river. *Journal of Environmental Management*, **90**(3): S224–S236. doi: 10.1016/j.jenvman.2008.07.022.

Burrows, A.D. and Steffler, P.M. (2005) Depth averaged modeling of flow around individual boulders. *Proceedings of the 17th Canadian Hydrotechnical Conference*, CSCE, Edmonton, Canada.

Caamaño, D., Goodwin, P. and Buffington, J.M. (2010) Flow structure through pool-riffle sequences and a conceptual model for their sustainability in gravel-bed rivers. *River Research and Applications*. doi:10.1002/rra.1463.

Capra, H., Breil, P. and Souchon, Y. (1995) A new tool to interpret magnitude and duration of fish habitat variations. *Regulated Rivers: Research and Management*, **10**: 281–289.

Carbonneau, P.E., Fonstad, M.A., Marcus, W.A. and Dugdale, S.J. (2012) Making riverscapes real. *Geomorphology*, **137**(1): 74–86. doi: 10.1016/j.geomorph.2010.09.030.

Cardenas, M.B. and Wilson, J.L. (2007) Hydrodynamics of coupled flow above and below a sediment–water interface with triangular bedforms. *Advances in Water Resources*, **30**: 301–313.

Carrivick, J.L. (2007) Hydrodynamics and geomorphic work of jökulhlaups (glacial outburst floods) from Kverkfjöll volcano, Iceland. *Hydrological Processes*, **21**: 725–740. doi: 10.1002/hyp.6248.

Casas, M.A., Lane, S.N., Hardy, R.J., Benito, G. and Whiting, P.J. (2010) Reconstruction of subgrid-scale topographic variability and its effect upon the spatial structure of three-dimensional river flow. *Water Resources Research*, **46**: W03519. doi: 10.1029/2009WR007756.

Castellarin, A., Di Baldassarre, G., Bates, P.D. and Brath, A. (2009) Optimal cross-sectional spacing in Preissmann scheme 1D hydrodynamic models. *Journal of Hydraulic Engineering*, **135**(2): 96–105.

Chaudhry, M.H. (1993) *Open-Channel Flow*, Prentice-Hall Inc., Englewood Cliffs, NJ.

Cheng, A.H.-D. and Cheng, D.T. (2005) Heritage and early history of the boundary element method. *Engineering Analysis with Boundary Elements*, **29**: 268–302.

Chow, V.T. (1959) *Open-Channel Hydraulics*, McGraw-Hill Book Co., New York, NY.

Clark, J.S., Rizzo, D.M., Watzin, M.C. and Hession, W.C. (2008) Spatial distribution and geomorphic condition of fish habitat in streams: an analysis using hydraulic modelling and geostatistics. *River Research and Applications*, **24**(7): 885–899. doi: 10.1002/rra.1085.

Clifford, N.J., Robert, A. and Richards, K.S. (1992) Estimation of flow resistance in gravel-bedded rivers: A physical explanation of the multiplier of roughness length. *Earth Surface Processes and Landforms*, **17**(2): 111–126.

Clifford, N.J., Soar, P.J., Harmar, O.P., Gurnell, A.M., Petts, G.E. and Emery, J.C. (2005) Assessment of hydrodynamic simulation results for eco-hydraulic and eco-hydrological applications: a spatial semivariance approach. *Hydrological Processes*, **19**: 3631–3648. doi: 10.1002/hyp.5855.

Conner, J.T. (2011) *Effect of cross-section interpolated bathymetry on 2D hydrodynamic results in a large river*, University of Idaho, Moscow, ID, 58 pp.

Cowan, W.L. (1956) Estimating hydraulic roughness coefficients. *Agricultural Engineering*, **37**(7): 473–475.

Crowder, D.W. and Diplas, P. (2000a) Evaluating spatially explicit metrics of stream energy gradients using hydrodynamic model simulations. *Canadian Journal of Fisheries and Aquatic Sciences*, **57**: 1497–1507.

Crowder, D.W. and Diplas, P. (2000b) Using two-dimensional hydrodynamic models at scales of ecological importance. *Journal of Hydrology*, **230**: 172–191.

Crowder, D.W. and Diplas, P. (2002) Vorticity and circulation: spatial metrics for evaluating flow complexity in stream habitats. *Canadian Journal of Fisheries and Aquatic Sciences*, **59**: 633–645. doi: 10.1139/F02-037.

Cunge, J.A., Holly, F.M. and Verwey, A. (1980) *Practical Aspects of Computational River Hydraulics*, Pitman Advanced Publishing Program, Boston.

Daraio, J.A., Weber, L.J., Newton, T.J. and Nestler, J.M. (2010) A methodological framework for integrating computational fluid dynamics and ecological models applied to juvenile freshwater mussel dispersal in the Upper Mississippi River. *Ecological Modelling*, **221**: 201–214.

Deltares (2011) User manual Delft3D-FLOW, Deltares, Delft, The Netherlands.

DHI Software (2001) *MIKE 3, Estuary and coastal hydraulics and oceanography*, Hørsholm, DK.

Dietrich, W.E. and Smith, D.J. (1983) Influence of the point bar on flow through curved channels. *Water Resources Research*, **19**(5): 1173–1192.

Dietrich, W.E. and Whiting, P.J. (1989) Boundary shear stress and sediment transport in river meanders of sand and gravel. In Ikeda, S. and Parker, G. (eds) *Water Resources Monograph 12: River Meandering*, American Geophysical Union, Washington DC, pp. 1–50.

Duan, J.G. (2004) Simulation of flow and mass dispersion in meandering channels. *Journal of Hydraulic Engineering*, **130**(10): 964–976. doi: 10.1061/(ASCE)0733-9429(2004)130:10(964).

Dugdale, S.J., Carbonneau, P.E. and Campbell, D. (2010) Aerial photosieving of exposed gravel bars for the rapid calibration of airborne grain size maps. *Earth Surface Processes and Landforms*, **35**(6): 627–639. doi: 10.1002/esp.1936.

Elkins, E.M., Pasternack, G.B. and Merz, J.E. (2007) The use of slope creation for rehabilitating incised, regulated, gravel-bed rivers. *Water Resources Research*, **43**: W05432. doi: 10.1029/2006WR005159.

Escobar-Arias, M.I. and Pasternack, G.B. (2010) A hydrogeomorphic dynamics approach to assess in-stream ecological functionality using the functional flows model, part 1-model characterstics. *River Research and Applications*, **26**: 1103–1128. doi:10.1002/rra.1316.

Fausch, K.D., Torgersen, C.E., Baxter, C.V. and Li, H.W. (2002) Landscapes to riverscapes: Bridging the gap between research and conservation of stream fishes. *Bioscience*, **52**(6): 483–498.

Ferziger, J.H. and Peric, M. (2002) *Computational Methods for Fluid Dynamics*, Springer, New York, NY.

Finlavson, B.A. (1972) *The Method of Weighted Residuals and Variational Principles: With Application in Fluid Mechanics, Heat and Mass Transfer*, Academic Press, Inc., New York, NY.

Fischer, H.B., List, E.J., Koh, R.C.Y., Imberger, J. and Brooks, N.H. (1979) *Mixing in Inland and Coastal Waters*, Academic Press, San Diego, CA.

Fluent Inc. (2003) *Fluent User's Guide*, Distributor: Fluent US, Lebanon, New Hampshire.

Fonstad, M.A. and Marcus, W.A. (2005) Remote sensing of stream depths with hydraulically assisted bathymetry (HAB) models. *Geomorphology*, **72**: 320–339. doi:10.1016/j.geomorph.2005.06.005.

Franz, D.D. and Melching, C.S. (1997) *Full Equations (FEQ) model for the solution of the full, dynamic equations of motion for one-dimensional unsteady flow in open channels and through control structures*, U.S. Geological Survey.

French, J.R. and Clifford, N.J. (2000) Hydrodynamic modelling as a basis for explaining estuarine environmental dynamics: some computational and methodological issues. *Hydrological Processes*, **14**(11–12): 2089–2108.

Frissell, C.A., Liss, W.J., Warren, C.E. and Hurley, M.D. (1986) A hierarchical framework for stream habitat classification. *Environmental Management*, **10**: 199–214.

Froehlich, D.C. (1989) *HW031.D – Finite Element Surface-Water Modeling System: Two-Dimensional Flow in a Horizontal Plane – Users' Manual*, Federal Highway Administration Report.

García, A., Jorde, K., Habit, E., Caamaño, D. and Parra, O. (2011) Downstream environmental effects of dam operations: changes in habitat quality for native fish species. *River Research and Applications*, **27**(3): 312–327. doi: 10.1002/rra.1358.

Geleynse, N., Storms, J.E.A., Walstra, D.-J.R., Jagers, H.R.A., Wang, Z.B. and Stive, M.J.F. (2011) Controls on river delta formation: insights from numerical modelling. *Earth and Planetary Science Letters*, **302**: 217–226.

Ghanem, A. (1995) *Two Dimensional Finite Element Modeling of Flow in Aquatic Habitats*, University of Alberta, Edmonton, Alberta, Canada.

Ghanem, A., Steffler, P., Hicks, F. and Katopodis, C. (1996) Two-dimensional hydraulic simulation of physical habitat conditions in flowing streams. *Regulated Rivers: Research and Management*, **12**(2–3): 185–200.

Glenn, J. (2011) *Effect of Transect Location, Transect Spacing and Interpolation Methods on River Bathymetry Accuracy*, University of Idaho, Moscow, ID.

Grace, J.R. and Taghipour, F. (2004) Verification and validation of CFD models and dynamic similarity for fluidized beds. *Powder Technology*, **139**: 99–110. doi: 10.1016/j.powtec.2003.10.006.

Grant, J.W. and Kramer, D.L. (1990) Territory size as a predictor of the upper limit to population density of juvenile salmonids in streams. *Canadian Journal of Fisheries and Aquatic Sciences*, **47**: 1724–1737.

Hardy, R.J., Lane, S.N., Ferguson, R.I. and Parsons, D.R. (2003) Assessing the credibility of a series of computational fluid dynamic simulations of open channel flow. *Hydrological Processes*, **17**(8): 1539–1560. doi: 10.1002/hyp.1198.

Hardy, T.B. (1998) The future of habitat modeling and instream flow assessment techniques. *Regulated Rivers: Research and Management*, **14**(5): 405–420.

Harlow, F.H. and Nakayama, P.I. (1968) *Transport of Turbulence Energy Decay Rate*, Los Alamos Science Lab.

Hassan, M.A., Gottesfeld, A.S., Montgomery, D.R., Tunnicliffe, J.F., Clarke, G.K.C., Wynn, G., Jones-Cox, H., Poirier, R., MacIsaac, E., Herunter, H. and Macdonald, S.J. (2008) Salmon-driven bed load transport and bed morphology in mountain streams. *Geophysical Research Letters*, **35**, L04405. doi:10.1029/2007GL032997.

Hayes, J.W., Hughes, N.F. and Kelly, L.H. (2007) Process-based modelling of invertebrate drift transport, net energy intake and reach carrying capacity for drift-feeding salmonids. *Ecological Modelling*, **207**(2–4): 171–188. doi: 10.1016/j.ecolmodel.2007.04.032.

Heritage, G.L., Milan, D.J., Large, A.R.G. and Fuller, I.C. (2009) Influence of survey strategy and interpolation model on DEM quality. *Geomorphology*, **112**: 334–344. doi:10.1016/j.geomorph.2009.06.024.

Hervouet, J.-M. and Van Haren, L. (1996) Recent advances in numerical methods for fluid flows. In Anderson, M.G. *et al.* (eds) *Floodplain Processes*, John Wiley & Sons, Ltd, Chichester, UK, pp. 183–214.

Hilldale, R.C. and Raff, D. (2007) Assessing the ability of airborne LiDAR to map river bathymetry. *Earth Surface Processes and Landforms*, **33**(5): 773–783.

Hirsch, C. (2007) *Numerical Computation of Internal and External Flows*, Butterworth-Heinemann, Oxford, UK.

Hirt, C.W. and Nichols, B.D. (1981) Volume of fluid (VOF) method for the dynamics of free boundaries. *Journal of Computational Physics*, **39**: 201–225.

Hodge, R., Brasington, J. and Richards, K.S. (2009) In situ characterization of grain-scale fluvial morphology using terrestrial laser scanning. *Earth Surface Processes and Landforms*, **34**(7): 954–968. doi: 10.1002/esp.1780.

Hodskinson, A. (1996) Computational fluid dynamics as a tool for investigating separated flow in river bends. *Earth Surface Processes and Landforms*, **21**: 993–1000.

Hodskinson, A. and Ferguson, R.I. (1998) Numerical modeling of separated flow in river bends: model testing and experimental investigation of geometric controls on the extent of flow separation at concave bank. *Hydrological Processes*, **12**: 1323–1338.

Holley, E.R. and Abraham, G. (1973) Laboratory studies on transverse mixing in rivers. *Journal of Hydraulic Research*, **11**(3): 219–253. doi: 10.1080/00221687309499775.

Horritt, M.S. (2002) Evaluating wetting and drying algorithms for finite element models of shallow water flow. *International Journal of Numerical Methods in Engineering*, **55**: 835–851. doi: 10.1002/nme.529.

Horritt, M.S., Bates, P.D. and Mattinson, M.J. (2006) Effects of mesh resolution and topographic representation in 2D finite volume models of shallow water fluvial flow. *Journal of Hydrology*, **329**: 306–314.

Huang, J.V. and Greimann, B. (2010) *User's Manual for SRH-1D 2.6 (Sedimentation and River Hydraulics – One Dimension, Version 2.6)*, Bureau of Reclamation, Denver, CO.

Hussaini, M.Y. and Zang, T.A. (1987) Spectral methods in fluid dynamics. *Annual Review of Fluid Mechanics*, **19**: 339–367.

Ibbeken, H. and Schleyer, R. (1986) Photo-sieving: A method for grain-size analysis of coarse-grained, unconsolidated bedding surfaces. *Earth Surface Processes and Landforms*, **11**(1): 59–77. doi: 10.1002/esp.3290110108.

Ingham, D.B. and Ma, L. (2005) Fundamental equations for CFD in river flow simulations. In Bates, P.D. *et al.* (eds) *Computational Fluid Dynamics: Applications in Environmental Hydraulics*, John Wiley & Sons, Ltd, Chichester, UK, p. 531.

Javan, M. and Eghbalzadeh, A. (2011) 2D width-averaged numerical simulation of density current in a diverging channel. *Proceedings of the 2nd International Conference on Environmental Science and Development*, IACSIT Press, Singapore.

Jenkins, G. (1993) Estimating eddy kinematic viscosity in compound channels. In Wang, S.Y. (ed.) *Advances in Hydro-Science and Engineering*, Center for Computational Hydroscience and Engineering, pp. 1277–1282.

Johannesson, H. and Parker, G. (1989) Linear theory of river meanders. In Parker, G. (ed.) *Monograph 12: River Meandering*, American Geophysical Union, pp. 181–213.

Johnson, R.W. (1998) *The Handbook of Fluid Dynamics*, CRC Press LLC, Boca Raton, FL.

Jorde, K. and Bratrich, C. (1998) Influence of river bed morphology and flow regulations in diverted streams: effects on bottom shear stress patterns and hydraulic habitat. In Bretschko, G. and Helesic, J. (eds) *Advances in River Bottom Ecology IV*, Backhuys Publishers, Leiden, The Netherlands, pp. 47–63.

Keefer, T.N. (1971) *The Relationship of Turbulence to Diffusion in Open-Channel Flows*, Colorado State University, Fort Collins, CO.

Kilsby, N.N. (2008) *Reach-Scale Spatial Hydraulic Diversity in Lowland Rivers: Characterisation, Measurement and Significance for Fish*, University of Adelaide, Adelaide, Australia.

Kinzel III, P.J., Wright, C.W., Nelson, J.M. and Burman, A.R. (2007) Evaluation of an experimental LiDAR for surveying a shallow, braided, sand-bedded river. *Journal of Hydraulic Engineering*, **133**(7): 838–842.

Kondolf, M.G. (1995) Geomorphological stream channel classification in aquatic habitat restoration: uses and limitations.

Aquatic Conservation: Marine and Freshwater Ecosystems, **5**: 127–141.

Kondolf, M.G., Larsen, E.W. and Williams, J.G. (2000) Measuring and modeling the hydraulics environment for assessing instream flows. *North American Journal of Fisheries Management*, **20**: 1016–1028.

Köse, Ö. (2011) Distribution of turbulence statistics in open-channel flow. *International Journal of the Physical Sciences*, **6**(14): 3426–3436.

Kristof, P., Benes, B., Krivanek, J. and St'ava, O. (2009) Hydraulic erosion using smoothed particle hydrodynamics. *Proceedings of Eurographics 2009*, **28**(2): 219–228.

Lacey, R.W.J. and Millar, R.G. (2004) Reach scale hydraulic assessment of instream salmonid habitat restoration. *Journal of the American Water Resources Association*, **40**(6): 1631–1644.

Lai, Y.G. (2009) *SRH2D: Two-Dimensional Depth-Averaged Flow Modeling with an Unstructured Hybrid Mesh.*

Lamouroux, N., Capra, H. and Pouilly, M. (1998) Predicting habitat suitability for lotic fish: linking statistical hydraulic models with multivariate habitat use models. *Regulated Rivers: Research and Management*, **14**: 1–11.

Lamouroux, N., Scouchon, Y. and Herouin, E. (1995) Predicting velocity frequency distribution in stream reaches. *Water Resources Research*, **31**(9): 2367–2375.

Lancaster, J. and Downes, B.J. (2010) Linking the hydraulic world of individual organisms to ecological processes: putting ecology into ecohydraulics. *River Research and Applications*, **26**: 385–403.

Lane, S.N. (1998) Hydraulic modeling in hydrology and geomorphology: a review of high resolution approaches. *Hydrological Processes*, **12**: 1131–1150.

Lane, S.N. and Bates, P.D. (1998) Special Issue: High Resolution Flow Modelling. *Hydrological Processes*, **12**(8): 1129–1396.

Lane, S.N. and Richards, K.S. (1998) High resolution, two-dimensional spatial modelling of flow processes in a multi-thread channel. *Hydrological Processes*, **12**(8): 1279–1298.

Lane, S.N. and Richards, K.S. (2001) The 'validation' of hydro-dynamic models: Some critical perspectives. In Bates, P.D. and Anderson, M.G. (eds) *Model Validation for Hydrological and Hydraulic Research*, John Wiley & Sons, Ltd, Chichester, UK, pp. 413–438.

Lane, S.N., Hardy, R.J., Elliott, L. and Ingham, D.B. (2004) Numerical modeling of flow processes over gravelly surfaces using structured grids and a numerical porosity treatment. *Water Resources Research*, **40**: W01302. doi:10.1029/2002WR001934.

Lane, S.N., Hardy, R.J., Ferguson, R.I. and Parsons, D.R. (2005) A framework for model verification and validation of CFD schemes in natural open channel flows. In Bates, P.D. *et al.* (eds) *Computational Fluid Dynamics: Applications in Environmental Hydraulics*, John Wiley & Sons, Ltd, Chichester, UK, p. 531.

Lane, S.N., Bradbrook, K.F., Richards, K.S., Biron, P.A. and Roy, A.G. (1999) The application of computational fluid dynamics to natural river channels: three-dimensional versus two-dimensional approaches. *Geomorphology*, **29**(1–2): 1–20.

Le Pichon, C., Gorges, G., Baudry, J., Goreaud, F. and Boët, P. (2009) Spatial metrics and methods for riverscapes: quantifying variability in riverine fish habitat patterns. *Environmetrics*, **20**(5): 512–526.

Le Pichon, C., Gorges, G., Boët, P., Baudry, J., Goreaud, F. and Faure, T. (2006) A spatial explicit resource-based approach for managing stream fishes in riverscapes. *Environmental Management*, **37**(3): 322–335.

Leclerc, M., Boudreault, A., Bechara, J.A. and Corfa, G. (1995) Two-dimensional hydrodynamic modeling: a neglected tool in the instream flow incremental methodology. *Transactions of the American Fisheries Society*, **124**(5): 645–662.

Leclerc, M., Capra, H., Valentin, S., Boudreault, A. and Cote, Y. (eds) (1996) *Ecohydraulique 2000: Proceedings of the 2nd IAHR symposium on habitat hydraulics*, IAHR, Quebec.

Legleiter, C.J. and Kyriakidis, P.C. (2008) Spatial prediction of river channel topography by kriging. *Earth Surface Processes and Landforms*, **33**: 841–867. doi: 10.1002/esp.1579.

Legleiter, C. J., Harrison, L.R. and Dunne, T. (2011) Effect of point bar development on the local force balance governing flow in a simple, meandering gravel bed river. *Journal of Geophysical Research*, **116**: F01005. doi:10.1029/2010JF001838.

Leonard, A. (1980) Vortex methods for flow simulation. *Journal of Computational Physics*, **37**: 289–335.

Logan, B.L., McDonald, R.R., Nelson, J.M., Kinzel III, P.J. and Barton, G.J. (2011) *Use of Multidimensional Modeling to Evaluate a Channel Restoration Design for the Kootenai River, Idaho*, U.S. Geological Survey, Reston, Virginia.

Löhner, R. (2001) *Applied Computational Fluid Dynamics Techniques*, John Wiley & Sons, Ltd, Chichester, UK.

MacWilliams, M.L.J., Wheaton, J.M., Pasternack, G.B., Street, R.L. and Kitanidis, P.K. (2006) Flow convergence routing hypothesis for pool-riffle maintenance in alluvial rivers. *Water Resources Research*, **42**, W10427. doi:10.1029/2005WR004391.

Marzadri, A.,Tonina, D., Bellin, A., Vignoli, G. and Tubino, M. (2010) Effects of bar topography on hyporheic flow in gravel-bed rivers. *Water Resources Research*, **46** W07531. doi:10.1029/2009WR008285.

McDonald, R.R., Nelson, J.M. and Bennett, J.P. (2005) *Multidimensional Surface-water Modeling System User's Guide*, U.S. Geological Survey.

McKean, J.A., Nagel, D., Tonina, D., Bailey, P., Wright, C.W., Bohn, C. and Nayegandhi, A. (2009) Remote sensing of channels and riparian zones with a narrow-beam aquatic-terrestrial LIDAR. *Remote Sensing*, **1**: 1065–1096. doi:10.3390/rs1041065.

Menter, F.R. (1994) Two-equation eddy-viscosity turbulence models for engineering applications. *American Institute of Aeronautics and Astronautics Journal*, **32**(8): 1598–1605.

Merwade, V., Cook, A. and Coonrod, J. (2008) GIS techniques for creating river terrain models for hydrodynamic modeling and flood inundation mapping. *Environmental Modelling and Software*, **23**(10–11): 1300–1311.

Miller, A.J. and Cluer, B.L. (1998) Modeling considerations for simulation of flow in bedrock channels. In Tinkler, K.J. and Wohl, E.E. (eds) *Rivers Over Rock: Fluvial Processes in*

Bedrock Channels, American Geophysical Union, Washington, DC.

Moin, P. and Mahesh, K. (1998) Direct numerical simulation: a tool in turbulence research. *Annual Review of Fluid Mechanics*, **30**: 539–578.

Moir, H.J. and Pasternack, G.B. (2009) Substrate requirements of spawning chinook salmon (*Oncorhynchus Tshawytscha*) are dependent on local channel hydraulics. *River Research and Applications*. doi: 10.1002/rra.1292.

Monaghan, J.J. (1988) An introduction to SPH. *Computer Physics Communications*, **48**: 89–96.

Morvan, H., Knighton, D., Wright, N., Tang, X. and Crossley, A. (2008) The concept of roughness in fluvial hydraulics and its formulation in 1D, 2D and 3D numerical simulation models. *Journal of Hydraulic Research*, **46**(2): 191–208.

Mouton, A., Meixner, H., Goethals, P.L.M., De Pauw, N. and Mader, H. (2007) Concept and application of the usable volume for modelling the physical habitat of riverine organisms. *River Research and Applications*, **23**(5): 545–558.

Munson, B.R., Young, D.F. and Okiishi, T.H. (2006) *Fundamentals of Fluid Mechanics*, John Wiley & Sons, Inc., Hoboken, NJ.

Nelson, J.M. and Smith, J.D. (1989) Flow in meandering channels with natural topography. In Parker, G. (ed.) *Monograph 12: River Meandering*, American Geophysical Union, pp. 69–102.

Nelson, J.M., Bennett, J.P. and Wiele, S.M. (2003) Flow and sediment-transport modeling. In Kondolf, M.G. and Piégay, H. (eds) *Tools in Fluvial Geomorphology*, John Wiley & Sons, Ltd, Chichester, UK.

Nestler, J.M., Goodwin, R.A., Smith, D.L., Anderson, J.J. and Li, S. (2008) Optimum fish passage and guidance designs are based on the hydrogeomorphology of natural rivers. *River Research and Applications*, **24**(2): 148–168.

Newson, M.D. and Newson, C.L. (2000) Geomoprhology, ecology and river channel habitat: mesoscale approaches to basin-scale challenges. *Progress in Physical Geography*, **24**(2): 195–217.

Nicholas, A.P. (2001) Computational fluid dynamics modelling of boundary roughness in gravel-bed rivers: an investigation of the effects of random variability in bed elevation. *Earth Surface Processes and Landforms*, **26**(4): 345–362.

Nicholas, A.P. and McLelland, S.J. (1999) Hydrodynamics of floodplain recirculation zone investigated by field monitoring and numerical simulation. In Marriott, S.B. and Alexander, J. (eds) *Floodplain: Interdisciplinary Approaches*, Geological Society, London, pp. 15–26.

Nicholas, A.P. and Mitchell, C.A. (2003) Numerical simulation of overbank processes in topographically complex floodplain environments. *Hydrological Processes*, **17**(4): 727–746. doi: 10.1002/hyp.1162.

Nicholas, A.P. and Sambrook Smith, G.H. (1999) Numerical simulation of three-dimensional flow hydraulic in a braided channel. *Hydrological Processes*, **13**: 913–929.

Nikora, V.I. Goring, D.G. and B.J.F. Biggs, B.J.F. (1998) On gravel-bed roughness characterization. *Water Resources Research*, **34**(3): 517–527.

Nikora, V.I. Goring, D.G., McEwan, I. and Griffiths, G. (2001) Spatially averaged open-channel flow over rough bed. *Journal of Hydraulic Engineering*, **127**(2):123–133.

Nikora, V., McEwan, I., McLean, S., Coleman, S.E., Pokrajac, D. and Walters, R. (2007) Double-averaging concept for rough-bed open-channel and overland flows: applications. *Journal of Hydraulic Engineering*, **133**(8): 884–895. DOI: 10.1061/(ASCE)0733-9429(2007)133:8(884).

Nikora, V.I. (2010) Hydrodynamics of aquatic ecosystems: An interface between ecology, biomechanics and environmental fluid mechanics. *River Research and Applications*, **26**(4): 367–384. doi: 10.1002/rra.1291.

Nuze, I., Tominaga, A. and Nakagawa, H. (1993) Field measurements of secondary currents in straight rivers. *Journal of Hydraulic Engineering*, **119**: 598–614.

Nystrom, E.A., Oberg, K.A. and Rehmann, C.R. (2002) Measurements of turbulence with acoustic Doppler current profilers: Source of error and laboratory results. *Proceedings of Hydraulic Measurements and Experimental Methods 2002*, ASCE, Estes Park, CO.

Oliveira, N.R.B., Fortunato, A.B. and Baptista, A.M. (2000) Mass balance in Eulerian–Lagrangian transport simulations in estuaries. *Journal of Hydrologic Engineering, ASCE*, **126**(8): 605–614.

Olsen, N.R.B. (1999) Two-dimensional numerical modelling of flushing processes in water reservoirs. *Journal of Hydraulic Research*, **37**(1): 3–16.

Olsen, N.R.B. (2010) *A Three-Dimensional Numerical Model for Simulation of Sediment Movements in Water Intakes with Multiblock Option SSIIM*, Department of Hydraulic and Envirnmental Engineering, The Norwegian University of Science and Technology.

Olsen, N.R.B. (2011) SSIIM User's Manual.

Olsen, N.R.B. and Kjellesvig, H.M. (1998) Three-dimensional numerical modeling of bed for estimation of maximum local scour depth. *Journal of Hydraulic Research*, **36**: 579–590.

Olsen, N.R.B. and Stokseth, S. (1995) 3-dimensional numerical modeling of water flow in a river with large-bed roughness. *Journal of Hydraulic Research*, **33**(1): 571–581.

OpenFOAM (2011) User's Manual.

Orr, H.G., Large, A.R.G., Newson, M.D. and Walsh, C.L. (2008) A predictive typology for characterising hydromorphology. *Geomorphology*, **100**: 32–40. doi:10.1016/j.geomorph.2007.10.022.

Osmundson, D.B., Ryel, R.J., Lamarra, V.L. and Pitlick, J. (2002) Flow–sediment-biota relations: implications for river regulation effects on native fish abundance. *Ecological Applications*, **12**(6): 1719–1739.

Packman, A.I. and Bencala, K.E. (1999) Modeling methods in study of surface–subsurface hydrological interactions. In Jones, J.B. and Mulholland, P.J. (eds) *Streams and Ground Waters*, Academic Press, San Diego, California, pp. 45–80.

Packman, A.I., Salehin, M. and Zaramella, M. (2004) Hyporheic exchange with gravel beds: basic hydrodynamic interactions and bedform-induced advective flows. *Journal of Hydraulic Engineering*, 130(7): 647–656.

Papanicolaou, A.N.T., Elhakeem, M. and Wardman, B. (2011) Calibration and verification of a 2D hydrodynamic model for simulating flow around emergent bendway weir structures. *Journal of Hydraulic Engineering*, 137(1): 75–89. doi: 10.1061/(ASCE)HY.1943-7900.0000280.

Parasiewicz, P. and Dunbar, M. (2001) Physical habitat modeling for fish – a developing approach. *Archiv für Hyrdobiologie*, 135(2–4): 239–268.

Pasternack, G.B. (2011) *2D Modeling and Ecohydraulic Analysis*, Davis, CA.

Pasternack, G.B. and Senter, A. (2011) *21st Century Instream Flow Assessment Framework for Mountain Streams*, Public Interest Energy Research (PIER), California Energy Commission.

Pasternack, G.B., Wang, C.L. and Merz, J.E. (2004) Application of a 2D hydrodynamic model to design of reach-scale spawning gravel replenishment on the Mokelumne river, California. *River Research and Applications*, 20(2): 205–225. doi: 10.1002/rra.748.

Pasternack, G.B., Gilbert, A.T., Wheaton, J.M. and Buckland, E.M. (2006) Error propagation for velocity and shear stress prediction using 2D models for environmental management. *Journal of Hydrology*, 328: 227–241.

Payne, B.A. and Lapointe, M.F. (1997) Channel morphology and lateral stability: effects on distribution of spawning and rearing habitat for Atlantic Salmon in a wandering cobbled-bed river. *Canadian Journal of Fisheries and Aquatic Sciences*, 54: 2627–2636.

Payne, T.R. (2003) The concept of weighted usable area as relative suitability index. In *International IFIM User's Workshop*, Fort Collins, CO.

Payne, T.R. and Allen, M.A. (2009) Application of the use-to-availability electivity ratio for developing habitat suitability criteria in PHABSIM instream flow studies. *Proceedings of the 7th International Symposium on Ecohydraulics*, IAHR, Concepcion, Chile, 12–16 January, 2009.

Payne, T.R., Eggers, S.D. and Parkinson, D.B. (2004) The number of transects required to compute a robust PHABSIM habitat index. *Hydroécologie Appliquée*, 14(1): 27–53.

Peiró, J. and Sherwin, S. (2005) Finite difference, finite element and finite volume methods for partial differential equations. In Sidney, Y. (ed.) *Handbook of Material Modeling*, Springer: The Netherlands, Chapter 8.

Petts, G.E. (1996) Water allocation to protect the river ecosystems. *Regulated Rivers: Research and Management*, 12: 353–365.

Plant, W.J., Keller, W.C. and Hayes, K. (2005) Measurement of river surface currents with coherent microwave systems. *IEEE Transactions on Geoscience and Remote Sensing*, 43(6): 1242–1257. doi: 10.1109/TGRS.2005.845641.

Poff, L.N. (2002) Ecological response to and management of increased flooding caused by climate change. *Philosophical Transactions of The Royal Society London*, 360: 1497–1510.

Poff, L.N., Allan, J.D., Bain, M.B., Karr, J.R., Prestegaard, K.L., Richter, B.D., Sparks, R.E. and Stromberg, J.C. (1997) The natural flow regime: A paradigm for river conservation and restoration. *Bioscience*, 47(11): 769–784.

Qian, L. and Vezza, M. (2001) A vorticity-based method for incompressible unsteady viscous flows. *Journal of Computational Physics*, 172(2): 515–542.

Railsback, S. (1999) Reducing uncertainties in instream flow studies. *Fisheries*, 24(4): 24–26.

Rehmel, M. (2007) Application of acoustic Doppler velocimeters for streamflow measurements. *Journal of Hydraulic Engineering*, 133(12): 1433–1438. doi: 10.1061/(ASCE)0733-9429(2007)133:12(1433).

Rice, S.P., Lancaster, J. and Kemp, P. (2009) Experimentation at the interface of fluvial geomorphology, stream ecology and hydraulic engineering and the development of an effective, interdisciplinary river science. *Earth Surface Processes and Landforms*. doi: 10.1002/esp.1838.

Rice, S.P., Little, S., Wood, P.J., Moir, H.J. and Vericat, D. (2010) Special Issue: Ecohydraulics at Scales Relevant to Organisms. Selected Papers from the British Hydrological Society National Meeting, Loughborough University, UK, June 2008. *River Research and Applications*, 26(4): 363–527.

Richter, B.D., Baumgartner, J.V., Powell, J. and Braun, D.P. (1996) A method for assessing hydrologic alteration within ecosystems. *Conservation Biology*, 10(4): 1163–1174.

Richter, B.D., Baumgartner, J.V., Wigington, R. and Braun, D.P. (1997) How much water does a river need? *Freshwater Biology*, 37: 231–249.

Roache, P.J. (1997) Quantification of uncertainty in computational fluid dynamics. *Annual Review of Fluid Mechanics*, 29: 123–160. doi: 10.1146/annurev.fluid.29.1.123.

Rodi, W. (2000) *Turbulence Models and Their Application in Hydraulics: A State of the Art Review*, International Association for Hydraulic Research, A.A. Balkema Publishers, Rotterdam, The Netherlands.

Rodi, W., Pavlovic, R.N. and Srivatsa, S.K. (1981) Prediction of flow and pollutant spreading in rivers. In Fisher, H.B. (ed.) *Transport Models for Inland and Coastal Waters*, Academic Press, New York, NY, p. 542.

Roy, M.L., Roy, A.G. and Legendre, P. (2009) The relations between 'standard' fluvial habitat variables and turbulent flow at multiple scales in morphological units of a gravel-bed river. *River Research and Applications*, 26(4): 439–455. doi: 10.1002/rra.1281.

Ruff, J.F. and Gelhar, L.W. (1972) Turbulent shear flow in porous boundary. *Journal of the Engineering Mechanics Division ASCE*, 98(EM4), 975–991.

Sagaut, P. (2006) *Large Eddy Simulation for Incompressible Flows: An Introduction*, Springer, New York.

Samuels, P.G. (1990) Cross-section location in 1-D models. In White, W.R. (ed.) *Proceedings of the International Conference on River Flood Hydraulics*, Wallingford, Wiley Paper K1: 339–350.

Schneider, M. (2001) *Habitat- und Abflussmodellierung für Fließgewässer mit unscharfen Berechnungsansätzen*. PhD dissertation, Mitteilungen des Instituts für Wasserbau, Heft 108, Universität Stuttgart, Eigenverlag.

Schneider, M., Noack, M. and Gebler, T. (2008) *Handbook for the Habitat Simulation Model: CASiMiR*, SJE Schneider & Jorde Ecological Engineering Ltd and University of Stuttgart, Stuttgart, Germany.

Shen, Y. and Diplas, P. (2008) Application of two- and three-dimensional computational fluid dynamics models to complex ecological stream flows. *Journal of Hydrology*, **348**: 195–214.

Shields, F.D.J. (2010) Aquatic habitat bottom classification using ADCP. *Journal of Hydraulic Engineering*, **136**(5): 336–342. doi: 10.1061/(ASCE)HY.1943-7900.0000181.

Shih, T.H., Liou, W.W., Shabbir, A., Yang, Z. and Zhu, J. (1995) A new κ-ε eddy-viscosity model for high Reynolds number turbulent flow – model development and validation. *Computers and Fluids*, **42**(3:) 832–722.

Simões, F.J.M. (2011) Finite volume model for two-dimensional shallow environmental flow. *Journal of Hydraulic Engineering*, **137**(2): 173–182. doi: 10.1061/(ASCE)HY.1943-7900.0000292.

Simões, F.J.M. and Wang, S. S.-Y. (1997) Numerical prediction of three-dimensional mixing in compound open channel. *Journal of Hydraulic Research*, **35**(5): 619–642.

SJE Schneider & Jorde Ecological Engineering (2007) *Sunk-Schwallbetrieb Kraftwerk Schiffenen, Jungfischhabitat-Untersuchungen mit dem Simulationsmodell CASiMiR in der Saane*, Stuttgart, Germany.

Smagorinsky, J. (1963) General circulation experiment with the primitive equations. I. The basic experiment. *Monthly Weather Review*, **91**(3): 99–164.

Smith, J.D. and McLean, S.R. (1984) A model for flow in meandering streams. *Water Resources Research*, **20**(9): 1301–1315.

Sotiropoulos, F. (2005) Introduction to statistical turbulence modelling for hydraulic engineering flows. In Bates, P.D. *et al.* (eds) *Computational Fluid Dynamics: Applications in Environmental Hydraulics*, John Wiley & Sons, Ltd, Chichester, UK, pp. 91–120.

Soulsby, R.L. (1980) Selecting record length and digitisation rate for near-bed turbulence measurements. *Journal of Physical Oceanography*, **10**(2): 208–218.

Spalding, D.B. (1985) The computation of flow around ships with allowance for free-surface and density-gradient effects. *Proceedings of the 1st International Maritime Simulation Symposium*, Munich.

Statzner, B. (2008) How views about flow adaptations of benthic stream invertebrates changed over the last century. *International Review of Hydrobiology*, **93**(4–5): 593–605.

Statzner, B., Fuchs, U. and Higler, L.W.G. (1996) Sand erosion by mobile predaceous stream insects: Implications for ecology and hydrology. *Water Resources Research*, **32**(7): 2279–2288.

Statzner, B., Sagnes, P., Champagne, J.-Y. and Viboud, S. (2003) Contribution of benthic fish to the patch dynamics of gravel and sand transport in streams. *Water Resources Research*, **39**(11): 1309. doi:10.1029/2003WR002270.

Statzner, B., Arens, M.-F., Champagne, J.-Y., Morel, R. and Herouin, E. (1999) Silk-producing stream insects and gravel ero-

sion: Significant biological effects on critical shear stress. *Water Resources Research*, **35**(11): 3495–3506.

Statzner, B., Fievet, E., Champagne, J.-Y., Morel, R. and Herouin, E. (2000) Crayfish as geomorphic agents and ecosystem engineers: Biological behavior affects sand and gravel erosion in experimental streams. *Limnology and Oceanography*, **45**(5): 1030–1040.

Steffler, P. and Blackburn, J. (2002) River2D: User Manual.

Stewart, G.B. (2000) *Two-Dimensional Hydraulic Modeling for Making Instream-Flow Recommendations*, Colorado State University, Fort Collins, CO.

Stockdale, R.J., McLelland, S.J., Middleton, R. and Coulthard, T.J. (2008) Measuring river velocities using GPS River Flow Tracers (GRiFTers). *Earth Surface Processes and Landforms*, **33**(8): 1315–1322. doi: 10.1002/esp.1614

Stone, M.C. (2005) *Natural Stream Flow Fields: Measurements and Implications for Periphyton*. PhD thesis, Washington State University, 201 pp.

Sukhodolov, A., Thiele, M. and Bungartz, H. (1998) Turbulence structure in a river reach with sand bed. *Water Resources Research*, **34**(5): 1317–1334.

Tennekes, H. and Lumley, J.L. (1972) *A First Course in Turbulence*, MIT Press, Cambridge, MA.

Tingsanchali, T. and Maheswaran, S. (1990) Two-dimensional depth-averaged flow computation near groynes. *Journal of Hydraulic Engineering*, **116**: 103–125.

Tolossa, H.G., Tuhtan, J.A., Schneider, M. and Wieprecht, S. (2009) Comparison of 2D hydrodynamic models in river reaches of ecological importance: HYDRO˙AS-2D and SRH-W. *Proceedings of the IAHR Congress on Water Engineering for a Sustainable Environment*, IAHR, Vancouver, Canada, 9–14 August, 2009.

Tonina, D. and Buffington, J.M. (2007) Hyporheic exchange in gravel-bed rivers with pool-riffle morphology: laboratory experiments and three-dimensional modeling. *Water Resources Research*, **43**, W01421.

Tonina, D. and Buffington, J.M. (2009) A three-dimensional model for analyzing the effects of salmon redds on hyporheic exchange and egg pocket habitat. *Canadian Journal of Fisheries and Aquatic Sciences*, **66**: 2157–2173. doi: 10.1139/F09-146.

Tonina, D. and McKean, J.A. (2010) Climate change impact on salmonid spawning in low-land streams in Central Idaho, USA. *Proceedings of the 9th International Conference on Hydroinformatics 2010*, Chemical Industry Press, Tianjin, China.

Tonina, D. and Buffington, J.M. (2011) Effects of stream discharge, alluvial depth and bar amplitude on hyporheic flow in pool-riffle channels. *Water Resources Research*, **47**, W08508. doi:10.1029/2010WR009140.

Tonina, D. (2012) Surface water and streambed sediment interaction: The hyporheic exchange. In Gualtieri, C. and Mihailović, D.T. (eds). *Fluid mechanics of environmental interfaces.*, CRC Press, Taylor & Francis Group, London, UK, pp. 255–294.

Tonina, D., McKean, J.A., Tang, C. and Goodwin, P. (2011) New tools for aquatic habitat modeling. *Proceedings of the 34th IAHR World Congress 2011*, IAHR, Brisbane, Australia.

USACE (1996) *Users guide to RMA2, Version 4.3*, US Army Corps of Engineers, Waterways Experiment Station Hydraulic Laboratory, Vicksburg, MS.

USACE (2002) *HEC-RAS River Analysis System: User's Manual, v3.1*, US Army Corps of Engineers, Hydrological Engineering Center, Davis, CA.

USBR (2011) *Final Biological Assessment and Final Essential Fish Habitat Determination for the Preferred Alternative of the Klamath Facilities Removal EIS/EIR*, U.S. Department of the Interior Bureau of Reclamation.

USGS (1992) *Policy Statements on Stage Accuracy*, U.S. Geological Survey, Washington, DC.

USGS (2001) *PHABSIM for Windows: User's Manual and Exercises*. Open File Report 01-340, U.S. Geological Survey.

Vallé, B.L. and Pasternack, G.B. (2006) Field mapping and digital elevation modelling of submerged and unsubmerged hydraulic jump regions in a bedrock step–pool channel. *Earth Surface Processes and Landforms*, **31**(6): 646–664. doi: 10.1002/esp.1293.

van Balen, W., Uijttewaal, W.S.J. and Blanckaert, K. (2010) Large-eddy simulation of a curved open-channel flow over topography. *Physics of Fluids*, **22**(7): 075108.

Vannote, R.L., Minshall, W.G., Cummins, K.W., Sedell, J.R. and Cushing, C.E. (1980) The river continuum concept. *Canadian Journal of Fisheries and Aquatic Sciences*, **37**: 130–137.

Vermeulen, B., Hoitink, A.J.F. and Sassi, M.G. (2011) Coupled ADCPs can yield complete Reynolds stress tensor profiles in geophysical surface flows. *Geophysical Research Letters*, **38**, L06406. doi:10.1029/2011GL046684.

Versteeg, H. and Malalasekera, W. (1995) *An Introduction to Computational Fluid Dynamics: The Finite Volume Method*, Longman Scientific & Technical, London, UK.

Vidal, J.-P., Moisan, S., Faure, J.-B. and Dartus, D. (2007) River model calibration, from guidelines to operational support tools. *Environmental Modelling and Software*, **22**(11): 1628–1640. doi: 10.1016/j.envsoft.2006.12.003.

Vionnet, C.A., Tassi, P.A. and Vide, J.P.M. (2004) Estimates of flow resistance and eddy viscosity coefficients for 2D modeling on vegetated floodplains. *Hydrological Processes*, **18**: 2907–2926.

Vollmer, S., de los Santos Ramos, F., Daebel, H. and Kühn, G. (2002) Micro scale exchange processes between surface and subsurface water. *Journal of Hydrology*, **269**: 3–10.

Waddle, T. (2009) Field evaluation of a two-dimensional hydrodynamic model near boulders for habitat calculation. *River Research and Applications*, **26**: 730–741. doi: 10.1002/rra.1278.

Walther, S.C., Marcus, W.A. and Fonstad, M.A. (2011) Evaluation of high resolution, true colour, aerial imagery for mapping bathymetry in a clear water river without ground-based depth measurements. *International Journal of Remote Sensing*, **32**(15): 4343–4363. doi: 10.1080/01431161.2010.486418.

Wasantha Lal, A.M. (1995) Calibration of riverbed roughness. *Journal of Hydraulic Engineering*, **121**(9): 664–671.

Weber, L.J., Goodwin, R.A., Nestler, J.M. and Anderson, J.J. (2006) Application of an Eulerian–Lagrangian-Agent method (ELAM) to rank alternative designs of a juvenile fish passage facility. *Journal of Hydroinformatics*, **8**(4): 271–295.

Wheaton, J.M., Pasternack, G.B. and Merz, J.E. (2004) Spawning habitat rehabilitation – 2. Using hypothesis development and testing in design, Mokelumne River, California, U.S.A. *International Journal of River Basin Management*, **2**(1): 21–37.

White, D.F., Stanford, J.A. and Kimball, J.S. (2002) Application of airborne multispectral digital imagery to quantify riverine habitats at different base flows. *River Research and Applications*, **18**(6): 583–594. doi: 10.1002/rra.695.

Whited, D.C., Kimball, J.S., Lorang, M.S. and Stanford, J.A. (2011) Estimation of juvenile salmon habitat in Pacific rim rivers using multiscalar remote sensing and geospatial analysis. *River Research and Applications*. doi: 10.1002/rra.1585.

Whiting, P.J. and Dietrich, W.E. (1991) Convective accelerations and boundary shear stress over a channel bar. *Water Resources Research*, **27**(5): 783–796.

Wiele, S.M., Wilcock, P.R. and Grams, P.E. (2007) Reach-averaged sediment routing model of a canyon river. *Water Resources Research*, **43**, W02425. doi:10.1029/2005WR004824.

Wilcox, D.C. (1988) Re-assessment of the scale-determining equation for advance models. *AIAA Journal*, **26**(11): 1299–1310.

Wilcox, D.C. (2006) *Turbulence Modeling for CFD*, DCW Industries Inc., La Cañada, CA.

Williams, J.G. (2010) Lost in space, the sequel: Spatial sampling issues with 1D PHABSIM. *River Research and Applications*, **26**(3): 341–352. doi: 10.1002/rra.1258.

Wilson, C.A.M.E., Bates, P.D. and Hervouet, J.-M. (2002) Comparison of turbulence models for stage–discharge rating curve prediction in reach-scale compound channel flows using two-dimensional finite element methods. *Journal of Hydrology*, **257**: 42–58.

Wilson, C.A.M.E., Stoesser, T. and Bates, P.D. (2005) Modelling of open channel flow through vegetation. In Bates, P.D. *et al.* (eds) *Computational Fluid Dynamics: Applications in Environmental Hydraulics*, John Wiley & Sons, Ltd, Chichester, UK.

Wilson, C.A.M.E., Yaggi, O., Rauch, H.-P. and Stoesser, T. (2006) Application of the drag force approach to model the flow-interaction of natural vegetation. *International Journal of River Basin Management*, **4**(2): 137–146.

Wolman, M.G. (1954) Method of sampling coarse river bed material. *Eos (Transactions, American Geophysical Union)*, **35**: 951–956.

Wolterstorff, G., Gray, T., Belmonte, E. and Milner, G. (2003) Holistic site development with floodplain, stream habitat and wetland mitigation. *Proceedings of the World Water and Environmental Resources Congress*, Philadelphia, Pennsylvania.

Woodhead, S., Asselman, N., Zech, Y., Soares-Frazao, S., Bates, P.D. and Kortenhaus, A. (2006) *Evaluation of Inundation Models: Limits and Capabilities of Models*, HR Wallingford, UK.

Yakhot, V. and Orszag, S.A. (1986) Renormalization group analysis of turbulence: I. Basic theory. *Journal of Scientific Computing*, **1**: 1–51.

Zienkiewicz, O. and Cheung, Y. (1965) Finite elements in the solution of field problems. *The Engineer*, pp. 507–510.

4

The Habitat Modelling System CASiMiR: A Multivariate Fuzzy Approach and its Applications

Markus Noack[1], Matthias Schneider[2] and Silke Wieprecht[3]

[1]Federal Institute of Hydrology, Department M3 – Groundwater, Geology, River Morphology, Mainzer Tor 1, D-56068 Koblenz, Germany
[2]Schneider & Jorde Ecological Engineering GmbH, Viereichenweg 12, D-70569 Stuttgart, Germany
[3]Institute for Modelling Hydraulic and Environmental Systems, Department of Hydraulic Engineering and Water Resources Management, University of Stuttgart, Pfaffenwaldring 61, D-70569 Stuttgart, Germany

4.1 Introduction

4.1.1 Background

For hundreds of years, throughout the history of human development, rivers have been diverted for agricultural irrigation, hydropower, navigation, provision of drinking water, removal of wastewater, etc. As a consequence, the state of worldwide aquatic biodiversity is far worse than for other ecosystems (Revenga *et al.*, 2000). In order to quantify and predict ecological impacts, aquatic habitat simulation tools have been used for decades in water resources management. One of the first widely available physical habitat models was PHABSIM (Bovee, 1982; Milhous *et al.*, 1989), a component of the Instream Flow Incremental Methodology (IFIM, Bovee, 1982; Stalnaker *et al.*, 1995) that was developed in North America in the 1970s. In the 1980s, physical habitat models became an important tool for river management and nowadays they are applied worldwide. Moreover, they are still the focus of ongoing debate and research. Today, a great variety of different model types and techniques has been developed, encompassing nearly all types of aquatic organisms.

These physical habitat models use so-called physical–biota relationships (Conallin *et al.*, 2010), which represent the core of predictive habitat modelling as they aim to assess how environmental factors control the distribution of available habitats for species and communities. Although many studies assess the biotic response to altered environmental conditions, there is still a clear need for quantifying these species–environmental relationships (Guisan and Zimmermann, 2000). From a more practical point of view, water resources managers require methods and tools to quantify the disturbance level of ecosystems compared to a reference state. To choose among the available modelling approaches, water managers have to strike a compromise regarding model performance, policy, finances, scale and data requirements (Conallin *et al.*, 2010).

This chapter focuses on the application of the habitat simulation model CASiMiR, with its fuzzy approach to the interface of biological responses to flow changes and physical habitat variables. After a brief introduction to the fuzzy approach, a comparison with the classical preference functions approach is presented together with two state-of-the-art model concepts and applications beyond classical physical habitat modelling applications, implying the adaptability of the fuzzy approach to present issues in ecohydraulic modelling.

4.1.2 Physical habitat modelling in general

As physical habitat is a key factor in evaluating the ecological status of rivers (Maddock, 1999). The classical approach to qualifying and quantifying habitat consists

Ecohydraulics: An Integrated Approach, First Edition. Edited by Ian Maddock, Atle Harby, Paul Kemp and Paul Wood.
© 2013 John Wiley & Sons, Ltd. Published 2013 by John Wiley & Sons, Ltd.

of estimating habitat indices defining an optimum range of abiotic conditions for indicator species (Leclerc *et al.*, 2003). Based on the similarity between existing conditions and preferred conditions of aquatic organisms, an estimate of habitat quality can be assigned for a specific location. The most common index to describe the biological response to abiotic attributes is the habitat suitability index (HSI), which ranges from 0.0 (unsuitable) to 1.0 (most suitable) and represents the suitability of a habitat for a target species or life stage. The linkage between biological responses and abiotic factors can be approached through a number of different techniques. They are mainly classified as univariate methods, considering individual habitat variables, or multivariate approaches, taking into account interactions between habitat variables to determine the target species or life stage response to the abiotic factors. Based on the integration of HSI values and the hydraulic characteristics of the stream, the weighted usable area (WUA) or the hydraulic habitat index (HHS, Stalnaker *et al.*, 1995) for a target species or life stage can be estimated as a function of the flow rate (Gore and Nestler, 1988). The HHS divides the WUA by the wetted area, leading to an index ranging from 0.0 to 1.0, thus eliminating the influence of the wetted area to facilitate model comparisons between study-sites.

According to Conallin *et al.* (2010), aquatic habitat modelling tools can be used to:

- identify individual biota–physical habitat relationships;
- assess the quality of the physical habitat variables through impact on the biota;
- predict likely biological responses if hydromorphological changes to a system occur.

Several reviews about existing habitat modelling techniques can be found in the literature (e.g. Rosenfeld, 2003; Harby *et al.*, 2004; Ahmadi-Nedushan *et al.*, 2006; Conallin *et al.*, 2010).

4.1.3 Fuzzy logic in ecohydraulic modelling

Fuzzy set theory, developed by Zadeh (1965), assumes that complex systems are characterized by imprecise transitions between different states of a system. Contrary to Boolean logic, fuzzy logic allows systems to be in intermediate states, which is especially germane and important in ecological modelling, as transitions in ecology are not crisp but gradual (Salski, 1992; Cadenasso *et al.*, 2003). Fuzzy logic has proven to be an excellent modelling technique to deal with ecological gradients, as the overlapping fuzzy set theory reflects these gradual transitions

between predefined classes (van Broekhoven *et al.*, 2006; Mouton, 2008). One major advantage of fuzzy logic modelling is that it allows the use of qualitative data for numerical processing (Schneider, 2001) and provides the ability to consider multivariate effects without assuming independence of input variables. This approach also accounts for numerous combinations of many descriptive habitat variables and allows easy model interpretability (no black box), which is crucial for communication and improves the basis for collaboration between scientists, river managers and decision-makers.

4.2 Theoretical basics of the habitat simulation tool CASiMiR

4.2.1 Background and development

The habitat simulation tool CASiMiR (Computer Aided Simulation Model for Instream Flow Riparia) is an MS Windows-based simulation model developed in the 1990s by the Institute of Hydraulic Engineering at the University of Stuttgart (Jorde, 1996; Schneider, 2001). The motivation for model development arose from the objective to develop more sophisticated and ecology-related minimum flow solutions for assessing hydropower plant diversions. The first version of CASiMiR (1993) includes habitat modelling for macroinvertebrates based on a preference function approach using FST-hemispheres (Statzner *et al.*, 1991) to account for the spatial distribution of the bed shear stress (Jorde, 1996). In 1996, CASiMiR advanced its applicability to fish preferences and in 1998, the multivariate fuzzy approach was implemented (Schneider, 2001). Further improvements to CASiMiR – in 2001 and 2003 – included interfaces to several 1D and 2D hydrodynamic models in order to integrate detailed information about river hydraulics for different discharge scenarios without extensive field measurements. Currently, CASiMiR allows integration with any multidimensional hydrodynamic model and actual applications encompass a wide range of river management strategies such as restoration measures, influence of sediment dynamics, vegetation on riverbanks and floodplains, impacts due to hydropeaking and the evaluation of mesohabitats. Moreover, an advanced macroinvertebrate simulation tool has been developed, allowing the calculation of FST numbers based on detailed hydrodynamic modelling instead of relying on time-consuming FST measurements (Kopecki, 2008). CASiMiR and its fuzzy approach continue to grow in popularity and are applied broadly at the international level to solve a wide variety of ecohydraulic issues. Examples are long-term fish habitat changes in floodplain channels in The Netherlands (Kerle *et al.*, 2002), dam operation and its effect on physical habitat conditions in Chile (Garcia *et al.*, 2010) and modelling sturgeon habitats in a Chinese river (Yi *et al.*, 2010).

4.2.2 Functional principle of CASiMiR

Fuzzification

The first step in fuzzy modelling is the fuzzification of the chosen habitat variables by defining overlapping membership functions in order to describe their parameter ranges. This process defines real numbers between 0.0 and 1.0, where 0.0 means that the habitat variable value does not belong to a membership function, while 1.0 means that it belongs entirely. Usually, a habitat variable is subdivided into several membership functions which are described by linguistic variables (e.g. low, medium or high flow velocities) forming a fuzzy set (see also Chapter 5, Section 5.2.1). The whole parameter range is not only defined by the physical values observed in the field but also by the range of how the target species uses this physical attribute. Although a membership function can be described by any functional form, they are usually represented by simple trapezoidal or triangular functions, as illustrated for the habitat variable flow velocity in Figure 4.1.

Assuming an overlapping membership function, which reflects gradual transitions between parameter classes, fuzzy theory allows an appropriate representation of eco-logical gradients. For example, a particular fish may not differentiate between 0.19 m/s or 0.21 m/s and the fuzzy description that uses partly 'LOW' and partly 'MEDIUM' flow velocities approximates more closely the ecological gradients that aquatic organisms tend to follow (Adriaenssens *et al.*, 2004).

Fuzzy rule system

After fuzzification, the physical–biota relationships have to be determined using IF–THEN rules. A fuzzy rule consists of several arguments (habitat variables), building a premise as the first part and a consequence (HSI) in the second part. This procedure has the significant advantage that expert knowledge can be transferred easily into preference datasets and combinations of relevant physical criteria can be addressed. Hence, the experts themselves define the conditions under which habitat quality is described as 'HIGH', 'MEDIUM' or 'LOW'. Table 4.1 gives an example of a rule set describing the habitat requirements of European grayling (*Thymallus thymallus*) for spawning in the River Aare, Switzerland.

For any possible combination of habitat characteristics the total number of rules that must be provided depends on the number of input variables (e.g. velocity, depth and substrate) and membership functions (e.g. Low (L), Medium (M) or High (H)). The ability to define complex physical–biota relationships as a series of simple IF–THEN rules fits with the simulation of imprecise problems that commonly arise in ecological modelling.

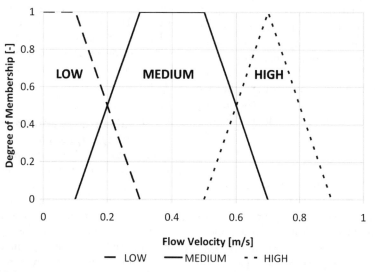

Figure 4.1 Example of defining triangular and trapezoidal membership functions for the habitat variable flow velocity (fuzzification).

Table 4.1 Example of a fuzzy rule set describing the habitat requirements of European grayling (*Thymallus thymallus*) for spawning in the River Aare, Switzerland.

	Velocity	Depth	Substrate	HSI	Examples:
1	L			L	rule 1 IF velocity 'Low' THEN HSI 'Low'
2		H		L	rule 2 IF depth 'High' THEN HSI 'Low'
3	M or H	L	M	M	IF velocity 'Medium' or 'High' AND rule 3 depth 'Low' AND substratum 'Medium' THEN HSI 'Medium'
4	M or H	L	L	L	vel = velocity L = Low
5	M or H	M	M	H	dep = water depth M = Medium
6	M or H	M	H	H	sub = substratum H = High
7	M or H	H	H	H	HSI = habitat suitability index

Inference system

For a fuzzy simulation, the inference processor runs systematically through the entire set of rules and determines the degree of fulfilment (DOF) depending on the combination of input variables. The better the rule reflects the habitat parameters, the higher the DOF. Usually, several rules have a DOF > 0, or, in other words, are partly true and become activated. They have to be combined in order to define a total consequence considering the different weights of these activated rules. This is illustrated for two input variables (flow velocity and water depth) and two rules in Figure 4.2. In this example, input data to compute the DOF are a flow velocity of 0.7 m/s and a water depth of 0.4 m.

For rule A in Figure 4.2, the DOF is 0.25 according to the minimum method applied for 'HIGH' flow velocities and 'LOW' water depths. The DOF of rule B is 0.5, as the degree of the membership for 'HIGH' flow velocities is lower compared to 'MEDIUM' water depth. As the DOF in rule B is higher than in rule A, it receives a higher weight, visualized in the total consequence of rule A and B. For an ecological interpretation, it can be stated that the inference method allows for an automatic weighting of the rules according to the rule-specific DOF. The rule that reflects the best input parameter combination receives the highest weight.

Defuzzification

The total consequence derived by the inference processor is a fuzzy set resulting from the DOFs of rule A and B using the maximum product method. In order to transform this fuzzy information back into a crisp number, a process called 'defuzzification' is applied. In CASiMiR, the method 'Centre of Gravity' (COG) is implemented, which calculates the COG of the area encompassing both

black triangles. Applying this method gives the result of defuzzification (the HSI value) using the *x*-coordinate of the COG value of the total consequence, as shown in Figure 4.2.

4.2.3 Calibration of the fuzzy approach

The model is calibrated by comparing observed fish locations with the predicted habitat quality. There are several ways to calibrate the fuzzy approach in CASiMiR:

1 by changing the numerical approaches, e.g. different inference methods;

2 by modifying the system of fuzzy rules which is part of the input dataset;

3 by altering the fuzzy sets of habitat parameters.

In most cases, option 1 does not have a significant effect on the final results (Schneider, 2001). Option 2 must only be chosen in agreement with the involved fish experts. The most commonly used method is option 3. While the definition of, for example, a 'medium' flow velocity is initially made from the human point of view, the calibration process can be seen as an adaption from the 'fish' point of view.

4.2.4 Advantages and limitations of the fuzzy approach

According to Schneider (2001) and Conallin *et al.* (2010), the multivariate fuzzy approach considers interactions of habitat variables without assuming independence. Moreover, qualitative knowledge-based data can be implemented in numerical processing, coinciding well with information about physical–biota relationships that is often available in a more qualitative than quantitative form. All modelling steps are comprehensible (no black box) and the linguistic formulation of physical–biota

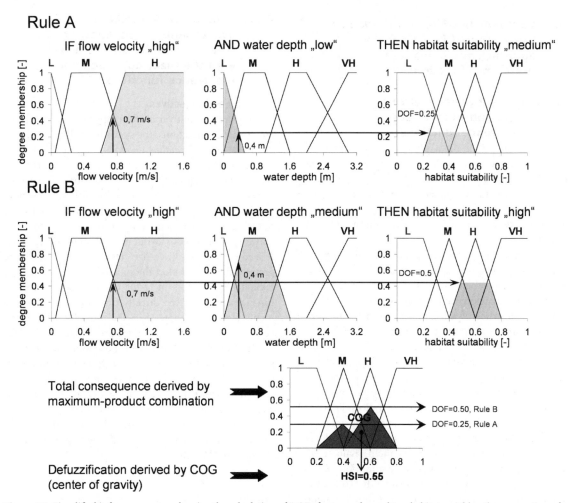

Figure 4.2 Simplified inference process showing the calculation of DOFs for two rules and two habitat variables, the aggregation of both rules to a total consequence using the maximum product method and the defuzzification to a crisp HSI value.

relationships allows a clear interpretation that is important for communicating results to river managers.

However, fuzzy logic modelling also has its limitations. According to Adriaenssens *et al.* (2004), the subjectivity of the fuzzy approach is one of the major weaknesses. As the definitions of fuzzy sets and fuzzy rules are the core of fuzzy models, the quality of results depends strongly on the expert knowledge used to define the rules and sets (see also Chapter 5, Section 5.1.4). Cornelissen *et al.* (2001) stress three aspects to evaluate expert knowledge: criteria specification to determine expert knowledge, proper elicitation of expert knowledge and methods to test the reliability of membership functions. Another limitation is the number of habitat variables considered and the membership functions which define the total number of

rules. Too many input variables lead to exploding proliferation of the number of fuzzy rules and can result in an inappropriate model application. Even for highly qualified and experienced experts it might be nearly impossible to estimate physical–biota relationships considering the combined effect of more than five or six variables.

The following three sections illustrate applications of the habitat model CASiMiR and its fuzzy approach. The first case study compares the multivariate fuzzy approach with univariate preference functions, while the second case study gives an example of how CASiMiR copes with morphodynamic processes, considering additional parameters in a multi-step fuzzy model. The third case study describes an application on the meso-scale, investigating habitat fragmentation and connectivity.

4.3 Comparison of habitat modelling using the multivariate fuzzy approach and univariate preference functions

4.3.1 Biota–physical relations: fuzzy approach versus preference functions

As the definition of biota–physical relationships is a basis component of most habitat models, this comparison aims to present the different modelling results using the multivariate fuzzy approach – the standard approach in CASiMiR – and univariate preference functions as linkages between abiotic habitat description and biotic response.

In univariate preference functions the suitability of a habitat is a function of a single variable, characterizing one of the physical characteristics of a habitat. Each of the univariate preference curves is compared with the physical properties of a location in the river to derive a one-parametrical HSI value (e.g. HSI_{Depth}, $HSI_{Velocity}$). To obtain a total habitat suitability index, the individual suitability indices are combined using mathematical operations like product, arithmetic mean or geometric mean (Vadas and Orth, 2001). Generally, there are two major assumptions in using such composite indices. Firstly, all variables are equally important to the aquatic organisms and secondly all environmental variables are independent, with no interaction between them (Bain, 1995). The product method is based on the assumption that fish select each particular variable independently of other variables (Bovee, 1986). Consequently, the product equation yields a zero total HSI for any given unsuitable habitat variable. The arithmetic mean method is based on the assumption that habitat variables are compensatory and good habitat conditions on one variable (e.g. velocity) can compensate for poor conditions on other variables (e.g. depth). The geometric mean method also implies some compensation but also leads to zero suitability for any zero-valued single HSI value (Ahmadi-Nedushan et al., 2006).

In contrast, the multivariate fuzzy approach does not require any mathematical operators, as the total HSI is obtained directly from the inference system. The link to the biotic response is defined in the fuzzy rule system, with different weights according to the DOFs of each activated rule. Hence, the input variables are neither considered independently nor do they have equal importance.

4.3.2 Case study: River Aare, Switzerland – simulation of spawning habitats for grayling with the fuzzy approach and preference functions

Situation and objectives

To compare the multivariate fuzzy approach with univariate preference functions, the spawning habitats of European grayling (Thymallus thymallus) in the River Aare, Switzerland were investigated. The comparison is based on a field campaign in 2001 where locations of spawning sites were mapped and their functionality was observed. To obtain information about grayling requirements, the water depth and flow velocity were directly measured above spawning redds, while the substrate was characterized considering three size classes (fine gravel, coarse gravel, cobbles) to reflect the substrate composition. The habitat requirements in terms of univariate preference functions were derived directly from the field measurements (see also Chapter 5, Section 5.3.1). After a first verification, it turned out that the preferences received for a side branch were different to those found in the main stream and had to be adapted by including the knowledge of local fish experts. The fuzzy sets and fuzzy rules were created using the measured data as well as expert knowledge of a local fish biologist. For the comparison, the habitat model CASiMiR was applied, as it facilitates working with both approaches, although the fuzzy approach is standard in CASiMiR.

Results

To calculate a total HSI using univariate preference functions, the model user has to decide upon a mathematical operator to combine the individual HSI values. In this example, the product, the arithmetic mean and the geometric mean method were considered. In Figure 4.3, the total HSI of the univariate preference functions for each mathematical method are visualized versus the HSI obtained by the multivariate fuzzy approach, all at a discharge of 100 m³/s, which is a common flow rate during the spawning season in the River Aare.

In Figure 4.3, considerable variability in all simulated HSI distributions for both the univariate preference function with its different mathematical combination methods and the multivariate fuzzy approach can be seen. In particular, the results for the univariate preference functions show great inconsistency, even though identical preference functions were applied. The different results are

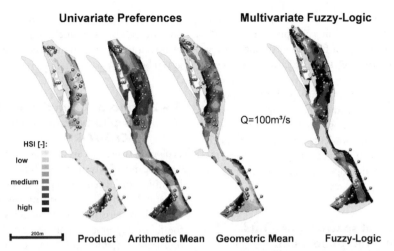

Figure 4.3 Spatial HSI distribution for univariate preference functions using different mathematical operations and for the multivariate fuzzy approach in comparison with information about observed spawning redds (pins).

produced solely by the different mathematical operations combining the individual HSI values.

Regarding the observed spawning areas (the pins in Figure 4.3 indicate the edges of the spawning areas), it was difficult to make an assessment just by visual comparison, therefore the percentages of observed spawning areas in three HSI classes were analysed to test model performance, as shown in Figure 4.4.

The comparison with observed spawning locations in Figure 4.4 confirms the enormous variability in the results obtained by the different approaches and combination operators in habitat modelling. In terms of model performance, Figure 4.4 gives the poorest performance for univariate preference functions using the product method. Here, 89% of observed spawning redds are located in areas where the model simulates a poor habitat quality and only 3% of observed redds are in areas with high HSI values. For the arithmetic and geometric means, the performance is improved, however: 68% for the arithmetic mean and 88% for the geometric mean of observed redds are in areas with low and medium HSI values. This is despite the adaption of the preference function

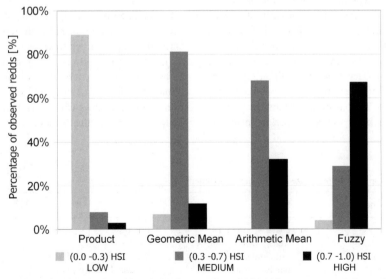

Figure 4.4 Comparison of simulated HSI values (classified in low, medium, high) with percentages of observed spawning areas.

mentioned above, which leads to a better result than a purely data-based function. The calibrated multivariate fuzzy approach (adaptation of fuzzy sets) simulates 67% of observed redds correctly (high HSI) and 4% incorrectly (low HSI). The results of this case study indicate that the multivariate fuzzy approach performs better in terms of the predictive accuracy than approaches using univariate preference functions. This might be attributed mainly to the biological knowledge that is incorporated in the form of fuzzy rules and fuzzy sets instead of using simple mathematical operations linking individual HSI indices.

4.4 Simulation of spawning habitats considering morphodynamic processes

4.4.1 Ecological relevance of morphodynamic processes

According to the European Water Framework Directive (WFD), preservation and establishment of fish spawning habitats should be considered as one of the major aims in successful river restoration (Hauer *et al.*, 2008). River dynamic processes such as flow alteration, sediment transport and seasonal flow events lead to mobilization of channel bed sediments and provide renewal of substrate conditions but can also affect early fish life stages or even destroy an entire generation (Schiemer *et al.*, 2003). Intrusion of fines or embeddedness contributes to a decrease in substrate permeability in the hyporheic interstitial and negatively impacts the development of early life stages. It may inhibit the emergence of fry from the interstitial spaces to the free water column and reduce the supply of dissolved oxygen and the transport of metabolic waste during egg incubation (Heywood and Walling, 2007). The significance of substrate characteristics is essential for the ecological assessment of spawning habitats. In particular, high flow events, their frequency and intensity, are

crucial in maintaining spawning habitats as these periodical events remove fine sediments and avoid clogging of interstitial spaces in suitable spawning habitats in a gravel bed.

4.4.2 Concept of implementing morphodynamic processes in CASiMiR

Ecological issues that are addressed by an approach implementing morphodyamics in habitat modelling include the identification of reduced HSI due to embeddedness and the simulation of mitigation measures to enhance substrate characteristics via morphodynamic modelling. In the following approach, an extended substrate description is implemented in CASiMiR in the form of embeddedness, packing and pore space, which constitute the most drastic effect on spawning habitat quality. Data for the extended substrate description were obtained by visual assessment during field observations. Table 4.2 provides information about the mapped parameters and their classification (Schälchli, 2002; modified by Eastman, 2004):

Based on the fuzzy logic method implemented in CASiMiR, these morphodynamic attributes were included as imprecise expert knowledge.

Due to the large amount of effort required to define fuzzy rules for more than four different parameters simultaneously, it is convenient to divide the approach into two subsequent steps, as illustrated in the flow chart in Figure 4.5.

The first step implements the conventional habitat modelling approach using basic habitat parameters (Figure 4.5), resulting in the calculation of HSI_{base}. If HSI_{base} is less than a predefined threshold value ($HSI_{threshold}$, e.g. 0.4), habitat suitability is defined as 'LOW' based on the conventional parameters. If this minimum requirement of habitat suitability is not fulfilled, no further investigation of embeddedness will be performed. For an estimated HSI_{base} higher than the threshold value in the second step,

Table 4.2 Mapping approach to recognize and evaluate embeddedness.

ID	Embeddedness	Substrate	Packing	Pore space
1	no	coarse-grained	very loose	coarse-pored
2	low	well-graded stones	loose	coarse–fine pored
3	moderate	sand and cohesives	slightly consolidated	fine-pored
4	high	partly cohesives	moderately consolidated	fine–no pores
5	complete	area-wide cohesives	strongly consolidated	no visible pores

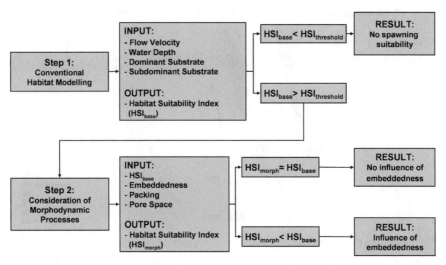

Figure 4.5 Flow chart to evaluate the influence of embeddedness on spawning habitat suitability.

additional morphodynamic processes are considered. The comparison between HSI_{base} and HSI_{morph} reflects the influence of the additional substrate characteristics and their importance for spawning habitat quality. An exemplary rule for the second step might be formulated as:

IF HSI_{base} is 'high' AND embeddedness 'medium' AND packing 'low' AND pore space 'high', THEN HSI_{morph} is 'high'

4.4.3 Case study: River Mur, Austria – morphodynamic processes and gravel-spawning fish habitat

Situation and objectives

The influence of morphodynamic processes in terms of embeddedness and dynamic substrate considerations was investigated at the River Mur in the Austrian Alps during the ALPRESERV project (Interreg IIIB, Eberstaller *et al.*, 2008). The study area is situated downstream of a power plant with a heavily impacted sediment regime, leading to strongly armoured layers and significant substrate embeddedness. Additionally, the flow velocities and water depths exceed the tolerable values of spawning suitability for European grayling (*Thymallus thymallus*) due to fixed river banks and the resulting narrowing of the river bed. In this study, two scenarios were investigated using the CASiMiR model: first, the influence of the current existing morphological conditions on spawning habitat suitability was analysed and secondly, the effects of a hypothetically proposed mitigation scenario were investigated. The mitigation scenario included the construction of groynes and a river bed widening to enhance the abiotic conditions, in particular the impaired substrate conditions. The mitigation scenario was implemented in a 2D morphodynamic model in order to simulate the sediment changes during a flushing hydrograph. The spawning habitat requirements for grayling were provided by Austrian fish experts based on electro-fishing, experience and a literature review.

Results

In order to analyse the influence of morphodynamic processes, the modelling results considering the extended description of substrate characteristics were compared with the conventional modelling approach. Figure 4.6 illustrates the mapping results for embeddedness, packing and pore space as well as their influence on spawning habitat suitability for grayling at a common discharge during the spawning period of 45 m^3/s.

Considering the parameters embeddedness, packing and pore space emphasizes that almost no suitable habitats for spawning grayling can be detected under existing conditions. The strong influence of the mapped parameters, particularly packing and pore space, has a direct impact on the extent of suitable areas. This leads to a significant reduction of spawning habitat suitability compared to conventional habitat modelling. As these results correspond to the biological observations for this specific river section, the need to consider an extended

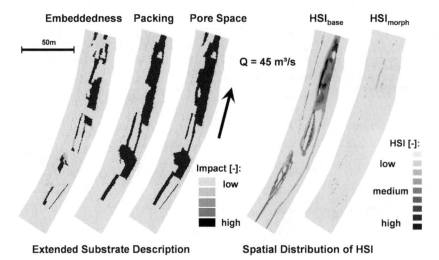

Figure 4.6 Results of mapping embeddedness, packing and pore space and their influence on spawning habitat suitability for grayling.

substrate description for assessing spawning habitat quality is proved by the simulation results of CASiMiR.

Figure 4.7 shows the proposed mitigation scenario with the implementation of two groynes plus river widening and the resulting bed level changes due to sediment flushing. Furthermore, we can also see the improved spawning habitat suitability compared to the conventional habitat modelling with a static representation of substrate characteristics.

In Figure 4.7, the bed elevation of the unmodified reach is shown on the left map, while the adjacent map represents the modified reach considering the changed bed topography due to the mitigation measures. The central

map illustrates the bed level changes (erosion and accumulation) after the flushing hydrograph, assumed to be a one-year flood event. Particularly on the left-hand side of the river, the sedimentation results in minor water depths and, combined with river bed widening, in a reduction of local flow velocities. Erosion at the downstream end of the study site results in significant substrate changes. Water depths in this section are still in a tolerable range for spawning activities. Comparing the resulting habitat suitability of both situations, significant improvements are detected in large areas on the right side of the river reach. As an integrated result, the habitat availability would almost be doubled by the proposed mitigation measures.

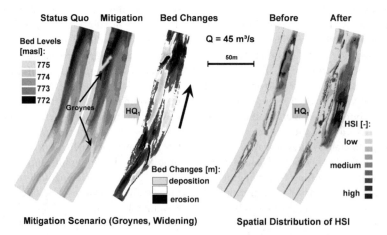

Figure 4.7 Results of the morphodynamic simulation applying a defined flushing hydrograph and its mitigation effect in terms of spawning habitat suitability.

Table 4.3 Classification of mesohabitats in MesoCASiMiR and collected parameters. Adapted from Maddock & Bird (1996).

Mesohabitat types											
Turbulent						Non-turbulent					
Very fast			Fast	Moderately fast		Moderately fast	Moderately slow		Slow		
Fall	Cascade	Chude	Rapid	Riffle		Run		Glide	Pool	Ponded	Other
Collected parameters in each mesohabitat											
Flow velocity			Flow diversity			Depth			Depth variability		
Dominant substrate			Subdominant substrate			Cover type			Embeddedness		
Packing			Pore space								

4.5 Habitat modelling on meso- to basin-scale

4.5.1 Requirements of habitat assessment on larger scales

As river management, according to the WFD, is conducted mostly on a basin scale, the upscaling of microhabitat models is an important task, but the extrapolation of micro-scale models to larger scales introduces a high level of uncertainty (Maddock, 1999). Moreover, habitat assessment at the micro-scale is time consuming and labour intensive. Therefore, intermediary approaches between micro- and macro-scale are required (Borsányi, 2002) to provide an objective representation of various characteristic morphological formations and habitat characteristics. These meso-scale approaches are based on so-called 'hydromorphological units' (HMUs), whereby each specific HMU is characterized by similar hydraulic patterns and morphological features. Given fishes' natural mobility, observations at the meso-scale are less affected by coincidence than at the micro-scale, providing relatively meaningful clues about species choice for habitats. Therefore, an ecological assessment on the meso-scale provides a suitable approach for good river basin management (Parasiewicz, 2003) in order to quantify impacts on habitat fragmentation and connectivity with a tolerable expenditure in time and costs. The meso-scale concept of CASiMiR is described briefly in the next section.

4.5.2 Concept of evaluation of mesohabitats using MesoCASiMiR

Providing suitable habitat-mapping methods is one of the crucial aspects in the development of mesohabitat models. For objective mapping in MesoCASiMiR, the parameters to be collected are classified with overlapping ranges in order to facilitate assignment to predefined parameters (Eisner *et al.*, 2005). Table 4.3 gives an overview of the mesohabitat types, sorted by degree of turbulence/flow velocity and the referring parameters.

As well as hydromorphological information, data are also collected on the current flow rate, the closest gauging station, migration barriers, etc. When performing habitat simulation, MesoCASiMiR first calculates the habitat suitability of each mesohabitat using the fuzzy approach of the micro-scale model. Input data are representative, averaged values of the conventional habitat parameters such as flow velocity, water depth, etc. for an average flow rate. The intermediate results are habitat maps for different species and life stages. To assess longer river sections or catchments with respect to their ecological functions, the spatial distribution of different mesohabitats, their extent and their suitability are taken into account.

The term 'habitat section' is introduced, defining a river section between two migration barriers. These habitat sections are evaluated using biological knowledge to consider the frequency distribution and portion of different habitat types. Another aspect of habitat networking is the passability of migration barriers. Even if the habitat section solely provides sufficient connectivity, it can hardly contribute to the survival of fish populations if spawning habitats are not accessible. Therefore, the functionality of migration facilities is used as an input parameter. Information about the attraction flow, the construction and hydraulics of the migration facility as well as the length of the upstream backwater is incorporated in a functional classification of migration barriers. As a result, the impact of each single barrier is part of the integrated assessment of habitat fragmentation and migratory potential in an investigated river reach.

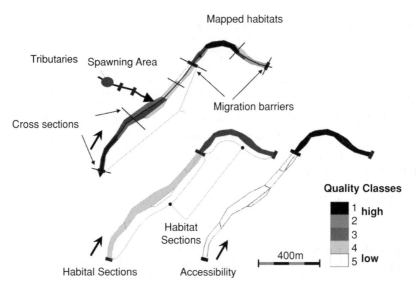

Figure 4.8 Results of multi-step fuzzy modelling in MesoCASiMiR in terms of connectivity and accessibility. The first step evaluates the mesohabitat quality, the second step the quality of a habitat section and the last step the accessibility of spawning areas.

Figure 4.8 provides an overview of the modelling steps implemented in MesoCASiMiR to evaluate mesohabitat distribution and habitat accessibility.

Based on the mapped mesohabitat distribution in Figure 4.8, the habitat quality of habitat sections is computed considering habitat connectivity, distances and mesohabitat types.

Given this concept, cumulative effects of migration barriers can be considered to locate migration bottlenecks and to prioritize mitigation measures. Hence, MesoCASiMiR allows a comprehensible method to investigate whole river systems based on rapid mapping and, further, on mesohabitat distribution, habitat connectivity and migration ability.

4.5.3 Case study: River Neckar, Germany – habitat fragmentation and connectivity

Situation and objectives

The first version of MesoCASiMiR was part of an integrated regional model within the project RIVERTWIN, funded by the European Commission, to assess the ecological status of rivers and to simulate the effects of different climate and management scenarios on river ecology (Schneider *et al.*, 2006). MesoCASiMiR was applied in Germany for the whole River Neckar (367 km), considering mainly the indicator species barbel (*Barbus barbus*). One goal was to assess the impact of navigation on habitat quality and the accessibility of spawning habitats, as most of the migration barriers in the River Neckar are actually not passable for fish. On the basis of a feasibility study provided by the regional fish administration, different scenarios were developed to improve the migratory potential. The mesohabitat characteristics and the functional capability of all migration barriers were mapped to evaluate the habitat connectivity and the accessibility of spawning habitats for river sections limited by migration barriers. The biological knowledge to formulate the underlying fuzzy rules was provided by fish experts familiar with the river's specific conditions.

Results

The river shows different characteristics in the navigable reach downstream of the city Plochingen versus the non-navigable part upstream. This is clearly reflected in the habitat availability for different life stages of the indicator species barbel. Figure 4.9 shows the results of the simulated habitat quality for adult, juvenile and spawning fish classified into five categories.

The results in Figure 4.9 emphasize the impact of navigation in the Neckar River. For all life stages, a higher habitat quality can be observed in the non-navigable river reach compared to the habitat sections affected by navigation. The navigable river reach provides only a few sections with moderate to high habitat quality, but they are separated by sections with low quality and tend to

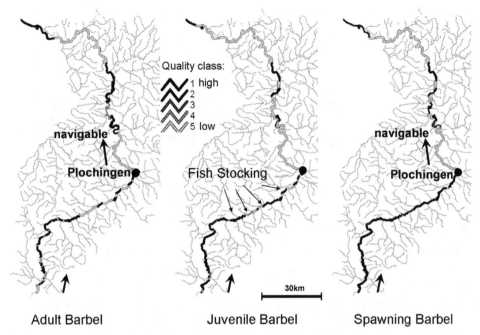

Figure 4.9 Results of MesoCASiMiR evaluating the quality of habitat sections for adult, juvenile and spawning barbel.

be isolated. The partly higher quality for adults in the navigable, regulated section reflects their higher tolerance to homogenous conditions and backwater compared to juveniles or spawning fish. For the non-navigable reach, the opposite conditions are prevalent, as hydromorphological conditions provide a wider spectrum of flow and morphological characteristics.

Investigations on the prioritization of measures to mitigate conditions for migratory fish and thus increase accessibility to spawning areas are illustrated by two scenarios in Figure 4.10.

The status quo in Figure 4.10 indicates the poor accessibility of the high quality spawning grounds in the non-navigable river reach given the high number of non-passable migration barriers. Reproduction areas are only accessible for short river reaches and are only used in combination with fish stocking (Figure 4.9). Scenario 1 aims to improve the accessibility by mitigating the most downstream barriers to enable upstream migration. However, since in the lower part of the River Neckar spawning grounds have almost disappeared, access to reproduction areas for barbel is not enhanced. Scenario 2 includes the rehabilitation of migration facilities in the upper part of the navigable section and is much more effective, since the non-navigable section with higher potential for reproduc-

tion is connected to the lower, less valuable parts of the river.

The model allows us to analyse which migration barriers should be mitigated first (e.g. rehabilitation of existing, non-functional fish bypasses) to achieve the most positive effect in terms of connectivity between habitat sections. The tributaries (only partly considered in the study) are of additional importance as they are hydromorphologically more heterogeneous compared to the main river and can provide more suitable spawning habitats if they are accessible.

4.6 Discussion and conclusions

Habitat model users need to know about the limitations of their tools, as models generally are simplified representations of nature. This is especially important in habitat modelling, as aquatic ecosystems are characterized by enormous complexity and ecological data often show a high bias, for example, important components of ecosystems like biological interactions (growth, predator pressure, energy budget, metabolic processes) are neglected in physical habitat modelling but can strongly influence habitat quality and habitat choice of aquatic species

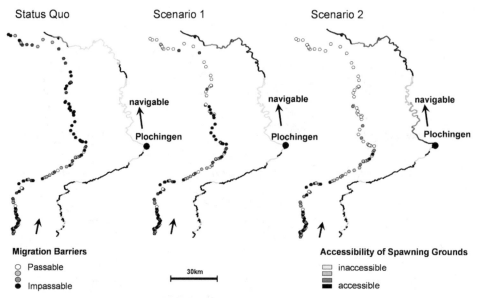

Figure 4.10 Results of MesoCASiMiR evaluating the accessibility of spawning grounds for barbel for the status quo and two mitigation scenarios. Passability of migration barriers (dots, left) and accessibility to spawning grounds (line, right).

(Gordon *et al.*, 2004). Another aspect in habitat modelling is the proper use of physical–biota relationships. A significant source of error is the transfer of habitat preferences between different ecosystems (Conallin *et al.*, 2010). Ideally, habitat preferences should be developed based on site-specific studies (Heggenes, 1990), as factors influencing fish habitat vary locally and fish respond spatially and temporally to these conditions. Furthermore, results of physical habitat modelling are often interpreted as direct indicators of abundance or population density, but physical habitat is a necessary but not a sufficient condition for the presence of organisms or ecological functionality. So far, little information is known about how much habitat suitability or WUA is required to represent population dynamics expressed in biomass or abundance (Gordon *et al.*, 2004).

The main strength of the fuzzy approach for fish habitat modelling is the ability to deal with imprecise and qualitative data that reflect ecological gradients and the available knowledge as well as the capability to consider the interaction of habitat parameters. The fuzzy rules representing expert knowledge are descriptors of general preferences of fish rather than precise indicators of preference. This leads to a robust method, especially in poor data situations, and allows the incorporation of semi-quantitative expert knowledge. If biological data (fish locations and abundances) are available, the physical–biota relationships in

terms of fuzzy rules and sets can become more specific by adapting them through a calibration process. Thus, it is highly recommended to collect field data for calibration and, wherever possible, to validate the initial chosen fuzzy sets and rules. However, a knowledge-driven approach includes some subjectivity and recent research proves the ability to use a data-driven approach in fuzzy logic to determine a biota–physical relationship if sufficient data are available for generating statistical distributions and training sets (Mouton, 2008).

As habitat modelling tools are applied for a wide range of water management topics, key issues are the formulation of biota–physical relationships and the adaptability of a model for specific ecohydraulic purposes. Therefore, this chapter contains a comparison between univariate preference functions and the multivariate fuzzy approach and two applications of the habitat model CASiMiR dealing with the inclusion of morphodynamic processes and the assessment of migratory aspects on the basin-scale.

The model comparison between univariate preference functions and the multivariate fuzzy approach yields an enhanced model performance for the multivariate fuzzy approach. The capability to incorporate qualitative and local biological knowledge in generating biota–physical relationships reflects the biotic response in a more robust way compared to univariate preference functions, which require mathematical operators to calculate the cumulative influence of habitat variables on biotic response. Furthermore, the mathematical combinations presume independency of habitat variables and consequently lack proper weighting of the applied habitat variables. With the experience of numerous projects worldwide and in rivers of varying size, the many fish experts involved in the development of fuzzy sets and expert fuzzy rules need some time to get accustomed to this way of defining habitat preferences, but once they become familiar with it, they find it easy to handle, more appropriate and closer to their kind of thinking than preference functions.

The case study on the River Mur, Austria reveals the importance of an extended description of substrate characteristics (embeddedness, packing and pore size) for simulating spawning habitat suitability for a gravel spawning fish. However, further intensive studies are required to develop an efficient operational interface between morphodynamic and habitat simulation models. Moreover, focused research is necessary to increase the biological knowledge about dependencies between sediment dynamics and their limitations on specific river organisms. In addition, the consideration of interactions between suspended load and bed material (e.g. sediment infiltration) is of great importance to enable the incorporation of parameters such as embeddedness into the simulation of mitigation scenarios.

The mapping methodology of MesoCASiMiR has been developed to meet the demands of a fast and efficient field data collection methodology on one hand and to be detailed enough for reliable fish habitat modelling on the other hand. In the case study on the Neckar River, the efficiency of single and cumulative mitigation measures related to migration barriers was evaluated by scenario simulations. However, the existing knowledge about habitat connectivity, diversity, networking and fragmentation is very limited. Little is known about the relations between mesohabitat distribution, habitat type-to-type distances and the final cumulative effects on habitat quality and ecosystem functions. A better understanding of these issues is crucial for future investigations of meso-habitats on a basin-scale.

The habitat modelling system CASiMiR and its multivariate fuzzy approach are based on an open, flexible structure that can be adapted to a wide range of ecohydraulic issues. Nevertheless, even though it allows linkage between arbitrary habitat parameters and their biological response, CASiMiR is based on physical–biota relationships and the quality of the results depends on the reliability of these relationships. As experienced in the past, the further development and improvement of habitat suitability models and their prediction capabilities will require close collaboration between fish biologists, river engineers and managers – a real interdisciplinary challenge for the future.

References

Acreman, M.C. and Dunbar, M.J. (2004) Defining environmental river flow requirements – a review. *Hydrology and Earth System Science*, **8**: 861–876. doi:10.5194/hess-8-861-2004.

Adriaenssens, V., DeBaets, B., Goethals, P.L.M. and DePauw, N. (2004) Fuzzy-rule based models for decision support in ecosystem management. *Science of the Total Environment*, **319**: 1–12. doi:10.1016/S0048-9697(03)00433-9.

Ahmadi-Nedushan, B., St-Hilaire, A., Bérubé, M., Robichaud, E., Thiémonge, N. and Bobée, B. (2006) A review of statistical methods for the evaluation of aquatic habitat suitability for instream flow assessment. *River Research and Applications*, **22**: 503–523. doi:10.1002/rra.918.

Bain, M.B. (1995) Habitat at the local scale: multivariate patterns for stream fishes. *Bulletin Français de la Pêche et de la Pisciculture*, **337/833/933:** 771–561.

Borsányi, P. (2002) *A Meso-Scale Habitat Classification Method for Production Modelling of Atlantic Salmon in Norway.* IFIM users workshop, Fort Collins, USA.

Bovee, K.D. (1982) *A Guide to Stream Analysis Using the Instream Flow Incremental Methodology.* Instream flow information paper 12, FWS/OBS 82/26, U.S. Fish and Wildlife Service.

Bovee, K.D. (1986) *Development and Evaluation of Habitat Suitability Criteria for Use in the Instream Flow Incremental Methodology.* U.S. Fish and Wildlife Service Biological Report, **86**(7): 1–235.

Cadenasso, M.L., Pickett, S.T.A., Weathers, K.C. and Jones, C.G. (2003) A framework for a theory of ecological boundaries. *BioScience*, **53**: 750–758. doi:10.1641/0006-3568(2003)053[0750:AFFATO]2.0.CO;2.

Conallin, J., Boegh, E. and Jensen, J. (2010) Instream physical habitat modeling types: an analysis as stream hydro-morphological modeling tools for EU water resource managers. *Journal of River Basin Management*, **8**: 93–107, doi:10.080/1571521003715123.

Cornelissen, A.M.G., Van den Berg, J., Koops, W.J., Grossman, M. and Udo, H.M.J. (2001) Assessment of the contribution of sustainability indicators to sustainable development: a novel approach using fuzzy set theory. *Agriculture, Ecosystems and Environment*, **86**: 173–185.

Eastman, K. (2004) *Effects of Embeddedness on Fish Habitats: An Approach for the Implementation in the Habitat Simulation Model CASiMiR.* Masters Thesis, Institute of Hydraulic Engineering, Universitaet Stuttgart.

Eberstaller, J., Pinka, P. and Schneider, M. (2008) *Impact Analysis and Recommendations Regarding Reservoir Flushing Strategies.*

ALPRESERV – Sustainable Sediment Management in Alpine Reservoirs Considering Ecological and Economical Aspects, report volume 6.

Eisner, A., Young, C., Schneider, M. and Kopecki, I. (2005) Meso-CASiMiR – new mapping method and comparison with other current approaches. In Harby, A., Baptist, M., Duel, H., Dunbar, M., Goethals, P., Huusko, A., Ibbotson, A., Mader, H., Pedersen, M.L., Schmutz, S. and Schneider, M. (eds) *Proceedings of COST-Action 626*, Silkeborg, Denmark.

García, A., Jorde, K., Habit, E., Caamaño, D. and Parra, O. (2010) Downstream environmental effects of dam operations: Changes in habitat quality for native fish species. *River Research and Applications*, published online, doi:10.1002/rra.1358.

Gordon, N.D., McMahon, T.A., Finlayson, B.L., Gippel, C.J. and Nathan, R.J. (2004) *Stream Hydrology: An Introduction for Ecologists*, 2nd edition, John Wiley & Sons, Ltd, Chichester, UK.

Gore, J.A. and Nestler, J.M. (1988) Instream flow studies in perspective. *Regulated Rivers*, **2**: 93–101. doi:10.1002/rrr.3450020204.

Guisan, A. and Zimmermann, N.E. (2000) Predictive habitat distribution models in ecology. *Ecological Modelling*, **135**: 147–186. doi:10.1016/S0304-3800(00)00354-9.

Harby, A., Baptist, M., Dunbar, M.J. and Schmutz, S. (2004) *COST Action 626: State-of-the-art in data sampling, modelling analysis and applications of river habitat modelling*, European Aquatic Modelling Network.

Hauer, C., Unfer, G., Schmutz, S. and Habersack, H. (2008) Morphodynamic effects on the habitat of juvenile cyprinids (*Chondrostoma nasus*) in a restored Austrian lowland river. *Environmental Management*, **42**(2): 279–296. doi: 10.1007/s00267-008-9118-2

Heggenes, J. (1990) Habitat utilisation and preferences of juvenile Atlantic salmon (*Salmo salar*) in streams. *Regulated Rivers*, **5**: 341–354.

Heywood, M.J.T. and Walling, D.E. (2007) The sedimentation of salmonid spawning gravels in the Hampshire Avon catchment, UK: implications for the dissolved oxygen content of intragravel water and embryo survival. *Hydrological Processes*, **21**: 770–788. doi:10.1002/hyp.6266.

Jorde, K. (1996) *Ökologisch begründete, dynamische Mindestwasserregelungen bei Ausleitungskraftwerken*. PhD thesis, Institute of Hydraulic Engineering, Universitaet Stuttgart.

Kerle, F., Zöllner, F., Schneider, M., Böhmer, J., Kappus, B. and Baptist, M.J. (2002) Modelling of long-term fish habitat changes in restored secondary floodplain channels of the river Rhine. In King, J. and Brown, C (eds) Environmental Flows for River Systems, *Proceedings of the 4th Ecohydraulics Symposium*, Cape Town, South Africa.

Kopecki, I. (2008) *Calculational Approach to FST-Hemispheres for Multiparametrical Benthos Habitat Modelling*. PhD thesis, Institute of Hydraulic Engineering, Universitaet Stuttgart.

Leclerc, M.A., St-Hilaire, A. and Bechara, J.A. (2003) State-of the-art and perspectives on habitat modelling. *Canadian Water Resources Journal*, **28**: 153–172.

Maddock, I. (1999) The importance of physical habitat assessment for evaluating river health. *Freshwater Biology*, **41**: 373–391. doi:10.1046/j.1365-2427.1999.00437.x.

Maddock, I. and Bird, D. (1996) The application of habitat mapping to identify representative PHABSIM sites on the River Tavy, Devon, UK. In Leclerc, M., Capra, H., Valentin, S., Boudreault, A. and Coté, Y. (eds) *Proceedings of the 2nd International Symposium on Habitats and Hydraulics*, Quebec, Canada, Vol. B, 203–214.

Milhous, R.T., Updike, M.A. and Schneider, D.M. (1989) *Physical Habitat Simulation Reference Manual Version II*. Instream flow information paper 26, U.S. Fish and Wildlife Service.

Mouton, A.M. (2008) *A Critical Analysis of Performance Criteria for the Evaluation and Optimisation of Fuzzy Species Distribution Models*. PhD thesis, University of Gent, Belgium.

Mouton, A.M., Schneider, M., Kopecki, I., Goethals, P.L.M. and DePauw, N. (2006) Application of MesoCASiMiR: Assessment of *Baetis rhodani* Habitat Suitability. *Proceedings of the 3rd iEMSs Biennial Meeting*, ISBN 1-4243-0852-6.

Parasiewicz, P. (2003) Upscaling: integrating habitat model into river management. *Canadian Water Resources Journal*, **28**: 1–17.

Revenga, C., Brunner, J., Henninger, N., Kassem, K. and Payne, R. (2000) *Pilot Analysis of Global Ecosystems: Freshwater Systems*, World Resources Institute, Washington, DC.

Rosenfeld, J. (2003) Assessing the habitat requirements of stream fishes: An overview and evaluation of different approaches. *Transactions of the American Fisheries Society*, **132**: 953–968.

Salski, A. (1992) Fuzzy knowledge-based models in ecological research. *Ecological Modelling*, **63**: 103–112. doi:10.1016/0304-3800(92)90064-L.

Schälchli, U. (2002) Die innere Kolmation von Fliessgewässersohlen – Eine neue Methode zur Erkennung und Bewertung. *Fischnetz-Info*, **9**: 5–6.

Schiemer, F., Keckeis, H. and Kammler, E. (2003) The early life history stages of riverine fish: ecophysiological and environmental bottlenecks. *Comparative Biochemistry and Physiology Part A*, **133**: 439–449.

Schneider, M. (2001) *Habitat- und Abflussmodellierung für Fließgewässer mit unscharfen Berechnungsansätzen*. PhD thesis, Institute of Hydraulic Engineering, Universitaet Stuttgart.

Schneider, M., Kopecki, I. and Eisner, A. (2006) *Entwicklung des Habitatmodells Mesocasimir und Anwendung im Neckareinzugsgebiet, Tagungsdokumentation, Zukunftsperspektiven für ein integriertes Wasserresourcen Management im Einzugsgebiet Neckar*.

Stalnaker, C.B., Lamb, B.L., Henriksen, J., Bovee, K. and Bartholow, J. (1995) The Instream Flow Incremental Methodology: A Primer for IFIM. *U.S. Geological Survey Biological Report*, **29**: 45.

Statzner, B., Kohmann, F. and Hildrew, A.G. (1991) Calibration of FST-hemispheres against bottom shear stress in a laboratory flume. *Freshwater Biology*, **26**: 227–231.

Vadas, R.L. and Orth, D.J. (2001) Formulation of habitat suitability models for stream fish guilds: do the standard methods work? *Transactions of the American Fisheries Society*, **130**: 217–235. doi: 10.1577/1548-8659(2001)130<0217:FOHSMF>2.0.CO;2.

van Broekhoven, E., Adriaenssens, V., De Baets, B. and Verdonschot, P.F.M. (2006) Fuzzy rule-based macroinvertebrate habitat suitability models for running waters. *Ecological Modelling*, **198**: 71–84. doi:10.1016/j.ecolmodel.2006.04.006.

Vilizzi, L., Copp, G.H. and Roussel, J.-M. (2004) Assessing variation in suitability curves and electivity profiles in temporal studies of fish habitat use. *River Research and Applications*, **20**: 605–618. doi: 10.1002/rra.767.

Yi, Y., Wang, Z. and Yang, Z. (2010) Two-dimensional habitat modeling of Chinese sturgeon spawning sites. *Ecological Modelling*, **221**: 864–875. doi:10.1016/j.ecolmodel.2009.11.018.

Zadeh, L.A. (1965) Fuzzy-Sets. *Information and Control*, **8**: 338–353.

5

Data-Driven Fuzzy Habitat Models: Impact of Performance Criteria and Opportunities for Ecohydraulics

Ans Mouton[1], Bernard De Baets[2] and Peter Goethals[3]

[1] Research Institute for Nature and Forest, Department of Management and Sustainable Use, Ghent University, Kliniekstraat 25, B-1070 Brussels, Belgium

[2] Department of Mathematical Modelling, Statistics and Bioinformatics, Ghent University, Coupure links 653, 9000 Gent, Belgium

[3] Aquatic Ecology Research Unit, Department of Applied Ecology and Environmental Biology, Ghent University, J. Plateaustraat 22, B-9000 Gent, Belgium

5.1 Challenges for species distribution models

5.1.1 Knowledge-based versus data-driven models

Although expert knowledge and data-driven methods have been described as two separate approaches by many authors, actually there is no sharp distinction between them because all expert knowledge has been derived from observations or measurements. Knowledge derived from data is often quantitative, while most sensory-perceived knowledge is qualitative. Furthermore, expert knowledge can describe environmental processes, but also the occurrence of species, which integrates several environmental processes. The subject and origin of the expert knowledge determine the niche that may be described by the knowledge.

Most aquatic expert knowledge that focuses on species occurrence is derived from sensory field perceptions and is thus qualitative. An example would be a fisherman saying, 'At a high flow velocity and a low depth you'll find many brown trout in this river'. Expert models translate this quantitative knowledge into linguistic rules (Zadeh, 1965;

Jorde *et al.*, 2000; van Broekhoven *et al.*, 2006) such as 'if flow velocity is high and depth is low, then the habitat suitability for brown trout is high'.

> Expert knowledge-based models may take into account most of the ecological knowledge available.

5.1.2 Ecological boundaries

To construct the linguistic expert rules, the range of the model variables depth, flow velocity and habitat suitability is divided into classes such as low, medium and high. Most habitat suitability models apply the crisp boundary approach, which states, for example, that a water column depth lower than 0.5 m is low, whereas a depth exceeding 0.5 m is high. Consequently, the boundary between two consecutive classes is crisp because it is situated at a single value of the variable. However, transitions in ecology are not crisp but gradual, resulting in ecological variables. Specifically, if the class boundary between a low and a high water column depth is set at 0.5 m, a depth of 0.49999 m

Ecohydraulics: An Integrated Approach, First Edition. Edited by Ian Maddock, Atle Harby, Paul Kemp and Paul Wood.
© 2013 John Wiley & Sons, Ltd. Published 2013 by John Wiley & Sons, Ltd.

will be classified as low, while a depth of 0.50001 m will be high, which does not match with ecological boundary theory (Strayer *et al.*, 2003). Fuzzy logic has proven to be an appropriate expert model technique to deal with these ecological variables because the boundaries between the classes of the input variables are overlapping in a fuzzy model and thus reflect these gradual transitions between classes (van Broekhoven *et al.*, 2006; Mouton *et al.*, 2007).

> Fuzzy logic models may be more appropriate for ecological modelling than classical modelling techniques because the overlap of the fuzzy classes reflects the ecological boundary concept.

5.1.3 Interdependence of variables

The fuzzy logic approach addresses another shortcoming of the widely used PHABSIM related habitat models, which is the application of independent habitat suitability curves. These curves describe HS inadequately since, in reality, physical habitat variables are not independent (Heggenes, 1996). Since the first application of PHABSIM, different methods have been developed to overcome this problem. Several authors have suggested that linking different habitat variables through simple mathematical operations is not adequate for the description of habitat suitability (Heggenes, 1996; Sekine *et al.*, 1997). Although sensitivity analysis with different sets of univariate preference curves (Bovee, 1986) could partially solve this problem, Sekine *et al.* (1997) proposed the use of weighting factors to combine different habitat preferences based on different variables. Another option, the multivariate species response curves which were described earlier, is hardly used in practical applications due to several mathematical limitations (Bovee *et al.*, 1998). Fuzzy habitat models, by contrast, take into account interactions between habitat variables in a multivariate habitat suitability analysis (Jorde *et al.*, 2000; Adriaenssens *et al.*, 2004; van Broekhoven *et al.*, 2006).

Other authors propose a more data-driven approach (Lamouroux *et al.*, 1998; Mouton *et al.*, 2010b; Thu Huong *et al.*, 2010) but many of these methods require large amounts of data or are not transferable to different river types (Lamouroux *et al.*, 1998). These data requirements significantly restrict applicability of these approaches, even if data collection is facilitated by new techniques such as fish tagging. Fuzzy habitat models allow the inclusion of expert knowledge in the calculation of habitat suitability, hence compensating for situations where few fish data could be collected in the field.

> Fuzzy logic models may be more appropriate for habitat suitability modelling than some PHABSIM-based approaches, since these fuzzy models include variable interdependence.
>
> Fuzzy logic models may be more appropriate for habitat suitability modelling than some data-driven models if field data are lacking.

5.1.4 The knowledge acquisition bottleneck

More recently, knowledge-based models have become a popular technique for ecological modelling, resulting in numerous applications (Adriaenssens *et al.*, 2004). However, the main bottleneck in the application of a merely knowledge-driven approach is the need for ecological expert knowledge. Although ecological research currently produces a wealth of knowledge about the habitat requirements of the various species, the formalisation of problem-relevant human expert knowledge is often difficult and tedious (Žnidaršic *et al.*, 2006).

First, most of the available knowledge is species specific, while, for example, habitat preference curves estimated for one species in a given reach can often not be extrapolated to other communities of which this species forms a part (Karr, 1991). A far greater threat for the application of expert models lies in the lack of consistency in the expert knowledge, which is reported by several authors (Acreman and Dunbar, 2004; Hernandez *et al.*, 2006). In a study on the transferability of species distribution models for butterflies, Vanreusel *et al.* (2007) observed some transferability of ecological expert knowledge, but their study only covers a small study area, which minimises the likelihood of different impacts on the realised niche. Boyce *et al.* (2002) describe a number of biological problems that interfere with the niche conservation concept. These results are in line with the environmental gradient categorisation which was discussed earlier. Due to their direct impact on the habitat suitability of aquatic species, species response curves to direct variables are expected to have constant shapes, whereas the shape of a species response curve to an indirect variable may depend on the situation (Hutchinson, 1957). Consequently, expert knowledge or species response curves to a specific variable can only be consistent over two situations if this variable is a direct variable for the species of interest in both situations. However, the same variable can be a direct variable in one river and an indirect one in another river, because a direct variable was defined as the most proximal variable in the chain of processes that link the variable to its impact. Flow velocity, for instance, has been widely accepted as a direct variable for fish presence in rivers with a good chemical water quality, whereas flow velocity may be farther down the chain of processes that link this variable to its impact in severely polluted rivers due to interactions with other variables such as oxygen concentration. Specifically, high flow velocities, for instance, may lead to a higher oxygen concentration due to water splashing. This oxygen concentration may thus be a direct variable for fish presence in polluted rivers, whereas flow velocity is an indirect variable in these rivers. A general assumption of most knowledge-based habitat suitability models is that only the most direct variables are included in the model, and the species response to these variables is thus consistent and transferable between different rivers. Since the distinction between direct and indirect variables is situation-dependent, however, species response curves could take any shape depending on the studied case (Guisan and Zimmermann, 2000; Acreman and Dunbar, 2004). This has important implications for the applicability of knowledge-based models (Hudson *et al.*, 2003). Particular care should be taken in interpreting model results beyond their original domain (Boyce *et al.*, 2002).

> The acquisition of expert knowledge is the key bottleneck for the application of fuzzy logic models in habitat suitability modelling.

5.1.5 Data-driven knowledge acquisition

Recent research has shown that complementing expert systems by data-driven techniques can solve this 'knowledge acquisition bottleneck' (Žnidaršic *et al.*, 2006). For example, the induction of fuzzy rule-based models by heuristic search algorithms is often used in the field of fuzzy rule learning (Hüllermeier, 2005). However, ecological data show some specific characteristics such as a low or high species prevalence (i.e. the frequency of occurrence), a high likelihood that false absences have been included in the dataset, etc. Since these characteristics should be taken into account when developing data-driven ecological models, some of these characteristics are addressed in this chapter, with a focus on the development of reliable models and on the evaluation of these models.

The development of a reliable data-driven model requires a sound training and validation procedure. Model training is the process in which the model parameters are iteratively adjusted to increase the agreement between the model predictions and the observations, which are

referred to as the training dataset. Since this agreement is assessed by the performance criterion, model training aims to optimise the performance criterion. Cross-validation is often applied to assess the robustness of the training results by randomly creating different training datasets. If the number of model parameters is relatively high compared to the number of training data points, model training may result in a model which predicts the observations too accurately. Specifically, the reliability and the robustness of the resulting model decrease during model training because the model increasingly reflects the training data. The model predictions would agree poorly with the observations from other datasets because it only describes the specific ecological processes included in the original training data. In such cases, model training is a trade-off between model specificity and model generality or robustness. This situation often occurs when neural networks or decision trees are applied and is called 'overfitting'. To avoid overfitting, the available dataset is often split into a training set and a test set. Model training is then stopped when the model performance on the test set decreases (Fielding and Bell, 1997).

Although several authors have emphasised the importance of correct model training and evaluation, both procedures are often neglected or applied erroneously (Fielding and Bell, 1997; Allouche *et al.*, 2006). This problem is often related to the application of the performance criterion which is used to evaluate the model performance. This criterion is the key component of model training and evaluation, and this chapter aims to incorporate both procedures into ecological modelling in an ecologically relevant way by analysing different performance criteria and suggesting challenges for their correct application.

> Ecologically relevant model training and evaluation is a crucial step in the development process of a data-driven species distribution model.

5.2 Fuzzy modelling

5.2.1 Fuzzy rule-based modelling

In fuzzy models, linguistic values such as 'low', 'moderate' and 'high' are assigned to the output variable (for instance, habitat suitability) and to the input variables (for instance, depth, flow velocity, width, substrate or cover). These linguistic values are defined by fuzzy sets (Zadeh, 1965) and not by conventional sets with crisp boundaries

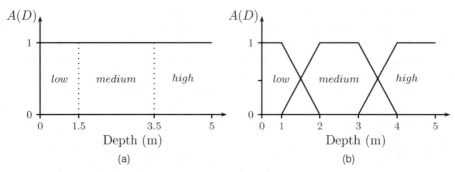

Figure 5.1 Definition of the three linguistic values assigned to depth (D) by means of (a) crisp and (b) fuzzy sets (A(D) = the membership degree of a depth value to the fuzzy set).

(hereafter called crisp sets). When using these crisp sets, for instance, depths below 1.5 m would be considered 'low', depths between 1.5 and 3 m 'moderate' and depths higher than 3 m 'high'. Hence, a given depth would either belong to a set (it has a membership degree of 1 to this set) or it would not. A membership function of a particular fuzzy set, however, indicates the degree to which an element belongs to this fuzzy set, with membership ranging from zero to one (see also Chapter 4, Section 4.2.2). Consequently, the membership functions of the fuzzy sets have overlapping boundaries and the linguistic statement 'the velocity is quite moderate but tending to be high' can be translated into a velocity which has a membership degree of 0.4 to the fuzzy set 'moderate' and of 0.6 to the fuzzy set 'high' (Figure 5.1). In this chapter, all membership functions are defined by four parameters (am, bm, cm and dm): the membership degree linearly increases between am and bm from 0 to 1, is equal to 1 between bm and cm and linearly decreases from 1 to 0 between cm and dm (Figure 5.2). A triangular membership function is obtained when bm equals cm (Figure 5.2).

The fuzzy rule base relates the input variables to the habitat suitability for a given species and consists of IF–THEN rules, such as 'IF depth is moderate AND flow velocity is high AND cover is high THEN habitat suitability is high'. The IF part of the rule, called the antecedent, describes in which situation this rule applies, while the THEN part, called the consequent, indicates whether the habitat in this situation is suitable or not for the species. Given crisp values of the input variables, the output of the fuzzy model is calculated as described by van Broekhoven *et al.* (2006). For each instance, the membership degrees to the fuzzy sets of each input variable are calculated. The degree of fulfilment of each rule is then calculated as the minimum of the membership degrees in its antecedent. Finally, to each linguistic output value, a fulfilment degree is assigned equal to the maximum of the fulfilment degrees of all rules with the output value under consideration in their consequent (see also Chapter 4, Section 4.2.2). The approach is similar to the Mamdani–Assilian procedure (Assilian, 1974; Mamdani, 1974) in which the fuzzy output is converted into a crisp one based on the fuzzy sets of the output variables. However, in presence–absence modelling, often a different type of model is applied: a fuzzy classifier (van Broekhoven *et al.*, 2006; 2007). Both observed and modelled values of the output variable are assigned to the fuzzy set with the highest fulfilment degree, which allows comparison of the modelled output with the observed output and calculation of performance measures. Specifically, if a value of habitat suitability has 0.6 membership in fuzzy set 'low' and 0.4 membership in fuzzy set 'moderate', then the value is assigned to 'low'.

5.2.2 Fuzzy rule base optimisation

Introduction
Although linguistic fuzzy rules were derived originally from expert knowledge, more recently various techniques have been developed to identify the rules and fuzzy sets from data, such as fuzzy clustering, neural learning

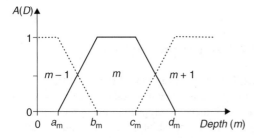

Figure 5.2 The parameters a_m, b_m, c_m, and d_m of the trapezoidal fuzzy set m of the variable 'Depth'.

methods and genetic algorithms (Hüllermeier, 2005). In this chapter, the accuracy of interpretable linguistic models will be improved, but we will also present a method which guarantees that the increase in accuracy is also ecologically relevant. This method consists of two steps: first, the fuzzy sets are created, based either on expert knowledge or by optimisation on data; then, the fuzzy rule base is created by optimising the fuzzy rule consequents. Within the latter optimisation method, different performance criteria will be applied and results will be compared.

Fuzzy sets optimisation

Fuzzy sets are of particular interest for ecological modelling since they allow definition of different environmental conditions and gradients between these conditions. For instance, a depth range between 0 and 0.5 m can be considered 100% 'low', the depth range between 1 and 2 m 100% 'moderate' and depth changes gradually from 'low' to 'moderate' between 0.5 and 1 m. In this scenario, the model will consider all depths between 0 and 0.5 m as equal environmental conditions because the ecologist developing the model defined this depth range as one environmental condition (e.g. one mesohabitat type). Consequently, the parameters of the membership functions corresponding to the fuzzy sets of the input variables have often been derived from expert knowledge. However, if no *a priori* knowledge on the fuzzy sets is available, data-driven sets optimisation followed by further fine tuning of the sets by the ecological expert may overcome this problem. Previous research showed that optimising sets to guarantee a uniform distribution of the input variables over the fuzzy sets may be a good starting point (Mouton *et al.*, 2008a). Specifically, if a fuzzy set of an input variable contains very few training instances, rules which involve this fuzzy set will be trained inadequately, whereas uniform distribution of the input variables over the fuzzy sets may reduce this problem. Therefore, before fuzzy rule base optimisation based on fixed fuzzy sets, fuzzy sets were optimised based on the Shannon–Weaver entropy (Shannon and Weaver, 1963), ranging between 0 and 1, with an entropy of 1 representing a perfectly uniform distribution. This entropy allows quantification of the uniformity if the fuzzy sets are converted into crisp ones whose boundaries are the points having a membership degree of 0.5 to the corresponding fuzzy set. This fuzzy sets optimisation method was not applied in this chapter but is described in detail in Mouton *et al.* (2008a).

The fuzzy sets optimisation method ensures that the distribution of the training data instances over the fuzzy sets is optimal. In a situation with n fuzzy sets, sets which contain less than $1/n$ of the data would be expanded as far as possible, whereas sets which contain more than $1/n$ of the data would be reduced to, or split into, smaller sets. This method avoids a situation where empty or poorly represented sets are included in the model and increases model efficiency by deleting redundant sets. Several authors have suggested similar approaches and demonstrated that such methods significantly improve model performance (Casillas *et al.*, 2003). However, a more uniform distribution of the input data over the fuzzy sets does not guarantee that each fuzzy rule is represented uniformly in the input data. Water depth values, for instance, can be uniformly distributed over a 'low' and a 'high' fuzzy set, but this does not imply that many sampling sites with a high flow velocity and a high depth are represented in the training set.

Hill-climbing algorithms for rule base training

The term 'hill-climbing' implies an iterative improvement technique, and thus describes both maximisation and minimisation problems. The technique is applied to a single solution, the current solution, in the search space. During each iteration, a new solution is selected from the neighbourhood of the current solution. If that solution provides a better value in light of the evaluation function, the new solution becomes the current solution. Otherwise, some other neighbour is selected and tested against the current point. The method terminates if no further improvement is possible. Hill-climbing methods are often started from a large variety of different starting solutions because these methods can only provide locally optimal values, and these values depend on the selection of the starting point. In this chapter, the initial points are chosen at random, and the algorithm is stopped if the best solution is found for the fifth time. For problems with many local optima, particularly those where these optima have large basins of attraction, it's often very difficult to locate a globally optimal solution. However, several authors have shown that there is no way to choose a single search method that can serve well in every case (Michalewicz and Fogel, 2000), which is referred to as the 'no free lunch theorem'. If poor consistency between different iterations had been observed, a solution could be to run the algorithm r times (with $r > 100$) and to select the model that was obtained in most iterations as the global optimum. However, in this chapter, the algorithm was consistent between the five different iterations and thus no more than five iterations were needed.

There are a few versions of hill-climbing algorithms, which differ mainly in the way a new solution is selected for comparison with the current solution. In this chapter, the steepest ascent hill-climbing algorithm is applied. Initially, all possible neighbours v_n of the current solution are considered, and the one v_n that returns the best model performance is selected to compete with the current solution, v_c. If the model performance of v_c is worse than the performance of v_n, then the new solution v_n becomes the current solution. Otherwise, no local improvement is possible and the algorithm has reached a local or a global optimum. In such a case, the next iteration of the algorithm is executed with a new current solution selected at random (Michalewicz and Fogel, 2000).

To generate a reliable habitat suitability model, the consequents of the fuzzy rules in this chapter were optimised using a nearest ascent hill-climbing algorithm. First, the fuzzy sets were optimised as mentioned earlier. Once the sets were optimised, they did not change during the further training procedure. Starting from these fixed fuzzy sets and a rule base with randomly selected rule consequents, the linguistic value in the consequent of one randomly selected rule was changed into its neighbouring linguistic value and the impact on model performance was calculated. If model performance increased, the algorithm continued with the adjusted rule base; if not, it continued with the original one. If a linguistic term had two neighbouring linguistic terms (e.g. the linguistic term 'moderate' may have the two neighbouring linguistic terms 'low' and 'high'), the impact of both neighbouring linguistic terms on model performance was calculated and compared. To indicate the robustness of the optimisation results, s-fold cross-validation was applied. First, s partitions were constructed by randomising the original dataset and assigning each data point to one partition without replacement, such that each partition contained $1/s$ of the total number of data points of the dataset. Ten sets of both a training and a test fold were then created by identifying one partition as the test fold and by grouping the $s-1$ other partitions into the training fold. This procedure resulted in s different training and test folds. The species prevalence and the fuzzy set configuration were constant for all partitions and, thus, for all training and test folds.

Different models were trained based on different performance measures, as described in the next sections. Different training scenarios were created to compare the results of optimisation based on the different performance measures (Table 5.2). Each training iteration was stopped when no further increase of the performance measure on the test fold was observed. Each training iteration was repeated and the obtained rule base was compared to each rule base obtained in previous iteration steps. The resulting rule base similarity indicated the percentage of rule consequents that was identical for two rule bases. If the rule base with the highest performance on the test fold was obtained five times, this rule base was selected as the final rule base and training continued on another fold as in the following algorithm:

Algorithm 1: Training algorithm

$t \leftarrow 0$
for each *fold* **do**
    stop $\leftarrow 0$
    while stop $\leftarrow 5$ **do**
        $t \leftarrow t + 1$
        Train rule base based on training fold
        Perftest$_t$ $\leftarrow$ the performance of the resulting
                rule base, *RB$_t$*, on the test fold
        if $t = 1$ **then**
            Maximal *Perftest* $\leftarrow$ *Perftest$_t$*
            Maximal *RB* $\leftarrow$ *RB$_t$*
        else if *Perftest$_t$* = Maximal *Perftest* **then**
            Maximal *Perftest* $\leftarrow$ *Perftest$_t$*
            Maximal *RB* $\leftarrow$ *RB$_t$*
            stop $\leftarrow 0$
        else if *Perftest$_t$* = Maximal *Perftest* **then**
            if similarity of *RB$_t$* and Maximal
                    RB = 100% **then**
                stop $\leftarrow$ *stop* + 1
        end
    end
    end
end

In this work, the fuzzy rule base contained fuzzy rules representing each possible combination of input variable sets. However, not every combination of input variable sets was present in the studied reach. To obtain an indication of which rules were relevant, the fuzzy sets were turned into crisp ones by assuming that an input value does not belong to a set if its membership degree to this set is < 0.5. Each data point could be assigned to one environmental condition, resulting in the distribution of the data points over the 'crisp' environmental conditions described by the rules. The distribution also gives an indication of the usefulness of the obtained rule base over a range of environmental conditions that can be found in the study area. The similarity between the rule base of

Table 5.1 The confusion matrix as a basis for evaluation of observed data. The table cross-tabulates observed values against predicted values: true-positives, *a*; false-positives, *b*; false-negatives, *c*; true-negative values, *d*.

		Observed	
		Present	Absent
Predicted	Present	*a*	*b*
	Absent	*c*	*d*

model A and model B was calculated as the % *CCI*, assuming that the rule base of model A equals the observed values and the one of model B the predicted output (Mouton *et al.*, 2008b).

Performance criteria

The key component of the model training and validation procedures is the performance criterion that evaluates the model performance. Performance criteria can deal with either continuous or discrete model outputs, or with both. If a model generates discrete predictions, these outputs can be summarised in a confusion matrix (Fielding and Bell, 1997), which compares the model predictions to the observations (Table 5.1). Several performance criteria have been derived from this confusion matrix, including overall predictive accuracy or the percentage of correctly

classified instances (*CCI*; Fielding and Bell, 1997), sensitivity, specificity, *Kappa* (Fielding and Bell, 1997) and the true skill statistic (*TSS*; Allouche *et al.*, 2006). The latter two criteria range from −1 to 1, whereas all other criteria range from 0, where models are completely inaccurate, to 1, where presence–absence is perfectly predicted.

For fuzzy modelling, the average deviation *AD* (van Broekhoven *et al.*, 2007) is often applied because it incorporates the specific characteristics of fuzzy classifiers with an ordered set of classes and can deal with the fuzzy outputs of these models. Specifically, several performance criteria have been developed to evaluate and train presence–absence models, but most of these criteria are based on the confusion matrix, which requires a threshold to distinguish between present and absent predictions. Since these criteria cannot deal with the fuzzy output of a fuzzy classifier, valuable information may be lost by transferring this fuzzy output to the crisp output which is needed to generate the confusion matrix. Performance measures which are derived from the confusion matrix, for instance, are not sensitive to the position of the classes where the wrong classification occurs (van Broekhoven *et al.*, 2007).

Therefore, van Broekhoven *et al.* (2007) introduced the *AD*, which returns the average deviation between the position of the output class obtained with the model and the position of the output class stored in the training set. Although the *AD* incorporates the fuzzy characteristics of fuzzy habitat models, it depends on the prevalence of the training set and it does not consider the ecological difference between overestimation and underestimation

Table 5.2 Measures of predictive accuracy calculated from a 2×2 error matrix (Table 5.1). The percentage of Correctly Classified Instances (*CCI*) is the rate of correctly classified cells. Sensitivity (*Sn*) is the probability that the model will correctly classify a presence. Specificity (*Sp*) is the probability that the model will correctly classify an absence. The *Kappa* statistic and *TSS* normalise the overall accuracy by the accuracy that might have occurred by chance alone. In all formulae, $n_{tot} = a + b + c + d$.

Measure	Formula	
CCI	$\dfrac{a+d}{n_{tot}}$	(5.1)
Sn	$\dfrac{a}{a+c}$	(5.2)
Sp	$\dfrac{d}{b+d}$	(5.3)
Kappa	$\dfrac{\left(\dfrac{a+d}{n_{tot}}\right) - \dfrac{(a+b)(a+c)+(c+d)(d+b)}{n_{tot}^2}}{1 - \dfrac{(a+b)(a+c)+(c+d)(d+b)}{n_{tot}^2}}$	(5.4)
TSS	$Sn + Sp - 1$	(5.5)

of the observations by the model. Therefore, the aAD was introduced (Mouton *et al.*, 2009a). This performance criterion includes a parameter α which ranges between 0 and 1 and allows stimulation of overprediction or underprediction, depending on the prevalence of the training set. To define the aAD, the cumulative deviation $D_{i,j}$ between the position of the class i obtained with the model and the position of the observed class i is described as:

$$D_{i,j} = \sum_{k=1}^{i} A_k(y_{\text{model},j}) - \sum_{k=1}^{i} A_k(y_{\text{data},j}) \quad (5.1)$$

with $A_k(y_{\text{data},j})$ the membership degree of the jth observed output to the kth output class and $A_k(y_{\text{model},j})$ the membership degree of the jth model output to the kth output class.

The aAD is then defined as:

$$aAD = \frac{1}{N} \cdot \sum_{j=1}^{N} \sum_{i=1}^{n-1} \left(\frac{|D_{i,j}| + D_{i,j}}{2} + \alpha \cdot \frac{|D_{i,j}| - D_{i,j}}{2} \right) \quad (5.2)$$

with N the number of data points and n the number of output classes. Given a set of ordered output classes, the model shows overprediction if $D_{i,j}$ is negative, whereas the model underestimates the observed outputs if $D_{i,j}$ is positive. Following this notation, AD equals $\frac{1}{N} \cdot \sum_{j=1}^{N} \sum_{i=1}^{n-1} |D_{i,j}|$. In contrast to the performance measures that are based on the confusion matrix, AD and aAD are zero if the model output equals the reference output and increase with increasing distance between the reference output and the model output.

The difference between AD, aAD and confusion matrix-based performance measures such as CCI is illustrated in Mouton *et al.* (2009a). Both AD and aAD can be applied to presence–absence model outputs and thus to the confusion matrix. Consequently, it can be shown that, for presence–absence predictions with crisp outputs,

$$AD = \frac{c + b}{N} = 1 - CCI, \quad (5.3)$$

while

$$aAD = \frac{c + \alpha \cdot b}{N}. \quad (5.4)$$

For fuzzy presence–absence models, however, a simplified version of these criteria could be applied because only two output classes, absent and present, are considered. This particular case is described in detail in Mouton *et al.* (2009a).

5.3 Case study

5.3.1 Study area and collected data

The study site is a 1300 m reach of the Aare River in the Bern department, Switzerland, and is situated along the city of Thun (Figure 5.3). Up to this point, the Aare River is draining an area of about 2490 km^2 and is classified as a 7th order stream. The average flow is 111 m^3/s, with

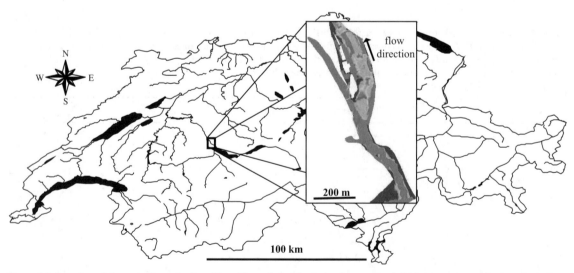

Figure 5.3 Location of the study site at the Aare River, Thun, Switzerland. A colour version of this image appears in the colour plate section of the book.

respective base and peak flows of 23 and 570 m³/s. The Aare River at the studied site was originally a braided river with large gravel banks. However, from the beginning of the 18th century, anthropogenic disturbances were introduced for flood control and hydropower generation (EAWAG, 2002). Hence, the flow regime is altered and controlled by flood control weirs. Nevertheless, the studied site contains some of the major spawning habitats for European grayling in Switzerland.

To develop the species distribution model, an intensive monitoring campaign was set up. In the studied reach, 50 cross-sections were defined and water depth was measured along each cross-section at equal distances of about 1 m using a Raytheon 760 depth measuring device (Raytheon, MA, USA). Flow velocity was measured with a Flo-Mate 2000 flow meter (Marsh–McBirney Inc., MD, USA) at 40% of the water column height in 14 of these cross-sections at equal distances of about 25 m, resulting in 63 measurements. The substrate composition of the river bed surface was assessed by underwater photography with DIN A4 frames. Hence, the dominating substrate of the different sampling points in the studied reach could be assessed visually. If a sampling point was covered by macrophytes, both substrate percentages were set to 99%. This substrate combination cannot be observed in the river reach and hence the definition of this specific situation will not affect optimisation results. All data were collected at a flow of ca. 100 m³/s and no significant flow changes were observed during the measurements.

A finite element grid of the studied reach with 5625 elements and 22 500 nodes was generated using SMS (surface water modelling system, Brigham Young University) software, while the size of the grid cells was adjusted depending on river geometry. Flow velocity and depth values were calculated at each node by a two-dimensional hydraulic model which was generated using FESWMS (Finite Element Surface Water Modelling System, U.S. Geological Survey). Additional measurements of depth and flow velocity were performed in the whole reach, more specifically in the spawning areas, to validate the hydraulic model. The hydraulic modelling was conducted by Schneider & Jorde Ecological Engineering in cooperation with the Swiss Federal Institute of Aquatic Science and Technology (EAWAG) and is described in detail in EAWAG (2002).

European grayling spawns in faster flowing patches (0.1–0.4 m/s) with fine to medium-sized gravel substrate. During egg deposition, the trembling female grayling is pushing its abdomen into the gravel substrate, thereby

creating small grooves (Fabricus and Gustafson, 1955). These light-coloured grooves can easily be distinguished from the substrate which is mostly covered with dark brown algae. Hence, the spawning grounds of grayling were visually identified and localised using GPS (Garmin 12X). Each grid cell in the studied reach was defined as suitable for spawning or not by combining the results of the hydraulic simulations with the observations of the spawning grounds. The resulting dataset contained 22 510 grid cells, one output variable indicating whether the habitat was suitable for spawning or not and four input variables characterising physical habitat in each grid cell (Table 5.3).

5.3.2 Fuzzy rule-based modelling and rule base training

Ten-fold cross-validation was applied to indicate the robustness of the optimisation results. The folds were constructed by randomising the original dataset and assigning each data point to one fold without replacement. The species prevalence (i.e. the frequency of occurrence) was constant for all ten folds and equal to the prevalence of the original dataset (0.203 = 4579/22 510).

The parameters of the membership functions corresponding to the fuzzy sets of the input variables were derived from an ecological study of spawning grayling in the Aare River (EAWAG, 2002). The parameters of these membership functions are shown in Table 5.3. A uniform distribution of the input variables over the fuzzy sets was needed to generate reliable rule bases. The Shannon–Weaver entropy (Shannon and Weaver, 1963) quantified the uniformity and was calculated for the different input variables to assess the quality of the fixed fuzzy sets used in this chapter.

Different models were trained based on five performance measures: the percentage of Correctly Classified Instances, *CCI*, and *Kappa*, the *TSS*, the *AD* and the *aAD*.

In this chapter, the fuzzy rule base contained 108 (= 3 × 4 × 3 × 3) fuzzy rules representing each possible combination of input variable sets. However, not every combination of input variable sets was present in the studied reach (Figure 5.4). To obtain an indication of which rules were relevant, the fuzzy sets were turned into crisp ones by assuming that an input value does not belong to a set if its membership degree to this set is < 0.5. Each data point could be assigned to one environmental condition, resulting in the distribution of the data points over the 108 'crisp' environmental conditions described by the rules (Figure 5.4). The distribution also gives an

Table 5.3 Input variables recorded and the corresponding fuzzy sets of the species distribution models. The entropy, indicating the uniformity of the distribution of the values of a variable over its fuzzy sets, was calculated for all input variables. The parameters represent a_m, b_m, c_m and d_m, as shown in Figure 5.2, respectively.

Input variable	Fuzzy set	Parameters	Entropy
Depth (m)	Shallow	(0.0,0.0,0.0, 0.5)	0.808
	Moderate	(0.0,0.5,1.0,2.0)	
	Deep	(1.0,2.0,2.0,3.0)	
	Very deep	(2.0,3.0,6.6,6.6)	
Flow velocity (m/s)	Low	(0.00,0.00,0.05,0.25)	0.818
	Moderate	(0.05,0.25,0.25,0.50)	
	High	(0.25,0.50,0.68,0.68)	
Percentage of fine gravel	Low	(0,0,10,50)	0.822
(2 mm–2 cm) (%)	Moderate	(10,50,50,90)	
	High	(50,90,100,100)	
Percentage of medium-sized gravel	Low	(0,0,10,50)	0.940
(2 cm–5 cm) (%)	Moderate	(10,50,50,90)	
	High	(50,90,100,100)	

indication of the usefulness of the obtained rule base over a range of environmental conditions that can be found in the Aare River.

5.3.3 Model application

All the aforementioned algorithms were included in a C# toolbox called FISH (freely available upon request Ans.Mouton@INBO.be). This toolbox allows selection of the performance criterion applied for model training, application of *s*-fold cross-validation and calculation of rule coverage of the optimised rule base (Mouton *et al.*, 2011). The presented fuzzy modelling approach

also enables variable selection, as described in Mouton *et al.* (2009c), although for a large number of variables (> 10), other data mining methods may be more appropriate (Mouton *et al.*, 2010b; Thu Huong *et al.*, 2010). Specifically, variables could first be selected based on these other methods, and then the fuzzy model could be created based on the remaining variables (Mouton *et al.*, 2009c).

5.3.4 Results

Comparison of the training based on *aAD* with training based on four other performance criteria revealed that the latter four criteria produced rule bases which

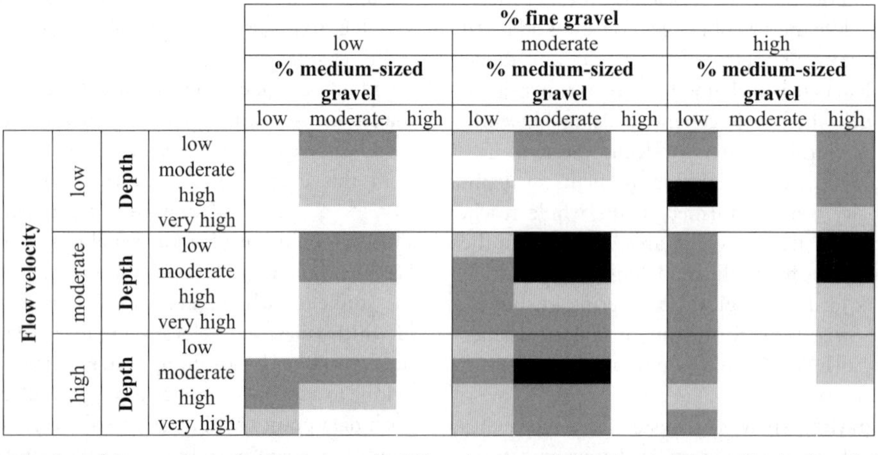

Figure 5.4 Distribution of the samples in the dataset over the 108 environmental situations considered in the fuzzy habitat suitability models. All environmental situations were represented by less than 22.9% of all data points (0% of instances (▭), 0–0.5% of instances (▭), 0.5–5% of instances (▬), > 5% of instances (▬)).

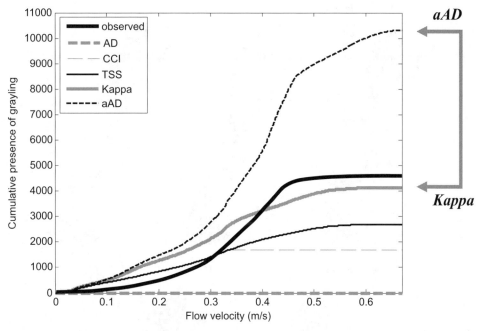

Figure 5.5 The cumulative predictions of grayling for the variable flow velocity obtained after training based on the percentage of correctly classified instances (*CCI*), *Kappa*, the average deviation (*AD*), the true skill statistic (*TSS*) and the adjusted average deviation (*aAD*) with $\alpha = 0.32$. Training was performed on the original dataset.

underestimated the observations (Figure 5.5). However, the shape of the cumulative predictions obtained by training based on *Kappa* and on the *TSS* was more similar to the shape of the cumulative observations than it was for the underestimating rule bases derived based on *CCI* and *AD*. This may indicate that the rule bases obtained from *Kappa* and *TSS* training may be ecologically more relevant than those obtained after training based on *aAD* with a value of α that leads to underestimating rule bases (Figure 5.5). Nevertheless, the shape of the cumulative predictions obtained from training based on *aAD* with a value of α of 0.32 (Figure 5.5) still corresponds the most to the observations. Moreover, other performance criteria and visualisation of the model predictions indicate that model performance is acceptable for this training scenario (Figure 5.6).

5.3.5 Discussion

In the presented study, the habitat preferences derived from field observations or expert knowledge (see also Chapter 4, Section 4.3.2) were very similar to those predicted by the fuzzy rule-based models. The predicted flow velocity preferences for spawning grayling were slightly higher than the 0.1–0.4 m/s range found by Fabricus

and Gustafson (1955). This was confirmed by Nykänen and Huusko (2002), who found flow velocity preferences between 0.5 and 0.6 m/s and suggested that these preferences might be transferable between rivers. The model results were also supported by the wider flow velocity preferences (0.2–0.9 m/s) observed in other studies (Gönczi, 1989; Sempeski and Gaudin, 1995).

Several authors describe depth preferences of spawning grayling which are significantly lower (0.1–0.5 m) than both the observed and modelled preferences in this study (Gönczi, 1989; Sempeski and Gaudin, 1995; Nykänen and Huusko, 2002). However, these deeper spawning habitats in the Aare River have been observed for more than 20 years (EAWAG, 2002). Possible explanations for the contrasting depth preferences may be differences in topology of the sampled rivers or ecological factors such as competition or predation. Specifically, the shallower spawning areas in the Aare River are regularly disturbed by swans, which might cause a shift in depth preferences (EAWAG, 2002).

Unlike the depth preferences, the modelled and observed substrate preferences strongly correspond to previous research, since several authors agree that spawning grayling prefers medium-sized gravel (Gönczi, 1989;

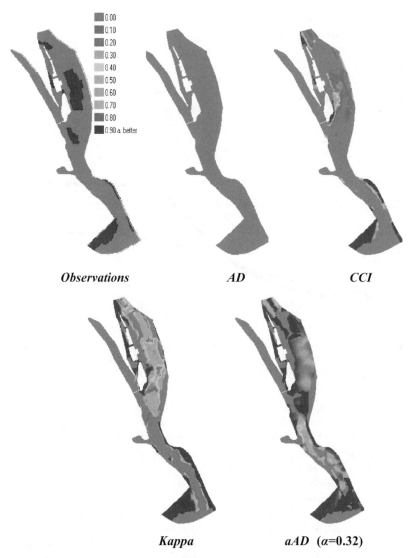

0.00	
0.10	
0.20	
0.30	
0.40	
0.50	
0.60	
0.70	
0.80	
0.90 a. better	

Observations   *AD*   *CCI*

Kappa   *aAD* (α=0.32)

Figure 5.6 Visualisation of the model predictions in the study area, showing the habitat suitability for spawning grayling calculated by the fuzzy rule base optimised based on the correctly classified instances (*CCI*), the average deviation (*AD*), *Kappa* and the adjusted average deviation (*aAD*) with α = 0.32. A colour version of this image appears in the colour plate section of the book.

Sempeski and Gaudin, 1995; Nykänen and Huusko, 2002). Gönczi (1989) suggests that gravel thickness may also affect spawning preferences. Although this variable was not considered in this chapter because field data were lacking, the presented fuzzy rule-based approach would allow fast and easy implementation of such expert knowledge in the fuzzy knowledge base.

Comparison of the results from different optimisation scenarios and folds indicated the hill-climbing method generated consistent results. This consistency is enhanced by the size of the dataset used with respect to the relatively small sampling area and by the uniform distribution of the training instances over the different rules. However, since only a few organisms were observed at a depth higher than 2.5 m and a flow velocity higher than 0.45 m/s, model predictions above these thresholds will be less reliable, as was discussed earlier (Figure 5.4). Specifically, some rules apply to environmental conditions which are not present in the dataset, which was also reflected in the lower entropy of the rule base compared to the entropy of the individual input variables. Since the consequent of these rules cannot be reliably predicted, the optimised rule bases only apply to environmental conditions which are covered by the data (van Broekhoven et al., 2006).

To evaluate the different models, two aspects were taken into account: the predictive accuracy and the shape of the cumulative predictions curve. This approach is based on the assumption that an ecologically relevant model should show an acceptable predictive accuracy and should minimise underprediction, given the high likelihood that the dataset contains false absences. If the shape of the cumulative predictions curve reflects the shape of the cumulative observations curve, this implies that no net underprediction was observed in the studied range. Consequently, these models may be ecologically more reliable than models that are underpredicting but may show better predictive accuracy (Mouton et al., 2010a). If more detailed information was available, α could, for instance, be further fine-tuned to increase predictive accuracy. Specifically, if data have been collected very accurately in a region of the study area, reducing the likelihood of false absences in the data for this region, α could be chosen to obtain the highest predictive accuracy in this region, whereas all other instances of the study area will be predicted based on the same α.

Although an optimal parameter value could be found by applying sensitivity analysis, a more important problem with these flexible performance criteria could be the difficulty in saying which models are better (Mouton et al., 2010a). The results presented in this chapter showed that comparison of the shapes of the cumulative prediction curves may provide an indication of the ecological relevance of the different optimised models. Specifically, observations may be more appropriate than expert knowledge to evaluate model performance because differences in species' dispersal patterns and associated gene flow may lead to subtle variations in habitat preferences of some species due to local adaptations (Holt, 2003). Even in the absence of genetically driven differences in habitat use, species could express different realised niches (Hutchinson, 1957) as a result of spatial variation in predators, competitors or other biotic factors (Hutchinson, 1957; Holt, 2003; Hernandez et al., 2006). Consequently, conservationists should distinguish between models which reliably or less reliably predict species distribution. Maggini et al. (2006) show that applying a training parameter or threshold that minimises the difference between over- and underprediction may be a suitable rule of thumb. Applied to the results of this chapter, this suggests that the

optimal rule base could be obtained by applying a value of α between 0.35 and 0.30. Further discussion on this issue is provided in Mouton et al. (2009a; 2009b).

This chapter confirms that the performance criteria applied for model training may significantly affect the results of the training process. Consequently, these findings suggest that model developers should use training scenarios based on different performance criteria or on an adjustable criterion such as the aAD. They should thus carefully select an appropriate training performance criterion or parameter which reflects the ecological model purpose (Mouton et al., 2010a). Although this approach is illustrated in this chapter by applying the adjusted average deviation to train a fuzzy habitat suitability for spawning grayling, the presented methods could be applied to any ecosystem or species. Therefore, this chapter may contribute to improving the reliability of ecological models in general and thus provide valuable insights for ecosystem management.

References

Acreman, M. and Dunbar, M.J. (2004) Defining environmental river flow requirements – a review. *Hydrology and Earth System Sciences*, **8**: 861–876.

Adriaenssens, V., De Baets, B., Goethals, P.L.M. and De Pauw, N. (2004) Fuzzy rule-based models for decision support in ecosystem management. *Science of the Total Environment*, **319**: 1–12.

Allouche, O., Tsoar, A. and Kadmon, R. (2006) Assessing the accuracy of species distribution models: prevalence, kappa and the true skill statistic (TSS). *Journal of Applied Ecology*, **43**: 1223–1232.

Assilian, S. (1974) *Artificial Intelligence in the Control of Real Dynamical Systems*. PhD thesis, London University, London, UK.

Bovee, K.D. (1986) *Development and evaluation of habitat suitability criteria for use in the instream flow incremental methodology*. Instream flow information paper 21, biological report 86, U.S. Fish and Wildlife Service, Washington, DC.

Bovee, K.D., Lamb, B.L., Bartholow, J.M., Stalnaker, C.B., Taylor, J. and Henriksen, J. (1998) *Stream Habitat Analysis Using the Instream Flow Incremental Methodology*. Information and Technology Report USGS/BRD/ITR-1998-0004, Fort Collins, Colorado.

Boyce, M.S., Vernier, P.R., Nielsen, S.E. and Schmiegelow, F.K.A. (2002) Evaluating resource selection functions. *Ecological Modelling*, **157**: 281–300.

Casillas, J., Cordón, O., Herrera, F. and Magdalena, L. (2003) *Interpretability Issues in Fuzzy Modelling*. Volume 128 of

Studies in Fuzziness and Soft Computing, Springer-Verlag, Heidelberg.

EAWAG (2002) *Fishereiliches Gutachten über die Aarebaggerung in Thun. Forschungsanstalt des ETH-Bereichs*, EAWAG (Fish Federal Institute for Environmental Science and Technology), Kastanienbaum, Switzerland (in German).

Fabricus, E. and Gustafson, K.J. (1955) Observation of the spawning behaviour of the grayling, *Thymallus thymallus. Report of the Institute of Freshwater Research, Drottningholm*, **36**: 75–103.

Fielding, A.H. and Bell, J.F. (1997) A review of methods for the assessment of prediction errors in conservation presence/absence models. *Environmental Conservation*, **24**: 38–49.

Gönczi, A.P. (1989) A study of physical parameters at the spawning sites of the European grayling (*Thymallus thymallus* L.). *Regulated Rivers: Research and Management*, **3**: 221–224.

Guisan, A. and Zimmermann, N.E. (2000) Predictive habitat distribution models in ecology. *Ecological Modelling*, **135**: 147–186.

Heggenes, J. (1996) Habitat selection by brown trout (*Salmo trutta*) and young Atlantic salmon (*S. salar*) in streams: Static and dynamic hydraulic modelling. *Regulated Rivers-Research and Management*, **12**: 155–169.

Hernandez, P.A., Graham, C.H., Master, L.L. and Albert, D.L. (2006) The effect of sample size and species characteristics on performance of different species distribution modeling methods. *Ecography*, **29**: 773–785.

Holt, R.D. (2003) On the evolutionary ecology of species' ranges. *Evolutionary Ecology Research*, **5**: 159–178.

Hudson, H.R., Byrom, A.E. and Chadderton, W.L. (2003) *A Critique of IFIM – Instream Habitat Simulation in the New Zealand Context*. Science for Conservation 231, Department of Conservation, Wellington, New Zealand.

Hüllermeier, E. (2005) Fuzzy methods in machine learning and data mining: Status and prospects. *Fuzzy Sets and Systems*, **156**: 387–406.

Hutchinson, G.E. (1957) Population Studies – Animal Ecology and Demography – Concluding Remarks. *Cold Spring Harbor Symposia on Quantitative Biology*, **22**: 415–427.

Jorde, K., Schneider, M. and Zöllner, F. (2000) Analysis of instream habitat quality – preference functions and fuzzy models. In Wang, Z.Y. and Hu, S.-X. (eds) *Stochastic Hydraulics 2000*, Balkema, Rotterdam, pp. 671–680.

Karr, J.R. (1991) Biological integrity – a long-neglected aspect of water-resource management. *Ecological Applications*, **1**: 66–84.

Lamouroux, N., Capra, H. and Pouilly, M. (1998) Predicting habitat suitability for lotic fish: Linking statistical hydraulic models with multivariate habitat use models. *Regulated Rivers-Research and Management*, **14**: 1–11.

Maggini, R., Lehmann, A., Zimmermann, N.E. and Guisan, A. (2006) Improving generalized regression analysis for the spatial prediction of forest communities. *Journal of Biogeography*, **33**: 1729–1749.

Mamdani, E. (1974) Application of fuzzy algorithms for control of a simple dynamic plant. *Proceedings IEE*, **121**: 1585–1588.

Michalewicz, Z. and Fogel, D.B. (2000) *How to Solve It: Modern Heuristics*, Springer-Verlag, Berlin, Heidelberg.

Mouton, A.M., De Baets, B. and Goethals, P.L.M. (2008a) Entropy-based fuzzy set optimisation for reducing ecological modelling complexity. In Sànchez-Marrè, M., Béjar, J., Comas, J., Rizzoli, A.E. and Guariso, G. (eds) *Proceedings of the iEMSs Fourth Biennial Meeting: 'International Congress on Environmental Modelling and Software'*. International Environmental Modelling and Software Society, July 2008, Barcelona, Catalonia.

Mouton, A.M., De Baets, B. and Goethals, P.L.M. (2010a) A theoretical analysis of performance criteria for presence/absence species distribution models. *Ecological Modelling*, **221**: 1995–2002.

Mouton, A.M., Dedecker, A.P. and Goethals, P.L.M. (2010b) Selecting variables for habitat suitability of Asellus (Crustacea, Isopoda) by applying input variable contribution methods to Artificial Neural Network models. *Environmental Modelling and Assessment*, **15**: 65–79. doi: 10.1007/s10666-009-9192-8.

Mouton, A.M., De Baets, B., van Broekhoven, E. and Goethals, P.L.M. (2009a) Prevalence-adjusted optimisation of fuzzy models for species distribution. *Ecological Modelling*, **220**: 1776–1786.

Mouton, A.M., Jowett, I.G., Goethals, P.L.M. and De Baets, B. (2009b) Comparison of generalised additive models and a data-driven fuzzy approach for the modelling of fish species in New Zealand. *Ecological Informatics*, **4**: 215–225. doi: 10.1016/j.ecoinf.2009.07.006

Mouton, A.M., van Broekhoven, E., Goethals, P.L.M. and De Baets, B. (2009c) Knowledge-based versus data-driven fuzzy habitat suitability models for river management. *Environmental Modelling and Software*, **24**: 982–993.

Mouton, A.M., Alcaraz-Hernández, J.D., De Baets, B., Goethals, P.L.M. and Martínez-Capel, F. (2011) Data-driven habitat suitability models for brown trout in Spanish Mediterranean rivers. *Environmental Modeling and Software*, **26**: 615–622.

Mouton, A.M., Schneider, M., Depestele, J., Goethals, P.L.M. and De Pauw, N. (2007) Fish habitat modelling as a tool for river management. *Ecological Engineering*, **29**: 305–315.

Mouton, A.M., Schneider, M., Armin, P., Holzer, G., Müller, R., Goethals, P.L.M. and De Pauw, N. (2008b) Optimisation of a fuzzy physical habitat model for spawning European grayling (*Thymallus thymallus* L.) in the Aare river (Thun, Switzerland). *Ecological Modelling*, **215**(1–3): 122–132.

Nykänen, M. and Huusko, A. (2002) Suitability criteria for spawning habitat of riverine European grayling. *Journal of Fish Biology*, **60**: 1351–1354.

Sekine, M., Imai, T. and Ukita, M. (1997) A model of fish distribution in rivers according to their preference for environmental factors. *Ecological Modelling*, **104**: 215–230.

Sempeski, P. and Gaudin, P. (1995) Habitat selection by grayling .1. Spawning habitats. *Journal of Fish Biology*, **47**: 256–265.

Shannon, C.E. and Weaver, W. (1963) *Mathematical Theory of Communication*, University of Illinois Press, Urbana, Illinois.

Strayer, D.L., Power, M.E., Fagan, W.F., Pickett, S.T.A. and Belnap, J. (2003) A classification of ecological boundaries. *Bioscience*, **53**: 723–729.

Thu Huong, H., Lock, K., Mouton, A.M. and Goethals, P.L.M. (2010) Application of classification trees and support vector machines to model the presence of macroinvertebrates in rivers in Vietnam. *Ecological Informatics*, **5**: 140–146. doi: 10.1016/j.ecoinf.2009.12.001.

van Broekhoven, E., Adriaenssens, V. and De Baets, B. (2007) Interpretability-preserving genetic optimization of linguistic terms in fuzzy models for fuzzy ordered classification: An ecological case study. *International Journal of Approximate Reasoning*, **44**: 65–90.

van Broekhoven, E., Adriaenssens, V., De Baets, B. and Verdonschot, P.F.M. (2006) Fuzzy rule-based macroinvertebrate habitat suitability models for running waters. *Ecological Modelling*, **198**: 71–84.

Vanreusel, W., Maes, D. and Van Dyck, H. (2007) Transferability of species distribution models: a functional habitat approach for two regionally threatened butterflies. *Conservation Biology*, **21**: 201–212.

Zadeh, L.A. (1965) Fuzzy sets. *Information and Control*, **8**: 338–353.

Žnidaršic, M., Jakulin, A., Džeroski, S. and Kampichler, C. (2006) Automatic construction of concept hierarchies: The case of foliage-dwelling spiders. *Ecological Modelling*, **191**: 144–158.

6

Applications of the MesoHABSIM Simulation Model

Piotr Parasiewicz[1,5], Joseph N. Rogers[1], Paolo Vezza[2], Javier Gortázar[3], Thomas Seager[1], Mark Pegg[4], Wiesław Wiśniewolski[5] and Claudio Comoglio[2]

[1]Rushing Rivers Institute, 592 Main Street, Amherst, MA 01002, USA
[2]Turin Polytechnic, Corso Duca degli Abruzzi 24, c/o DITAG, 10129 Torino, Italy
[3]Ecohidráulica S.L., Calle Rodríguez San Pedro 13, 4°7, 28015 Madrid, Spain
[4]University of Nebraska, 402 Hardin Hall, Lincoln, NE 68583-0974, USA
[5]The Stanisław Sakowicz Inland Fisheries Institute, ul. Oczapowskiego 10, 10-719 Olsztyn 4, Poland

6.1 Introduction

The MesoHABSIM approach is a physical habitat modelling system created for the purpose of instream habitat management in applications such as hydro-power and water withdrawals mitigation, as well as river channel restoration planning. It has been developed and tested between 2000 and 2010 at Cornell University, the University of Massachusetts and the Rushing Rivers Institute. The first concept of the model was published in Parasiewicz (2001). The latest description of the method was presented through a series of papers in 2007 and 2008, which established a procedural benchmark of the model (Parasiewicz, 2007a; 2007b; 2008a; 2008b). The MesoHABSIM approach has been applied in over 30 rivers and the methodology has been refined and adapted to the particular circumstances of each project. The current software implementation of MesoHABSIM (Sim-Stream) includes a number of tools facilitating the interpretation and presentation of the results for use in regulatory environments. The purpose of this chapter is to present the current state of the methodology as well as to demonstrate the utility of the model in different environments and for

varied applications. We provide a short description of key methodological steps and discuss variations that can be supported with examples of their application. For details of each methodological step, the reader should refer to Parasiewicz (2001; 2007a; 2007b; 2008a; 2008b).

6.2 Model summary

The process of model development consists of the following procedures:

1 Identifying biological targets and indicators.
2 Establishing habitat suitability criteria.
3 Mapping and developing an evaluation of instream habitats.
4 Adjusting biophysical templates to reflect reference habitat.
5 Time series analysis.
6 Interpretation and application.

Mesohabitat types are defined by hydromorphological units (HMUs), such as pools and rapids. Mesohabitats are mapped under multiple flow conditions at chosen representative sites along the river. The sites and their

---

Ecohydraulics: An Integrated Approach, First Edition. Edited by Ian Maddock, Atle Harby, Paul Kemp and Paul Wood.
© 2013 John Wiley & Sons, Ltd. Published 2013 by John Wiley & Sons, Ltd.

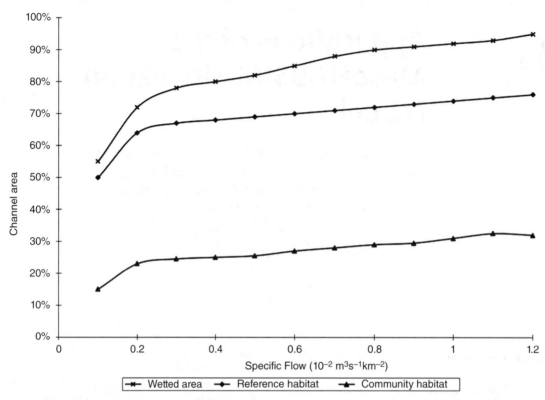

Figure 6.1 An example of habitat rating curves for generic fish representing the entire amount of suitable habitat and for community habitat, which is a sum of habitat for individual species weighted by their expected proportions in the community. The flows (*x*-axis) are standardized to the watershed area. The suitable habitat at reference conditions, expressed as a proportion of river channel area (*y*-axis), is available across most of the wetted area, however the community habitat is much lower, indicating that more habitat is available to less common species.

quantitative representativeness are defined during an extensive reconnaissance phase. Fish and/or invertebrate data are collected in randomly distributed mesohabitats where habitat surveys are also conducted. These data are used for developing mathematical models that describe which mesohabitats are used by animals more frequently and hence are assumed to be more or less suitable. This allows the evaluation of habitat availability at a range of flows using suitable area as a metric.

Habitat rating curves represent changes in the area of suitable habitat for species and communities in response to flow and allow for the determination of habitat quantity at any given flow within the range of surveyed discharges (Figure 6.1). These rating curves can be developed for river units of any size, making them useful for drawing conclusions about the suitability of channel patterns or habitat structures for specific river sections as well as for the entire river.

In combination with hydrologic time series data, habitat rating curves are used to create Uniform Continuous-Under-Threshold (UCUT) curves for the analysis of frequency, magnitude and duration of significant habitat events (Figure 6.2). UCUT curves evaluate continuous durations of events when available habitat is less than a specified quantity and help to select probabilistic thresholds from the frequency of these events. UCUT curves serve as a basis for the development of ACTograms, which managers can use to determine habitat bottlenecks (Bovee *et al.*, 1998) (Figure 6.3). These procedures are described in more detail below.

6.2.1 Identifying biological targets and indicators

In this step, we define the aquatic resource elements for which the model will be developed. We select seasonal assemblages of these resources as indicators of habitat

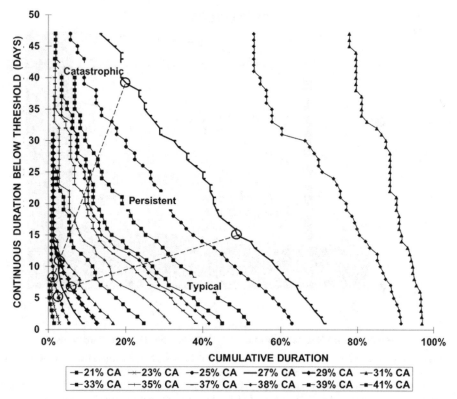

Figure 6.2 An example of Uniform Continuous Under-Threshold curves for determination of HSTs. Each curve on the diagram represents the cumulative duration of events when habitat is lower than a threshold (x-axis) for a continuous duration of days depicted on the y-axis. The reduction in slope as well as the increase of spacing between two curves indicates an increase in the frequency of 'under-threshold' events. We select the most outstanding curves to identify the rare, critical and common (lines with circles) thresholds and their inflection points (circles) to demarcate associated persistent and catastrophic durations of events with less habitat than indicated by the threshold (see Parasiewicz, 2007b).

use that will help guide the assessment of altered flow regimes or potential restoration actions. Seasonal aquatic resource elements selected can be fish, invertebrates, species-specific life stages, species groups, species guilds or entire aquatic resource communities. The most comprehensive approach is to establish a model of an expected or desired community consisting of a species list that includes the proportions of each species in the community. As described in Parasiewicz (2007b), we most commonly use the Target Fish Community (TFC) approach described by Bain and Meixler (2008) for this purpose. By comparing the proportional structure of the observed fish community with the expected structure and with available habitat, we can determine if the habitat is a limiting factor for some species or for a particular life stage and a potential reason for their low numbers (see Figure 6 in Parasiewicz, 2008b). This may serve as a basis for adjustment to the

habitat template (i.e. modification of channel morphology) by increasing the habitat proportions for specific species. It may also be used as an end-point restoration model by which restoration success can be measured.

We sometimes develop a Reference Fish Community (RFC), which, in contrast to the TFC, represents the seasonal estimate of natural fish fauna composition. This includes species that are currently underrepresented in the recent stream surveys because they were extirpated or impacted by anthropogenic factors. The RFC gives better insight into the expected natural community structure, which allows modelling habitat structure that would support such a community in the target river. The estimates of expected proportions of these species have been established in two possible ways: either by approximating the species' expected percentage within the community with the assistance of expert opinion or by calculating the

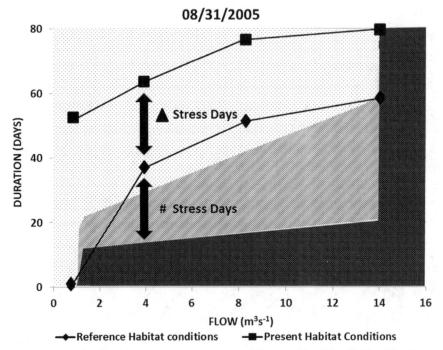

Figure 6.3 Example of an ACTogram for the Summer Rearing and Growth bioperiod for Eightmile River. The durations on the y-axis represent time in days for which flows have been below the level indicated on the x-axis. The coloured areas indicate if the event duration should be considered typical (black), persistent (grey) or catastrophic (spotted). The squares and diamonds indicate the period of flows under a specific value (e.g. $4\,\mathrm{m^3s^{-1}}$) on August 31, 2005 for two different scenarios (reference and present conditions). The increase in the number of stress days represents the impact on habitat at any given flow level.

average density of these species with data from remediation projects or historical monitoring (e.g. Parasiewicz *et al.*, 2007a; 2007b).

The community composition will vary between seasons (or bioperiods *sensu* Parasiewicz, 2007b), especially in rivers with high levels of seasonal migration by specific species. To develop a model of habitat required to sustain the community structure in a bioperiod, we usually select representative species consisting of the most common species or a species of particular interest to use as indicators. This is a pragmatic approach because for rare species, establishing habitat suitability criteria offers a particular challenge due to the lack of observational data. Frequently, for the purpose of the determination of instream flows, the selected species belong to the macro-habitat guild of fluvial specialists and fluvial dependents, as defined by Kinsolving and Bain (1993).

Another option for indicator species selection is to divide the community into habitat-use guilds and select one or more species as guild representatives (Leonard and Orth, 1988; Vadas and Orth, 2001). One benefit of this

approach is that the guilds are considered to be more universal in their application at regional scales. Welcomme *et al.* (2006) developed a set of 'environmental guilds' that group riverine fish species based on their response to hydrologic and geomorphologic changes in the ecosystem. This approach is particularly useful in rivers with a distinct hydrologic and geomorphic separation of habitat or where a large number of species or species groups with common habitat needs are present. Additionally, the guild approach provides the ability to use information from more abundant, representative species within a guild to help characterize habitat suitability information for a rare species that, on its own, would be too rare to gather adequate data for. The approach was applied on the Niobrara River, Nebraska, USA, where the species compositions vary longitudinally within the Niobrara River (Wanner *et al.*, 2009), but common habitat requirements exist that allow creation of species subsets (Table 6.1). These commonalities then allow for an assessment of habitat availability using information from a collective suite of species rather than individual species where data

Table 6.1 Habitat guild definitions proposed for the Niobrara River. Guilds are defined using Welcomme *et al.* (2006) classifications.

Guild name	Definition
Eupotamonic benthic	Inhabit benthic habitats and typically found in the main channel. Generally intolerant of low dissolved oxygen.
Eupotamonic phytophilic	Longitudinal migrants that also use the floodplain (lateral movements). Juveniles found in or near floodplain.
Eupotamonic pelagophilic	Main channel residents that migrate long distances.
Parapotamonic	Generally species that prefer semi-lotic habitat and are intermediate between migrants and sedentary species.
Plesiopotamonic	Typically found in open water or along stream edges or in flooded floodplain. Tolerant to lower, but not anoxic, dissolved oxygen concentrations

may be limited. For example, pallid sturgeon (*Scaphirhyncus albus*) is a federally endangered species in the United States, which likely means this species is not abundant and information on its habitat use in the Niobrara River would be sparse. We do know that pallid sturgeons are in the eupotamonic pelagophil guild and that several other species have common mesohabitat requirements. Collectively, the information used to model and assess habitat information for the entire guild in this example could also be used to infer similar habitat availability for pallid sturgeon.

Therefore it is possible to identify the assemblages of mesohabitat types utilized by the guilds and potentially to define habitat-based groupings. Such an approach was utilized in the Powder River, Wyoming, USA, in which cluster analysis of mapped mesohabitats was applied to define habitat use guilds (Senecal, 2009).

At this point, individual target species can be modelled (e.g. Ballesterro *et al.*, 2006) either by selecting the species with the most flow-dependent habitat rating curve or by the development of a community habitat rating curve derived from proportional weighting of individual curves where the weights are derived from the species community level proportions.

The least comprehensive option, but also commonly applied in fish habitat studies, is to determine the indicators using a list of expected species without defining their expected proportions. This makes the model more coarse and insensitive to detecting changes in community structure, which is frequently the consequence of anthropogenic impacts. However, determining indicators is simplified in its approach, as qualitative data are much more frequently available. Vezza (2010) utilized this approach, where the target fish species expected to be found in small

streams were identified using regionalized ichthyic zonation (Carta Ittica Regionale, 1992–2004).

Finally, the investigator or resource agency may decide *a priori* which species are of the greatest interest and develop the model for only these species. This is also a common habitat modelling approach as it directly addresses the management needs of resource agencies or public interest. However, this approach must be exercised with caution as recommendations do not explicitly consider potential benefits or impacts to other components of the aquatic community.

These ideas are applicable not only for fish, but also for macroinvertebrates or other aquatic resources. Since the identification of invertebrates at the species level may be prohibitively expensive, frequently family-level models are developed instead and have been used in other habitat assessment approaches for decades. In the Lamprey River study, we created a collective model for Odonata, Ephemeroptera, Plecoptera, Trichoptera and Generic EPT taxa, which shows the available habitat as a function of channel area for these four families of macroinvertebrates (Parasiewicz *et al.*, 2008) (Figure 6.4).

Models developed for freshwater mussels have also been successful in defining suitable habitat and recommended flow regimes. In one application, a model was developed for Unioids as a group on the Souhegan River in New Hampshire, USA. The developed habitat rating curves did not indicate a change in available habitat as a function of flow. This lack of sensitivity in available habitat versus changes in discharge is attributed to the community-level model that was based on a wide range of species-specific relationships. A second model created for the freshwater pearl mussel (*Margaritifera margaritifera*) in Wekepeke Brook indicated, for similar flow range, increased habitat

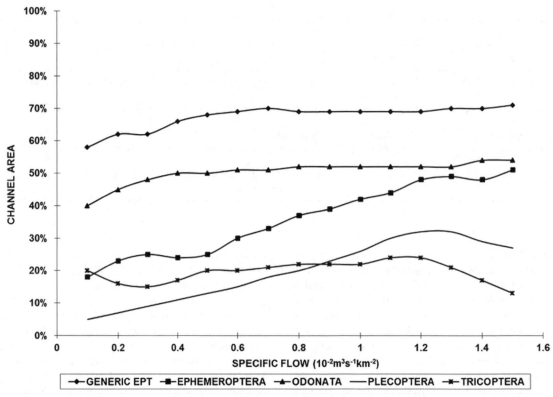

Figure 6.4 Habitat rating curves for selected invertebrate families developed for the Lamprey River, NH. Generic EPT indicates collective habitat for ephemeropterans, plecopterans and trichopterans (see Parasiewicz *et al.*, 2008).

area as a function of increasing flow rates. This confirmed what is known from literature reviews regarding the life history and habitat use of this species (Parasiewicz and Rogers, 2010).

6.2.2 Establishing habitat suitability criteria

The next analysis step is to establish habitat suitability criteria for the selected indicator species. The criteria describe the combination of physical habitat attributes that correlate with the species' presence, indicating suitable habitats, or high abundance, indicating optimal habitats. As described by Parasiewicz (2001; 2007a), every HMU is associated with categorical variables describing presence, absence or abundance of cover types such as woody debris or boulders, as well as the relative distribution of depth, velocities and substrate classes occurring within the units. MesoHABSIM allows for the use of different approaches for determining suitable combinations of these attributes as long as they are compatible with the above data structure.

The simplest approach is to use the information obtained from literature studies to specify the range of velocities, depths, substrate conditions, types of HMU and cover attributes that have been determined as adequate for the species presence. Each of these five HMU descriptive categories can then be defined as preferable or critical to a species presence. When HMU attributes from the field surveys fall within the specified ranges of the developed suitability values, then the HMU is determined to be suitable for a particular species. The number of fulfilled HMU descriptive categories is used as a factor separating suitable from optimal habitats (i.e. 3 is suitable, ≥4 is optimal). We typically use this method for bioperiods where more detailed empirical data are not easily obtainable and the model can be less precise.

One option is to calibrate literature-based habitat suitability criteria with fish observations at the reach level. For example, in the Tajuña River study in Spain, habitat suitability criteria were adjusted based on linear regression analysis between observed brown trout density in the electrofished sites (ca. 100m-long reaches) and the amount of suitable habitat within the reach. Development of habitat suitability criteria followed a stepwise iterative process: at each step, one class of one variable (a single depth, velocity or substrate class, or a single HMU or cover type) was included in the model or excluded from it and the linear regression analysis between trout density and suitable habitat was calculated. If the regression analysis reflected a better fit, the change was included in the next model; if not, it was rejected (Gortazar et al., 2011).

The most precise criteria for a target species or guild can be developed using empirical data collected from one or more rivers. Data collected from multiple rivers provide a wider range of habitat availability and utilization than those occurring in one river. Therefore, the model better captures the species-specific response to environmental variability. In such a case, numerous HMUs are sampled for target species or guilds, using the methodologies described by Parasiewicz (2007a). In recent years, the Rushing Rivers Institute has established a large database, with well over 1000 samples of HMUs obtained from over 15 rivers across the Northeastern USA. So far, over 36 fish species, 30 species of odonates and 3 invertebrate families are included in the database, which is growing continuously. This allows for the establishment of a more robust criterion, which can be transferred between rivers in the region and helps to limit the fish sampling effort on each project to those required for the model's validation. These data serve as a basis for the calculation of multivariate probabilistic models such as logistic regression. One important element in this process is to isolate the habitat attributes that have a significant influence on fish presence, such as the use of the Akaike information criterion (Sakamoto, 1991) instead of stepwise regression. Furthermore, in the cross-validation procedure, we apply the computed formula to the validation data (e.g. 20% of available data) and compare the number of fish observations with the predictions of suitable habitat. This procedure is repeated 20 times and each time a new randomly selected dataset is retained for validation purposes. After 20 runs, the model generates a list of parameters that were selected in at least two of the runs and computes another model using only these parameters as input attributes. To further improve model quality, we investigate the standard errors of each final model and remove the attributes with high standard errors. The remaining attributes are then used in the calculations of the probability of presence or high abundance. Receiver Operational Characteristics (ROC) curves serve as the basis for the identification of probability cutoff values that distinguish between not suitable, suitable and optimal habitats (Metz,1986; Pearce

and Ferrier, 2000). HMUs with probability of presence higher than the selected cutoff are considered to be suitable habitats. Out of those, the HMUs with probability of abundance higher than the cutoff are considered optimal. This analytical framework is currently supported within the Sim-Stream 8 modelling system (Rushing Rivers Inc., 2010).

The methods described above for development of habitat suitability criteria are illustrative of common approaches; the use of alternative approaches for criteria development external to the MesoHABSIM system are not precluded, as long as the criteria meet the basic input format for criteria curves.

6.2.3 Mapping and evaluation of instream habitat

The application of MesoHABSIM can be accommodated across a wide array of spatial scales from complete delineation to subsampling of representative river sections within longer homogeneous river reaches. The approach can be based on expert opinion or sophisticated analysis using statistical approaches. Time, cost and logistics constraints play an important role in selection of the appropriate method. The section below demonstrates a few examples of selected sampling strategies.

The Stony Clove Creek, New York, USA study mapped the entire 16 km length of river multiple times (Parasiewicz et al., 2003). This effort was time and cost intensive. Each survey by multiple parallel teams took two weeks to complete, during which time flows frequently changed, thus creating additional data-processing complications. In contrast, the reconnaissance survey of the Little River, CT indicated that it would be most effective to map the entire length of the study area (5 km) because of its short length and streamlined processing.

In the Eightmile River, Connecticut, USA study (28 km), watershed maps and aerial photographs were used by local fish biologists and residents to conduct a preliminary reconnaissance. This allowed the selection of representative sites (smaller portions of a stream segment that are proportionally representative) with a reasonable level of confidence (Parasiewicz et al., 2007b), but without the ability to quantifiably justify our choice. In contrast, in studies on the Quinebaug (34 km, Connecticut, USA), Pomperaug (21 km, Connecticut, USA) and Mill (20 km, Massachusetts, USA) Rivers, the reconnaissance consisted of a detailed mapping of all HMUs in the study area and the representative sites were selected with the help of a sensitivity analysis of HMU distributions. Since this effort was

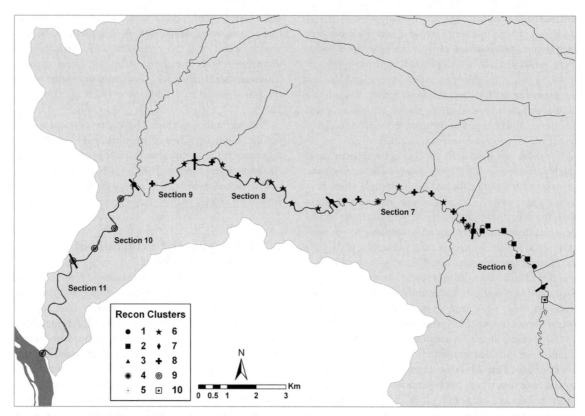

Figure 6.5 A map of the Świder River, Poland with locations of identified clusters and section boundaries. Each symbol is located at the end of a homogenous river reach and indicates the cluster classification. The thick section markers are placed in the locations where cluster patterns are changing.

intensive, we developed a more effective protocol for gathering the necessary information. This protocol involves hiking or boating the entire project area river length while estimating proportions of hydromorphologic units and mesohabitat characteristics for homogenous sections of river instead of mapping each individual HMU. Usually, such an on-the-ground survey overestimates the number of sections, which are later grouped with the help of cluster analysis. For example, the Świder River study in Poland, a 4th order river 75 km long, was delineated into 11 sections. For each section, we recorded the estimated proportions of different HMU types and habitat cover categories while excluding depth and velocity data collection. After the survey, the team identified the major observed breakpoints in river morphology. The cluster analysis supported these delineations and allowed us to identify locations where the habitat conditions changed

by grouping the sections that were similar to each other (see Figure 6.5).

The Niobrara River, Nebraska, USA study required a slightly different approach for identifying homogenous sections and representative study sites due to its long study length (530 km) and small number of access locations. A helicopter equipped with mounted video was used to view the proposed project area. During the flight, a tablet PC loaded with aerial photos and basic GIS data layers was used to annotate observed points of interest (e.g. sandbars, islands, impoundments, tributaries, bank and valley characteristics, etc). Observations were spot-checked by landing at several locations and then later by ground truthing.

Using these GPS points along with the aerial photos, survey video and photo documentation, an initial section delineation was developed. Major considerations for section breaks were related to morphological changes, including the presence/absence of sandbars and islands, diversity of perceived hydromorphologic units, sinuosity and bank characteristics as well as the location of dams and major tributaries.

Alexander *et al.* (2010) developed a segment-scale geomorphic classification system for the Niobrara River and divided the study area into 25 distinct geomorphic segments. Our initial delineation resulted in 21 reconnaissance-based sections. The statistical analysis of data developed by Alexander and the qualitative analysis of reconnaissance data resulted in many similar proposed section breaks, which, in the end, were merged to form 16 project sections.

Using river access locations and aerial photo observations, we chose two- to three-mile representative sites in each section. We created whisker diagrams for each of Alexander's morphometric attributes at the section and site levels. These plots were compared to ensure that the means, first and third quartiles were statistically similar. The representative sites were lengthened or shortened to create a better fit and, in some cases, a new location was chosen altogether.

MesoHABSIM was also applied for the determination of minimum environmental flows in small first order streams in the entire region of Piedmont in NW Italy. Twenty-five streams were selected for their natural conditions with respect to flow regime, fish community composition and homogeneous spatial distribution across the region. Within each stream, the representative site was defined by its proximity to the drainage basin outlet, the absence of human impacts and the possibility to survey from 5 to 10% of the stream length in one day (Vezza *et al.*, 2011).

From this experience we conclude that, for the river sections up to 5 km, the best approach is a complete delineation of the study site. For longer rivers we recommend a representative site approach; however, the sophistication of the site selection methods increases with the length of the river. For rivers up to 100 km in length, we recommend on-foot or boat reconnaissance surveys such as the one conducted on the Świder River and aerial data analysis for longer sections. For the regional scale, GIS data can be used for site selections.

6.2.4 Habitat survey

The habitat survey describes all mesohabitats within the selected representative study sites. The purpose is to delineate the distribution and area of habitat types at each target flow. The number of surveyed flows strongly depends on the range of discharges being evaluated. If the assessment targets low flow conditions, then the mapping effort can be limited to a minimum of three flows distributed strategically (e.g. more surveys at the conditions where more dramatic changes are expected, i.e. lower flows). Obviously, the more surveys can be afforded, the greater resolution in the shape of the rating curve can be expected and the inflection points are defined more precisely. If higher flows need to be evaluated, additional surveys may be necessary; however, it should be noted that the habitat suitability criteria may change in response to flow increases (i.e. shifting between search for shelter versus foraging). This would require development of additional suitability curves for high flows. However, the ability to observe the species in such conditions may not be practical.

In general, most MesoHABSIM studies focus on low flow conditions. The survey consists of two processes: mapping of HMUs and the collection of hydraulic data in random locations. The following steps detail a basic MesoHABSIM surveying procedure:

1 Monitor the river for target survey flow occurrence (typically three to five) identified using flow time series analysis. Typically, the highest survey flow targets highest summer low flows and the lowest flow targets the annual minimum flow experienced on that river during the rearing and growth period. The other surveys are distributed between the two, with a preference for capturing low flow events if more than three surveys are possible. This helps to define observed habitat availability changes typical at the lower flows. Surveys can be conducted within 10% of the target flow. It is imperative that the flows remain constant during the one-day survey to avoid complicated post-processing of field data that may compromise the development of habitat versus flow relationships.

2 Determine the extent of the first HMU as follows. Walk or canoe the river, depending on river depth and accessibility, and note water surface characteristics (ripples, slope), river bottom morphology and uniformity of bank and shore-use characteristics. Continue moving downstream until a noticeable change in one of these characteristics occurs (see Parasiewicz 2007a for details).

3 Note the location of this change and draw a polygon on a field computer to delineate that HMU. After the sketch is completed, record the observed characteristics including: the mesohabitat type, dominate substrate and wetted/bankfull width. Next, indicate the

absence (<5% of area), presence (≤50%) or abundance (>50%) of instream habitat attributes such as: boulders, woody debris and undercut banks. Finally, note shoreline attributes, which include information on land use, erosion and irregular shores and comments (see Parasiewicz, 2007a for a list of all attributes).

4 A second team records depth, velocity and substrate information from the HMU. The HMU is divided into estimated zones of similar hydraulic and substrate conditions to better classify the range of observed characteristics. The HMU is sampled at a minimum of seven random locations distributed proportionally in each stratum to best characterize the conditions within the HMU.

These steps can be accomplished using any number of simple mapping techniques or with integrated GPS and field-based laptop computer systems.

6.2.5 Upscaling

The data collected during the habitat surveys serve as a basis for the development of habitat rating curves for each site. Using the developed suitability criteria, each HMU is evaluated to determine if it offers suitable or optimal habitat for each species at each surveyed flow (for details, see Parasiewicz, 2007a). The area of HMUs with suitable (or optimal) habitats is summarized for each site and plotted against a constant unit of area such as the wetted area at the highest measured flow, or the channel area of the site. Effective habitat is calculated as an aggregation of suitable and optimal habitat with different weights, to assure the high contribution of optimal habitat. Typically, we use 0.25 of suitable and 0.75 of optimal habitat as weights to define effective habitat.

Alternatively, composite habitat suitability indices can be used to weight the area of each HMU and create Weighted Usable Area (WUA). Although it is used widely in other studies, we do not recommend this method due to the fact that units with large areas and low suitabilities could produce the same WUA value as small units with high suitabilities. This could potentially lead to restoration efforts that create large, sub-standard, instead of high quality, habitats.

In addition to curves for individual species, habitat-rating curves for generic fish (the total amount of habitat available for the chosen fish community), as described in Parasiewicz (2007a), and community habitat-rating curves are calculated. The community habitat-rating curve is constructed by weighting the suitable habitat area of each species by its expected proportion in the Target or Reference Fish Community. Since a generic fish habitat

approach represents the habitat area that is suitable for any of the species in the investigated community, it represents the total amount of habitat available. In contrast, the community habitat-rating curve takes into account the habitat availability that supports the desired structure of the fish community. Frequently, the habitat-rating curves for generic fish habitat are plotted together with the community habitat curves and the curve representing the change in the wetted area (Figure 6.1). This allows one to determine whether there is a lot of habitat available (generic curve) and whether the habitat structure does not reflect the community structure. This can be concluded if there is a substantial vertical distance between both habitat curves in the diagram. This diagram can also be used to assist in the evaluation of potential habitat improvements associated with potential restoration measures and can be used for planning purposes.

The habitat-rating curves for each site are upscaled by a length-weighted sum to represent river segments. The river segment length is usually defined as a portion of the river where we would expect a specific structure of the fish community. The change may be due to natural factors such as the confluence of a major tributary or waterfall, or anthropogenic factors such as dams or flow withdrawals. On the Souhegan River, for example, the river segment division was due to a major change in gradient and geology that coincided with a change between two Level III ecoregions (Omernik, 1987; Ballesterro et al., 2006). In the case of the Pomperaug and Eightmile Rivers, the change in stream order from third to fourth at the confluence of two major river branches was the reason for developing multiple Reference Fish Communities (Parasiewicz et al., 2007a; 2007b).

A different river segmentation took place in the regional application in Piedmont, Italy. The environmental flow requirements of fish communities were upscaled from the local level to the entire region of interest, integrating the MesoHABSIM results within the regional water planning process. The reference streams were grouped according to the Classification and Regression Trees (CART) algorithm, defining homogenous sub-regions distinct from both environmental flow needs of aquatic fauna and catchment/reach characteristics. Building the tree, CART split the learning sample (i.e. 21 catchment/reach characteristics as independent variables and the environmental flow needs as the dependent variable) by using a binary recursive partitioning algorithm (see Vezza et al., 2011). Based on the resulting four groups of catchments represented by the terminal nodes of the regression tree, the resulting classification assigned the minimum environmental flow value to each group. First, latitude of the catchment centroid, then longitude and the maximum elevation were used for partitioning, identifying four sub-regions characterized by homogeneous hydro-ecological features (i.e. climate, flow regime, topography and fish community composition).

6.2.6 Adjusting biophysical templates to reflect reference habitat

The next step recommended in the MesoHABSIM approach is to consider and simulate structural improvements of the riverbed to create a habitat that would better support the targeted fish community. These simulations take place through appropriate modification of GIS maps and the information gathered in the project-specific database. As described in Parasiewicz (2007b), the simplest approach is to begin with the simulation of removing the most obvious anthropogenic factors, such as impoundments and dams or restoring the connectivity based on historic data and aerial imagery. More recently, we developed an approach directly investigating habitat needs of indicator species. It begins with comparing fish and habitat structure, as described above, and the identification of species that either lack or have a surplus in available habitat. In the subsequent step, we investigate multivariate criteria and compare HMUs that were predicted to be suitable and not suitable for the species. The purpose of this screening is to isolate physical attributes which, if modified at the river scale, would change the habitat structure to better support the expected community. The simulation takes place in an iterative process, where we introduce proposed changes to the project database one at a time, to determine model sensitivity. For example, when comparing the target fish community structure with the current habitat structure on the Wekepeke River, MA, we noticed a particularly important lack of habitat for brook trout (*Salvelinus fontinalis*), which historically should dominate the fish community. At the same time, the habitat available for blacknose dace (*Rhinichthys atratulus*) appeared to be excessive. The analysis of habitat suitability criteria for these species pointed to the need for a greater area of pool, riffle and run HMUs, abundance of shallow margins and undercut banks as well as phytal (submerged plants, floating stands, etc.) and cobbles with a variable percentage of gravel and sand substrate. Six different simulations were carried out in an iterative process, varying the variables noted above by modifying HMU areas and abundance of cover attributes. This allowed us to calculate a reference habitat structure and to

identify measures that would lead to these improvements (Parasiewicz and Rogers, 2010).

6.2.7 Reference flow time series

The final element necessary to determine reference conditions, in addition to the Reference Fish Community and reference habitat structure, is to develop a reference flow time series. This can be accomplished by a number of techniques based on the context of the study. Techniques range from the estimation of flow regime characteristics at an ungauged site based on an index gauge (e.g. Fennessey and Vogel, 1990; Fennessey, 1994) to distributed parameter catchment-level rainfall-runoff modelling (e.g. Leavesley *et al.*, 1983).

6.2.8 Habitat time series analysis

The reference flow time series and reference habitat structure are eventually used to develop a reference habitat time series, which describes the expected amount of habitat that would exist given the reference flow time series. The habitat time series are investigated with the help of UCUT curves to establish natural habitat stressor thresholds (HSTs) (see Parasiewicz, 2007b). The purpose of this analysis is to investigate habitat duration patterns and to identify conditions that could create pulse and press disturbances, as described by Niemi *et al.* (1990). A pulse stressor causes an instantaneous alteration in fish densities, while a press disturbance causes a sustained alteration of species composition. In the habitat analysis, this can be caused either by extreme habitat limitation regardless of duration or by catastrophically long duration events with critically low habitat availability. Press disturbances can be caused by frequent occurrence of persistent-duration events with critically low habitat availability. Therefore, identifying HSTs requires taking into account habitat magnitude as well as the duration and frequency of non-exceedance events, as described below.

To identify an HST, a habitat time series and the UCUT curves are developed (see Parasiewicz, 2007b for detail). As documented by Capra *et al.* (1995), the curves are a good tool to predict the impact of the frequency and duration of biological conditions. The curves evaluate the continuous duration and frequency of continuous non-exceedance events for different habitat magnitudes. Rapid changes in the frequency pattern are used to distinguish between typical and unusual events and to identify HSTs for rare versus common events. Rare habitat events happen infrequently or for only a short period of time. The

Table 6.2 Flow management criteria developed for two flow levels on the Saugatuck River, CT. Base flow is equivalent to common habitat conditions, subsistence is equivalent to rare habitat levels and absolute minimum is the lowest flow on record.

Bioperiod Approximate dates	Rearing and growth July–Sept	Fall spawning Oct–Nov	Overwintering Dec–Feb	Spring flood Mar–Apr	Spring spawning May–June
Base flow ($10^{-2} m^3 s^{-1} km^{-2}$)	0.74	0.40	2.09	2.09	1.10
Allowable duration under (days)	34	13	20	19	14
Catastrophic duration (days)	85	56	47	35	42
Subsistence flow ($10^{-2} m^3 s^{-1} km^{-2}$)	0.06	0.06	0.44	1.10	0.40
Allowable duration under (days)	14	8	18	10	10
Catastrophic duration (days)	49	26	33	15	20
Abs. minimum flow ($10^{-2} m^3 s^{-1} km^{-2}$)	0.002	0.006	0.052	0.204	0.051

common habitat threshold divides normal conditions that occur frequently from uncommon events.

The HST captures rare and common habitat characteristics together with their durations. The method specifies two duration thresholds: persistent and catastrophic based on the frequency of occurrence. Exceedance of those durations causes habitat stress days (HSDs). The cumulative frequency of events that are longer than the threshold value captures natural limitations shaping the aquatic community. Anthropogenic factors (e.g. flow diversions) often increase the frequency of such events, ergo the number of HSDs.

The HST can be used to develop criteria for ecological flow management. These criteria include the magnitude of rare, critical and common flows as well as the durations of persistent and catastrophic events and are used for the development of flow pulsing strategies (see Table 6.2), as described in Parasiewicz (2008b).

Eventually, these flow criteria are summarized in the form of ACTograms. The ACTogram approach attempts to capture all essential parameters (flow, habitat, duration and function) in a single set of graphs. The boundaries demarcating the black, striped and spotted areas (e.g. Figure 6.3) are defined by the flow–habitat relationship. Where boundary lines slope upward to the right, greater flows are indicative of greater habitat quantity. In such cases, persistent low flows may endanger ecological resources.

ACTograms plot the number of consecutive days that flows have persisted below a chosen threshold, typically the rare, critical and common thresholds mentioned above. Unlike traditional hydrographs, which plot flow on the ordinate (i.e. y-axis) and time (e.g. return period or event duration) on the abscissa (i.e x-axis), the ACTogram

reverses this relationship. ACTograms are designed to answer: 'How long can the current flow condition persist before creating press or pulse stressor?', whereas hydrographs are designed to solve for flow at a particular time.

To plot flow data on the ACTogram, it is necessary to track the number of consecutive days that flows have remained below a threshold of interest. For example, in Figure 6.3, two curves are presented, each representing flow-duration conditions on one day. The reference line indicates that the flow has been less than $1 \ m^3 s^{-1}$ for 0 days, less than $4 \ m^3 s^{-1}$ for 40 days, less than $8 \ m^3 s^{-1}$ for 50 days and less than $14 \ m^3 s^{-1}$ for 60 days. Note that these flow-duration conditions persist simultaneously on that day. A theoretically infinite number of flow thresholds may be plotted, but as a practical matter it is likely that three or four thresholds will be sufficient. It is sensible to track thresholds that are in the flow range in which changes in slope of between black/striped/spotted areas occur. To complete the plot, each flow/consecutive-day data point is connected with a line. The result shows a flow-duration frontier that begins on an x-axis intercept at the left edge of the ACTogram and generally slopes higher to the right as flows increase. Intrusion of any part of the frontier into the striped zone of the ACTogram is indicative of an anticipated stressed ecosystem. As dry days continue, the frontier will creep upwards. Upon entering the spotted zone, the ACTogram indicates that habitat quality has potentially suffered critical damage and the bioperiod function has been seriously impaired. However, an increase in flow will break the consecutive day streak at all thresholds less than the new, higher flow. In this case, the frontier to the left of the new flow will be returned to zero, but will remain high to the right of this flow. The flow/duration frontier is dynamic and new flows

Table 6.3 Summary of number of stress days calculated for current conditions and simulated scenarios. P refers to persistent and C to catastrophic events. Moderate changes in stress days are lightly shaded; severe changes are darkly shaded.

Current habitat structure

Event duration	P	C	P	C	P	C	P	C	P	C
Withdrawal	0		0.001		0.014		0.028		0.15	
Mitigation	no		no		no		no		Dynamic augmentation	
Common events NSD	111%	102%	97%	100%	129%	160%	167%	402%	98%	0%
Rare events NSD	176%	388%	243%	525%	243%	525%	243%	525%	86%	0%

Improved habitat structure

Event duration	P	C	P	C	P	C	P	C
Withdrawal $m^3 s^{-1}$	0		0.014		0.014		0.014	
Mitigation	no		Minimum flow		Static augmentation		Dynamic augmentation	
Common events NSD	127%	122%	127%	122%	29%	0%	99%	0%
Rare events NSD	111%	188%	93%	97%	0%	0%	53%	0%

must be plotted each day to monitor the river condition accurately.

6.2.9 Scenario comparison

The first step in this process is to define a list of viable scenarios that should be investigated. For example, the objectives of the Wekepeke Brook project were to define possible flow and habitat augmentation scenarios to compensate for planned water withdrawal sites in the brook's headwaters. To compare various flow scenarios for their impact on fish fauna, we simulated the modification of two factors: flow time series and habitat structure. Multiple flow time series were available: historical flows and simulated flows, which modelled three volumes of water withdrawals (0.001, 0.014 and 0.028 $m^3 s^{-1}$). The water withdrawals could also potentially be mitigated by imposition of minimum flows and by flow augmentations from an upstream reservoir. Flow augmentation could take the form of continuous and pulsed releases (dynamic augmentation). There were two options for spatial habitat distribution patterns: reference and present morphology.

For each scenario, habitat time series were developed and UCUT analysis described by Parasiewicz (2007b; 2008a) was used for comparison. Plotting the UCUT for selected rare and common thresholds provides an insight into the change in frequency of such events. To further compare the change in persistent and catastrophic events, the HSD was computed. The cumulative duration for the lowest persistent and the shortest catas-

trophic events are related to those of the reference conditions and presented as proportions of reference durations. If the proportion is between 50% and 200%, the durations are considered to be similar (i.e. HSD) or the count of days that the current or simulated conditions are above the common level is less than twice the original. Cases where durations exceed 200% or are shorter than 50% are considered remarkable, and those exceeding 300% or less than 5%, severe. The results are presented by colour-shaded tables to highlight the difference (see example in Table 6.3).

6.2.10 Interpretation and application

MesoHABSIM has been applied on over 30 rivers throughout the USA and Europe (www.MesoHABISM.org/projects). The majority of applications and model development took place on streams and small rivers of high to moderate gradient in the Northeastern USA. However, applications have been conducted on a variety of different systems such as prairie streams (Powder River, WY), a large coastal lowland river (Santee River, SC), a large braided river (Niobrara River, NE), alpine first order streams (Piedmont, Italy), a lowland meandering river (Świder River, Poland) as well as a Mediterranean river in the plateau of Castile (Tajuña River, Spain). These attest to the fact that there are no apparent limitations to MesoHABSIM's application in terms of river size and character. The main difference in application of MesoHABSIM across spatial gradients is related to field data acquisition, where larger systems are sampled using both photography and boats while smaller systems are sampled using ground-based techniques. A particular strength of MesoHABSIM is data collection in areas of complex and diverse habitat structure, where the precise collection of microhabitat data is very tedious. This is as much the case for dynamic, braided rivers as for small, high-gradient streams and has been documented clearly during the study of Piedmont's mountainous streams and the braided Niobrara River.

MesoHABSIM is often used for the analysis of a specific flow range. We consider this a strength of the model, because it prevents application of the developed habitat suitability criteria to high flow conditions. Furthermore, repeated surveys of representative sites at chosen flow thresholds allow for a detailed understanding of site specifics. Although this approach may require additional field effort and logistics, it offers returns and insights different from more remote techniques. In our experience, time spent in the field is very valuable for project understanding and outweighs the greater post-processing efforts of some hydrodynamic models.

Although MesoHABSIM has been developed with hydropower in mind, to date it has been most frequently applied in questions of instream flow management as related to industrial and municipal water withdrawals. Most notably, it serves as a method for determining Protected Instream Flows standards for the state of New Hampshire, where the flow regime developed by use of the approach was the basis for the implementation of the Water Quality Act on the Souhegan and Lamprey Rivers. The criteria developed by the model described flows protective to fish, invertebrates and rare and endangered riparian species. They are currently applied in the management of the Souhegan River.

Similarly, MesoHABSIM has been included as one potential tool for examination of instream flow regimes in The Nature Conservancy's Ecologically Sustainable Water Management Framework (ESWM). This approach consists of six consecutive steps that allow for the development of a proper management strategy, using available data supported by models such as those described in this chapter (Richter *et al.*, 2003). MesoHABSIM was used for the implementation of step 1 of this approach ('developing initial numerical estimates of key aspects of river flow necessary to sustain native species and natural ecosystem functions') on the Saugatuck/Aspetuck River in Connecticut. We considered multiple aspects of river ecology and defined flow needs for instream and riparian ecological targets in different seasons (Parasiewicz *et al.*, 2010).

In another, similar project on the Fenton River, CT, the flow criteria served as a regulatory basis for limiting well water withdrawals (Nadim *et al.*, 2007; Jacobson *et al.*, 2008). On the Pomperaug River, CT, the HSTs were used as a basis for developing a Habitat Meter, which signals in real time the habitat status on the Pomperaug River Watershed Association's website (www.Pomperaug.org). In this project, we used habitat time series to simulate future watershed development scenarios, demonstrating the applicability of the model to the questions of global climate change.

On several projects, our focus was on river channel restoration. The model can select restoration measures that are the most beneficial to fish fauna by offering focused habitat improvements. Such recommendations have, to date, been general in nature, as the model has a limited ability to address the detail necessary for construction planning. We recommend that for detailed site-specific designs, a micro-scale analysis is needed.

Overall, the ten years of MesoHABSIM's development and application have proved the utility of the model for river management and restoration. It has been applied on a variety of river sizes and types without major difficulty and can be used as an ecological status assessment tool as well as for planning instream flow management and channel restorations. It successfully addresses issues associated with hydropower generation, water supplies, channelization and river restoration. Compared to other meso-scale approaches such as MesoCASiMiR or the Norwegian Mesohabitat method (Borsanyi *et al.*, 2004; Eisner *et al.*, 2005), or mesohabitat typing used within the PHABSIM framework, the model incorporates scientific rigour in data collection as well as in analysis because:

1 The approach integrates expert habitat mapping with hydraulic measurements to characterize meso-scale features conducted at multiple flow conditions based on the analysis of detailed reconnaissance surveys.

2 It uses multivariate statistical models.

3 It uses multiple cross-validations for model calibration.

It offers not only advantages in data collection effort and sampling intensity, but also incorporates a number of innovative analytical possibilities such as quantification of habitat run-length (durations under threshold) and HSD analysis. The system provides a wide array of analytical possibilities and sophisticated approaches to the integration of habitat availability and flow regime assessments.

The approach offers a useful tool for river management planning that is critically needed for the implementation of modern water laws such as the European Water Framework Directive or the South African National Water Act.

It complements the micro- and macro-scale modelling of fish habitat, closing an important gap that impaired river management in the past. It also presents opportunities for the advancement of river science in general. Through the process of model development alone, we have learned a lot about habitat and ecological processes in riverine environments that may not have come easily without using this research tool.

Acknowledgements

The authors of this chapter would like to acknowledge the numerous colleagues, technicians and students who participated in development of the method. We would like to thank all those who funded and reviewed our work in the past, and those who contributed advice and data for this chapter.

References

Alexander, J.S., Zelt, R.B. and Schaepe, N.J. (2010) Hydrogeomorphic and hydraulic habitats of the Niobrara River, Nebraska—with special emphasis on the Niobrara National Scenic River. U.S. Geological Survey Scientific Investigations Report, 2010–5141: 62.

Bain, M.B. and Meixler, M.S. (2008) A target fish community to guide river restoration. *River Research and Application*, **24**: 453–458. doi: 10.1002/rra.1065.

Ballestero, T., Kretchmar, D., Carboneau, L., Parasiewicz, P., Legros, J., Rogers, J.N., Steger, T. and Jacobs, J. (2006) *Souhegan River Protected Instream Flow Report*. Report NHDES-R-WD-06-50 for New Hampshire Department of Environmental Services, Concord, NH. http://mesohabsim .org/projects/projects.html accessed 2007.

Borsanyi, P., Alfredsen, K., Harby, A., Ugedal, O. and Kraxner, K. (2004) A meso-scale habitat classification method for production modeling of Atlantic Salmon in Norway. *Hydroecologie Appliquee*, **14**: 119–138. doi:10.1051/hydro:2004008.

Bovee, K.D., Lamb, B.L., Bartholow, J.M., Stalnaker, C.B., Taylor, J. and Henriksen, J. (1998) *Stream habitat analysis using the Instream Flow Incremental Methodology*. Information and Technical Report USGS/BRD-1998-0004, U.S. Geological Survey, Biological Resources Division, Fort Collins, CO.

Capra, H., Breil, P. and Souchon, Y. (1995) A new tool to interpret magnitude and duration of fish habitat variations. *Regulated Rivers: Research and Management*, **10**: 281–289.

Carta Ittica Regionale (1992–2004) *Carta ittica relativa al territorio della Regione Piemontese – The ichthyic zonation for Piedmont region*. Volume: Bibliografia faunistica. Regione Piemonte, Italia.

Eisner, A., Young, C., Schneider, M. and Kopecki, I. (2005) Meso-CASIMIR – new mapping method and comparison with other current approaches. *Proceedings from the final COST 26 meeting*, Silkeborg, Denmark.

Fennessey, N.M. (1994) *A Hydro-Climatological Model of Daily Streamflow for the Northeast United States*. PhD dissertation, Tufts University, Medford, MA.

Fennessey, N. and Vogel, R.M. (1990) Regional flow duration curves for ungaged sites in Massachusetts. *Journal of Water Resources Planning and Management, ASCE*, **116**(4): 530–549.

Gortázar, J., Parasiewicz, P., Alonso-González, C. and García de Jalón, D.G. (2011) Physical habitat assessment in the Tajuña River (Spain) by means of the MesoHABSIM approach. *Limnetica*, **30**(2): 379–392.

Jacobson, R.A., Warner, G., Parasiewicz, P., Bagtzoglou, R. and Ogden, F. (2008) An interdisciplinary study of the effects of groundwater extraction on freshwater fishes. *International Journal of Ecological Economics and Statistics*, **12**(F08): 7–26.

Kinsolving, A.D. and Bain, M.B. (1993) Fish assemblage recovery along a riverine disturbance gradient. *Ecological Applications*, **3**(3): 531–544.

Leavesley, G.H., Lichty, R.W., Troutman, B.M. and Saindon, L.G. (1983) *Precipitation-runoff modeling system – Users manual*. U.S. Geological Survey Water-Resources Investigations Report, 83-4238: 207.

Leonard, P.M. and Orth, D.J. (1988) Use of habitat guilds of fishes to determine instream flow requirements. *North American Journal of Fisheries Management*, **8**: 399–409.

Metz, C.E. (1986) ROC methodology in radiologic imaging. *Investigative Radiology*, **21**: 720–733.

Nadim, F., Bagtztoglo, A.C., Baun, S.A., Warner, G., Jacobson, R.A. and Parasiewicz, P. (2007) Management of adverse impact of a public water supply well-field on the aquatic habitat of a stratified drift stream in eastern Connecticut. *Water Environment Research*, **79**(1): 43–56. doi: 10.2175/ 106143006X136801.

Niemi, G.J., Devore, P., Detenbeck, N., Taylor, D., Lima, A., Pastor, J., Yount, J.D. and Naimar, R.J. (1990) Overview of case studies on recovery of aquatic systems from disturbance. *Environmental Management*, **14**(5): 571–588.

Omernik, J.M. (1987) Ecoregions of the coterminous United States. Map (scale 1:7,500,000). *Annals of the Association of American Geographers*, **77**(1): 118–125.

Parasiewicz, P. (2001) MesoHABSIM – a concept for application of instream flow models in river restoration planning. *Fisheries*, **29**(9): 6–13. doi: 10.1577/1548-8446(2001)026<0006:M>2.0. CO;2

Parasiewicz, P. (2007a) The MesoHABSIM model revisited. *River Research and Application*, **23**(8): 893–903. doi: 10.1002/ rra.1045

Parasiewicz, P. (2007b) Developing a reference habitat template and ecological management scenarios using the MesoHABSIM model. *River Research and Application*, **23**(8): 924–932. doi: 10.1002/rra.1044

Parasiewicz, P. (2008a) Habitat time-series analysis to define flow-augmentation strategy for the Quinebaug River, Connecticut and Massachusetts, USA. *River Research and Application*, **24**: 439–452. doi: 10.1002/rra.1066

Parasiewicz, P. (2008b) Application of MesoHABSIM and target fish community approaches for selecting restoration measures of the Quinebaug River, Connecticut and Massachusetts, USA. *River Research and Application*, **24**: 459–471. doi: 10.1002/rra.1064

Parasiewicz, P. and Rogers, J.N. (2010) *Development of a Mitigation Method and Ecological Impacts Offset for Spring Water Withdrawals on Downstream Habitats for Fish and Invertebrates*. Report by Rushing Rivers Institute, Amherst, MA. http://mesohabsim.org/projects/projects.html.

Parasiewicz, P., Ehman, S.B. and Corp, P. (2003) *Fish Habitat Assessment on Stony Clove Creek, NY using MesoHABSIM*. Technical report, Cornell University, Ithaca, NY, pp. 99. http://mesohabsim.org/projects/projects.html.

Parasiewicz, P., Legros, J.D., Rogers, J.N. and Wirth, M.J. (2007a) *Assessment and restoration of instream habitat for the Pomperaug, Nonnewaug and Weekeepeemee Rivers of Connecticut*. Report for Pomperaug Watershed Coalition, Northeast Instream Habitat Program, University of Massachusetts, Amherst. http://mesohabsim.org/projects/projects.html.

Parasiewicz, P., Rogers, J.N., Legros, J.D. and Wirth, M.J. (2007b) *Assessment and restoration of instream habitat of the Eightmile River in Connecticut: Developing a MesoHABSIM model*. Northeast Instream Habitat Program, University of Massachusetts, Amherst. http://mesohabsim.org/projects/projects.html.

Parasiewicz, P., Thompson, D., Walden, D., Rogers, J.N. and Harris, R. (2010) *Saugatuck River Watershed Environmental Flow Recommendations*. Report for The Nature Conservancy and Aquarion, Rushing Rivers Institute, Amherst, MA. http://mesohabsim.org/projects/projects.html.

Parasiewicz, P., Rogers, J.N., Larson, A., Ballesterro, T., Carboneau, L., Legros, J. and Jacobs, J. (2008) *Lamprey River Protected Instream Flow Report*. Report NHDES-R-WD-08-26 for New Hampshire Department of Environmental Services, Concord, NH. http://mesohabsim.org/projects/projects.html.

Pearce, J. and Ferrier, S. (2000) Evaluating the predictive performance of habitat models developed using logistic regression. *Ecological Modelling*, **133**(3): 225–245.

Richter, B.D., Mathews, R., Harrison, D.L. and Wigington, R. (2003) Ecologically sustainable water management: managing river flows for ecological integrity. *Ecological Applications*, **13**: 206–224. doi:10.1890/1051-0761(2003)013[0206:ESWMMR]2.0.CO;2

Rushing Rivers Inc. (2010) *Sim-Stream 8.0 Users Manual*. Amherst, MA. www.Sim-Stream.com.

Sakamoto, Y. (1991) Categorical Data Analysis by AIC, Kluwer Academic Publishers, Dordrecht.

Senecal, A.C. (2009) *Fish Assemblage Structure and Flow Regime of the Powder River, Wyoming: An Assessment of the Potential Effects of Flow Augmentation Related to Energy Development*. Masters thesis, Department of Zoology and Physiology, University of Wyoming, Laramie, WY.

Vadas, R.L. and Orth, D.J. (2001) Formulation of habitat suitability models for stream-fish guilds: do the standard methods work? *Transactions of the American Fisheries Society*, **130**: 217–235.

Vezza, P. (2010) *Regional Meso-Scale Models for Environmental Flows Assessment*. PhD thesis, Polytechnic University of Turin. Turin, Italy.

Vezza, P., Parasewicz, P., Rosso, M. and Comoglio, C. (2011) Defining environmental flow requirements at regional scale by using meso-scale habitat models and catchments classification. *Rivers Research and Application*, doi: 10.1002/rra.1571.

Wanner, G.A., Pegg, M.A., Shuman, D.A. and Klumb, R.A. (2009) *Niobrara River fish community downstream of Spencer Dam, Nebraska*. Progress report to U.S. Fish and Wildlife Service, Pierre, South Dakota.

Welcomme, R.L., Winemiller, K.O. and Cowx, I.G. (2006) Fish environmental guilds as a tool for assessment of ecological condition of rivers. *River Research and Applications*, **22**: 377–396. doi: 10.1002/rra.914

7

The Role of Geomorphology and Hydrology in Determining Spatial-Scale Units for Ecohydraulics

Elisa Zavadil[1] and Michael Stewardson[2]

[1]Alluvium Consulting Australia, 21–23 Stewart Street, Richmond, Victoria 3121, Australia
[2]Department of Infrastructure Engineering, Melbourne School of Engineering, The University of Melbourne, Melbourne 3010, Australia

7.1 Introduction

Over the last two decades considerable attention has been focused on links between ecology and the physical in-stream environment, where abiotic processes have been widely recognized as key drivers of aquatic ecosystem structure and function. This includes the influence of both chemical and physical processes on the biotic community, including water temperature (e.g. Ward, 1985; Hawkins *et al.*, 1997), discharge and hydraulics (e.g. Statzner and Hilger, 1986; Beisel *et al.*, 1998; Jowett, 2003; Gillette *et al.*, 2006) and channel structure (e.g. Roussel and Bardon-net, 1996; Thomson *et al.*, 2001; Maddock *et al.*, 2004). The combined influence of these factors defines the living space or 'physical habitat' of in-stream biota (Maddock, 1999).

The discipline of eco-geomorphology is founded on this habitat-centred view for ecological systems, whereby physical structure is considered to be a dominant driver of ecological patterns. Eco-geomorphology has been defined as:

• The ecological aspects of fluvial geomorphology (Thorp *et al.*, 2006).

• An interdisciplinary approach to the study of river systems that integrates hydrology, fluvial geomorphology and ecology (Thoms and Parsons, 2002).
• Linking the physical form and processes of rivers with ecological response through adoption of an ecosystem perspective (Hudson, 2002).

One of the main focus points in eco-geomorphic research has been to improve our conceptual understanding of the spatial organization of riverine ecosystems (e.g. Fisher *et al.*, 2007; Post *et al.*, 2007; Renschler *et al.*, 2007), with a particular focus on the variability of physical habitat. This has included proposals for several hierarchical scale frameworks, incorporating spatial scales at the catchment, segment, reach, meso and micro levels (e.g. Frissell *et al.*, 1986; Maddock and Bird, 1996; Rowntree, 1996; Newson and Newson, 2000; Thoms and Parsons, 2002; Zavadil, 2009). These frameworks are expected to support prediction and appraisal of river condition throughout catchments, and enable integration and coordination of river research and management across different scales. This approach relies on identifying thresholds in physical heterogeneity to define boundaries between units at all scales.

In practice, however, the selection of appropriate spatial scales remains problematic, particularly in the field of ecohydraulics where the scale of reference has a strong influence on study outcomes (e.g. measuring species abundance, dispersal and other population dynamics). While many scale frameworks have been proposed, there has been little attempt to compare existing frameworks and review and advance their practicality for research and management. In this chapter we outline how our fundamental geomorphic and hydrologic understanding of channel networks has contributed to our perception of how stream systems are structured, and ultimately to the development of spatial scale hierarchy frameworks. We present a synthesis and comparison of selected frameworks and identify some of the challenges associated with defining scale units. For several of the more common scale units, the segment, reach and meso scales, we examine the geomorphic significance of the boundary criteria used to define them. A synthesis of these boundary criteria is proposed to provide a more practical and standardized approach to defining these spatial scales for research and management in ecohydraulics and related fields.

7.2 Continuum and dis-continuum views of stream networks

Stream networks have long been perceived as hierarchical in structure, with each tributary in a catchment having its own sub-catchment contributing runoff, and so larger catchments consist of a hierarchy of smaller ones (Summerfield, 1991). Horton (1945) laid the foundation for modern network analysis with the concepts of stream order and drainage density. These quantitative concepts replaced previous qualitative descriptions of drainage basins and their constituent networks, and explicitly recognize hierarchical structure (Knighton, 1998). Modifications by others such as Strahler (1952) and Shreve (1967) produced the more commonly used stream ordering methods of today. Subsequently, stream networks are typically perceived to consist of a series of links (lengths of stream) and nodes (confluences) within a hierarchical framework.

Yet, despite the link–node structure, downstream channel change within a stream network has generally been treated as continuous. The pioneers of continuous downstream hydraulic geometry, Leopold and Maddock (1953), introduced quantitative relationships to examine variations in channel dimensions (morphology) throughout river networks. Leopold and Maddock (1953) related

bankfull width (W), depth (D), and velocity (U) to mean discharge (Q) along a river channel in the form of a power function, $X = aQ^b$ (where X is W, D or U and b is the regional exponent). The smooth, continuous adjustment in channel morphology implied by this power function is useful for defining regional trends. For example, downstream hydraulic geometry relationships model how rivers get progressively wider and deeper (from headwaters to the mouth). There are, of course, limits to the use of these relationships; for example, where river systems may experience variable geologic control. However, these relationships are useful for establishing broad-scale trends associated with downstream channel change through catchments.

It is the case, then, that channel change across stream networks can be perceived as having both a continuum of change (e.g. Leopold and Maddock, 1953) and also a dis-continuum of change (e.g. link–node structure: Strahler, 1952; Shreve, 1967). Both of these views of stream networks are valid depending on the scale and resolution at which the channel network is examined. For example, smooth downstream continuum relationships would typically emerge when observing broad-scale data scattered across the whole catchment, whereas a step-change in channel morphology might be observed with finer-scale data along a shorter segment of stream that includes a major tributary junction. The recent trend in river research, and particularly in the field of eco-geomorphology, has been to focus more attention on the definition of discrete scales within channel networks, scales that typically represent step-changes in channel form (when viewed at finer resolutions) within the wider downstream continuum.

Principles of fluvial geomorphology and hydrology, including continuum and dis-continuum views of how stream networks adjust with increasing discharge, play a key role in the development and refinement of conceptual models of ecosystem structure. Early geomorphic principles of downstream hydraulic geometry relationships (Leopold and Maddock, 1953) inspired initial views on ecosystem structure such as the River Continuum Concept (RCC; Vannote et al., 1980). The RCC framework predicts gradual adjustments of biota and ecosystem processes in accordance with perceived gradual downstream changes in channel dimensions and dynamics. More recent advances in eco-geomorphology (commencing with Frissell et al., 1986) have initiated a shift towards (or rather back to) a more discontinuous link–node or 'patch' view of stream systems, through the development of a scale hierarchy for ecosystem structure. For example, contrasting to a continuum view, the recent Riverine Ecosystem Synthesis (Thorp *et al.*, 2006, Thorp, 2009) portrays rivers as having downstream arrays of large hydrogeomorphic patches (channel and floodplain types) and associated ecological functional process zones. The geomorphic scale hierarchy, and associated discrete geomorphic scales or patches, has become a fundamental framework in the field of ecohydraulics and related research areas.

7.3 Evolution of the geomorphic scale hierarchy

7.3.1 Origins

Frissell *et al.* (1986) presented an early geomorphic approach to a 'hierarchical framework for stream habitat classification.' The framework was designed to allow more systematic interpretation and descriptions of watershed–stream relationships. In developing their framework, Frissell *et al.* (1986) highlighted the need for an integrative, systematic approach for understanding the considerable variability within and among stream systems and stream communities. In particular, three specific needs were outlined for stream management (Frissell *et al.*, 1986, p. 199), which are re-iterated here and are undoubtedly still three big questions facing researchers and managers today:

1 How do we select representative or comparable sites in such diverse environments?

2 How can we interpret in a broader context, or how far can we reasonably extrapolate, information gathered at specific sites?

3 How do we assess past, present and possible future states of a stream?

The authors consider that small-scale physical habitat systems develop within constraints set by the larger-scale systems of which they are a part, and therefore advocate that a spatially nested hierarchical model provides a useful classification, noting in particular several benefits of using a hierarchical structure, outlined by Godfrey (1977):

• classification at higher levels narrows the set of variables needed at lower levels;

• it provides for integration of data from diverse sources and of different levels of resolution; and

• it allows the scientist or manager to select the level of resolution most appropriate to his or her objectives.

On this understanding, the initial scale hierarchy was established (Figure 7.1; Table 7.1), incorporating, on successively lower levels, stream segment, reach, pool/riffle and microhabitat subsystems (Frissell *et al.*, 1986).

The hierarchy model relies on the key assumption that structure, operation and development of stream communities is determined largely by the physical stream habitat (together with the pool of species available), and that the structure and dynamics of stream habitat itself is determined by the surrounding watershed (Frissell *et al.*, 1986). It is important to note that the physical environment is not the only influence on stream biota. Competition, predation and other processes regulating population

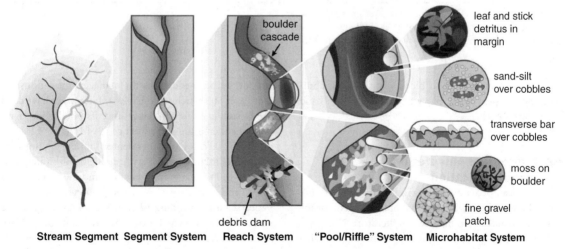

Figure 7.1 Hierarchical organization of a stream system and its subsystems: the original model. FISRWG, 1998; after Frissell *et al.*, 1986, p. 202. With kind permission from Springer Science and Business Media.

Table 7.1 Habitat subsystems: spatial boundaries and general descriptions. Reproduced from Frissell *et al.*, 1986, p. 203. With kind permission from Springer Science and Business Media.

Spatial scale (m)	Vertical boundaries[a]	Longitudinal boundaries[b]	Lateral boundaries[c]	Descriptive notes
Stream system 10^3	Total initial basin relief, sea level or other base level	Drainage divides and seacoast, or chosen catchment area	Drainage divides, bedrock faults, joints controlling ridge valley development	All surface waters within a watershed. A scale relevant to basin-wide assessments of management activities.
Segment system 10^2	Bedrock elevation, tributary junction or falls elevation	Tributary junctions, major falls, bedrock lithologic or structural discontinuities	Valley side slopes or bedrock outcrops controlling lateral migration	A portion of stream flowing through a single bedrock type, relatively uniform in slope. A scale most relevant to assessing regional impacts from dams, diversions, channelization, etc.
Reach system 10^1	Bedrock surface, relief of major sediment-storing structures	Slope breaks, structures capable of withstanding < 50-year flood	Local side slopes or erosion-resistant banks, 50-year floodplain margins	Potentially the least physically discrete unit in the hierarchy. A length of a stream segment lying between breaks in slope, bank material, vegetation and the like. Most common field scale for biologists and geomorphologists.
Pool/riffle system 10^0	Depth of bedload subject to transport in < 10-year flood; top of water surface	Water surface and bed profile slope breaks, location of genetic structures	Mean annual flood channel, mid-channel bars, other flow-splitting obstructions	A sub-section of a reach, having characteristic bed topography, water surface slope, depth, and velocity patterns (riffles, glides, cascades, pools, etc). Potentially useful in determining habitat suitability for various organisms.
Microhabitat system 10^{-1}	Depth to particles immovable in mean annual flood, water surface	Zones of differing substrate type, size, arrangement, water depth, velocity	Same as longitudinal	Patches within pool/riffle systems that have relatively homogenous substrate type, water depth and velocity. Useful units for the investigation of fish and invertebrate behaviour.

[a]Vertical dimension refers to upper and lower surfaces
[b]Longitudinal dimension refers to upstream–downstream extent
[c]Lateral dimension refers to cross-channel or equivalent horizontal extent

dynamics will contribute to biotic distributions, abundance and dispersal (e.g. Flory and Milner, 1999; Lauer and Spacie, 2004).

7.3.2 Adoption, adaptation and application

Since publication of the initial Frissell *et al.* (1986) framework, there has been widespread adoption of geomorphic scale hierarchies across the relevant literature. Numerous authors have adopted, incorporated, adapted and/or cited the Frissell *et al.* (1986) approach as a conceptual basis for structuring ecohydraulic and related research, and also to

support recommendations for management actions. To date, over 600 articles in the relevant literature have cited the Frissell *et al.* (1986) paper since its publication, with citations increasing steadily since 1994 (Figure 7.2). The main journal categories represented in those citations are predominantly associated with the fields of biology and ecology (40%), geography, geology and the geo-sciences (10%) and a diversity of other environmental and conservation sciences including fisheries, limnology, engineering and conservation (combined to give the remaining 50%).

In addition to the widespread adoption of the Frissell *et al.* (1986) framework, there have been several adaptations and alternate frameworks published since 1996 (e.g.

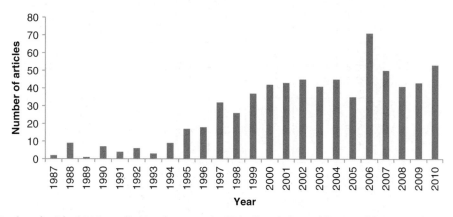

Figure 7.2 Number of articles (680 in total) citing the approach of Frissell *et al.* since publication of their seminal paper in 1986. Data from Web of Science, November 2010.

Table 7.2). These scale hierarchies have been applied to a range of ecohydraulic and related research and management; examples of these include:

• demonstrating the working scales of habitat mapping and Physical Habitat Simulation (PHABSIM) (Maddock and Bird, 1996);

• targeting restoration measures for fish habitat improvement relevant to various scales (Muhar, 1996);

• providing a broader framework for the application of more commonly used meso-scale units in ecohydraulic studies, such as biotopes and hydraulic units (Rowntree, 1996; Padmore, 1997; Newson and Newson, 2000; Taylor *et al.*, 2000; Thomson *et al.*, 2001);

• providing a broader framework for upscaling and downscaling between scales (Habersack, 2000) and providing guidance on differences between process-relevant scales and those most appropriate for practical measurements (Newson and Newson, 2000);

• linking regional variables (e.g. climate, geology, topography, lithology and sediment transport) to different structures and functioning of stream ecosystems at lower spatial scales (Cohen *et al.*, 1998) and linking geomorphic form with ecosystem function at a suite of scales for a specific catchment (Fisher *et al.*, 2007);

• advancing environmental flow applications through considering the impact of environmental flow manipulation on geomorphology and ecology for a suite of functional scale units that have direct relevance to ecological communities (Thoms and Parsons, 2002).

7.3.3 Difficulties

The number of scale framework adaptations (Table 7.2) highlights the widespread acceptance of geomorphic scale

hierarchies for characterizing stream systems, and the ongoing efforts to develop and refine these frameworks. However, the discrepancies in scale units and scale boundaries across the adaptations highlight one of the main difficulties in the practical use of these frameworks – that is the high level of subjectivity involved in the definition of scale units.

In most cases, the definition of boundaries for geomorphic units at all scales relies heavily on visual assessments of 'significant' scale boundaries. For example, slope breaks, zones of differing substrate type and structural discontinuities (Table 7.1) are all based on subjective assessments of what constitutes a point of significant physical difference. In this way, scales are determined largely based on the grouping of perceived thresholds in channel form and function. For example, obvious geomorphic features (e.g. stream confluences and topographic breaks in channel gradient) are typically used to define segment boundaries, based on the assumption that they represent a threshold change in physical form and the composition of finer-scale units (reach and meso scale). Similarly, uniform patches of flow and substrate are used to define many meso-scale units, on the assumption that these units are geomorphically distinct. The subjective nature of these distinctions makes the definition of discrete geomorphic scales difficult, and application of these frameworks challenging. One avenue of research currently being undertaken to reduce the subjectivity of meso-scale unit survey (biotopes) is the use of a terrestrial laser scanner (TLS) to collect high-resolution spatial data of water surface roughness (Milan *et al.*, 2010). From these data, hydraulic variability is expressed by assigning a local standard deviation value to a set of

Table 7.2 Comparison of spatial scales used in selected geomorphic scale hierarchies.

Spatial dimensions (approx)	Scale	Frissell et al. (1986)	Maddock and Bird (1996)	Muhar (1996)	Rowntree (1996)	Cohen et al. (1998)	Habersack (2000)	Newson and Newson (2000)	Thomson et al. (2001)	Thoms and Parsons (2002)	Fisher et al. (2007)	Zavadil (2009)
10^3–10^4	Catchment scale	Stream system	Drainage basin type	Stream system	Catchment Zones	Basin Valleys	Catchment	Catchment Subcatchment	Catchment Landscape unit	River system	Catchment	Catchment Zone
10^2–10^3 m	Segment scale	Segment system	Sector	Segment	Segment		Section	segment		Functional process zone	Segment	Segment Sub-segment
10^1–10^2 m	Reach scale	Reach system	Reach	Reach	Reach			Reach Site	River style	River reach	Reach	Reach
10^0–10^1 m	Meso scale	Pool/riffle system	Site/transect	Macro-habitat	Morphological unit	Meso-habitats	Local	Morphological unit	Geomorphic unit	Functional channel set	Channel sub-unit	Morphological unit
10^{-1}–10^0 m					Biotope		Point	Transect Biotope Cell	Hydraulic unit	Functional unit		Geotope Biotope
10^{-2}–10^{-1} m	Micro scale	Micro-habitat system	Patch	Micro-habitat				Patch		Meso-habitat		Point/patch

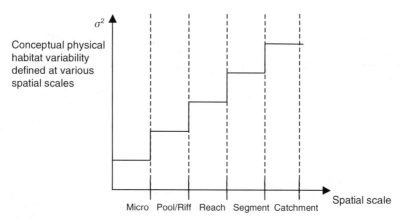

Figure 7.3 Conceptual diagram of hierarchy theory: discrete spatial scales and increasing physical variability with a perceived threshold change (or step-change) between scales.

adjacent water surface values in order to map biotope types.

A second difficulty in applying geomorphic scale hierarchies is the possibility that different spatial thresholds will be associated with different physical variables. This concept is best understood by first examining what 'hierarchy theory' means in practice. Hierarchy theory implies that, for a given parameter distributed in space, at a defined level of resolution, thresholds in physical variability exist for a discrete set of spatial scales, with the variability at smaller scales being constrained by that at larger scales. A conceptual illustration of this hierarchical scenario for stream systems is shown in Figure 7.3. For any given physical parameter, one would expect increased variability from small to larger scales, and some threshold change (or step-change) in variability between the scales. For example, channel area at the micro scale would be typical of one point (cross-section), but then would be more variable at the pool/riffle scale (e.g. encompassing area variation over a pool unit), and then more variable again at the reach scale (e.g. encompassing area variation over several pool and riffle units) and so on.

It follows that each variable will have its own inherent pattern of variability in time and space, with differing scale thresholds (i.e. changes in variability with scale). There are a wide range of geomorphic, hydrologic and ecologic parameters and processes considered in echydraulics and related fields of research and management. Scale frameworks such as Frissell *et al.* (1986) appear to be largely based on changes in channel morphology. However, there are many other physical parameters contributing to the condition of the in-stream environment, including hydraulics, substrate and vegetation. Thresh-

olds of change for these parameters will not necessarily correlate with those relevant to channel morphology. The conceptual diagram in Figure 7.4 demonstrates this idea, illustrating how different parameters that influence the physical form of the in-stream environment may differ in where their step-changes in variability occur with scale (not necessarily coinciding with channel morphology). This further complicates the goal of defining a set of discrete scales which represent a threshold change for overall physical habitat variability and associated ecosystem structure.

These difficulties of scale unit boundary subjectivity, and the potential for varying thresholds in different aspects of physical habitat, are not insurmountable, but do need to be considered when defining spatial scales for practical research and management purposes. When applying scale hierarchies, it is important to have an appreciation of the significance of scale unit boundaries, and associated limitations, for various physical and ecological parameters.

7.4 Defining scale units

The variation in what scales are included across the existing frameworks (Table 7.2) is largely a result of uncertainty in the significance of scale unit boundary criteria. However, there are three scales which are relatively common across the frameworks, the meso-, reach and segment scales. These three scales are also the base scales from which further subdivisions are typically made. Definitions of these three scales are discussed here to promote a more standard framework, incorporating recent advances

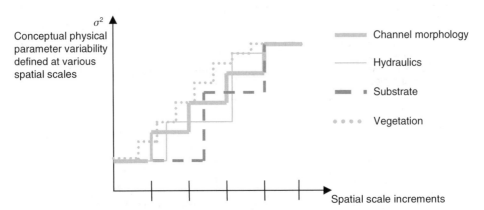

Figure 7.4 Conceptual diagram illustrating potential variability of different habitat parameters (related to hydraulics, substrate and vegetation) with increasing spatial scale, each according to its own threshold/step-change patterns and potentially independent of step-changes associated with channel morphology.

in geomorphic research on improved scale unit boundary definition.

7.4.1 Meso scale

The meso scale includes morphological units and biotope/cell (or equivalent) units, i.e. from 10^{-1} to 10^{1} m (Table 7.2). Terminology for meso-scale units is often used interchangeably (e.g. morphological unit, biotope unit, cell/flow-type); however, there are distinct differences in the practical definition of these units.

Morphological/geomorphic units

Morphological or geomorphic units are generally considered to be sedimentological features such as pools, riffles, glides and waterfalls (e.g. Wadeson, 1996; Parsons *et al.*, 2002). However, the identification of morphological/geomorphic units still often requires some observation of flow characteristics to distinguish them (Table 7.3).

Physical habitat units (biotopes, hydraulic units, functional units)

Other, smaller units at the meso scale, like biotopes, hydraulic or functional units (Table 7.2), vary with discharge. These flow-dependent meso-scale units have become increasingly popular over the last decade as the finest-scale unit of the hierarchy. This is in part due to a perception that these units have potential to reflect changes in geomorphic characteristics at a scale most relevant to in-stream ecology, i.e. the potential to reflect changes in physical habitat (Padmore, 1998; Newson and Newson, 2000).

The characterization of meso-scale physical habitat units has progressed from earlier classifications such as pool, riffle, run (e.g. Jowett, 1993), into a more standardized set of biotopes, hydraulic or functional units (e.g. Wadeson, 1995; Padmore *et al.*, 1996; Padmore, 1997, 1998; Wadeson and Rowntree, 1999, 2001, 2002; Clifford *et al.*, 2006). These biotope (and similar) units are determined primarily by flow type, plus some consideration of substrate and/or channel setting (e.g. Table 7.4). Flow-types refer to the dominant surface flow characteristics and provide information on morphologic in-channel variability at the micro scale. While the terms 'flow type' and 'biotope' are often used interchangeably, the biotope generally refers to a larger spatial unit (or patch) of the channel, and flow type generally refers to point- or micro-scale assessments of flow that make up a biotope unit (Padmore, 1998; Newson and Newson, 2000). For example, a pool biotope may be identified by a dominance of 'no perceptible flow' across an area of the channel.

Habitat mapping is a popular application of flow-types and biotopes in the ecohydraulic literature. Mapping of biotope units over river reaches at various discharges can provide an understanding of habitat diversity across space and time. For example, biotope mapping has been used in South Africa for more than a decade to assist with habitat assessment as part of water resource allocation and ecological flow assessments (Wadeson and Rowntree, 2001; 2002); it is also part of the River Habitat Survey in the UK (Fox *et al.*, 1998), and has been used as an approach to managing environmental flows in Australia (Dyer and Thoms, 2006).

Table 7.3 Morphological/geomorphic units defined by sedimentological features: examples from the Buffalo River, South Africa (Wadeson, 1996) and AUSRIVAS assessment protocol, Australia (Parsons *et al.*, 2002).

Unit	Description
Morphological units (Wadeson, 1996)	
Plunge pool	Erosional feature below a resistant stratum (waterfall)
Bedrock pool	Topographic low point formed behind resistant stratum lying across the channel
Alluvial pool	Topographic low point formed by scour within an alluvial bed, often associated with fines in the bed, especially at low flows
Step	Occurs within headwater streams, characterized by large clasts organized into discrete channel-spanning accumulations
Plane bed	Consists of large clasts relative to flow depth and lack of well-defined bedforms
Riffle	Topographic high points in an undulating bed-long profile composed of coarser sediments
Geomorphic units (Parsons *et al.*, 2002)	
Waterfall	Height >1m, gradient >60 degrees
Cascade	Step height <1m, gradient 5–60 degrees, strong currents
Rapid	Gradient 3–5 degrees, strong currents, rocks break surface
Riffle	Gradient 1–3 degrees, moderate currents, surface unbroken but unsmooth
Glide	Gradient 1–3 degrees, small currents, surface unbroken and smooth
Run	Gradient 1–3 degrees, small but distinct and uniform current, surface unbroken
Pool	Area where stream widens or deepens and current declines
Backwater	A reasonable sized (>20% of channel width) cutoff section away from main channel

Physical habitat units and flow types have largely superseded the use of morphological/geomorphic units in river research and management (e.g. Padmore, 1998; Newson and Newson, 2000). This may largely be associated with the convenience of working with the smallest practical scale unit, as smaller units can be aggregated up to larger units when required. However, there is the added complication of these physical habitat units being flow-dependent, which can complicate assessment and comparisons.

Hydraulic and geomorphic significance

Recent research at the meso scale has focused on defining the hydraulic character of physical habitat units defined by flow type. Investigations from New Zealand, the UK and South Africa have shown good relationships between flow types and channel hydraulics, with Froude number (ratio of velocity to square root of depth, or ratio of kinematic to potential energy) having the most distinct differences between flow types (Jowett, 1993; Padmore, 1998; Wadeson and Rowntree, 1999; Zavadil *et al.*, 2011). However, the Froude number and flow type relationship is likely to be a complicated one, as very different combinations of water depth and velocity can produce a similar Froude number (Clifford *et al.*, 2006). Flow-type composition has been shown to be sensitive to variations in discharge, channel bed topography (depth diversity) and cross-sectional channel geometry (Zavadil *et al.*, 2011).

Wallis *et al.* (2010) demonstrated that the proportion, size, shape and relative location of hydraulic patches (combinations of depth and mean column velocity) all varied with discharge, and that intermediate flows appeared to support the most diverse hydraulic patch composition and configuration, indicating that the spatial influence of meso-scale bedforms on hydraulic patches is mediated by temporal variations in discharge.

The geomorphic significance of surface flow types has been further demonstrated at the cross-sectional, or 'geotope' scale. Cross-sections of the channel that are dominated by riffle-run (unbroken standing waves or rippled surface), glide (smooth boundary turbulent flow) or deadwater/pool (no perceptible flow) biotopes have been shown to have significant distinctions in their channel form (based on statistical tests) (Zavadil *et al.*, 2011). Geotopes may be considered to be at a smaller scale than morphological units, but a coarser scale than biotope units (Table 7.5). In this way, a series of similar cross-sectional geotopes can be used to define the extent of morphological units. This also provides a less subjective way to identify morphological units.

Table 7.4 Example classification of surface flow types and associated biotopes (after Padmore, 1998, p. 27 and Newson and Newson, 2000, p. 204).

Flow type	Description	Associated biotope(s)
FF Free fall	Water falls vertically and without obstruction from a distinct feature, generally more than 1m high and often across the full channel width.	*Waterfall*
CH Chute	Fast, smooth boundary turbulent flow over boulders or bedrock. Flow is in contact with the substrate and exhibits upstream convergence and downstream divergence.	*Spill* – over exposed bedrock *Cascade* – over individual boulders
BSW Broken standing waves	White water 'tumbling' waves with the crest facing in an upstream direction. Associated with 'surging' flow.	*Cascade* – downstream of boulder flow diverges or 'breaks' *Rapid*
USW Unbroken standing waves	Undular standing waves in which the crest faces upstream without 'breaking'.	*Riffle*
Rip Rippled	Surface turbulence does not produce waves, but symmetrical ripples which move in a general downstream direction.	*Run*
UP Upwelling	Secondary flow cells visible at the water surface by vertical 'boils' or circular horizontal eddies.	*Boil*
SBT Smooth boundary turbulent	Flow in which relative roughness is sufficiently low that very little surface turbulence occurs. Very small turbulent flow cells are visible, reflections are distorted and surface 'foam' moves in a downstream direction. A stick placed vertically in the flow creates an upstream-facing 'V'.	*Glide*
NPF No (or scarcely) perceptible flow	Surface foam appears to be stationary and reflections are not distorted. A stick placed on the water surface will remain still.	*Pool* – full channel width *Marginal deadwater* – does not occupy the full channel width

7.4.2 Reach scale

The reach scale is perhaps the most frequently employed scale across river research and management. Typically, the reach scale is of larger spatial extent than the meso scale, and smaller spatial extent than the segment scale (Table 7.2); however, the term is often used to refer to river lengths from tens of metres up to tens of kilometres. Unlike meso and segment scales, the reach scale does not have the advantage of visually or measurably discrete geomorphic boundary criteria (e.g. biotopes, confluences), and is, therefore, far more difficult to define.

Reaches are often treated as both physically discrete scales and as sampling units, which can be conflicting views. At the reach scale, a reach can be viewed as having a pattern of channel form that is distinct in some way from neighbouring reaches, such as variations in slope or sequences of meso-scale units (e.g. pool-riffle sequence). However, the reach is often also used as a sample of conditions along the entire segment. Although somewhat distinct from other reaches, a 'representative reach' is considered to be representative of what conditions are like along that segment, as viewed at the segment scale. In this way, the reach scale is a popular sampling unit as it is often considered to be a spatial extent that is most practical for survey efforts. When time and budget are limited, a representative reach is typically sought, from which results can be inferred or extrapolated for the wider region (e.g. segment). However, identifying representative reaches can be difficult and highly subjective, as criteria used to define the reach scale are often variable and ill-defined. The remaining discussion here focuses on the definition of representative reaches.

Representative reaches have commonly been defined by a range of criteria including some multiple of channel width (e.g. Angermeier and Schlosser, 1989; Simonson et al., 1994; Montgomery and Buffington, 1997; Myers and Swanson, 1997) and/or an extent which includes two or three pool-riffle sequences or one meander wavelength (e.g. Chee et al., 2006) and/or an extent defined by similar land use or other channel setting features. Reach length

Table 7.5 Summary of meso-scale unit explanations derived from Newson and Newson, 2000; Padmore, 1997, 1998; Padmore *et al.*, 1996; Wadeson, 1995; Wadeson and Rowntree, 1999, 2001, 2002; Zavadil, 2009; Zavadil *et al.*, 2011.

Meso-scale unit	Description
Morphologic/geomorphic unit	• A sedimentological unit delineated by relative longitudinal changes (breaks) in channel gradient and substrate size. • These units are not considered to be flow-dependent, however can often be distinguished by observing dominant surface flow. • Units can contain a set of different physical habitat sub-units, although the dominant physical habitat sub-unit will be the same as the overall morphological unit. For example, an alluvial pool morphological unit will have a dominance of 'pool' biotopes, but may also contain 'deadwater' or 'boil' biotopes. • Can also be defined as a unit encompassing a series of similar cross-sectional geotopes.
Cross-sectional geotope unit	• Cross-sectional units representing significantly distinct depth–velocity environments as defined by dominant cross-sectional flow type.
Physical habitat unit (biotope, hydraulic unit, functional unit)	• A flow-dependent unit delineated by dominant surface flow type, plus channel setting and substrate. • These units are at a finer scale than morphological units and are able to reflect more detail in the depth–velocity environment, which is useful for characterizing physical habitat variability in both space and time. • Units are primarily defined by the dominant surface flow type. For example, a 'pool' biotope will have a dominant flow type of 'no perceptible flow'. • Multiple points of the same flow type within the channel denote a biotope cell.
Flow type	• Visible surface flow characteristics at a given discharge and point in time. • A finer scale than physical habitat units, defined at the cell or point scale and reflect the depth–velocity environment at that location.

can, therefore, range from metres to kilometres. In practical terms, the definition and extent of representative reaches is often determined by logistical constraints such as time, budget and accessibility to the river. In the absence of any other tangible geomorphic criteria, these logistical constraints often become the only criteria for reach scale extent.

The selection of representative reaches in both upland and lowland environments has proven to be difficult and generally unreliable due to the considerable reach-to-reach variation in channel morphology as measured by hydraulic parameters (Zavadil, 2009). This may be in part due to variations in channel gradient or other sedimentological variations between reaches (at the reach scale), resulting in relatively discrete reaches within the broader channel segments.

Further results from recent research (Zavadil, 2009) indicate that the magnitude of variation in channel change between reaches is likely to be greater in lowland environments compared to upland environments. In lowland environments it is likely that we can have less confidence

in selecting a reach that is representative of depth-related parameters (mean and variation) compared to width (Zavadil, 2009). Conversely, in upland environments it is likely that we can have less confidence in selecting a reach that is representative of width-related parameters (mean and variation) compared to depth. Furthermore, in upland environments, we can have greater confidence that a selected reach is representative of both the mean and variation in channel form parameters simultaneously, compared to lowland environments where reaches which are representative of mean values are not as likely to be representative of the variability in those parameters (Zavadil, 2009). Therefore, it is likely that, in general, representative reaches can be selected with more confidence in an upland environment compared to a lowland environment.

Confidence in selecting representative reaches can potentially be increased by the initial inspection of aerial photographs of the river length, particularly in lowland environments (Zavadil, 2009). Aerial photographs can be used to assist in identifying features of the channel planform, including valley confinement, channel position, stream length, sinuosity, relict channels and cutoffs, number of channels, land use, human impacts and vegetation condition. However, this approach has been shown to be generally unreliable in isolation of follow-up, on-ground assessments (Zavadil, 2009).

7.4.3 Segment scale

Segment scale units (Table 7.2) are typically defined based on broad-scale changes or thresholds in topography (valley confinement, slope), geology and hydrology within the catchment (e.g. Table 7.1). Important changes in geology and topography are more easily identifiable; for example, waterfalls, gorge constrictions and the interface between confined valleys and unconfined alluvial floodplains. These landscape changes are sometimes defined as a slightly broader scale than the segment scale; for example, 'types', 'zones' or 'landscape units'. Finer-scale changes

to hydrology and channel morphology are then attributed to the 'segment' scale, which is further discussed here.

Boundary criteria

At the segment scale, stream confluences (or tributary junctions) are most commonly employed as scale unit boundary criteria, within the context of more obvious landscape features, as confluences are often associated with more subtle changes in the physiographic setting (topography, geology, hydrology). Table 7.6 provides a comparison of the criteria used to define the segment scale across a selection of scale hierarchies. Definition criteria are relatively similar across the frameworks; however, they are also predominantly subjective. For example, the definition of a 'significant' change in topography, geology and hydrology is relatively undefined. Stream confluences are both explicitly and implicitly recognized as boundary criteria in these frameworks.

Table 7.6 Comparison of segment scale boundary criteria.

Framework	Unit name	Definition criteria
Frissell *et al.*, 1986	Segment	Vertical boundaries: Bedrock elevation, tributary junction or falls elevation
		Longitudinal boundaries: Tributary junctions, major falls, bedrock lithologic or structural discontinuities
		Lateral boundaries: Valley side slopes or bedrock outcrops controlling lateral migration
		Description: A portion of stream flowing through a single bedrock type, relatively uniform in slope. A scale most relevant to assessing regional impacts from dams, diversions, channelization, etc.
Rowntree, 1996	Segment	A length of channel along which there is no significant change in the imposed flow discharge or sediment load.
Habersack, 2000	Section	Planview river morphology, slopes, sediment regime, etc.
Thomson *et al.*, 2001	Landscape unit	Topographic unit determined on the basis of local relief, valley slope and morphology. Defines the valley setting.
Newson and Newson, 2000	Segment	Gradient/geology/discharge – from maps
Thoms and Parsons, 2002	Functional process zone	Lengths of the river system that have similar discharge and sediment regimes; can be defined by major breaks in slope and from style of river channel or floodplain.
Zavadil, 2009	Segment	Stream length within a sub-catchment, bound by obvious geologic features (e.g. geologic intrusions) or shifts in valley confinement or river style AND/OR confluences of geomorphic significance (bankfull symmetry ratio > 0.5 or sub-catchment symmetry ratio > 0.2) with evidence of morphological features (bars etc.).
	Sub-segment	Stream length within a segment, bound by confluences of hydrological significance (bankfull symmetry ratio < 0.5, or sub-catchment symmetry ratio < 0.2) and absence of morphological features (bars, etc.).

Stream confluences

Stream confluences have long been used to divide catchments into more manageable scale units for ecohydraulic and related studies. This approach is based on the assumption that the hydrological input at confluence zones effects an immediate physical change to the in-stream environment, and by comparison, the channel morphology between confluences is relatively homogeneous (i.e. the step-change principle). Research focused on understanding and predicting morphologic adjustment at confluences has highlighted the complexity of these changes (e.g. Miller, 1958; Gippel, 1985a; Roy and Woldenberg, 1986; Rhoads, 1987; Roy and Roy, 1988; Rice and Church, 1998; Rice, 1998, 1999; Benda *et al.*, 2004a, 2004b). While it would be reasonable to expect that confluence characteristics (size, location, symmetry, etc.) would influence the degree and significance of morphologic adjustment, gaining an understanding of these relationships has proven difficult. Research to date has been approached from both *qualitative* and *quantitative* perspectives.

The *qualitative* approach focuses on visible confluence characteristics and features to infer the geomorphic significance of a confluence. For example, from a qualitative perspective, Benda *et al.* (2004a; 2004b) assessed the geomorphic significance of confluences by noting the presence of gradient steepening/lowering, upstream sediment deposition, changing substrate size, boulders and rapids, terraces, floodplains, side channels, mid-channel bars, ponds, logjams, meanders and channel instability. Type, form and age distribution of fluvial landforms can then be used as a way of assigning some measure of geomorphic significance to confluences, to reflect the degree of morphologic adjustment in relation to the landscape.

Quantitative predications of morphological adjustment at confluences have been more difficult to explore, and in the past have been considered largely unachievable due to low resolution and limited data (Gippel, 1985b; Benda *et al.*, 2004b) and the complexity of riverine environments (Rhoads, 1987). Quantitative observations and predictions can theoretically be made for channel morphology in terms of both sedimentology and geometry (size and variability). Research on sedimentology at confluences (Rice, 1998, 1999; Rice and Church, 1998) has attempted to differentiate between significant and insignificant tributaries by examining their association with grain size discontinuities and the initiation or termination of a fining trend in the main stream at confluences. Important outcomes of this research highlighted that sedimentological networks and hydrological networks do not necessarily correspond, and subsequently that models based solely on hydrological network order may be fundamentally inappropriate for understanding sediment fluxes and sediment characteristics within fluvial systems (Rice, 1998).

Quantifying the nature and magnitude of morphologic adjustment in terms of channel geometry at confluences has received much less attention than sedimentology. Rhoads (1987) summarized that if channel characteristics are adjusted primarily to bankfull flow, where channel-forming discharges occur simultaneously on the major and minor tributaries, then continuity of flow implies that downstream hydraulic geometry should adequately describe changes at confluences (based on the work by Miller, 1958; Gippel, 1985a; Roy and Woldenberg, 1986; Roy and Roy, 1988). This assumes that changes in channel form are governed primarily by changes in hydrology. On this basis, Rhoads (1987) states that empirical analyses (e.g. Leopold *et al.*, 1964; Richards, 1982) indicate that the expected relationships among morphologic variables are:

$$W_0 > W_1 > W_2 \qquad (7.1)$$

$$D_0 > D_1 > D_2 \qquad (7.2)$$

$$C_0 > C_1 > C_2 \qquad (7.3)$$

$$F_0 > F_1 > F_2 \qquad (7.4)$$

$$S_0 < S_1 < S_2 \qquad (7.5)$$

where W is channel width, D is depth, C is cross-sectional area, F is the form ratio, S is the channel gradient and the subscripts 0, 1 and 2 refer to the receiving stream, major tributary and minor tributary, respectively. This means we would expect the receiving (downstream) channel to be wider, deeper, of greater capacity and form ratio and with a flatter gradient than either of the upstream merging channels.

However, demonstrating these theoretical relationships (Equations (7.1)–(7.5)) has proven difficult in practice. This was highlighted by Rhoads's (1987) review of previous datasets (datasets included Richards, 1980; Gippel, 1985a; Rhoads, 1986; Roy and Woldenberg, 1986). Rhoads (1987) found that the expected relationships occurred only at relatively symmetrical junctions, where bankfull symmetry ratio is greater than or equal to 0.7 ($Q_2/Q_1 \geq 0.7$ where Q represents the bankfull or channel-forming flow of the small tributary (2) and larger main stem (1)). Recent research (Zavadil, 2009) supports this notion of a symmetry ratio threshold for determining where predictable step-changes in channel morphology occur at confluence zones, and further proposes that a bankfull symmetry ratio as low as 0.5 may be associated with significant step-changes in channel morphology at confluences. This 0.5 bankfull symmetry ratio (ratio of bankfull channel areas)

Table 7.7 Revised geomorphic scale hierarchy based on a synthesis of advances in boundary criteria.

Extent	Scale	Unit	Description and criteria
$10^3 - 10^4$ m	*Catchment scale*	*Catchment/basin*	Area lying within topographical boundaries. Ultimate drainage basin boundary, within which the channel network forms.
		Zone	Upland – Midland – Lowland as determined by relative shifts in topography within a catchment. May be synonymous with erosion, transfer and deposition zones.
		Sub-catchment	Area lying within local topographical boundaries. Smaller drainage area within the broader channel network.
$10^2 - 10^3$ m	*Segment scale*	*Segment*	Confluences associated with a significant step-change in bankfull channel morphology (Zavadil, 2009). Stream length within a sub-catchment, bound by obvious geologic features (e.g. geologic intrusions) or shifts in valley confinement or river style (e.g. Brierley and Fryirs, 2000) AND/OR confluences of geomorphic significance as defined by bankfull symmetry ratio ≥ 0.5 or sub-catchment symmetry ratio > 0.2 AND evidence of morphological features (bars, etc.; Benda *et al.*, 2004a, 2004b).
		Sub-segment	Confluences not associated with significant geomorphic adjustment to the channel, yet still constituting a hydrological input to the in-stream environment (Zavadil, 2009). Stream length within a segment, bound by smaller confluences of bankfull symmetry ratio < 0.5 or sub-catchment symmetry ratio < 0.2 AND absence of morphological features (bars, etc.; Benda *et al.*, 2004a, 2004b).
$10^1 - 10^2$ m	*Reach scale*	*Reach*	Length of stream within a segment, comprising more than one meso-scale unit. Representative reach should encompass more than two pool/riffle sequences, and/or one meander wavelength. Avoid encompassing any local discontinuities or disturbances to the channel. Note: channel form is highly variable over the reach scale. Identification of representative reaches likely to be more problematic in lowland environments and will likely require detailed site inspections (Zavadil, 2009).
$10^0 - 10^1$ m	*Meso scale*	*Morphological unit*	A unit encompassing a series of the same geotopes. Basic units of pool, riffle, glide. An extension of geotopes whereby several pool geotopes (several cross-sections dominated by no perceptible flow) constitute a pool morphological unit.
		Geotope (cross-section)	Cross-section of the stream. Geotope is defined by dominance of a statistically significant flow type (significant by Froude number) along the cross-section (Zavadil, 2009). Basic units of pool (no perceptible flow), riffle (rippled flow or unbroken standing waves), glide (smooth boundary turbulent flow).
$10^{-1} - 10^0$ m		*Biotopes (cell)*	Multiple points of the same flow type within the channel denote a biotope cell. Provides a framework for delineating hydraulic variability and for assessing depth diversity across a given reach. The range of biotopes and associated flow types as specified by Padmore (1998) or similar.
$10^{-2} - 10^{-1}$ m	*Micro scale*	*Point/patch*	Individual points of flow and bed or bank interaction. For example, flow–substrate interaction at a given point.

was also found to be associated with sub-catchment symmetry ratios (ratio of sub-catchment areas) as low as 0.2 (Zavadil, 2009).

Geomorphic significance

In combining outcomes from qualitative and quantitative research to date on confluence zone dynamics, Zavadil (2009) proposed a distinction between confluences of likely geomorphic significance and those of hydrologic significance (Table 7.6). Confluences of geomorphic significance are likely to be associated with bankfull symmetry ratios >0.5 (or estimated by sub-catchment symmetry ratios >0.2) and may be used in conjunction with observations of morphologic features (Benda *et al.* 2004a; 2004b) and geologic and topographic changes in the landscape such as channel confinement, waterfalls and river style (Brierley and Fryirs, 2000) as criteria for defining segment scale units. Other, more asymmetrical confluences which may be considered to have more of a hydrological input only (no significant step-change in channel form) are used to denote 'sub-segments'.

7.5 Advancing the scale hierarchy: future research priorities

As highlighted throughout this chapter, advances in geomorphic and hydrologic understanding have contributed to the definition of more meaningful spatial scale units for ecohydraulics and related fields, particularly at the meso and segment scales. A synthesis of practical scale descriptions and associated geomorphic boundary criteria is presented in Table 7.7.

There are a range of future research opportunities to further advance the geomorphic and hydrologic rigour underpinning spatial scales for river research and management, including testing and refinement of scale unit boundary criteria at all scales. At the meso scale, an improved understanding of the temporal sensitivity of geotopes, biotopes and flow types to varying stage heights would assist in unit definition and also the pursuit of important flow–ecology links. At the segment scale, refinement of the proposed symmetry ratio threshold for geomorphically significant confluences is required before bankfull or sub-catchment symmetry ratios can confidently be used more widely for defining geomorphically significant confluences as segment unit boundaries. Improved understanding of reach scale variability and the definition of representative reaches would also be beneficial for all research and management areas that frequently employ this scale for investigations.

More generally, as outlined early on in this chapter, scale hierarchies have been widely adopted across the fields of ecology, biology, geo sciences and other environmental sciences and management fields, yet with little critical review of the conceptual premise behind them. More critical reviews of the concept of how a scale hierarchy is useful for understanding stream system organization, and associated implications for ecosystem structure, should be pursued. Given the widespread applications of these largely untested scale frameworks, it is possible that hierarchy models of discrete spatial scales may actually be limiting our view of stream systems to this conceptual format. Furthermore, geomorphic scale hierarchies have only been applied in the context of dendritic stream systems, and in limited landscape settings and climatic zones. Further research could consider the applicability of these scale frameworks and scale unit boundary criteria to other types of drainage systems (e.g. distributary, radial, parallel) and a diversity of landscapes. Efforts to refine and test geomorphic scale hierarchies and scale unit boundary criteria will continue to complement ongoing research to test the ecological significance of various scales (e.g. Newson *et al.*, 1998; Harper *et al.*, 2000), and will contribute to a more practical and meaningful scale framework for ecohydraulics and other fields.

References

Angermeier, P.L. and Schlosser, I.J. (1989) Species-area relationships for stream fishes. *Ecology,* **70**(5): 1450–1462.

Beisel, J., Ussegliao-Polatera, P., Thomas, S. and Moreteau, J. (1998) Stream community structure in relation to spatial variation: the influence of mesohabitat characteristics. *Hydrobiologia,* **389**: 73–88.

Benda, L., Andras, K., Miller, D. and Bigelow, P. (2004a) Confluence effects in rivers: Interactions of basin scale, network geometry, and disturbance regimes. *Water Resources Research,* **40**: W05402. doi: 10.1029/2003WR002583).

Benda, L., Poff, L., Miller, D., Dunne, T., Reeves, G., Pess, G. and Pollock, M. (2004b) The network dynamics hypothesis: How channel networks structure riverine habitats. *Bioscience,* **54**(5): 413–427.

Brierley, G.J. and Fryirs, K. (2000) River styles, a geomorphic approach to catchment characterization: Implications for river rehabilitation in Bega catchment, New South Wales, Australia. *Environmental Management,* **25**(6): 661–679.

Chee, Y., Webb, A., Stewardson, M. and Cottingham, P. (2006) *Victorian environmental flows monitoring and assessment program: Monitoring and evaluation of environmental flow releases in the Thomson River.* Report prepared for the West Gippsland Catchment Management Authority and the Department of Sustainability and Environment. e-Water Cooperative Research Centre, Melbourne.

Clifford, N.J., Harmar, O.P., Harvey, G. and Petts, G. (2006) Physical habitat, eco-hydraulics and river design: a review and re-evaluation of some popular concepts and methods. *Aquatic Conservation : Marine and Freshwater Ecosystems,* **16**(4): 389–408.

Cohen, P., Andriamahefa, H. and Wasson, J. (1998) Towards a regionalization of aquatic habitat: distribution of mesohabitats at the scale of a large basin. *Regulated Rivers: Research and Management,* **14**: 391–404.

Dyer, F. and Thoms, M. (2006) Managing river flows for hydraulic diversity: an example of an upland regulated gravel-bed river. *River Research and Applications,* **22**: 257–267.

Fisher, S.G., Heffernan, J.B., Sponseller, R.A. and Welter, J.R. (2007) Functional ecomorphology: Feedbacks between form and function in fluvial landscape ecosystems. *Geomorphology,* **89**(1–2): 84–96.

Flory, E. and Milner, A.M. (1999) The role of competition in invertebrate community development in a recently formed stream in Glacier Bay National Park, Alaska. *Aquatic Ecology,* **33**(2): 175–184.

Fox, P.J.A., Naura, M. and Scarlett, P. (1998) An account of the derivation and testing of a standard field method, River Habitat Survey. *Aquatic Conservation: Marine and Freshwater Ecosystems,* **8**: 455–475.

Frissell, C.A., Liss, W.J., Warren, C.E. and Hurley, M.D. (1986) A hierarchical framework for stream habitat classification: Viewing streams in a watershed context. *Environmental Management,* **10**(2): 199–214.

Gillette, D.P., Tiemann, J.S.E., Edds, D.R. and Wildhaber, M.L. (2006) Habitat use by a Midwestern U.S.A riverine fish assemblage: effects of season, water temperature and river discharge. *Journal of Fish Biology,* **68**: 1494–1512.

Gippel, C.J. (1985a) Changes in stream channel morphology at tributary junctions, Lower Hunter Valley, New South Wales. *Australian Geographical Studies,* **23**: 291–307.

Gippel, C.J. (1985b) *Limitations of the downstream hydraulic geometry approach to the study of stream channel morphology.* N.S.W Geography Students Conference, University of Newcastle.

Godfrey, A.E. (1977) A physiographic approach to land use planning. *Environmental Geology,* **2**: 43–50.

Habersack, H.M. (2000) The river-scaling concept (RSC): a basis for ecological assessments. *Hydrobiologia,* **422/423**: 49–60.

Harper, D.M., Kemp, J.L., Vogel, B. and Newson, M.D. (2000) Towards the assessment of 'ecological integrity' in running waters of the United Kingdom. *Hydrobiologia,* **422/423**: 133–142.

Hawkins, C.P., Hogue, J.N., Decker, L.M. and Feminella, J.W. (1997) Channel morphology, water temperature, and assemblage structure of stream insects. *Journal of the North American Benthological Society,* **16**(4): 728–749.

Horton, R.E. (1945) Erosional development of streams and their drainage basins: hydrophysical approach to quantitative morphology. *Bulletin of the Geological Society of America,* **56**: 275–370.

Hudson, H.R. (2002) Linking the physical form and processes of rivers with ecological response. In Dyer, F., Thoms, M.C. and Olley, J.M. (eds) *The Structure, Function and Management Implications of Fluvial Sedimentary Systems,* IAHS, Alice Springs, Australia, pp. 121–139.

Jowett, I.G. (1993) A method for objectively identifying pool, run, and riffle habitats from physical measurements. *New Zealand Journal of Marine and Freshwater Research,* **27**: 241–248.

Jowett, I.G. (2003) Hydraulic constraints on habitat suitability for benthic invertebrates in gravel-bed rivers. *River Research and Applications,* **19**: 495–507.

Knighton, D. (1998) *Fluvial Forms and Processes: A New Perspective,* Arnold, London.

Lauer, T.E. and Spacie, A. (2004) Space as a limiting resource in freshwater systems: competition between zebra mussels (*Dreissena polymorpha*) and freshwater sponges (*Porifera*). *Hydrobiologia,* **517**(1–3): 137–145.

Leopold, L.B. and Maddock, T.M. (1953) *The Hydraulic Geometry of Stream Channels and Some Physiographic Implications.* Geological Survey Professional Paper **252**: 118–127.

Leopold, L.B., Wolman, G.M. and Miller, J.P. (1964) *Fluvial Processes in Geomorphology,* W.H. Freeman and Company, San Francisco.

Maddock, I. (1999) The importance of physical habitat assessment for evaluating river health. *Freshwater Biology,* **41**: 373–391.

Maddock, I. and Bird, D. (1996) The application of habitat mapping to identify representative PHABSIM sites on the River Tavy, Devon, UK. *Proceedings of the 2nd International Symposium on Habitat Hydraulics,* **Vol. B**: 149–162.

Maddock, I., Thoms, M., Jonson, K., Dyer, F. and Lintermans, M. (2004) Identifying the influence of channel morphology on physical habitat availability for native fish: application to the two-spined blackfish (*Gadopsis bispinosus*) in the Cotter River, Australia. *Marine and Freshwater Research,* **55**: 173–184.

Milan, D.J., Heritage, G.L., Large, A.R.G. and Entwistle, N.S. (2010) Mapping hydraulic biotopes using terrestrial laser scan data of water surface properties. *Earth Surface Processes and Landforms,* **35**(8): 918–931.

Miller, J.B. (1958) *High Mountain Streams: Effects of Geology on Channel Characteristics and Bed Material.* New Mexico State Bureau of Mines and Mineral Resources Memoir 4.

Montgomery, D.R. and Buffington, J.M. (1997) Channel-reach morphology in mountain drainage basins. *Geological Society of America Bulletin,* **109**: 596–611.

Muhar, S. (1996) Habitat improvement of Austrian rivers with regard to different scales. *Regulated Rivers: Research and Management,* **12**: 471–482.

Myers, T.J. and Swanson, S. (1997) Precision of channel width and pool area measurements. *Journal of the American Water Resources Association*, **33**(3): 647–659.

Newson, M.D. and Newson, C.L. (2000) Geomorphology, ecology and river channel habitat: mesoscale approaches to basin-scale challenges. *Progress in Physical Geography*, **24**(2): 195–217.

Newson, M.D., Harper, D.M., Padmore, C.L., Kemp, J.L. and Vogel, B. (1998) A cost-effective approach for linking habitats, flow types and species requirements. *Aquatic Conservation: Marine and Freshwater Ecosystems*, **8**: 431–446.

Padmore, C.L. (1997) Biotopes and their hydraulics: A method for defining the physical component of freshwater quality. In Boon, P.J. and Howell, D.L. (eds) *Freshwater Quality: Defining the Indefinable?* HMSO, London, pp. 251–257.

Padmore, C.L. (1998) The role of physical biotopes in determining the conservation status and flow requirements of British rivers. *Aquatic Ecosystem Health and Management*, **1**: 25–35.

Padmore, C.L., Newson, M.D. and Charlton, M.E. (1996) Instream habitat: Geomorphological guidance for habitat identification and characterisation. In Rowntree, K.M. (ed.) *The Hydraulics of Physical Biotopes – Terminology, Inventory and Calibration*. Report of a workshop held at Citrusdal 4–7 February 1995. WRC Report KV84/96, Citrusdal, pp. 27–41.

Parsons, M., Thoms, M. and Norris, R. (2002) *Australian River Assessment System: AusRivAS Physical Assessment Protocol*. Monitoring River Health Initiative Technical Report no 22, University of Canberra, Australia.

Post, D.M., Doyle, M.W., Sabo, J.L. and Finlay, J.C. (2007) The problem of boundaries in defining ecosystems: A potential landmine for uniting geomorphology and ecology. *Geomorphology*, **89**(1–2): 111–126.

Renschler, C.S., Doyle, M.W. and Thoms, M. (2007) Geomorphology and ecosystems: Challenges and keys for success in bridging disciplines. *Geomorphology*, **89**(1–2): 1–8.

Rhoads, B.L. (1986) Process and Response in Desert Mountain Fluvial Systems. Unpublished PhD dissertation, Arizona State University, Tempe.

Rhoads, B.L. (1987) Changes in stream channel characteristics at tributary junctions. *Physical Geography*, **8**(4): 346–361.

Rice, S. (1998) Which tributaries disrupt downstream fining along gravel-bed rivers? *Geomorphology*, **22**: 39–56.

Rice, S. (1999) The nature and controls on downstream fining within sedimentary links. *Journal of Sedimentary Research*, **69**(1): 32–39.

Rice, S. and Church, M. (1998) Grain size along two gravel-bed rivers: Statistical variation, spatial pattern and sedimentary links. *Earth Surface Processes and Landforms*, **23**: 345–363.

Richards, K.S. (1980) A note on changes in channel geometry at tributary junctions. *Water Resources Research*, **16**(1): 241–244.

Richards, K.S. (1982) *Rivers: Form and Process in Alluvial Channels*, Methuen and Co., London.

Roussel, J. and Bardonnet, A. (1996) Spatiotemporal use of the riffle/pool unit by fish in a Brittany brook. In Leclerc, M.,

Capra, H., Valentin, S., Boudreault, A. and Cote, I. (eds) *Proceedings of the 2nd International Symposium on Habitat Hydraulics*, INRS, Quebec City, Canada, pp. 341–352.

Rowntree, K.M. (1996) *The Hydraulics of Physical Biotopes – Terminology, Inventory and Calibration*. WRC Report KV84/96 SBN 1 86845 229 8, Citrusdal.

Roy, A.G. and Roy, R. (1988) Changes in channel size at river confluences with coarse bed material. *Earth Surface Processes and Landforms*, **13**: 77–84.

Roy, A.G. and Woldenberg, M.J. (1986) A model for changes in channel form at a river confluence. *Journal of Geology*, **94**: 402–411.

Shreve, R.L. (1967) Infinite topologically random channel networks. *Journal of Geology*, **75**: 178–186.

Simonson, T.D., Lyons, J. and Kanehl, P.D. (1994) Quantifying fish habitat in streams: Transect spacing, sample size, and a proposed framework. *North American Journal of Fisheries Management*, **14**: 607–615.

Statzner, B. and Hilger, B. (1986) Stream hydraulics as a major determinant of benthic zonation patterns. *Freshwater Biology*, **16**: 127–139.

Strahler, A.N. (1952) Hyposometric (area-altitude) analysis of erosional topography. *Bulletin of the Geological Society of America*, **63**: 1117–1142.

Summerfield, M. (1991) *Global Geomorphology*, Longman/Wiley, London/New York.

Taylor, M.P., Thomson, J.R., Fryirs, K. and Brierley, G.J. (2000) Habitat assessment using the River Styles methodology. *Ecological Management and Restoration*, **1**(3): 223–226.

Thoms, M.C. and Parsons, M. (2002) Eco-geomorphology: an interdisciplinary approach to river science. In Dyer, F., Thoms, M.C. and Olley, J.M. (eds) *The Structure, Function and Management Implications of Fluvial Sedimentary Systems*, IAHS, Alice Springs, Australia, pp. 113–119.

Thomson, J.R., Taylor, M.P., Fryirs, K.A. and Brierley, G.J. (2001) A geomorphological framework for river characterization and habitat assessment. *Aquatic Conservation: Marine and Freshwater Ecosystems*, **11**: 373–389.

Thorp, J.H. (2009) Models of ecological processes in riverine ecosystems. *Encyclopaedia of Inland Waters*, **1**: 448–455.

Thorp, J.H., Thoms, M.C. and Delong, M.D. (2006) The riverine ecosystem synthesis: Biocomplexity in river networks across space and time. *River Research and Applications*, **22**: 123–147.

Vannote, R.L., Minshall, G.W., Cummins, K.W., Sedell, J.R. and Cushing, C.E. (1980) The River Continuum Concept. *Canadian Journal of Fisheries and Aquatic Sciences*, **37**: 130–137.

Wadeson, R.A. (1995) *The Development of the Hydraulic Biotope Concept Within a Catchment Based Hierarchical Geomorphological Model*. PhD thesis, Rhodes University, Grahamstown, South Africa.

Wadeson, R.A. (1996) The biotope concept: A geomorphological perspective. In Rowntree, K.M. (ed.) *The Hydraulics of Physical Biotopes – Terminology, Inventory and Calibration*, Citrusdal.

Wadeson, R.A. and Rowntree, K.M. (1999) Application of the hydraulic biotope concept to the classification of instream habitats. *Aquatic Ecosystem Health and Management*, **1**(2): 143–157.

Wadeson, R.A. and Rowntree, K.M. (2001) The application of a hydraulic biotope matrix to the assessment of available habitat: Potential application to IFRs and river health monitoring. *African Journal of Aquatic Science*, **26**: 67–73.

Wadeson, R.A. and Rowntree, K.M. (2002) Mapping hydraulic biotopes for ecological flow assessments. *Proceedings of the Environmental Flows for River Systems incorporating the Fourth International Ecohydraulics Symposium: An international working conference on assessment and implementation*, Cape Town, South Africa, 3–8 March, 2002.

Wallis, C., Maddock, I., Visser, F. and Acreman, M. (2010) A framework for evaluating the spatial configuration and temporal dynamics of hydraulic patches. *River Research and Applications*, in press (doi: 10.1002/rra.1468).

Ward, J.V. (1985) Thermal characteristics of running waters. *Hydrobiologia*, **125**: 31–46.

Zavadil, E. (2009) *On the Geomorphic Criteria Underpinning Spatial Scale Hierarchies in River Research and Management*. PhD thesis, The University of Melbourne, Melbourne.

Zavadil, E., Stewardson, M., Turner, M. and Ladson, A. (2011) An evaluation of surface flow types as a rapid measure of channel morphology for the geomorphic component of river condition assessments. *Geomorphology*, in press (doi: 10.1016/j.geomorph.2011.10.034).

8 Developing Realistic Fish Passage Criteria: An Ecohydraulics Approach

Andrew S. Vowles, Lynda R. Eakins, Adam T. Piper, James R. Kerr and Paul Kemp

International Centre for Ecohydraulics Research, University of Southampton, Highfield, Southampton, SO17 1BJ, UK

8.1 Introduction

Globally, anthropogenic structures placed within streams and rivers have fragmented the fluvial continuum, causing population declines and in some instances extirpation of aquatic organisms unable to effectively disperse between habitats (Odeh, 1999; Lucas and Baras, 2001; Agostinho *et al.*, 2005). While wider negative environmental effects of disrupting lotic environments with anthropogenic structures are understood (Ward and Stanford, 1983), mitigating the impacts of river fragmentation on the movements of fish has been a central focus of rehabilitation efforts, due to their economic, ecological and cultural significance. River fragmentation is frequently associated with large hydroelectric dams (Odeh, 1999; Larinier, 2000; Figure 8.1), yet even low-head structures, such as weirs and culverts, can constitute a major obstruction to the movements of fish (Ovidio and Philippart, 2002; O'Hanley and Tomberlin, 2005). Therefore, management of anthropogenic development of river environments must provide efficient fish passage across all scales of impoundment (Odeh, 1999).

Engineered fish passes (or fishways) have been employed since at least the 17th century and are commonly used to restore habitat connectivity for fishes (Clay, 1995). Historically, fish passage research focused on facilitating the upstream movements of migratory fish, such as adult salmonids (Calles and Greenberg, 2005;

Williams *et al.*, 2012), for which determining physiological swimming capabilities was essential for establishing water velocity criteria (Stringham, 1924). Non-salmonid species of low commercial importance and downstream migrating life stages received less attention (Roscoe and Hinch, 2010; Williams *et al.*, 2012). Accordingly, anthropogenic barriers in conjunction with fish passage facilities designed primarily for salmonid movements can prevent or limit many non-salmonid migrations, necessitating research on multi-species and multi-life stage swimming performance (Moser *et al.*, 2002; Cheong *et al.*, 2006; Mallen-Cooper and Brand, 2007). Furthermore, many fish passes still have low efficiency for the target species, reflecting a need to understand not only fish swimming performance but also behaviour in response to associated hydraulic and other environmental factors (Laine *et al.*, 1998; Lucas *et al.*, 1999; Bunt *et al.*, 2012).

Linking biological and physical features of aquatic systems necessitates an interdisciplinary research approach (Lancaster and Downes, 2010). In modern river management, the importance of this is increasingly acknowledged with 'ecohydraulics', a sub-discipline of ecohydrology, gaining popularity (Wood *et al.*, 2008; Rice *et al.*, 2010; Towler *et al.*, 2012). Ecohydraulics combines expertise from ecology and biology with hydraulic engineering to answer mutually important questions, using accepted methodologies to advance river management (Nestler *et al.*, 2008; Rice *et al.*, 2010). Adopting an ecohydraulics

Ecohydraulics: An Integrated Approach, First Edition. Edited by Ian Maddock, Atle Harby, Paul Kemp and Paul Wood.
© 2013 John Wiley & Sons, Ltd. Published 2013 by John Wiley & Sons, Ltd.

(a)   (b)

Figure 8.1 (a) Lower Monumental Dam on the Snake River, a major tributary of the Columbia River (USA), where decades of fish passage research has been conducted as hydropower dams block historic salmon spawning sites (photograph: Andrew Vowles); (b) a dam under construction in China, the country with the greatest number of large dams. The development of fish passage criteria for Chinese species is in its infancy (photograph: Lynda Eakins).

approach to fish passage research would advance the methods used to define suitable design criteria, and could be used to identify, quantify and understand responses of fish to the hydraulic environment at relevant scales (Roy *et al.*, 2010). Through ecohydraulics, the development of fish passes can be approached in a more holistic and inter-disciplinary way.

This chapter will critically review the traditional methods of obtaining criteria used in fish pass design. The ecohydraulic approach will be described in the context of fish passage, the benefits will be discussed and future challenges identified.

8.2 Developing fish passage criteria

8.2.1 The traditional approach

Although some of the early studies on locomotion and swimming ability were prompted by the need to facilitate free passage of fish at anthropogenic barriers (Stringham, 1924), much of the information on which current fish pass design is based was obtained during studies intended for other purposes (Webb, 1975; Beamish, 1978). The traditional approach to developing fish passage criteria has been biased towards forced swimming experiments conducted under controlled and uniform hydraulic conditions. These techniques allowed classification of swimming speed and endurance into: (1) sustained, (2) pro-

longed and (3) burst swimming (Brett, 1964; Beamish, 1978).

Sustained swimming speed describes that which can be maintained for hours without fatigue and is fuelled only by aerobic, red myotomal muscles (Brett *et al.*, 1958). For convenience, during laboratory tests, sustained swimming is commonly defined as the speed which can be maintained for over 200 minutes (Beamish, 1978; Hammer, 1995). Burst swimming, used for low-duration, highly energetic activities such as passing short velocity barriers, is fuelled by the anaerobic metabolism of white myotomal muscles and has commonly been considered to be that maintained for less than an arbitrary value of 20 seconds before fatigue (Beamish, 1978). Within this category, the exceptionally high velocity reached when swimming for only one or two seconds, important in ambush prey capture and predator escape, is termed the fast start performance or sprint swimming speed of the fish (Domenici and Blake, 1997; Nelson *et al.*, 2002). Prolonged swimming results in fatigue between 20 seconds and 200 minutes and is fuelled mainly by red muscles, with white muscles contributing at higher speeds (Beamish, 1978).

The 'fixed velocity' or 'endurance' test defines the range of the three aforementioned categories of swimming. Fish are swum at a fixed velocity until fatigue. By repeating trials over a range of velocities, those that can be maintained for at least 200 minutes and those that result in fatigue within this time can be calculated and endurance curves produced (Brett, 1964, 1967; Beamish, 1966; and, more recently, Peake *et al.*, 1997; Langdon and Collins, 2000). The relationship between swimming speed and endurance can be used to estimate the maximum water velocity through culverts or fish passes of different lengths that do not result in fatigue (Peake *et al.*, 1997). Maximum swimming distance is calculated from the fish's ground speed (the difference between swimming velocity (U) and water velocity (V)) and endurance time (t) at the swimming velocity (Bainbridge, 1960):

$$D = (U - V)\, t \qquad (8.1)$$

According to this model, distance of ascent is maximised by swimming at a constant optimum speed, deviations from which reduce performance (Castro-Santos, 2005; 2006).

Within the prolonged performance category is the 'critical swimming speed' (U_{crit}), the most common measure of fish swimming capability (Brett, 1964). Fish are swum at incremental velocity levels for predetermined time intervals (traditionally 30–60 minutes; Hammer, 1995) until fatigue (normally impingement on a downstream screen which may be electrified to encourage fish to swim to exhaustion). U_{crit} is calculated as:

$$U_{crit} = U_1 + \left[\Delta U \left(T_1 / \Delta T \right) \right] \qquad (8.2)$$

where U_1 is the highest velocity maintained for the entire time interval (cm s^{-1}), ΔU is the velocity increment (cm s^{-1}), T_1 is the time elapsed at the velocity of fatigue (min) and ΔT is the time interval used (min). The U_{crit} is an easily comparable measure of performance between species or under varying environmental conditions (for a review, see Hammer, 1995) and provides velocity values that fish can swim against for a period of time equivalent to ΔT. U_{crit} values have been commonly used to define suitable water velocities in culverts and fish passes of various lengths (e.g. Jones *et al.*, 1974; Santos *et al.*, 2007).

In Japan, a similar method, the swimming ability index (SAI), was developed as a method of comparing performance between species and sizes of fish (Tsukamoto *et al.*, 1975). A swimming curve derived from endurance times at different velocities was created and swimming ability assumed proportional to the area below the curve (see Tsukamoto *et al.*, 1975). The SAI allows a comparison of overall swimming ability, taking into consideration both sustained and burst speeds and providing an effective method of comparing overall swimming performance. However, information gained from the SAI simply provides an overview of performance, rather than the necessary detail on ability to pass through known flow conditions, and as such is unlikely to have a direct application

to fish pass design. The authors are unaware of the use of this index outside of Asia.

Early research into fish swimming performance used apparatus such as the fish wheel (see Bainbridge and Brown, 1958) and swim chamber (see Blazka *et al.*, 1960; Beamish, 1978) to quantify swimming speeds and the effects of body length and water temperature. Advances in respirometry allowed oxygen consumption and therefore metabolic rates during swimming to be calculated (Brett, 1965), while studies of muscle kinematics led to the formalisation of the relationship between tail beat frequency, tail beat amplitude, stride length and swimming speed (Bainbridge, 1958; Wardle, 1975). This innovative research and the ensuing equations of Zhou (1982) relating maximum swimming speed and endurance, dependent on limited energy stores, formed the basis of Beach's (1984) swimming endurance curves, which have subsequently been reproduced widely in fish passage manuals (e.g. Bell, 1991; Clay, 1995). More recently, research has cautioned against the application of these data due to the unrealistic or flawed assumptions on which these swimming endurance curves were based (see Section 8.2.2).

Concurrent with forced swimming studies, a more direct approach to obtaining fish pass design criteria and estimating the maximum distance fish could ascend against prescribed water velocities was achieved using large open channels (Castro-Santos and Haro, 2006). Prototype fish passes constructed at a research facility at Bonneville Dam on the Columbia River (USA) enabled the determination of swimming performance of actively migrating fish (Collins and Elling, 1960), including Pacific salmonids (*Oncorhynchus* spp.) and American shad (*Alosa sapidissima*) (Weaver, 1963; 1965). In addition to ability to pass velocity barriers of various magnitudes, impacts of changes in fish pass hydraulics (e.g. streaming or plunging flow) on fish movements were tested (Collins and Elling, 1960). Although this research has received little recognition in the published literature (Castro-Santos and Haro, 2006), it has inspired recent work using open channel flumes, enabling more realistic measures of swimming performance, behaviour and hydraulic conditions to be quantified (see Section 8.2.2).

Downstream fish migrations are frequently assumed to be largely passive (Thorpe and Morgan, 1978), and while in-river barriers can physically block upstream movements, impacts on downstream migrants are less frequently considered (Kemp and Williams, 2009). Downstream migrating fish face potential injury and mortality during passage over high dams, through hydroelectric turbines (Čada, 2001) or other water off-takes (e.g. pumping

Figure 8.2 A 30m-long, near-vertical bar screen (12.5 mm bar spacing) installed diagonally across a mill channel leading to a small-scale (0.34 megawatt) hydropower facility on the River Tay, Scotland, UK. A fish bypass system is located at the end of the screen (photograph: Paul Kemp).

stations; Piper *et al.*, 2012), as well as increased predation risk, energetic cost, risk of disease and propensity to pass through suboptimal routes due to delay at anthropogenic barriers to migration (Poe *et al.*, 1991; Venditti *et al.*, 2000; Castro-Santos and Haro, 2003; Garcia de Leaniz, 2008). To mitigate for such impacts it is essential to understand the swimming capabilities and behaviours of downstream migrating life stages.

To divert fish away from hazardous locations (e.g. hydropower turbines) and towards bypasses which provide an alternative route that reduces risk of injury and delay, a range of physical and behavioural barriers have been developed (Clay, 1995; Taft, 2000). Physical screens (Figure 8.2) are the most common method of diverting fish and, with the correct bar spacing and when angled across the flow, can efficiently guide some life stages towards more benign routes (Larinier and Travade, 1999). However, if the velocity upstream and running perpendicular to the screen face (referred to as the escape velocity, Figure 8.3) is excessive relative to swimming capability, then fish can become damaged by mechanical abrasion on contact, and potentially become impinged, often resulting in high rates of mortality due to suffocation (Hadderingh and Jager, 2002; Calles *et al.*, 2010). The escape velocity (measured 10 cm upstream of the screen) has traditionally been based on laboratory measurements of fish swimming ability, e.g. lower than the upper limit of sustained swimming speeds (Larinier and Travade, 2002; Turnpenny and O'Keeffe, 2005). The escape velocity should remain

low even when axial velocities (Figure 8.3) are high and should be lower than the sweeping velocity (Figure 8.3) along the screen face, which helps deflect fish towards bypass channels (Turnpenny and O'Keeffe, 2005). In the UK, fish screen guidelines recommend escape velocities not in excess of 25, 30, 50 and 60 cm s^{-1} for coarse fish (in a UK context considered to be cyprinids, percids and

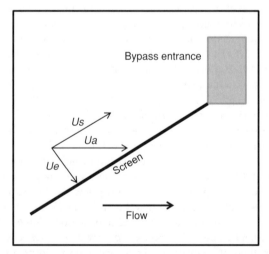

Figure 8.3 Flow velocity components in front of an angled bar screen. *Ua* is the axial velocity, *Ue* is the escape velocity and *Us* is the sweep velocity along the screen face. Adapted from Turnpenny & O'Keeffe, 2005. Contains Environment Agency information © Environment Agency and database right.

pike, *Esox lucius*) and shad, lamprey (*Lampetra* sp. and *Petromyzon marinus*), eels, and salmonids, respectively (Environment Agency, 2009). Higher sweeping velocities, in some instances, could be used to reduce exposure time of fish to the screen face (Swanson *et al.*, 2004). For larval life stages, screening can be problematic due to their poor swimming ability and small size, often resulting in easy entrainment through bar spaces (Carter and Reader, 2000).

As an alternative to (or in combination with) physical screens, a range of behavioural barriers, including light, sound, electricity and hydrodynamic (louvre) devices, have been developed to deter downstream moving fish from intakes (Taft, 2000). The effectiveness of a behavioural barrier relies on the fish detecting and avoiding an artificial stimulus or suite of stimuli and as a consequence being repelled from hazardous areas (Čada and Szluha, 1978). Behavioural barriers have had varying degrees of success, with physical screens generally more effective, largely because fish response to environmental stimuli can be variable both between and within species (Popper and Carlson, 1998). Importantly, site hydraulics can impact the fish's ability to detect and respond to environmental stimuli. If velocities are excessive, then fish may not be able to exhibit a response, even if the stimulus is detected.

8.2.2 Criticisms of the traditional approach

Measures of fish swimming ability obtained from forced swimming experiments have received criticism for their universal application to multispecies fish passage design (Castro-Santos and Haro, 2006; Kemp *et al.*, 2011). In part this is because fish passage design is considered to be biased towards upstream movements of anadromous salmonids, but also because swim chambers generate unnatural hydraulic conditions (Enders *et al.*, 2003) and prevent fish from expressing natural, performance-enhancing behaviours (Peake and Farrell, 2006; Russon and Kemp, 2011).

Unnatural hydraulic conditions

The uniform flow profiles generated within swim chambers are rare in nature, where flows are inherently complex (Lacey *et al.*, 2012). By conducting swim chamber respirometry experiments under different turbulent flow conditions, Enders *et al.* (2003) demonstrated a positive relationship between turbulence and cost of swimming. Reduced swimming efficiency in turbulent environments could stem from the movements required to maintain stability (e.g. changes in speed and direction) being more energetically costly than under the steady swimming (maintenance of speed and direction) required during forced swimming experiments (Enders *et al.*, 2003).

Turbulence scale, vorticity (angular velocity) and eddy orientation can also negatively impact swimming performance and stability (e.g. Tritico and Cotel, 2010). An eddy diameter greater than 2/3 the body length of perch (*Perca fluviatilis*) reduces critical swimming speeds (Lupandin, 2005). In a similar study, Tritico and Cotel (2010) exposed creek chub (*Semotilus atromaculatus*) to a distribution of eddy sizes, vorticities and orientations to assess their influence on stability and swimming speed. Only the largest eddies, of similar diameter to fish length, cause stability challenges, termed 'spills'. These are characterised by spins in an orientation consistent with the rotation of eddies and subsequent loss of ability to maintain station against the flow (Tritico and Cotel, 2010). Eddy diameter alone is not the causative factor, as, during low-velocity, large-eddy-diameter treatments, spills are not observed; stability challenges are associated with higher vorticity. Larger eddies also have the greatest impact on swimming performance when rotating around a horizontal, rather than vertical axis, as fish have insufficient flexibility in this plane to counter the rotational forces experienced (Tritico and Cotel, 2010). Within the fluvial environment, turbulence has the potential to influence habitat selection, station holding and migratory abilities of fishes (Smith *et al.*, 2005; Cotel *et al.*, 2006). Yet, for fish passes, turbulence is necessary to dissipate energy so that water velocity is reduced to suitable levels relative to fish swimming capabilities, creating contradictory pressures for those designing such structures. Recent research illustrates there may be a trade-off between energy dissipation using turbulence and the disadvantages of increased complexity of flow. As such, turbulence is a key hydraulic component that requires quantification and consideration during fish pass design, a factor overlooked during traditional swim chamber experiments.

Although the cost of swimming can increase due to turbulence, it has long been suggested that it can also be reduced when turbulence is well structured; for example, fish may exploit the predictable eddies generated by the propulsive movements of other fish (Weihs, 1973). More recently, Liao *et al.* (2003) have demonstrated that juvenile rainbow trout (*Oncorhynchus mykiss*) are able to voluntarily alter their body kinematics when exposed to predictably shed eddies generated in the wake of a structure (a vertically oriented, D-shaped cylinder). Further experiments confirm that muscle activity is reduced during this form of locomotion (termed the Kármán gait), suggesting that the energy expenditure for fish utilising these eddies is lower than in more uniform flows of comparable velocity (Liao, 2004). Vision, in addition to mechanosensory perception, is also important when considering behavioural responses to turbulence. Trout prefer to Kármán gait when visual cues are available but entrain behind cylinders under dark conditions (Liao, 2006). In natural rivers, eddy shedding is often unpredictable and variable in scale, vorticity and orientation, unlike the controlled laboratory conditions (Tritico, 2009). Despite this, the potential to apply this knowledge to fish pass design (i.e. modify turbulent characteristics in an unnatural channel located in the field to facilitate more efficient passage) presents an interesting avenue for future research.

Prohibition of performance-enhancing behaviour

Traditional swim chambers confined fish in areas of limited size as a necessity for researchers aiming to manipulate test conditions (velocity) while controlling for confounding variables (e.g. temperature, oxygen, turbulence). Such constraints hamper natural swimming behaviours, ignoring their positive influence on performance. A recent comparison of the critical swimming speeds of common carp (*Cyprinus carpio*) shows that longer swim chambers are less restrictive to burst-coast swimming and more conducive to higher performance (Tudorache *et al.*, 2007), resulting in higher swimming speeds and duration of burst-coast locomotion. This indicates that swimming in constrained space limits exhibition of performance-enhancing behaviour and tests in shorter chambers may not accurately represent maximum locomotory capability (Tudorache *et al.*, 2007). In a similar study, smallmouth bass (*Micropterus dolomieu*) exhibited lower critical swimming speeds when confined to respirometers in comparison to free-swimming fish in a long open channel (Peake, 2004). Fatigue during respirometer tests occurred at a velocity that closely matched the gait transitional speed to burst-coast swimming in the open channel, suggesting that fatigue in respirometers is a behavioural response to energetically inefficient or biomechanically difficult locomotion (Peake and Farrell, 2006). The swimming ability of migratory anguilliform fish with elongated morphologies has also been observed in large, open-channel flumes. When ascending experimental weirs, river lamprey (*Lampetra fluviatilis*) and European eel (*Anguilla anguilla*), species of conservation concern in Europe, attain burst velocities greater than $1.75–2.12 \text{ m s}^{-1}$ (Russon and Kemp,

2011). These values are higher than previously reported, further supporting the observation that greater performance is obtained for a range of species swimming in large flumes compared to swim chambers.

Swimming performance metrics used to define fish passage criteria, but which ignore the role played by behaviour, underestimate performance. Increasing velocity tests designed to establish U_{crit} values traditionally employ time intervals of 30–60 minutes (Hammer, 1995) and, although shorter intervals have been used (e.g. 2, 5 and 10 minutes; Peake and McKinley, 1998), it is often rapid bursts of a few seconds (e.g. through vertical slots, orifices or notches) that determine successful fishway passage (Gowans *et al.*, 2003; Kemp *et al.*, 2011). Performance data for fish in large open channels, motivated to burst through short-velocity barriers may, therefore, provide more realistic fish pass criteria in some instances. The use of swimming endurance models based on data obtained using swim chambers has also been questioned. When six migratory non-salmonids attempted to pass velocity barriers in an open channel flume, no biochemical or behavioural evidence (e.g. based on attempt rate) suggested that these fish volitionally swam to exhaustion (Castro-Santos, 2004; Haro *et al.*, 2004). Furthermore, the extent to which maximum distance traversed was affected by water velocity varied among species, as did effects of other covariates, such as temperature (Haro *et al.*, 2004). Interestingly, Castro-Santos (2005) observed only migratory species (anadromous clupeids) to adapt their swimming strategy (an increase in ground speed above the optimal level for prolonged swimming) to maximise the distance travelled up an open channel. Ability to successfully traverse velocity barriers, therefore, is influenced by swimming performance and behaviour acting in combination (Castro-Santos and Haro, 2006). This research suggests the use of traditional swimming endurance models to form fish passage criteria should be used with caution. Many assumptions of these models, such as fish glycogen stores (used during prolonged and burst swimming) being similar between individuals and species, fish swimming at a constant optimal speed to physiological exhaustion and negligible variation in swimming performance within and between species, are unfounded (Castro-Santos, 2006), but, as previously mentioned (Section 8.2.1), formed the basis of swimming endurance curves commonly reproduced in fish passage manuals. This provides a potential explanation for the variable efficiency of many fishways (Bunt *et al.*, 2012). For downstream migrants, it has been reported that some species contact physical screens at velocities much lower than their measured swimming capabilities, suggesting that application of traditional swimming performance data to the development of screen criteria may also be insufficient, and that behaviour plays a prominent role (Swanson et al., 2004; 2005).

8.2.3 The ecohydraulic approach

To advance fish passage design to improve efficiency for multiple species, there is a need to quantify swimming performance and behaviour under realistic hydraulic conditions for a range of locomotory guilds. To achieve this there is a need to: (1) create and quantify hydraulic conditions at biologically relevant scales and (2) quantify swimming performance under conditions where natural behaviours can be expressed, using appropriate metrics.

Quantifying hydraulic conditions

Quantifying the hydraulic environment at biologically relevant scales remains a key challenge. Historically, studies were constrained to investigating biological interactions with time-averaged velocity fluctuations, owing to the unidirectional nature of available flow-measuring devices (e.g. impeller current meters). Accordingly, a limited number of hydraulic metrics, such as longitudinal turbulence intensity (K_u), could be calculated:

$$K_u = \sigma_u / \bar{u} \qquad (8.3)$$

where σ_u is the standard deviation of the time-averaged longitudinal velocity and $\bar{u}$ is the average velocity (m s^{-1}). This oversimplified the inherent complexity of natural flows. However, with the application of more sophisticated measurement and flow visualisation techniques to fish passage research, more accurate hydraulic information is being collected (e.g. Wang et al., 2010; Silva et al., 2012). Acoustic Doppler velocimeters (ADVs) measure a small sample of the flow at discrete points. The ADV probe emits short pairs of acoustic pulses, and receivers measure the change in pitch or frequency of the returned sound, giving a three-dimensional (3D) velocity reading (Voulgaris and Trowbridge, 1998). ADVs provide excellent temporal resolution, but arrays of ADVs that collect velocity data simultaneously are required to improve spatial resolution (Tritico et al., 2007). A similar technology, the acoustic Doppler current profiler (ADCP), has the potential to map bathymetry while simultaneously measuring 3D velocities throughout the water column (Shields et al., 2003). Whilst mainly intended to provide discharge measurements for commercial purposes, ADCPs have been applied to study turbulence in open channels (Stacey et al., 1999) and map river habitat (Shields et al., 2003),

offering great potential for ecohydraulics research, particularly in complex field settings.

Three-dimensional hydraulic information is enabling the influence of more complex measures of turbulence on fish swimming performance and stability to be considered (e.g. Tritico and Cotel, 2010). Eddy length scale (L_u), a common measure of the scale of turbulent flow within rivers, is defined as the spatial extent over which a region of fluid is correlated, and is calculated as:

$$L_u \cong \bar{u}L_t \qquad (8.4)$$

where L_t, the eddy time scale (s), is the time taken for a self-correlated eddy to pass through a sample volume of water. Within laboratory settings, eddy diameter (a measure of the spatial extent of the rotating fluid) is the more frequently reported metric of turbulent scale (Lacey et al., 2012). Through the quantification of such metrics, recent research (e.g. that of Lupandin, 2005; Tritico and Cotel, 2010 discussed in Section 8.2.2) highlights the biological relevance of turbulent scale. Eddies smaller than a threshold size impart an even distribution of rotational forces along a fish's body without impairing balance. When encountering an eddy similar in size to their length, fish are unbalanced by hydrodynamic forces pushing in opposite directions. This negatively influences swimming performance (Tritico and Cotel, 2010) and therefore ability to ascend areas of high velocity, such as fish passes.

In addition to turbulent scale, the 3D nature of ADVs has allowed a multitude of additional metrics to be calculated, such as relative turbulence intensity, turbulent kinetic energy and Reynolds stresses (see Chapter 2). These measures of turbulence have been reported from ADV data in numerous recent studies on fish spatial preferences, swimming ability and passage through prototype fishways (Nikora et al., 2003; Enders et al., 2005; Smith et al., 2005; Cotel et al., 2006; Enders et al., 2009; Wang et al., 2010; Wang and Hartlieb, 2011; Silva et al., 2011, 2012).

Advances in measuring hydraulic conditions at biologically relevant scales were made through the use of ADVs. However, ADVs can perform poorly during conditions of very high turbulence and where air is entrained in the water column (MacVicar et al., 2007); in such instances, flow visualisation methods may be more appropriate. Particle image velocimetry (PIV) uses small seeding particles to visualise the hydraulic environment (Okamoto et al., 2000). Typically, multiple photographic or high-speed video methods record the particles as they pass through a laser sheet (laser PIV) that illuminates the hydraulic area of interest (Green *et al.*, 2011). Knowledge of the time duration (or period) between consecutive images, together with particle travel distance, and auto-correlation methods allows velocity and directional data to be attained (Grant, 1997; Adrian, 2005). Such information is valuable for quantifying instantaneous measures directly from the flow field, such as the circulation about an eddy (angular momentum per unit mass), Γ_e:

$$\Gamma_e = \omega_e a_e \qquad (8.5)$$

where ω_e is vorticity (twice the angular velocity) and a_e is the eddy size, i.e. the area circumscribed by the point of maximum circulation.

For seeding particles to follow the fluctuations of turbulent flow accurately, they must be small (tens of microns in diameter) and the flow field sufficiently illuminated (e.g. using lasers) to generate clear images (Adrian, 2005). The higher spatial resolution obtained without using apparatus that disturbs the flow field (Okamoto *et al.*, 2000) makes it an attractive technique for mapping fluid dynamics within controlled flume environments. However, the financial costs of using this technique can be substantially greater than those accrued using ADVs.

Links between fluid dynamics and fish swimming performance, stability and behaviour are starting to emerge through the adoption of techniques that allow accurate quantification of hydraulics. The intensity, periodicity, orientation and scale of turbulence are all considered to be biologically relevant (see Lacey *et al.*, 2012). However, it is not clear which metrics researchers should measure, as several have been proposed (e.g. turbulence intensity, Enders *et al.*, 2003; turbulent kinetic energy,

Smith *et al.*, 2006; shear stress; Silva *et al.*, 2011, 2012; eddy size, Lupandin, 2005; eddy orientation and vorticity, Tritico and Cotel, 2010). Frameworks that help shape which hydraulic parameters are quantified and integrate laboratory and field research have been developed (Lacey *et al.*, 2012), which should aid the application of an ecohydraulics approach to fish passage research.

Quantifying swimming performance and behaviour

The use of large, open-channel flumes enables the study of volitional fish movement and performance-enhancing behaviours in response to conditions relevant to fish passage. Direct observation of fine-scale behaviours using filming techniques and tracking software allows analysis and quantification of spatial distributions, trajectories and speed of movement, and interactions between fish, and for this information to be linked to empirical maps or models (e.g. Computational Fluid Dynamic simulations) of the hydraulic environment (Figure 8.4). For example, on approaching abrupt accelerations of water velocity, downstream-moving brown trout (*Salmo trutta*) exhibit avoidance responses as a threshold velocity gradient across the body is reached, and when encountering an additional light stimulus, avoidance is enhanced (Vowles and Kemp, 2012). The ecohydraulics approach provides increasingly detailed information (e.g. volitional burst swimming abilities and threshold hydraulic values) that contribute to the development of generic behavioural rules for fish passage.

Integrating techniques across disciplines via the ecohydraulics approach yields benefits to fish passage research. Until recently, those attempting to improve the efficiency

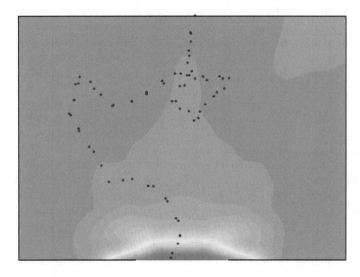

Figure 8.4 The path of a Pacific salmon smolt (*Oncorhynchus tshawytscha*) video tracked during passage through an orifice weir in a large, open-channel flume (red and blue dots represent head and tail positions, respectively); fish movements have been overlain onto the accelerating velocity profile (0–1.53 m s^{-1}) created by the weir. A colour version of this image appears in the colour plate section of the book.

of fish passes or screens, and who were enlightened enough to consider behaviour, often relied on relatively simplistic metrics such as the proportion of a population that shows avoidance or attraction to a particular stimulus (e.g. accelerating velocity). Tests often involved naïve wild fish that experienced the stimulus only once (Haro *et al.*, 1998; Kemp *et al.*, 2005), and thus were not trained or 'conditioned' in the classical sense. Such an approach ignores the bias exhibited by individuals, in which some fish may not respond to stimuli they detect, while others will. By adopting a psychometric method commonly employed by experimental psychologists and described in Signal Detection Theory, it is now possible to dissociate the influence of bias from the ability to discriminate to obtain more realistic measures of behavioural performance in ethological studies (Kemp *et al.*, 2012). Future fish passage research that adopts the ecohydraulic approach should not be based only on advances in novel technology that enable the complexity of the fluid environment to be determined at everfiner resolution, but must also strive to improve methods by which animal behaviour is appropriately described.

Historically, the field-based application of the ecohydraulics approach has been restricted by the resolution, accuracy and frequency at which both fish movements and complex hydraulic environments can be measured. Conventional fish-tracking techniques are best suited to reach-scale studies of movements, quantifying the location of tagged individuals within a general area rather than their absolute position, or confirmed passage at a fixed point, e.g. dam, weir or associated fish pass. Whilst offering insight into rates and timings of movements, life-history strategies and physical capabilities of many species, in addition to quantitative evaluation of fish passage efficiency, such studies arguably result in subjective interpretation of behaviours and correlative factors. Recent advances in acoustic telemetry now provide the potential for 2- or 3D fish movements to be tracked at near-continuous, sub-metre resolution. Movement trajectories are obtained from calculated positions of tagged fish, based on the differences in arrival times of transmitted signals at multiple hydrophones, which are typically positioned around the perimeter of the study site. Brown *et al.* (2009) used acoustic telemetry to generate 3D distribution and movement data for silver-phase American eels (*A. rostrata*) within a dam forebay. Diurnal movement patterns, depth preferences and circling behaviours were identified, with eels rarely found to utilise a surface bypass as a route of downstream passage. Although acoustic telemetry provided a more comprehensive measure of movement patterns than conventional radio teleme-

try techniques, the sample size was small and behaviour was highly variable, making it difficult to form robust conclusions. Piper *et al.* (in preparation) were able to build on this research by employing acoustic telemetry to quantify the approach and downstream passage of silverphase European eel (*A. anguilla*) at an array of water regulation structures. Flow fields and channel bathymetry within the study site were mapped concurrently using a raft-mounted ADCP. Integration of datasets allowed fish behaviours such as milling and rapid upstream rejection to be linked to physical and hydrodynamic components with sub-metre accuracy. Acoustic telemetry techniques can suffer from similar disadvantages to radio telemetry, such as site geometry, negative post-operative effects on fish, restricted tag life and imposition of a minimum size of fish that may be used. These may pose potential limitations to the wide-scale application of this method in developing fish passage design criteria. Despite this, by adopting a range of techniques in the field and laboratory, the ecohydraulics approach continues to close the knowledge gap in understanding the behavioural and physiological capabilities of multiple fish species and life stages.

8.3 Conclusions

Fish swimming capabilities determined using swim chambers (e.g. Brett, 1964) and endurance models based on unrealistic assumptions (e.g. Beach, 1984) have formed the main biological information used in fish pass design, and are still widely used where velocity criteria are required (e.g. Santos *et al.*, 2007). However, these traditional approaches can underestimate the locomotory capacity of fish volitionally swimming under more natural conditions (Castro-Santos, 2005, 2006; Peake and Farrell, 2006; Russon and Kemp, 2011). Further, the unidirectional flows generated within swim chambers rarely occur in nature, where flows are characterised by varying levels of turbulence, which affects habitat selection, station holding and migratory abilities of fish. Due to these limitations, we recommend that future research should be based on an interdisciplinary approach to advance fish passage through the development of realistic, multi-species and multi-life stage design criteria.

Ecohydraulics aims to link the physical properties of flowing water with biological and ecological processes (Lancaster and Downes, 2010). For fish passage, ecohydraulics allows hydraulic features of interest to be quantified and linked to the swimming performance and behavioural response of fish. Research should continue to adopt an ecohydraulics approach and conduct studies across a range of spatial scales and combine the advantages of both laboratory and field-based techniques. Technological developments in telemetry and hydraulic profiling allow this to take place in the field, while the ever-advancing techniques employed within flumes enable the direct observation and quantification of behaviour and hydraulic parameters at much finer scales, under conditions in which test variables are manipulated while confounding factors are controlled.

Understanding the fundamental reasons why fish reject or progress through fish passes, be it due to physiological ability or behaviour, will greatly improve our capacity to facilitate more efficient passage or to deter fish from entering potentially hazardous locations. With this in mind, ecohydraulics must not be constrained to linking hydraulic and ecological processes, but should focus on bridging gaps between disciplines. For fish passage, psychometric theories could advance the understanding of mechanisms governing migrant behaviour (Kemp *et al.*, 2012), while other environmental stimuli may also be influential (e.g. Vowles and Kemp, 2012). Interdisciplinary interactions will likely benefit fish passage and other areas of river science in tackling unanswered questions. Adopting an ecohydraulics approach throughout all aspects of river management should, therefore, be encouraged.

8.4 Future challenges

Current fish passes are not achieving their primary goal of restoring habitat connectivity. A recent meta-analysis reported mean up- and downstream fish passage efficiencies of around 40 and 70%, respectively – well below target levels (see Noonan *et al.*, 2012). The evaluation and monitoring of fish passage facilities post construction must, therefore, be a future research priority. Quantification of both attraction (proportion of fish that are guided towards and enter the fish pass) and passage (proportion of fish ascending or descending having entered the fish pass) efficiency for a range of fishway types and configurations and for a multitude of species is required to aid the optimisation and design of current and future fish passes (Bunt *et al.*, 2012). Post-passage effects on survival and reproduction form another aspect of barrier passage often ignored in evaluations (Roscoe and Hinch, 2010), yet they should also be considered when quantifying impacts of river fragmentation and mitigation measures on the population viability of fish (Pompeu *et al.*, 2012).

For ecohydraulics research, there remains much to be done. Recent studies described in this chapter focus on quantifying swimming capability and behaviour under realistic, complex flows. Yet more work is required to verify which hydraulic metrics are most appropriate from a biological perspective (Lacey *et al.*, 2012), and how fish collect, process and respond to environmental stimuli (Liao, 2007; Kemp *et al.*, 2012). Although such research has clear value for fish pass development, examples of successful application remain scarce, with the majority of fish passes still found to be functioning poorly (Bunt *et al.*, 2012; Noonan *et al.*, 2012). In this respect, an ecohydraulics approach to fish passage research could be considered in its infancy. Advances in technology will likely facilitate future field research to assess swimming and behaviour in situ and enable better integration of experimental, modelling and field-based approaches. However, ultimately a shift in scientific culture to one that is less conservative and which embraces interdiscipinarity is needed.

References

Adrian, R.J. (2005) Twenty years of particle image velocimetry. *Experiments in Fluids*, **39**: 159–169.

Agostinho, A.A., Thomas, S.M. and Gomes, L.C. (2005) Conservation of the biodiversity of Brazil's inland waters. *Conservation Biology*, **19**: 646–652.

Bainbridge, R. (1958) The speed of swimming of fish as related to size and to the frequency and amplitude of the tail beat. *The Journal of Experimental Biology*, **35**: 109–133.

Bainbridge, R. (1960) Speed and stamina in three fish. *The Journal of Experimental Biology*, **37**: 129–153.

Bainbridge, R. and Brown, R.H.J. (1958) An apparatus for the study of the locomotion of fish. *Journal of Experimental Fish Biology*, **35**: 134–137.

Beach, M.H. (1984) *Fish Pass Designs – Criteria for the Design and Approval of Fish Passes and Other Structures to Facilitate the Passage of Migratory Fish in Rivers*. Fisheries Research Technical Report, Lowestoft, England, No. 78.

Beamish, F.W.H. (1966) Swimming endurance of some Northwest Atlantic fishes. *Journal of the Fisheries Research Board of Canada*, **23**: 341–347.

Beamish, F.W.H. (1978) Swimming capacity. In Hoar, W.S. and Randall, D.J. (eds) *Fish Physiology*, Volume 7, Academic Press, Inc., New York.

Bell, M. (1991) *Fisheries Handbook of Engineering Requirements and Biological Criteria*. Portland, OR: U.S. Army Corps of Engineers, Fish Passage Development and Evaluation Program, North Pacific Division.

Blazka, P., Volf, M. and Ceplea, M. (1960) A new type of respirometer for determination of the metabolism of fish in an active state. *Physiologia Bohemoslovaca*, **9**: 553–560.

Brett, J.R. (1964) The respiratory metabolism and swimming performance of young sockeye salmon. *Journal of the Fisheries Research Board of Canada*, 21: 1183–1226.

Brett, J.R. (1965) The relation of size to rate of oxygen consumption and sustained swimming speed of sockeye salmon (*Oncorhynchus nerka*). *Journal of the Fisheries Research Board of Canada*, 23: 1491–1501.

Brett, J.R. (1967) Swimming performance of sockeye salmon (*Oncorhynchus nerka*) in relation to fatigue time and temperature. *Journal of the Fisheries Research Board of Canada*, 21: 1731–1741.

Brett, J.R., Hollands, M. and Alderice, D.R. (1958) The effect of temperature on the cruising speed of young sockeye and coho salmon. *Journal of the Fisheries Research Board of Canada*, 15: 587–605.

Brown, L., Haro, A. and Castro-Santos, T. (2009) Three-dimensional movement of silver-phase American eels in the forebay of a small hydroelectric facility. In Casselman, J.M. and Cairns, D.K. (eds) *Eels at the Edge*, American Fisheries Society, Symposium 58, Bethesda, Maryland, pp. 277–291.

Bunt, C.M., Castro-Santos, T. and Haro, A. (2012) Performance of fish passage structures at upstream barriers to migration. *River Research and Applications*, 28: 457–478.

Čada, G.F. (2001). The development of advanced hydroelectric turbines to improve fish passage survival. *Fisheries*, 26: 14–23.

Čada, G.F. and Szluha, A.T. (1978) A biological evaluation of devices used for reducing entrapment and impingement losses at thermal power plants. *International Symposium on the Environmental Effects of Hydraulic Engineering Works*, Knoxville, TN, USA, 12 September, 1978.

Calles, E.O. and Greenberg, L.A. (2005) Evaluation of nature-like fishways for re-establishing connectivity in fragmented salmonid populations in the River EMÅN. *River Research and Applications*, 21: 951–960.

Calles, E.O., Olsson, I.C., Comoglio, C., Kemp, P.S., Blunden, L., Schmitz, M. and Greenberg, L.A. (2010) Size-dependent mortality of migratory silver eels at a hydropower plant, and implications for escapement to the sea. *Freshwater Biology*, 55: 2167–2180.

Carter, K.L. and Reader, J.P. (2000) Patterns of drift and power station entrainment of 0+ fish in the River Trent, England. *Fisheries Management and Ecology*, 7: 447–464.

Castro-Santos, T. (2004) Quantifying the combined effects of attempt rate and swimming capacity on passage through velocity barriers. *Canadian Journal of Fisheries and Aquatic Science*, 61: 1602–1615.

Castro-Santos, T. (2005) Optimal swim speeds for traversing velocity barriers: an analysis of volitional high-speed swimming behavior of migratory fishes. *The Journal of Experimental Biology*, 208: 421–432.

Castro-Santos, T. (2006) Modeling the effect of varying swim speeds on fish passage through velocity barriers. *Transactions of the American Fisheries Society*, 135: 1230–1237.

Castro-Santos, T. and Haro, A. (2003) Quantifying migratory delay: a new application of survival analysis methods. *Canadian Journal of Fisheries and Aquatic Science*, 60: 986–996.

Castro-Santos, T. and Haro, A. (2006) Biomechanics and fisheries conservation. In Shadwick, R.E. and Lauder, G.V. (eds) *Fish Physiology*, Volume 23, Fish Biomechanics, Academic Press, New York, pp. 469–523.

Cheong, T.S., Kavvas, M.L. and Anderson, E.K. (2006) Evaluation of adult white sturgeon swimming capabilities and applications to fishway design. *Environmental Biology of Fishes*, 77: 197–208.

Clay, C.H. (1995) *Design of Fishways and Other Fish Facilities*, 2nd edition, Lewis Publishers, Boca Raton.

Collins, G.B. and Elling, C.H. (1960) *Fishway Research at the Fisheries-Engineering Research Laboratory*. U.S. Department of the Interior, Fish and Wildlife Service, Bureau of Commercial Fisheries, Circular 98, Washington DC.

Cotel, A.J., Webb, P.W. and Tritico, H. (2006) Do brown trout choose locations with reduced turbulence? *Transactions of the American Fisheries Society*, 135: 610–619.

Domenici, P. and Blake, R.W. (1997) The kinematics and performance of fish fast-start swimming. *The Journal of Experimental Biology*, 200: 1165–1178.

Enders, E.C., Boisclair, D. and Roy, A.G. (2003) The effects of turbulence on the cost of swimming for juvenile Atlantic salmon (*Salmo salar*). *Canadian Journal of Fisheries and Aquatic Science*, 60: 1149–1160.

Enders, E.C., Boisclair, D. and Roy, A.G. (2005) A model of total swimming costs in turbulent flow for juvenile Atlantic salmon (*Salmo salar*). *Canadian Journal of Fisheries and Aquatic Science*, 62: 1079–1089.

Enders, E.C., Roy, M.L., Ovidio, M., Hallot, E.J., Boyer, C., Petit, F. and Roy, A.G. (2009) Habitat choice by Atlantic salmon parr in relation to turbulence at a reach scale. *North American Journal of Fisheries Management*, 29: 1819–1830.

Environment Agency (2009) *Good Practice Guidelines to the Environment Agency Hydropower Handbook: The Environmental Assessment of Proposed Low Head Hydropower Developments*, Environment Agency, Almondsbury, Bristol, UK.

Garcia de Leaniz, C. (2008) Weir removal in salmonid streams: implications, challenges and practicalities. *Hydrobiologia*, 609: 83–96.

Gowans, A.R.D., Armstrong, J.D., Priede, I.G. and Mckelvey, S. (2003) Movements of Atlantic salmon migrating upstream through a fish-pass complex in Scotland. *Ecology of Freshwater Fish*, 12: 177–189.

Grant, I. (1997) Particle image velocimetry: a review. *Proceedings of the Institute of Mechanical Engineers*, 211(C): 55–76.

Green, T.M., Lindmark, E.M., Lundström, T.S. and Gustavsson, L.H. (2011) Flow characterization of an attraction channel as entrance to fishways. *River Research and Applications*, 27: 1290–1297.

Hadderingh, R.H. and Jager, Z. (2002) Comparison of fish impingement by a thermal power station with fish populations in the Ems Estuary. *Journal of Fish Biology*, **61**: 105–124.

Hammer, C. (1995) Fatigue and exercise tests with fish. *Comparative Biochemistry and Physiology Part A: Physiology*, **112**: 1–20.

Haro, A., Castro-Santos, T., Noreika, J. and Odeh, M. (2004) Swimming performance of upstream migrant fishes in open-channel flow: a new approach to predicting passage through velocity barriers. *Canadian Journal of Fisheries and Aquatic Science*, **61**: 1590–1601.

Haro, A., Odeh, M., Noreika, J. and Castro-Santos, T. (1998) Effect of water acceleration on downstream migratory behavior and passage of Atlantic salmon smolts and juvenile American shad at surface bypasses. *Transactions of the American Fisheries Society*, **127**: 118–127.

Jones, D.R., Kiceniuk, J.W. and Bamford, O.S. (1974) Evaluation of the swimming performance of several fish species from the Mackenzie River. *Journal of the Fisheries Research Board of Canada*, **31**: 1641–1647.

Kemp, P.S., Anderson, J.J. and Vowles, A.S. (2012) Quantifying behaviour of migratory fish: Application of signal detection theory to fisheries engineering. *Ecological Engineering*, **41**: 22–31.

Kemp, P.S., Gessel, M.H. and Williams, J.G. (2005) Fine-scale behavioural responses of Pacific salmonid smolts as they encounter divergence and acceleration of flow. *Transactions of the American Fisheries Society*, **134**: 390–398.

Kemp, P.S., Russon, I.J., Vowles, A.S. and Lucas, M.C. (2011) The influence of discharge and temperature on the ability of upstream migrant adult river lamprey (*Lampetra fluviatilis*) to pass experimental overshot and undershot weirs. *River Research and Applications*, **27**: 488–498.

Kemp, P.S. and Williams, J.G. (2009) Illumination influences the ability of migrating juvenile salmonids to pass a submerged experimental weir. *Ecology of Freshwater Fish*, **18**: 297–304.

Lacey, R.W.J., Neary, V.S., Liao, J.C., Enders, E.C. and Tritico, H.M. (2012) The IPOS framework: linking fish swimming performance in altered flows from laboratory experiments to rivers. *River Research and Applications*, **28**: 429–443.

Laine, A., Kamula, R. and Hooli, J. (1998) Fish and lamprey passage in a combined Denil and vertical slot fishway. *Fisheries Management and Ecology*, **5**: 31–44.

Lancaster, J. and Downes, B.J. (2010) Linking the hydraulic world of individual organisms to ecological processes: putting ecology into ecohydraulics. *River Research and Applications*, **26**: 385–403.

Langdon, S.A. and Collins, A.L. (2000) Quantification of the maximal swimming performance of Australasian glass eels, *Anguilla australis* and *Anguilla reinhardtii*, using a hydraulic flume swimming chamber. *New Zealand Journal of Marine and Freshwater Research*, **34**: 629–636.

Larinier, M. (2000) Dams and fish migration. Contributing paper for *Environmental Issues, Dams and Fish Migration*, World Commission on Dams.

Larinier, M. and Travade, F. (1999) The development and evaluation of downstream bypasses for juvenile salmonids at small hydroelectric plants in France. In Odeh, M. (ed.) *Innovations in Fish Passage Technology*, American Fisheries Society, Bethesda, Maryland.

Larinier, M. and Travade, F. (2002) Downstream migration: problems and facilities. *Bulletin Francais de la Peche et de la Pisciculture*, **346**: 181–207.

Liao, J.C. (2004) Neuromuscular control of trout swimming in a vortex street: implications for energy economy during Kármán gait. *Journal of Experimental Biology*, **207**: 3495–3506.

Liao, J.C. (2006) The role of the lateral line and vision on body kinematics and hydrodynamic preference of rainbow trout in turbulent flow. *The Journal of Experimental Biology*, **209**: 4077–4090.

Liao, J.C. (2007) A review of fish swimming mechanics and behaviour in altered flows. *Philosophical Transactions of the Royal Society B – Biological Sciences*, **362**: 1973–1993.

Liao, J.C., Beal, D.N., Lauder, G.V. and Triantafyllou, M.S. (2003) The Kármán gait: novel body kinematics of rainbow trout swimming in a vortex street. *Journal of Experimental Biology*, **206**: 1059–1073.

Lucas, M.C. and Baras, E. (2001) *Migration of Freshwater Fishes*, Blackwell Science Ltd, Oxford.

Lucas, M.C., Mercer, T., Armstrong, J.D., McGinty, S. and Rycroft, P. (1999) Use of a flat-bed passive integrated transponder antenna array to study the migration and behaviour of lowland river fishes at a fish pass. *Fisheries Research*, **44**: 183–191.

Lupandin, A.I. (2005) Effects of flow turbulence on swimming speed of fish. *Biology Bulletin*, **32**: 461–466.

MacVicar, B.J., Beaulieu, E., Champagne, V. and Roy, A.G. (2007) Measuring water velocity in highly turbulent flows: field tests of an electromagnetic current meter (ECM) and an acoustic Doppler velocimeter (ADV). *Earth Surface Processes and Landforms*, **32**: 1412–1432.

Mallen-Cooper, M. and Brand, D.A. (2007) Non-salmonids in a salmonid fishway: what do 50 years of data tell us about past and future fish passage? *Fisheries Management and Ecology*, **14**: 319–332.

Moser, M.L., Matter, A.L., Stuehrenberg, L.C. and Bjornn, T.C. (2002) Use of an extensive radio receiver network to document Pacific lamprey (*Lampetra tridentata*) entrance efficiency at fishways in the Lower Columbia River, USA. *Hydrobiologia*, **483**: 45–53.

Nelson, J.A., Gotwalt, P.S., Reidy, S.P. and Webber, D.M. (2002) Beyond U_{crit}: Matching swimming performance tests to the physiological ecology of the animal, including a new fish "drag strip". *Comparative Biochemistry and Physiology: Part A*, **133**: 289–302.

Nestler, J.M., Goodwin, R.A., Smith, D.L. and Anderson, J.J. (2008) A mathematical and conceptual framework for ecohydraulics. In Wood, P.J., Hannah, D.M. and Sadler, J.P. (eds)

Hydroecology and Ecohydrology: Past, Present and Future, John Wiley & Sons, Ltd, Chichester, UK, pp. 205–224.

Nikora, V.I., Aberle, J., Biggs, B.J.F., Jowett, I.G. and Sykes, J.R.E. (2003) Effects of fish size, time-to-fatigue and turbulence on swimming performance: a case study of *Galaxias maculatus*. *Journal of Fish Biology*, **63**: 1365–1382.

Noonan, M.J., Grant, J.W.A. and Jackson, C.D. (2012) A quantitative assessment of fish passage efficiency. *Fish and Fisheries*, **13**: 450–464.

Odeh, M. (1999) *Innovations in Fish Passage Technology*, American Fisheries Society, Bethesda, Maryland.

O'Hanley, J.R. and Tomberlin, D. (2005) Optimizing the removal of small fish passage barriers. *Environmental Modeling and Assessment*, **10**: 85–98.

Okamoto, K., Nishio, S., Saga, T. and Kobayashi, T. (2000) Standard images for particle-image velocimetry. *Measurement Science and Technology*, **11**: 685–691.

Ovidio, M. and Philippart, J-C. (2002) The impact of small physical obstacles on upstream movements of six species of fish. *Hydrobiologia*, **483**: 55–69.

Peake, S. (2004) An evaluation of the use of critical swimming speed for determination of culvert water velocity criteria for smallmouth bass. *Transactions of the American Fisheries Society*, **133**: 1472–1479.

Peake, S.J. and Farrell, A.P. (2006) Fatigue is a behavioural response in respirometer-confined smallmouth bass. *Journal of Fish Biology*, **68**: 1742–1755.

Peake, S. and McKinley, R.S. (1998) A re-evaluation of swimming performance in juvenile salmonids relative to downstream migration. *Canadian Journal of Fisheries and Aquatic Sciences*, **55**: 682–687.

Peake, S., Beamish, F.W.H., McKinley, R.S., Scruton, D.A. and Katopodis, C. (1997) Relating swimming performance of lake sturgeon, *Acipenser fulvescens*, to fishway design. *Canadian Journal of Fisheries and Aquatic Sciences*, **54**: 1361–1366.

Piper, A.T., Wright, R.M. and Kemp, P.S. (2012) The influence of attraction flow on upstream passage of European eel (*Anguilla anguilla*) at intertidal barriers. *Ecological Engineering*, **44**: 329–336.

Poe, T.P., Hansel, H.C., Vigg, S., Palmer, D.E. and Prendergast, L.A. (1991) Feeding of predacious fishes on out-migrating juvenile salmonids in John Day reservoir, Columbia River. *Transactions of the American Fisheries Society*, **120**: 405–420.

Pompeu, P.S., Agostinho, A.A. and Pelicice, F.M. (2012) Existing and future challenges: the concept of successful fish passage in South America. *River Research and Applications*, **28**: 504–512.

Popper, A.N. and Carlson, T.J. (1998) Application of sound and other stimuli to control fish behavior. *Transactions of the American Fisheries Society*, **127**: 673–707.

Rice, S.P., Little, S., Wood, P.J., Moir, H.J. and Vericat, D. (2010) The relative contributions of ecology and hydraulics to ecohydraulics. *River Research and Applications*, **26**: 363–366.

Roscoe, D.W. and Hinch, S.G. (2010) Effectiveness monitoring of fish passage facilities: historical trends, geographic patterns and future directions. *Fish and Fisheries*, **11**: 12–33.

Roy, M.L., Roy, A.R. and Legendre, P. (2010) The relationship between 'standard' fluvial habitat variables and turbulent flow at multiple scales in morphological units of a gravel-bed river. *River Research and Applications*, **26**: 439–455.

Russon, I.J. and Kemp, P.S. (2011) Experimental quantification of the swimming performance and behaviour of spawning run river lamprey *Lampetra fluviatilis* and European eel *Anguilla anguilla*. *Journal of Fish Biology*, **78**: 1965–1975.

Santos, H.A., Pompeu, P.S. and Martinez, C.B. (2007) Swimming performance of the neotropical fish *Leporinus reinhardti* (Characiformes: Anastomidae). *Neotropical Icthyology*, **5**: 139–146.

Shields, F.D., Knight, S.S., Testa, S. and Cooper, C.M. (2003) Use of Acoustic Doppler Current Profilers to describe velocity distributions at the reach scale. *Journal of the American Water Resource Association*, **39**: 1397–1408.

Silva, A.T., Katopodis, C., Santos, J.M., Ferreira, M.T. and Pinheiro, A.N. (2012) Cyprinid swimming behaviour in response to turbulent flow. *Ecological Engineering*, **44**: 314–328.

Silva, A.T., Santos, J.M., Ferreira, M.T., Pinheiro, A.N. and Katopodis, C. (2011) Effects of water velocity and turbulence on the behaviour of Iberian barbel (*Luciobarbus bocagei*, Steindachner 1864) in an experimental pool-type fishway. *River Research and Applications*, **27**: 360–373.

Smith, D.L., Brannon, E.L. and Odeh, M. (2005) Response of juvenile rainbow trout to turbulence produced by prismatoidal shapes. *Transactions of the American Fisheries Society*, **134**: 741–753.

Smith, D.L., Brannon, E.L., Shafii, B. and Odeh, M. (2006) Use of the average and fluctuating velocity components for estimation of volitional rainbow trout density. *Transactions of the American Fisheries Society*, **135**: 431–441.

Stacey, M.T., Monismith, S.G. and Burau, J.R. (1999) Observations of turbulence in a partially stratified estuary. *Journal of Physical Oceanography*, **29**: 1950–1970.

Stringham, E. (1924) The maximum speed of fresh-water fishes. *The American Naturalist*, **58**(655): 156–161.

Swanson, C., Young, P.S. and Cech Jr, J.J. (2004) Swimming in two-vector flows: performance and behavior of juvenile Chinook salmon near a simulated screened water diversion. *Transactions of the American Fisheries Society*, **133**: 265–278.

Swanson, C., Young, P.S. and CechJr, J.J. (2005) Close encounters with a fish screen: integrating physiological and behavioral results to protect endangered species in exploited ecosystems. *Transactions of the American Fisheries Society*, **134**: 1111–1123.

Taft, E.P. (2000) Fish protection technologies: a status report. *Environmental Science and Policy*, **3**: 349–359.

Thorpe, J.E. and Morgan, R.I.G. (1978) Periodicity in Atlantic salmon, *Salmo salar*, smolt migration. *Journal of Fish Biology*, **12**: 541–548.

Towler, B., Hoar, A. and Ahlfeld, D. (2012) Ecohydrology and fish-passage engineering: legacy of Denil and the call for a more inclusive paradigm. *Journal of Water Resources Planning and Management*, **138**: 77–79.

Tritico, H.M. (2009) *The Effects of Turbulence on Habitat Selection and Swimming Kinematics of Fishes*. PhD dissertation, University of Michigan, USA.

Tritico, H.M. and Cotel, A.J. (2010) The effects of turbulent eddies on the stability and critical swimming speed of creek chub (*Semotilus atromaculatus*). *The Journal of Experimental Biology*, **213**: 2284–2293.

Tritico, H.M., Cotel, A.J. and Clarke, J.N. (2007) Development, testing and demonstration of a portable submersible miniature particle imaging velocimetry device. *Measurement Science and Technology*, **18**: 2555–2562.

Tsukamoto, K., Kajihara, T. and Nishiwaki, M. (1975) Swimming ability of fish. *Bulletin of the Japanese Society of Scientific Fisheries*, **41**: 167–174.

Tudorache, C., Viaenen, P., Blust, R. and De Boeck, G. (2007) Longer flumes increase critical swimming speeds by increasing burst-glide swimming duration in carp *Cyprinus carpio*, L. *Journal of Fish Biology*, **71**: 1630–1638.

Turnpenny, A.W.H. and O'Keeffe, N. (2005) *Screening for Intake and Outfalls: A Best Practice Guide*. Science Report SC030231, Environment Agency, Almondsbury, Bristol, UK.

Venditti, D.A., Rondorf, D.W. and Kraut, J.M. (2000) Migratory behaviour and forebay delay of radio-tagged juvenile fall Chinook salmon in a lower Snake River impoundment. *North American Journal of Fisheries Management*, **20**: 41–52.

Voulgaris, G. and Trowbridge, J.H. (1998) Evaluation of the acoustic Doppler velocimeter (ADV) for turbulence measurements. *Journal of Atmospheric and Oceanic Technology*, **15**: 272–289.

Vowles, A.S. and Kemp, P.S. (2012) Effects of light on the behaviour of brown trout (*Salmo trutta*) encountering acceler-ating flow: Application to downstream fish passage. *Ecological Engineering*, **47**: 247–253.

Wang, R.W. and Hartlieb, A. (2011) Experimental and field approach to the hydraulics of nature-like pool-type fish migration facilities. *Knowledge and Management of Aquatic Ecosystems*, **400-05**. doi: 10.1051/kmae/2011001.

Wang, R.W., David, L. and Larinier, M. (2010) Contribution of experimental fluid mechanics to the design of vertical slot fish passes. *Knowledge and Management of Aquatic Ecosystems*, **396-02**. doi: 10.1051/kmae/2010002.

Ward, J.V. and Stanford, J.A. (1983) The serial discontinuity concept of lotic ecosystems. In Fontaine, T.D. and Bartell, S.M. (eds) *Dynamics of Lotic Ecosystems*, Ann Arbor Science, Ann Arbor, Michigan, pp. 29–42.

Wardle, C.S. (1975) Limit of fish swimming speed. *Nature*, **255**: 725–727.

Weaver, C.R. (1963) Influence of water velocity upon orientation and performance of adult migrating salmonids. *Fishery Bulletin*, **63**: 97–121.

Weaver, C.R. (1965) Observations on the swimming ability of adult American shad (*Alosa sapidissima*). *Transactions of the American Fisheries Society*, **94**: 382–385.

Webb, P.W. (1975) Hydrodynamics and energetics of fish propulsion. *Bulletin of the Fisheries Research Board of Canada*, **190**: 1–159.

Weihs, D. (1973) Hydrodynamics of fish schooling. *Nature*, **241**: 290–291.

Williams, J.G., Armstrong, G., Katopodis, C., Larinier, M. and Travade, F. (2012) Thinking like a fish: A key ingredient for development of effective fish passage facilities at river obstructions. *River Research and Applications*, **28**: 407–417.

Wood, P., Hannah, D. and Sadley, J. (eds) (2008) *Hydroecology and Ecohydrology: Past, Present and Future*, John Wiley & Sons, Ltd, Chichester, UK.

Zhou, Y. (1982) *The Swimming Behaviour of Fish in Towed Gears: A Re-examination of the Principles*. Working paper 4, Department of Agriculture and Fisheries of Scotland.

Species–Habitat Interactions

9

Habitat Use and Selection by Brown Trout in Streams

Jan Heggenes[1] and Jens Wollebæk[2]

[1]Telemark University College, Department of Environmental Sciences, Hallvard Eikas Plass, N-3800 Bø i Telemark, Norway

[2]The Norwegian School of Veterinary Science, Department of Basic Sciences and Aquatic Medicine, Box 8146, Dep. 0033 Oslo, Norway

9.1 Introduction

Spatial environmental heterogeneity is a primary factor that may influence reproductive success and growth of plants and animals (Andrewartha and Birch, 1954; Pulliam, 1988). Decreased habitat complexity may negatively impact the growth and survival of freshwater organisms, e.g. fish (Miller *et al.*, 1989; Finstad *et al.*, 2007; Brockmark *et al.*, 2010). Freshwater fish tend to exhibit size-structured ecological communities (Hutchings and Morris, 1985; Elliott, 1994). Therefore, selective differences and asymmetric competition for habitat may provide a mechanism that allows the coexistence of size classes and species (L'Abée-Lund *et al.*, 1993; Bremset and Heggenes, 2001; Armstrong and Nislow, 2006). Conversely, disturbance of the spatial environment, e.g. by human intervention, may also disrupt fish communities. The widely distributed salmonids have evolved life history strategies that may drive ecosystems (Cederholm *et al.*, 1999; Gende *et al.* 2002; Janetski *et al.*, 2009) and are culturally important species. Thus, integration of hydraulics and physical habitat use and selection by salmonids is critical to ecohydraulics to provide environmentally sustainable management solutions. In a key paper, Chapman (1966) suggested that competition for space and food regulates salmonid populations in streams. He further suggested that trout and salmon were niche generalists, since they often inhabit unstable and unpredictable environments, where generalist strategies will increase fitness (Andrewartha and Birch, 1954; Slobodkin and Rapoport, 1974).

In temperate streams, competition for space among salmonid fish is particularly important, because it may also substitute for competition for food, i.e. energy intake, in summer (the growing season). Stream salmonids, in particular the young, are drift-feeding, and may therefore compete for holding stations based on their net energy gain value, which is a result of a trade-off between drifting food availability and swimming costs (Fausch, 1984; Hughes and Dill, 1990; Hayes *et al.*, 2007). Net energy intake is often considered a link between habitat selection and fitness (Hill and Grossman, 1993), although risk may also modify habitat selection patterns (Dill and Fraser, 1984; Gilliam and Fraser, 1987; Forrester and Steele, 2004). Therefore, habitat selection in stream salmonids is considered ecologically important and closely linked to ecohydraulics, as a result it has attracted much scientific and management interest. In brown trout (*Salmo trutta*) inhabiting natural temperate streams, habitat selection at low temperatures may principally be a risk-avoidance strategy, but it appears to be context-dependent (Huusko *et al.*, 2007). Brown trout habitat use and selection at higher temperatures during typical summer low flow conditions have been studied extensively both in the field and laboratory, and reviewed (Heggenes, 1989; Heggenes *et al.*, 1999; Armstrong *et al.*, 2003). This knowledge is the basis for ecohydraulic applications. There is,

---

Ecohydraulics: An Integrated Approach, First Edition. Edited by Ian Maddock, Atle Harby, Paul Kemp and Paul Wood.
© 2013 John Wiley & Sons, Ltd. Published 2013 by John Wiley & Sons, Ltd.

however, less systematic knowledge as to how brown trout may respond to varying hydraulic conditions, i.e. natural floods or hydropeaking (Halleraker *et al.*, 2003; Flodmark *et al.*, 2004, 2006; Heggenes *et al.*, 2007).

As the use of habitat-hydraulic models in environmental impact analysis and fisheries management has become more common, the translation of habitat behaviours into statistical models has become an important research and management subject (Ahmadi-Nedushan *et al.*, 2006; Mouton *et al.*, 2009; Ayllon *et al.*, 2010). Habitat-hydraulic models serve three main purposes: to quantify habitat requirements, to predict fish occurrences from abiotic and biotic habitat variables, and to improve our understanding of the complex species–habitat interactions. Models are, however, simplified representations of reality; e.g. based on optimal habitat suitability models per variable, simplified or no interaction among habitat variable selection and use, and/or consistent habitat use across different stream flows, temperatures and light conditions (Ahmadi-Nedushan *et al.*, 2006; Ayllon *et al.*, 2009; Mouton *et al.*, 2009). These models may, therefore, provide potential for better understanding of variables and processes at work, rather than realistic predictions. This also stresses the need for more research on spatial and temporal dynamics in fish habitat selection. An important advance in this context with considerable relevance for ecohydraulics has been the development and testing of different energy-based foraging models for brown trout and other stream salmonids (Hayes *et al.*, 2000, 2007; Booker *et al.*, 2004). Another recent focus in fish habitat research is on the importance of the hyporheic zone and groundwater flow in habitat selection (Brunke and Gonser, 1997; Boulton, 2007; Olsen *et al.*, 2009). In this chapter we use brown trout as a model species and discuss some basic issues, i.e. observation methods with their associated potential bias, the concept of habitat, and key hydraulic and biotic factors relevant to fish, before we summarize aspects of habitat use and selection for different fish life stages. Bioenergetic models and the understudied importance of the hyporheic zone are additionally briefly reviewed.

9.2 Observation methods and bias

It is an unfortunate fact that to do field habitat studies, the observer often has to be on and/or in the stream. This implies some sort of observation bias for most methods applied hitherto (Conallin *et al.*, in press). Furthermore, because brown trout habitat requirements change throughout their lifecycle (fish size, activity; Bardonnet

and Heland, 1994; Roussel *et al.*, 1999; Wollebaek *et al.*, 2008), in time (season, day–night; Heggenes *et al.*, 1993; Roussel and Bardonnet, 2002; Bardonnet *et al.*, 2006) and space (Greenberg *et al.*, 1997; Heggenes *et al.*, 2002; Ayllon *et al.*, 2010), effective sampling methods must be applicable for the entire year as well as night and day and high–low flows. This may require a systematic combination of observation methods to obtain representative habitat use and selection data, but few studies do so (Conallin *et al.*, in press). Sampling methods may be (1) more suitable in certain areas (spatial scale issues) (Durance *et al.*, 2006), (2) more suitable at certain times of the year or diel period (temporal scale issues) (Bardonnet *et al.*, 2006; Hickey and Closs, 2006), (3) more suitable for certain life stages (sampling bias) (Heggenes *et al.*, 1990, 1991; Enders *et al.*, 2007) or (4) observing a small number of fish per sampling effort (statistical power) (Conallin *et al.*, in press). Whereas statistical power can be calculated in retrospect, potential sampling bias cannot. Depending on observation methods, bias may come in three forms: (1) bias in sampling probability across spatial and temporal habitat variation, (2) bias across fish life history stages, e.g. size class and behaviours and (3) intrusiveness or disturbance of fish position (fright bias). Ideally, these biases should be quantified and considered during the data analysis, but in practice this is limited to cautionary notes in data interpretation.

The methods available to determine habitat requirements of brown trout (and stream salmonids in general) differ in their applicability to varying stream conditions (Cunjak *et al.*, 1988; Heggenes *et al.*, 1990; 1991; Hickey and Closs, 2006). Underwater observations (snorkelling) and riverbank observation following specific protocols may have low intrusiveness (rarely inducing observable fright reaction) (Bovee and Cochnauer, 1977; Bachman, 1984; Heggenes *et al.*, 1991) and give the least biased observations of actual fish positions and habitat use in flowing waters (Heggenes *et al.*, 1991; Bonneau *et al.*, 1995). Snorkelling is applicable for both diurnal and nocturnal observations (although requiring the use of a hand-held torch at night; Heggenes *et al.*, 1991). However, it requires sufficient depths for snorkelling, which may be infrequent in smaller streams, and snorkelling may not be suitable in slow water, or during floods when there is reduced visibility. Riverbank observations have been used since the 1960s as a diurnal method (Jenkins, 1969; Bachman, 1984), but the method is sensitive to surface turbulence, overhead cover, aquatic macrophytes, light conditions, turbidity and fish size (Heggenes *et al.*, 1990, 1991; Bonneau *et al.*, 1995).

Other, much-used methods include electrofishing such as single-pass electrofishing (Bohlin *et al.*, 1989) where preferably the entire stream bed is fished, or certain stream areas enclosed by nets (Price and Peterson, 2010). To reduce the fright bias considered a problem using single-pass electrofishing (Bovee, 1982), systematic point-abundance electrofishing has been used (Persat and Copp, 1989; Copp and Garner, 1995). However, both methods are prone to fright bias. The operator is in the stream and highly visible, and the electric field alerts fish at a distance (Bohlin *et al.*, 1989). Electrofishing as a nocturnal method or during floods is difficult and impractical (Heggenes *et al.*, 1990; Conallin *et al.*, in press).

Radio and acoustic telemetry do not require instream observers and are methods with low or no fright bias (Lucas and Baras, 2000), but bias may be associated with stress due to handling, surgical insertion of transmitters into the body cavity, hydrodynamic 'load' of external transmitters or behavioural changes, although dummy tests suggest such bias may be limited (Connors *et al.*, 2002; Zale *et al.*, 2005). Cost and logistics typically limit the number of fish tagged (Enders *et al.*, 2007), reducing statistical power in most habitat studies. The method is not applicable to small fish, e.g. juvenile trout, due to the size of transmitters (transmitter weight should preferably be less than 2% of body weight, Zale *et al.*, 2005), thus leaving a substantial proportion of the population in most streams unsampled. An alternative to radio and acoustic telemetry is the use of passive integrated transponder (PIT) tags (Enders *et al.*, 2007). A portable or stationary instream detecting device is used to locate PIT-tagged fish to work out habitat requirements (Roussel *et al.*, 2004; Cucherousset *et al.*, 2005, 2010). Detection distance is, however, short (approximately 70 cm), and therefore most habitat studies involve an operator wading the stream with the portable device, inducing fright bias. PIT detecting can be used as both a diurnal and nocturnal method in any type of habitat, and may be used on smaller fish than radio or acoustic tags (Acolas *et al.*, 2007). A promising recent development is the use of antennae grid systems laid out in the bottom substrate, e.g. the Kraken system (Johnston *et al.*, 2009), which gives excellent temporal data, but for a limited stream area.

Portable or fixed underwater cameras may be used as an alternative to diving to determine habitat use (Bovee, 1982; Moore and Scott, 1988; Marchand *et al.*, 2002). An advantage compared to most other methods is the continual monitoring of fish activity (e.g. foraging, resting, hiding and social interactions), allowing detailed analysis at a later time both of habitat use and behaviours. It may

also function as both a diurnal and nocturnal method, depending on the portable setup. However, similar to some other methods, it tends to give relatively few observations (Conallin *et al.*, in press) within a limited spatial field. Spotlighting has been used for gaining abundance estimates of brown trout and may be comparable to electrofishing abundance estimates for larger fish (Hickey and Closs, 2006), but it appears to be inefficient and biased for smaller trout (Conallin *et al.*, in press).

9.3 Habitat

The term habitat is often subdivided into macrohabitat, mesohabitat and microhabitat. Macrohabitat may describe the general type of space in which an animal lives, and applies to a scale larger than the animal's normal daily range (Kramer *et al.*, 1997), typically spatial scales of 10^1 metres (Frissell *et al.*, 1986; Allan, 1995) or often more. Data on geomorphology, hydrology and climate on a scale of stream reaches or subcatchments, combined with fish habitat observations, may be referred to as macrohabitat studies (Stanford, 1996; Cunjak and Therrien, 1998). Mesohabitat studies typically quantify ecohydraulic characteristics of stream reaches (habitat types) within the normal daily range used by fish, i.e. patches from a few to tens of m^2 (Heggenes *et al.*, 1999; Armstrong *et al.*, 2003). Microhabitat is a term used for describing small spaces within mesohabitats and usually refers to spatial scales of 10^{-1} metres (Frissell *et al.*, 1986; Allan, 1995). Microhabitat studies in trout may quantify ecohydraulic characteristics in one or a number of stream points (e.g. at the snout position of the individual fish) with additional information on the immediate surroundings of the fish (i.e. patches up to several cm^2; Marchildon *et al.*, 2011).

9.4 Abiotic and biotic factors

Habitat includes both abiotic and biotic factors, in complex interaction. Brown trout habitat use is affected by a number of factors and at multiple scales, e.g. the landscape, population and individual level, and may therefore be variable and flexible in time and space (Heggenes *et al.*, 1999; Armstrong *et al.*, 2003; Ayllon *et al.*, 2010). On the individual level, choice of holding station (microhabitat) is often thought to be best described by optimal foraging theory, i.e. the fish attempts to maximize net energy intake (Bachman, 1984; Hughes, 1998; Hayes *et al.*, 2007), modified, however, by (predation) risk and competition

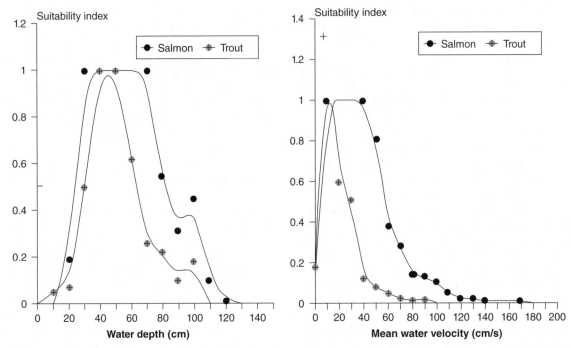

Figure 9.1 Habitat suitability curves and segregation between sympatric brown trout and young Atlantic salmon for water depth and mean water velocity in Gjengedalselva River, Norway. From Heggenes, Brabrand and Saltveit, 1990, reprinted by permission of the publisher (Taylor & Francis Ltd, http://www.tandf.co.uk/journals).

(see Section 9.8 below; Dill and Fraser, 1984; Gilliam and Fraser, 1987). Individual trout exhibit some variation in habitat preferences (Johnsson *et al.*, 2000). Specific habitat requirements for different trout instream life stages must be known for the entire year and lifecycle so that important habitat bottlenecks and other events (e.g. swim-up, spawning) can be identified and managed (Elliott *et al.*, 1997; Armstrong and Nislow, 2006). Habitat requirements and use may change over a diel period (Johnson and Covich, 2000; Hansen and Closs, 2005; Bardonnet *et al.*, 2006). Brown trout may be mostly crepuscular in their foraging habits in winter (Heggenes *et al.*, 1993; Johnson and Douglass, 2009), and spend the day hiding under cover. By contrast, at summer temperatures, trout may be more active at dawn and dusk (Bardonnet and Heland, 1994; Bardonnet *et al.*, 2006). Abiotic landscape factors other than temperature and light that may affect habitat use are the frequency and amplitude of water flow and the availability and distribution of different habitats (e.g. Kocik and Ferreri, 1998; Clark *et al.*, 2008).

At the population level, intra- and interspecific competition may affect habitat use by brown trout (Kennedy and Strange, 1986; Bremset and Heggenes, 2001; Armstrong *et al.*, 2003). Intraspecific dominance is usually deter-

mined by fish size (Jenkins, 1969; Bachman, 1984; Stradmeyer *et al.*, 2008), and the larger fish tend to occupy the more profitable stream (micro) habitats (Hughes, 1998; Hayes *et al.*, 2007; Stradmeyer *et al.*, 2008). There may also be a 'prior residence' or owners' dominance advantage (Brännäs, 1995; Johnsson *et al.*, 1999; Harwood *et al.*, 2003), probably particularly important at the swim-up stage when the young establish territories (Armstrong and Nislow, 2006). Laboratory studies indicate that trout density *per se* may affect behavioural skills (Brockmark *et al.*, 2010; Sloman and Baron, 2010) and habitat use (Greenberg, 1994), but it is less clear how this translates to actual field conditions (Sloman *et al.*, 2002; 2008).

Brown trout tend to dominate over Atlantic salmon (*Salmo salar*), with which they often occur in sympatry (Figure 9.1). Brown trout suppress feeding by salmon (Stradmeyer *et al.*, 2008) or displace salmon to the faster areas (Kennedy and Strange, 1986; Bremset and Heggenes, 2001; Heggenes *et al.*, 2002). It is, however, difficult to separate niche preferences from active segregation due to competition, and the two species also display adaptations to different habitats (e.g. larger pectoral fins in salmon). Thus, interactions may be offset by different habitat preferences. The Biotic–Abiotic Constraining Hypothesis

(BACH) suggests that such biotic interactions as intra- and interspecific competition and predation risk can have an overriding negative influence on a species, even when abiotic conditions are suitable (Quist and Hubert, 2005). Habitat suitability and the linked abundance of a fish species is predicted to be largely regulated by abiotic habitat factors when densities of competitors or predators are low. However, when these densities are high, the habitat use of the fish species is changed and abundance suppressed by biotic interactions regardless of environmental conditions. Although corroborated by laboratory experiments (e.g. Brockmark et al., 2010; Sloman and Baron, 2010), it is unclear to what extent this affects habitat use by brown trout in natural streams. Complex biotic–abiotic interactions are difficult to disentangle in field studies. In trout they may be particularly important during the 'early life critical period' during swim-up, i.e. the transition from dependence on maternal yolk reserves to independent feeding (Armstrong and Nislow, 2006).

9.5 Key hydraulic factors

A wide range of physical habitat variables have been included in studies of spatial niche selection in brown trout (Wesche et al., 1987; Jowett, 1990; 1992). The more frequently measured variables for trout are water depth, water velocity (or stream gradient), substrate particle size and cover. These correlated instream variables have been shown to affect in situ distribution of trout in streams (Gibson, 1988; Heggenes et al., 1999; Armstrong et al., 2003; Ahmadi-Nedushan et al., 2006). Nevertheless, the explanatory power of these basic key physical factors, influenced by fluid dynamics, is variable (Heggenes, 2002; Armstrong et al., 2003; Milner et al., 2003), indicative of the complex and dynamic habitat–fish relationships. Other hydraulic and physical factors (often more difficult to measure) or derived variables may be important and higher resolution in data sampling may be required (Marchildon et al., 2011). Different physical variables may also influence fish position choice at different spatial scales. For example, water velocity defines both mesohabitat and microhabitat use in brown trout (Table 9.1). Temporal issues have attracted more attention during recent years. Temporally variable light and temperatures affect behaviour and spatial niche selection in trout (e.g. Rimmer et al., 1984; Heggenes et al., 1993; Johnson and Douglass, 2009). Effects of varying water flows, e.g. hydropeaking, on brown trout stream habitat appear to be largely negative due to stranding and reduced availability of slow-flowing stream margin habitats, in particular for the smallest young, although more research is needed (Halleraker et al., 2003; Flodmark et al., 2004, 2006; Armstrong and Nislow, 2006; Korman et al., 2009). One conclusion is that these key hydraulic factors interact in dynamic and complex ways to determine trout holding position. Combinations of depth and water velocity may often be selected for (Shirvell and Dungey, 1983; Heggenes, 2002; Ayllon et al., 2009), to some extent depending on fish size (Table 9.1).

9.6 Habitat selection

In an ecohydraulic context, habitat suitability has often been quantified as univariate habitat suitability curves (HSCs), constructed by measuring these key physical characteristics of patches occupied by fish. To establish habitat selection criteria, however, it is also necessary to measure habitat availability, i.e. to include patches not occupied by fish (Figure 9.2). Ideally, we have a 'cafeteria situation' (sensu Krebs, 1989), meaning that individual fish at any point in space and time can select their preferred habitat. Trout are distributed between patches in a non-uniform way. Thus, if a stream is sampled at random and represents a 'cafeteria situation' for each observed fish individual, generalized probability functions for in situ habitat selection can be constructed directly from such ecohydraulic data. Great amounts of data are necessary to generalize, and it is a problem to extrapolate the data outside the stream and range of habitat conditions sampled, to establish 'universals' (Morantz et al., 1987). The 'cafeteria situation' is, of course, rare or absent in most field studies, and data represent a semi-selective situation at best, where for example Liebig's law of the minimum, i.e. control by scarcest resource rather than the total amount of resources available, comes into play, making transferability of data a dubious exercise (Tharme, 2003; Strakosh et al., 2003; Hedger et al., 2004). Furthermore, questions about predictability (e.g. suitability of habitat) depend on the scale at which they are asked (e.g. Bult et al., 1996). In the spatially heterogeneous ecosystems of natural streams, designated patch size is likely to affect the form of the resulting probability functions. Events within any one patch may also be variable and unpredictable, whereas events on a larger scale may be quite stable (Baran et al., 1995). Adding to this, brown trout is also a highly variable species (Elliott, 1994), reflected in its wide distribution, and exhibits flexibility and individual variation in habitat selection (Johnsson et al., 2000; Heggenes, 2002;

Table 9.1 Stream habitats used by brown trout for daytime feeding in summer and spawning.

Habitat variable	Measure	Values	References
Small parr (< 7 cm)			
Depth mesohabitat	General range	5–35 cm	Heggenes *et al.*, 1999; Ayllon *et al.*, 2010
	Preference	< 30–40 cm	Bohlin, 1977; Kennedy and Strange, 1982; Bardonnet and Heland, 1994; Ayllon *et al.*, 2010
Velocity mesohabitat	General range	10–60 cms^{-1}	Heggenes *et al.*, 1999; Ayllon *et al.*, 2010
	Range for fry	0–20 cms^{-1}	Bardonnet and Heland, 1994
	Range for 0+	20–50 cms^{-1}	Heggenes, 1996
Velocity microhabitat	General range	0–10 cms^{-1}	Heggenes *et al.*, 1999
Substrate particle size	General range	16–256 mm	Heggenes *et al.*, 1999
	Range	50–70 mm	Heggenes, 1988b
	Range	10–90 mm	Bardonnet and Heland, 1994
Larger (> 7 cm)			
Depth mesohabitat	General range	Positively related to size	Heggenes *et al.*, 1999; Ayllon *et al.*, 2010
	Preference	> 50 cm	Heggenes, 1988a, 2002; Ayllon *et al.*, 2010
	Mean preference	65 cm	Shirvell and Dungey, 1983
	Range	14–122 cm	Shirvell and Dungey, 1983
	Range	40–75 cm	Mäki-Petays *et al.*, 1997
	Range	9–305 cm	Heggenes, 2002
	Mean	69 cm	Heggenes, 2002
Velocity mesohabitat	General range	0–50 cms^{-1}	Heggenes *et al.*, 1999; Ayllon *et al.*, 2010
	Preference	0–60 cms^{-1}	Ayllon *et al.*, 2010
	Range	10–70 cms^{-1}	Heggenes, 1988a
	Range	0–65 cms^{-1}	Shirvell and Dungey, 1983
	Range	0–142 cms^{-1}	Heggenes, 2002
	Range	0–100 cms^{-1}	Ayllon *et al.*, 2010
	Mean	26.7 cms^{-1}	Shirvell and Dungey, 1983
	Mean	24 cms^{-1}	Heggenes, 2002
Velocity microhabitat	General range and preference	0–20 cms^{-1}	Heggenes *et al.*, 1999
	Preference	0–10 cms^{-1}	Heggenes, 2002
	Mean	14 cms^{-1}	Heggenes, 2002
	Range	0–81cms^{-1}	Heggenes, 2002
Substrate particle size	General range	Weakly positive to size	Heggenes *et al.*, 1999
	Range	8–128 cm	Eklöv *et al.*, 1999
	Preferred range	128–256 mm	Heggenes, 2002
Spawning			
Depth mesohabitat	Range	15–45 cm	Louhi *et al.*, 2008
	Range	6–82 cm	Shirvell and Dungey, 1983
	Range	23–215 cm	Wollebaek *et al.*, 2008
	Mean	25.5 cm	Witzel and MacCrimmon, 1983
	Mean	31.7 cm	Shirvell and Dungey, 1983
	Mean	103 cm	Wollebaek *et al.*, 2008
Velocity mesohabitat	Range	20–55 cms^{-1}	Louhi *et al.*, 2008
	Range	11–80 cms^{-1}	Witzel and MacCrimmon, 1983
	Range	15–75 cms^{-1}	Shirvell and Dungey, 1983
	Range	2–124 cms^{-1}	Wollebaek *et al.*, 2008
	Mean	46.7 cms^{-1}	Witzel and MacCrimmon, 1983
	Mean	39.4 cms^{-1}	Shirvell and Dungey, 1983
	Mean	47 cms^{-1}	Wollebaek *et al.*, 2008

Table 9.1 (*Continued*)

Habitat variable	Measure	Values	References
Substrate particle size	Range	1.6–6.4 cm	Louhi *et al.*, 2008
	Range	2–37 cm	Wollebaek *et al.*, 2008
	Mean	0.69 cm	Witzel and MacCrimmon, 1983
	Mean	7 cm	Wollebaek *et al.*, 2008
	Critical percentage fines < *c.* 2 mm	> 10%	Crisp and Carling, 1989; Louhi *et al.*, 2008
Depth in gravel of egg burial	Mean	15.2 cm	Crisp and Carling, 1989
	Minimum	14 cm	Witzel and MacCrimmon, 1983

Ayllon *et al.*, 2010). Therefore, it is, in this general *ex situ* context, useful to define in broad ecohydraulic terms the (meso) habitats successfully used by brown trout for the life stages we have some knowledge about: small parr (< 7 cm), summer rearing (> 7 cm) and spawning habitat (Table 9.1; Heggenes *et al.*, 1999; Armstrong *et al.*, 2003). Information on optimal habitat and relative qualities of different habitats should be obtained by *in situ* studies.

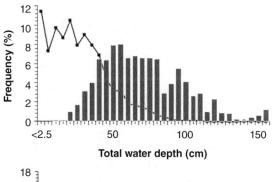

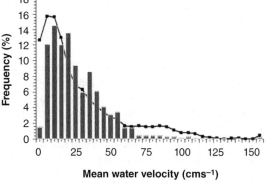

Figure 9.2 Differences between water depth and mean water velocity use (bars) and availability (line) for wild brown trout in eight boreal streams. From Heggenes, 2002, reprinted by permission of the publisher (Taylor & Francis Ltd, http://www.tandf.co.uk/journals).

In general, the smaller fish, and in particular at the swim up stage, tend to hold positions close to the bottom, often in more slow, shallow water, i.e. typically along the stream margins. With increasing size and swimming ability, brown trout successively use deeper and midstream or preferably pool areas. In particular, the deep pool areas are habitat for the larger trout. However, substantial niche overlaps between size groups are indicative of a habitat generalist strategy. Both between and within-stream variability in habitat selection may be generated by different spatial patterns, i.e. habitat availabilities, and spatial scales per key hydraulic factor. Water depths and substrate tend to have patchy distributions on large scales relative to the fish (several metres), and compensatory behaviours may involve the energetic cost of relatively long movements between few patches. On the other hand, water velocities, and in particular focal water velocities, may more often constitute a small-scale spatial mosaic with high connectivity, e.g. in the water column, presenting a 'cafeteria situation' to the fish, requiring less movement. Results reported in the literature, with considerable spatial variation in selection of depths, but less in particular for snout water velocities, support the inference that spatial scale and pattern are important for habitat selection in brown trout (Heggenes, 2002; Ayllon *et al.*, 2010). However, different hydraulic factors may be limiting or less favourable in different streams and reaches, and the relative importance of variables will vary.

The rather wide range of habitat used by brown trout is likely an adaptive response to the often unpredictable spatio-temporal heterogeneity in many temperate streams. Rather than narrow optima, which will tend to be time- and site-specific, the tactics of individual brown trout are to be selective within wide tolerance limits. These tactics depend critically on the mobility of individual fish. Brown trout is a mobile species, exhibiting large individual and temporal variation in movement patterns and

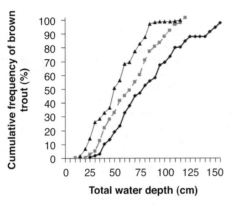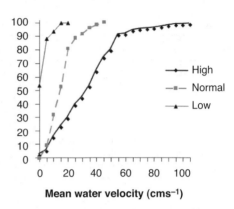

Figure 9.3 Temporal changes in use of water depths and mean water velocities, depending on water flows, for wild brown trout in a small boreal stream. From Heggenes, 2002, reprinted by permission of the publisher (Taylor & Francis Ltd, http://www.tandf.co.uk/journals).

propensity to settle or reside in a home area or explore (Armstrong *et al.*, 1997; Heggenes *et al.*, 2007; Zimmer *et al.*, 2010). They have the capacity to respond to environmental variation and impacts on virtually any natural temporal and spatial scale by behavioural movements to shift habitat. Movement is expected to be a typical response to (local) adversity. The ideal free distribution model (e.g. Fretwell and Lucas, 1970) states that movement may be a response to adjust fitness in a spatially and temporally variable environment, i.e. trout move to better habitat patches.

This resilience combined with the complex fish–habitat interactions, poses some challenges for ecohydraulic habitat modelling (Ayllon *et al.*, 2009; 2010). Within these habitat ranges, biotic interactions, such as intra- and interspecific competition, food availability, social interactions and predation, may play important roles during periods or sites of relative environmental stability, both in narrowing the spatial niche of brown trout and producing local site-dependent shifts. The connection between individual habitat selection and population regulation becomes even more complex (Armstrong *et al.*, 2003; Milner *et al.*, 2003).

9.7 Temporal variability: light and flows

Habitat holding capacity may be affected by limiting ecological events (Orth, 1987; Elliott, 1994). Important in an ecohydraulic context are temporal changes in habitat conditions due to water flow, but also temperature and light. Few studies quantify habitat use at night, but some studies, mostly on juvenile trout, indicate that trout are

more spread out at night with more fish occupying the shallow, slow stream areas, e.g. stream margins (Harris *et al.*, 1992; Hubert *et al.*, 1994; Roussel and Bardonnet, 1995, 1999; Heggenes, 2002).

9.7.1 Case study: varying water flows and habitat use

Water flows may cause major temporal changes in habitat availabilities on a seasonal, diurnal or shorter time scale. And changes in water flow may indeed induce considerable changes in spatial niche use by brown trout (Figure 9.3; Heggenes, 2002; see also Scruton *et al.*, 2008). In a natural river, with natural, increasing flows and concurrent changes in habitat availability, trout exhibited a flexible response by occupying areas providing successively greater depths and water velocities (Figure 9.3). This response is not, however, flow conditional, i.e. proportional with flow, but rather, to a limited extent, flow compensatory.

Hydropeaking operations generate shifts in downstream flows over short time periods much more extreme than under natural conditions. Obviously, changes in wetted area affect habitat availability, but how are trout able to respond to such an unnatural regime of habitat variation? Water drawdown causes physiological stress and behavioural aggression (Flodmark *et al.*, 2002; Stradmeyer *et al.*, 2008) and may be so rapid that 0+ brown trout in particular strand (Figure 9.4; Halleraker *et al.*, 2003), increasing population mortality. Drawdown velocity *per se* is therefore an important ecological consideration. There is some indication that at least some individuals may 'learn' and habituate to such an unnatural flow regime

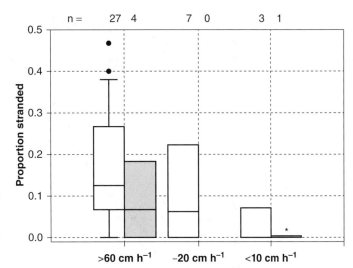

Figure 9.4 Proportion of stranded 0+ brown trout for three dewatering rates (*x*-axis) in summer (water temperatures 10–12 °C) during day (white boxes) and night (grey boxes). Asterisk indicates no stranded trout. Reproduced from Halleraker *et al.*, 2003, with permission from John Wiley & Sons.

(Figure 9.5; Saltveit *et al.*, 2001; Halleraker *et al.*, 2003; Flodmark *et al.*, 2002, 2006).

The shifts in response to flow may override other causal factors, e.g. biotic interactions, in determining habitat selection. For example, considerably higher niche overlap has been reported between sympatric and presumably competing brown trout and Atlantic salmon (Bremset and Heggenes, 2001; Armstrong *et al.*, 2003) than niche overlaps found for allopatric brown trout at high, normal and low flows. It is particularly interesting that selection

of focal water velocities, generally narrow, presumably for energetic reasons, and therefore considered 'universal', also appear to shift depending on water flows (Figure 9.6). These shifts may suggest stress. Extreme low flows may indeed be population bottlenecks for trout (Elliott, 1994; Elliott *et al.*, 1997; James *et al.*, 2010). Low flows involve unfavourable habitat conditions, combined with increased competition and predation risk and reduced food resources. Extreme floods, in particular when temperatures are low, may also reduce salmonid population

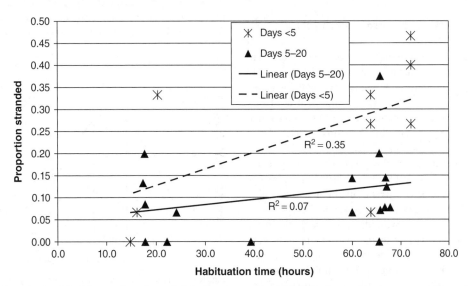

Figure 9.5 Learning by doing? Proportion of stranded 0+ brown trout (rapid daylight dewatering at water temperatures 10–12 °C) versus habituation time at stable flow. A higher proportion of trout < 5 days in the experimental stream (first dewatering episode) stranded, than for trout 5–20 days in the stream (i.e. experiencing 2–5 dewatering episodes). Reproduced from Halleraker *et al.*, 2003, with permission from John Wiley & Sons.

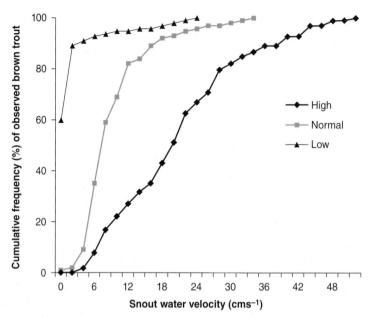

Figure 9.6 Temporal changes in use of focal water velocities, depending on water flows, for wild brown trout in a small boreal stream. From Heggenes, 2002, reprinted by permission of the publisher (Taylor & Francis Ltd, http://www.tandf.co.uk/journals).

numbers (Elwood and Waters, 1969; Erman *et al.*, 1988; Elliott, 1994; Jensen and Johnsen, 1999) probably because of the hostile environment during floods.

9.8 Energetic and biomass models

The principal foraging strategy for trout in summer may be described as an energy intake-maximizing sit-and-wait-strategy. This is because the cost curve with increased water velocity is rather flat, whereas the drift availability peaks due to reduced capture success at higher water velocities (Figure 9.7; Hill and Grossman, 1993; Hayes *et al.*, 2000, 2007; Booker *et al.*, 2004). Trout reduce energy expenditure by holding positions in low-velocity microniches (Bachman, 1984; Fausch, 1984), presumably close to higher-velocity currents in order to maximize drift-based energy intake (Wankowski and Thorpe, 1979; Bachman, 1984, see also 'hydraulic strain' in Nestler *et al.*, 2008). This rather narrow spatial microniche, indicative of an energy-conserving strategy for feeding, has formed the functional basis for microhabitat holding position models, rather successful in predicting 'snap-shot' feeding positions (Fausch, 1984; Hughes and Dill, 1990; Hayes *et al.*, 2000, 2007; Booker *et al.*, 2004). However, the models have been rather less successful in predicting fish num-

bers. Many temperate streams during normal flows will provide ample niches with, for example, suitable focal velocities and correlated heights above the bottom, and thus rarely be a limiting factor. That may, of course, be different during extreme ecological events. Habitat-based bioenergetic models may also capture important environmental determinants of fish growth and survival (e.g.

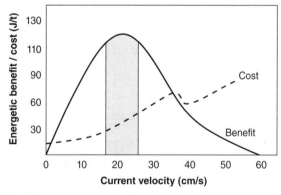

Figure 9.7 Principal effect of water velocity on drift availability (benefit) and swimming cost for a stream fish (after Hill and Grossman, 1993; Hayes *et al.*, 2000; Piccolo *et al.*, 2008). An optimum foraging window can be identified with respect to maximizing net energy gain.

Kennedy *et al.*, 2008) and thus improve our understanding of the complex fish–habitat interactions. However, such models have met with more mixed success with respect to quantifying more than rather broad habitat requirements and related fish growth potential (Nislow *et al.*, 2000; Kennedy *et al.*, 2008) and are, therefore, indirectly, still limited as predictors of actual population numbers. Modelling the spatial and temporal hydraulic dynamics in temperate streams, for example, is a daunting task.

9.9 The hyporheic zone

The apparent change in focal water velocities with varying flows may also be influenced by trout actively using the hyporheic zone. Unfortunately, groundwater and rivers have traditionally been treated as distinct entities (Brunke and Gonser, 1997; Boulton *et al.*, 1998). Although a number of review papers over the last decade have pointed out the vertical connection and functional role and significance of the hyporheic zone (Brunke and Gonser, 1997; Malard *et al.*, 2002; Boulton, 2007), this vertical dimension remains understudied compared to longitudinal and lateral connectivity (Power *et al.*, 1999), perhaps for logistic and conceptual reasons. Recent empirical studies indicate the complexity and spatial patchiness of the hyporheic habitat (Malard *et al.*, 2002; Malcolm *et al.*, 2003; Finstad *et al.*, 2007). This patchiness is dynamic, depending on the relative contributions of fluxes of surface water and groundwater through the hyporheic zone, producing a fine-scale spatio-temporal habitat mosaic. The functional importance of the hyporheic zone for fish is largely controlled by the interstitial spaces (substratum composition), flow rates (residence time) and direction, sediment stability and hydrological exchange processes determining temperatures, dissolved oxygen and nutrients (Brunke and Gonser, 1997; Malcolm *et al.*, 2003; Boulton, 2007; Calles *et al.*, 2007). The hyporheic zone may provide its inhabitants with low-velocity microniches (energy, sheer stress), protection against extreme temperatures (winter ice formation, physiologically dangerous summer temperatures) and desiccation, protection from large predators and stable substrates during bedload movement. Environmental impacts such as flow and temperature extremes are, therefore, likely to induce active use of the hyporheic zone. High and low flows and the associated restriction of suitable habitat, and temperature extremes, are environmental stressors in trout (Elliott, 1994; Borgstrøm and Museth, 2005; James *et al.*, 2010). The hyporheic zone likely provides refuge during such events. Young brown

trout certainly tend to seek refuge on, and probably in, the bottom, also during normal foraging (Gustafsson *et al.*, 2010), likely as a response to predation risk and higher water velocities.

Data on if and how trout may actively use the hyporheic zone are, however, largely lacking (Power *et al.*, 1999). The common method of direct (underwater) observation will not observe fish that are actually in the substrate, and electrofishing cannot separate fish that were in the substrate. Comparative summer studies of both methods suggest that Arctic char (*Salvelinus alpinus*; Halselva and Komagelva Rivers, northern Norway) may spend considerable time concealed in the substrate (Heggenes and Saltveit, 2007; see also Holierhoek and Power, 1995), compared with trout and Atlantic salmon. It is also known that several salmonid species may take advantage of groundwater upwellings in winter and summer and actively seek out such patches (*Thymallus arcticus*, Scotland, West *et al.*, 1992; *Oncorhynchus clarkii*, Canada, Brown *et al.*, 1994; *O. clarkii*, USA, Harper and Farag, 2004; *Salmo salar*, Scotland, Malcolm *et al.*, 2005).

In contrast, the importance of hyporheic habitat characteristics during spawning and early development has, for obvious reasons, been studied extensively (Table 9.1; Hansen, 1975; Calles *et al.*, 2007; Greig *et al.*, 2007; Soulsby *et al.*, 2009).

9.10 Spatial and temporal complexity of redd microhabitat

Reproductive habitats are a necessity for population sustainability (Pulliam, 1988), and scarcity of suitable spawning habitat may affect brown trout populations directly. Brown trout deposit their eggs in fluvial waters in the substrate in constructions called redds (White, 1942). Published studies on redd site selection typically examine substrate particle sizes, water velocities and water depths, generally considered the most critical microhabitat variables for redds, in addition to vorticity (Crowder and Diplas, 2002). A systematic review of brown trout redd habitat indicated optimal depths, velocities and substrate size of 15–45 cm, 20–55 cm s^{-1} and 16–64 mm, respectively (Table 9.1; Louhi *et al.*, 2008). There have been some attempts to develop suitability criteria indexes for brown trout spawning habitat (Raleigh *et al.*, 1986; Louhi *et al.*, 2008; Wollebaek *et al.*, 2008). However, site-specific habitat criteria should be implemented where possible. Relatively wide ranges of spawning habitats are utilized, related to available habitat and fish size (Champigneulle *et al.*, 2003; Zimmer and Power, 2006; Wollebaek *et al.*, 2008). For example, brown trout require little fine sediments and good oxygen for their redds (Table 9.1; Acornley and Sear, 1999), but may select substrate particle sizes from c. 0.1 mm (Louhi *et al.*, 2008) to about 10% of their body length (Kondolf and Wolman, 1993) for spawning. Barlaup *et al.* (2008) documented flexibility also in the selection of water flow. Similar to non-mature trout, there appears to be a positive correlation between the size of spawning trout (and other salmonids) and water depth, velocity and substrate size (Louhi *et al.*, 2008; Wollebaek *et al.*, 2008; Johnson *et al.*, 2010). Brown trout, therefore, show some intra- and inter-stream flexibility in redd site selection and use (Wollebaek *et al.*, 2008). Habitat criteria for spawning may, however, be narrower than for recruitment (Louhi *et al.*, 2008).

Environmental constraints, both natural and anthropogenic, generate variation in (micro) habitat use across streams and year-to-year variation in available habitat. For example, reduced availability of spawning habitat may slow population growth (Sabaton *et al.*, 2008). Brown trout inhabiting unstable habitats may be migratory and display flexible habitat use to cope with habitat fluctuations (Ostergaard *et al.*, 2003; Wollebaek *et al.*, 2010). Indeed, habitat-specific population structuring within tributaries (Carlsson *et al.*, 1999) may be genetically founded. Therefore, spawning habitat conservation may be critical to retain the integrity of brown trout populations.

9.11 Summary and ways forward

Observation methods may be biased and should be chosen carefully depending on study objectives and instream spatial and temporal heterogeneity. Habitat–fish interactions are extremely complex. Spatial and temporal hydraulic stream heterogeneity strongly influence summer habitat use and selection by brown trout. Ecological theory suggests niche generalist strategies are favoured in such unstable environments, and brown trout may be considered a (meso) habitat generalist. Habitat studies indicate that key physical factors are water depth, velocities, substrates and cover. They are correlated and interact to determine habitat suitability. Trout favour slower-flowing run and pool areas, in particular with increasing fish size, but selection ranges are wide. Great overlaps in habitat used by differently sized trout indicate flexibility in habitat selection, e.g. in response to varying hydraulic conditions. Considerable within and between stream variation in habitat use

and also selection, largely depending on habitat availabilities, is common. Temporal variation, related to changes in water flows, temperatures and light, may produce shifts in habitat use. The flexibility in summer habitat use by brown trout poses some challenges for ecohydraulic modelling, and indicates that the common use of habitat-hydraulic models based on static univariate habitat suitability curves across stream flows, diel and seasonal periods is flawed. Energetic models for microhabitat position choice may exhibit predictive power during stable summer flow conditions, but are of less use to model spatial and temporal heterogeneity and impact scenarios.

The responses in habitat selection of the trout must be studied along axes of scaled temporal and spatial variation in habitat conditions. More studies are needed on the importance of the hyporheic zone. Trout habitat use responses to hydropeaking operations, which are currently an increasing concern, are understudied. Biotic confounding effects related to intra and interspecific competitive interactions and (predation) risk must be considered in such studies. A substantial proportion of the variability in habitat selection is not explained by the key hydraulic factors usually studied, perhaps because they are not studied at different spatial scales. Hydraulic factors not studied, or derived variables, may be of importance.

References

Acolas, M.L., Roussel, J.M., Lebel, J.M. and Bagliniere, J.L. (2007) Laboratory experiment on survival, growth and tag retention following PIT injection into the body cavity of juvenile brown trout (*Salmo trutta*). *Fisheries Research*, **86**: 280–284.

Acornley, R.M. and Sear, D.A. (1999) Sediment transport and siltation of brown trout (*Salmo trutta* L.) spawning gravels in chalk streams. *Hydrological Processes*, **13**: 447–458.

Ahmadi-Nedushan, B., St-Hilaire, A., Berube, M., Robichaud, E., Thiemonge, N. and Bobee, B. (2006) A review of statistical methods for the evaluation of aquatic habitat suitability for instream flow assessment. *River Research and Applications*, **22**: 503–523.

Allan, J.D. (1995) *Stream Ecology: Structure and Function of Running Waters*, Chapman & Hall, London.

Andrewartha, H.G. and Birch, L.C. (1954) *The Distribution and Abundance of Animals*, University of Chicago Press, Chicago.

Armstrong, J.D. and Griffiths, S.W. (2001) Density-dependent refuge use among over-wintering wild Atlantic salmon juveniles. *Journal of Fish Biology*, **58**: 1524–1530.

Armstrong, J.D. and Nislow, K.H. (2006) Critical habitat during the transition from maternal provisioning in freshwater fish, with emphasis on Atlantic salmon (*Salmo salar*) and brown trout (*Salmo trutta*). *Journal of Zoology*, **269**: 403–413.

Armstrong, J.D., Braithwaite, V.A. and Huntingford, F.A. (1997) Spatial strategies of wild Atlantic salmon parr: exploration and settlement in unfamiliar areas. *Journal of Animal Ecology*, **66**: 203–211.

Armstrong, J.D., Kemp, P.S., Kennedy, G.J.A., Ladle, M. and Milner, N.J. (2003) Habitat requirements of Atlantic salmon and brown trout in rivers and streams. *Fisheries Research*, **62**: 143–170.

Ayllon, D., Almodovar, A., Nicola, G.G. and Elvira, B. (2009) Interactive effects of cover and hydraulics on brown trout habitat selection patterns. *River Research and Applications*, **25**: 1051–1061.

Ayllon, D., Almodovar, A., Nicola, G.G. and Elvira, B. (2010) Ontogenetic and spatial variations in brown trout habitat selection. *Ecology of Freshwater Fish*, **19**: 420–432.

Bachman, R.A. (1984) Foraging behavior of free-ranging wild and hatchery brown trout in a stream. *Transactions of the American Fisheries Society*, **113**: 1–32.

Baran, P., Delacoste, M., Poizat, G., Lascaux, J.M., Lek, S. and Belaud, A. (1995) Multi-scales approach of the relationships between brown trout (*Salmo trutta* L.) populations and habitat features in the central Pyrenees. *Bulletin Francais de la Peche et de la Pisciculture*, **337/833/933**: 604–993.

Bardonnet, A. and Heland, M. (1994) The influence of potential predators on the habitat preferenda of emerging brown trout. *Journal of Fish Biology*, **45**: 131–142.

Bardonnet, A., Poncin, P. and Roussel, J.M. (2006) Brown trout fry move inshore at night: a choice of water depth or velocity? *Ecology of Freshwater Fish*, **15**: 309–314.

Barlaup, B.T., Gabrielsen, S.E., Skoglund, H. and Wiers, T. (2008) Addition of spawning gravel – A means to restore spawning habitat of Atlantic salmon (*Salmo salar* L.) and anadromous and resident brown trout (*Salmo trutta* L.) in regulated rivers. *River Research and Applications*, **24**: 543–550.

Bohlin, T. (1977) Habitat selection and intercohort competition of juvenile sea-trout *Salmo trutta*. *Oikos*, **29**(1): 112–117.

Bohlin, T., Hamrin, S., Heggberget, T.G., Rasmussen, G. and Saltveit, S.J. (1989) Electrofishing – theory and practice with special emphasis on salmonids. *Hydrobiologia*, **173**: 9–43.

Bonneau, J.L., Thurow, R.F. and Scarnecchia, D.L. (1995) Capture, marking, and enumeration of juvenile bull trout and cutthroat trout in small, low-conductivity streams. *North American Journal of Fisheries Management*, **15**: 563–568.

Booker, D.J., Dunbar, M.J. and Ibbotson, A. (2004) Predicting juvenile salmonid drift-feeding habitat quality using a three-dimensional hydraulic-bioenergetic model. *Ecological Modelling*, **177**: 157–177.

Borgstrom, R. and Museth, J. (2005) Accumulated snow and summer temperature – critical factors for recruitment to high mountain populations of brown trout (*Salmo trutta* L.). *Ecology of Freshwater Fish*, **14**: 375–384.

Boulton, A.J. (2007) Hyporheic rehabilitation in rivers: restoring vertical connectivity. *Freshwater Biology*, **52**: 632–650.

Boulton, A.J., Findlay, S., Marmonier, P., Stanley, E.H. and Valett, H.M. (1998) The functional significance of the hyporheic zone in streams and rivers. *Annual Review of Ecology and Systematics*, **29**: 59–81.

Bovee, K.D. (1982) *A Guide to Stream Habitat Analysis Using the Instream Flow Incremental Methodology*. Instream flow information paper 12, FWS/OBS-82/86. U.S. Fish and Wildlife Service, Washington, DC.

Bovee, K.D. and Cochnauer, T. (1977) *Development and Evaluation of Weighted Criteria, Probability of Use Curves for Instream Flow Assessment: Fisheries*. Instream flow information paper 21, FWS/OBS-77/63. U.S Fish and Wildlife Service, Washington, DC.

Brännäs, E. (1995) First access to territorial space and exposure to strong predation pressure – a conflict in early emerging Atlantic salmon (*Salmo salar* L) fry. *Evolutionary Ecology*, **9**: 411–420.

Bremset, G. and Heggenes, J. (2001) Competitive interactions in young Atlantic salmon (*Salmo salar* L) and brown trout (*Salmo trutta* L) in lotic environments. *Nordic Journal of Freshwater Research*, **75**: 127–142.

Brockmark, S., Adriaenssens, B. and Johnsson, J.I. (2010) Less is more: density influences the development of behavioural life skills in trout. *Proceedings of the Royal Society B – Biological Sciences*, **277**: 3035–3043.

Brown, R.S., Stanislawski, S.S. and Mackay, W.C. (1994) Effects of frazil ice on fish. In Prowse, T.D. (ed.) *Proceedings of the Workshop in Environmental Aspects of River Ice*, National Hydrology Research Institute, Saskatoon, pp. 261–278.

Brunke, M. and Gonser, T. (1997) The ecological significance of exchange processes between rivers and groundwater. *Freshwater Biology*, **37**: 1–33.

Bult, T.P., Haedrich, R.L. and Schneider, D.C. (1996) Description of a technique for multiscale analyses: spatial scaling and habitat selection in riverine habitats. In Leclerc, M., Capra, H., Valentin, S., Boudreault, A. and Cote, Y. (eds) *Proceedings of the second IAHR Symposium on habitat hydraulics, Ecohydraulics 2000*, Vol. B, INRS-Eau, Quebec, pp. 67–80.

Calles, O., Nyberg, L. and Greenberg, L. (2007) Temporal and spatial variation in quality of hyporheic water in one unregulated and two regulated boreal rivers. *River Research and Applications*, **23**: 829–842.

Carlsson, J., Olsen, K.H., Nilsson, J., Overli, O. and Stabell, O.B. (1999) Microsatellites reveal fine-scale genetic structure in stream-living brown trout. *Journal of Fish Biology*, **55**: 1290–1303.

Cederholm, C.J., Kunze, M.D., Murota, T. and Sibatani, A. (1999) Pacific salmon carcasses: Essential contributions of nutrients and energy for aquatic and terrestrial ecosystems. *Fisheries*, **24**: 6–15.

Champigneulle, A., Largiader, C.R. and Caudron, A. (2003) Reproduction of the brown trout (*Salmo trutta* L.) in the torrent of Chevenne, Haute-Savoie. An original functioning? *Bulletin Francais de la Peche et de la Pisciculture*, 41–69.

Chapman, D.W. (1966). Food and space as regulators of salmonid populations in streams. *American Naturalist*, **100**: 345–357.

Clark, J.S., Rizzo, D.M., Watzin, M.C. and Hession, W.C. (2008) Spatial distribution and geomorphic condition of fish habitat in streams: An analysis using hydraulic modelling and geostatistics. *River Research and Applications*, **24**: 885–899.

Conallin, J. (2009) Instream Physical Habitat Suitability Modelling in Danish Small Lowland Streams: The Development of Habitat Suitability Indices for Juvenile Brown Trout (*Salmo trutta*). PhD thesis, Roskilde Universitet, Roskilde.

Conallin, J., Olsen, M., Boegh, E., Kristensen, A., Pedersen, S., Heggenes, J. and Jensen, J.K. (2013) Determination of juvenile brown trout habitat use in small lowland streams: a comparison of sampling methods. *In press*.

Connors, K.B., Scruton, D., Brown, J.A. and McKinley, R.S. (2002) The effects of surgically-implanted dummy radio transmitters on the behaviour of wild Atlantic salmon smolts. *Hydrobiologia*, **483**: 231–237.

Copp, G.H. and Garner, P. (1995) Evaluating the microhabitat use of fresh-water fish larvae and juveniles with point abundance sampling by electrofishing. *Folia Zoologica*, **44**: 145–158.

Crisp, D.T. and Carling, P.A. (1989) Observations on siting, dimensions and structure of salmonid redds. *Journal of Fish Biology*, **34**(1): 119–134.

Crowder, D.W. and Diplas, P. (2002) Vorticity and circulation: spatial metrics for evaluating flow complexity in stream habitats. *Canadian Journal of Fisheries and Aquatic Sciences*, **59**: 633–645.

Cucherousset, J., Roussel, J.M., Keeler, R., Cunjak, R.A. and Stump, R. (2005) The use of two new portable 12-mm PIT tag detectors to track small fish in shallow streams. *North American Journal of Fisheries Management*, **25**: 270–274.

Cucherousset, J., Britton, J.R., Beaumont, W.R.C., Nyqvist, M., Sievers, K. and Gozlan, R.E. (2010) Determining the effects of species, environmental conditions and tracking method on the detection efficiency of portable PIT telemetry. *Journal of Fish Biology*, **76**: 1039–1045.

Cunjak, R.A. and Therrien, J. (1998) Inter-stage survival of wild juvenile Atlantic salmon, *Salmo salar* L. *Fisheries Management and Ecology*, **5**: 209–223.

Cunjak, R.A., Prowse, T.D. and Parrish, D.L. (1998) Atlantic salmon (*Salmo salar*) in winter: "the season of parr discontent"? *Canadian Journal of Fisheries and Aquatic Sciences*, **55**: 161–180.

Dill, L.M. and Fraser, A.H.G. (1984) Risk of predation and the feeding behavior of juvenile coho salmon (*Oncorhynchus kisutch*). *Behavioral Ecology and Sociobiology*, **16**: 65–71.

Durance, I., Lepichon, C. and Ormerod, S.J. (2006) Recognizing the importance of scale in the ecology and management of riverine fish. *River Research and Applications*, **22**: 1143–1152.

Eklov, A.G., Greenberg, L.A., Bronmark, C., Larsson, P. and Berglund, O. (1999) Influence of water quality, habitat and species richness on brown trout populations. *Journal of Fish Biology*, **54**: 33–43.

Elliott, J.M. (1994) *Quantitative Ecology and the Brown Trout*, Oxford University Press, Oxford, UK.

Elliott, J.M., Hurley, M.A. and Elliott, J.A. (1997) Variable effects of droughts on the density of a sea-trout *Salmo trutta* population over 30 years. *Journal of Applied Ecology*, **34**: 1229–1238.

Elwood, J.W. and Waters, T.F. (1969) Effects of floods on food consumption and production rates of a stream brook trout population. *Transactions of the American Fisheries Society*, **98**: 253–262.

Enders, E.C., Clarke, K.D., Pennell, C.J., Ollerhead, L.M.N. and Scruton, D.A. (2007) Comparison between PIT and radio telemetry to evaluate winter habitat use and activity patterns of juvenile Atlantic salmon and brown trout. *Hydrobiologia*, **582**: 231–242.

Erman, D.C., Andrews, E.D. and Yoderwilliams, M. (1988) Effects of winter floods on fishes in the Sierra Nevada. *Canadian Journal of Fisheries and Aquatic Sciences*, **45**: 2195–2200.

Fausch, K.D. (1984) Profitable stream positions for salmonids: relating specific growth rate to net energy gain. *Canadian Journal of Zoology–Revue Canadienne de Zoologie*, **62**: 441–451.

Finstad, A.G., Einum, S., Forseth, T. and Ugedal, O. (2007) Shelter availability affects behaviour, size-dependent and mean growth of juvenile Atlantic salmon. *Freshwater Biology*, **52**: 1710–1718.

Flodmark, L.E.W., Vollestad, L.A. and Forseth, T. (2004) Performance of juvenile brown trout exposed to fluctuating water level and temperature. *Journal of Fish Biology*, **65**: 460–470.

Flodmark, L.E.W., Forseth, T., L'Abee-Lund, J.H. and Vollestad, L.A. (2006) Behaviour and growth of juvenile brown trout exposed to fluctuating flow. *Ecology of Freshwater Fish*, **15**: 57–65.

Flodmark, L.E.W., Urke, H.A., Halleraker, J.H., Arnekleiv, J.V., Vollestad, L.A. and Poleo, A.B.S. (2002) Cortisol and glucose responses in juvenile brown trout subjected to a fluctuating flow regime in an artificial stream. *Journal of Fish Biology*, **60**: 238–248.

Forrester, G.E. and Steele, M.A. (2004) Predators, prey refuges, and the spatial scaling of density-dependent prey mortality. *Ecology*, **85**: 1332–1342.

Fretwell, S.D. and Lucas Jr, H.L. (1970) On territorial behavior and other factors influencing habitat distribution in birds. I. Theoretical development. *Acta Biotheoretica*, **19**: 16–36.

Frissell, C.A., Liss, W.J., Warren, C.E. and Hurley, M.D. (1986) A hierarchical framework for stream habitat classification – viewing streams in a watershed context. *Environmental Management*, **10**: 199–214.

Gende, S.M., Edwards, R.T., Willson, M.F. and Wipfli, M.S. (2002) Pacific salmon in aquatic and terrestrial ecosystems. *Bioscience*, **52**: 917–928.

Gibson, R.J. (1988) Mechanisms regulating species composition, population structure, and production of stream salmonids: a review. *Polskii Archivum Hydrobiologie*, **35**: 469–495.

Gilliam, J.F. and Fraser, D.F. (1987) Habitat selection under predation hazard: test of a model with foraging minnows. *Ecology*, **68**: 1856–1862.

Greenberg, L.A. (1994) Effects of predation, trout density and discharge on habitat use by brown trout, *Salmo trutta*, in artificial streams. *Freshwater Biology*, **32**: 1–11.

Greenberg, L.A., Bergman, E. and Eklov, A.G. (1997) Effects of predation and intraspecific interactions on habitat use and foraging by brown trout in artificial streams. *Ecology of Freshwater Fish*, **6**: 16–26.

Greig, S.M., Sear, D.A. and Carling, P.A. (2007) A review of factors influencing the availability of dissolved oxygen to incubating salmonid embryos. *Hydrological Processes*, **21**: 323–334.

Gustafsson, P., Bergman, E. and Greenberg, L.A. (2010) Functional response and size-dependent foraging on aquatic and terrestrial prey by brown trout (*Salmo trutta* L.). *Ecology of Freshwater Fish*, **19**: 170–177.

Halleraker, J.H., Saltveit, S.J., Harby, A., Arnekleiv, J.V., Fjeldstad, H.P. and Kohler, B. (2003) Factors influencing stranding of wild juvenile brown trout (*Salmo trutta*) during rapid and frequent flow decreases in an artificial stream. *River Research and Applications*, **19**: 589–603.

Hansen, E.A. (1975) Some effects of groundwater on brown trout redds. *Transactions of the American Fisheries Society*, **104**: 100–110.

Hansen, E.A. and Closs, G.P. (2005) Diel activity and home range size in relation to food supply in a drift-feeding stream fish. *Behavioral Ecology*, **16**: 640–648.

Harper, D.D. and Farag, A.M. (2004) Winter habitat use by cutthroat trout in the Snake River near Jackson, Wyoming. *Transactions of the American Fisheries Society*, **133**: 15–25.

Harris, D.D., Hubert, W.A. and Wesche, T.A. (1992) Habitat use by young-of-year brown trout and effects on weighted usable area. *Rivers*, **3**: 99–105.

Harwood, A.J., Griffiths, S.W., Metcalfe, N.B. and Armstrong, J.D. (2003) The relative influence of prior residency and dominance on the early feeding behaviour of juvenile Atlantic salmon. *Animal Behaviour*, **65**: 1141–1149.

Hayes, J.W., Hughes, N.F. and Kelly, L.H. (2007) Process-based modelling of invertebrate drift transport, net energy intake and reach carrying capacity for drift-feeding salmonids. *Ecological Modelling*, **207**: 171–188.

Hayes, J.W., Stark, J.D. and Shearer, K.A. (2000) Development and test of a whole-lifetime foraging and bioenergetics growth model for drift-feeding brown trout. *Transactions of the American Fisheries Society*, **129**: 315–332.

Hedger, R.D., Dodson, J.J., Bergeron, N.E. and Caron, F. (2004) Quantifying the effectiveness of regional habitat quality index models for predicting densities of juvenile Atlantic salmon (*Salmo salar* L.). *Ecology of Freshwater Fish*, **13**: 266–275.

Heggenes, J. (1988a) Substrate preferences of brown trout fry (*Salmo trutta*) in artificial stream channels. *Canadian Journal of Fisheries and Aquatic Sciences*, **45**: 1801–1806.

Heggenes, J. (1988b) Effect of experimentally increased intraspecific competition on sedentary adult brown trout (*Salmo trutta*) movement and stream habitat choice. *Canadian Journal of Fisheries and Aquatic Sciences*, **45**: 1163–1172.

Heggenes, J. (1989) Physical habitat selection by brown trout (*Salmo trutta*) in riverine systems. *Nordic Journal of Freshwater Research*, **64**: 74–90.

Heggenes, J. (1996) Habitat selection by brown trout (*Salmo trutta*) and young Atlantic salmon (*S. salar*) in streams. Static and dynamic hydraulic modelling. *Regulated Rivers: Research and Management*, **12**(2–3): 155–169.

Heggenes, J. (2002) Flexible summer habitat selection by wild, allopatric brown trout in lotic environments. *Transactions of the American Fisheries Society*, **131**: 287–298.

Heggenes, J. and Dokk, J.G. (2001) Contrasting temperatures, waterflows, and light: Seasonal habitat selection by young Atlantic salmon and brown trout in a boreonemoral river. *Regulated Rivers: Research and Management*, **17**: 623–635.

Heggenes, J. and Saltveit, S.J. (1990) Seasonal and spatial microhabitat selection and segregation in young Atlantic salmon, *Salmo salar* L. and brown trout, *Salmo trutta*, L. in a Norwegian river. *Journal of Fish Biology*, **36**(5): 707–720.

Heggenes, J. and Saltveit, S.J. (2007) Summer stream habitat partitioning by sympatric Arctic charr, Atlantic salmon and brown trout in two sub-Arctic rivers. *Journal of Fish Biology*, **71**: 1069–1081.

Heggenes, J., Bagliniere, J.L. and Cunjak, R.A. (1999) Spatial niche variability for young Atlantic salmon (*Salmo salar*) and brown trout (*S. trutta*) in heterogeneous streams. *Ecology of Freshwater Fish*, **8**: 1–21.

Heggenes, J., Brabrand, A. and Saltveit, S.J. (1990) Comparison of 3 methods for studies of stream habitat use by young brown trout and Atlantic salmon. *Transactions of the American Fisheries Society*, **119**: 101–111.

Heggenes, J., Brabrand, A. and Saltveit, S.J. (1991) Microhabitat use by brown trout, *Salmo trutta* L. and Atlantic salmon, *S. salar* L., in a stream – A comparative-study of underwater and river bank observations. *Journal of Fish Biology*, **38**: 259–266.

Heggenes, J., Saltveit, S.J., Bird, D. and Grew, R. (2002) Static habitat partitioning and dynamic selection by sympatric young Atlantic salmon and brown trout in south-west England streams. *Journal of Fish Biology*, **60**(1): 72–86.

Heggenes, J., Krog, O.M.W., Lindas, O.R., Dokk, J.G. and Bremnes, T. (1993) Homeostatic behavioral responses in a changing environment: brown trout (*Salmo trutta*) become nocturnal during winter. *Journal of Animal Ecology*, **62**: 295–308.

Heggenes, J., Omholt, P.K., Kristiansen, J.R., Sageie, J., Okland, F., Dokk, J.G. and Beere, M.C. (2007) Movements by wild brown trout in a boreal river: response to habitat and flow contrasts. *Fisheries Management and Ecology*, **14**: 333–342.

Hickey, M.A. and Closs, G.P. (2006) Evaluating the potential of night spotlighting as a method for assessing species composition and brown trout abundance: a comparison with electrofishing in small streams. *Journal of Fish Biology*, **69**: 1513–1523.

Hill, J. and Grossman, G.D. (1993) An energetic model of micro-habitat use for rainbow trout and rosyside dace. *Ecology*, **74**: 685–698.

Holierhoek, A.M. and Power, G. (1995) Responses of wild juvenile arctic char to cover, light and predator threat. *Nordic Journal of Freshwater Research*, **71**: 296–308.

Hubert, W.A., Harris, D.D. and Wesche, T.A. (1994) Diurnal shifts in use of summer habitat by age-0 brown trout in a regulated mountain stream. *Hydrobiologia*, **284**: 147–156.

Hughes, N.F. (1998) A model of habitat selection by drift-feeding stream salmonids at different scales. *Ecology*, **79**: 281–294.

Hughes, N.F. and Dill, L.M. (1990) Position choice by drift-feeding salmonids: model and test for Arctic grayling (*Thymallus arcticus*) in sub-Arctic mountain streams, interior Alaska. *Canadian Journal of Fisheries and Aquatic Sciences*, **47**: 2039–2048.

Hutchings, J.A. and Morris, D.W. (1985) The influence of phylogeny, size and behavior on patterns of covariation in salmonid life histories. *Oikos*, **45**: 118–124.

Huusko, A., Greenberg, L., Stickler, M., Linnansaari, T., Nykänen, M., Vehanen, T., Koljonen, S., Louhi, P. and Alfredsen, K. (2007) Life in the ice lane: The winter ecology of stream salmonids. *River Research and Applications*, **23**: 469–491.

James, D.A., Wilhite, J.W. and Chipps, S.R. (2010) Influence of drought conditions on brown trout biomass and size structure in the Black Hills, South Dakota. *North American Journal of Fisheries Management*, **30**: 791–798.

Janetski, D.J., Chaloner, D.T., Tiegs, S.D. and Lamberti, G.A. (2009) Pacific salmon effects on stream ecosystems: a quantitative synthesis. *Oecologia*, **159**: 583–595.

Jenkins, T.M. (1969) Social structure, position choice and microdistribution of two trout species (*Salmo trutta* and *Salmo gairdneri*) resident in mountain streams. *Animal Behavior Monographs*, **2**: 56–123.

Jensen, A.J. and Johnsen, B.O. (1999) The functional relationship between peak spring floods and survival and growth of juvenile Atlantic salmon (*Salmo salar*) and brown trout (*Salmo trutta*). *Functional Ecology*, **13**: 778–785.

Johnson, J.H. and Douglass, K.A. (2009) Diurnal stream habitat use of juvenile Atlantic salmon, brown trout and rainbow trout in winter. *Fisheries Management and Ecology*, **16**: 352–359.

Johnson, J.H., Nack, C.C. and McKenna, J.E. (2010) Migratory salmonid redd habitat characteristics in the Salmon River, New York. *Journal of Great Lakes Research*, **36**: 387–392.

Johnson, S.L. and Covich, A.P. (2000) The importance of night-time observations for determining habitat preferences of stream biota. *Regulated Rivers: Research and Management*, **16**: 91–99.

Johnsson, J.I., Carlsson, M. and Sundstrom, L.F. (2000) Habitat preference increases territorial defence in brown trout (*Salmo trutta*). *Behavioral Ecology and Sociobiology*, **48**: 373–377.

Johnsson, J.I., Nobbelin, F. and Bohlin, T. (1999) Territorial competition among wild brown trout fry: effects of ownership and body size. *Journal of Fish Biology*, **54**: 469–472.

Johnston, P., Berube, F. and Bergeron, N.E. (2009) Development of a flatbed passive integrated transponder antenna grid for continuous monitoring of fishes in natural streams. *Journal of Fish Biology*, **74**: 1651–1661.

Jowett, I.G. (1990) Factors related to the distribution and abundance of brown and rainbow trout in New-Zealand clear-water rivers. *New Zealand Journal of Marine and Freshwater Research*, **24**: 429–440.

Jowett, I.G. (1992) Models of the abundance of large brown trout in New Zealand rivers. *North American Journal of Fisheries Management*, **12**: 417–432.

Kennedy, B.P., Nislow, K.H. and Folt, C.L. (2008) Habitat-mediated foraging limitations drive survival bottlenecks for juvenile salmon. *Ecology*, **89**: 2529–2541.

Kennedy, G.J.A. and Strange, C.D. (1982) The distribution of salmonids in upland streams in relation to depth and gradient. *Journal of Fish Biology*, **20**(5): 579–591.

Kennedy, G.J.A. and Strange, C.D. (1986) The effects of intra- and inter-specific competition on the distribution of stocked juvenile Atlantic salmon, *Salmo salar* L., in relation to depth and gradient in an upland trout, *Salmo trutta* L., stream. *Journal of Fish Biology*, **29**: 199–214.

Kocik, J.F. and Ferreri, C.P. (1998) Juvenile production variation in salmonids: population dynamics, habitat, and the role of spatial relationships. *Canadian Journal of Fisheries and Aquatic Sciences*, **55**: 191–200.

Kondolf, G.M. and Wolman, M.G. (1993) The sizes of salmonid spawning gravels. *Water Resources Research*, **29**: 2275–2285.

Korman, J., Yard, M., Walters, C. and Coggins, L.G. (2009) Effects of fish size, habitat, flow, and density on capture probabilities of age-0 rainbow trout estimated from electrofishing at discrete sites in a large river. *Transactions of the American Fisheries Society*, **138**: 58–75.

Kramer, D.L., Rangeley, R.W. and Chapman, L.J. (1997) Habitat selection: patterns of spatial distribution from behavioural decisions. In Godin, J.G.J. (ed.) *Behavioural Ecology of Teleost Fishes*, Oxford University Press, New York, pp. 37–80.

Krebs, C.J. (1989) *Ecological Methodology*, Harper Collins Publications, New York.

L'Abee-Lund, J.H., Langeland, A., Jonsson, B. and Ugedal, O. (1993) Spatial segregation by age and size in Arctic charr: a trade-off between feeding possibility and risk of predation. *Journal of Animal Ecology*, **62**: 160–168.

Louhi, P., Maki-Petays, A. and Erkinaro, J. (2008) Spawning habitat of Atlantic salmon and brown trout: General criteria and intragravel factors. *River Research and Applications*, **24**: 330–339.

Lucas, M.C. and Baras, E. (2000) Methods for studying spatial behaviour of freshwater fishes in the natural environment. *Fish and Fisheries*, **1**: 283–316.

Maki-Petays, A., Muotka, T., Huusko, A., Tikkanen, P. and Kreivi, P. (1997) Seasonal changes in habitat use and preference by juvenile brown trout, *Salmo trutta*, in a northern boreal river. *Canadian Journal of Fisheries and Aquatic Sciences*, **54**: 520–530.

Malard, F., Tockner, K., Dole-Olivier, M.J. and Ward, J.V. (2002) A landscape perspective of surface–subsurface hydrological exchanges in river corridors. *Freshwater Biology*, **47**: 621–640.

Malcolm, I.A., Soulsby, C., Youngson, A.F. and Hannah, D.M. (2005) Catchment-scale controls on groundwater–surface water interactions in the hyporheic zone: Implications for salmon embryo survival. *River Research and Applications*, **21**: 977–989.

Malcolm, I.A., Soulsby, C., Youngson, A.F. and Petry, J. (2003) Heterogeneity in ground water–surface water interactions in the hyporheic zone of a salmonid spawning stream. *Hydrological Processes*, **17**: 601–617.

Marchand, F., Magnan, P. and Boisclair, D. (2002) Water temperature, light intensity and zooplankton density and the feeding activity of juvenile brook charr (*Salvelinus fontinalis*). *Freshwater Biology*, **47**: 2153–2162.

Marchildon, M.A., Annable, W.K., Imhof, J.G. and Power, M. (2011) A high-resolution hydrodynamic investigation of brown trout (*Salmo trutta*) and rainbow trout (*Oncorhynchus mykiss*) redds. *River Research and Applications*, **27**(3): 345–359.

Miller, R.R., Williams, J.D. and Williams, J.E. (1989) Extinctions of North American fishes during the past century. *Fisheries*, **14**: 22–38.

Milner, N.J., Elliott, J.M., Armstrong, J.D., Gardiner, R., Welton, J.S. and Ladle, M. (2003) The natural control of salmon and trout populations in streams. *Fisheries Research*, **62**: 111–125.

Moore, A. and Scott, A. (1988) Observations of recently emerged sea trout, *Salmo trutta* L., fry in a chalk stream, using a low-light underwater camera. *Journal of Fish Biology*, **33**: 359–360.

Morantz, D.L., Sweeney, R.K., Shirvell, C.S. and Longard, D.A. (1987) Selection of microhabitat in summer by juvenile Atlantic salmon (*Salmo salar*). *Canadian Journal of Fisheries and Aquatic Sciences*, **44**: 120–129.

Mouton, A.M., Jowett, I., Goethals, P.L.M. and de Baets, B. (2009) Prevalence-adjusted optimisation of fuzzy habitat suitability models for aquatic invertebrate and fish species in New Zealand. *Ecological Informatics*, **4**: 215–225.

Nestler, J.M., Goodwin, R.A., Smith, D.L., Anderson, J.J. and Li, S. (2008) Optimum fish passage and guidance designs are based in the hydrogeomorphology of natural rivers. *River Research and Applications*, **24**: 148–168.

Nislow, K.H., Folt, C.L. and Parrish, D.L. (2000) Spatially explicit bioenergetic analysis of habitat quality for age-0 Atlantic salmon. *Transactions of the American Fisheries Society*, **129**: 1067–1081.

Olsen, M., Boegh, E., Pedersen, S. and Pedersen, M.F. (2009) Impact of groundwater abstraction on physical habitat of brown trout (*Salmo trutta*) in a small Danish stream. *Hydrology Research*, **40**: 394–405.

Orth, D.J. (1987) Ecological considerations in the development and application of instream flow-habitat models. *Regulated Rivers: Research and Management*, **1**: 171–181.

Ostergaard, S., Hansen, M.M., Loeschcke, V. and Nielsen, E.E. (2003) Long-term temporal changes of genetic composition in brown trout (*Salmo trutta* L.) populations inhabiting an unstable environment. *Molecular Ecology*, **12**: 3123–3135.

Persat, H. and Copp, G.H. (1989) Electrofishing and point abundance sampling for ichthyology of large rivers. In Cowx, I.G. (ed.) *Developments in Electrofishing*, Fishing News Books, Blackwell Scientific Publications, Oxford, pp. 203–215.

Piccolo, J.J., Hughes, N.F. and Bryant, M.D. (2008) Water velocity influences prey detection and capture by drift-feeding juvenile coho salmon (*Oncorhynchus kisutch*) and steelhead (*Oncorhynchus mykiss irideus*). *Canadian Journal of Fisheries and Aquatic Sciences*, **65**(2): 266–275.

Power, G., Brown, R.S. and Imhof, J.G. (1999) Groundwater and fish – insights from northern North America. *Hydrological Processes*, **13**: 401–422.

Price, A.L. and Peterson, J.T. (2010) Estimation and modeling of electrofishing capture efficiency for fishes in wadeable warmwater streams. *North American Journal of Fisheries Management*, **30**: 481–498.

Pulliam, H.R. (1988) Sources, sinks, and population regulation. *American Naturalist*, **132**: 652–661.

Quist, M.C. and Hubert, W.A. (2005) Relative effects of biotic and abiotic processes: A test of the Biotic–Abiotic Constraining Hypothesis as applied to cutthroat trout. *Transactions of the American Fisheries Society*, **134**: 676–686.

Raleigh, R.F., Zuckerman, L.D. and Nelson, P.C. (1986) *Habitat Suitability Index Models and Instream Flow Suitability Curves: Brown Trout*. Report 82, U.S Department of the Interior, Fish and Wildlife Service, National Ecology Center, Washington.

Rimmer, D.M., Paim, U. and Saunders, R.L. (1984) Changes in the selection of microhabitat by juvenile Atlantic salmon (*Salmo salar*) at the summer autumn transition in a small river. *Canadian Journal of Fisheries and Aquatic Sciences*, **41**: 469–475.

Roussel, J.M. and Bardonnet, A. (1995) Diel activity and use of a riffle/pool unit by one-year-old brown trout (*Salmo trutta* L.). *Bulletin Francais de la Peche et de la Pisciculture*, **338**: 221–230.

Roussel, J.M. and Bardonnet, A. (1999) Ontogeny of diel pattern of stream-margin habitat use by emerging brown trout, *Salmo trutta*, in experimental channels: influence of food and predator presence. *Environmental Biology of Fishes*, **56**: 253–262.

Roussel, J.M. and Bardonnet, A. (2002) The habitat of juvenile brown trout (*Salmo trutta* L.) in small streams: Preferences, movements, diel and seasonal variations. *Bulletin Francais de la Peche et de la Pisciculture*: 435–454.

Roussel, J.M., Bardonnet, A. and Claude, A. (1999) Microhabitats of brown trout when feeding on drift and when resting in a lowland salmonid brook: effects on Weighted Usable Area. *Archiv Fur Hydrobiologie*, **146**: 413–429.

Roussel, J.M., Cunjak, R.A., Newbury, R., Caissie, D. and Haro, A. (2004) Movements and habitat use by PIT-tagged Atlantic salmon parr in early winter: the influence of anchor ice. *Freshwater Biology*, 49: 1026–1035.

Sabaton, C., Souchon, Y., Capra, H., Gouraud, V., Lascaux, J.M. and Tissot, L. (2008) Long-term brown trout population responses to flow manipulation. *River Research and Applications*, 24: 476–505.

Saltveit, S.J., Halleraker, J.H., Arnekleiv, J.V. and Harby, A. (2001) Field experiments on stranding in juvenile Atlantic salmon (*Salmo salar*) and brown trout (*Salmo trutta*) during rapid flow decreases caused by hydropeaking. *Regulated Rivers: Research and Management*, 17(4): 609–622.

Scruton, D.A., Pennell, C.J., Bourgeois, C.E., Goosney, R.F., King, L., Booth, R.K., Eddy, W., Porter, T.R., Ollerhead, L.M.N. and Clarke, K.D. (2008) Hydroelectricity and fish: a synopsis of comprehensive studies of upstream and downstream passage of anadromous wild Atlantic salmon, *Salmo salar*, on the Exploits River, Canada. *Hydrobiologia*, 609: 225–239.

Shirvell, C.S. and Dungey, R.G. (1983) Microhabitats chosen by brown trout for feeding and spawning in rivers. *Transactions of the American Fisheries Society*, 112: 355–367.

Slobodkin, L.B. and Rapoport, A. (1974) An optimal strategy of evolution. *Quarterly Review of Biology*, 49: 181–200.

Sloman, K.A. and Baron, M. (2010) Conspecific presence affects the physiology and behaviour of developing trout. *Physiology and Behavior*, 99: 599–604.

Sloman, K., Baker, D. and Wilson, R. (2008) Are there physiological correlates of dominance in natural trout populations? *Animal Behaviour*, 76(4): 1279–1287.

Sloman, K.A., Wilson, L., Freel, J.A., Taylor, A.C., Metcalfe, N.B. and Gilmour, K.M. (2002) The effects of increased flow rates on linear dominance hierarchies and physiological function in brown trout, *Salmo trutta*. *Canadian Journal of Zoology–Revue Canadienne de Zoologie*, 80: 1221–1227.

Soulsby, C., Malcolm, I.A., Tetzlaff, D. and Youngson, A.F. (2009) Seasonal and inter-annual variability in hyporheic water quality revealed by continuous monitoring in a salmon spawning stream. *River Research and Applications*, 25: 1304–1319.

Stanford, J.A. (1996) Landscapes and catchment basins. In Hauer, F.R. and Lamberti, G.A. (eds) *Methods in Stream Ecology*, Academic Press, San Diego, pp. 3–23.

Stradmeyer, L., Hojesjo, J., Griffiths, S.W., Gilvear, D.J. and Armstrong, J.D. (2008) Competition between brown trout and Atlantic salmon parr over pool refuges during rapid dewatering. *Journal of Fish Biology*, 72: 848–860.

Strakosh, T.R., Neumann, R.M. and Jacobson, R.A. (2003) Development and assessment of habitat suitability criteria for adult brown trout in southern New England rivers. *Ecology of Freshwater Fish*, 12: 265–274.

Tharme, R.E. (2003) A global perspective on environmental flow assessment: Emerging trends in the development and application of environmental flow methodologies for rivers. *River Research and Applications*, 19: 397–441.

Wankowski, J.W.J. and Thorpe, J.E. (1979) Spatial distribution and feeding in Atlantic salmon, *Salmo salar* L. juveniles. *Journal of Fish Biology*, 14: 239–247.

Wesche, T.A., Goertler, C.M. and Hubert, W.A. (1987) Modified habitat suitability index model for brown trout in southeastern Wyoming. *North American Journal of Fisheries Management*, 7: 232–237.

West, R.L., Smith, M.W., Barber, W.E., Reynolds, J.B. and Hop, H. (1992) Autumn migration and overwintering of Arctic grayling in coastal streams of the Arctic national wildlife refuge, Alaska. *Transactions of the American Fisheries Society*, 121: 709–715.

White, H.C. (1942) Atlantic salmon redds and artificial spawning beds. *Journal of the Fisheries Research Board of Canada*, 6: 37–45.

Witzel, L.D. and Maccrimmon, H.R. (1983) Redd site selection by brook trout and brown trout in southwestern Ontario streams. *Transactions of the American Fisheries Society*, 112: 760–771.

Wollebaek, J., Heggenes, J. and Røed, K.H. (2010) Disentangling potential stocking introgression and natural migration in brown trout: survival success and recruitment failure in populations with semi-supportive breeding. *Freshwater Biology*, 55: 2626–2638.

Wollebaek, J., Thue, R. and Heggenes, J. (2008) Redd site microhabitat utilization and quantitative models for wild large brown trout in three contrasting boreal rivers. *North American Journal of Fisheries Management*, 28: 1249–1258.

Zale, A.V., Brooke, C. and Fraser, W.C. (2005) Effects of surgically implanted transmitter weights on growth and swimming stamina of small adult westslope cutthroat trout. *Transactions of the American Fisheries Society*, 134: 653–660.

Zimmer, M.P. and Power, M. (2006) Brown trout spawning habitat selection preferences and redd characteristics in the Credit River, Ontario. *Journal of Fish Biology*, 68: 1333–1346.

Zimmer, M., Schreer, J.F. and Power, M. (2010) Seasonal movement patterns of Credit River brown trout (*Salmo trutta*). *Ecology of Freshwater Fish*, 19: 290–299.

10 Salmonid Habitats in Riverine Winter Conditions with Ice

Ari Huusko[1], Teppo Vehanen[2] and Morten Stickler[3]

[1] Finnish Game and Fisheries Research Institute, Manamansalontie 90, 88300 Paltamo, Finland
[2] Finnish Game and Fisheries Research Institute, Paavo Havaksen tie 3, 90014 Oulun yliopisto, Finland
[3] Statkraft AS, Lilleakerveien 6, 0216 Oslo, Norway

10.1 Introduction

Boreal areas are characterized by strong alterations in environmental conditions between seasons. In winter, natural fluvial systems are generally associated with low water temperatures, various ice phenomena, low discharge rates, short daylight time and low heat radiation (Prowse, 2001a, 2001b). Except for low discharges these conditions are in contrast with summer conditions. In fluvial environments, where marked seasonal environmental changes occur, adaptive changes in fish habitat use, behaviour and activity are expected (Chapman, 1966). In summer conditions, the interaction between physical river habitats and the habitat selection of fish has been shown to be complex (Klemetsen et al., 2003). In winter, this complexity increases with the formation of various types of ice and ice-related processes even on a short temporal scale, leading to a dynamic environment (Cunjak, 1996; Huusko et al., 2007; Stickler et al., 2008a; Linnansaari, 2009). Winter conditions challenge the possibilities for specific descriptions of both riverine habitat conditions and habitats preferred by stream-dwelling fish. Notwithstanding the general belief that conditions in winter strongly influence fish activity, the overwintering ecology of stream-dwelling fish has not been studied extensively (Hubbs and Trautman, 1935; Cunjak, 1996; Reynolds, 1997; Huusko et al., 2007; Linnansaari, 2009). In fact, the majority of previous work conducted in winter has been carried out at above-freezing water temperatures without the presence of ice (Huusko et al., 2007). Thus, relatively little is known about the behaviour of fish in relation to ice (Roussel et al., 2004; Stickler et al., 2008a; Linnansaari, 2009), and little experimental work has been conducted on the impact of different ice conditions on fish (but see Finstad et al., 2004a; Linnansaari et al., 2008; Huusko et al., 2011). To this end, technological developments, such as active and passive telemetry, are improving the ability to study fish in cold climate fluvial environments (Greenberg and Giller, 2000; Alfredsen and Tesaker, 2002; Robertson et al., 2004; Linnansaari et al., 2007).

A prerequisite for successful management of fish and their habitats in boreal rivers is to identify and quantify the critical habitat factors limiting fish production that can be influenced by management (Cunjak, 1996; Huusko et al., 2007). This requires a comprehensive understanding of fish ecology in all relevant life history stages, and of fluvial geomorphology and river hydraulics. Ecohydraulics integrates information from ecology and fluid dynamics to analyze and predict ecological responses to hydrological and hydraulic changes taking place in lotic ecosystems. This chapter takes a novel ecohydraulics perspective by considering salmonid ecology during the winter, and particularly interactions with temperatures of water and, in its solid state, as ice. The relationship between winter habitat and salmonid performance is investigated as the information obtained may provide an important addition

Ecohydraulics: An Integrated Approach, First Edition. Edited by Ian Maddock, Atle Harby, Paul Kemp and Paul Wood.
© 2013 John Wiley & Sons, Ltd. Published 2013 by John Wiley & Sons, Ltd.

177

Table 10.1 Main ice processes over the course of winter outlined by three types of rivers (adapted from Huusko *et al.*, 2007).

Ice regimes	River type		
	Small, steep rivers	Large rivers	Regulated rivers
Freeze-up	Border ice	Border ice	Border ice
	Dynamic ice formation	Ice over formation	Dynamic ice formation
Main winter	Extended dynamic ice	Stable ice cover	Less surface ice
	formation	Dynamic ice formation	Local ice runs
	Anchor ice dams	in open riffles	Increased dynamic ice formation
	Local ice runs		
Ice break-up	Thermal and mechanical	Thermal ice break-up	Repeated mechanical ice break-ups
	ice break-up		throughout winter

to the toolbox of both ecologists and resource managers using an ecohydraulics approach to understand and predict fish responses to hydrological and hydraulic changes taking place in winter. The information will help us to understand the relationships between potential bottlenecks in fish production and winter conditions in specific stream systems (Cunjak and Therrien, 1998). Here, winter is considered as a period of ice formation and freezing water temperatures. Further, winter is divided into periods based on distinctive ice conditions (ice regimes) (Huusko *et al.*, 2007), which are described in terms of their basic processes: freeze-up (the channel starts to freeze and dynamic ice processes mostly prevail), main winter (the surface of a stream is mainly, or totally, covered with solid ice) and break-up (the ice cover starts to break up and water temperature rises). Due to fish sampling difficulties in conditions with ice phenomena, cold temperatures and variable flow dynamics, specific descriptions of riverine habitat conditions and which habitats fish are assumed to prefer are limited in number in the scientific literature (Huusko *et al.*, 2007). To this end, almost all winter habitat criteria of salmonids are based on frequency analysis of habitat conditions utilized by different life stages and species (Huusko *et al.*, 2007). These 'utilization functions' depict the conditions that were being used by the fish when the fish observations were made, and are discussed here as habitat use. Habitat use may not always accurately describe a species' habitat preferences, where habitat use is related to prevailing habitat conditions, i.e. to habitat availability. Unfortunately, only a few fish winter habitat studies have utilized this kind of design (for example, Mäki-Petäys *et al.*, 2004). Basic knowledge of lifecycles, life stages and general migration patterns of salmonids is not covered in this chapter, but can be found in Milner *et al.* (2003), Klemetsen *et al.* (2003) and Armstrong *et al.* (2003).

10.2 Ice processes in running waters

Ice formation in running waters may lead to both dynamic and stable conditions depending on channel morphology, gradient and discharge (Alfredsen and Tesaker, 2002; Stickler and Alfredsen, 2009). While ice production in small, steep rivers, as well as in regulated rivers, can be dynamic with both thawing and freezing throughout the entire winter (Lindström and Hubert, 2004; Stickler *et al.*, 2010), ice conditions in lowland rivers are generally more stable (Brown *et al.*, 2011) (Table 10.1). In small headwater streams, snow accumulation can bridge over the channel and also provide stable conditions.

Ice formation (freeze-up regime) begins in late autumn when low air temperatures and negative energy balances cool the water until it reaches freezing point. Ice first appears along river margins (border ice), emergent boulders and in low water velocity areas, for example in pools (Matousek, 1984). In low-gradient rivers and river sections (gradient < 0.4%; Tesaker, 1994), ice continues to grow laterally until a complete ice cover is established, referred to as static ice formation (Carstens, 1966). In steep rivers or river sections (gradient > 0.4%), for example in riffles and rapids, the ice regime is characterized by dynamic ice formation (Figures 10.1 and 10.2). Dynamic ice formation occurs only in turbulent super-cooled water (T_{Water} −0.005 to −0.04 °C; Stickler and Alfredsen, 2009), and consists of the formation of tiny ice particles in the water column, termed 'frazil' (Altberg, 1936; Devik, 1944). These particles are carried along in the water column and readily adhere to any submerged object, or they agglomerate and form floating slush (Figure 10.1). The formation process may take place within a brief period of time with growth at night and thawing during the day, thus creating a highly unstable river environment.

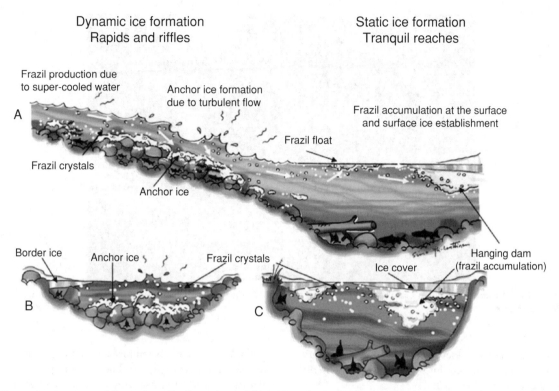

Figure 10.1 A conceptual illustration of ice formation and fish habitat selection in streams and rivers during the freeze-up regime. A: longitudinal profile of a rapid and a pool section of a stream, showing dynamic ice formation in the rapids and static ice formation in the pool section. White arrows indicate the pathways of frazil crystals. B: a cross-section profile of a rapid. C: a cross-section profile of a pool. Juvenile fish are sheltering in the rapid section while adults and large-sized immature fish prefer habitats in tranquil river sections. Updated from Power *et al.*, 1993. Drawings by Simo Yli-Lonttinen, generated from the original work of Environmental Aspects of River Ice. 1993. T.D. Prowse & N.C. Gridley (eds), NHRI Science Report No. 5, 155 p., Environment Canada (NHRI), Saskatoon, SK. © Environment Canada (1993).

Figure 10.2 A view over a small, steep river during a freeze-up. Frazil is adhered to boulders forming anchor ice patches. The shoreline on the right has some border ice. Photo by Morten Stickler.

Figure 10.3 An anchor ice dam in a small, steep river during freeze-up. A large pool was established upstream of the anchor ice dam. Photo by Morten Stickler.

However, in periods of severe cold, dynamic ice formation has also been observed during daytime (Benson, 1955; Stickler and Alfredsen, 2009).

Dynamic ice formation may lead to significant in-stream changes. In pools and slow-flowing river sections with low levels of turbulence, frazil is transported in the water surface or under prevailing ice cover, and may accumulate in large quantities, leading to the formation of 'hanging dams' (Figure 10.1). Large hanging dams may lead to local flooding upstream and reduce wetted areas downstream. Once a hanging dam is formed, it may last until spring (Brown *et al.*, 2000; Kylmänen *et al.*, 2001). In areas with sufficient turbulence, frazil ice may be transported from the water surface down to the river bed, forming anchor ice (Figures 10.1 and 10.3; Ashton, 1986). Anchor ice typically consists of loose ice crystals (with a density similar to water) that may form extensive, porous blankets of ice up to 50 cm thick (Benson, 1955; Power *et al.*, 1999; Stickler and Alfredsen, 2009). However, recent studies by Stickler and Alfredsen (2009) have reported observations of anchor ice with a density similar to solid ice armouring the river bed. During events with anchor ice formation, flow conditions may be substantially altered (Prowse and Gridley, 1993; Beltaos, 1995; Kerr *et al.*, 2002; Stickler *et al.*, 2010). Anchor ice increases both the water level, by its accumulation on the river bed, and the water velocity by smoothing out irregularities on the river bottom (Prowse and Gridley, 1993). Anchor ice rafting (i.e.

anchor ice that removes sediments from the bottom as it moves) together with increased flows can transport substantial amounts of coarse sediment from the river bed (Kempema and Ettema, 2010). Although anchor ice is usually observed in highly turbulent, shallow rapids with rough substrates, it has also been observed in both deep (up to 20 m; Ashton, 1986) and low-gradient areas (Stickler and Alfredsen, 2009). In river sections where the river gradient increases (for example, at the top of rapids) or in riffles with large substrate (boulders), anchor ice may establish 'anchor ice dams', causing similar effects to hanging dams (Figure 10.3). However, riffles and highly turbulent river sections usually remain open due to exchange of energy between releases of frictional heat and ice formation. After the freeze, during the mid-winter ice regime, ice dynamics in flowing water are controlled by local meteorology, topography and the physical characteristics of the river. Continued dynamic subsurface ice formation and the lateral growth of border ice finally lead to channel bridging by surface ice (Figure 10.4). The period dominated by surface ice is considered to be relatively stable in terms of physical habitat variability in comparison with the early winter subsurface ice period (Prowse, 2001a, 2001b). Large rivers in particular, as well as stream and river sections with a low gradient, can generally be characterized as stable and ice-covered. In contrast, small, steep rivers can undergo an extended period of dynamic ice formation before reaching stable winter conditions (Tesaker,

Rapids and riffles

Tranquil reaches

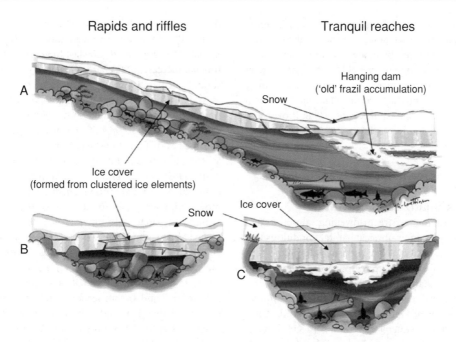

Figure 10.4 A conceptual illustration of ice conditions and fish habitat selection in streams and rivers during the main winter regime. A: longitudinal profile of a rapid and a pool section of a stream, showing stabilized ice cover both in the rapids and in the pool section. In this case, the rapid section has typically become ice covered by clustered ice elements of different origins (such as border ice, anchor ice and frazil slush), although the rapids may also have only a partial ice cover over the whole winter. Ice cover prevents water temperature from super-cooling in the rapids and frazil formation has stopped. Hanging dams, developed during freeze-up, may exist throughout the main winter. B: a cross-section profile of a rapid. C: a cross-section profile of a pool. Juvenile fish are sheltering in the rapid section while adults and large-sized immature fish prefer habitats in tranquil river sections. Reduced flow may decrease the amount of habitat suitable for juvenile salmonids in the rapids (see also Figure 10.1). Updated from Power *et al.*, 1993. Drawings by Simo Yli-Lonttinen, generated from the original work of Environmental Aspects of River Ice. 1993. T.D. Prowse & N.C. Gridley (eds), NHRI Science Report No. 5, 155 p., Environment Canada (NHRI), Saskatoon, SK. © Environment Canada (1993).

1994). In these systems, surface ice formation is prevented by significant turbulent water flow and such conditions can be dominated by dynamic ice formation throughout the winter. However, partial or full ice cover may develop either as a result of frazil ice accumulation – where frazil ice adheres to solid objects such as boulders, and border ice clustering to form larger structures – or from upstream anchor ice dams (Stickler *et al.*, 2010).

In man-modified river systems, such as flow-regulated streams, the stable mid-winter ice regime may be altered significantly (Alfredsen and Tesaker, 2002). Hydropower production, particularly in high-head systems, changes the discharge and the water temperature, causing repeated ice break-ups and increased ice production. Regulated river systems during winter can be characterized as unstable with reduced periods of stable ice cover and an increased degree of dynamic ice production. Ice cover in regulated rivers is also typically susceptible to breaking. Local ice runs and ice jams may even occur in mid-winter.

During the ice break-up regime in spring, the conditions that lead to an ice break-up fall between two extremes: thermal break-up and mechanical break-up (Prowse and Gridley, 1993; Prowse, 1994). Thermal ice break-ups are initiated as solar radiation increases, thereby reducing ice strength before the main spring runoff occurs. As the water level increases, weakened ice cover breaks up and moves downstream without any significant effect on river flood levels and with only modest disturbance to the river channel (Prowse, 1994). However, a decrease in water quality may occur if locally polluted snow and ice melt (Prowse and Culp, 2004; Laudon *et al.*, 2005). In some cases, for example in regulated rivers or rivers experiencing spates of warm weather with heavy rainfall, ice break-up may occur before the ice cover is thermally weakened, i.e. mechanical break-up. Mechanical ice break-up has the potential to result in rapid break-up with ice jamming and ice-induced flooding (Prowse, 1994). In small, steep streams with several anchor ice dams, mechanical ice break-up may lead to a 'domino effect' where a flood wave builds up and results in severe flooding when dams are breaking. Furthermore, during mechanical ice break-up, considerable changes to the river bed may occur due to increased scouring (caused both by high water velocity and the transported ice) and sediment transport (Ashton, 1986; Cunjak *et al.*, 1998; Prowse and Culp, 2004).

10.3 Salmonids in winter ice conditions

10.3.1 Acclimatization in winter

In boreal environments, stream-dwelling salmonids face seasonal changes in their physical habitat conditions that challenge their capacity to cope with such alterations (e.g. Fretwell, 1972). The freeze-up regime is considered to be the most drastic change due to the ever-changing hydro-physical conditions associated with low water temperatures (Cunjak, 1996; Hurst, 2007; Huusko *et al.*, 2007; Linnansaari, 2009). Water temperature affects fish significantly because they are poikilothermal and their body temperature varies according to the surrounding environment (Diana, 1995). When water temperatures decline in the autumn, the metabolism of fish is reduced and their ability to swim, feed, avoid predators and defend their position decline (Cunjak, 1996; Huusko *et al.*, 2007). Fish rely, to a large degree, on energy stored in their bodies to survive, and winter performance is therefore dictated by the need to minimize energy expenditure (Heggenes *et al.*, 1993; Cunjak, 1996).

Conditions that tax energy reserves require fish to make appropriate behavioural decisions to ensure their survival. Acclimation to winter conditions involves physiological and behavioural adaptation, which strongly affect performance and habitat selection (Somero *et al.*, 1996; Cunjak, 1996; Reynolds, 1997). One of the most distinctive behavioural changes that takes place when water temperatures cool is the search for suitable winter habitats. The autumnal shift from summer habitats to overwintering habitats in stream-dwelling fish is triggered by cooling water temperatures, prompting them to find a habitat where they can conserve energy (Fausch and Young, 1995; Cunjak, 1996), shelter from harsh conditions (Huusko *et al.*, 2007) and be protected from endothermic predators and piscivores (Valdimarson and Metcalfe, 1998). This shift typically occurs at water temperatures between 3 and 6 °C (Hillman *et al.*, 1987; Jakober *et al.*, 1998; Nykänen *et al.*, 2001; Bramblet *et al.*, 2002), but other studies have reported shifts at temperatures of 10 °C for

Atlantic salmon (*Salmo salar)* (Rimmer *et al.*, 1983) and up to 14 °C for adult European grayling (*Thymallus thymallus)* (Nykänen *et al.*, 2004a). In addition to temperature, an increase in discharge (Peterson, 1982; Tschaplinski and Harman, 1983; Youngson *et al.*, 1983), and possibly even changes in day length are thought to play a role in the timing of autumnal habitat shifts (Huusko *et al.*, 2007). The scale of the habitat shift may vary, depending on the fish species or their life stage, from a few metres within a riffle up to tens of kilometres of river length (Rimmer *et al.*, 1983; Gowan *et al.*, 1994; Young, 1998; Nykänen *et al.*, 2004a, 2004b). Young, small-sized fish often only move within a river section towards large cobble–boulder dominated patches (Rimmer *et al.* 1984; Cunjak *et al.* 1998; Heggenes and Dokk, 2001; Mäki-Petäys *et al.*, 1997, 2004). Large juvenile and adult salmonids cease to defend their summer feeding territories in rapids and runs, and emigrate from such fast velocity sites with a high potential for frazil and anchor ice formations to more tranquil habitats or river margins with more stable ice conditions (Heggenes *et al.*, 1993; Brown and Mackay, 1995; Simpkins *et al.*, 2000; Nykänen *et al.*, 2001, 2004a; Saraniemi *et al.*, 2008). However, such shifts are not mandatory if the summer habitat area also contains suitably sheltered locations for overwintering. For example, Linnansaari *et al.* (2005, 2009) and Stickler *et al.* (2008a) reported that during the freeze-up regime, salmonids do not necessarily avoid frazil and anchor ice in rivers where coarse substrates exist. It is also suggested that salmonids utilize anchor ice as cover (Whalen and Parrish, 1999; Roussel *et al.*, 2004; Stickler *et al.*, 2008a). Therefore, even in the event of severe anchor ice, increased habitat shift may not occur, especially amongst juveniles (Roussel *et al.*, 2004; Stickler *et al.*, 2008b; Linnansaari and Cunjak, 2010).

Another distinctive behaviour that occurs towards winter is the suppression of daytime activity to simulate nocturnal activity (Eriksson, 1978; Cunjak, 1988; Heggenes *et al.*, 1993). Such behaviour could have evolved in response to adverse physical conditions associated with autumnal flooding or ice formation. It has been suggested that leaving the river bed before nightfall may reduce the risk of being displaced by, or entrapped in, ice (Hartman, 1965; Heggenes *et al.*, 1993; Whalen *et al.*, 1999). However, as the switch to nocturnal behaviour takes place long before ice formation commences – at a water temperature threshold of about 8–10 °C (Rimmer *et al.*, 1983; Fraser *et al.*, 1993), displacement and entrapment cannot be the sole reasons for this. Balancing decreased mortality risk when sheltering with increased feeding rate when exposed is believed to be a key determinant of diel patterns of sheltering in many animals. Orpwood *et al.* (2010; see also Orpwood *et al.*, 2006) reported that, both in summer and winter, food availability did not affect the extent to which young Atlantic salmon parr were nocturnal, revealing that the risk of predation is a more likely explanation for nocturnal behaviour.

At low temperatures, fish are less able to avoid endothermic predators, such as mink (*Mustela vison*) and otter (*Lutra lutra*) as these animals are not encumbered by the same metabolic constraints as fish (Fraser *et al.*, 1993; Heggenes *et al.*, 1993; Valdimarson and Metcalfe, 1998; Metcalfe *et al.*, 1999; Hiscock *et al.*, 2002). Nocturnal activity is assumed to prevail throughout the winter, although assessment of the effects of different ice regimes on the activity of fish has not been considered in previous studies. To this end, Linnansaari *et al.* (2008) observed that although juvenile Atlantic salmon were nocturnal throughout winter, daytime activity increased after surface ice formed. Similar observations were made by Gregory and Griffith (1996), who showed that juvenile rainbow trout (*Oncorhynchus mykiss*) did not conceal themselves in coarse substrates as often during the daytime under surface ice. A river channel with complete ice-cover presumably creates a lower risk of predation from endothermic animals (Valdimarson and Metcalfe, 1998) by making it more difficult to access in-stream habitats, for example.

After acclimatization to winter conditions before and during the freeze-up regime, stream-dwelling salmonids have been observed to be highly sedentary during the main winter regime with stable ice cover (Meyer and Griffith, 1997; Komadina-Douthwright *et al.*, 1997; Jakober *et al.*, 1998, Muhfeld *et al.*, 2001; Nykänen *et al.*, 2004a, 2004b; Linnansaari *et al.*, 2008, 2009; Saraniemi *et al.*, 2008). Due to the primary concern of minimizing energy expenditure, fish seem to minimize their movements, and if required to do so by local changing environmental conditions, they adjust by moving relatively short distances (Linnansaari *et al.*, 2008, 2009; Brown *et al.*, 2000; Saraniemi *et al.*, 2008). In fact, stable ice cover, as such, may increase the use of a particular habitat thought to be unsuitable, e.g. during the freeze-up period. Linnansaari *et al.* (2008) reported that during the open water and freezing period, no Atlantic salmon parr were observed in a stream section with small substrate sizes, while once the section was covered in surface ice, the parr dispersed in this area and remained there until ice break-up. Habitat stability in the form of stable ice cover seems to be important to fish (Finstad *et al.*, 2004b, 2004c). On the other hand, as winter progresses, mechanical ice break-up due to warm weather,

increased discharge or flow regulation can return freeze-up like conditions, even forcing fish to make multiple long-range movements as new additional ice formation occludes habitats used earlier (Brown *et al.*, 2000; 2001). In addition, Cunjak *et al.* (1998) showed evidence that mechanical ice break-ups triggered by rain-on-snow in midwinter resulted in severe scouring of the stream bed in Catamaran Brook (New Brunswick, Canada), decreasing the apparent survival rates of young Atlantic salmon.

In spring, ice cover deteriorates due to an increase in temperature and discharge. The break-up of stable ice cover transforms a stream or river back into an open system. For salmonids, the vernal period of warm acclimation is not as physiologically demanding as cold acclimation in autumn, with the exception that the fish have to re-acclimatize when weak from having endured the winter (Cunjak *et al.*, 1998). Improved environmental conditions and feeding opportunities following the ice cover period typically ensure a rapid increase in body lipids and subsequent growth (Cunjak *et al.*, 1998; Koljonen *et al.*, 2012). At the same time, fish behavioural and habitat use patterns typical in summer conditions are restored.

Vernal habitat shifts by stream-dwelling salmonids often take place in a reverse direction to the autumnal shift. For example, European grayling abandoned over-wintering sites in lentic pool-like river sections during and just after the thermal ice break-up, as flow rates and water temperature increased; they were then found in their summer habitats of rapids and runs when flooding had started to decrease slightly and water temperature had reached 5 °C (Nykänen *et al.*, 2004a, 2004b). In addition, anadromous and adfluvial salmonid adults, having overwintered in pool sections of rivers, leave the river just after ice break-up and when high discharges still prevail (Komadina-Douthwright *et al.*, 1997; Saraniemi *et al.*, 2008).

10.3.2 Winter habitat criteria

There are essential differences between the way salmonids use summer and winter habitats (Mäki-Petäys *et al.*, 1997, 2004; Enders *et al.*, 2007). Habitat use by salmonids has been shown to be size-related in summer, with larger fish using deeper and swifter-flowing stream areas with coarser substrates and often also dominating profitable feeding positions (Schlosser, 1987; Hughes and Dill, 1990). The 'bigger fish–deeper habitat' relationship frequently documented for stream salmonids is mainly related to the risk of predation from aquatic and terrestrial predators (e.g. Schlosser, 1987) or to inter-cohort competition, whereby larger individuals restrict smaller conspecifics to less favourable habitats (e.g. Hughes and Dill, 1990; Mäki-Petäys *et al.*, 1997). During acclimatization to winter with cooling water temperatures, i.e. before and during the freeze-up regime, such size-specific habitat use appears to be disrupted, resulting in relatively similar habitat criteria for all salmonid size classes throughout the winter period (Mäki-Petäys *et al.*, 1997, 2004).

Size-related habitat use is generally regarded as a result of avoiding intraspecific overlap in resource use (e.g. Mäki-Petäys *et al.*, 1997). Thus, size-independent habitat use in winter, if it proves to be more common than previously believed (Mäki-Petäys *et al.*, 1997, 2004), may have far-reaching effects on wintering salmonid assemblages. This is especially applicable in severe freeze-up and main winter conditions, where ice formation and reduced flow may decrease the amount of habitats suitable for juvenile salmonids. Closely similar habitat use by different-sized salmonids may create intensive intra- and interspecific competition (Cunjak *et al.*, 1998; Whalen and Parrish, 1999). Size-independent habitat use by juvenile salmonids in winter documented by Jakober *et al.* (2000) indicated a high degree of both intra- and interspecific overlap in daytime habitat use by bull trout (*Salvelinus confluentus*) and cutthroat trout (*Oncorhynchus clarki*) in two Rocky Mountain streams. In contrast, Jakober *et al.* (2000) showed strict habitat partitioning, both among and within species, during winter nights. Close associations of different sized fish have been typically explained by decreased territorial behaviour and generally reduced activity due to the need to save energy at lower water temperatures (Heggenes *et al.*, 1993; Brown, 1999; Vehanen and Huusko, 2002). For example, Cunjak and Power (1986) concluded that aggregation of two trout species in winter results from the cessation of territorial behaviour. Harwood *et al.* (2001), however, suggested that aggregation results from salmonids defending smaller territories in winter rather than from a total lack of territorial behaviour. Salmonids are also known to compete for shelter during wintertime (Harwood *et al.*, 2002). Based on these findings, competition for suitable winter habitat is more probable during daylight hours, when salmonids are documented to shelter among the substrate (e.g. Heggenes *et al.*, 1999; Heggenes and Dokk, 2001).

Suitable winter habitats for salmonids in streams and rivers are sites that allow fish to minimize energy expenditure and, at the same time, maximize protection from environmental variation and predation (Heggenes *et al.*, 1993; Cunjak, 1996; Armstrong *et al.*, 2003; Mäki-Petäys *et al.*, 2004). Juvenile salmonids, regardless of size, use

large substrate sizes for their wintering habitat (from large pebbles to boulders) (Rimmer *et al.*, 1984; Cunjak, 1988; Heggenes and Dokk, 2001; Mitro and Zale, 2002; Mäki-Petäys *et al.*, 1997, 2004; Enders *et al.*, 2007; Linnansaari *et al.*, 2008). Large, unembedded substratum (i.e. loose structure with plenty of interstitial sites among cobble and boulder formations) plays a key role, especially during the periods of subsurface ice formation (Stickler *et al.*, 2008a; 2008b). Mäki-Petäys *et al.* (2004) reported that all size classes of young Atlantic salmon preferred cobbles or boulders in autumn during freeze-up and slightly finer substrates (large pebbles–boulders) in main winter. Cunjak (1988) found that young Atlantic salmon of 5–15 cm in length sheltered under cobbles and boulders of 11–41 cm in diameter, with young-of-year juveniles taking advantage of rocks of 16–19 cm and elder juveniles using slightly larger rocks of 21–23 cm in diameter, on average (Figure 10.1; Figure 10.4). Rimmer *et al.* (1984) concluded that over 68% of the rocks used for sheltering by young Atlantic salmon were over 20 cm in diameter, and this was clearly larger than the stones used in summer conditions (6–7 cm in diameter, on average). Young brown trout (*Salmo trutta*) selected coarse rocks (6–51 cm in diameter) during daytime concealment in the substratum, compared to dark hours (from gravel to 12 cm rocks), when fish were out of shelters (Heggenes *et al.*, 1993). The importance of the availability of shelters is stressed by observations that the number of fish overwintering in a particular area goes hand in hand with the availability of shelters in the substratum (Tschaplinski and Harman, 1983; Meyer and Griffith, 1997; Harvey *et al.*, 1999; Finstad *et al.* 2007). In addition, according to Stickler *et al.* (2008a, 2008b), Atlantic salmon parr could remain in the stream reaches regardless of anchor ice deposition, as they were able to find suitable shelter in the interstices of rocks and thus could avoid any harmful effects of accumulating ice (Figure 10.1). Besides coarse substrate, several other features have been reported as providing suitable shelter, including large woody debris, undercut banks, vegetation and ice (Baltz *et al.*, 1991; Heggenes *et al.*, 1993; Cunjak, 1996; Muhlfeld *et al.*, 2003; Linnansaari *et al.*, 2008; Johnson and Douglass, 2009). In large rivers, juveniles can also use deep, slow-flowing waters as shelter (Bonneau and Scarnecchia, 1998; Hiscock *et al.*, 2002) both in winter and summer conditions (Linnansaari *et al.*, 2010).

The importance of slow velocity refugia among the stream bed structures to overwintering salmonids is widely recognized, with Brown *et al.* (2011) concluding that suitable water velocities within these sites should be less than one body length per second. However, in fish habitat studies more commonly measured mean water column velocity at the positions of young salmonids reportedly indicate large variation, although there seems to be a trend towards lower velocities compared to corresponding measurements conducted in summer habitats. Mäki-Petäys et al. (1997, 2004) reported that overwintering young Atlantic salmon and brown trout were located in deeper habitats farther from the shore and larger fish seemed to prefer areas with slower velocities than their summer habitats. This resulted in relatively similar habitat criteria of water velocity and depth for all size classes of these salmonids during winter, whereby stream areas with slow mean water column velocity (< 20 cm·s^{-1}) and moderate depth (30–50 cm) were preferred by all juvenile fish, regardless of size (see also Jakober et al., 2000). Many other previous studies, conducted either in summer or winter, indirectly support these results. In winter, the typical values of mean water column velocities reported are < 40 cm s^{-1} for juvenile Salmo spp. (Rimmer et al., 1984; Heggenes et al., 1993; Whalen and Parrish, 1999; Heggenes et al., 1999; Heggenes and Dokk, 2001), and < 30 cm s^{-1} for Oncorhynchus spp. (Baltz et al., 1991; Quinn and Peterson, 1996; Harper and Farag, 2004). However, the consistency of species assemblage in a certain stream reach seems to have implications on the use of velocity fields. Johnson and Douglass (2009) found that Atlantic salmon juveniles occupied sites of relatively swift water velocities (mean water column values 32–68 cm s^{-1}) when in sympatry with brown trout and rainbow trout. Enders et al. (2007) reported that brown trout juveniles in winter preferred slower water velocities than Atlantic salmon juveniles (mean water column velocity: brown trout 30 vs. salmon 49 cm s^{-1}; refugia velocities 9 and 13 cm s^{-1}, respectively), but found no preference for depth for either species. Indeed, depth, as such, may be less important in winter habitat selection because salmonids have been reported to use a large variation of depths, from > 1-m pools to < 20-cm riffles in winter (e.g. Cunjak and Power, 1987; Baltz et al., 1991; Heggenes et al. 1993; Whalen and Parrish, 1999; Heggenes and Dokk, 2001; Harper and Farag, 2004; Enders et al., 2007). The variability in both the observed depth and mean water column velocity use by young salmonids is obviously a consequence of the site-specific availability of sheltered locations in relation to spatial distribution of depth and velocity fields.

Adult and large-sized immature salmonids generally adopt a different behavioural strategy to facilitate survival during freeze-up and main winter regimes, using slow-flowing, deep sections of rivers (Heggenes et al.,

1993; Brown and Mackay, 1995; Jakober et al., 1998; Brown, 1999; Nykänen et al., 2004a, 2004b; Östergren and Rivinoja, 2008; Saraniemi et al., 2008). While juvenile fish are able to use crevices within the substratum or bury themselves under layers of stones, adult fish, due to their size, are not often able to find suitable cover enabling them to shelter against velocity and ice events (Figure 10.1; Figure 10.4). Many kinds of lentic habitats, such as pools, off-channel ponds (or alcoves), logjams and beaver ponds have been described as suitable overwintering areas for adult salmonids (Tschaplinski and Harman, 1983; Swales and Levings, 1989; Nickelson et al., 1992; Harper and Farag, 2004; Lindström and Hubert, 2004). However, during the onset and progress of winter, dynamic ice conditions can make habitat conditions unsuitable for fish, even in low-gradient habitats like pools. For example, Komadina-Douthwright et al. (1997), Kylmänen et al. (2001), Lindstrom and Hubert (2004) and Barrineau et al. (2005) have reported that hanging dams can occupy up to 80% of pool volume and cause dramatic changes in habitat conditions within pools, including reduced physical space and swift local current velocities. Adult salmonids have been observed to be more mobile in areas with unstable ice conditions than in areas with stable ice conditions (Jacober et al., 1998; Brown et al., 1994, 2000; Simpkins et al., 2000; Saraniemi et al., 2008). Because areas with stable ice conditions, and related stable habitats, are often limited in streams and rivers in winter, it is common for overwintering larger-sized salmonids to be found in relatively dense groups or aggregations within those habitats suitable for overwintering (Cunjak, 1996).

Although the basic pattern of habitat selection, i.e. strong selection of slow-velocity habitats by overwintering adult and large-sized immature salmonids, seems uniform between different species and rivers (Heggenes et al., 1993; Brown and Mackay, 1995; Cunjak, 1996; Nykänen et al., 2004b; Saraniemi et al., 2008), there are few field data on the exact conditions prevailing in overwintering sites. Much of this is due to the extreme field conditions necessary for carrying out surveys when rivers are covered with ice. Radiotelemetry studies carried out during ice formation and the surface ice cover period have found that radio-tagged, large (over 60 cm in length) brown trout in a sub-Arctic river (with winter discharges between 3 and 10 m^3 s^{-1}) in northern Finland consistently used pool habitats (96% of observations) with water depths from 2 m up to 30 m and estimated water velocity < 10 cm s^{-1} (Saraniemi et al., 2008). European grayling in another sub-Arctic river in northern Finland also selected pool habitats with sand and gravel beds; radio-tagged fish being located mainly at water depths of 1–2.5 m and mean water column velocities < 30 cm s^{-1} during the freeze-up ice regime (discharge about 10 m^3 s^{-1}) (Nykänen et al., 2004a). In the main winter, with a clearly lower discharge of about 3 m^3 s^{-1}, grayling were located at water depths of < 1 m, and mean water column velocities < 5 cm s^{-1} under 70–80 cm thick ice cover (Nykänen et al., 2004a, 2004b). Others studies, based on diving observations in the ice-free sections of rivers (Cunjak and Power, 1986; Calkins, 1989; Heggenes et al., 1993), have found similar results on habitat use by adult salmonids: fish consistently used deep, slow-velocity areas (mean water column velocity < 40 cm s^{-1}, but < 5 cm s^{-1} at the focal sites of fish) prone to surface ice formation.

10.4 Summary and ways forward

The specific descriptions of riverine habitat conditions and the preferences of salmonids and other fish in winter conditions form a challenging issue. In-stream habitats of streams and rivers are extremely variable due to dynamic ice formation both during the autumn freeze-up and potential mid-winter ice break-ups. In addition, inter- and intra-winter variability in habitat conditions seems to be highly river-specific (Stickler and Alfredsen, 2009). According to present knowledge, salmonids seem to be seasonally local fish with seasonally differing habitat requirements. Well before freeze-up, stream-dwelling fish acclimatize to winter by behavioural adaptation and move to suitable overwintering habitats (Figure 10.5; Cunjak,

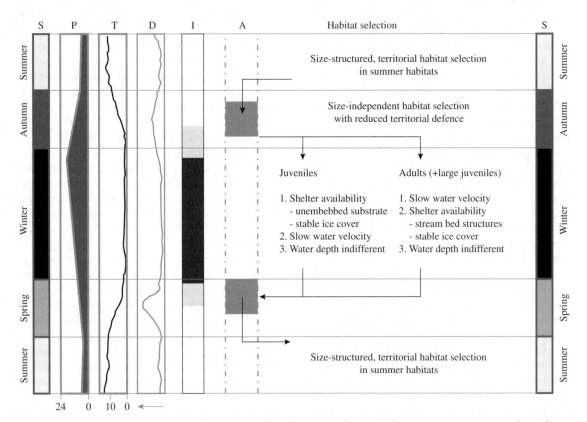

Figure 10.5 A chart summarizing the typical environmental conditions prevailing in northern temperate streams together with corresponding responses in salmonid physical habitat selection. S = season; P = photoperiod (dark shading = night length (hours)); T = water temperature (°C); D = discharge pattern; I = ice processes (white = no ice, light grey = freeze-up regime, grey = ice cover regime, light grey = ice break-up regime); A = acclimatization to changing conditions (grey box = main acclimatization period between warm and cold environmental conditions, with habitat shifts and behavioural changes (see text), dashed line = minor scale adjustments to prevailing conditions); habitat selection = main modes of habitat selection during summer and winter with stream ice. Main habitat characteristics and their weight in habitat selection for juvenile and adult salmonids during and after acclimatization to the cold water period are given.

1996; Huusko *et al.*, 2007; Linnansaari, 2009). In general, salmonids are reportedly day-active, sit-and-wait predators in summer (Hughes and Dill, 1990; Heggenes *et al.*, 1993), whereas at the onset of winter they adopt a shelter-and-move strategy, avoiding the risks of predation and ice damage by being active at night and seeking shelter in the streambed during the day (e.g. Heggenes *et al.*, 1993; Heggenes and Dokk, 2001). Complex mixes of habitat features can provide suitable winter habitats for salmonids. To this end, shelter availability seems to be a key variable that may alone govern the habitat selection and performance of salmonids irrespective of the age/size of the fish. Fish can tolerate a wide range of depths and water column velocities if the river bed substratum, as such, provides shelter, offering protection from predation, displacement and physical damage (Figure 10.5; Stickler *et al.*, 2008a, 2008b; Finstad *et al.*, 2009).

Notwithstanding the recent improved understanding of winter ecology of stream fish (e.g. Finstad *et al.*, 2007; Stickler *et al.*, 2008a, 2008b; Linnansaari, 2009), there remains much to be learned about the winter habitat requirements of salmonids (Cunjak, 1996; Huusko *et al.*, 2007). There is a clear need for innovative ways to describe habitat availability and selection in winter conditions. To better define the winter habitat requirements of salmonids, more research needs to be conducted not only in winter but also in other seasons, on various scales (e.g. microhabitat vs. macrohabitat; Rabeni and Sowa, 1996) and in different kinds of streams and rivers, as well as under experimental conditions. Although habitat complexity is considered to be good for successful performance of overwintering fish (Cunjak *et al.*, 1998), the functional understanding of how the physical structures of the river bed and ice processes interact with fish performance and production is unclear. Because abiotic and biotic factors change over different spatio-temporal scales in stream environments, it will be a complicated task to analyze and predict which particular habitats fish should occupy under varying stream conditions. There is a need to understand the role of the spatial relationships of different habitats for each life stage and season affecting the overall production of stream-dwelling salmonids. The functional habitat unit concept, launched by Kocik and Ferreri (1998), could serve as a template for more comprehensive analyses of seasonal habitat shifts and the spatial ecology of salmonids. Any valid management scheme aiming to assess or enhance stream habitats for salmonids in northern areas needs to rest on the knowledge of season- and size-specific habitat requirements of salmonid species at various scales.

With respect to the spatio-temporally constrained nature of river ice dynamics during the winter, river ice processes are not conventionally considered in lotic environment assessment and modelling tools, despite the importance of ice dynamics for cold-climate stream systems. Existing hydraulic models are not only weakly suited to model low-gradient river systems in winter but they also cannot be used to forecast ice processes in rivers that are affected by dynamic ice formation. The use of hydraulic models or various spatial mapping tools that do not consider both static and dynamic ice formation are usually incomplete when considering in-stream dynamics in steep rivers affected by ice formation. In addition, most of the biological information on salmonid habitat selection applied so far in habitat-hydraulic models is derived from measurements made during summer (Cunjak, 1996; Huusko *et al.*, 2007) and is, therefore, not appropriate for winter habitat simulations (but see, for example, Mäki-Petäys *et al.*, 1997, 2004; Nykänen *et al.*, 2004a, 2004b). To this end, recent developments in the acquisition and analysis of airborne- or shore-sensed images and video imagery have the potential to improve habitat analysis during different ice regimes (e.g. Bradley *et al.*, 2002). Using ground-penetrating radar, it is possible to monitor anchor ice and frazil ice development, and to gain accurate measurements of surface ice thickness, frazil ice float and hanging dam dimensions, and their temporal durations (Kylmänen *et al.*, 2001). Recent technological developments to monitor fish behaviour and movements (Linnansaari and Cunjak, 2007; Johnston *et al.*, 2009) by active and passive telemetry seem to offer promising tools to gather individual-based information on fish performance under icy conditions. All of the above examples of methodological developments represent a step forward, but they should be used more routinely in the context of stream habitat studies to reveal their usefulness in practical stream management.

Challenges and increased pressure due to anthropogenic disturbances (hydro-regulation, land-use activities) and predicted climate change scenarios (Corell, 2006) pose a threat to northern fluvial ecosystems and challenge the future management of salmonid populations. Remedial actions, such as river habitat enhancement and minimum environmental flow assessments, are today common procedures used in freshwater fisheries management to increase the habitat quality of degraded rivers. However, the impact of these actions on the river ice regime has only attracted limited attention, despite its importance, not to mention its subsequent potential impact on the performance of overwintering fish (but see Scruton *et al.*,

2005). Studies focusing on changes in the ice regime and ice formation due to remedial action should have, as a high priority, a focus on ensuring that suitable habitats are created when making habitat improvement efforts. Correspondingly, there is a need to investigate habitat selection, growth and survival rates of both juvenile and adult salmonids in man-modified rivers that experience extensive dynamic ice formation during winter.

References

Alfredsen, K. and Tesaker, E. (2002) Winter habitat assessment strategies and incorporation of winter habitat in the Norwegian habitat assessment tools. *Hydrological Processes*, **16**: 927–936.

Altberg, W.J. (1936) Twenty years of work in the domain of underwater ice formation, 1915–35. *International Union of Geodesy and Geophysics, International Association of Scientific Hydrology*, **23**: 373–407.

Armstrong, J.D., Kemp, P.S., Kennedy, G.J.A., Ladle, M. and Milner, N.J. (2003) Habitat requirements of Atlantic salmon and brown trout in rivers and streams. *Fisheries Research*, **62**: 143–170.

Ashton, G.D. (1986) *River and Lake Ice Engineering*, Water Resources Publications, Littleton, Colorado, USA.

Baltz, D.M., Vondracek, B., Brown, L.R. and Moyle, P.B. (1991) Seasonal changes in microhabitat selection by rainbow trout in a small stream. *Transactions of the American Fisheries Society*, **120**: 166–176.

Barrineau, C.E., Hubert, W.A., Dey, P.D. and Annear, T.C. (2005) Winter ice processes and pool habitat associated with two types of constructed instream structures. *North American Journal of Fisheries Management*, **25**: 1022–1033.

Beltaos, S. (1995) *River Ice Jams*, Water Research Publications, Ontario, Canada.

Benson, N.G. (1955) Observations on anchor ice in a Michigan trout stream. *Ecology*, **36**: 529–530.

Bonneau, J.L. and Scarnecchia, D.L. (1998) Seasonal and diel changes in habitat use by juvenile bull trout (*Salvelinus confluentus*) and cutthroat trout (*Oncorhynchus clarki*) in a mountain stream. *Canadian Journal of Zoology*, **76**: 783–790.

Bradford, M.J. (1997) An experimental study of stranding of juvenile salmonids on gravel bars and in side channels during rapid flow decreases. *Regulated Rivers: Research and Management*, **13**: 295–401.

Bradley, A.A., Kruger, A., Meselhe, E.A. and Muste, M.V.I. (2002) Flow measurement in streams using video imagery. *Water Resources Research*, **38**: 1315.

Bramblett, R.G., Bryant, M.D., Wright, B.E. and White, R.G. (2002) Seasonal use of small tributary and main-stem habitats by juvenile steelhead, coho salmon, and Dolly Varden in a southeastern Alaska drainage basin. *Transactions of the American Fisheries Society*, **131**: 498–506.

Brown, R.S. (1999) Fall and early winter movements of cutthroat trout, *Oncorhynchus clarki*, in relation to water temperature and ice conditions in Dutch Creek, Alberta. *Environmental Biology of Fishes*, **55**: 359–368.

Brown, R.S. and Mackay, W.C. (1995) Fall and winter movements of and habitat use by cutthroat trout in the Ram River, Alberta. *Transactions of the American Fisheries Society*, **124**: 873–885.

Brown, R.S., Hubert, W.A. and Daly, S.F (2011) A primer on winter, ice, and fish: What fisheries biologists should know about winter ice processes and stream-dwelling fish. *Fisheries*, **36**: 8–26.

Brown, R.S., Power, G. and Beltaos, S. (2001) Winter movements and habitat use of riverine brown trout, white sucker and common carp in relation to flooding and ice break-up. *Journal of Fish Biology*, **59**: 1126–1141.

Brown, R.S., Stanislawski, S.S. and Mackay, W.C. (1994) Effects of frazil ice on fish. In Prowse, T.D. (ed.) *Proceedings of the Workshop on Environmental Aspects of River Ice*, National Hydrology Research Institute, Saskatoon, Saskatchewan, NHRI Symposium series **12**: 261–278.

Brown, R.S., Power, G., Beltaos, S. and Beddow, T.A. (2000) Effects of hanging ice dams on winter movements and swimming activity of fish. *Journal of Fish Biology*, **57**: 1150–1159.

Calkins, D.J. (1989) *Winter Habitats of Atlantic Salmon, Brook Trout, Brown Trout and Rainbow Trout*. Special Report 89-34, U.S. Army Corps of Engineers, Cold regions Research and Engineering Laboratory, Hanover, NH, USA.

Carstens, T. (1966) Experiments with supercooling and ice formation in flowing water. *Geophysica Norwegica*, **XXVI**(9): 1–18.

Chapman, D.W. (1966) Food and space as regulators of salmonid populations in streams. *American Naturalist*, **100**: 345–357.

Corell, R.W. (2006) Challenges of climate change: An Arctic perspective. *Ambio*, **35**:148–152.

Cunjak, R.A. (1988) Behavior and microhabitat of young Atlantic salmon (*Salmo salar*) during winter. *Canadian Journal of Fisheries and Aquatic Sciences*, **45**: 2156–2160.

Cunjak, R.A. (1996) Winter habitat of selected stream fishes and potential impacts from land-use activity. *Canadian Journal of Fisheries and Aquatic Sciences*, **53**: 267–282.

Cunjak, R.A. and Power, G. (1986) Winter habitat utilization by stream resident brook trout (*Salvelinus fontinalis*) and brown trout (*Salmo trutta*). *Canadian Journal of Fisheries and Aquatic Sciences*, **43**: 1970–1981.

Cunjak, R.A. and Power, G. (1987) The feeding and energetics of stream-resident trout in winter. *Journal of Fish Biology*, **31**: 493–511.

Cunjak, R.A. and Therrien, J. (1998) Inter-stage survival of wild juvenile Atlantic salmon, *Salmo salar* L. *Fisheries Management and Ecology*, **5**: 209–224.

Cunjak, R.A., Prowse, T.D. and Parrish, D.L. (1998) Atlantic salmon (*Salmo salar*) in winter: The season of parr 'discontent'? *Canadian Journal of Fisheries and Aquatic Sciences*, **55**: 161–180.

Devik, O. (1944) Ice formation in lakes and rivers. *The Geographical Journal*, **103**: 193–203.

Diana, J.S. (1995) *Biology and Ecology of Fishes*, Cooper Publishing Group, Carmel, Indiana, USA.

Enders, E.C., Clarke, K.D., Pennell, C.J., Ollerhead, L.M.N. and Scruton, D.A. (2007) Comparison between PIT and radio telemetry to evaluate winter habitat use and activity patterns of juvenile Atlantic salmon and brown trout. *Hydrobiologia*, **582**: 231–242.

Eriksson, L.-O. (1978) Nocturnalism versus diurnalism – dualism within fish individuals. In Thorpe, J.E. (ed.) *Rhythmic Activity of Fishes*, Academic Press, New York, USA.

Erkinaro, J., Shchurov, I.L., Saari, T. and Niemelä, E. (1994) Occurrence of Atlantic salmon parr in redds at spawning time. *Journal of Fish Biology*, **45**: 899–900.

Fausch, K.D. and Young, M.K. (1995) Evolutionary significant units and movement of resident stream fishes: a cautionary tale. *American Fisheries Society Symposium*, **17**: 360–370.

Finstad, A.G., Næsje, T.F. and Forseth, T. (2004a) Seasonal acclimatization of thermal performance in juvenile Atlantic salmon (*Salmo salar* L.). *Freshwater Biology*, **49**: 1459–1467.

Finstad, A.G., Einum, S., Forseth, T. and Ugedal, O. (2007) Shelter availability affects behaviour, size-dependent and mean growth of juvenile Atlantic salmon. *Freshwater Biology*, **52**: 1710–1718.

Finstad, A.G., Einum, S., Ugedal, O. and Forseth, T. (2009) Spatial distribution of limited resources and local density regulation in juvenile Atlantic salmon. *Journal of Animal Ecology*, **78**: 226–235.

Finstad, A.G., Forseth, T., Naesje, T.F. and Ugedal, U. (2004b) The importance of ice cover for energy turnover in juvenile Atlantic salmon. *Journal of Animal Ecology*, **73**: 959–966.

Finstad, A.G., Ugedal, O., Forseth, T. and Næsje, T.F. (2004c) Energy-related juvenile winter mortality in a northern population of Atlantic salmon (*Salmo salar*). *Canadian Journal of Fisheries and Aquatic Sciences*, **61**: 2358–2368.

Fraser, N.H.C., Metcalfe, N.B. and Thorpe, J.E. (1993) Temperature-dependent switch between diurnal and nocturnal foraging in salmon. *Proceedings of the Royal Society B*, **252**: 135–139.

Fretwell, S.D. (1972) *Populations in a Seasonal Environment*, Princeton University Press, Princeton, NJ, USA.

Gowan, C., Young, M.K., Fausch, K.D. and Riley, S.C. (1994) Restricted movement in resident stream salmonids: a paradigm lost? *Canadian Journal of Fisheries and Aquatic Sciences*, **51**: 2626–2637.

Greenberg, L.A. and Giller, P. (2000) The potential of the flat-bed passive integrated transponder antennae for studying habitat use by stream fishes. *Ecology of Freshwater Fish*, **9**: 74–80.

Gregory, J.S. and Griffith, J.S. (1996) Winter concealment by subyearling rainbow trout: space size selection and reduced concealment under surface ice and in turbid water conditions. *Canadian Journal of Zoology*, **74**: 451–455.

Harper, D.D. and Farag, A.M. (2004) Winter habitat use by cutthroat trout in the Snake River near Jackson, Wyoming. *Transactions of the American Fisheries Society*, **133**: 15–25.

Hartman, G.F (1965) The role of behaviour in the ecology and interaction of underyearling coho salmon (*Oncorhynchus kisutch*) and steelhead trout (*Salmo gairdneri*). *Journal of the Fisheries Research Board of Canada*, **22**: 1035–1081.

Harvey, B.C., Nakamoto, R.J. and White, J.L. (1999) Influence of large woody debris and a bankfull flood on movement of adult resident coastal cutthroat trout (*Onchorynchus clarki*) during fall and winter. *Canadian Journal of Fisheries and Aquatic Sciences*, **56**: 2161–2166.

Harwood, A.J., Metcalfe, N.B., Armstrong, J.D. and Griffiths, S.W. (2001) Spatial and temporal effects of interspecific competition between Atlantic salmon (*Salmo salar*) and brown trout (*Salmo trutta*) in winter. *Canadian Journal of Fisheries and Aquatic Sciences*, **58**: 1133–1140.

Harwood, A.J., Metcalfe, N., Griffiths, S.W. and Armstrong, J.D. (2002) Intra- and inter-specific competition for winter concealment habitat in juvenile salmonids. *Canadian Journal of Fisheries and Aquatic Sciences*, **59**: 1515–1523.

Heggenes, J. and Dokk, J.G. (2001) Contrasting temperatures, water flows, and light: seasonal habitat selection by young Atlantic salmon and brown trout in a boreonemoral river. *Regulated Rivers: Research and Management*, **17**: 623–635.

Heggenes, J., Bagliniere, J.L. and Cunjak, R.A. (1999) Spatial niche variability for young Atlantic salmon (*Salmo salar*) and brown trout (*Salmo trutta*) in heterogeneous streams. *Ecology of Freshwater Fish*, **8**: 1–21.

Heggenes, J., Krog, O.M.W., Lindås, O.R., Dokk, J.G. and Bremnes, T. (1993) Homeostatic behavior responses in a changing environment: brown trout (*Salmo trutta*) becomes nocturnal during winter. *Journal of Animal Ecology*, **62**: 295–308.

Hillman, T.W., Griffith, J.S. and Platss, W.S. (1987) Summer and winter habitat selection by juvenile Chinook salmon in a highly sedimented Idaho stream. *Transactions of the American Fisheries Society*, **116**: 185–195.

Hiscock, M.J., Scruton, D.A., Brown, J.A. and Clarke, K.D. (2002) Winter movement of radio-tagged juvenile Atlantic salmon in Northeast Brook, Newfoundland. *Transactions of the American Fisheries Society*, **131**: 577–581.

Hubbs, C.L. and Trautman, M.B. (1935) The need for investigating fish conditions in winter. *Transactions of the American Fisheries Society*, **65**: 51–56.

Hughes, N.F. and Dill, L.M. (1990) Position choice of drift-feeding salmonids: model and test for Arctic grayling (*Thymallus arcticus*) in subarctic mountain streams, interior Alaska. *Canadian Journal of Fisheries and Aquatic Sciences*, **47**: 2039–2048.

Hurst, T.P. (2007) Causes and consequences of winter mortality in fishes. *Journal of Fish Biology*, **71**: 315–345.

Huusko, A., Mäki-Petäys, A., Stickler, M. and Mykrä, H. (2011) Fish can shrink under harsh living conditions. *Functional Ecology*, 25: 628–633.

Huusko, A., Greenberg, L., Stickler, M., Linnansaari, T., Nykänen, M., Vehanen, T., Koljonen, S., Louhi, P. and Alfredsen, K. (2007) Life in the ice lane: The winter ecology of stream salmonids. *River Research and Applications*, 23: 469–491.

Jakober, M.J., McMahon, T.E. and Thurow, R.F. (2000) Diel habitat portioning by bull char and cutthroat trout during fall and winter in Rocky mountain streams. *Environmental Biology of Fishes*, 58: 79–89.

Jakober, M.J., McMahon, T.E., Thurow, R.F. and Clancy, C.G. (1998) Role of stream ice on fall and winter movements and habitat use by bull trout and cutthroat in Montana headwater streams. *Transactions of the American Fisheries Society*, 127: 223–235.

Johnson, J.H. and Douglass, K.A. (2009) Diurnal stream habitat use of juvenile Atlantic salmon, brown trout and rainbow trout in winter. *Fisheries Management and Ecology*, 16: 352–359.

Johnston, P., Berube, F. and Bergeron, N.E. (2009) Development of a flatbed passive integrated transponder antenna grid for continuous monitoring of fishes in natural streams. *Journal of Fish Biology*, 74: 1651–1661.

Kempema, E.W. and Ettema, R. (2010) Anchor ice rafting: observations from Laramie River. *River Research and Applications*, 27(8): 1073–1116. doi: 10.1002/rra.1450.

Kerr, D.J., Shen, H.T. and Daly, S.F. (2002) Evolution and hydraulic resistance of anchor ice on gravel bed. *Cold Regions Science and Technology*, 35: 101–114.

Klemetsen, A., Amundsen, P.-A., Dempson, J.B., Jonsson, B., Jonsson, N., O'Connell, M.F. and Mortensen, E. (2003) Atlantic salmon *Salmo salar* L., brown trout *Salmo trutta* L. and Arctic charr *Salvelinus alpinus* (L.): a review of aspects of their life histories. *Ecology of Freshwater Fish*, 12: 1–59.

Kocik, J.F. and Ferreri, C.P. (1998) Juvenile production variation in salmonids: population dynamics, habitat, and the role of spatial relationships. *Canadian Journal of Fisheries and Aquatic Sciences*, 55(1): 191–200.

Koljonen, S., Huusko, A., Mäki-Petäys, A., Mykrä, H. and Muotka, T. (2012) Body mass and growth of overwintering brown trout in relation to stream habitat complexity. *River Research and Applications*, 28: 62–70. doi: 10.1002/rra.1435.

Komadina-Douthwright, S.M., Caissie, D. and Cunjak, R.A. (1997) Winter movement of radio-tagged Atlantic salmon (*Salmo salar*) kelts in relation to frazil ice in pools of the Miramichi River. *Canadian Technical Report of Fisheries and Aquatic Sciences*, 2161: 1–66.

Kylmänen, I., Huusko, A., Vehanen, T. and Sirniö, V.-P. (2001) Aquatic habitat mapping by the ground-penetrating radar. In Nishida, T., Kailola, P.J. and Hollingworth, C.E. (eds) *Proceedings of the first international symposium on geographic information system (GIS) in fishery sciences*, Fishery GIS research group, Saitama, Japan, pp. 186–194.

Laudon, H., Poleo, A.B.S., Voellestad, L.A. and Bishop, K. (2005) Survival of brown trout during spring flood in DOC-rich streams in northern Sweden: the effect of present acid deposition and modeled pre-industrial water quality. *Environmental Pollution*, 135: 121–130.

Lindström, J.W. and Hubert, W.A. (2004) Ice processes affect habitat use and movements of adult cutthroat trout and brook trout in a Wyoming foothills stream. *North American Journal of Fisheries Management*, 24: 1331–1342.

Linnansaari, T. (2009) *Effects of Ice Conditions on Behaviour and Population Dynamics of Atlantic Salmon (Salmo salar L.) Parr*. PhD thesis, Department of Biology, University of New Brunswick, Fredericton, NB, Canada.

Linnansaari, T. and Cunjak, R.A. (2007) The performance and efficacy of a two-person operated portable PIT-antenna for monitoring spatial distribution of stream fish populations. *River Research and Applications*, 23: 559–564.

Linnansaari, T. and Cunjak, R.A. (2010) Patterns in apparent survival of Atlantic salmon (*Salmo salar*) parr in relation to variable ice conditions throughout winter. *Canadian Journal of Fisheries and Aquatic Sciences*, 67: 1744–1754.

Linnansaari, T., Cunjak, R.A. and Newbury, R. (2008) Winter behavior of juvenile Atlantic salmon *Salmo salar* L. in experimental stream channels: effect of substratum size and full ice cover on spatial distribution and activity pattern. *Journal of Fish Biology*, 72: 2518–2533.

Linnansaari, T., Roussel, J.-M., Cunjak, R.A. and Halleraker, J.H. (2007) Efficacy and accuracy of portable PIT-antennae when locating fish in ice-covered streams. *Hydrobiologia*, 582: 281–287.

Linnansaari, T., Keskinen, A., Romakkaniemi, A., Erkinaro, E. and Orell, P. (2010) Deep habitats are important for juvenile Atlantic salmon *Salmo salar* in large rivers. *Ecology of Freshwater Fish*, 19: 618–626.

Linnansaari, T., Alfredsen, K., Stickler, M., Arnekleiv, J.V., Harby, A. and Cunjak, R.A. (2009) Does ice matter? Site fidelity and movements by Atlantic salmon (*Salmo salar* L.) parr during winter in a substrate enhanced river reach. *River Research and Applications*, 25: 773–787.

Linnansaari, T., Stickler, M., Alfredsen, K., Arnekleiv, J.V., Cunjak, R.A., Fjeldstad, H.-P., Halleraker, J.H. and Harby, A. (2005) Movements and behavior by juvenile Atlantic salmon in relation to ice conditions in small rivers in Canada and Norway. In Anon. (ed.) *Proceedings from the13th Workshop on the Hydraulics of Ice Covered Rivers*, Committee on River Ice Processes and the Environment, Hanover, New Hampshire, USA, pp. 83–101.

Mäki-Petäys, A., Erkinaro, J., Niemelä, E., Huusko, A. and Muotka, T. (2004) Spatial distribution of juvenile Atlantic salmon (*Salmo salar*) in a subarctic river: size-specific changes in a strongly seasonal environment. *Canadian Journal of Fisheries and Aquatic Sciences*, 61: 2329–2338.

Mäki-Petäys, A., Muotka, T., Huusko, A., Tikkanen, P. and Kreivi, P. (1997) Seasonal changes in habitat use and preference by juvenile brown trout, *Salmo trutta*, in a northern boreal river. *Canadian Journal of Fisheries and Aquatic Sciences*, **54**: 520–530.

Matousek, V. (1984) Types of ice run and conditions for their formation. In *IAHR International Symposium on Ice*, volume 1, Hamburg, Germany, pp. 315–327.

Metcalfe, N.B., Fraser, N.H.C. and Burns, M.D. (1999) Food availability and the nocturnal vs. diurnal foraging trade-off in juvenile salmon. *Journal of Animal Ecology*, **68**: 371–381.

Meyer, K.A. and Griffith, J.S. (1997) Effects of cobble–boulder substrate configuration on winter residency of juvenile rainbow trout. *North American Journal of Fisheries Management*, **17**: 77–84.

Milner, N.J., Elliott, J.M., Armstrong, J.D., Gardiner, R., Welton, J.S. and Ladle, M. (2003) The natural control of salmon and trout populations in streams. *Fisheries Research*, **62**: 111–125.

Mitro, M.G. and Zale, A.V. (2002) Seasonal survival, movement, and habitat use of age-0 rainbow trout in the Henrys Fork of the Snake River, Idaho. *Transactions of the American Fisheries Society*, **131**: 271–286.

Muhlfeld, C.C., Bennett, D.H. and Marotz, B. (2001) Fall and winter habitat use and movement by Columbia River redband trout in a small stream in Montana. *North American Journal of Fisheries Management*, **21**: 170–177.

Muhlfeld, C.C., Glutting, S., Hunt, R., Daniels, D. and Marotz, B. (2003) Winter diel habitat use and movement by subadult bull trout in the Upper Flathead River, Montana. *North American Journal of Fisheries Management*, **23**: 163–171.

Nichelson, T.E., Solazzi, M.F., Johnson, S.L. and Rodgers, J.D. (1992) Effectiveness of selected stream improvement techniques to create suitable winter rearing habitat for juvenile coho salmon in Oregon coastal streams. *Canadian Journal of Fisheries and Aquatic Sciences*, **49**: 790–794.

Nykänen, M., Huusko, A. and Lahti, M. (2004a) Changes in movement, range and habitat preferences of adult grayling from late summer to early winter. *Journal of Fish Biology*, **64**: 1386–1398.

Nykänen, M., Huusko, A. and Lahti, M. (2004b) Movements and habitat preferences of adult grayling (*Thymallus thymallus* L.) from late winter to summer in a boreal river. *Archiv fur Hydrobiologie*, **161**: 417–432.

Nykänen, M., Huusko, A. and Mäki-Petäys, A. (2001) Seasonal changes in the habitat use and movements of adult European grayling in a large subarctic river. *Journal of Fish Biology*, **58**: 506–519.

Orpwood, J.E., Armstrong, J.D. and Griffiths, S.W. (2010) Interactions between riparian shading and food supply: a seasonal comparison of effects on time budgets, space use and growth in Atlantic salmon *Salmo salar*. *Journal of Fish Biology*, **77**: 1835–1849.

Orpwood, J.E., Griffiths, S.W. and Armstrong, J.D. (2006) Effects of food availability on temporal activity patterns and growth of Atlantic salmon. *Journal of Animal Ecology*, **75**: 677–685.

Östergren, J. and Rivinoja, P. (2008) Overwintering and downstream migration of sea trout (*Salmo trutta* L.) kelts under regulated flows – northern Sweden. *River Research and Applications*, **24**: 551–563.

Peterson, N.P. (1982) Immigration of juvenile coho salmon (*Oncorhynchus kisutch*) into riverine ponds. *Canadian Journal of Fisheries and Aquatic Sciences*, **39**: 1308–1310.

Power, G., Brown, R.S. and Imhof, J.G. (1999) Groundwater and fish – insights from northern North America. *Hydrological Processes*, **13**: 401–422.

Power, G., Cunjak, R.A., Flannagan, J. and Katopodis, C. (1993) Biological effects of river ice. In Prowse, T.D. (ed.) *River-Ice Ecology*, National Water Research Institute, Environment Canada, Saskatoon, SK, Canada.

Prowse, T.D. (1994) Environmental significance of ice to stream flow in cold regions. *Freshwater Biology*, **32**: 241–259.

Prowse, T.D. (2001a) River ice ecology. Part A: hydrologic, geomorphic, and water-quality aspects. *Journal of Cold Regions Engineering*, **15**: 1–16.

Prowse, T.D. (2001b) River-ice ecology. Part B: Biological aspects. *Journal of Cold Regions Engineering*, **15**: 17–33.

Prowse, T.D. and Culp, J.M. (2004) Ice breakup: a neglected factor in river ecology. *Canadian Journal of Civil Engineering*, **39**: 128–144.

Prowse, T.D. and Gridley, N.C. (1993) *Environmental Aspects of River Ice*. Science Report No. 5, National Hydrology Research Institute, Environment Canada, Saskatoon.

Quinn, T.P. and Peterson, N.P. (1996) The influence of habitat complexity and fish size on over-winter survival and growth of individually marked juvenile coho salmon (*Oncorhynchus kisutch*) in Big Beef Creek, Washington. *Canadian Journal of Fisheries and Aquatic Sciences*, **53**: 1555–1564.

Rabeni, C.F. and Sowa, S.P. (1996) Integrating biological realism into habitat restoration and conservation strategies for small streams. *Canadian Journal of Fisheries and Aquatic Sciences*, **53**: 252–259.

Reynolds, J.B. (1997) Ecology of overwintering fishes in Alaskan freshwaters. In Milner, A.M. and Oswood, M.W. (eds) *Freshwaters of Alaska – Ecological Synthesis*, Ecological studies 119, Springer-Verlag, Inc, New York, pp. 281–308.

Rimmer, D.M., Paim, U. and Saunders, R.L. (1983) Autumnal habitat shifts of juvenile Atlantic salmon (*Salmo salar*) in a small river. *Canadian Journal of Fisheries and Aquatic Sciences*, **40**: 671–680.

Rimmer, D.M., Saunders, R.L. and Paim, U. (1984) Changes in the selection of microhabitat by juvenile Atlantic salmon (*Salmo salar*) at the summer–autumn transition in a small river. *Canadian Journal of Fisheries and Aquatic Sciences*, **41**: 469–475.

Robertson, M.J., Pennell, C.J., Scruton, D.A., Robertson, G.J. and Brown, J.A. (2004) Effects of increased flow on the behaviour of Atlantic salmon parr in winter. *Journal of Fish Biology*, **65**: 1070–1079.

Roussel, J.-M., Cunjak, R.A., Newbury, R., Caissie, D. and Haro, A. (2004) Movements and habitat use by PIT-tagged Atlantic salmon parr in early winter: the influence of anchor ice. *Freshwater Biology*, **49**: 1026–1035.

Saraniemi, M., Huusko, A. and Tahkola, H. (2008) Spawning migration and habitat use of adfluvial brown trout, *Salmo trutta*, in a strongly seasonal boreal river. *Boreal Environment Research*, **13**: 121–132.

Schlosser, I.J. (1987) The role of predation in age- and size-related habitat use by stream fishes. *Ecology*, **68**: 651–659.

Scruton, D.A., Pennell, C.J., Robertson, M.J., Ollerhead, L.M.N., Clarke, K.D., Alfredsen, K., Harby, A. and McKinley, R.S. (2005) Seasonal response of juvenile Atlantic salmon to experimental hydropeaking power generation in Newfoundland, Canada. *North American Journal of Fisheries Management*, **25**: 964–974.

Simpkins, D.G., Hubert, W.A. and Wesche, T.A. (2000) Effects of fall-to-winter changes in habitat and frazil ice on the movements and habitat use of juvenile rainbow trout in a Wyoming tailwater. *Transactions of the American Fisheries Society*, **129**: 101–118.

Somero, G.N., Dahlhoff, E. and Lin, J.J. (1996) Stenotherms and eurytherms: mechanisms establishing thermal optima and tolerance ranges. In Johnston, I.A. and Bennett, A.F. (eds) *Animals and Temperature*, Cambridge University Press, Cambridge, UK, pp. 53–78.

Stickler, M. and Alfredsen, K. (2009) Anchor ice formation in streams: a field study. *Hydrological Processes*, **23**: 2307–2315.

Stickler, M., Alfredsen, K.T., Linnansaari, T. and Fjelstad, H.-P. (2010) The influence of dynamic ice formation on hydraulic heterogeneity in steep streams. *River Research and Applications*, **26**: 1187–1197.

Stickler, M., Alfredsen, K., Scruton, D.A., Pennell, C., Harby, A. and Økland, F. (2007) Mid-winter activity and movement of Atlantic salmon parr during ice formation events in a Norwegian regulated river. *Hydrobiologia*, **582**: 81–89.

Stickler, M., Enders, E.C., Pennell, C.J., Cote, D., Alfredsen, K. and Scruton, D.A. (2008a) Stream gradient-related movement and growth of Atlantic salmon parr during winter. *Transactions of the American Fisheries Society*, **137**: 371–385.

Stickler, M., Enders, E.C., Pennell, C.J., Cote, D., Alfredsen, K. and Scruton, D.A. (2008b) Habitat use of Atlantic salmon *Salmo salar* in a dynamic winter environment: the influence of anchor-ice dams. *Journal of Fish Biology*, **73**: 926–944.

Swales, S. and Levings, C.D. (1989) Role of off-channel ponds in the life cycle of coho salmon (*Oncorhynchus kisutch*) and other juvenile salmonids in the Coldwater River, British Columbia. *Canadian Journal of Fisheries and Aquatic Sciences*, **46**: 232–242.

Tesaker, E. (1994) Ice formation in steep rivers. In Anon (ed.) *Proceedings of the International Ice Symposium*, IAHR, Trondheim, Norway, pp. 630–638.

Tschaplinski, P.J. and Harman, G.F. (1983) Winter distribution juvenile coho salmon (*Oncorhynchus kisutch*) before and after logging in Carantion Creek, British Columbia, and some implications for overwintering survival. *Canadian Journal of Fisheries and Aquatic Sciences*, **40**: 452–461.

Valdimarson, S.K. and Metcalfe, N.B. (1998) Shelter selection in juvenile Atlantic salmon, or why do salmon seek shelter in winter? *Journal of Fish Biology*, **52**: 42–49.

Vehanen, T. and Huusko, A. (2002) Behaviour and habitat use of young-of-the-year Atlantic salmon (*Salmo salar*) at the onset of winter in artificial streams. *Archiv für Hydrobiologie*, **154**: 133–150.

Whalen, K.G. and Parrish, D.L. (1999) Nocturnal habitat use of Atlantic salmon parr in winter. *Canadian Journal of Fisheries and Aquatic Sciences*, **56**: 1543–1550.

Whalen, K.G., Parrish, D.L. and Mather, M.E. (1999) Effect of ice formation on selection of habitats and winter distribution of post-young-of-the-year Atlantic salmon parr. *Canadian Journal of Fisheries and Aquatic Sciences*, **56**: 87–96.

Young, M.K. (1998) Absence of autumnal changes in habitat use and location of adult Colorado River cutthroat trout in a small stream. *Transactions of the American Fisheries Society*, **127**: 147–151.

Youngson, A.F., Buck, R.J.G., Simpson, T.H. and Hay, D.W. (1983) The autumn and spring emigrations of juvenile Atlantic salmon, *Salmo salar* L., from Girnock Burn, Aberdeenshire, Scotland: environmental release of migration. *Journal of Fish Biology*, **23**: 625–639.

11
Stream Habitat Associations of the Foothill Yellow-Legged Frog (*Rana boylii*): The Importance of Habitat Heterogeneity

Sarah Yarnell

Center for Watershed Sciences, University of California, Davis, One Shields Avenue, Davis, CA 95616, USA

11.1 Introduction

In the emerging field of ecohydraulics, which explores the interdisciplinary boundary between hydraulics and ecology in river systems (Rice *et al.*, 2010), research has focused almost exclusively on the instream habitat requirements of fish and benthic macroinvertebrates (e.g. Jowett *et al.*, 1991; Guay *et al.*, 2000 among many others). However, there are other aquatic species that are uniquely adapted to the hydraulic complexity of riverine environments. For example, some amphibian species have evolved to take advantage of seasonally available habitat provided by rivers in Mediterranean climates (e.g. Western spotted frog, *Heleioporus albopunctatus*, in southwestern Australia), while others have adapted their body morphology to withstand high velocities associated with flowing stream environments (e.g. Tailed frog, *Ascaphus truei*, in the northwestern USA). Given the direct connection these amphibian species have with their hydraulic environment, the methods, tools and conceptual insights emerging from the study of ecohydraulics can further the understanding and conservation of these sensitive species.

The Foothill yellow-legged frog (*Rana boylii*) is one such stream-dwelling frog native to California and southern Oregon (USA) that has received attention in recent decades as one of several native amphibians in rapid decline (Nussbaum *et al.*, 1983; Jennings and Hayes, 1994). Currently occupying less than 50% of its historic range (Davidson *et al.*, 2002), *R. boylii* is listed as a Species of Special Concern in California and is afforded priority consideration in resource management decisions. Largely confined to moderate gradient streams at mid-elevations, *R. boylii* has been particularly vulnerable to introduced species and large-scale land use disturbances, notably those associated with flow regulation (Kupferberg, 1996; Lind, 2005). Altered stream flow regimes, resulting changes in sediment supply and instream habitat and establishment of non-native species are some of the primary factors contributing to the decline of the species (Lind *et al.*, 1996; Yarnell, 2000; Kupferberg *et al.*, 2009).

Research on the life history and habitat requirements of *R. boylii* has greatly increased in the last decade. Various studies have provided an improved understanding of *R. boylii* behaviour (Van Wagner, 1996; Wheeler, 2007), the development of early life stages (Kupferberg, 1997), phylogeny and genetic structure (Lind, 2005; Peek, 2010) and individual movement (Bourque, 2008). *R. boylii* is strongly tied to the physical characteristics of streams (Kupferberg, 1996; Van Wagner, 1996; Yarnell, 2000; Lind, 2005; Haggarty, 2006). Breeding and rearing sites are typically limited to channel locations with asymmetric channel shape, coarse substrate and stable hydraulic conditions, while post-metamorphic life stages are found in a wide variety of instream habitats. While these findings indicate that different age classes have distinctly different geomorphic and hydraulic habitat requirements, information on which physical parameters may be linked to habitat selection across multiple life stages and at multiple spatial scales is lacking.

Variation and diversity in physical stream habitat, defined as the combination of structural and hydraulic features that are dynamic both in time and space, have been shown to increase aquatic species diversity (Beisel *et al.*, 2000; Brown, 2003) and individual species success and survival (Power, 1992; Torgersen and Close, 2004). In particular, greater variability in hydraulic habitat patches through space and time has been associated with higher species diversity in fish and macroinvertebrate assemblages (Dyer and Thoms, 2006; Pastuchova *et al.* 2008). *R. boylii*, with its distinctive variability in life stage-specific habitat requirements, may respond to, and even directly benefit from, increased hydraulic habitat heterogeneity. In response to annual fluctuations in discharge common to California's Mediterranean climate, seasonal habitat requirements by *R. boylii* may change as fluctuating flows change hydraulic habitat conditions (Lind, 2005; Yarnell *et al.*, 2010). As a result, physical habitat associations may be evident at multiple spatial scales as individuals move between breeding, foraging and overwintering habitats. Stream reaches with greater habitat heterogeneity may provide the diversity of hydraulic and geomorphic habitat needed annually by all life stages without the risk associated with long migrations between suitable habitats.

This study focused on habitat use by *R. boylii* in several small streams in the Yuba River watershed in the northern Sierra Nevada mountains, California, USA. To test the hypothesis that *R. boylii* is associated with stream reaches of higher habitat heterogeneity in terms of hydraulic and geomorphic characteristics, two scales of habitat use were assessed: (1) mesohabitat characteristics for each life stage throughout one biologically active season (spring–fall), where mesohabitat is defined on the scale of 1–10 m², and (2) reach-scale associations with reach morphology, locations near tributaries and locations

with greater valley width, where a reach is defined on the scale of 100–1000 m².

11.2 Methods for quantifying stream habitat

11.2.1 Study area

Four streams in the northern Sierra Nevada mountains in California were selected for study, each known to sustain populations of *R. boylii*: Shady Creek, Rush Creek and Humbug Creek on the South Yuba River and Oregon Creek on the Middle Yuba River (Figure 11.1). These creeks vary both in terms of stream habitat availability and occupation by *R. boylii*. Shady Creek has the highest population density and individuals reside year-round within the creek. The remaining creeks have been known to occasionally support breeding, but the majority of individuals observed in each creek are juveniles and adults, as most breeding likely takes place in the tributary confluences.

The four study creeks are similar to most mid-elevation Sierran drainages, having moderate to steep slopes, confined valleys with occasional bedrock outcrops, narrow disconnected riparian zones, coarse substrates and cascade, step-pool, and riffle-pool morphologies (after the classification of Montgomery and Buffington, 1997) (Table 11.1). All four creeks have been subject to various land uses including mining (in-stream, hydraulic and high-banking), logging and development. Shady Creek differs from the other study creeks in that it continues to recover from extensive past aggradation of hydraulic mining debris. Some reaches with steeper slopes have recovered to the original bedrock surfaces, but the majority of the stream continues to degrade through vast piles of tailings, leaving remnant terraces behind. The four creeks represent common stream habitats that are known to support *R. boylii* populations elsewhere in the northern Sierra Nevada (Van Wagner, 1996; Peek, 2010).

Survey segments were selected in areas known to consistently support *R. boylii* including those sites where breeding was observed previously. Segments ranged in length from 1.2 km to 3.2 km, depending on accessibility, and incorporated as many reach morphologies and habitat types common to that creek as possible. Because site fidelity is high for *R. boylii* (Kupferberg, 1996), and individuals rarely migrate greater than several hundred metres unless necessary (Bourque, 2008), each creek was considered independently in regard to

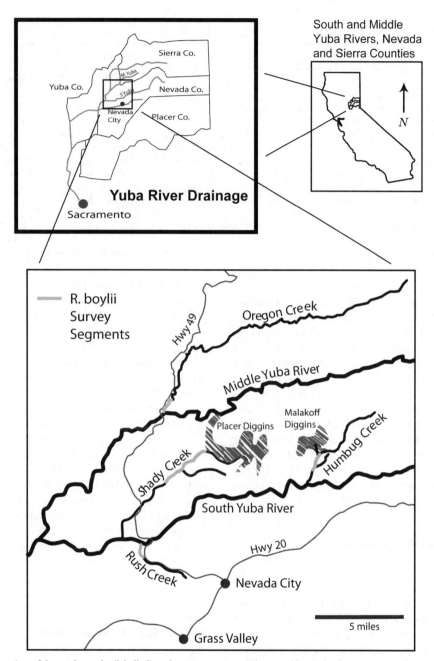

Figure 11.1 Location of the study creeks (labelled) and survey segments. The term 'diggins' refers to an area of exposed hydraulic mining debris, remnant from hydraulic mining practices in the late 19th century.

the diversity and availability of reach types available for occupation. Thus, the goal was to select segments with enough length and diversity to adequately assess reach-scale habitat associations, but short enough to be surveyed in one 10-hour day.

11.2.2 *Rana boylii* sampling

R. boylii visual encounter surveys (Heyer *et al.*, 1994) were completed in 2003 on each creek – once per month from the start of the breeding season in May through to the end of the foraging season in October. Surveys consisted

Table 11.1 Habitat characteristics of survey segments for each creek.

	Shady Creek	Humbug Creek	Rush Creek	Oregon Creek
Total drainage area (km²)	37.58	27.75	14.56	91.08
Total elevational range (m)	660–1470	450–1400	340–940	310–890
Survey segment length (km)	3.2	1.8	1.2	1.5
Survey segment slope (m/m)	0.017	0.034	0.086	0.039
Reach morphologies	Riffle-pool, Plane bed, Braided	Cascade, Step-pool, Forced riffle-pool	Cascade, Step-pool, Bedrock	Cascade, Step-pool, Bedrock
Dominant channel morphology	Riffles, Runs, Gravel bars, Shallow pools	Plunge pools, Boulder steps, Coarse bars, High-gradient riffles	Plunge pools, Boulder steps, Cascades, Coarse bars	Plunge pools, Boulder steps, Cascades
Dominant riparian type	Willow thickets with open gravel bars	Mature alders	Mature alders	Mature alders
Dominant substrate types	Cobble, Gravel, Sand	Cobble, Boulder	Boulder, Cobble	Boulder, Bedrock, Cobble

of two observers walking each survey segment upstream during the middle part of the day, visually identifying frogs and documenting mesohabitat data for each observed use location. Individual encounters were categorized by life stage (egg mass, tadpole, juvenile, adult), with details on their size, location and habitat associations recorded. Juveniles were identified based on snout-to-vent length (35 mm or less), while adults were identified by standard sexual dimorphism traits such as size and presence or absence of swollen joints (Zweifel, 1955). *R. boylii* are strictly stream inhabitants and are usually found within 1–2 metres from the stream margin (Nussbaum *et al.*, 1983). Individuals are often sighted perched on the edge of rocks protruding from the water surface or sitting partially submerged in the water at the stream margin.

11.2.3 Habitat associations

At each frog observation point, a series of physical habitat characteristics at the mesohabitat scale was recorded (Table 11.2). A variety of quantitative and qualitative variables were chosen in an effort to determine how well each performed in describing habitat associations. Although most previous research on *R. boylii* has been qualitative in nature (Zweifel, 1955; Van Wagner, 1996), more recent studies have used discrete quantitative variables to assess hydraulic characteristics such as depth, velocity and shear stress (Kupferberg, 1996; Yarnell, 2000; Lind, 2005). While the use of quantitative variables is undoubtedly more objective and repeatable, there is some value to assessing the applicability of subjective qualitative variables (such as 'pool') in describing aspects of the habitat that may not be accounted for in a single quantitative

variable (Hawkins *et al.*, 1993). Therefore, an assessment of both qualitative and quantitative variables was completed to provide some insight into both the applicability of the measures and the mesohabitat associations of the different life stages.

Reach types in each creek survey segment did not noticeably shift during the survey season (high flows occur during winter), and thus were delineated in the field during summer when survey time was greatest. Reach-type boundaries were recorded using a handheld GPS and a 1:24 000 scale topographic map, and the point data were incorporated into a GIS (ArcGIS 8.3; ESRI, 2002) as a continuous line overlay, where each line segment represented the stream length of the mapped reach type. The location of each frog observation was also recorded in the field using a handheld GPS and a topographic map, and incorporated into GIS as point data overlays. The associated reach type for each frog observation was recorded in the field, verified in the GIS data and used for subsequent statistical analyses.

11.2.4 Statistical analyses

Logistic regression was used to determine which mesohabitat variables were significant in predicting the probability of observing certain life stages, while reach-scale habitat associations were assessed using chi-square and paired mean testing. The data were evaluated both seasonally (subsets of spring: May–June, summer: July–August, and fall: September–October) and annually (full dataset: May–October) using the statistical program SPSS v12.0 (SPSS, 2003).

Table 11.2 Habitat variables measured or assessed at the location of each individual life stage observation.

Variable	Variable type	Categories	Measurement description
Local depth	Quantitative	N/A	Depth taken at or adjacent to where individual was observed, representing the local average depth of the surrounding mesohabitat
Local velocity	Quantitative	N/A	Velocity measured using a Marsh–McBirney FloMate, and taken at or adjacent to where individual was observed, representing the local average velocity of the surrounding mesohabitat
Channel width	Quantitative	N/A	Average width of water surface where individual was observed
Dominant substrate	Categorical	Bedrock, Boulder, Cobble, Gravel, Sand, Silt	Size of dominant substrate individual was observed attached to, perched on or in the surrounding mesohabitat
Degree of substrate sorting	Categorical	Well-sorted, Moderately sorted, Poorly sorted	Degree of sorting of particles in the surrounding mesohabitat (Compton, 1985)
Cover class	Categorical	0–25%, 25–50%, 50–75%, 75–100%	Measured using a densiometer and defined as any vegetation occurring in the vertical space located immediately above the water surface
Riparian type	Categorical	Cobble bar, Pure willow, Willow/alder mix, Mature riparian	Categories represent stages of riparian succession adjacent to the surrounding mesohabitat; following methods described in Lind (2005)
Geomorphic unit	Categorical	Pool, Run, Riffle, Rapid, Bar, Step, Bedrock chute	Units identified based on common definitions found in the literature and representing the geomorphic structure of the surrounding mesohabitat (e.g. Hawkins et al., 1993)
Hydraulic unit	Categorical	Standing water, Scarcely perceptible flow, Smooth surface flow, Upwelling, Rippled, Unbroken standing wave, Broken standing wave, Chute, Freefall	Units identified based on definitions described in Thomson et al. (2001) and representing the hydraulic structure (water surface specifically) of the surrounding mesohabitat
Reach type	Categorical	Cascade, Step-pool, Plane bed, Riffle-pool, Braided, Bedrock	Types identified based on definitions described by Montgomery and Buffington (1997), typically ranging in stream length from 25–250 m

Logistic regression

Unlike linear regression, logistic regression requires limited assumptions regarding normality or homoscedasticity (Hosmer and Lemeshow, 1989, Trexler and Travis, 1993). Specifically, logistic regression only requires that observations are independent and that explanatory variables are linearly related to a log transform of the response. Because the heterogeneous distribution of organisms across the environment often does not meet the standard assumptions of normality or linearity, the logistic modelling approach has been used to predict species–habitat relationships at a variety of scales (Knapp and Preisler, 1999; Torgersen and Close, 2004).

To determine the probability of observing a certain life stage given the mesohabitat characteristics, each life stage was modelled in relation to the others as a binary response (Table 11.3) using the following logistic model:

$$\log(P/(1-P)) = B_0 + B_{1 \times 1} + \cdots + B_n X_n \quad (11.1)$$

where P is the probability of observing a life stage and n is the number of independent mesohabitat variables. Each life stage was modelled in succession of age (e.g. 'tadpole versus older than tadpole' rather than 'tadpole versus not tadpole' as would be done for habitat use versus non-use data) in order to reflect the seasonal variation in life stage assemblage (egg masses only exist in spring; tadpoles only exist in summer). To validate the results from these seasonally based models, a series of models in reverse succession (e.g. adult versus younger than adult, etc.) was also run using the full annual dataset.

To determine which mesohabitat variables were suitable for inclusion in the logistic regression, a non-parametric cross-correlation analysis was completed, and the least descriptive variable of any correlated pair of variables (based on Spearman's Rho values > 0.3) was excluded. Riparian type was significantly correlated with cover class, hydraulic unit was strongly correlated with local velocity and geomorphic unit, and geomorphic unit was significantly correlated with local velocity. Riparian type was a less specific descriptor than cover class, and local velocity as a quantitative variable was a preferable measure for the modelling analysis; therefore, riparian type, hydraulic unit and geomorphic unit were excluded from the statistical analyses. As a result, a total of six mesohabitat variables were included in the logistic regression (velocity, depth, dominant substrate size, degree of sorting, cover class and stream channel width), with channel width excluded from the full year dataset analysis and the May/June analysis due to errors in field measurements of width during the egg mass surveys (Table 11.3).

Reach-scale statistical analyses

Reach type, distance to nearest tributary and valley width were incorporated into ArcGIS 8.3 (ESRI, 2002) to facilitate comparison with the observed frog distributions from each monthly survey. Reach type was a continuous line segment overlay based on data collected in the field during the August frog surveys. Tributary junctions were mapped in ArcGIS as a point overlay based on identification from a 1:24 000 scale topographic map and verification in the field. Valley width was calculated from 1:24 000 scale topographic maps and imported as a point overlay, where each

point represented the average valley width for 50 m of stream length. 50 m was chosen as a reasonable length for determining differences in width on the maps as well as a length appropriate to the typical home range of an adult frog (Bourque, 2008). Frog location points were then compared to these three mapped overlays using the geoprocessing tools in ArcGIS.

A chi-square test was used to test the null hypothesis that frogs were equally spaced throughout each stream survey segment relative to reach type (i.e. the number of frogs observed in each reach type was proportional to the relative abundance of each reach type). The availability of each reach type was determined using ArcGIS and statistically compared to the observed percentage of frogs in each reach type. Similarly, a one-sample t-test was used to test the null hypotheses that frogs were equally spaced throughout each stream survey segment with respect to tributary confluences (i.e. expected mean distance to tributary equalled the observed mean distance) and valley width (i.e. expected mean valley width equalled observed mean valley width). Observed distances and widths were determined for each mapped frog point in ArcGIS, and expected distances and widths were calculated from an overlay of evenly distributed points spaced 50 m apart in each survey segment. Statistical tests were completed on the annual dataset for each stream and on the seasonal datasets for Shady and Humbug Creeks. The sample sizes on Rush and Oregon Creeks were too small to allow for seasonal analyses.

11.3 Observed relationships between *R. boylii* and stream habitat

Population densities varied widely between the four study creeks (Table 11.4). With 90% of the total number of

Table 11.3 Summary of logistic regression models tested. Parameters entered into each model were based on the cross-correlation analysis.

Life stage	Dataset evaluated	Parameters entered into stepwise algorithm
Egg masses vs. older	Spring	Velocity, Depth, Dominant substrate, Degree of sorting, Cover
Tadpoles vs. older	Summer	Velocity, Depth, Width, Dominant substrate, Degree of sorting, Cover
Tadpoles vs. egg masses	Annual	Velocity, Depth, Dominant substrate, Degree of sorting, Cover
Juveniles vs. older (a)	Fall	Velocity, Depth, Width, Dominant substrate, Degree of sorting, Cover
Juveniles vs. older (b)	Fall	Velocity, Depth, Width, Dominant substrate, Cover
Juveniles vs. younger	Annual	Velocity, Depth, Dominant substrate, Degree of sorting, Cover
Adults vs. younger	Annual	Velocity, Depth, Dominant substrate, Degree of sorting, Cover
Adult males vs. adult females	Annual	Velocity, Depth, Width, Dominant substrate, Degree of sorting, Cover

Table 11.4 Summary of *R. boylii* population density for each survey on each study stream.

Survey date (2003)	Creek	Total observations	Distance surveyed (km)	Population density (#/km)
May 20	Shady	40	3.2	12.5
June 3	Shady	130	3.2	40.6
July 3	Shady	62	3.2	19.4
August 15	Shady	90	3.2	28.1
September 12	Shady	142	3.2	44.4
October 15	Shady	114	3.2	35.6
June 10	Humbug	3	1.8	1.7
July 9	Humbug	5	1.8	2.8
August 14	Humbug	10	1.8	5.6
September 18	Humbug	13	1.8	7.2
October 13	Humbug	12	1.8	6.7
June 13	Rush	1	1.2	0.8
July 11	Rush	5	1.2	4.2
August 12	Rush	1	1.2	0.8
September 17	Rush	3	1.2	2.5
October 14	Rush	3	1.2	2.5
June 12	Oregon	3	1.5	2.0
July 10	Oregon	1	1.5	0.7
August 13	Oregon	0	1.5	0.0
September 16	Oregon	0	1.5	0.0
October 9	Oregon	2	1.5	1.3

observations, Shady Creek had much higher population densities than any of the other study creeks. As discussed above, Shady Creek supports all life stages throughout the season, while predominantly juveniles and adults occur on the other creeks. As a result, statistical analyses of life stage–mesohabitat associations were only completed on Shady Creek. However, the mesohabitat associations on Shady Creek were qualitatively compared with the adult and juvenile mesohabitat associations on the other study creeks in order to see if patterns were similar across creek systems.

11.3.1 Mesohabitat associations in Shady Creek

Frequency analysis

An initial analysis of the frequency of each life stage association with each categorical mesohabitat variable revealed a few general patterns (Figure 11.2). Frogs were observed with the highest frequency in the most prevalent habitats on Shady Creek: habitats with open cover canopies (cover class = 1), mixed alder/willow riparian types and riffles with poorly sorted gravel and cobble substrates. However, egg masses were most often found in open riparian areas with little to no vegetation, attached to boulders or cobbles in pools with scarcely perceptible flow. Tadpoles were

found at riffle margins and in pools with smooth surface flow or scarcely perceptible flow, while juveniles were observed with the highest frequencies in shallow riffles with poorly sorted gravel substrates. In addition to the prevalent habitats, adults were observed with moderate frequencies in higher velocity habitats such as riffles and cascades with cobble substrates.

Logistic regression

Hydraulic habitat characteristics were the most significant predictors of life stage–habitat associations, with velocity as the most significant predictor, present in every model except that which distinguished egg masses from tadpoles (Table 11.5). As velocity increased, the likelihood of observing successively older life stages increased. Depth was the next most significant predictor, present in every model except that which distinguished tadpoles from older life stages. As depth increased, the likelihood of observing older life stages increased, except for juveniles, which were negatively associated with increasing depth. The combination of velocity and depth indicated general patterns of association for each life stage, particularly for pre-metamorphic life stages (Figure 11.3).

Substrate was an important predictor for distinguishing habitats used by younger life stages from those of older life stages. Egg masses had a positive association with

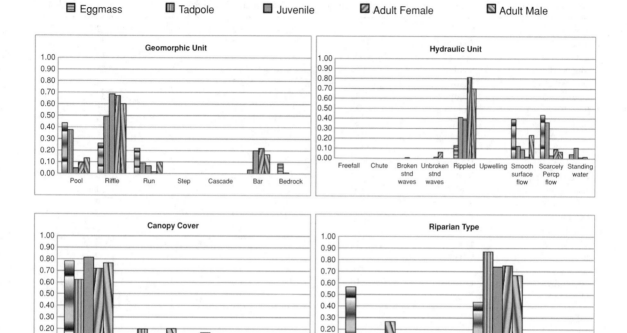

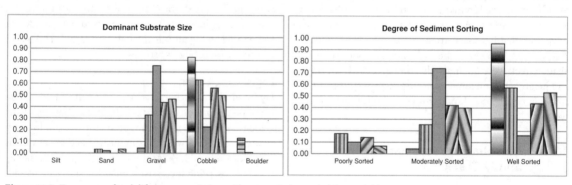

Figure 11.2 Frequency of each life stage association with categorical mesohabitat variables.

boulders, while tadpoles and juveniles were positively associated with gravel. In some models, dominant substrate size was a better predictor than degree of substrate sorting (e.g. egg masses vs. older, juveniles vs. younger), while in others, degree of substrate sorting was a better predictor (e.g. juveniles vs. older, adults vs. younger). In all models except tadpoles versus older, some measure of substrate increased the predictive power of

the model when combined with one or both hydraulic variables.

11.3.2 Mesohabitat associations on other study creeks

The available habitats on Humbug, Oregon and Rush Creeks were more varied than those on Shady Creek, often with large boulder and bedrock substrates, partial

Table 11.5 Results of logistic regression modelling. Variables are listed in order of inclusion in the stepwise algorithm. Positive standardized coefficients indicate a positive association between that variable and the life stage of interest, while negative coefficients indicate a negative association. Wald chi-square tests the significance of each variable within the model, and Omnibus chi-square tests the significance of the resulting final model.

Life stage	Variables included in stepwise model	Standardized coefficients	Wald chi-square	p-value	Omnibus test chi-square	Nagerkelke R^2	% correctly reclassified
Egg masses vs older	Dominant substrate (gravel/cobble/boulder relative to sand)	19.227/ 23.086/ 43.472	11.104	0.063			
	Velocity	−21.893	12.496	< 0.001			
	Depth	−22.817	3.462	0.011			
model summary				*< 0.001*	*142.231*	*0.832*	*92.9*
Tadpoles vs older	Velocity	−13.544	44.42	< 0.001			
model summary				*< 0.001*	*78.925*	*0.548*	*82.9*
Tadpoles vs egg masses	Dominant substrate (gravel/cobble/boulder relative to sand)	−19.355/ −21.133/ −26.822	21.664	< 0.001			
	Degree of sorting (poorly/ moderately sorted relative to well sorted)	3.683/4.025	13.292	0.001			
	Depth	24.389	6.902	0.009			
model summary				*< 0.001*	*87.957*	*0.597*	*83.1*
Subadults vs older (a)	Degree of sorting (poorly/ moderately sorted relative to well sorted)	2.564/2.793	22.847	< 0.001			
	Depth	−25.66	15.698	< 0.001			
	Velocity	−5.869	15.192	< 0.001			
	Dominant substrate (gravel relative to cobble)	−1.328	8.883	0.003			
model summary				*< 0.001*	*131.619*	*0.592*	*81.2*
Subadults vs older (b)	Dominant substrate (gravel relative to cobble)	2.329	37.897	< 0.001			
	Depth	−23.048	16.548	< 0.001			
	Velocity	−5.851	17.682	< 0.001			
model summary				*< 0.001*	*102.645*	*0.489*	*79.0*
Subadults vs younger	Velocity	10.106	42.728	< 0.001			
	Dominant substrate (gravel/cobble/boulder relative to sand)	−0.902/ −2.07/ −22.5	19.304	< 0.001			
	Degree of sorting (poorly/ moderately sorted relative to well sorted)	2.002/0.815	30.077	< 0.001			
	Depth	−16.415	15.237	< 0.001			
model summary				*< 0.001*	*203.317*	*0.555*	*78.4*
Adults vs younger	Velocity	5.974	77.214	< 0.001			
	Depth	13.32	34.384	< 0.001			
	Degree of sorting (poorly/ moderately sorted relative to well sorted)	−1.064/ −0.632	15.149	0.001			
	Dominant substrate (gravel/cobble/boulder relative to sand)	0.330/1.000/ −0.120	9.384	0.025			
model summary				*< 0.001*	*177.84*	*0.366*	*77.3*

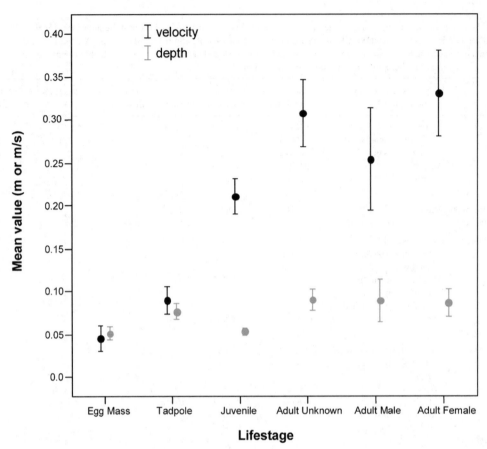

Figure 11.3 Annual life stage associations with local depth and velocity on Shady Creek. Points represent mean velocity (m/s) and depth (m); error bars represent 95% confidence interval.

to full canopy cover and a mature riparian zone. Steeper reach morphologies provided a variety of mesohabitats that were rare or absent from Shady Creek, such as steep boulder cascades, deep plunge pools and coarse boulder bars and steps. As a result, individuals were observed in coarser substrates and in denser canopy cover. However, with respect to hydraulic parameters, most life stages were observed in generally similar habitat conditions. As velocity increased, the number of observations of adults increased (Figure 11.4).

11.3.3 Reach-scale habitat associations

Three reach-scale characteristics were analyzed in detail using ArcGIS to determine associations on each study stream with reach morphology (measured as reach type), valley morphology (measured as valley width) and tributary proximity (measured as distance to nearest tributary). Results from the chi-square and one-sample

t-tests revealed statistically significant habitat associations on Shady, Humbug and Rush Creeks. Due to the small sample size on Oregon Creek, statistical analyses could not be completed.

Analysis of the annual datasets revealed high habitat use relative to availability in certain reach types, while the seasonal datasets provided some insight into the movements of adults throughout the season. On Shady Creek, frogs of all life stages throughout the year were consistently observed in braided reaches more often than would be expected if individuals were evenly disbursed along the creek, and in plane bed reaches less than expected (Figure 11.5). Individuals were also observed less than expected in cascade reaches, but were generally neutral with regard to riffle-pool reaches. Unlike the other study creeks, these habitat associations were consistent across life stages and were significant to highly significant for every life stage ($p = 0.002 - < 0.001$; $\chi^2 = 6.25 - 36.9$).

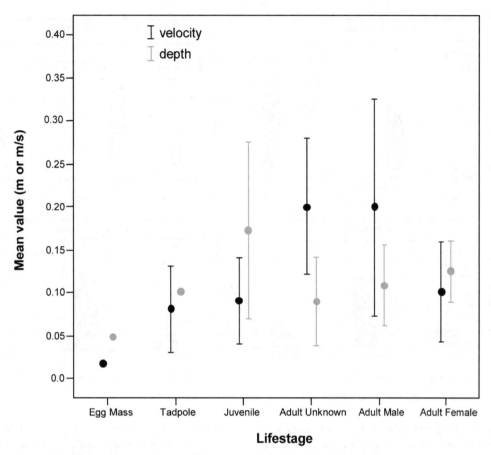

Figure 11.4 Annual life stage associations with local depth and velocity on Humbug, Oregon and Rush Creeks. Points represent mean velocity and depth; error bars represent 95% confidence interval.

On the remaining creeks, adults were observed more frequently in some reach types than others (e.g. riffle-pool and step-pool reaches) (Figure 11.5), but relative occupancy in each reach type varied between creeks.

During the spring breeding season, both adults and egg masses on Shady Creek were observed in braided reaches much more than expected and in plane bed reaches less than expected (adults: $p = 0.10$, $\chi^2 = 6.25$; egg masses: $p = < 0.001$, $\chi^2 = 23.68$), while juveniles were observed in all reach types proportional to the availability (Figure 11.6a). By summer, however, while tadpoles remained in the braided reaches ($p = < 0.001$, $\chi^2 = 20.49$), adults showed no preference for reach type, and juveniles were observed more often in plane bed and riffle-pool reaches than braided reaches ($p = 0.037$, $\chi^2 = 8.46$) (Figure 11.6b). This change in reach type association indicates movement by both the adults and juveniles throughout the summer foraging season. In fall, relative occupancy

in braided reaches versus plane bed reaches was significantly higher for all life stages once again (tadpole: $p = 0.03$, $\chi^2 = 8.65$; juvenile: $p = < 0.001$, $\chi^2 = 56.9$; adults: $p = 0.005$, $\chi^2 = 13.02$) (Figure 11.6c).

With regard to tributary proximity, frogs on Shady Creek were observed farther than expected from tributary junctions, while frogs on the other study creeks were observed closer than expected to tributaries (Figure 11.7). The seasonal dataset for Shady and Humbug Creeks revealed the same significant trends as the annual dataset with each season. Individuals on Oregon Creek were observed closer to tributaries than the expected average mean distance; however, the sample size was too small for statistical analysis.

Frogs on Humbug Creek did not show an association with valley width ($p > 0.10$), but individuals on both Shady and Rush Creeks were observed more often than expected in stream locations with wider than average

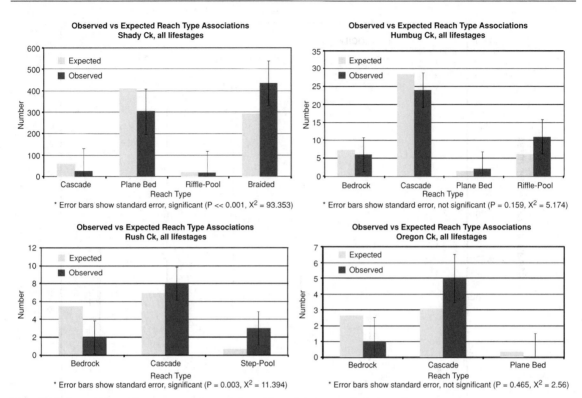

Figure 11.5 Annual observed versus expected association with reach type on each study creek.

valley width throughout the year (Figure 11.8). Associations with valley width were not significant in the Humbug Creek seasonal datasets or the summer and fall datasets for Shady Creek. In spring on Shady Creek, however, individuals did show a higher relative occupancy in stream locations with wider than average valley width ($p = 0.001$, $t = 3.296$).

11.4 Discussion

11.4.1 Mesohabitat associations

The results from the statistical analyses indicate that quantitative parameters related to hydraulics, specifically velocity, depth and substrate size, were the most useful predictors of mesohabitat associations by life stage for *R. boylii*. Velocity was the single most useful parameter, distinguishing all life stages and indicating a strong relationship between aging life stage and progressively higher velocities. When velocity was evaluated in combination with other hydraulically related parameters, such as depth and substrate, the ability to predict individual life stage–habitat associations increased greatly. The combination

of velocity and depth was useful in predicting habitat associations for pre-metamorphic life stages; evaluating the combination of velocity, depth and substrate through logistic regression modelling resulted in an average of 82% of life stage–habitat associations being predicted correctly. These results indicate that habitat hydraulics are likely a key component of habitat suitability for all *R. boylii* life stages. Further study on the relationship between life stage and small-scale microhabitat hydraulic parameters (e.g. point velocity (versus local velocity in this study), point depth, shear stress, roughness, Froude number, Reynolds number, among other hydraulic variables) may provide additional insights into *R. boylii* habitat suitability.

The observed mesohabitat associations on all four study creeks generally agree with results from previously published studies and, in sum, indicate specific hydraulic habitats are utilized throughout the *R. boylii* lifecycle. Kupferberg (1996) found that *R. boylii* selected breeding sites with higher than average width to depth ratios and lower than average velocities, such that negative effects from fluctuating discharges were minimized. Lind (2005), Wheeler (2007) and Lind and Yarnell (2008) furthered this study by examining breeding sites across the northern

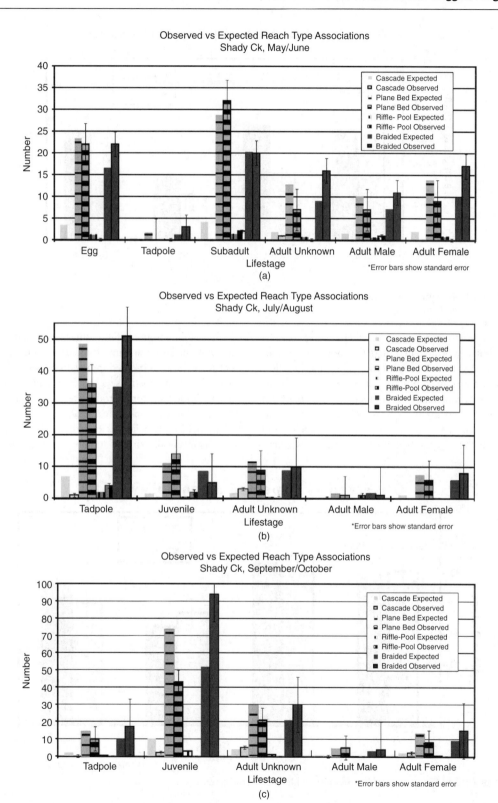

Figure 11.6 Seasonal observed versus expected association with reach type on Shady Creek: (a) spring; (b) summer; (c) fall.

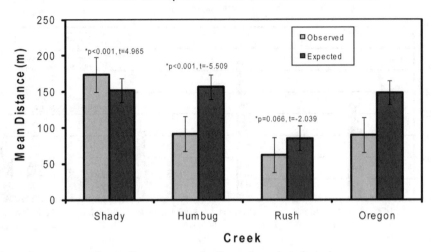

Figure 11.7 Observed versus expected mean distance to nearest tributary for each study creek.

California coast range and the northern Sierra Nevada. Their results indicated that *R. boylii* optimizes reproductive success by selecting breeding sites that have low velocity and hydraulic stability (e.g. wide, shallow cross-sections or downstream of boulders) with large-sized substrate. The results from Shady Creek show a similar relationship between egg masses, low velocity, and coarse substrate. While Shady Creek, due to its exten-

sive hydraulic mining tailings, differs in habitat availability from the other streams studied previously, the fact that *R. boylii* was observed in breeding sites with similar hydraulic characteristics suggests these habitat conditions may be important for reproductive success regardless of the nature or location of the stream.

The habitat associations of tadpoles in Shady Creek were hydraulically similar to those of egg masses (low

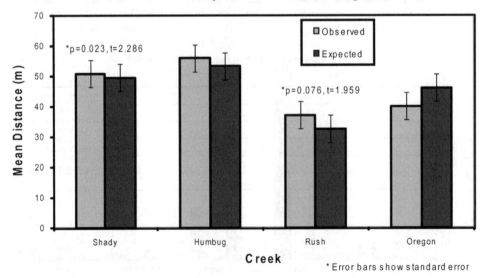

Figure 11.8 Observed versus expected mean valley width for each study creek.

velocity, shallow depth), and agreed with habitat observations from other studies (Lind and Yarnell, 2008). *R. boylii* tadpoles are adapted to the stable, low-flow stream conditions common to the summer dry season in California, and are not morphologically adapted to withstand high stream flow and associated high velocities (Kupferberg *et al.*, 2011). Their apparent preference for low-velocity, shallow stream habitats confers a growth advantage due to high algal resources often associated with low-velocity locations (Kupferberg, 1996) and the limited energy expenditure required to maintain stream position. As a result, successful breeding sites have been observed in stream locations that are not only hydraulically stable during the egg development period in spring, but are also directly connected to stream locations with preferred hydraulic habitat conditions for tadpoles in summer (Yarnell *et al.*, 2012).

Data on the habitat associations of older life stages are more limited; however, the results from this study complement previous work examining *R. boylii* life history. Van Wagner (1996) and Haggarty (2006) found juveniles most commonly occupied shallow, moderate-velocity run and glide habitats, while adults occupied a wide range of habitats. With a more quantitative approach using hydraulic geometry relationships to assess hydraulic conditions as flow fluctuated, Yarnell (2000) found that juveniles were most often observed in channels with shallow overbank areas that maintained moderate velocities as discharge fluctuated, while adults were observed in narrower channels where depth and velocity fluctuated. The results from Shady Creek support these findings, indicating that adults were most likely to be associated with habitats with higher velocity and depth, while shallow habitats were more likely to support juveniles.

The study results show that a variety of hydraulic conditions are utilized throughout the *R. boylii* lifecycle, indicating that habitat diversity in both time and space is important. Although depth and velocity are usually correlated with substrate size in most stream environments, life stage associations with substrate, depth and velocity on Shady Creek were not correlated. Rather, life stages were often observed in association with habitats where substrate size did not reflect associated depth and velocity conditions. For example, egg masses were often found attached to coarse substrates in low velocities and depths, and juveniles were often observed on smaller substrates such as gravel and small cobble, but in shallow, high-velocity flows. These associations suggest that *R. boylii* may require specific mesohabitat conditions that are the result of previous high flows. High velocities scour and transport finer material downstream, leaving coarse cobbles and boulders behind. Only when flows subsequently decrease will a mesohabitat of low velocity, low depth and coarse substrate required for successful oviposition emerge. As a result, streams with high hydraulic habitat diversity both in space and time are more likely to provide the mesohabitat conditions required for successful *R. boylii* populations.

The results from this and other studies (as discussed above) show that hydraulically related parameters, such as velocity, depth and substrate, are clearly associated with suitable *R. boylii* egg mass and tadpole habitat. Additionally, Kupferberg *et al.* (2009; 2010) showed that negative impacts from hydraulic conditions inducing scour can decrease survival and recruitment, resulting in decreased frog population size. However, hydraulic habitat conditions alone are not sufficient to define *R. boylii* habitat suitability. Lind and Yarnell (2008) emphasized the fact that other ecological factors such as biological influences, water quality (particularly temperature), human influences and reach-scale influences also contribute to *R. boylii* habitat suitability. Lancaster and Downes (2010) argued a similar position, cautioning against inferring habitat preferences based on observed habitat–abundance relationships alone. Therefore, while *R. boylii* habitat requirements can be usefully characterized using hydraulic habitat parameters, which, in turn, can be readily evaluated using instream flow assessment techniques, an ecohydraulics-based management tool widely utilized to assess flow impacts in regulated stream systems (Yarnell *et al.*, 2012), a full assessment of habitat suitability must include ecological influences as well as hydraulic conditions.

11.4.2 Reach-scale associations

Statistically significant associations of *R. boylii* with specific reach types on each creek revealed not only associations with larger-scale habitat characteristics, but an association with stream locations with a wide variety of hydraulic habitats. On Shady Creek, braided reaches had a higher relative abundance of all life stages than other available reach types. The multi-channelled nature of the braided reaches provided a number of exposed channel bars, scour pools in locations where channels converged or moved around vegetation, shallow gently flowing runs and higher gradient riffles. Within this diversity of habitat types, hydraulic habitat conditions for each life stage occurred. Similarly, on Humbug and Rush Creeks, riffle-pool and step-pool reaches were the most heavily used reach types, each potentially providing a greater amount of hydraulically variable pool and bar habitat than either bedrock or cascade reaches.

Unlike the mesohabitat analyses, which were based on habitat use data only, the reach-scale data incorporated both abundance and absence of individuals (use and non-use data) and therefore results could be described as habitat 'preferences' as is commonly the case in studies of species–habitat relationships (e.g. Ahmadi-Nedushan *et al.*, 2006; Guay *et al.*, 2000, among others). While these data do reflect relative differences in abundance that are likely linked to the various physical habitat characteristics available in the different reach types given the strong association *R. boylii* has with hydraulic habitat conditions, additional study would be needed to determine what other ecological and geomorphic conditions might be driving observed reach-scale preferences (Lancaster and Downes, 2010).

The analysis of reach type use data within the seasonal datasets provided additional insight regarding movement of individuals between habitats throughout the year. While mesohabitat associations shifted with progressively older life stages, the use of specific reach types remained consistent among the younger life stages. Tadpoles and juveniles on Shady Creek preferred braided reaches throughout the year, while adults showed varying associations with reach type depending on season. The seasonal variability in adult habitat likely indicates movement while foraging (Van Wagner, 1996; Bourque, 2008), while the lack of seasonal variability in reach type association with younger life stages indicates decreased movement, either by choice or necessity. Braided reaches appear to provide the variety of hydraulic habitats required for the younger life stages; consequently, movement upstream or downstream is minimized. Decreased movement is inherent to egg masses and young tadpoles, simply due to their limited mobility, but limited movement by older tadpoles and newly metamorphed juveniles may help limit energy expenditure and reduce the risk of exposure to predators. Thus, braided reaches would provide a significant benefit towards increased survival over less diverse reaches, such as cascades and plane bed reaches, where the costs of movement between habitats would be greater.

The results from Humbug, Rush and Oregon Creeks agree with data from other studies that suggest a preference by *R. boylii* for stream habitats closer to tributary junctions (Kupferberg, 1996; Peek, 2010). Tributaries often provide overwintering habitat and refuge from high winter and spring flows on larger streams. However, Lind (2005) did not find a preference by *R. boylii* for near-tributary locations on Hurdygurdy Creek, a small northern California coastal stream. Rather, breeding sites were commonly located in braided reaches where water temperature and algal growth were greater. She surmised the lack of association with tributaries was due to the relatively small stream size and the limited number of perennial tributaries. Similar conditions exist on Shady Creek – surface flow input from tributaries is often insignificant in late summer and fall due to low flows, and locations with high valley width have braided channels at high flows, providing an abundance of refuge habitat in the adjacent floodplains and low terraces. Therefore, like Hurdygurdy Creek, the open conditions of braided reaches on Shady Creek may provide an increased benefit for frogs over the potential refuge of tributaries.

11.4.3 Habitat heterogeneity

Physical habitat heterogeneity (the spatial variation of physical properties between habitats) has long been believed to influence biologic processes in stream ecosystems, both directly in terms of species richness and diversity (Palmer *et al.*, 1996; Downes *et al.*, 2000) and indirectly as the result of watershed disturbance processes (Townsend *et al.*, 1997; Poole, 2002). Specifically, greater heterogeneity in stream habitat can increase both aquatic species diversity (Beisel *et al.*, 2000; Brown, 2003) and individual species success and survival (Power, 1992; Torgersen and Close, 2004). For *R. boylii*, increased habitat heterogeneity at the reach scale can provide substantial benefits for each life stage, particularly from egg masses through metamorphosis. Reaches with higher heterogeneity provide a greater diversity of hydraulic habitat suitable to each of the life stages, provide a greater variety of geomorphic habitats with multiple functions (e.g. cover, forage, basking and breeding) and provide a greater variety of refugia as flows fluctuate.

Results from a concurrent study (Yarnell, 2008) showed that diversity in stream habitat type and spatial variation in these habitats can be quantified using a variety of habitat heterogeneity indices. These indices were shown not only to reflect the spatial diversity of channel structures at the reach scale, but to be ecologically meaningful. Using data from the same four study creeks within the Yuba River basin, reach types with the highest habitat heterogeneity had the highest abundance of *R. boylii*. Interestingly, reach types were not consistent in their level of measured heterogeneity across each of the tributaries, and thus did not correlate with *R. boylii* abundance across creeks. Rather, the reach type with the highest heterogeneity of those available within that creek was the reach type most populated by *R. boylii* in that drainage (Yarnell, 2008). Given the diverse hydraulic habitat needs of individuals across life stages, as indicated in this study, *R. boylii* appears to be associating with reaches with higher physical habitat heterogeneity, where a greater diversity of hydraulic habitats is available in a relatively short section of stream.

The association of all life stages with stream locations near tributary confluences and greater valley width also supports the hypothesis that reaches with greater physical heterogeneity provide optimal habitat for *R. boylii*. Benda *et al.* (2003) found that heterogeneity in channel morphology increased near tributary confluences prone to debris flows. Debris and alluvial inputs from tributaries are forms of watershed disturbance and act to increase habitat diversity by depositing large woody debris, inducing sediment deposition behind larger boulders and forming wider valley floors. Just as increased complexity in habitat types induces increased species richness by providing a greater variety of niches (Ward *et al.*, 2002), so too does increased habitat heterogeneity provide the variety of hydraulic habitat characteristics needed by *R. boylii* throughout its lifecycle.

The results from this study have direct implications for conservation and restoration practices. Because *R. boylii* habitat requirements vary across multiple spatial and temporal scales, streams where natural processes create a high diversity of physical habitats within a stream reach will have the highest benefit for long-term reproductive success. Specifically, reaches where a variety of hydraulic habitat conditions exist provide an increased likelihood that, in any given year, regardless of discharge, the required habitat conditions will exist for multiple life stages. For example, stream reaches with a greater geomorphic and hydraulic diversity are more likely to contain the low-velocity, shallow-depth, coarse-substrate habitat required for breeding in both low flow and high flow years than a uniform stream reach where the right habitat may exist only at certain flows. Therefore, restoration practices that maximize the diversity of hydraulic habitats available through space and time will be most likely to benefit *R. boylii*.

References

Ahmadi-Nedushan, B., St-Hilaire, A., Berube, M., Robichaud, E., Thiemonge, N. and Bobee, B. (2006) A review of statistical methods for the evaluation of aquatic habitat suitability for instream flow assessment. *River Research and Applications*, **22**(5): 503–523.

Beisel, J.N., Usseglio-Polatera, P. and Moreteau, J.C. (2000) The spatial heterogeneity of a river bottom: a key factor determining macroinvertebrate communities. *Hydrobiologia*, **422**: 163–171.

Benda, L., Veldhuisen, C. and Black, J. (2003) Debris flows as agents of morphological heterogeneity at low-order confluences, Olympic Mountains, Washington. *Geological Society of America Bulletin*, **115**(9): 1110–1121.

Bourque, R. (2008) *Spatial Ecology of an Inland Population of the Foothill Yellow-legged Frog (Rana boylii) in Tehama County, California*. Masters thesis, Humboldt State University.

Brown, B.L. (2003) Spatial heterogeneity reduces temporal variability in stream insect communities. *Ecology Letters*, **6**(4): 316–325.

Compton, R.R. (1985) *Geology in the Field*, John Wiley & Sons, Inc., New York.

Davidson, C., Shaffer, H.B. and Jennings, M.R. (2002) Pesticide drift, habitat destruction, UV-B, and climate-change hypotheses for California amphibiana. *Conservation Biology*, **16**(6), 1588–1601.

Downes, B.J., Lake, P.S., Schreiber, E.S.G. and Glaister, A. (2000) Habitat structure, resources and diversity: the separate effects of surface roughness and macroalgae on stream invertebrates. *Oecologia*, **123**(4): 569–581.

Dyer, F.J. and Thoms, M.C. (2006) Managing river flows for hydraulic diversity: An example of an upland regulated gravel-bed river. *River Research and Applications*, **22**(2): 257–267.

ESRI (2002) ESRI ArcGIS ver.8.3. www.esri.com, ESRI Inc.

Guay, J.C., Boisclair, D., Rioux, D., Leclerc, M., Lapointe, M. and Legendre, P. (2000) Development and validation of numerical habitat models for juveniles of Atlantic salmon (*Salmo salar*). *Canadian Journal of Fisheries and Aquatic Sciences*, **57**(10): 2065–2075.

Haggarty, M. (2006) *Habitat Differentiation and Resource Use Among Different Age Classes of Post Metamorphic Rana boylii on Red Bank Creek, Tehama County, California*. Masters thesis, Humboldt State University.

Hawkins, C.P., Kershner, J.L., Bisson, P.A., Bryant, M.D., Decker, L.M., Gregory, S.V., McCullough, D.A., Overton, C.K., Reeves, G.H., Steedman, R.J. and Young, M.K. (1993) A hierarchical approach to classifying stream habitat features. *Fisheries*, **18**(6): 3–12.

Heyer, W.R., Donnelly, M.A., McDiarmid, R.W., Hayek, L.C. and Foster, M.S. (eds) (1994) *Measuring and Monitoring Biological Diversity: Standard Methods for Amphibians*. Biological Diversity Handbook Series, Smithsonian Institution Press, Washington DC.

Hosmer, D.W. and Lemeshow, S. (1989) *Applied Logistic Regression*, John Wiley & Sons, Inc., New York.

Jennings, M.R. and Hayes, M.P. (1994) *Amphibian and Reptile Species of Special Concern in California*. Final report, Rancho Cordova, California Department of Fish and Game Inland Fisheries Division.

JowettI.G., Richardson, J., Biggs, B.J.F., Hickey, C.W. and Quinn, J.M. (1991) Microhabitat preferences of benthic invertebrates and the development of generalized *Deleatidium* spp. habitat suitability curves, applied to four New Zealand rivers. *New Zealand Journal of Marine and Freshwater Resources*, **25**: 187–199.

Knapp, R.A. and Preisler, H.K. (1999) Is it possible to predict habitat use by spawning salmonids? A test using California golden trout (*Oncorhynchus mykiss aguabonita*). *Canadian Journal of Fisheries and Aquatic Sciences*, **56**: 1576–1584.

Kupferberg, S.J. (1996) Hydrologic and geomorphic factors affecting conservation of a river-breeding frog (*Rana boylii*). *Ecological Applications*, **6**(4): 1332–1344.

Kupferberg, S.J. (1997) Bullfrog (*Rana catesbeiana*) invasion of a California river: the role of larval competition. *Ecology*, **78**: 1736–1751.

Kupferberg, S.J., Lind, A.J. and Palen, W.J. (2010) *Pulsed Flow Effects on the Foothill Yellow-legged Frog* (Rana boylii): *Population Modeling*. Final report, California Energy Commission, PIER. Publication number TBD.

Kupferberg, S.J., Lind, A.J., Mount, J.F. and Yarnell, S.M. (2009) *Pulsed Flow Effects on the Foothill Yellow-legged Frog* (Rana boylii): *Integration of Empirical, Experimental and Hydrodynamic Modeling Approaches*. Final report, California Energy Commission, PIER. Publication number 500-2009-002. Available at: http://animalscience.ucdavis.edu/PulsedFlow/Kupferberg%20et%20al%202009.pdf

Kupferberg, S.J., Lind, A.J., Thill, V. and Yarnell, S.M. (2011) Water velocity tolerance in tadpoles of the foothill yellow-legged frog (*Rana boylii*): swimming performance, growth, and survival. *Copeia*, **2011**(1): 141–152.

Lancaster, J. and Downes, B.J. (2010) Linking the hydraulic world of individual organisms to ecological processes: Putting ecology into ecohydraulics. *River Research and Applications*, **26**: 385–403.

Lind, A.J. (2005) *Reintroduction of a Declining Amphibian: Determining an Ecologically Feasible Approach for the Foothill Yellow-legged Frog (Rana boylii) Through Analysis of Decline Factors, Genetic Structure, and Habitat Associations*. PhD dissertation, University of California, Davis.

Lind, A.J. and Yarnell, S.M. (2008) *Habitat Suitability Criteria for the Foothill Yellow-legged Frog* (Rana boylii) *in the Northern Sierra Nevada and Coast Ranges of California*. Final report, DeSabla Centerville Hydroelectric Project (FERC Project No. 803). Available at: http://www.eurekasw.com/DC/relicensing/Study%20Plan%20Implementation/Fish%20and%20Aquatic/HSC%20final%20report%2015Feb08.pdf

Lind, A.J., Welsh, H.H. and Wilson, R.A. (1996) The effects of a dam on breeding habitat and egg survival of the foothill yellow-legged frog (*Rana boylii*) in northwestern California. *Herpetological Review*, **27**(2): 62–67.

Montgomery, D.R. and Buffington, J.M. (1997) Channel-reach morphology in mountain drainage basins. *Geological Society of America Bulletin*, **109**(5): 596–611.

Nussbaum, R.A., Brodie, E.D. and Storm, R.M. (1983) *Amphibians & Reptiles of the Pacific Northwest*, Univeristy Press of Idaho.

Palmer, M.A., Arensburger, P., Martin, A.P. and Denman, D.W. (1996) Disturbance and patch specific responses: The interactive effects of woody debris and floods on lotic invertebrates. *Oecologia*, **105**(2): 247–257.

Pastuchova, Z., Lehotsky, M. and Greskova, A. (2008) Influence of morphohydraulic habitat structure on invertebrate communities (Ephemeroptera, Plecoptera and Trichoptera). *Biologia*, **63**: 720–729.

Peek, R. (2010) *Landscape Genetics of Foothill Yellow-legged Frogs* (Rana boylii) *in Regulated and Unregulated Rivers: Assessing Connectivity and Genetic Fragmentation*. Masters thesis, University of San Francisco.

Poole, G.C. (2002) Fluvial landscape ecology: addressing uniqueness within the river discontinuum. *Freshwater Biology*, **47**(4): 641–660.

Power, M.E. (1992) Habitat heterogeneity and the functional significance of fish in river food webs. *Ecology*, **73**(5): 1675–1688.

Rice, S.P., Little, S., Wood, P.J., Moir, H.J. and Vericat, D. (2010) The relative contributions of ecology and hydraulics to ecohydraulics. *River Research and Applications*, **26**: 1–4.

SPSS (2003) SPSS 12.0 for Windows: Base, Regression Models and Advanced Models. Chicago, IL, SPSS Inc.

Thomson, J.R., Taylor, M.P., Fryirs, K.A. and Brierley, G.J. (2001) A geomorphological framework for river characterization and habitat assessment. *Aquatic Conservation-Marine and Freshwater Ecosystems*, **11**(5): 373–389.

Torgersen, C.E. and Close, D.A. (2004) Influence of habitat heterogeneity on the distribution of larval Pacific lamprey (*Lampetra tridentata*) at two spatial scales. *Freshwater Biology*, **49**(5): 614–630.

Townsend, C.R., Scarsbrook, M.R. and Doledec, S. (1997) The intermediate disturbance hypothesis, refugia, and biodiversity in streams. *Limnology and Oceanography*, **42**(5): 938–949.

Trexler, J.C. and Travis, J. (1993) Nontraditional regression analysis. *Ecology*, **74**: 1629–1637.

Van Wagner, T.J. (1996) *Selected Life-history and Ecological Aspects of a Population of Foothill Yellow-legged Frogs* (Rana boylii) *from Clear Creek, Nevada County, California*. Masters thesis, California State University, Chico.

Ward, J.V., Tockner, K., Arscott, D.B. and Claret, C. (2002) Riverine landscape diversity. *Freshwater Biology*, **47**: 517–539.

Wheeler, C. (2007) *Temporal Breeding Patterns and Mating Strategy of the Foothill Yellow-legged Frog* (Rana boylii). Masters thesis, Humboldt State University.

Yarnell, S.M. (2000) *The Influence of Sediment Supply and Transport Capacity on Foothill Yellow-legged Frog Habitat, South Yuba River, California*. Masters thesis, University of California, Davis.

Yarnell, S.M. (2008) Quantifying physical habitat heterogeneity in an ecologically meaningful manner: A case study of the habitat preferences of the Foothill yellow-legged frog (*Rana boylii*). In Dupont, A. and Jacobs, H. (eds) *Landscape Ecology Trends*, Nova Science Publishers, Inc., pp. 89–112.

Yarnell, S.M., Viers, J.H. and Mount, J.F. (2010) Ecology and management of the spring snowmelt recession. *BioScience*, **60**(2): 114–127.

Yarnell, S.M., Lind, A.J. and Mount, J.F. (2012) Dynamic flow modeling of riverine amphibian habitat with application to regulated flow management. *River Research and Applications*, **28**(2): 177–191.

Zweifel, R.G. (1955) *Ecology, Distribution and Systematics of Frogs of the Rana boylii Group*, University of California Publications in Zoology, Berkeley, CA.

12 Testing the Relationship Between Surface Flow Types and Benthic Macroinvertebrates

Graham Hill[1], Ian Maddock[1] and Melanie Bickerton[2]

[1] Institute of Science and the Environment, University of Worcester, Henwick Grove, Worcester, WR2 6AJ, UK
[2] Geography, Earth and Environmental Sciences, University of Birmingham, Edgbaston, Birmingham, B15 2TT, UK

12.1 Background

Surface Flow Types (SFTs) are defined by patterns produced on the surface of running water as a result of the interaction between channel morphology and flow hydraulics. They are related to Channel Geomorphic Units (CGUs) (Hawkins *et al.*, 1993) and were first defined by Wadeson (1994) and investigated by Wadeson and Rowntree (1998). The national system for assessing river habitats in England and Wales, the River Habitat Survey (Environment Agency, 2003), uses SFTs as surrogates for CGUs whilst some habitat-modelling tools, e.g. MesoCASiMiR (Eisner *et al.*, 2005) require surveyors to map habitat areas that 'look different', which implies a reference to surface flow patterns. Surface Flow Types have been used to define habitat patches to investigate habitat diversity–discharge relationships (Dyer and Thoms, 2006) and the distribution of macroinvertebrate communities (Reid and Thoms, 2008).

Habitat classification schemes based on physical conditions (Hawkins *et al.*, 1993; Wadeson, 1994; Padmore, 1997; Wadeson and Rowntree, 1998; Tickner *et al.*, 2000) link in-stream hydraulics and morphology, whilst classifications of macroinvertebrate communities based on biotic variables, e.g. Functional Feeding Groups (Cummings, 1974) focus on the biological/ecological aspects. Together they address current ecohydraulic understanding, with 'habitat' being defined as the sum of the environmental conditions within which an individual organism or community lives. At the meso-scale of analysis, habitat has become the focus of hydroecological research. Maddock (1999) considers that the term 'habitat' implies a biological dimension, although that biological relevance is still not always clear. It is widely accepted that there are differences between types of physical habitat, and also perhaps their faunal composition, but there is also variety within them (Lancaster and Belyea, 2006). This study may, therefore, go some way to address the suggestion that flow biotope diversity could be used as a cost-effective way of assessing potential diversity (Harper *et al.*, 2000).

The biological relevance of mesohabitats, here defined as areas of river channel > 1 m^2 with the same SFT, is poorly understood. This chapter presents a case study examining the relationship between mesohabitats defined by their SFT, and the macroinvertebrate community associated with them in English lowland streams.

12.2 Ecohydraulic relationships between habitat and biota

Rivers comprise a highly variable mosaic of hydraulic microhabitats with varying water depth, velocity and

---

Ecohydraulics: An Integrated Approach, First Edition. Edited by Ian Maddock, Atle Harby, Paul Kemp and Paul Wood.
© 2013 John Wiley & Sons, Ltd. Published 2013 by John Wiley & Sons, Ltd.

Table 12.1 Channel Geomorphic Units (Hawkins *et al.*, 1993) compared with SFT mesohabitat descriptions (Padmore, 1997) and the SFT code used here.

Channel Geomorphic Unit (Hawkins *et al.*, 1993)	Surface Flow Types Padmore (1997)	SFT Code
Fall – Vertical drops of water	Free Fall – Vertically falling water, generally > 1 m high	FF
Cascade – Highly turbulent series of short falls and small scour basins		
Chute – Narrow, steep slots or slides in bedrock	Chute – Fast, smooth flow over boulders/bedrock	CH
Rapid – Moderately steep channel units with coarse substrate and planar profile	Broken standing wave – White water tumbling wave facing upstream	BW
Riffle – Shallow, fast flow with some substrate breaking the surface	Unbroken Standing Wave – Undular standing waves, upstream-facing unbroken crest	UW
Run – Moderately fast and shallow gradient with ripples on the water surface	Rippled – Surface turbulence produces symmetrical ripples moving downstream	RP
Boil – Strong vertical flow producing a characteristic 'boil' at the surface	Upwelling – Strong vertical flow producing a characteristic 'boil' at the surface	UP
Glide – Smooth, 'glass-like' surface, with visible flow movement along the surface	Smooth Boundary Turbulent – Little surface turbulence, reflections distorted	SM
Pool – Deep and slow flowing with fine substrate, usually little surface water movement visible	No Perceptible Flow (Scarcely Perceptible Flow) – Surface stationary and reflections not distorted	NP

substrate sizes and types (Marchildon *et al.*, 2011). These can be grouped into areas with similar characteristics based on geomorphic and hydraulic properties. Hawkins *et al.* (1993) devised a hierarchical scheme based on increasingly fine descriptions of morphological and hydraulic properties of CGUs (Table 12.1), which has been widely used for research (e.g. Maddock and Bird, 1996; Parasiewicz, 2001). Hydraulic properties were used by Wadeson (1994) and Wadeson and Rowntree (1998) to define mesohabitats by their SFT. Padmore (1997) developed detailed descriptions of SFTs (Table 12.1) and advocated that they could provide a standard description of in-stream habitat that was visually identifiable. Table 12.1 defines SFTs used in this study and the acronyms used herein; it also provides a comparison of descriptions between the CGUs of Hawkins *et al.* (1993) and SFTs. The Environment Agency in England and Wales adopted SFTs in the River Habitat Survey Method (Environment Agency, 2003).

Surface Flow Types result from the interaction between riverbed morphology, bed roughness and local hydraulics to produce a series of distinct patterns on the water surface. Turbulent flow is likely to be present in natural channels (Wadeson, 1994) and forms the upper part of the water column with a viscous boundary layer close to the bed caused by friction with the bed substrate. It is within the viscous layer that many macroinvertebrates live, although

the physical nature of hydraulic biotopes and the classification of in-stream habitats using substrate and SFT is complex (Wadeson and Rowntree, 1998).

Wadeson and Rowntree (1998) examined a range of hydraulic indices with respect to several South African rivers, encompassing a wide range of hydraulic biotopes at several discharges. They demonstrated that there was considerable overlap of index values between several hydraulic biotopes (SFTs). Wadeson and Rowntree (1998) indicated that there is an increase in index value through Pool (NP), Run (RP) and Riffle (UW), which appears to follow the increasing energy gradient. However, Glide (SM) was shown to have the highest index values. They conclude that, apart from rapid and cascade, which can be grouped together (BW), the hydraulic biotopes can be considered hydraulically distinct.

Turbulence increases with greater substrate size and higher flow velocity, and is associated with reduced water depth in several hydraulic biotope classes (Figure 12.1). With the exception of chute and upwelling flow, water column turbulence generates a series of distinct water surface patterns – upwelling flow is a plume of water directed towards the surface by a bed or bank form and chute flow is laminar in nature and occurs over flat surfaces (e.g. bedrock) or through narrow gaps between large substrate clasts and is principally found in upland areas.

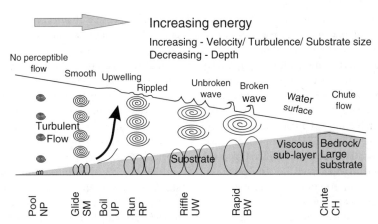

Figure 12.1 Conceptual diagram showing the interaction of substrate and water column hydraulics producing distinct patterns on the water surface – Surface Flow Types.

The nature of flow within hydraulic biotopes is complex. Harvey and Clifford (2009) concluded that burst–sweep turbulence generation, vortex shedding from individual clasts and bed microtopography, together with larger structures associated with tree roots and larger-scale forms, are responsible for the internal flow patterns in each hydraulic biotope. Glides (SM – see Table 12.1 for description) are considered to have the simplest flow structure, possibly related to burst–sweep structures. Riffles (UW) have a more complex form with vortices being shed from microform roughness, whilst pools (NP) are the most hydraulically complex biotope, characterised by burst–sweep structures and vortex shedding from small-scale grain roughness and larger-scale forms.

Benthic macroinvertebrate taxa are a convenient biological focus in ecohydraulic studies, as they are relatively large (> 1 mm in size), numerous and well studied, and less mobile than other faunal groups, such as fish. Both physical and behavioural adaptations help macroinvertebrate groups exploit a wide range of niches. Overcoming the hydraulic forces on and near the stream bed is important if macroinvertebrates are to remain within appropriate niches (Hynes, 1970). Adaptations of macroinvertebrate taxa to running waters are varied. For example, Heptageniidae mayfly larvae have a flattened dorso-ventral form, which may be an adaptation limiting the effects of high velocity (Hynes, 1970), although this shape also allows it to squeeze between stones to avoid the current altogether (Lancaster and Belyea, 2006). Adult Elmidae (Coleoptera) use their powerful claws to cling to the substrate, whilst the larvae of some Elmidae species have a flattened form with spines around the outer edges which engage with substrate roughness to prevent dislodgement (Hynes, 1970; Elliott, 2008). Snails can be seen to have large, soft contact areas by which they attach to the substratum, whilst Ancylidae also have the benefit of a 'streamlined' shell to prevent dislodgement in the high velocities where it is often found (Conchological Society of Great Britain and Ireland, 2011). Amongst the Trichoptera, those adapted to life in fast-flowing water, for example Goeridae, frequently build cases from dense materials, using large ballast stones (Wallace *et al.*, 2003), although this may also be due to the local availability of building materials. Limnephilidae, some species of which are more common in slow-flowing waters, have been observed using vegetation to build their cases, which would be less resilient than stone in turbulent areas. However, camouflage is also an important consideration for these animals and perhaps by using 'local' materials they are less susceptible to predation.

Physical habitat conditions at the micro-scale are important in determining the niche/s which macroinvertebrates inhabit. Lancaster and Belyea (2006) recorded a remarkable diversity between 99 macroinvertebrate samples collected from a single riffle. These results support Wadeson's (1994) view that geomorphologists are [still] not able to provide the level of detail required by ecologists. Lancaster and Belyea (2006) suggest that relationships between macroinvertebrate abundance and near-bed velocity are better informed by examination of velocity range, rather than the central tendency, and advocate the use of Quantile Regression.

The relationship between SFTs, near-bed hydraulics and stream macroinvertebrates in the regulated upland Cotter River, Australia were investigated by Reid and Thoms (2008), Surface Flow Types (NP, SM, RP, UW,

BW, CH) were used to identify habitats. They found that the SFTs had distinct hydraulic, substratum and macroinvertebrate community characteristics, although the distinction of the macroinvertebrate community between adjacent SFTs was not strong. Further, they demonstrated that the less strongly separated SFTs were adjacent to each other in an energy gradient: NP–SM; RP–UW; UW–BW and BW–CH, although NP and SM differed from RP, UW, BW and CH (see Table 12.1 for SFT descriptions). They contended that near-bed hydraulics and substratum were closely interlinked and together had the strongest relationship with the macroinvertebrate community. They found that downstream velocity was the most important variable influencing macroinvertebrate assemblage composition and taxa richness. Turbulence was also shown to have a strong relationship with macroinvertebrate community structure, although the reasons were unclear. They concluded that the relationships between SFTs, near-bed hydraulics and substrate character in upland regulated rivers were sufficient to support the hypothesis that SFTs were an *'effective way of characterising the physical habitat template controlling macroinvertebrate distributions'* (Reid and Thoms, 2008, p. 1054).

Surface Flow Types can also provide spatial information about habitat diversity at different flows (Dyer and Thoms, 2006), which suggests that SFTs could be used to define mesohabitats and provide an important tool for river managers.

Despite significant advances, the relationship between SFTs and macroinvertebrate communities is not well established, particularly in lowland rivers. To further this understanding, an investigation was undertaken in six lowland English rivers, addressing three questions:

1 Can Surface Flow Types in English lowland rivers be identified and recorded effectively?
2 Which hydraulic variables are the most appropriate to characterise Surface Flow Types in English lowland rivers: No perceptible flow, Smooth, Ripple, Unbroken wave or Upwelling?
3 What is the relationship between benthic macroinvertebrate communities and Surface Flow Types in English lowland rivers: No perceptible flow, Smooth, Ripple, Unbroken wave or Upwelling?

12.3 Case study

12.3.1 Site details and method

The relationship between SFT and macroinvertebrate communities was examined in six English lowland rivers

between 1 April and 30 June 2006 (Figure 12.2). A range of physical and biological data was determined for each site during three surveys at different discharges. Each survey estimated and mapped the extent of in-stream mesohabitats defined by five SFTs – NP, SM, RP, UW and UP. To characterise the physical environment of each SFT identified, measurements were made at five locations within the SFT. These included mean column velocity, water depth, substrate size class based on the Wentworth scale (Wentworth, 1922) (categorised as dominant, sub-dominant and 'other'), and percentage cover of algae, filamentous algae, bryophyte and macrophyte. During each survey three replicate macroinvertebrate samples were obtained by kick-sampling from one representative example of each SFT present at each site. Immediately downstream of each macroinvertebrate sample, a velocity profile was recorded, mean velocity was determined using a Valeport 801 Electronic Current Meter, from 20 readings taken at 1-second intervals – on the bed, 0.05 m and 0.10 m above the bed and at 0.10 m intervals to the surface. Macroinvertebrate samples were identified to family level in the laboratory.

12.3.2 Results

Can Surface Flow Types in English lowland rivers be identified and recorded effectively?

Eighteen surveys were conducted; three surveys at six sites located on six different streams. Stream channel edges at each site were checked against Ordnance Survey Mapping (surveyed scale 1:2500) using differential Global Positioning Satellite (dGPS) equipment with positional accuracy < 1 m and channel edges amended where necessary. The extents of 341 SFT mesohabitats were estimated and their positions recorded on large-scale channel plans. The mesohabitats comprised NP, SM, RP, UW, UP, BW, CH (Table 12.1) and 'Confused' (areas with a mixture of SFTs). 323 SFTs, comprising 42 NP, 97 SM, 119 RP, 55 UW and 10 UP, were investigated further. The remaining mesohabitats (BW, CH and Confused) were not investigated because they were rare in the lowland English rivers studied.

Which hydraulic variables are the most appropriate to characterise Surface Flow Types in English lowland rivers: No perceptible flow, Smooth, Ripple, Unbroken wave or Upwelling?

Depth and velocity The Kruskal–Wallis test (nonparametric ANOVA) was used to examine the variance of depth and mean column velocity across SFTs. There was a significant difference ($P < 0.05$) between depth in SFTs

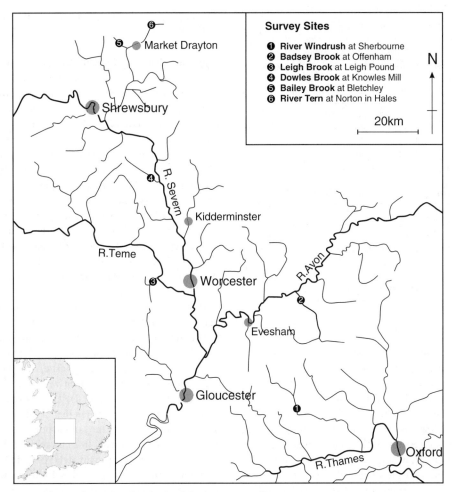

Survey Sites

❶ **River Windrush** at Sherbourne
❷ **Badsey Brook** at Offenham
❸ **Leigh Brook** at Leigh Pound
❹ **Dowles Brook** at Knowles Mill
❺ **Bailey Brook** at Bletchley
❻ **River Tern** at Norton in Hales

20km

Figure 12.2 Location of six sites on six English lowland rivers investigated in 2006.

and a significant difference in velocities. Pair-wise analysis of depth and velocity data, using the Mann–Whitney U Test, also showed that 90% of combinations of depth were significantly different between SFTs (Critical $P < 0.005$ using Bonferroni correction; Abdi, 2007) and that there was a significant difference (Critical $P < 0.005$ using Bonferroni correction) between 80% of mean column velocity combinations (Table 12.2). Of the non-significant depth pairs, only NP and UP were similar, whilst velocities in UP SFT were similar to SM and RP SFTs. This suggests that depth was significantly different between all SFT classes and that flow velocity in all SFTs except UP was significantly different.

HydroSignature analysis (Le Coarer, 2007) was used to identify the distinctiveness of SFTs. For a given flow,

HydroSignature quantifies the hydraulic variety of an area of aquatic habitat by calculating the surface area percentages in a depth and current velocity cross-classification. Each cell in the HydroSignature output shows the percentage of surface area containing that depth/velocity class. Calculations were made for each category of SFT using data from all surveys.

In NP SFT ∼ 90% of velocity records were < 0.1 m/s (Table 12.3, columns 1 and 2) and ∼ 82% of depth records were between 0.3 and 0.9 m (these values plot to the left of the matrix). SM and RP SFTs spread across the depth/velocity matrix, whilst the central group of UW SFT cells plot farther right (higher velocity) and higher (shallow); UP SFTs occupy a central area of the matrix. Summing the proportion of SFT area assigned to cells

Table 12.2 Contingency table showing significance of analysis of water depth and mean column velocity data from all surveys in six English lowland rivers using five Surface Flow Type mesohabitats using the Mann–Whitney U Test. Significance: b = significant using Bonferroni correction;.n.s. = not significant.

Depth	NP	SM	RP	UW
Smooth (SM)	< 0.001 b			
Rippled (RP)	< 0.001 b	< 0.001 b		
Unbroken wave (UW)	< 0.001 b	< 0.001 b	< 0.001 b	
Upwelling (UP)	0.011 n.s.	< 0.001 b	< 0.001 b	< 0.001 b
Velocity				
Smooth (SM)	< 0.001 b			
Rippled (RP)	< 0.001 b	< 0.001 b		
Unbroken wave (UW)	< 0.001 b	< 0.001 b	< 0.001 b	
Upwelling (UP)	< 0.001 b	0.133 n.s.	0.057 n.s.	< 0.001 b

(grouped by SFT) indicates that 90% of NP area was identified, 80% of UP, 73% of UW, 44% of SM and 33% of RP. HydroSignature analysis suggests that the depth/velocity cross-classes occupied by SFTs were distinctive in NP, UW and UP, but less so in SM and RP. Based on these data, Figure 12.3 shows depth and velocity range in each of the five SFTs investigated, the shaded areas show the inter-quartile range and the outer lines the overall data spread. The inter-quartile ranges of NP and UW SFTs were largely separated from other SFTs, whilst SM and RP overlapped, as did UP, although it was deeper than either SM or RP.

Substrate size The Kruskal–Wallis test (non-parametric ANOVA) was used to examine the differences between substrate sizes among SFTs. The analysis showed that there were significant differences ($P < 0.05$) in all categories of dominant substrate and sub-dominant substrates. Pairwise analysis using the Mann–Whitney U Test (Table 12.4) indicated that, for the dominant substrate, 70% of pairs differ (UP was not significantly different in three cases – NP, SM and RP) and in sub-dominant substrate, 80% of pairs differ (Critical $P < 0.005$ using Bonferroni correction; Abdi, 2007). These analyses indicated that dominant substrate size was significantly different between SFTs except between UP and NP, UP and SM and between UP and RP. Sub-dominant substrate size was also significantly different between SFTs, except between UP and NP and between RP and SM.

Principal Component Analysis (PCA) of the physical variables recorded in five classes of SFTs indicated that axes 1 and 2 accounted for 67% of variance. Both velocity and depth had strong relationships, with vector lengths exceeding 99%, and were opposite each other on the ordination plot (Figure 12.4, in which vectors are plotted against PCA axes 1 and 2). The dominant substrate vector was slightly longer than that of sub-dominant substrate. In the PCA plot, the sample points were plotted in bands running diagonally from bottom-left to top-right; this direction was at approximately 90° to the vector direction for both Dominant and Sub-dominant Substrate size. The stripes were, therefore, interpreted as relating to

Table 12.3 Distribution of depth and velocity classes by Surface Flow Type based on HydroSignature analysis. Key: NP – No Perceptible SFT; SM – Smooth SFT; RP – Rippled SFT; UW – Unbroken wave SFT; UP – Upwelling SFT.

Depth (m)										
0	np	NP	np	UW	**UW**	UW	UW	uw	Lower case	< 1% of total
0.1	NP	NP	UW	**UW**	UW	UW	**UW**	UW	UPPER CASE	1–5%
0.2	NP	SM	**SM**	SM	RP	**UP**	UW	uw	**UPPER BOLD**	5–10%
0.3	NP	SM	**SM**	UP	UP	**RP**	RP	uw	INVERSE	>10%
0.5	NP	**NP**	SM	UP	**RP**	RP	RP	uw		
0.7	NP	**NP**	SM	UP	UP	RP	RP			
0.9	**NP**	NP	SM	UP	UP	rp	rp			
1.4	NP									
	0	0.1	0.2	0.3	0.5	0.7	0.9	1.5		
	Velocity (m/sec)									

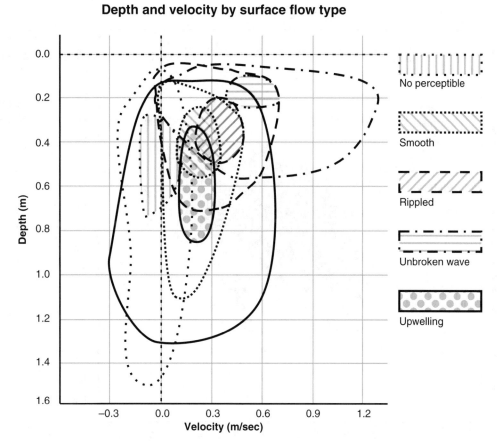

Figure 12.3 The range of depth and velocity values from all surveys by Surface Flow Type. The shaded areas show the inter-quartile range; the lines the range of values.

Table 12.4 Contingency table showing significance of analysis of using the Mann–Whitney U Test of substrate class size from five Surface Flow Types using data collected in six English lowland rivers. Significance: b = significant using Bonferroni correction; n.s. = Not significant.

Dominant substrate	NP	SM	RP	UW
Smooth (SM)	< 0.001 b			
Rippled (RP)	< 0.001 b	0.003 b		
Unbroken wave (UW)	< 0.001 b	< 0.001 b	< 0.001 b	
Upwelling (UP)	0.018 n.s.	0.332 n.s	0.023 n.s.	< 0.001 b
Sub-dominant substrate				
Smooth (SM)	< 0.001 b			
Rippled (RP)	< 0.001 b	0.115 n.s.		
Unbroken wave (UW)	< 0.001 b	< 0.001 b	< 0.001 b	
Upwelling (UP)	0.722 n.s.	0.001 b	< 0.001 b	< 0.001 b

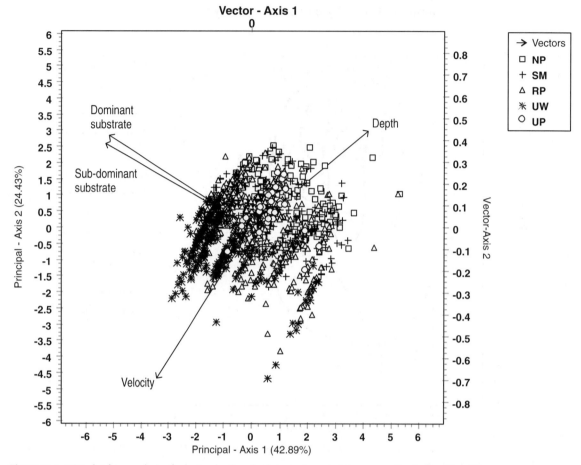

Figure 12.4 PCA plot from analysis of velocity, depth and substrate characteristics grouped by Surface Flow Type.

substrate size which was recorded in ten classes using the Wentworth Scale (Wentworth, 1922).

What is the relationship between benthic macroinvertebrate communities and No perceptible flow, Smooth, Ripple, Unbroken wave and Upwelling Surface Flow Types in English lowland rivers?

During each survey, one representative example of each SFT present was selected for macroinvertebrate sampling. Within this SFT unit, three macroinvertebrate samples were taken using a one-minute kick sample with a 500 μm mesh 'D' net. A total of 200 macroinvertebrate samples from 73 SFTs were collected (Table 12.5). Forty-three macroinvertebrate taxa were identified, most groups being identified to family level. Seven taxa containing < 5 individuals were omitted from further analysis. For the

Table 12.5 Macroinvertebrate sample points by Surface Flow Type mesohabitat recorded in six English lowland rivers between 1st April and 30th June 2006.

	Macroinvertebrate samples	Surface Flow Types sampled
No perceptible (NP)	46	19
Smooth (SM)	49	17
Ripple (RP)	54	18
Unbroken wave (UW)	41	14
Upwelling (UP)	10	5
Total	200	73

Table 12.6 Mean values of relative abundance, macroinvertebrate group richness, ASPT score, Shannon–Wiener diversity, LIFE score grouped by Surface Flow Type mesohabitats, from samples collected from six English lowland rivers. Key: NP – No perceptible; SM – Smooth; RP – Rippled; UW – Unbroken wave; UP – Upwelling; ASPT – Average Score per Taxon.

Means	NP	SM	RP	UW	UP	SFT rank (High–Low)	ANOVA P value
Relative abundance	83	109	126	270	92	UW RP SM UP NP	< 0.001
Richness	6.5	6.9	10.2	10.4	9.6	UW RP UP SM NP	< 0.001
ASPT	4.14	5.34	5.36	5.34	5.51	UP RP UW/SM NP	< 0.001
H diversity	1.04	1.30	1.57	1.42	1.91	UP RP UW SM NP	< 0.001
LIFE score	6.95	7.60	7.69	7.94	7.60	UW RP SM/UP NP	< 0.001

remaining taxa, abundance ranged from the minimum of 5 to the maximum of > 6700 individuals. Average abundance for five SFTs was NP = 83 (standard deviation (SD = 123); SM = 109 (SD = 130); RP = 126 (SD = 173); UW = 270 (SD = 182) and UP = 92 (SD = 96) (Table 12.6).

A series of macroinvertebrate metrics comprising relative abundance, macroinvertebrate taxa richness, Average Score per Taxon (ASPT), Shannon–Wiener diversity and The Lotic-invertebrate Index for Flow Evaluation (LIFE) score (Extence *et al.*, 1999) was used to examine the relationships between macroinvertebrate communities in five SFTs. For each metric, Kruskal–Wallis non-parametric ANOVA was used to examine variance between SFTs (Table 12.6) and the Mann–Whitney *U* Test was used to examine pair-wise differences.

Pair-wise analysis with the Mann–Whitney *U* Test (significance $P < 0.005$ with Bonferroni correction) showed that:

- Using relative abundance, 40% of pairs were significantly different ($P < 0.005$). UW significantly differed from all other SFTs.
- Using macroinvertebrate taxa richness, 50% of pairs were significantly different ($P < 0.005$). NP was significantly different from all other SFTs, as was SM/RP.
- Using ASPT, 40% of pairs were significantly different ($P < 0.005$). Only NP differed from all other SFTs.
- Using the Shannon–Wiener diversity index, 40% of pairs were significantly different ($P < 0.005$). This shows that NP differs from all other SFTs, and that SM differs from UP.
- Using the LIFE Score, 60% of pairs were significantly different ($P < 0.005$). The results show that NP and UW were most consistently different.

The results of pair-wise analysis indicated that the number of occasions (maximum five) when a significant difference existed between SFT pairs over the range of

macroinvertebrate metrics was variable (Table 12.7). NP SFT differed for all five metrics from SM, RP and UW, and from UP for four metrics. UW differed from NP for all five metrics and from SM, RP and UW for three each. SM and RP differed significantly for only one metric (H diversity). UP differed from SM for one metric (H diversity) and was not significantly different from RP for any metric.

Although a wide range of values were recorded, mean relative abundance and macroinvertebrate taxa richness increased from NP SM RP UW, reflecting the increase in mean velocity, whilst ASPT peaked at RP, increasing from NP SM UW RP. NP SFT was most consistently different from the other SFTs investigated. There were significant differences in relative abundance, richness and ASPT between some pairs of SFT, suggesting that adjacent pairs were more similar than pairs separated by one or more SFT on the velocity gradient. UP SFT did not fit easily into this pattern, suggesting that not only is it infrequent, but biologically variable.

Canonical Correspondence Analysis (CCA) was used to examine the relationship between macroinvertebrate communities and measured environmental variables. Collinearity between environmental variables was not appreciable (Variable Inflation Factor < 9) (Table 12.8). CCA based on all samples from 18 surveys of six rivers

Table 12.7 Summary table showing the number (maximum = five) of significantly different macroinvertebrate metrics based on pair-wise Mann–Whitney *U* Test results.

Summary metrics	NP	SM	RP	UW
Smooth (SM)	3			
Rippled (RP)	4	0		
Unbroken wave (UW)	All 5	3	2	
Upwelling (UP)	3	1	0	2

Table 12.8 Comparison of vector lengths (determined by CCA) for SFT and physical variables. NP – No perceptible; SM – Smooth; RP – Rippled; UW – Unbroken wave; UP – Upwelling, VIF – Variable inflation factor.

SFT	Vector length	VIF	Variable	Vector length	VIF
UW	0.80	5.7	Bed velocity	0.53	4.0
UP	0.70	1.4	Surface velocity	0.51	2.8
RP	0.68	5.9	Dominant substrate size	0.48	1.3
NP	0.56	5.0	Water depth	0.46	1.8
SM	0.52	4.9	Sub-dominant substrate size	0.40	1.3
			Velocity at bed + 0.05 m	0.33	4.8

was undertaken to establish the relative importance of SFTs and individual hydraulic and substrate variables. Examination of first and second ordination axis scores (Figure 12.5) showed very weak grouping of SFT sample communities, with only NP being obviously

distinguished from other SFTs. However, vector lengths determined from the first three ordination axis scores demonstrated that SFT, velocity, depth and substrate size were most strongly related to the macroinvertebrate communities present in each unit. The variance explained by CCA was low; axis 1 was 6.07%, axis 2: 2.53% and axis 3: 1.85%, totalling 10.45%.

Comparison of vector lengths (Table 12.8) showed that amongst SFTs, UW had the strongest association with macroinvertebrate communities, followed by UP, RP, NP and SM. Within the primary physical variables, velocity on the bed was ranked highest followed by surface velocity, dominant substrate size, water depth, sub-dominant substrate size and velocity 0.05 m above the bed.

Following the basic 'metrics' and community-level analyses, chi-square tests were used to identify to what extent individual macroinvertebrate taxa differed significantly in their frequency of occurrence between SFTs. Ten macroinvertebrate taxa (all families) showed significant differences between observed and expected frequencies and were therefore considered to be strongly associated with particular SFTs. Table 12.9 displays the observed numbers of these groups in each SFT.

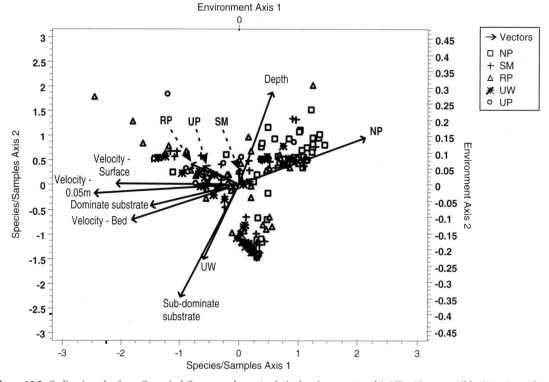

Figure 12.5 Ordination plot from Canonical Correspondence Analysis showing axes 1 and 2. NP – No perceptible; SM – Smooth; RP – Rippled; UW – Unbroken wave; UP – Upwelling.

Table 12.9 Chi-square analysis of macroinvertebrate data using data collected from six English lowland rivers showing taxa significantly associated with Surface Flow Type. P – Significance: ($P < 0.05$), D.F. – degrees of freedom; NP – No perceptible; SM – Smooth; RP – Rippled; UW – Unbroken wave; UP – Upwelling; LIFE group – Lotic-invertebrate Index for Flow Evaluation flow group.

	χ^2	D.F.	P	NP ($n = 46$)	SM ($n = 49$)	UP ($n = 10$)	RP ($n = 54$)	UW ($n = 41$)	LIFE group
				No of samples containing each taxa					
Elmidae	31.13	4	< 0.001	12	21	7	36	31	2
Ancylidae	20.83	4	< 0.001	0	3	1	10	12	2
Hydrobiidae	11.44	4	0.022	12	16	6	25	22	4
Heptageniidae	25.54	4	< 0.001	1	8	3	11	18	1
Hydropsychidae	17.59	4	0.001	1	2	2	7	11	2
Baetidae	33.33	4	< 0.001	5	24	5	28	30	2
Gammaridae	14.99	4	0.005	15	30	8	28	26	2
Ephemerellidae	12.54	4	0.014	7	21	4	19	19	2
Caenidae	17.95	4	0.001	0	7	4	5	9	4
Chironomidae	16.93	4	0.002	45	37	5	38	34	—

Nine of the ten macroinvertebrate taxa displaying a significant relationship with SFTs were most strongly associated with the highest-energy SFT (UW), and negatively associated with the lowest-energy SFT (NP).

LIFE scores, which categorise macroinvertebrates by velocity class (Extence *et al.*, 1999), are shown against each taxon in Table 12.9: Category 1 (Rapid) with BW or CH, Category 2 (Moderate/Fast flow) with UW or RP, Category 3 (Slow/Sluggish) with SM or RP, Category 4 (Flowing/Standing) with NP or SM and Category 5 (Standing) with NP. The taxa displaying the most significant associations (Elmidae, Ancylidae, Heptageniidae and Baetidae) all belong to LIFE score flow groups 1 or 2, indicating they were expected to occur in high-velocity habitats, and all were positively associated with UW, the highest-energy SFT sampled. Chironomidae differed from the other groups with highest relative frequency in NP; this family has no LIFE score flow group assigned. To investigate further the distribution of these ten taxa, Hydro-Signature matrices were used to display the range of depths and velocities over which these groups were recorded (Figure 12.6). In the matrix, the dark-coloured cell shows the median value, the heavy horizontal stripes show the interquartile range and the light vertical stripes the range of values. The velocity range suggested by LIFE flow velocity classes (Extence *et al.*, 1999) is shown (in grey) below the *x*-axis. This is a novel use of HydroSignature for ecological data, and usefully illustrates not only the dominant hydraulic conditions in which individual taxa were recorded, but also the relative ranges of distribution. For example, Elmidae and Ancylidae were both found in locations with a median velocity of 0.5 ms^{-1}, and median depth of 0.30 m, however Elmidae is found in a wider range of both velocity and depth than Ancylidae.

12.4 Discussion

This research demonstrates that five SFTs examined in six English lowland rivers can be effectively recorded by estimation on large-scale channel plans. The SFTs are physically distinct, with depth significantly different between SFT pairs in 90% of cases (Table 12.2) and velocity significantly different between SFT pairs in 80% of cases. Macroinvertebrate associations with SFTs, hydraulic and substrate variables are more subtle, but analyses of simple metrics, community type and individual taxa showed evidence of distinct relationships.

Macroinvertebrate richness and abundance were found to be highest in the Unbroken wave (UW) and lowest in No Perceptible flow (NP) SFT. Flow velocity was shown to be a key variable in the physical characterisation of SFTs; it was also shown to have the strongest relationship with macroinvertebrate community structure through the LIFE score (Extence *et al.*, 1999), suggesting that flow velocity is an important driver of macroinvertebrate community structure in SFT mesohabitats. Flow velocity was also shown to be a key variable in the physical characterisation of SFTs. A similar trend from low- to high-energy flow types was found by Pastuchova *et al.* (2010) in the Teplicka Brook, Slovenia. These authors also found a positive association between flow type and passive filter-feeding taxa. More energetic SFTs provide higher current velocities which are important for transporting food to

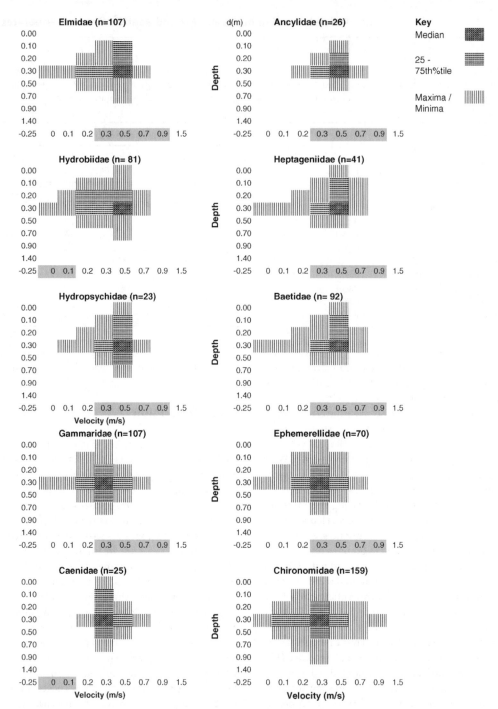

Figure 12.6 Matrix showing depth/velocity classes associated with ten macroinvertebrate groups from six English lowland rivers. Median values are shown in dark lines, the extents of the 25th and 75th quartiles with horizontal lines and the extents of maxima and minima with vertical lines. The velocity ranges of flow groups from LIFE are shown in mid grey below the *x*-axis. (*n* refers to the number of samples containing the macroinvertebrate group.)

filter-feeding invertebrates such as the net-spinning cad-disfly Hydropsyche and some Diptera fly larva such as Simuliidae (Wallace and Merrritt, 1980; Georgian and Thorp, 1992). Although the relationship between flow velocity and food capture efficiency is not simple due to behavioural adaptations (Brown et al., 2005) and larval size (Sagnes et al., 2008), higher current velocity habitats will favour a variety of filter-feeding taxa.

River sites with higher levels of macroinvertebrate taxa richness (Leigh Brook and River Windrush) had a wider range of substrates in the study reaches, whilst Bailey Brook and Badsey Brook had fewer. Low-energy SFTs (NP and SM) were dominated by small substrate size whilst higher-energy SFTs (RP and UW) had larger substrate sizes. Coarse and more heterogeneous substrates have been shown to promote greater macroinvertebrate rich-ness through increased topographical complexity alter-ing the pattern of near-bed flows (Cardinale et al., 2002; Brown et al. 2005, Death and Winterbourn, 1995). Coarse, stable substrates provide refugia from high velocities; retain organic matter which is a food source for shred-der species such as Gammaridae, and allow for the devel-opment of benthic algae which support grazers such as Ancylidae, Hydrobiidae and Elmidae. Pastuchova et al. (2010) also reported greater numbers of shredders and grazer-scraper species in high-energy flow types with coarse substrates.

The lowest energy SFT (NP) was associated with habi-tats dominated by fine substrates, which were less stable and suitable for a smaller range of macroinvertebrate taxa. Chironomidae, the only taxon found in this study to pre-fer the No Perceptible flow (NP) SFT, has been shown to increase in other studies of anthropogenic siltation of stream habitats (Wood and Armitage, 1997).

Potentially greater turbulence in high-energy SFTs may also be associated with increased dissolved oxygen avail-ability for benthic macroinvertebrates. This would explain the trend in ASPT scores (a measure of sensitivity to organic pollution which is strongly linked to dissolved oxygen), found to be highest in the Unbroken wave SFT (UW), which were characterised by higher water velocity and coarser substrate, resulting in higher water column turbulence. Elmidae were found to be significantly more frequent in UW in this study and are a good example of a group known to be associated with high dissolved oxygen, stable substrates and turbulent flow. Elmidae are assumed to be grazers of algae and bryophytes and rarely found in deep water where their food source does not grow well; larvae live between stones where velocity is lower (Elliott, 2008), whilst the adults cling to stones or vegetation. It is thought that Elmidae 'breathe' through pores in their exoskeleton as they have no gills, and will therefore benefit from high oxygen levels in turbulent SFTs.

This study showed the relationship between macroin-vertebrate communities and SFTs to be weak compared to between-site differences, and highlights the importance of scale. Whilst ecologists and hydrologists are increas-ingly recognising the importance of mesohabitats, these fall within a hierarchy of scales from microhabitat to catchment (Newson and Newson, 2000). Single stream studies of the upland regulated Cotter River, Australia (Reid and Thoms, 2008) and the Teplicka Brook, Slovenia (Pastuchova et al, 2010) found clearer separations between macroinvertebrate communities in the SFTs investigated. In the study of English lowland rivers presented here, macroinvertebrate community composition in SFTs was less distinct, and this may, in part, be due to the larger number of rivers investigated (200 samples taken from 73 SFT patches in six rivers) and the importance of catch-ment scale factors. Water quality, in particular, may be sig-nificant in restricting communities (Harper et al., 2000). Only indirect assessment of water quality was included in this study, using Biological Monitoring Working Party (BMWP) scores (Hawkes, 1997). These varied consider-ably from 80 to 21, suggesting that water quality may warrant further investigation. Direct physico-chemical measurements associated with each SFT sample could potentially be used to 'partial-out' the effects of water quality in a Canonical Correspondence Analysis (Hurl-bert, 1984) to separate between-site factors, and could also be used to elucidate between-SFT patterns in dis-solved oxygen related to turbulence.

Whilst larger-scale factors such as water quality may be important, hydraulic and sedimentary conditions also vary at smaller, microhabitat scales within SFTs, which are highly relevant to macroinvertebrates. Heino et al. (2004) suggested that many replicate samples were needed to establish macroinvertebrate community character, whilst Lancaster and Belyea (2006) found considerable variabil-ity in 99 macroinvertebrate samples obtained from one riffle. Greater taxa richness has been linked to increas-ing habitat patch heterogeneity (Tickner et al., 2000) and attributed to the range of niches available to macroin-vertebrates and to the relatively short distances between suitable niches which allow organisms to move from one to another.

In a study of the Kangaroo River, New South Wales, Brooks et al. (2005) demonstrated strong relationships between hydraulic variables within a single mesohabi-tat type – riffles. Interestingly, that study also reported

Table 12.10 Physical characteristics and macroinvertebrate communities associated with five Surface Flow Types in English lowland rivers. NP – No perceptible; SM – Smooth; RP – Rippled; UW – Unbroken wave; UP – Upwelling.

	Physical properties			Macroinvertebrate community		
SFT	Depth	Velocity	Substrate	Abundance	Richness	Dominant macroinvertebrate groups
NP	Deep	Slow, includes eddy (upstream) flow	Fine – silts or clays – or coarse highly embedded with fine materials	Low	Low	Chironomidae
SM	Less deep than NP	Low downstream	Sand and/or gravel	Slightly higher than in NP	Slightly higher than in NP	Ephemerellidae, Gammaridae, Baetidae, Chironomidae and Caenidae
RP	Moderately shallow	Moderate downstream	Gravel	Greater than NP and slightly higher than SM	Greater than NP and slightly higher than SM	Baetidae, Elmidae, Gammaridae, Ephemerellidae and Hydropsychidae
UW	Shallow	High downstream	Pebble or Cobble	Greatest	Greatest	Hydropsychidae, Ephemerellidae, Gammaridae, Heptageniidae, Ancylidae, Hydrobiidae and Chironomidae
UP	Deep	Moderate downstream with a strong vertical component	Sand and/or gravel	Moderate	Moderate	Gammaridae, Elmidae, Caenidae, Hydrobiidae, Baetidae and Ephemerellidae

negative associations between invertebrate abundance and richness with velocity. Maximum velocities in the Australian riffle sites were higher than those encountered in the English lowland rivers; this highlights the fact that the range of conditions, as well as scale, is an important consideration. The range of samples in the lowland rivers also included few vegetated habitats. Although benthic algae have been considered as an influence on the distribution of grazing macroinvertebrates, macrophytes are known to provide habitats with distinct communities (Harper *et al.*, 2000) and may alter near-bed hydraulic conditions within SFTs. Pastuchova *et al.* (2010) attribute greater macroinvertebrate abundance and richness in 'complex organic' habitats to heterogeneous and turbulent conditions similar to those usually associated with more energetic flow types. In summary, this case study suggests that mesohabitats, in English lowland rivers, defined by one of five SFTs are likely to have the physical characteristics and contain macroinvertebrate communities described in Table 12.10 (see Table 12.1 for a description of SFTs).

12.5 Wider implications

Mesohabitats have been used since the 1990s to describe homogeneous areas of river channel. Some methods, for example Rapid Habitat Mapping (Maddock and Bird, 1996) and MesoHABSIM (Parasiewicz, 2001) take no account of lateral habitat diversity, whilst the Norwegian Mesohabitat Classification Method (Borsányi, 2004) allows up to three habitats laterally across the channel and MesoCASiMiR (Eisner *et al.*, 2005) allows the surveyor to determine the habitats 'as seen'. In this research, the 'as seen' approach was followed as it allows a more complete representation of SFTs to be recorded. This research has demonstrated that mapping SFTs in this manner was effective and that the use of SFTs to define mesohabitats was appropriate to characterise the extent of such habitats

(Newson and Newson, 2000; Dyer and Thoms, 2006) and physical/biological relevance (Reid and Thoms, 2008).

The biological relevance of in-stream habitat assessments has yet to be fully understood, although they are clearly important. The use of such methods for investigation of macroinvertebrate communities has potential; however, there is a need for further research. Although the mesohabitat scale is widely used as an appropriate scale to study rivers (Newson and Newson, 2000), adopting this scale necessarily generalises the microhabitats present which are of importance to macroinvertebrates. Lancaster and Belyea (2006) found a great deal of variety in microhabitats in one riffle, and whilst they raise issues regarding generalising velocity–abundance relationships using central tendencies (Lancaster and Downes, 2010a; 2010b), others have argued that existing methods are appropriate in many instances (Lamouroux *et al.*, 2010).

HydroSignature (Le Coarer, 2007) has been shown to discriminate between some SFTs (over several discharges) quite well, although SM and RP are less distinct. This usefully allows the range of velocities/depths to be quantified and could potentially provide a tool for examining velocity–abundance relationships in the SFT and perhaps used/adapted to address some of the concerns raised by Lancaster and Downes (2010a; 2010b) in modelling habitat abundance of macroinvertebrate communities. Further, the use of SFTs may provide a route to remote sensing of in-stream habitats with potential cost benefits. Assessing river habitats at the mesoscale is advocated by several (e.g. Wadeson and Rowntree, 1998; Newson and Newson, 2000) and is the focus of several habitat assessment methods which use physical descriptors to identify habitats. SFTs have been used as surrogates for Channel Geomorphic Units in the UK River Habitat Survey and the basis for areas of research (Dyer and Thoms, 2006; Reid and Thoms, 2008). The biological relevance of such mesohabitats is still being debated, and although the diversity of microhabitats within a mesohabitat is challenging (Lancaster and Belyea, 2006; Lancaster and Downes, 2010a, 2010b) this is an area that requires further study.

12.6 Conclusion

This case study has attempted to identify associations between mesohabitats, defined by their Surface Flow Type, in English lowland rivers and the macroinvertebrate community contained within them. The common physical properties of the five SFTs have been shown to be more robust than the composition of the macroinvertebrate assemblages within them, although abundance and rich-

ness appear to increase with velocity. Ten macroinvertebrate taxa were associated with one or more of the SFTs, although CCA suggests that the community structure is complex and warrants further investigation. Eliminating variable water quality between rivers and identification to a lower taxonomic level (species perhaps) would be useful. However, whilst there are still questions to be answered, mesohabitats defined by their SFT, recorded 'as seen' in the channel could provide a useful method for river managers to better understand the complexities of in-stream habitats. Although there is still some way to go, this study has moved towards the challenge of finding a cost-effective measure of ecological process at the mesohabitat scale (Harper *et al.*, 2000).

References

Abdi, H. (2007) Bonferroni and Šidák corrections for multiple comparisons. In Salkind, N.J. (ed.) *Encyclopedia of Measurement and Statistics*, Sage, Thousand Oaks, CA.

Borsányi, P. (2004) Norway: using mesohabitats for upscaling: method development on the Nidelva and other rivers. In Harby, A., Baptist, M., Dunbar, M.J. and Schmutz, S. (eds) *State-of-the-Art in Data Sampling, Modelling Analysis and Applications of River Habitat Modelling*. COST Action 626 report.

Brooks, A.J., Haeusler, T., Reinfelds, I. and Williams, S. (2005) Hydraulic microhabitats and the distribution of macroinvertebrate assemblages in riffles. *Freshwater Biology*, 50: 331–344.

Brown, S.A., Ruxton, G.D., Pickup, R.W. and Humphries, S. (2005) Seston capture by *Hydropsyche siltalai* and the accuracy of capture efficiency estimates. *Freshwater Biology*, 50: 113–126.

Cardinale, B.J., Palmer, J.A. and Collins, S.L. (2002) Species diversity enhances ecosystem functioning through interspecific facilitation. *Nature*, 415: 426–429.

Conchological Society of Great Britain and Ireland (2011) FAMILY: ANCYLIDAE. [Online] http://www.conchsoc.org/aids_to_id/Ancylidae.php [Accessed 28/6/2011].

Cummings, K.W. (1974) Structure and function of stream ecosystems. *BioScience*, 24: 631–641.

Death, R.G. and Winterbourn, M.J. (1995) Diversity patterns in stream benthic invertebrate communities: The influence of habitat stability. *Ecology*, 76(5): 1446–1460.

Dyer, F.J. and Thoms, M.C. (2006) Managing river flows for hydraulic diversity: an example of an upland regulated gravel-bed river. *River Research and Applications*, 22: 257–267.

Eisner, A., Young, C., Schneider, M. and Kopecki, I. (2005) Meso-CASiMiR – new mapping method and comparison with other current approaches. In Harby, A. *et al.* (eds) *Proceedings of the Final COST 626 meeting*, Silkeborg, Denmark.

Elliott, J. (2008) The ecology of Riffle Beetles (Coleoptera: Elmidae). *Freshwater Reviews*, 1(2):189–203.

Environment Agency (2003) *River Habitat Survey in Britain and Ireland, Field Survey Guidance Manual*, 2003 version, Environment Agency, Bristol, UK.

Extence, C.A., Balbi, D.M. and Chadd, R.P. (1999) River flow indexing using British benthic macroinvertebrates: A framework for setting hydroecological objectives. *Regulated Rivers: Research and Management*, **15**: 543–574.

Georgian, T. and Thorp, J.H. (1992) Effects of microhabitat selection on feeding rates of net-spinning caddisfly larvae. *Ecology*, **73**: 229–240.

Harper, D.M., Kemp, J.L., Vogel, B. and Newson, M.D. (2000) Towards the assessment of 'ecological integrity' in running waters of the United Kingdom. *Hydrobiologia*, **422–423**: 133–142.

Harvey, G.L. and Clifford, N.J. (2009) Microscale hydrodynamics and coherent flow structures in rivers: implications for the characterisation of physical habitat. *River Research and Applications*, **25**: 160–180.

Hawkes, H.A. (1997) Origin and development of the Biological Monitoring Working Party score system. *Water Research*, **32**(3): 964–968.

Hawkins, C.P., Kershner, J.L., Bisson, P.A., Bryant, M.D., Decker, L.M., Gregory, S.V., McCullough, D.A., Overton, C.K., Reeves, G.H., Steedman, R.J. and Young, M.K. (1993) A hierarchical approach to classifying stream habitat features. *Fisheries*, **18**: 3–12.

Heino, J., Loughi, P. and Muotka, T. (2004) Identifying the scales of variability in stream macroinvertebrate abundance, functional composition and assemblage structure. *Freshwater Biology*, **49**: 1230–1239.

Hurlbert, S.H. (1984) Pseudoreplication and the design of ecological field experiments. *Ecological Monographs*, **54**(2): 187–211.

Hynes, H.N.B. (1970) *The Ecology of Running Waters*, Liverpool University Press, Liverpool, UK.

Lamouroux, N., Mérigoux, S., Capra, H., Dolédec, S., Jowett, I. and Statzner, B. (2010) The generality of abundance–environment relationships in microhabitats: a comment on Lancaster and Downes (2009). *River Research and Applications*, **26**: 915–920. doi: 10.1002/rra.1366.

Lancaster, J. and Belyea, L.R. (2006) Defining the limits to local density: alternative views of abundance–environment relationships. *Freshwater Biology*, **51**: 783–796.

Lancaster, J. and Downes, B. (2010a) Linking the hydraulic world of individual organisms to ecological processes: putting ecology into Ecohydraulics. *River Research and Applications*, **26**: 385–403. doi: 10.1002/rra.1274.

Lancaster, J. and Downes, B. (2010b) Ecohydraulics needs to embrace ecology and sound science, and to avoid mathematical artefacts. *River Research and Applications*, **26**: 921–929. doi: 10.1002/rra.1425.

Le Coarer, Y. (2007) Hydraulic signatures for ecological modelling at different scales. *Aquatic Ecology*, **41**(3): 451–459.

Maddock, I. (1999) The importance of physical habitat assessment for evaluating river health. *Freshwater Biology*, **41**: 373–391.

Maddock, I. and Bird, D. (1996) The application of habitat mapping to identify representative PHABSIM sites on the River Tavy, Devon, UK. In Leclerc, M. (ed.) *International Symposium on Habitat Hydraulics*, IAHR, Quebec, Canada.

Marchildon, M.A., Annable, W.K., Imhof, J.G. and Power, M. (2011) A high-resolution hydrodynamic investigation of brown trout (*Salmo trutta*) and rainbow trout (*Oncorhynchus mykiss*) redds. *River Research and Applications*, **27**: 345–359. doi: 10.1002/rra.1362.

Newson, M.D. and Newson, C.L. (2000) Geomorphology, ecology and river channel habitat; mesoscale approaches to basin-scale challenges. *Progress in Physical Geography*, **24**(2): 195–217.

Padmore, C.L. (1997) Biotopes and their hydraulics: a method for determining the physical component of freshwater habitat quality. In Boon, P.J. and Howell, D.L. (eds) *Freshwater Quality: Defining the Indefinable*, HMSO, Edinburgh, pp. 251–257.

Parasiewicz, P. (2001) MesoHABSIM: A concept for application of instream flow models in river restoration planning. *Fisheries*, **26**(9): 6–13.

Pastuchova, Z., Greskova, A. and Lehotsky, M. (2010) Spatial distribution pattern of macroinvertebrates in relation to morphohydraulic habitat structure: Perspectives for ecological stream assessment. *Polish Journal of Ecology*, **58**(2): 347–360.

Reid, M.A. and Thoms, M.C. (2008) Surface Flow Types, near-bed hydraulics and the distribution of stream macroinvertebrates. *Biogeosciences Discussions*, **5**: 1175–1204. Available at: www.biogeosciences-discuss.net/5/1175/2008.

Sagnes, P., Merigoux, S. and Peru, N. (2008) Hydraulic habitat use with respect to body size of aquatic insect larvae: Case of six species from a French Mediterranean type stream. *Limnologica*, **38**: 23–33.

Tickner, D., Armitage, P.D., Bickerton, M.A. and Hall, K.A. (2000) Assessing stream quality using information on mesohabitat distribution and character. *Aquatic Conservation: Marine and Freshwater Ecosystems*, **10**: 179–196.

Wadeson, R.A. (1994) A geomorphological approach to the identification and classification of instream flow environments. *South African Journal of Aquatic Sciences*, **20**: 1–24.

Wadeson, R.A. and Rowntree, K.M. (1998) Application of the hydraulic biotope concept to the classification of instream habitats. *Aquatic Ecosystem Health and Management*, **1**: 143–157.

Wallace, J.B. and Merritt, R.W. (1980) Filter-feeding ecology of aquatic insects. *Annual Review of Entomology*, **25**: 103–132.

Wallace, I.D., Wallace, B. and Phillipson, G.N. (2003) *Keys to the case-bearing Caddis larvae of Britain and Ireland*, FBA, Ambleside, UK.

Wentworth, C.K. (1922) A scale of grade and class terms for clastic sediments. *Journal of Geology*, **30**: 377–392.

Wood, P.J. and Armitage, P.D. (1997) Biological effects of fine sediment in the lotic environment. *Environmental Management*, **21**(2): 203–217.

13 The Impact of Altered Flow Regime on Periphyton

Nataša Smolar-Žvanut[1] and Aleksandra Krivograd Klemenčič[2]

[1] Institute for Water of the Republic of Slovenia, Hajdrihova 28c, SI-1000 Ljubljana, Slovenia
[2] University of Ljubljana, Faculty of Health Sciences, Department of Sanitary Engineering, SI-1000 Ljubljana, Slovenia

13.1 Introduction

The term 'periphyton' applies to microbiota living on any substratum: mineral, plant or animal (living or dead) (Wetzel, 1983, 2001). This word has been internationally accepted, despite the imprecision of the term 'microbiota', which applies to any kind of attached living organism, including bacteria, algae and animals. According to Nikora et al. (2002) 'periphyton' is a collective term for the micro-organisms that grow on stream beds, including algae, bacteria and fungi, with algae usually the dominant and most conspicuous component. According to many authors, 'periphyton' refers to a community of microscopic algae and cyanobacteria that grow on the surface of a variety of submerged substrata (Saravia et al., 1998; Godillot et al., 2001); in some cases it is also applied to macroscopic algae such as Cladophora, Audouinella, Vaucheria and other taxa (Vrhovšek et al., 2006). According to the EU Water Framework Directive (Directive of the European Parliament and of the Council, 2000/60/EC), the term 'phytobenthos' is used for microscopic algae, including cyanobacteria, although macroscopic algae are dealt with separately. In this chapter, the term 'periphyton' is used for microscopic and macroscopic algae, including cyanobacteria.

Periphyton is an essential component of lotic ecosystems; it is responsible for most primary production (McIntire, 1973; Apesteguia and Marta, 1979), forms the autotrophic base of stream food webs and is especially important in regulated tailwaters in which upstream sources of detritus are interrupted (Blinn et al., 1998). Periphyton also plays a major role in the metabolic conversion and partial removal of biodegradable material in streams (Lau and Liu, 1993), thereby helping to purify stream waters (Vymazal, 1988). Due to its rapid response to environmental changes, periphyton is also a useful indicator of stream water quality (Biggs, 1996a).

There are many factors important for the growth and development of periphyton including light, temperature, the nature of the substrate, water current/turbulence and flow velocity, pH, alkalinity, hardness, nutrients and other dissolved substances, salinity, oxygen and carbon dioxide (Hynes, 1979). Many of these factors are interrelated, with flow velocity and nutrient concentrations being among the most important factors influencing periphyton in streams (Stevenson, 1983; Reiter, 1986; Suren and Riis, 2010). These factors can fluctuate rapidly as a result of rainfall inputs and human activity. This generates the high spatial and temporal variability in periphyton biomass observed in field investigations (Morin and Cattaneo, 1992). Seasonal oscillations are driven largely by changes in the abiotic environment during the year, but short-term dynamics can be influenced by hydraulic disturbances resulting from floods (McCormick and Stevenson, 1991).

Periphyton is present in all habitats in rivers, from pools with standing water to high-velocity waterfalls. Differences in periphyton biomass and species composition can occur among different river habitats. These differences reflect spatial differences in shear stress, nutrient mass transfer and substratum type (Biggs, 1996a). In

Ecohydraulics: An Integrated Approach, First Edition. Edited by Ian Maddock, Atle Harby, Paul Kemp and Paul Wood.
© 2013 John Wiley & Sons, Ltd. Published 2013 by John Wiley & Sons, Ltd.

nutrient-rich streams, high algal biomass often develops in low flow-velocity runs and pools with filamentous green algae dominant, while in nutrient-poor streams, the highest algal biomass develops in high flow-velocity riffles where diatoms dominate (Biggs, 1996a).

The structure of the periphyton community depends on the characteristics of the river and its ecology. According to the EU Water Framework Directive (Directive of the European Parliament and of the Council, 2000/60/EC), type-specific biological reference conditions for diatoms represent a key biological quality element for each river type of high ecological status across all EU member states. However, human modification of the flow regime can degrade the ecological status of the river and this can be observed and detected through changes in the structure and biomass of the entire periphyton community. Where there has been a significant deviation or degradation of the community, Good Ecological Status (GES) of the water body will not be achieved.

The changes in the structure and biomass of the periphyton community due to altered flow regimes (frequently resulting from human activities) can influence higher trophic levels including invertebrates and fish, but may also lead to changes in habitat characteristics (Suren and Riis, 2010). In stable-flowing enriched streams, periphyton can proliferate, causing eutrophication and water management problems. Suren and Riis (2010) contend that if the aim of resource managers is to minimize the potential adverse effect of low flows, habitat changes caused by the indirect effects of reduced flow on aquatic plant growth must be considered as seriously as the direct effects of low flow on stream hydraulics. The aim of this chapter is to provide an overview of impacts on periphyton communities caused by modified hydrological regimes in rivers below dams.

13.2 Modified flow regimes

Flow is one of the most important variables and key drivers of river ecosystems (Bunn and Arthington, 2002; Malard et al., 2006). Key habitat parameters, such as flow, depth, velocity and habitat volume are dependent on stream flow. Flow is usually closely associated with other environmental conditions, such as water temperature, dissolved oxygen, channel morphology and substrate particle sizes (Richter et al., 1997). The natural flow regime of many rivers is characterized by regular floods, which can strongly influence the distribution and abundance of aquatic organisms (Poff and Hart, 2002). For instance,

the effect of floods has been evaluated for algal communities (Biggs and Close, 1989; Biggs et al., 1998b) and benthic macroinvertebrates (Robinson and Minshall, 1998; Lamouroux et al., 2004; Suren and Jowett, 2006). In general, artificial floods caused by the management of reservoirs can have the same physical impact on river ecosystems as natural floods. However, the impact of artificial floods on the structure and function of periphyton communities may be crucial if they coincide with the natural periods of low flow.

The majority of the rivers around the world are regulated to some extent, with over 45 000 large dams (> 15 m high) currently in operation and many others being built or planned in the future (World Commission on Dams, 2000; Nilsson et al., 2005). Reservoirs are used for a variety of functions including hydropower production, water supply, irrigation, recreation, navigation and flood control. Depending on their use and operation, they can change the river's natural flow regime and lead to both increases and decreases in flow variability. Dams profoundly affect the river hydrology and seasonal flow variability (O'Reilly and Silberblatt, 2009), resulting in changes to the timing, magnitude and frequency of high and low flows (Ligon et al., 1995; Power et al., 1996; Magilligan and Nislow, 2005); consequently, they produce a hydrologic regime differing significantly from the pre-impoundment natural flow regime (Poff et al., 1997). Many large impoundments were built to store water for hydro-electrical production, and they cause reduced or constant low flows in rivers downstream from the dam, with subsequent changes to instream habitats (Graf, 2006) or they eliminate sudden natural floods and droughts (Zimmermann and Ward, 1984). Dams built to control floods, for instance, will reduce peak flows, disconnect the river from its floodplain and reduce its capacity to purify the water as it moves through the watershed (Postel and Richter, 2003). The highest demand for water often coincides with seasonal low flows, when water is required for purposes such as irrigation. In addition, at low flow, any abstraction represents a higher proportion of total stream discharge than during periods of higher flow (Dewson et al., 2007).

The timing and duration of pulse releases below a dam are potentially important and must be undertaken in such a way as to reduce or avoid harm to biotic communities where possible (Petts and Maddock, 1995). Periphyton communities, like all other groups of aquatic flora and fauna, are adapted to the natural flow regime including periodic spates, which are important in regulating the abundance and structure of all aquatic communities.

River flow controls the frequency and magnitude of sediment transport and hence controls disturbance to the stream bed. Another major physical change within regulated systems is that flow may be kept constant at a low discharge for long periods, causing high proportions of boulders and other large substrata to project above the water's surface for unnaturally long time periods (Downes *et al.*, 2003).

Ecologists still struggle to predict and quantify biotic responses to altered flow regimes. According to Bunn and Arthington (2002), one obvious difficulty is the ability to distinguish the direct effects of modified flow regimes from impacts associated with land-use change, often linked with water resource development. They proposed four guiding principles regarding the influence of flow regimes on aquatic biodiversity:

Principle 1: Flow is a major determinant of physical habitat in streams.
Principle 2: Aquatic species have evolved life history strategies primarily in direct response to their natural flow regimes.
Principle 3: Maintenance of natural patterns of longitudinal and lateral connectivity is essential to the viability of populations of many species.
Principle 4: The invasion and success of exotic and introduced species in rivers is facilitated by the alteration of flow regimes.

Flow velocity and its variability have an impact on species composition, succession of species colonization, habitat characteristics and physiology of higher organisms and the metabolism of periphyton. Algae that developed under the influence of low flows have larger cells and are often filamentous algae, which grow more slowly than diatoms (Biggs, 1996b). Reduced flow velocities are associated with greater proportions of fine sediments on the river bed (Bundi and Eichenberger, 1989), which will be reflected in the periphyton community composition. Changes in flow velocities have a major influence on the development of the periphyton communities (Biggs and Stokseth, 1996; Biggs, 1996b). Relatively small changes in local flow velocity may change the hydraulic boundary conditions and have a significant impact on the occurrence of periphyton in rivers (Biggs and Stokseth, 1996). To understand periphyton community changes influenced by modified hydrological conditions, attention must be given to the complex hydraulic interactions that occur near to and at the substratum surface.

13.3 The impact of altered flow regime on periphyton

13.3.1 Species composition and abundance

The modified flow regime below a dam typically results in changes to the periphyton communities (Biggs, 2000). Altered flow regimes can impact species composition or the abundance of individual species, causing their numbers to increase or decrease (in addition, species replacement may occur over time). Furthermore, regulated conditions may provide new habitats or conditions which are suitable for species that did not occur previously in the stream (Growns and Growns, 2001).

Suren and Riis (2010) suggest that the longer the duration of low flow, the more the algal community will change and, in turn, the more the habitat quality will change. This may have significant consequences for benthic invertebrate communities and higher organisms. The pattern of succession in periphyton communities as a result of low flow is linked closely to both a river's enrichment status and the effects of invertebrate grazing, rather than to small temporal or spatial changes in hydraulic habitat during low flows (Suren *et al.*, 2003a). In non-enriched streams, green algae such as *Ulothrix zonata*, *Stigeoclonium* sp., *Spirogyra* sp. and cyanobacteria *Phormidium* sp. often form communities of low biomass. However, in enriched streams, algal communities are usually dominated by large filamentous taxa such as *Cladophora glomerata* and *Rhizoclonium* sp., which can form a high biomass (Biggs, 1996a). Filamentous green algae or cyanobacteria-dominated habitats will experience greater changes during low flows than diatom-dominated habitats, especially when they cover the entire stream bed (Suren, 2005; Suren and Riis, 2010).

Flow variation associated with hydropower production (hydropeaking) typically results in periodic aerial exposure of zones along the banks of regulated rivers. Exposure time can vary greatly, ranging from a few hours to several days. When combined with the unpredictability of flow variability, conditions are too extreme for most freshwater algae. To survive in the wet–dry transition zone, algae must tolerate frequent drying and rewetting, in addition to instream conditions and exposure to grazing organisms. The conditions that promote the most tolerant algae are long wet periods, minimal aerial exposure during daylight hours and riparian shade (Bergey *et al.*, 2010).

The health of the periphyton community in periodically aerial-exposed river margins affects the local stream food web (Blinn *et al.*, 1998). If algae can't tolerate drying conditions, banks that are periodically exposed will contribute little to aquatic productivity (Bergey *et al.*, 2010). Blinn *et al.* (1998) reported that periodically exposed zones in the Colorado River had a ten-fold lower productivity than permanently wet areas. According to Bergey *et al.* (2010), frequent cycles of exposure and submersion might be more detrimental to algae than longer, continuous exposure. No freshwater macroalgae appear to be common or pre-adapted to conditions in tailwaters and the wet–dry zone of rivers. Thus, no macroalgae are available to provide the food and shelter resources to sustain a wet–dry zone community. Shaver *et al.* (1997) reported that the filamentous cyanobacteria *Oscillatoria* sp. persisted better in the wet–dry zone than green algae *Cladophora* sp., especially if fine sediments were present. The reason hypothesized for this was that *Oscillatoria* sp. is motile and can migrate into the sediment to avoid desiccation, and the mucilage-coated filaments of *Oscillatoria* retain water, reducing the effect of desiccation.

Bergey *et al.* (2010) reported that filamentous algae form dense growths in regulated rivers, which can provide favourable habitat and cover for benthic invertebrates (Power, 1990; Shannon *et al.*, 1994). In low-velocity pools and runs in enriched streams (Biggs, 1996a), *Spirogyra* sp., *Cladophora* sp. and *Oedogoonuim* sp. are usually dominant, while in the Soča River in Slovenia, *Cladophora glomerata*, *Mougeotia* sp. and *Ulotrix zonata* are often dominant downstream from the Podsela Dam (Smolar-Žvanut, 2001).

In general, it is difficult to influence the hydrological disturbance regime of streams and rivers because of naturally occurring high-magnitude flows during floods (Biggs, 2000). Species accrual in periphyton after extensive flood disturbance is usually a biphasic process with different timescales for the individual phases. Phase I, which is common to all streams, occurs a day to a week after the flood and is dominated by taxa with high rates of immigration and reproduction (*r* strategists; Grime, 1979) and/or by taxa with high resistance to disturbance derived from local refugia (e.g. low-growing diatoms such as *Achnanthes minutisssima*, *Cocconeis placentula* and *Fragilaria vaucheriae*). Phase II leads to greater species richness and requires a considerably longer time – it is more variable in duration between streams (one month to many months) and is dominated by taxa with less disturbance resistance and much slower immigration/growth rates (*C* and *S* strategists such as cyanobacteria and filamentous green algae; Grime, 1979) derived from more distant refugia (Biggs and Smith, 2002). The frequency of floods dictates the time available for benthic algal accrual. During inter-flood periods, resource availability (light and nutrients), loss by grazing, spatial differences in water velocity and turbulence and the growing strategies of individual species become important in determining net community development (Biggs, 1996a). Under medium to low flood disturbance frequency and grazing, accrual processes tend to dominate, and where resource supply is moderate to high, it results in communities that are dominated by erect, stalked diatoms and/or filamentous green algae (e.g. *Cladophora glomerata* and *Melosira varians*). Where disturbance frequency and resource supply are moderate to low, the community tends to be dominated by filamentous cyanobacteria and red algae, with limited numbers of diatoms (e.g. *Nostoc* sp., *Tolypothrix* sp., *Schizothrix* sp., *Phormidium* sp., *Audouinella* sp. and *Epithemia* sp.). With moderate to high flood disturbance frequency, or heavy grazing, loss processes dominate and tend to result in communities dominated by low-growing diatoms that adhere tightly to the substratum or are capable of rapid colonization (particularly diatoms) (Biggs, 1996a). A number of studies have reported successional changes in the periphyton community from diatoms to filamentous green algae during periods of stable flows after floods (e.g. Peterson and Stevenson, 1992; Dent and Grimm, 1999) and particularly a reduction in species richness. A study on the Okuku River in New Zealand (Biggs and Stokseth, 1996) indicated an exponential increase in biomass and a shift from low-growing diatoms to stalked and filamentous diatoms followed by a transition to filamentous green algae over a three-month period following a major spring flood. In a study performed by Suren *et al.* (2003b) on the same river, these findings were not replicated. The reason hypothesized for the absence of successional changes in the periphyton community during the latter study was top-down invertebrate grazing control.

13.3.2 Biomass

Three main long-term temporal patterns in biomass can be distinguished among streams: (1) relatively constant, low biomass; (2) cycles of accrual and sloughing; and (3) seasonal cycles. These patterns are predominantly the result of interactions between the processes of biomass accrual and loss, with the hydraulic disturbance regime being the fundamental determinant of the overall balance between these processes (Biggs, 1996a).

The biomass of stream periphyton is controlled by spatial variations in flow velocity and forces (Biggs *et al.*, 1998a). A reduction in flow velocity, shear stress and erosive forces potentially leads to: (a) an increase in filamentous green algae biomass; (b) an increase or decrease in stalked diatoms and short filamentous algae biomass; and (c) a reduction in mucilaginous algae biomass (Biggs, 2000). Hydrological disturbance due to human activity can potentially change periphyton biomass. Flow regulation associated with reduced flow variability and increased bed stability can increase biomass, whereas increased flow variability typically decreases biomass, although this may depend on the pre-regulation conditions. Gravel abstraction from within the wetted channel can cause bed destabilization and therefore decrease biomass. In the case of intensification of land use, the impact on periphyton is typically associated with hydrological changes, especially if the catchment has a steep gradient. Ultimately, human activities that increase nutrient supply, light and base flow temperature may lead to an increase in periphyton biomass (Biggs, 2000).

Riverine impoundment alters the downstream flow/transport of bed sediments, resulting in bed armouring (i.e. the beds become paved with very stable, large cobbles and boulders comprising the surface layers). This provides an ideal substratum for periphyton to attain a high biomass. Most small- and medium-sized floods are prevented from flowing down the river (unless the reservoir is at full storage capacity), which means that normal flow variability is reduced and the natural ability of the system to remove excess accumulations of biomass is also impeded. Furthermore, the reduction in flow usually results in a reduction in flow velocities and hydraulic forces, which then allows a higher biomass of filamentous green algae to develop if nutrient levels are sufficient (Biggs, 2000). If flood disturbances are frequent and/or levels of enrichment are very low, then there will be low instream mean biomass. High levels of enrichment result in some algal accrual at high frequencies of disturbance because of more rapid inter-flood regeneration; some biomass can also accrue with low levels of enrichment, provided that the disturbance frequency is low (Biggs, 1996a).

Maximum biomass usually occurs at low flow velocities (Biggs et al., 1998a). Low flows enhance plant biomass through changes to hydraulic properties, light and temperature conditions, although these are dependent on stream type (Figure 13.1). During periods of hydrological quiescence, such as low flows, algae biomass accrual processes dominate, because loss processes associated with high velocities and shear stress, substrate movement and abrasion do not occur (Suren and Riis, 2010). Low periphyton biomass in regulated rivers can be maintained

with the long-term use of sequential floods (Mannes et al., 2008). Streams that experience frequent floods tend to have low average biomass, with communities dominated by low-growing, highly shear-resistant taxa (Biggs and Close, 1989; Biggs, 1996a). Conversely, streams with long intervals between disturbances, little resource stress and no heavy grazing tend to have communities with higher biomass and greater architectural complexity (Biggs, 1996a). Ward (1976) reported 3 to 20 times more epilithic algae in the regulated section of South Platte River in Colorado, below the dam, due to the stable substrate, increased nutrient availability and higher winter water temperatures. Research undertaken by Mannes et al. (2008) reported that floods in a Swiss pre-alpine river reduced periphyton standing stocks and although the periphyton recovered between floods, it was scoured by each new flood. Small floods (10–15 m^3/s) had a greater impact on periphyton biomass at the beginning of the experiment than in later years, which may be the result of periphyton adaptation on flood stress. High periphyton biomass, expressed as chlorophyll-a and organic matter, measured downstream from dams, has also been reported in other studies (Lowe, 1979; Bundi and Eichenberger, 1989; Smolar, 1997; Koudelkova, 1999; Smolar-Žvanut, 2001). This may be attributed primarily to minor fluctuations in the water temperature, flows lacking distinct seasonal variability, an increase in the concentration of nutrients and their absorption by algae (Koudelkova, 1999) and to large amounts of immobile sediment (Biggs et al., 2001).

The plant community that develops in the absence of floods largely determines the indirect effects of low flows on instream habitats (Figure 13.1) (Suren and Riis, 2010). Suren and Riis (2010) suggested that a three-dimensional habitat template can be developed to predict the plant communities in non-shaded streams (Figure 13.2A), although there are many interlinked factors responsible for controlling aquatic vegetation (Biggs, 1996b; 2000). The three-dimensional habitat template represents resource supply, ranging from oligotrophic systems (where nutrients and light may limit plant growth) to eutrophic streams (where nutrients and light do not limit plant growth) on the x-axis and interflood flow velocity, which influences both the dominant substrate size in the stream and the type of plant community, on the y-axis (Biggs, 1996b; Suren and Riis, 2010). Streams with low interflood flow velocities will generally be composed of smaller substrates (sand, pebble, gravel) and, consequently, plant communities adjusted to lower flow velocities. In contrast, streams with higher interflood flow

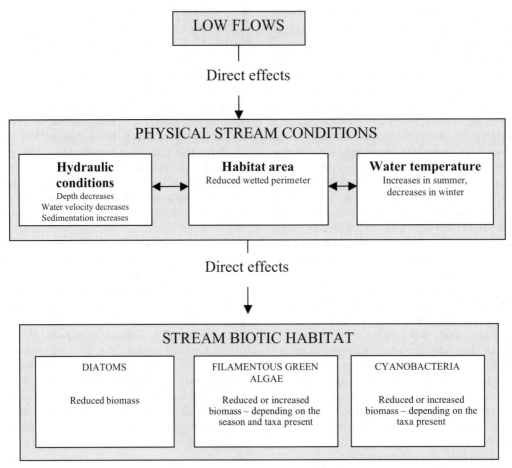

Figure 13.1 Conceptual model indicating the direct effects of low flow on periphyton biomass in streams. Changes to the physical conditions in a stream will have direct effects on the algae community and, as a result of changes to the pre-existing plant community, also direct effects on instream habitats. Modified with permission from Journal of the North American Benthological Society (2010), 29(2), 711–724. Copyright (2010) North American Benthological Society.

velocities will be dominated by bedrock or boulders and will develop plant communities that can tolerate high shear stress. The z-axis of the habitat template represents the overall substrate stability during floods (from fine substrates to large, stable bedrock and boulders) (Suren and Riis, 2010).

According to Suren and Riis (2010), diatoms dominate stream habitats with medium-to-high substrate stability during floods, low-to-high nutrient supply and low-to-high interflood flow velocities. In streams with moderately stable substrates, diatom biomass is controlled by a combination of resource supply and interflood flow velocity, with higher biomass being recorded in more resource-rich streams with reduced flow velocities (Figure 13.2B). Biggs and Hickey (1994) reported that the

composition of diatom communities is a reflection of both interflood flow velocity and nutrients, with small, prostrate species such as *Cocconeis* being found in fast-flowing water, and larger, stalked species, such as *Gomphonema*, being found in faster-flowing, nutrient-rich waters.

Filamentous green algae or cyanobacterial growths can become dominant in streams with gravel substrates, low interflood flow velocity and a moderate-to-high resource supply during periods of low flow (Biggs, 1990). During long periods of stable interflood flow, a gradual succession occurs from diatoms to filamentous green algae or thick cyanobacterial mats that can form extremely high biomass and cover the entire stream bed (Stevenson *et al.*, 1996; Suren *et al.*, 2003a; Suren and Riis, 2010). A summary

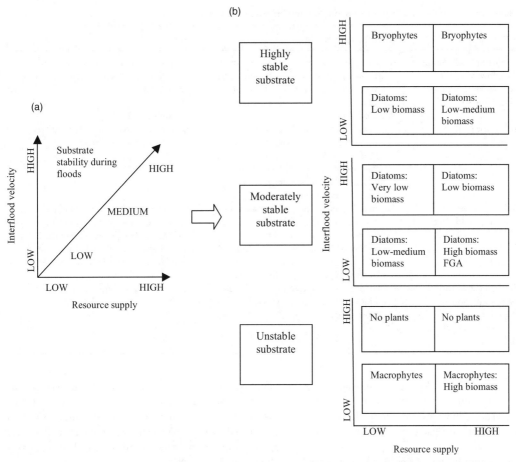

Figure 13.2 Aquatic plant community types in relation to stream physical condition according to Suren and Riis (2010) (modified with permission from Journal of the North American Benthological Society (2010), 29(2), 711–724. Copyright (2010) North American Benthological Society) in non-shaded streams. A: The three components of a stream's physical condition (resource supply, interflood velocity and substrate stability) interact to produce 12 potential stream types. B: Each cell represents one of the 12 potential stream types and shows the expected plant community (or characteristic plant community) before the onset of a low-flow period. FGA = filamentous green algae.

of the impact of altered flow regime on periphyton communities is given in Table 13.1.

13.3.3 Periphyton proliferations

Periphyton proliferations in rivers are a secondary effect of low flows (low shear stress) and can potentially alter habitat quality, resulting in degradation of ecosystem structure (Suren *et al.*, 2003a); a widely recognized problem for water resource managers (Biggs, 1988). Periphyton proliferations potentially affect abstraction for water supply, aesthetic appeal of the water and instream recreation activities (Biggs, 1996a), industrial use, instream biodiver-

sity, stock and domestic animal health (Biggs, 2000) and damage the structure and the function of the entire river ecosystem. Most severe cases occur under high nutrient loadings, which usually originate from intensive agricultural development or nutrient-rich wastewaters (Biggs, 1996a). Periphyton mats dominated by filamentous green algae are far more conspicuous than diatom-dominated mats with a similar biomass (Biggs, 2000). Filamentous green algae blooms may result in potentially deleterious changes to invertebrate and fish habitat (Suren *et al.*, 2003a). In rivers with a low level of enrichment and in heavily shaded forested streams where algae growth is

Table 13.1 Summary of periphyton community responses to modified flow regimes.

Flow variables affected	Periphyton responses	Sources
Increased stability of baseflow and reduction of flow variability below reservoirs or in reaches with regulated flows	Increased biomass; enhanced algal growth attributed to the absence of floods/high flows, and to an increase in nutrient availability and water transparency	Ward, 1976; Marcus, 1980; Ward and Stanford, 1983; Skulberg, 1984; Dufford et al., 1987
	Reduced biomass	Dewson et al., 2007
Low instream flow velocities	Increased biomass; enhanced growths of long filamentous green algae (*Cladophora*, *Mougeotia*, *Spirogyra*) and decreased growths of stalked, short filamentous or thick mucilaginous diatoms and cyanobacteria	Biggs et al., 1998a; Suren et al., 2003a; James and Suren, 2009
Floods/high flows below hydroelectric impoundment/ dam	Temporary biomass reduction; enhanced temporal variation in biomass; increased biomass variability; successional changes in periphyton community from diatoms to filamentous green algae after floods	Fisher et al., 1982; Peterson and Stevenson, 1992; Dent and Grimm, 1999; Biggs and Stokseth, 1996; Biggs and Smith, 2002; Uehlinger et al., 2003; Robinson and Uehlinger, 2008; Mannes et al., 2008
Flow variation associated with hydropower production (hydropeaking) causes periodic aerial exposure of zones along the river bank	Loss of algal cover; reduced algal biomass and lower productivity in aerial exposed zones	Blinn et al., 1998; Bergey et al., 2010

constrained by insufficient light (e.g. Keithan and Lowe, 1985), periphyton communities may not change as dramatically as a result of low flows, and reductions in flow will not lead to enhanced algae growth (Suren and Riis, 2010).

Since 1996, the monitoring of periphyton in the Soča River in Slovenia has been performed upstream and downstream of impoundments. Impoundment of the Soča River reduced sediment transport below the dams and, consequently, coarse-grained gravel dominates the substratum in these areas compared to river reaches upstream of the dams (Smolar-Žvanut, 2001). Together with nutrient enrichment and increased direct light on the channel as a result of reduced shading, limited sediment transport below the dams is the main reason for periphyton proliferation. These occur in summer in river sections with constant low flow and with low-frequency floods. Filamentous green algae (e.g. *Ulotrix, Cladophora, Mougeotia, Spirogyra*) usually proliferate in temperate region rivers, including Slovenia. Horner et al. (1983) suggested that values of chlorophyll-*a* higher than 100–150 mg m^{-2} or a cover of more than 20% by filamentous algae were unacceptable, while Biggs and Price (1987) reported that when filamentous algae cover was more than 40% of the river bed, it became very conspicuous from the river

bank. In the Soča River (Figure 13.3), the average values of chlorophyll-*a* were higher than 100 mg m^{-2} below the dam (Smolar-Žvanut, 2001), and the community was dominated by *Cladophora* sp., *Mougeotia* sp. and *Ulotrix* sp., resulting in extensive smothering of bed sediments (Biggs and Price, 1987).

13.4 Case studies from Slovenia

Although Slovenia is a small country, there are more than 600 small hydropower plants and 17 large hydropower plants, with many more planned. Most of these plants utilize all available water during the summer and winter low flow periods. The influence of reduced flows and modified flow regime (up to 98% reduction in flow) on periphyton communities in river reaches downstream of dams in the Tržiška Bistrica River, the Mojstranska Bistrica River and the Soča River have been documented previously (Smolar et al., 1998; Smolar-Žvanut, 2001; Smolar-Žvanut et al., 2005). Periphyton sampling was performed during winter and summer minimum flows. Reaches downstream from dams differed considerably in community composition, relative abundance and periphyton biomass from the sections upstream of dams and reaches where diverted water

Figure 13.3 Proliferation of filamentous green algae downstream of the dam on the Soča River, Slovenia.

was returned to the river. The abstraction of water from the Mojstranska Bistrica River (Smolar-Žvanut *et al.*, 2005), and the resulting low flow velocity below the dam, resulted in the absence of *Hydrurus foetidus* (Chrysophyceae) at these sites in winter period. During periods of constant low flow (more than two months) downstream of dams on the Soča River, periphyton usually attained a higher biomass on large pebbles and stones, primarily due to the limited mobility of bed sediments in the lower reaches of the watercourse (Smolar-Žvanut, 2001).

13.4.1 The Sava Dolinka River

The Moste Hydropower Plant (HPP Moste) was the first large power plant built on the Sava Dolinka River in Slovenia. The impoundment services a power plant, generating electricity during peak energy demand and, together with a dam on the left tributary (the Završnica stream), constitutes a uniform hydropower system. Initially, HPP Moste had a design flow of 34.5 m^3s^{-1} and operated during peak energy demand conditions. Due to the impoundment, a 2470 m-long section of the Sava Dolinka River below the dam is seriously affected by water abstraction for HPP Moste, and for most of the year, river flow is continually low. The river section at the outflow of the Moste power plant is affected by hydropeaking, which

causes rapid increases and reductions in flow downstream of the Sava Dolinka River. The operation of HPP Moste depends on the quantity of the inflow into the reservoir. It is operational depending on electricity requirements, usually during the daytime. If inflow into the reservoir is 14 m^3s^{-1}, the plant operates from 6 a.m. to 10 p.m., with abstractions of 20 m^3s^{-1}, except from 1 p.m. to 6 p.m. when abstractions range from 4 to 10 m^3s^{-1}. When it starts operating in the morning, the abstraction increases, on average, from 0 to 10 m^3s^{-1} in two minutes, or from 0 to 20 m^3s^{-1} in five minutes. HPP Moste is not in operation, as a rule, on Saturdays and Sundays, when water recharges the reservoir (Smolar-Žvanut, 2001).

Sampling of periphyton, determination of periphyton biomass and measurement of hydrological parameters were undertaken during periods of low flow in 1998 and 1999 within five river reaches (Figure 13.4): upstream (SA1, SA2) and downstream (SA3, SA4) from the dam, and at river reach subject to hydropeaking (SA5). Sampling at SA5 was always undertaken when HPP Moste was not in operation (Smolar-Žvanut, 2001).

Results of the analyses of the hydrological data indicate substantial changes to the flow regime of the Sava Dolinka River downstream of the Moste Dam. The changes include reduced mean discharge (stable low flow), a decrease in flow velocity and the stagnation and deposition of

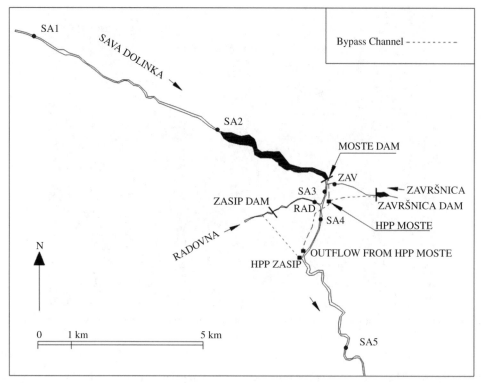

Figure 13.4 Map of hydropower system and sampling sites in the Sava Dolinka River (SA1–SA5).

sediments upstream of the dams, resulting in bed armouring downstream. At the time of periphyton sampling, river flow upstream of the dam was between 5.9 and 8.2 m^{3}s^{-1} (SA1, SA2), but was between 0.18 and 0.42 m^{3}s^{-1} below the impoundment (SA3, SA4).

Analysis of the impacts of the modified flow regime below the Moste Dam indicates substantial changes to the periphyton community. Low values of dry weight (average 56 g m^{-2}) and organic matter (average 19 g m^{-2}), but relatively high values of chlorophyll-a (average 107 mg m^{-2}) were recorded upstream of the Moste Dam at the reference site SA1. This may have resulted from reduced organic matter availability on the river bed due to high flow velocities and hydraulic forces. In addition, there was a higher proportion of living cells in the community recorded, since dead cells were swept away. The proliferation of algae in the Sava Dolinka River below the Moste Dam at SA3 and SA4 may be attributed to the long period of low flow and also the level of nutrients, light and sediment structure. The average values of dry–wet periphyton biomass (SA3 = 219 g m^{-2}, SA4 = 208 g m^{-2}), organic matter (SA3 = 56 g m^{-2}, SA4 = 48 g m^{-2}) were much

higher in comparison to the river upstream of the dam (dry–wet periphyton biomass at SA1 = 56 g m^{-2} and at SA2 = 139 g m^{-2}; organic matter at SA1 = 19 g m^{-2} and at SA2 = 34 g m^{-2}). But in the river reach farthest downstream (SA5) subject to hydropeaking, the average values of dry weight (78 g m^{-2}) and organic matter (32 g m^{-2}) were similar to the reference river section, while average values of chlorophyll-a were much higher (207 mg m^{-2}) in comparison to other river sections. Due to hydropeaking (flows downstream of hydropower plants may fluctuate daily or sub-daily due to hydropower operation), all organic particles and dead cells were eroded. However, it must also be acknowledged that the river section subject to hydropeaking was totally shaded. This may also result in higher quantities of chlorophyll-a within plant cells (Smolar-Žvanut, 2001).

During the period from 1998 to 1999, a total of 128 periphyton taxa were identified in the Sava Dolinka River and its tributaries (the Završnica and the Radovna). The highest number of species belonged to *Bacillariophyta*, similar to other Slovenian rivers (Kosi, 1988; Smolar, 1997; Smolar-Žvanut 2001; Peroci *et al.*, 2009; Krivograd

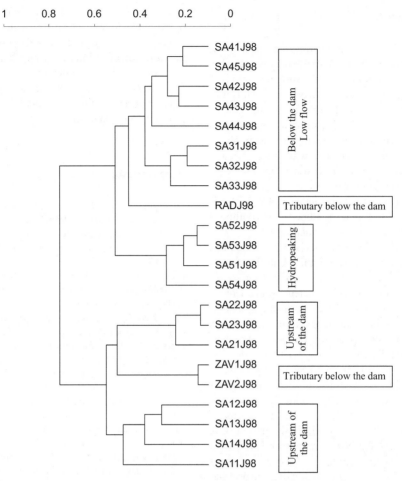

Figure 13.5 Bray–Curtis coefficient of similarity for the Sava Dolinka River. Modified from Smolar-Žvanut, 2001, with permission from Acta hydrotechnica.

Klemenčič and Toman, 2010). The stability of the river system downstream of the Sava Dolinka River dams (prolonged periods of constant flow, rare occurrence of high flows, low to moderate current velocities and adequate light conditions) facilitated the colonization of green algae, particularly *Bacillariophyta*. Environmental stability is the primary reason that more periphyton taxa were recorded below the dams compared to river reaches upstream. The effect of impounding and modifying the flow regime on the occurrence of *Hydrurus foetidus* was observed directly in the Sava Dolinka River. The species proliferated only during the winter months at sample site SA1. This may be ascribed to the low water temperatures recorded during the winter and to the constant flows (Ward, 1974; Valentin *et al.*, 1995; Smolar, 1997).

Traaen and Lindstrom (1983) reported that 90% of the algae *Hydrurus foetidus* were recorded occurred at velocities over 80 cm s^{-1} (measured 1 cm above the bed), and this is comparable to the results recorded in this study.

The Bray–Curtis coefficient of similarity indicates qualitative changes to the structure of the periphyton community for the Sava Dolinka River (Figure 13.5), providing a fast and effective analysis of similarities and differences among sites and their communities. The greatest similarities were recorded in river reaches influenced by water diversion for HPP Moste with a group of sites subject to hydropeaking sharing some similarities at the first level of division. A second group, separated at the first level of division, comprised all sample sites upstream of the Moste Dam.

13.5 Conclusions

The importance of periphyton in river management was considered in detail by Biggs (2000). He defined 'periphyton' as the primary transducer of the sun's rays into biologically based energy for stream ecosystems. Thus, this community is the 'grass' of streams for aquatic grazing animals (Biggs, 2000, p. 25). The structure and function of the periphyton community is highly dependent on the flow regime, but its inter-relatedness with other biotic and abiotic parameters in the river ecosystem must be considered. In some instances, invasive exotic (non-native) species (e.g. *Didymosphenia geminata* in New Zealand) can be more successful under modified flow conditions than indigenous taxa (e.g. Miller *et al.*, 2009).

Among the negative impacts associated with modified flow regimes on periphyton communities, algal proliferation is the primary effect recognized by water resource managers. Periphyton mats dominated by filamentous green algae are far more conspicuous than diatom-dominated mats with a comparable biomass. Therefore, the proposed limits on periphyton biomass were determined in terms of diatom/cyanobacterial mats and filamentous green algal mats in New Zealand rivers (Biggs, 2000). These limits should also be determined for other river types around the world to help guide river managers.

In the EU, the role of periphyton in the monitoring of river ecosystems has been more widely recognized since the adoption of the EU Water Framework Directive in 2000. However, among all groups of algae, only diatoms are used as indicators for the evaluation of ecological status in many EU countries; although they are also used as an indicator of potential impacts of impoundment and modified flow regimes. As a result, the EU Water Framework Directive does not take into account other groups of algae as equal elements of the river ecosystem. For example, the filamentous green algae *Cladophora* may be indicative of rivers with stable substrata and high light intensities downstream of the dams (Smolar, 1997) and cyanobacteria such as *Oscillatoria limosa* may be a good indicator of organically polluted rivers. We propose that all groups of algae should be considered in the evaluation of ecological status in rivers and that without it, significant gaps in knowledge regarding river status remain.

In the future, more research centred on the impact of altered flow regimes on periphyton should be directed towards the impacts of hydropeaking and the role of periphyton in the evaluation of environmental flows in rivers. Restoration of river habitats for periphyton with modified flow regime should include sustainable management of riverine sediments, the maintenance of river flows that mimic the natural magnitude and frequency of events and the reduction of pollution. Periphyton is the principal primary producer in fast-flowing rivers and given that they form the basis of most aquatic food chains, much more research is required to understand their role in instream processes and dynamics.

References

Apesteguia, C. and Marta, J. (1979) Producción de biofilm en ambientes acuáticos del río Paraná Medio. II: Medición de la velocidad de produccio;n media e instantánea. *Revista de la Asociacion de Ciencias Naturales del Litoral*, **10**: 39–48.

Bergey, E.A., Bunlue, P., Silalom, S., Thapanya, D. and Chantaramongkol, P. (2010) Environmental and biological factors affect desiccation tolerance of algae from two rivers (Thailand and New Zealand) with fluctuating flow. *Journal of the North American Benthological Society*, **29**(2): 725–736.

Biggs, B.J.F. (1988) Algal proliferations in New Zealand's shallow stony foothills-fed rivers: Towards a predictive model. *Verhandlungen des Internationalen Verein Limnologie*, **23**: 1405–1411.

Biggs, B.J.F. (1990) Periphyton communities and their environments in New Zealand rivers. *New Zealand Journal of Marine and Freshwater Research*, **24**: 367–386.

Biggs, B.J.F. (1996a) Patterns in benthic algae of streams. In Stevenson, R.J., Bothwell, M.L., and Lowe, R.L. (eds), *Algal Ecology: Freshwater Benthic Ecosystems*, Aquatic ecology series, Academic Press, San Diego, pp. 31–56.

Biggs, B.J.F. (1996b) Hydraulic habitat of plants in streams. *Regulated Rivers: Research and Management*, **12**: 131–144.

Biggs, B.J.F. (2000) *New Zealand Periphyton Guideline: Detecting, monitoring and managing enrichment of streams*, Ministry for the Environment, Wellington, New Zealand. Available at: http://www.mfe.govt.nz/publications/water/nz-periphyton-guide-jun00.html [Accessed 5 May 2011].

Biggs, B.J.F. and Close, M.E. (1989) Periphyton biomass dynamics in gravel bed rivers: the relative effects of flows and nutrients. *Freshwater Biology*, **22**: 209–231.

Biggs, B.J.F. and Price, G.M. (1987) A survey of filamentous algal proliferations in New Zealand rivers. *New Zealand Journal of Marine and Freshwater Research*, **21**: 175–192.

Biggs, B.J.F. and Hickey, C.W. (1994) Periphyton responses to a hydraulic gradient in a regulated river in New Zealand. *Freshwater Biology*, **32**: 49–59.

Biggs, B.J.F. and Smith, R. (2002) Taxonomic richness of stream benthic algae: Effects of flood disturbance and nutrients. *Limnology and Oceanography*, **47**(4): 1175–1186.

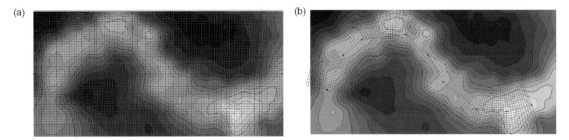

Plate 1 (Figure 3.4a, right and 3.4b, right) Structured (a) Cartesian and (b) curvilinear meshes for a natural reach. The averaged cell area is 1 m² and the averaged topographic survey is resolution 0.25pt/m². The right column shows the topography and the mesh and the left column the numerical domain for the three different mesh types.

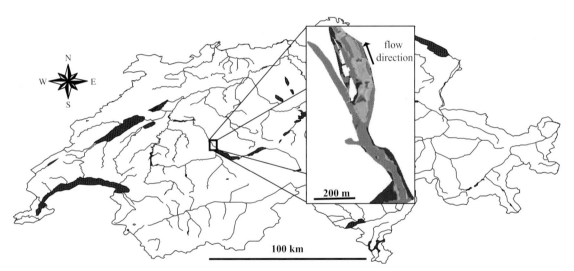

Plate 2 (Figure 5.3) Location of the study site at the Aare River, Thun, Switzerland.

Ecohydraulics: An Integrated Approach, First Edition. Edited by Ian Maddock, Atle Harby, Paul Kemp and Paul Wood.
© 2013 John Wiley & Sons, Ltd. Published 2013 by John Wiley & Sons, Ltd.

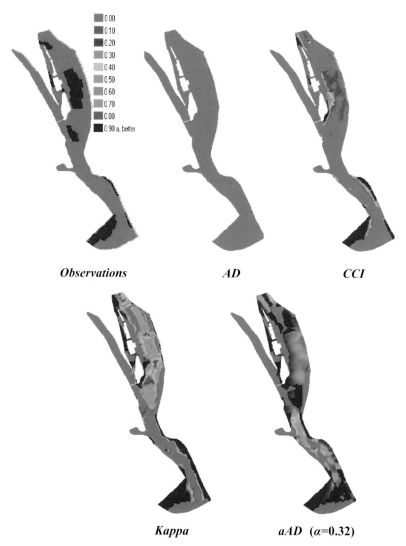

Observations   AD   CCI

Kappa   aAD (α=0.32)

Plate 3 (Figure 5.6) Visualisation of the model predictions in the study area, showing the habitat suitability for spawning grayling calculated by the fuzzy rule base optimised based on the correctly classified instances (*CCI*), the average deviation (*AD*), *Kappa* and the adjusted average deviation (*aAD*) with $\alpha = 0.32$.

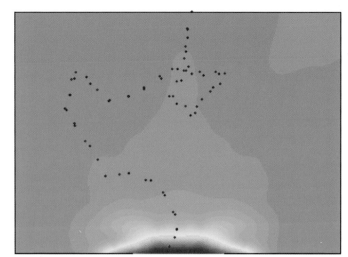

Plate 4 (Figure 8.4) The path of a Pacific salmon smolt (*Oncorhynchus tshawytscha*) video tracked during passage through an orifice weir in a large, open-channel flume (red and blue dots represent head and tail positions, respectively); fish movements have been overlain onto the accelerating velocity profile (0–1.53 m s^{-1}) created by the weir.

Plate 5 (Figure 19.4) Restoration of the River Drau in Austria has introduced refuges and delayed the rate of change in water level during hydropeaking (right) compared to the channelised situation to the left. Credit: Regional Government of Carinthia/S. Tichy.

Plate 6 (Figure 19.7) (Left) Retention basin downstream of the Alberschwende power plant in the Bregenzerach River in Austria (from S. Schmutz, personal communication); and (right) sketch of planned retention basin in the Hasliaare River where micro-turbines may also use the head difference between the basin and the river to produce power (Schweizer *et al.*, 2009).

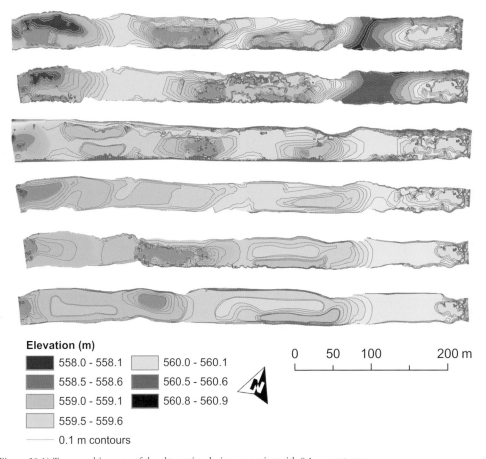

Elevation (m)

■ 558.0 - 558.1	■ 560.0 - 560.1	
■ 558.5 - 558.6	■ 560.5 - 560.6	
■ 559.0 - 559.1	■ 560.8 - 560.9	
■ 559.5 - 559.6		

— 0.1 m contours

0   50   100   200 m

Plate 7 (Figure 20.1) Topographic maps of the alternative design scenarios with 0.1-m contours.

Plate 8 (Figure 22.2) Area E showing inlets and areas compartmentalised by roads and culverts.

Plate 9 (Figure 22.9) Distribution of substrates used in the model. Resistance coefficients for each substrate are shown in Table 22.5.

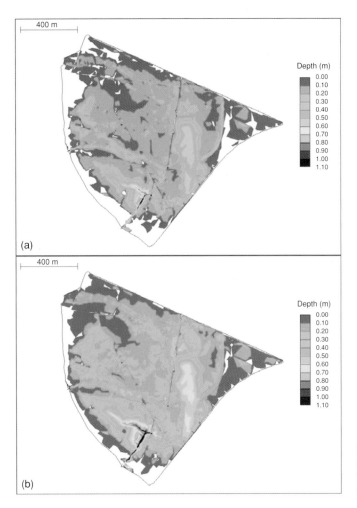

(a)

(b)

Depth (m)
0.00
0.10
0.20
0.30
0.40
0.50
0.60
0.70
0.80
0.90
1.00
1.10

Plate 10 (Figure 22.13) Modelled inundation depths for bed of Fish Fry Creek at 0.20 mAHD and Wader Creek closed (run f) for (a) spring tide low water and (b) spring tide high water.

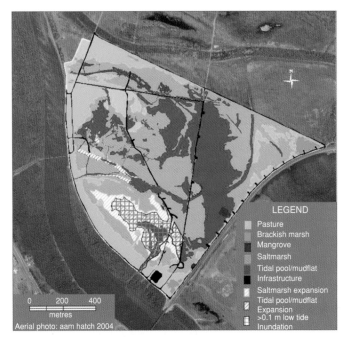

LEGEND

Pasture
Brackish marsh
Mangrove
Saltmarsh
Tidal pool/mudflat
Infrastructure
Saltmarsh expansion
Tidal pool/mudflat Expansion
>0.1 m low tide Inundation

0   200   400
metres
Aerial photo: aam hatch 2004

Plate 11 (Figure 22.14) Predicted future expansion of saltmarsh and tidal pool habitats as a result of raising the inlet of Fish Fry Creek to 0.20 mAHD and closing the Wader Creek inlet.

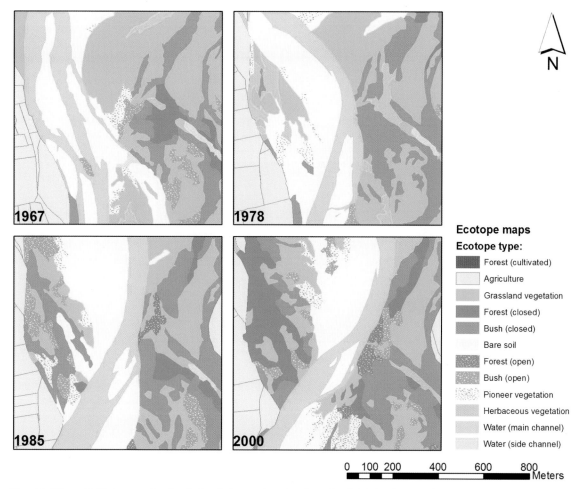

Ecotope maps

Ecotope type:

- Forest (cultivated)
- Agriculture
- Grassland vegetation
- Forest (closed)
- Bush (closed)
- Bare soil
- Forest (open)
- Bush (open)
- Pioneer vegetation
- Herbaceous vegetation
- Water (main channel)
- Water (side channel)

0   100 200   400   600   800
Meters

Plate 12 (Figure 23.2) An example of ecohydraulic interactions that shape floodplain landscapes. The 1967–1978 channel migration rejuvenated older ecotopes on the right river bank and created niches for forest regeneration/succession in the abandoned channels. By 1985, these channels had been colonised by vegetation that developed into forest during the 1985–2000 period. Reproduced from Geerling (2006), with kind permission from Springer Science + Business Media.

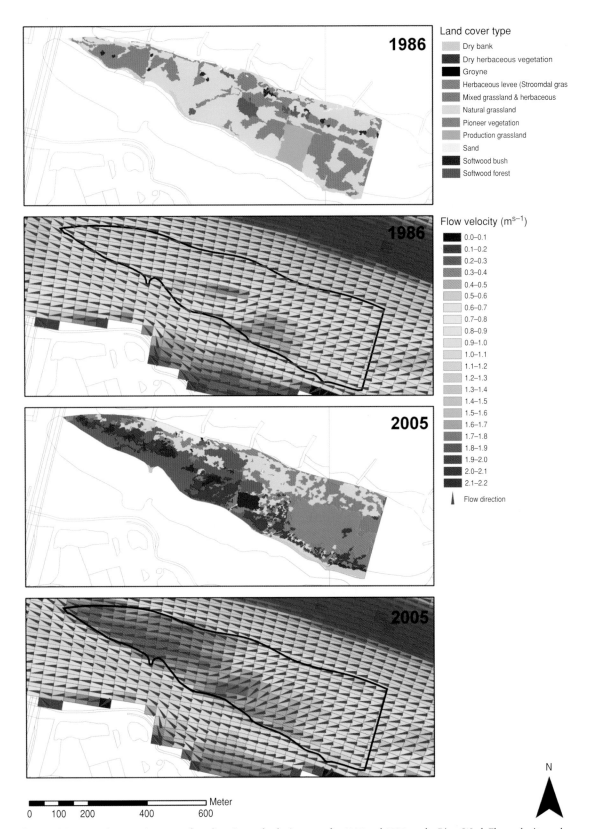

Plate 13 (Figure 23.6) Vegetation cover, flow direction and velocity maps for 1986 and 2005 on the River Waal. Flow velocity and direction vectors were computed for levels recorded for the 1995 flood event. Flow velocities were generally reduced in the study area between surveys. Reprinted from Geerling *et al.*, Copyright 2008, with permission from Elsevier.

Biggs, B.J.F. and Stokseth, S. (1996) Hydraulic habitat preferences for periphyton in rivers. *Regulated Rivers: Research and Management*, **12**: 251–261.

Biggs, B.J.F., Goring, D.G. and Nikora, V.I. (1998a) Subsidy and stress responses of stream periphyton to gradients in water velocity as a function of community growth form. *Journal of Phycology*, **34**: 598–607.

Biggs, B.J.F., Stevenson, R.J. and Lowe, R.L. (1998b) A habitat matrix conceptual model for stream periphyton. *Archiv für Hydrobiologie*, **143**: 21–56.

Biggs, B.J.F, Duncan, M.J., Suren, A.M. and Holomuzki, J.R. (2001) The importance of bed sediment stability to benthic ecosystems of streams. In Mosley, M.P. (ed.) *Gravel-Bed Rivers 5*. New Zealand Hydrological Society, Wellington, New Zealand, pp. 423–450.

Blinn, D.W., Shannon, J.P., Benenati, P.L. and Wilson, K.P. (1998) Algal ecology in tailwater stream communities: the Colorado River below Glen Canyon Dam. *Arizona Journal of Phycology*, **34**: 734–740.

Bundi, U. and Eichenberger, E. (1989) *Wasserentnahme aus Fliessgewässern: Gewässerökologische Anforderungen an die Restwasserführung (Water Abstraction from Running Waters: Water Ecology Requirements for Minimum Flows)*. Schriftenreihe Umweltschutz 10, Bundesamt für Umwelt, Wald und Landschaft Bern.

Bunn, S.E. and Arthington, A.H. (2002) Basic principles and ecological consequences of altered flow regimes for aquatic biodiversity. *Environmental Management*, **30**(4): 492–507.

Dent, C.L. and Grimm, N.B. (1999) Spatial heterogeneity of river water nutrient concentrations over successional time. *Ecology*, **80**: 2283–2298.

Dewson, Z.S., JamesA.B.W. and Death, R.G. (2007) Invertebrate community responses to experimentally reduced discharge in small streams of different water quality. *Journal of the North American Benthological Society*, **26**(4): 754–766.

Directive of the European Parliament and of the Council 2000/60/EC of 23 October 2000 on establishing a framework for community action in the field of water policy. *Official Journal of the European Communities*, L 327/1.

Downes, B.J, Entwisle, T.J. and Reich, P. (2003) Effects of flow regulation on disturbance frequencies and in-channel bryophytes and macroalgae in some upland streams. *River Research and Applications*, **19**: 27–42.

Dufford, R.G., Zimmermann, H.J., Cline, L.D. and Ward, J.V. (1987) Responses of epilithic algae to regulation of Rocky Mountain streams. In Craig, J.F. and Kemper, J.B. (eds) *Regulated Streams: Advances in Ecology*, Plenum Press, New York, pp. 383–390.

Fisher, S.G., Gray, L.J. Grimm, N.B. and Busch, D.E. (1982) Temporal succession in a desert river ecosystem following flash flooding. *Ecological Monographs*, **52**: 93–110.

Godillot, R., Caussade, B., Amezaine, T. and Capblancq, J. (2001) Interplay between turbulence and periphyton in rough open-channel flow. *Journal of Hydraulic Research*, **39**: 227–239.

Graf, W.L. (2006) Downstream hydrologic and geomorphic effects of large dams on American rivers. *Geomorphology*, **79**: 336–360.

Grime, J.P. (1979) *Plant Strategies and Vegetation Processes*, John Wiley & Sons, Ltd, Chichester, UK.

Growns, I.O. and Growns, J.E. (2001) Ecological effects of flow regulation on macroinverterbrate and periphytic diatom assemblages in the Hawkesbury-Nepean River, Australia. *Regulated Rivers: Research and Management*, **17**(3): 275–293.

Horner, R.R., Welch, E.B. and Veenstra, R.B. (1983) Development of nuisance periphytic algae in laboratory streams in relation to enrichment and velocity. In Wetzel, R.G. (ed.) *Periphyton of Freshwater Ecosystems*, Dr W. Junk Publishers, The Hague, pp. 121–134.

Hynes, H.B.N. (1979) *The Ecology of Running Waters*, 4th edition, Liverpool University Press, Liverpool, UK.

James, A.W. and Suren, A.M. (2009) The response of invertebrates to a gradient of flow reduction – an instream channel study in a New Zealand lowland river. *Freshwater Biology*, **54**: 2225–2242.

Keithan, E.D. and Lowe, R.L. (1985) Primary productivity and spatial structure of phytolithic growth in streams in the great Smoky Mountains National Park, Tennessee. *Hydrobiologia*, **123**: 59–67.

Kosi, G. (1988) *Vpliv hipolimnijske vode Blejskega jezera iz natege in kanalizacije na primarno produkcijo perifitona v Savi Bohinjki (The influence of hypolimnian water from Lake Bled on primary productivity of periphyton in the river Sava Bohinjka)*. Masters thesis, Univerza v Ljubljani, Biotehniška fakulteta v Ljubljani.

Koudelkova, B. (1999) *Effects of the Hydropower Peaking on Distribution of Periphyton in the Cross section of a Regulated River*. Dissertation abstract, Department of Zoology and Ecology, Faculty of Sciences, Masaryk University in Brno.

Krivograd Klemenčič, A. and Toman, M.J. (2010) Influence of environmental variables on benthic algal associations from selected extreme environments in Slovenia in relation to the species identification. *Periodicum Biologorum*, **112**(2): 179–191.

Lamouroux, N., Dolédec, S. and Gayraud, S. (2004) Biological traits of stream macroinvertebrate communities: effects of microhabitat, reach, and basin filters. *Journal of the North American Benthological Society*, **23**: 449–466.

Lau, Y.L. and Liu, D. (1993) Effect of flow rate on biofilm accumulation in open channels. *Water Research*, **27**: 335–360.

Ligon, F.K., Dietrich, W.E. and Trush, W.J. (1995) Downstream ecological effects of dams. *BioScience*, **45**(3): 183–192.

Lowe, R.L. (1979) Phytobenthic ecology and regulated streams. In Ward, J.V. and Stanford, J.A. (eds) *The Ecology of Regulated Streams*, Plenum Press, New York, pp. 25–34.

Magilligan, F.J. and Nislow, K.H. (2005) Changes in hydrologic regime by dams. *Geomorphology*, **71**: 61–78.

Malard, F., Uehlinger, U., Zah, R. and Tockner, K. (2006) Flood-pulse and riverscape dynamics in a braided glacial river. *Ecology*, **87**: 704–716.

Mannes, S., Robinson, C.T., Uehlinger, U., Scheurer, T., Ortlepp, J., Mürle, U. and Molinari, P. (2008) Ecological effects of a long-term flood program in a flow-regulated river. *Journal of Alpine Research*, **1**: 125–134.

Marcus, M.D. (1980) Periphytic community response to chronic nutrient enrichment by a reservoir discharge. *Ecology*, **61**: 387–399.

McCormick, P.V. and Stevenson, R.J. (1991) Mechanisms of benthic algal succession in lotic environments. *Ecology*, **72**: 1835–1848.

McIntire, C.D. (1973) Periphyton dynamics in laboratory streams: a simulation model and its implications. *Ecological Monographs*, **43**: 399–420.

Miller, M.P., McKnight, D.M., Cullis, J.D., Greene, A., Vietti, K. and Liptzin, D. (2009) Factors controlling streambed coverage of *Didymosphenia geminata* in two regulated streams in the Colorado Front Range. *Hydrobiologia*, **630**: 207–218.

Morin, A. and Cattaneo, A. (1992) Factors affecting sampling variability of freshwater periphyton and the power of periphyton studies. *Canadian Journal of Fisheries and Aquatic Sciences*, **49**: 1695–1703.

Nikora, V.I., Goring, D.G. and Biggs, B.J.F. (2002) Some observations of the effects of micro-organisms growing on the bed of an open channel on the turbulence properties. *Journal of Fluid Mechanics*, **450**: 317–341.

Nilsson, C., Reidy, C.A., Dynesius, M. and Revenga, C. (2005) Fragmentation and flow regulation of the world's large river systems. *Science*, **308**: 405–408.

O'Reilly, C. and Silberblatt, R. (2009) *Reservoir Management in Mediterranean Climates through the European Water Framework Directive*, Water Resources Center Archives, Hydrology, University of California.

Peroci, P., Smolar-Žvanut, N. and Krivograd Klemenčič, A. (2009) Ocena vpliva odvzema vode iz vodotoka Oplotnica na hidromorfološke in fizikalno-kemijske dejavnike ter na združbo perifitona. *Natura Sloveniae*, **11**(1): 5–23.

Peterson, C.G. and Stevenson, R.J. (1992) Resistance and resilience of lotic algal communities: importance of disturbance timing and current. *Ecology*, **73**(4): 1445–1461.

Petts, G.E. and Maddock, I. (1995) Flow allocation for in-river needs. In Calow, P. and Petts, G.E. (eds) *The Rivers Handbook: Hydrological and Ecological Principles*, Volume **2**, Blackwell Scientific Publications, Oxford, pp. 289–307.

Poff, N.L. and Hart, D.D. (2002) How dams vary and why it matters for the emerging science of dam removal. *BioScience*, **52**: 659–668.

Poff, N.L., Allan, J.D., Bain, M.B., Karr, J.R., Prestegaard, K.L., Richter, B.D., Sparks, R.E. and Stromberg, J.C. (1997) The natural flow regime. *BioScience*, **47**(11): 769–784.

Postel, S. and Richter, B. (2003) *Rivers for Life: Managing Water for People and Nature*, Island Press, Washington, Covelo, London.

Power, M.E. (1990) Benthic turfs vs floating mats of algae in river food webs. *Oikos*, **58**: 67–79.

Power, M.E., Dietrich, W.E. and Finlay, J.C. (1996) Dams and downstream aquatic biodiversity: potential food web consequences of hydrologic and geomorphic change. *Environmental Management*, **20**(6): 887–895.

Reiter, M.A. (1986) Interactions between the hydrodynamics of flowing water and the development of a benthic algal community. *The Journal of Freshwater Ecology*, **3**: 511–517.

Richter, B.D., Baumgartner, J.V., Wigington, R. and Braun, D.P. (1997) How much water does a river need? *Freshwater Biology*, **37**: 231–249.

Robinson, C.T. and Minshall, G.W. (1998) Macroinvertebrate communities, secondary production, and life history patterns in two adjacent streams in Idaho, USA. *Archiv für Hydrobiologie*, **142**: 257–281.

Robinson, C.T. and Uehlinger, U. (2008) Experimental floods cause ecosystem regime shift in a regulated river. *Ecological Applications*, **18**(2): 511–526.

Saravia, L.A., Momo, F. and Boffi Lissin, L.D. (1998) Modelling periphyton dynamics in running water. *Ecological Modelling*, **114**: 35–47.

Shannon, J.P., Blinn, D.W. and Stevens, L.E. (1994) Trophic interactions and benthic animal community structure in the Colorado River, Arizona, U.S.A. *Freshwater Biology*, **31**: 213–220.

Shaver, M.L., Shannon, J.P., Wilson, K.P., Benenati, P.L. and Blinn, D.W. (1997) Effects of suspended sediment and desiccation on the benthic tailwater community in the Colorado River, USA. *Hydrobiologia*, **357**: 63–72.

Skulberg, O.M. (1984) Effects of stream regulation on algal vegetation. In Lillehammer, A. and Saltveit, S.J. (eds) *Regulated Rivers*, University of Oslo Press, Oslo, pp. 107–124.

Smolar, N. (1997) *Ocena vpliva odvzema vode iz različnih tipov vodotokov na perifiton v času nizkih pretokov (An estimation of influence of water abstraction from different types of running waters on periphyton under low flow conditions)*. MS thesis, University of Ljubljana, Biotechnical Faculty.

Smolar-Žvanut, N., Mikoš, M. and Breznik, B. (2005) The impact of the dam in the Bistrica River on the aquatic ecosystem. *Acta Hydrotechnica*, **23**(39): 99–115.

Smolar, N., Vrhovšek, D. and Kosi, G. (1998). Effects of low flow on periphyton in three different types of stream in Slovenia. In Bretschko, G. and Helešic, J. (eds) *Advances in River Bottom Ecology*, Backhuys Publishers, Leiden, The Netherlands, pp. 107–116.

Smolar-Žvanut, N. (2001) The role of periphytic algae in the determination of the ecologically acceptable flow in running waters. *Acta Hydrotechnica*, **19**(30): 65–89.

Stevenson, R.J. (1983) Effects of currents and conditions simulating autogenically changing microhabitats on benthic diatom immigration. *Ecology*, **64**: 1514–1524.

Stevenson, R.J., Bothwell, M.L. and Lowe, R.L. (1996) *Algal Ecology. Freshwater Benthic Ecosystems*, Academic Press, San Diego.

Suren, A.M. (2005) Effects of deposited sediment on patch selection by two grazing stream invertebrates. *Hydrobiologia*, **549**: 205–218.

Suren, A.M. and Jowett, I.G. (2006) Effects of floods versus low flows on invertebrates in a New Zealand gravel-bed river. *Freshwater Biology*, **51**: 2207–2227.

Suren, A.M. and Riis, T. (2010) The effects of plant growth on stream invertebrate communities during low flow: a conceptual model. *Journal of the North American Benthological Society*, **29**(2): 711–724.

Suren, A.M., Biggs, B.J.F., Kilroy, C. and Bergey, L. (2003a) Benthic community dynamics during summer low-flows in two rivers of contrasting enrichment 1. Periphyton. *New Zealand Journal of Marine and Freshwater Research*, **37**: 53–70.

Suren, A.M., Biggs, B.J.F., Kilroy, C. and Bergey, L. (2003b) Benthic community dynamics during summer low-flows in two rivers of contrasting enrichment 2. Invertebrates. *New Zealand Journal of Marine and Freshwater Research*, **37**: 71–83.

Traaen, T.S. and Lindstrøm, E.A. (1983) Influence of current velocity on periphyton distribution. In Wetzel, R.G. (ed.) *Periphyton of Freshwater Ecosystems*, Dr W. Junk Publishers, Boston, pp. 97–99.

Uehlinger, U., Kawecka, B. and Robinson, C.T. (2003) Effects of experimental floods on periphyton and stream metabolism below a high dam in the Swiss Alps (River Spöl). *Aquatic Sciences*, **65**: 199–209.

Valentin, S., Wasson, J.G. and Phillipe, M. (1995) Effects of hydropower peaking on epilithon and invertebrate community trophic structure. *Regulated Rivers: Research and Management*, **10**: 105–119.

Vrhovšek, D., Kosi, G., Krivograd Klemenčič, A. and Smolar-Žvanut, N. (2006) *Monograph on Freshwater and Terrestrial Algae in Slovenia*, ZRC SAZU, Ljubljana.

Vymazal, J. (1988) The use of periphyton communities for nutrient removal from polluted streams. *Hydrobiologia*, **166**: 225–237.

Ward, J.V. (1974) A temperature-stressed stream ecosystem below a hypolimnial release mountain reservoir. *Archiv für Hydrobiologie*, **74**: 247–275.

Ward, J.V. (1976) Comparative limnology of differentially regulated sections of a Colorado mountain river. *Archives of Hydrobiology*, **78**: 319–342.

Ward, J.V. and Stanford, J.A. (1983) The serial discontinuity concept of lotic ecosystems. In Fontaine, T.D. and Bartell, S.M. (eds) *Dynamics of Lotic Ecosystems*, Ann Arbor Science, Ann Arbor. Michigan, pp. 29–42.

Wetzel, R.G. (1983) Attached algal–substrata interactions: fact or myth and when and how? In Wetzel, R.G. (ed.) *Periphyton of Freshwater Ecosystems*, Dr. W. Junk Publishers, Weinheim, pp. 207–215.

Wetzel, R.G. (2001) *Limnology: Lake and River Ecosystems*, 3rd edition, Academic Press, San Diego.

World Commission on Dams (2000) *Dams and Development: A New Framework for Decision-making*. Available at: http://www.internationalrivers.org/dams-and-development-new-framework-decision [Accessed 4. May 2011].

Zimmermann, H.J. and Ward, J.V. (1984) A survey of regulated streams in the Rocky Mountains of Colorado, U.S.A. In Lillehammer, A. and Saltveit, S.J. (eds) *Regulated Rivers*, Oslo University Press, Oslo, pp. 251–262.

14

Ecohydraulics and Aquatic Macrophytes: Assessing the Relationship in River Floodplains

Georg A. Janauer[1], Udo Schmidt-Mumm[1] and Walter Reckendorfer[2]

[1]Department of Limnology, University of Vienna, Althanstrasse 14, A-1090 Vienna, Austria
[2]WasserCluster Lunz – Biologische Station GmbH, Dr Carl Kupelwieser Promenade 5, A-3293 Lunz am See, Austria

14.1 Introduction

Aquatic macrophytes are, like any other organism, dependent on the properties of their environment. Water has a much higher density than air, the ambient medium of terrestrial plants, and the mechanical stress exerted by the movement of water in streams and rivers, or in lentic waters by wave action, determines the occurrence, propagation, physiology and competitiveness of aquatic plants. Like terrestrial plants, aquatic species rely on light, CO_2, nutrients and water, but true aquatic macrophytes are rare due to the demanding conditions of aquatic environments.

At its most fundamental level, ecohydraulic research examines the effects of water movement on organisms in their habitat. A growing literature is available on this topic, although ecohydraulic research has largely centred on fish, which are important indicators of environmental quality, conservation status and economic value (e.g. Haefner and Bowen, 2002; Pasternak et al., 2008; Schwartz and Herricks, 2008; Daraio et al., 2010; van de Wolfshaar et al., 2010; U.S. Army Corps of Engineers, 2010). Rice et al. (2010) highlighted that contributions within the field of ecohydraulics provided by engineers have outnumbered those from biologists and ecologists. As a consequence,

habitat modelling, at varying scales, has been dominated by technical and physically based views of hydraulic–organism relationships (Harvey and Clifford, 2009; Habersack et al., 2010).There are significant gaps in the spatial resolution of hydraulic models with regard to the spatial characteristics of aquatic plant stands. Addressing these gaps would significantly help in the development of strategies for environmentally sensitive planning and management of aquatic ecosystems (Janauer, 2000; Kemp et al., 2002; Bockelmann et al. 2004). Schnauder and Moggridge (2009) started to address this issue, but the results from a highly dynamic alpine river system with braided channel morphology and coarse bed material can't be easily transferred to other localities. Rice et al. (2010) also highlighted the need to address this research gap and called upon biologists, in particular, to contribute the relevant expertise required to upscale biological data into hydraulic models. This chapter aims to contribute to this field of study by outlining the general relationships between aquatic macrophytes and their hydraulic environment, and by examining how aquatic plants react to the conditions experienced within their habitat. This is demonstrated with direct reference to research undertaken in the Danube River corridor between Vienna and Bratislava.

Ecohydraulics: An Integrated Approach, First Edition. Edited by Ian Maddock, Atle Harby, Paul Kemp and Paul Wood.
© 2013 John Wiley & Sons, Ltd. Published 2013 by John Wiley & Sons, Ltd.

14.2 Macrophytes

Macrophytes were characterised by Westlake (1973) and Vollenweider *et al.* (1974) as aquatic plants of different taxonomic groups, comprising large algae (*Chara, Cladophora*), bryophytes (*Marchantia, Fontinalis*), pteridophytes (*Isoetes, Equisetum*) and vascular angiosperms (*Ranunculus, Nymphaea, Lemna*). In a more general definition, Wetzel (2001, p. 528) characterised this group of primary producers as 'the macroscopic forms of aquatic vegetation'. Riparian and instream aquatic plants influence channel bed roughness constraining run-off to the extent that it may lead to over-bank flow and may impede flow in channels used for hydro-power generation (Thomas and Schanz, 1976). Other ecohydraulic studies have considered the effect of ecologically sensitive macrophyte management (e.g. Pitlo, 1982; Kaenel and Uehlinger, 1998; Rohde *et al.*, 2005), and the effects of aquatic plants on their own microhabitat characteristics (Gregg and Rose, 1982; Bockelmann-Evans *et al.*, 2008; De Doncker *et al.*, 2009).

An extensive review of the controls of flow on macrophytes in lowland rivers demonstrated the importance of discharge and flow velocity on macrophyte colonisation and the complications resulting from the feedback of macrophyte growth on local flow velocity (Franklin *et al.*, 2008). However, Demars and Harper (2005) found no relationship between growth form and physical features of channels in lowland rivers.

In many rivers, flood frequency has been found to be a major influence on macrophyte abundance (Riis and Biggs, 2003). The optimum development of plant cover in Danish streams occurred at flow velocities between 0.3 and 0.5 m s^{-1} (Riis and Biggs, 2003). The dominance of flow has also been established for sections of Austrian river floodplains (Janauer, 2005; Barta *et al.*, 2009), with *Potamogeton filiformis* Pers. and *Potamogeton nodosus* Poir. associated with still water, and *Myriophyllum spicatum* L., *Potamogeton pusillus* L. and *Potamogeton natans* L. indicative of flow velocities less than 30 cm s^{-1}. Intermediate flow velocities (30–65 cm s^{-1}) were characterised by the occurrence of *Potamogeton pectinatus* L. At higher flow velocities there were no statistically significant occurrences of vascular species, although the absence of *Ceratophyllum demersum* L. and *Potamogeton perfoliatus* L. was significant ($p = 0.05$). In smaller rivers, *Ranunculus* species (e.g. *R. trichophyllus*) are often dominant in fast-flowing water. This dominance may be reduced where dense shading occurs and when shade-tolerant *Berula erecta* becomes the dominant taxa, e.g. in Austrian lowland streams (Janauer, 1981). However, in the centre of the main stem of large rivers, high flow velocities and sediment transport/erosion prevent the establishment and growth of macrophytes. In contrast, closer to the banks of large rivers, flow velocities are reduced but plants rarely grow in water deeper than 1.5 m. This phenomenon results from shading due to either suspended inorganic sediment or phytoplankton growth (Westlake, 1973). Permanent stands of plants even without roots, like *Ceratophyllum demersum*, occur in the main stem of lotic systems such as the River Danube in sheltered habitats, e.g. behind groynes (Janauer and Stetak, 2003).

Modern observations and research have built upon the seminal research undertaken by Butcher (1927), Gessner (1955) and Hynes (1970), leading to the development of theoretical approaches considering the effect of large-scale run-off events on macrophytes (Biggs *et al.*, 2005). The importance of flow velocity and sediment grain size to macrophyte development has been underlined by qualitative information indicating that fine-grained substratum patches typically support denser growth of macrophytes (Gessner, 1955, pp. 295 and 296; Sculthorpe, 1967, p. 59).

A variable geographical response to flow conditions has been recorded for macrophytes within different regions of Europe. For example, Characeans, which usually avoid fast flow in Northern Germany, are frequently found in running waters in Bavaria (Gessner, 1955), indicating a wide range of flow tolerances for many macrophyte species. Rapid flows are typically favoured, but not exclusively, by bryophytes (Figure 14.1; see also Janauer and Pall, 2003) and it has also been demonstrated that some aquatic plant species recorded in Southern Germany as emergents occur predominantly as submerged forms in Swedish rivers (Kohler *et al.*, 2000).

Aquatic community composition and the presence of individual species have been related to flow and substrate type, hydraulic resistance and anchoring strength, shear stress, abrasion erosion and longitudinal changes along a river course (see Haslam, 2006, for overview). Hydraulic resistance to flow increases with more bushy plant morphology. Anchoring strength and sensitivity to shear stress are hydraulic determinants of macrophyte species presence in running water environments and have been reviewed by Pitlo and Dawson (1993), Haslam (2006) and considered by Janauer and Jolankai (2008).

Flow velocity is also a primary factor influencing aquatic plant morphology (Table 14.1) and anatomy. The biometric characteristics of *Potamogeton perfoliatus* stem anatomy were studied in relation to variable flows ranging from 0.55 to 0.97 m s^{-1} (Autherith, 2008). When the

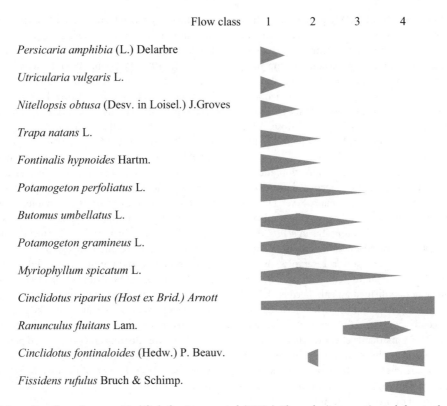

Figure 14.1 Macrophyte flow tolerances. Modified after Janauer *et al.* (2010a). Flow velocity was estimated close to the macrophyte stands: 1 = no flow; 2 = slow ($5 \leq 30$ cm s^{-1}); 3 = moderate (31–69 cm s^{-1}); 4 = fast (≥ 70 cm s^{-1}). Estimates were validated against direct flow velocity measurements (10 cm depth, next to plant stand on the open river side using a vane wheel cylinder probe ZS 18.5 mm, Höntzsch, Waiblingen, Germany) in the Austrian Danube (Schaffer, 1996; Hrivnak *et al.* 2009).

morphology of *P. perfoliatus* was examined, the length of intermodal sections of the main axis and the branches, as well as the area and length of leaves on the main axis, were significantly reduced under higher flow velocities. The anatomical parameters of *P. perfoliatus* significantly reduced at high flow velocities included diameter of the main stem and the central cylinder, thickness of cortex and the number of lacunae (the internal air spaces).

Lift forces, pressure forces and acceleration forces have been identified as primary hydraulic factors influencing submerged macrophytes (Wetzel, 2001). Submerged plants cope with these environmental stressors by developing flexible stems and leaves. Even so, flow velocities greater than 0.7 m s^{-1} are tolerated by few species (Janauer *et al.*, 2010b). Water level fluctuations are also encountered by macrophytes in floodplain water bodies

Table 14.1 Morphological response of aquatic plants to water flow. After Gessner, 1955, p. 302.

Species	Flow (cm s^{-1})	Leaf stalk (cm)	Leaf length (cm)
Berula erecta (Huds.) Coville	10	14.0	10–12
(waterparsnip)	60	1.7	3–5
		Shoot length (cm)	Leaf length (cm)
Myriophyllum sp.	20	100	2.5
(watermilfoil)	70	50	1.2–1.5

open to the main river channel (Wetzel, 2001). In these hydraulically connected areas, the abundance of aquatic macrophytes is significantly reduced compared to oxbows and cut-off channels at a greater distance to the main channel (Janauer, 2003). Aquatic plant stands also significantly attenuate flow velocity in the area immediately adjacent to them. The lateral extension of this 'boundary mantle' can be up to c. 20 cm (Gregg and Rose, 1982; Machata-Wenninger and Janauer, 1991; Sand-Jensen and Mebus, 1996)

Significant differences in species composition and the number of protected species were recorded from side channels directly connected to the main River Danube channel compared with water bodies only hydraulically connected during periods of flooding (Barta *et al.*, 2009). Janauer *et al.* (2008) reported lower macrophyte abundance in run-of-river hydro-power reservoirs on the Austrian River Danube than in semi- or near-natural floodplain waters, but confirmed an enhanced ecological value related to macrophytes and fish in artificial habitats of the most recently constructed reservoir 'Freudenau' (finished 1998). This remarkable effect resulted from the high abundance of macrophytes utilising suitable habitat in small artificial side channels. River channel widening may be an appropriate measure in some instances to facilitate the development of riparian vegetation (Rohde *et al.*, 2005), although the final channel shape will also influence its development.

14.3 Life forms of macrophytes in running waters

Aquatic and wetland plants are generally classified into broad categories based on their life forms. This scheme is independent of phylogenetic relationships and is based solely on the way in which the plants grow in relation to their environment. Such categories describe the general environmental conditions of plant habitats, especially with respect to hydraulic flow conditions.

Present classifications derive from the parallel concepts of life form and growth form. These were first developed by Raunkiaer (1934) and later expanded by Danserau (1959), merging plant traits with the structural description of the vegetation, based on Du Rietz's original research (1931). The first approach divides aquatic macrophytes according to their association with depth of water and substrate composition; the second defines plant groups of comparable structure (Hutchinson, 1975). Other variants have classified aquatic plants according to

their life- or growth form (e.g. Luther, 1949; Hejný, 1957; den Hartog and Segal, 1964; Hogeweg and Brenkert, 1969; Hutchinson, 1975; Mäkirinta, 1978; den Hartog and van der Velde, 1988; Wiegleb, 1991). However, there may be overlap between groups because individual species change from one form to another during their life history. Based on the primary habitat of occurrence, Sand-Jensen *et al.* (1992), Riis *et al.* (2001) and Bowden *et al.* (2006) followed a simplified scheme, classifying stream macrophytes into: (1) obligate submerged plants, (2) amphibious plants and (3) terrestrial plants. In contrast, Willby *et al.* (2000) classified aquatic macrophytes into groups of plants sharing the same attributes of traits and habitat characteristics. Their multivariate analysis, utilising 58 trait attributes resulted in a non-hierarchical classification of European hydrophytes into 20 different groups.

In order to avoid confusion, we follow the terminology of Sculthorpe (1967) and Cook (1996) to characterise the macrophytes growing in the channel and in the riparian zone of rivers and streams into the following groups: (i) rooted emergent, (ii) rooted submerged, (iii) floating leaved, (iv) free floating and (v) attached plants. The effect of different life forms on flow (Figure 14.2) is dependent on the density and shape of the macrophyte leaves, stem and root system.

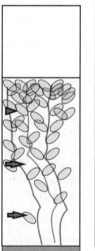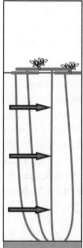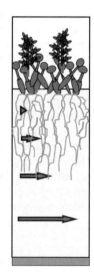

Figure 14.2 Impact of aquatic macrophyte life forms on flow conditions (schematic). Left: Rooted submerged – flow significantly reduced in dense stands. Middle: Rooted floating-leaf – flow moderately reduced by petioles, although this life form only occurs in habitats characterised by reduced flow velocities. Right: Free floating – flow significantly reduced in the root zone, but may be higher at greater depths. Image created by and copyright of Georg A. Janauer.

Emergent plants (including amphiphytes/amphibious plants and helophytes) are rooted in the substratum with their basal parts typically growing beneath the surface of the water, but leaves, stems and reproductive organs are airborne. Most of these plants are confined to riparian wetland areas of rivers and streams. 'Fringing herbs' are an important group of emergent vegetation on the banks of many tropical and subtropical slow-flowing rivers. These plants may encroach onto the water surface from the margins of the river and form floating mats or islands of vegetation. These are frequently secondarily colonised by non-floating plants (Junk, 1970). These mats of vegetation may cover the whole water surface, causing local oxygen depletion, reducing surface and subsurface flow velocities and subsequently leading to increased sedimentation rates.

To explain the development of vegetated islands in highly dynamic, fluvial alpine rivers, the feedback mechanisms between willow seedlings, hydraulic and geomorphological processes have been investigated (Schnauder and Moggridge, 2009). Riparian vegetation also increases flow resistance, modifies sediment transport and deposition processes and backwater channel morphology (Yen, 2002). The contribution of different vegetative roughness types to the total flow resistance depends on the type of vegetation and individual taxa, which may vary considerably temporally and spatially (Järvelä, 2002; 2004). However, emergent macrophytes are considered an important component of the ecohydraulic 'functional habitat' (mesohabitat) concept used in the UK as part of the River Habitat Survey methodology (Harper et al., 1998; Newson et al., 1998).

Floating plants spread their leaves on the water surface, but are typically rooted in the underlying substratum. Sculthorpe (1967) divided floating-leaved plants into two groups: (i) those with rhizomatous or cormose stems, and (ii) those with runners or stolons. Strong currents limit the presence and distribution of floating-leaved species, making them a less important ecohydraulic component of rivers and streams. Using Indicator Species Analysis and predetermined flow classes, Janauer et al. (2010a) found floating-leaved *Potamogeton nodosus* Poir. and *Trapa natans* L. to be frequently present in slow-flowing waters of the River Danube and its floodplain water bodies.

Rooted submerged plants typically spend their entire lifecycle beneath the surface of the water and flowering takes place in submerged, floating or emergent conditions. Rooted submerged macrophytes effectively divide river flow, creating micro-channels between their stands. With increasing abundance and a reduction in the conveyance of water, river level may rise and result in over-bank flow on to adjacent land, a feature that has been documented extensively (Dawson, 1988; Champion and Tanner, 2000; Pott and Remy, 2000; Haslam, 2006; Franklin et al., 2008; Bornette and Puijalon, 2011). Significant effects of aquatic plants on stream hydraulics, flow velocity and turbulence patterns around branching submerged macrophytes have also been examined by Marshall and Westlake (1990), Machata-Wenninger and Janauer (1991), Sand-Jensen and Mebus (1996), Sand-Jensen (1998), Sand-Jensen and Pedersen (1999), Green (2005a; 2005b) and Luhar et al. (2008). The Froude number, a dimensionless velocity/depth ratio, has been the most widely used hydraulic descriptor for the classification of running water biotopes and 'functional habitats' (Newson et al., 1998; Kemp et al., 2000; Chen et al., 2009). Despite its ecological utility being justified by its relationship with the density of invertebrate species (Jowett, 1993), it remains uncertain whether such hydraulic parameters can provide a comprehensive classification of all running-water habitat patches (Buffagni et al., 2000; Clifford et al., 2006).

14.4 Application of ecohydraulics for management: a case study on the Danube River and its floodplain

River–floodplain systems and their associated biodiversity have been significantly influenced by anthropogenic modification, resulting in long-term trends towards terrestrialisation and habitat fragmentation (Dynesius and Nilson, 1994; Ward, Tockner and Schiemer, 1999). In order to re-establish semi-natural conditions, rehabilitation programmes with a focus on enhancing lateral connectivity between river and floodplain have become increasingly important (Ormerod, 2003; Giller, 2005). Side-arm reconnection allows natural hydraulic forces to reshape water bodies and re-establish 'natural' channel heterogeneity. Reconnecting relict river branches increases fine sediment removal, enhances natural geomorphic processes and increases the turnover rate of aquatic and terrestrial habitats. Other goals of rehabilitation activities include compensation for structural engineering shortfalls and the reduction of bed scour in the main river stem (Reckendorfer et al., 2005). Ecologically orientated planning of side-arm reconnection requires an understanding of the relationship between hydraulic connectivity and the requirements of the organisms inhabiting floodplain ecosystems, many of which are highly regulated.

This understanding is also necessary for predicting future trajectories of change and sustaining floodplain and wetland ecosystem services and species biodiversity.

The case study outlined below analysed the composition of macrophyte assemblages in a river–floodplain system, based on hydrological connectivity and water depth. We relate these parameters to the aquatic macrophyte community, with the aim of identifying the mechanisms underlying community responses to ecohydraulic and hydromorphological change.

14.4.1 Study site

The River Danube is the second largest river in Europe (2850 km) and drains an area of 817 000 km^2. At Vienna, the Danube is a ninth-order river with a mean annual discharge of about 1950 m^3 s^{-1}. The bank-full discharge (recurrence time of approximately 1 year) is 5800 m^3 s^{-1} and the average slope in the study reach is 0.45‰. In 1996, the river-floodplain system east of Vienna was designated a National Park according to IUCN criteria ('Donau Auen National Park', DANP). The size of the National Park is about 93 km^2 and it comprises 65% floodplain forest, 20% open water and 15% meadows and fields.

Historically, the study reach of the Danube was braided, but a major river regulation scheme commenced in 1875, constraining the floodplain considerably. Restoration measures in the reach were first implemented in the mid 1990s. Reactivation of former inflowing sections improved the connectivity of cut-off river branches. Check-dams that regulated side arms and confined them into single sections were removed and additional outlet channels were created. Details regarding these restoration works were described by Tockner and Schiemer (1997) and Schiemer et al. (1999). Today, the connectivity between the main channel and the floodplain water bodies is partially intact (Baranyi et al., 2002; Reckendorfer et al., 2006) and the DANP comprises sections with varying lateral connectivity. However, hydraulically disconnected units remain the dominant habitat type (40%). Areas of intermediate and high connectivity (> 30 days) account for only 11% and 18% of total habitat, respectively.

14.4.2 Sampling

A total of 1271 quantitative samples of aquatic macrophytes, spanning a connectivity range from 0 to 171 days year^{-1}, were collected between August 1991 and October 2006. Sampling encompassed various water bodies located on floodplain sections on the right and left banks of a 40 km stretch of the Danube River (Figure 14.3). The dataset analysed comprised 601 survey units, a subsample from the whole dataset, plus additional data from the main stem.

14.4.3 Data analysis

Connectivity (*Cd*) is defined as the average annual duration (mean days per year 1961–90) of the upstream connection of floodplain water bodies with the main stem of the River Danube. This integrates several features of the different water bodies, such as current velocity, substratum composition and shear stress (for details, see Reckendorfer et al., 2006). Connectivity primarily depends on the flow pattern of the river and the position of the individual water bodies relative to river height (Hillman and Quinn, 2002). This was derived from data on: (i) the stage–discharge relationship of the river at the upstream end of each channel; (ii) the flow duration curve; and (iii) the stage at which water flows into the channel. The altitude of the inflow area and water level data were provided by the Austrian Federal Waterway Agency (Status: 1997). *Cd* was calculated for each water body located within the study area and water bodies were delineated from aerial photographs (provided by the National Park Donau-Auen GmbH, Orth/D, Austria) and based on inflow areas and transverse check-dams using ArcView 3.1. The maximum depth (d_{max}) was calculated as the distance between the theoretical groundwater level at mean water level and the deepest point of the water body. Five Habitat Types (HTs) were distinguished, based on the aggregated connectivity parameters within the main channel (*Cd*) and water depth:

HT-1: connectivity > 355 days/year (permanent lotic conditions; main stem)

HT-2: connectivity > 120 and < 355 days/year

HT-3: connectivity > 3.65 and < 120 days/year

HT-4: connectivity < 3.65 days/year (connected at the mean annual high water level)

HT-5: temporary water bodies: $d_{max} < 0$ m (at mean water in the Danube River).

We used Indicator Species Analysis (ISA) to look for significant indicator life forms and indicative species among different habitat types. The ISA method can be applied to calculate indicator values (IVs) for (i) life forms, as defined above, by multiplying the relative abundance of each life form in a specific HT-group by its relative frequency in that HT-group (McCune and Grace, 2002), or (ii) by following the same process for individual species. The IV reaches its maximum value when all abundances of a life form or a species are found in a single HT and

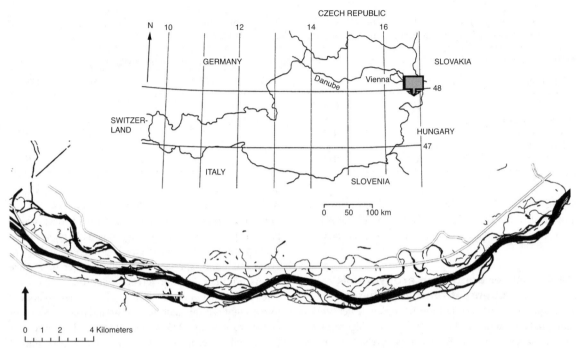

Figure 14.3 The Austrian Danube study reach between Vienna (left limit of the map section) and the Morava River (right limit of the map section), which forms the Austrian–Slovak border near Bratislava. This reach comprises the water bodies of the 'Donau-Auen National Park' (DANP, 'Danube River and Floodplain National Park' by IUCN appointment). Black lines: open water; grey open lines: flood protection dykes.

when the life form or species occurs in all survey units of that HT. A Monte Carlo simulation test with 1000 randomised runs was used to determine the significance ($P \leq 0.05$) of the indicator value (McCune and Mefford, 2006).

14.4.4 Ecohydraulic habitat–macrophyte relationship

Indicator Species Analysis was first used to statistically characterise the assemblage of aquatic vegetation in different habitats based on life forms. It was then used to identify species associated with individual survey units of the Habitat Types. For each of the life forms recorded, ISA resulted in significant indicator values (IVs, Table 14.2), but the allocation to Habitat Types was different ($P < 0.001$). Species numbers per life form were: attached – 6; rooted with floating-leaves – 10; rooted submerged – 39; free floating – 8; and rooted emergent – 56.

The attached life forms consisted primarily of rheophilic mosses and reached their highest values in the main river channel (HT-1). This Habitat Type had the highest hydraulic connectivity of those examined (Fig-

ure 14.4). River mosses tend to live in fast and shallow waters (Janauer et al., 2010b), partially due to their need for constant replenishment of CO_2 and nutrients directly from the water (Madsen et al., 1993). Habitat type HT-2 was also closely associated with the main channel, typically side branches with permanent flow. Habitat Type HT-3 sites were characterised by more transient flowing conditions and were located near the confluence with the main channel. No life forms were significantly associated with or absent from these two environments, making discrimination between them almost impossible at the life form level.

In contrast, floating-leaved, rooted submerged and free-floating plants were all significantly associated with HT-4 (Figure 14.4). This HT was characterised by greater water depths and reduced duration of through-flow originating from the main channel and was also reported by Schratt-Ehrendorfer (1999). Finally, rooted emergent plants were significantly associated with HT-5 (Figure 14.4) and represent habitats at greater distances from the main stem and with shallow water depth. This habitat type typically experiences short-duration flooding and

Table 14.2 Indicator life forms by ISA for Habitat Types (HT) 1–5. The Maximum Indicator Value (IV, **bold**, significance: $P \leq 0.05$) allocates a life form to a Habitat Type. Data source: Janauer *et al.* 2010a, unpublished analysis.

Life form	Maximum IV in HT	*P*-value	Indicator Value (%)				
			HT-1	HT-2	HT-3	HT-4	HT-5
Attached	1	0.0002	**73**	0	3	0	0
Floating leaved	4	0.0002	0	0	0	**43**	5
Root submerged	4	0.0004	0	8	14	**37**	20
Free-floating	4	0.0074	0	5	11	**33**	16
Root emergent	5	0.0002	0	1	2	41	**53**

is characterised by lentic/still water conditions through-out most of the year. The characterisation of the HTs is strongly supported at the species level (see Table 14.3).

14.4.5 Water depth – an ecohydraulic requirement for macrophytes

Most species in the River Danube floodplain water bodies exhibited a wide range of depth preference (Figure 14.5). Several species, such as *Potamogeton berchtoldii* Fieber, *Ranunculus sceleratus* L., three stoneworts and *Berula erecta* (Hudson) Coville showed a preference for water less than 2 m deep. In contrast, *Najas minor* Allioni, *Nitellopsis obtusa* J. Groves and some *Potamogeton* species such as *Potamogeton perfoliatus* L., *Potamogeton nodosus* Poir. and *Potamogeton pectinatus* L. dominated in deep water.

At 3.6–3.3 m depth, a group of species (*Potamogeton pectinatus* to *Chara delicatula* J.Agardh) forms a 'deep water' plateau, followed by a distinctive gradient (*Nymphaea alba* L. to *Fontinalis antipyretica* Hedw.) between 2.8 and 1.5 m. From this point, species with shallow water preference follow, but their depth preferences vary to a great extent. The two distinct gradient sections of the diagram appear to represent typical patch/habitat boundaries or ecotones (Janauer and Kum, 1996), whereas the two plateaus resemble central patch/habitat regions where a larger group of species can withstand the stress of greater submersion and the attenuation of incident radiation. Species with a 5-percentile value below zero were recorded either on wet ground (e.g. amphiphytes, growing either within or out of water) or in close-by deeper depressions, when water levels in the main water body were lower.

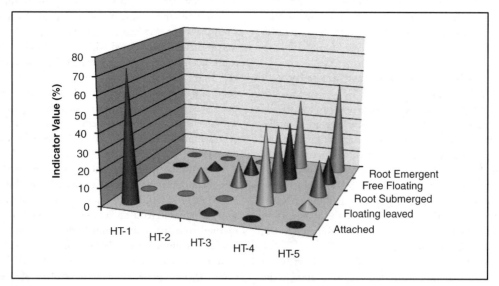

Figure 14.4 Indicator Values of each life form for Habitat Types (HT) 1–5. Only the highest value of each life form makes it 'indicative' of a Habitat Type. Data source: Janauer *et al.* (2010b).

Table 14.3 Two-way table of Indicator Species Analysis (ISA) for Habitat Types (HT) 1–5, Indicator Values, significance (**bold** denotes $P \leq 0.05$) and cumulative abundance. Non-significant species are not included in this table (Data source: Janauer *et al.* 2010a, unpublished analysis). Taxa Moss sp., *Vaucheria* sp., *Rumex* sp., *Carex* sp. and *Callitriche* sp. were not determined to species level in the dataset.

				Abundance				
Species name	Species code	Observed Indicator Value (IV)	P*	HT-1	HT-2	HT-3	HT-4	HT-5
Cinclidotus riparius	Cin rip	100.0	**0.0002**	**132**	0	0	0	0
Cinclidotus mucronatus	Cin muc	5.9	**0.0308**	**1**	0	0	0	0
Elodea canadensis	Elo can	17.4	**0.0068**	0	**295**	333	133	0
Polygonum hydropiper	Pol hyd	25.8	**0.0008**	0	5	**74**	3	1
Butomus umbellatus	But umb	22.4	**0.0010**	0	5	**195**	65	8
Elodea nuttallii	Elo nut	21.1	**0.0034**	0	0	**454**	219	0
Myriophyllum spicatum	Myr spi	28.6	**0.0044**	0	114	**876**	1865	1
Potamogeton perfoliatus	Pot per	20.2	**0.0106**	0	77	**438**	691	0
Moss sp.	Mos spe	9.4	**0.0160**	0	0	**132**	0	0
Eleocharis palustris	Ele pal	12.2	**0.0194**	0	1	**49**	41	0
Potamogeton pectinatus	Pot pec	22.0	**0.0204**	1	140	**1004**	2906	8
Vaucheria sp.	Vau spe	6.3	**0.0310**	0	0	**254**	0	0
Spirodela polyrhiza	Spi pol	14.8	**0.0404**	0	74	**361**	87	8
Nuphar lutea	Nup lut	38.3	**0.0004**	0	0	0	**5332**	0
Myriophyllum verticillatum	Myr ver	32.4	**0.0004**	0	16	83	**2987**	0
Sagittaria sagittifolia	Sag sag	28.8	**0.0004**	0	1	29	**2229**	1
Hippuris vulgaris	Hip vul	28.7	**0.0006**	0	0	6	**2514**	9
Najas marina	Naj mar	21.6	**0.0008**	0	0	0	**1264**	0
Carex elata	Car ela	28.3	**0.0018**	0	0	0	**3240**	170
Potamogeton lucens	Pot luc	25.0	**0.0018**	0	0	32	**2217**	8
Nymphaea alba	Nym alb	16.6	**0.0042**	0	0	1	**307**	0
Utricularia vulgaris	Utr vul	15.3	**0.0056**	0	0	0	**1096**	1
Ranunculus x glueckii	Ran x gl	14.4	**0.0082**	0	0	0	**997**	1
Rumex sp.	Rum spe	14.6	**0.0098**	0	0	0	**193**	1
Sparganium erectum	Spa ere	14.0	**0.0106**	0	0	0	**516**	1
Schoenoplectus lacustris	Sch lac	19.6	**0.0122**	0	0	0	**1507**	108
Potamogeton pusillus	Pot pus	14.6	**0.0134**	0	0	0	**1081**	0
Lythrum salicaria	Lyt sal	13.1	**0.0176**	0	0	1	**127**	1
Mentha aquatica	Men aqu	18.9	**0.0178**	0	1	5	**1114**	91
Riccia fluitans	Ric flu	16.9	**0.0200**	0	0	0	**1068**	43
Potamogeton nodosus	Pot nod	10.8	**0.0230**	0	0	0	**773**	0
Sparganium emersum	Spa eme	18.9	**0.0418**	0	2	155	**1871**	138
Veronica anagallis-aquatica	Ver ana	15.0	**0.0436**	0	0	0	**1355**	94
Iris pseudacorus	Iri pse	45.5	**0.0002**	0	0	3	241	**79**
Alisma plantago-aquatica	Ali pla	43.2	**0.0002**	0	0	69	485	**215**
Phragmites australis	Phr aus	42.8	**0.0002**	0	8	166	18834	**1568**
Alisma lanceolatum	Ali lan	39.4	**0.0002**	0	2	79	357	**98**
Ranunculus sceleratus	Ran sce	39.1	**0.0002**	0	0	0	169	**144**
Phalaris arundinacea	Pha aru	38.5	**0.0002**	0	36	214	1174	**387**
Carex sp.	Car spe	37.2	**0.0002**	0	1	120	2213	**592**
Oenanthe aquatica	Oen aqu	32.9	**0.0002**	0	0	0	471	**127**
Myosotis palustris	Myo pal	32.2	**0.0002**	0	17	54	1585	**167**
Callitriche sp.	Cal spe	27.7	**0.0008**	0	49	461	1723	**589**
Lycopus europaeus	Lyc eur	16.4	**0.0010**	0	0	0	21	**19**
Scirpus sylvaticus	Sci syl	7.1	**0.0034**	0	0	0	0	**16**
Ranunculus trichophyllus	Ran tri	19.3	**0.0042**	0	0	0	817	**229**
Polygonum mite	Pol mit	19.7	**0.0050**	0	0	0	1726	**174**
Alopecurus aequalis	Alo aeq	12.8	**0.0100**	0	0	0	817	**171**
Hydrocharis morsus-ranae	Hyd mor	13.0	**0.0240**	0	0	0	626	**79**
Equisetum palustre	Equ pal	5.9	**0.0266**	0	0	1	3	**2**

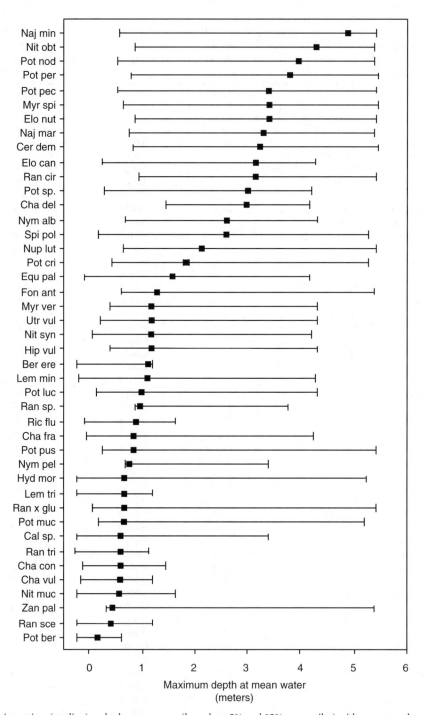

Figure 14.5 Species optima (median) and tolerance ranges (bars show 5% and 95% percentiles) with respect to the maximum depth of floodplain water bodies during periods of mean water level in the Danube (for details, see text). Full species names and authorities are given in Appendix 14.A at the end of this chapter.

The water depth preferences (medians) of the species identified in this 40 km reach of the River Danube and its numerous floodplain water bodies may differ from situations found in other geographical regions. However, based on knowledge of other reaches in the River Danube (e.g. Sarbu *et al.*, 2011) and the MIDCC dataset of the macrophyte survey of the full river length, including its floodplain waters (Janauer, 2005), this study represents a valuable contribution to the understanding of ecohydraulic factors influencing the mosaic of aquatic vegetation patches recorded.

14.5 Conclusion

Important aspects of the relationship between aquatic organisms and their environment have been explored in this chapter, and we have demonstrated that they can be described and quantified by ecohydraulic methods. Our examination of the literature indicates that there is a growing interest regarding the hydraulic–aquatic macrophyte relationship and that this historically neglected field of research is gaining importance and a wider interest. The hydraulic properties of aquatic macrophytes can be characterised by their morphology and growth form. However, there is a need for further extensive research to examine the influence of hydraulics on riparian and instream vegetation and this should be undertaken drawing upon the interdisciplinary expertise of biologists, hydrologists and hydraulicists. This holistic approach is necessary for the future implementation of river restoration schemes, flood control measures and to assist in the development of water resource and environmental protection legislation (e.g. EU Water Framework and Habitats Directive). This also demonstrates the key role that ecohydraulic research can play in the management and conservation of riverine ecosystems.

Acknowledgements

The research on macrophyte communities and hydraulic and hydrological parameters in the Danube River floodplain and the main channel was supported, in part, by an EU Life Project (LIFE98NAT/A/005422), the Austrian Federal Waterway Agency, the National Park Donau-Auen GmbH and the Municipality of Vienna (MA 45). Data on water level fluctuations were provided by the Austrian Federal Waterway Agency. We are grateful to our colleagues who contributed to the field work and species determination, especially Karin Pall, Georg Kum and Ulrike Wychera. Eva Lanz assisted the authors in screening recent literature. We thank the reviewers for comments made on the contents and structure of this chapter.

Appendix 14.A: Abbreviations used in Figure 14.5, including full plant names and authorities

Abbreviation	Species name	Authorities
Ber ere	*Berula erecta*	(Hudson) Coville
Cal sp.	*Callitriche* sp.	no Author
Cer dem	*Ceratophyllum demersum*	L.
Cha con	*Chara contraria*	A. Braun ex Kütz.
Cha del	*Chara delicatula*	A.N. Desv.
Cha fra	*Chara fragilis*	Desv. in Loisel.
Cha vul	*Chara vulgaris*	L.
Elo can	*Elodea canadensis*	Michx.
Elo nut	*Elodea nuttallii*	(Planch.) H.St.John
Equ pal	*Equisetum palustre*	L.
Fon ant	*Fontinalis antipyretica*	Hedw.
Hip vul	*Hippuris vulgaris*	L.
Hyd mor	*Hydrocharis morsus-ranae*	L.
Lem min	*Lemna minor*	L.
Lem tri	*Lemna trisulca*	L.
Myr spi	*Myriophyllum spicatum*	L.
Myr ver	*Myriophyllum verticillatum*	L.
Naj mar	*Najas marina*	L.
Naj min	*Najas minor*	All.
Nit muc	*Nitella mucronata*	(A.Braun) Miq.
Nit obt	*Nitellopsis obtusa*	(Desv. in Loisel.) J.Groves
Nit syn	*Nitella syncarpa*	(Thuill.) Chevall.
Nup lut	*Nuphar lutea*	(L.) Sibth. & Sm.
Nym alb	*Nymphaea alba*	L.
Nym pel	*Nymphoides peltata*	(S.G.Gmel.) Kuntze
Pot ber	*Potamogeton berchtoldii*	Fieber
Pot cri	*Potamogeton crispus*	L.
Pot luc	*Potamogeton lucens*	L.
Pot muc	*Potamogeton mucronatus*	Sond.
Pot nod	*Potamogeton nodosus*	Poir.
Pot pec	*Potamogeton pectinatus*	L.
Pot per	*Potamogeton perfoliatus*	L.
Pot pus	*Potamogeton pusillus*	L.
Pot sp.	*Potamogeton* sp.	no author
Ran cir	*Ranunculus circinatus*	Sibth.
Ran sce	*Ranunculus sceleratus*	L.
Ran sp.	*Ranunculus* sp.	no author
Ran tri	*Ranunculus trichophyllus*	Chaix
Ran x glu	*Ranunculus x glueckii*	A.Félix
Ric flu	*Riccia fluitans*	L. emend Lorb.
Spi pol	*Spirodela polyrhiza*	(L.) Schleid.
Utr vul	*Utricularia vulgaris*	L.
Zan pal	*Zannichellia palustris*	L.

References

Autherith, B. (2008) *Morphologische und anatomische Aspekte von* Potamogeton perfoliatus *und* Myriophyllum spicatum *bei unterschiedlichen Strömungsgeschwindigkeiten: Wiener Neustädter Kanal, Biedermannsdorf.* Diploma thesis, University of Vienna.

Baranyi, C., Hein, T., Holarek, C., Keckeis, S. and Schiemer, F. (2002) Zooplankton biomass and community structure in a Danube River floodplain system. Effects of hydrology. *Freshwater Biology*, **47**: 473–482.

Barta, V., Schmidt-Mumm, U. and Janauer, G.A. (2009) Adapting floodplain connectivity conditions – a prerequisite for sustaining aquatic macrophyte diversity in the UNESCO Biosphere Reserve Lobau (Austria). *Ecohydrology and Hydrobiology*, **9**: 73–81.

Biggs, B.J.F., Nikora, V.I. and Snelder, T.H. (2005) Linking scales of flow variability to lotic ecosystem structure and function. *River Research and Applications*, **21**: 283–298.

Bockelmann, B.N., Fenrich, E.K., Lin, B.K. and Falconer, R.A. (2004) Development of an ecohydraulics model for stream and river restoration. *Ecological Engineering*, **22**: 227–235.

Bockelmann-Evans, B.N., Davies, R. and Falconer, R.A. (2008) Measuring bed shear stress along vegetated river beds using FST-hemispheres. *Journal of Environmental Management*, **88**: 627–637.

Bornette, G. and Puijalon, S. (2011) Response of aquatic plants to abiotic factors: a review. *Aquatic Sciences*, **73**: 1–14.

Bowden, W.B., Glime, J.M. and Riis, T. (2006) Macrophytes and bryophytes. In Hauer, F.R. and Lamberti, G.A. (eds) *Methods in Stream Ecology*, 2nd edition, Academic Press/Elsevier, Amsterdam, pp. 381–414.

Buffagni, A., Crosa, G.A., Harper, D.M. and Kemp, J.L. (2000) Using macroinvertebrate species assemblages to identify river channel habitat units: an application of the functional habitats concept to a large, unpolluted Italian river (River Ticino, northern Italy). *Hydrobiologia*, **435**: 213–225.

Butcher, R.W. (1927) A preliminary account of the vegetation of the River Itchen. *Journal of Biology*, **15**: 55–65.

Champion, P.D. and Tanner, C.C. (2000) Seasonality of macrophytes and interaction with flow in a New Zealand lowland stream. *Hydrobiologia*, **441**: 1–12.

Chen, Y.C., Kao, S.P., Lin, J.Y. and Yang, H.C. (2009) Retardance coefficient of vegetated channels estimated by the Froude number. *Ecological Engineering*, **35**: 1027–1035.

Clifford, N.J., Harmar, O.P., Harvey, G. and Petts, G.E. (2006) Physical habitat, eco-hydraulics and river design: a review and re-evaluation of some popular concepts and methods. *Aquatic Conservation: Marine and Freshwater Ecosystems*, **16**: 389–408.

Cook, C.D.K. (1996) *Aquatic Plant Book*, SPB Academic Publishing, Amsterdam.

Danserau, P. (1959) Vascular aquatic plant communities of southern Quebec: A preliminary analysis. *Transactions of the Northeast Wildlife Conference*, **10**: 27–54.

Daraio, J.A., Weber, L.J., Netwon, T.J. and Nestler, J.M. (2010) A methodological framework for integrating computational fluid dynamics and ecological models applied to juvenile freshwater mussel dispersal in the Upper Mississippi River. *Ecological Modelling*, **221**: 201–214.

Dawson, F.H. (1988) Water flow and the vegetation of running water. In Symoens, J.J. (ed.) *Vegetation of Inland Waters*, Kluwer Academic Publishers, Dordrecht, pp. 283–309.

De Doncker, L., Troch, P., Verhoeven, R., Bal, K., Meire, P. and Quintellier, J. (2009) Determination of the Manning roughness coefficient influenced by vegetation in the river As and Biebrza river. *Environmental Fluid Mechanics*, **9**: 549–567.

Demars, B.O.L. and Harper, D.M. (2005) Water column and sediment phosphorus in a calcareous lowland river and their differential response to point source control measures. *Water, Air and Soil Pollution*, **167**: 273–293.

Den Hartog, C. and Segal, S. (1964) A new classification of the water-plant communities. *Acta Botanica Neerlandica*, **13**: 367–393.

Den Hartog, C. and Van der Velde, G. (1988) Structural aspects of aquatic plant communities. In Symoens, J.J. (ed.) *Vegetation of Inland Waters*, Kluwer Academic Publishers, Dordrecht, pp. 113–153.

Du Rietz, G.E. (1931) Life-forms of terrestrial flowering plants. *Acta Phytogeographica Suecica*, **3**: 1–95.

Dynesius, M. and Nilson, C. (1994) Fragmentation and flow regulation in the northern third of the world. *Science*, **266**: 753–762.

Franklin, P., Dunbar, M. and Whitehead, P. (2008) Flow controls on lowland river macrophytes: a review. *Science of the Total Environment*, **400**: 369–378.

Gallant, M.J. and Haralampides, K. (2005) An ecohydraulics approach to increasing production of juvenile wild Atlantic salmon (*Salmo salar*). *Proceedings of the Annual Conference of the Canadian Society for Civil Engineering.*

Gessner, F. (1955) Hydrobotanik. *Die physiologischen Grundlagen der Pflanzenverbreitung im Wasser. I. Energiehaushalt.* VEB Deutscher Verlag der Wissenschaften, Berlin, pp. 286–297.

Giller, P.S. (2005) River restoration: seeking ecological standards. Editor's introduction. *Journal of Applied Ecology*, **42**: 201–207.

Green, J.C. (2005a) Velocity and turbulence distribution around lotic macrophytes. *Aquatic Ecology*, **39**: 1–10.

Green, J.C. (2005b) Modelling flow resistance in vegetated streams: review and development of new theory. *Hydrological Processes*, **19**: 1245–1259.

Gregg, W.W. and Rose, F.L. (1982) The effects of aquatic macrophytes on the stream microenvironment. *Aquatic Botany*, **14**: 309–324.

Habersack, H., Tritthart, M., Liedermann, M. and Hauer, C. (2010) Micro- and Mesoscale habitat modelling. *8th International Symposium on Ecohydraulics*, Korean Water Resources Association, Seoul, pp. 900–907.

Haefner, J.W. and Bowen, M.D. (2002) Physical-based model of fish movement in fish extraction facilities. *Ecological Modelling*, **152**: 227–245.

Harper, D.M., Smith, C.D., Kemp, J.L. and Crosa, G.A. (1998) The use of "functional habitats" in the conservation, management and rehabilitation of rivers. In Bretschko, G. and Helesic, J. (eds) *Advances in River Bottom Ecology*, Backhuys Publishers, Leiden, pp. 315–326.

Harvey, G.L. and Clifford, N.J. (2009) Microscale hydrodynamics and coherent flow structures in rivers: Implications for the characterization of physical habitat. *River Research and Applications*, **25**: 160–180.

Haslam, S.M. (2006) *River Plants: The macrophytic vegetation of watercourses*, 2nd revised edition, Forrest Text, Cardigan.

Hejný, S. (1957) Ein Beitrag zur ökologischen Gliederung der Makrophyten der tschechoslowakischen Niederungsgewässer. *Preslia*, **29**: 349–368.

Hillman, T.J. and Quinn, G.P. (2002) Temporal changes in macroinvertebrate assemblages following experimental flooding in permanent and temporary wetlands in an Australian floodplain forest. *River Research and Applications*, **18**: 137–154.

Hogeweg, P. and Brenkert, A.L. (1969) Structure of aquatic vegetation: A comparison of aquatic vegetation in India, The Netherlands and Czechoslovakia. *Tropical Ecology*, **10**: 139–162.

Hrivnák, R., Otahelova, H. and Valachovič, M. (2009) Macrophyte distribution and ecological status of the Turiec River (Slovakia): changes after seven years. *Archive of Biological Science*, **61**: 297–306.

Hutchinson, G.E. (1975) *A Treatise on Limnology, Volume III: Limnological Botany*, John Wiley & Sons, Inc., New York.

Hynes, H.B.N. (1970) *The Ecology of Running Waters*, Liverpool University Press, Liverpool, p. 89.

Janauer, G.A. (1981) Die Zonierung submerser Wasserpflanzen und ihre Beziehung zur Gewässerbelastung am Beispiel der Fischa (Niederösterreich). *Verhandlungen der Zoologisch-Botanischen Gesellschaft in Österreich*, **120**: 73–98.

Janauer, G.A. (2000) Ecohydrology: fusing concepts and scales. *Ecological Engineering*, **16**: 9–16.

Janauer, G.A. (2003) Makrophyten der Augewässer. In Janauer, G.A. and Hary, N. (eds) *Ökotone – Donau – March*, Universitätsverlag Wagner, Innsbruck.

Janauer, G.A. (2005) *MIDCC – Macrophytes, river corridor, land use, habitats: A multifunctional study in the Danube catchment based on a GIS-approach*. Final project report, The Federal Ministry of Education, Science and Sports, Vienna.

Janauer, G.A. and Jolankai, G. (2008) Lotic vegetation processes. In Harper, D.M., Zalewski, M. and Pacini, N. (eds) *Ecohydrology: Processes, Models and Case Studies*, CAB International, Wallingford, pp. 46–61.

Janauer, G.A. and Kum, G. (1996) Aquatic macrophytes: indicators for ecotones in backwaters of the River Danube (Austria). *Archiv für Hydrobiologie*, **113**: 477–483.

Janauer, G.A. and Pall, K. (2003) Impoundment Abwinden-Asten, Austria (river-km 2136–2119.5): species distribution features and aspects of historical river status. *Archiv für Hydrobiologie Supplement Large Rivers*, **14**: 87–95.

Janauer, G.A. and Stetak, D. (2003) Macrophytes of the Hungarian lower Danube valley (1498–1468 rkm). *Archiv für Hydrobiologie*, **147**: 167–180.

Janauer, G.A., Schmidt-Mumm, U. and Reckendorfer, W. (2010a) The macrophyte–floodplain habitat relationship: indicator species, diversity and dominance. *Proceedings of The International Association for Danube Research*, Dresden.

Janauer, G.A., Schmidt-Mumm, U. and Schmidt, B. (2010b) Aquatic macrophytes and current velocity in the Danube River. *Ecological Engineering*, **36**: 1138–1145.

Janauer, G.A., Lanz, E., Schmidt-Mumm, U., Schmidt, B. and Waidbacher, H. (2008) Aquatic macropyhtes and hydroelectric power station reservoirs in regulated rivers: man-made ecological compensation structures and the "ecological potential". *Ecohydrology and Hydrobiology*, **8**: 149–157.

Järvelä, J. (2002) Flow resistance of flexible and stiff vegetation: a flume study with natural plants. *Journal of Hydrology*, **269**: 44–54.

Järvelä, J. (2004) Determination of flow resistance caused by non-submerged woody vegetation. *International Journal of River Basin Management*, **2**: 61–70.

Jowett, I.G. (1993) A method for objectively identifying pool, run, and riffle habitats from physical measurements. *New Zealand Journal of Marine and Freshwater Research*, **27**: 241–248.

Junk, W. (1970) Investigations on the ecology and production-biology of the "floating meadows" (Paspalo-Echinochloetum) on the Middle Amazon. Part I: The floating vegetation and its ecology. *Amazoniana*, **11**: 449–495.

Kaenel, B.R. and Uehlinger, U. (1998) Effect of plant cutting and dredging on habitat conditions in streams. *Archive of Hydrobiology*, **143**: 257–273.

Kemp, J.L., Harper, D.M. and Crosa, G.A. (2000) The habitat-scale ecohydraulics of rivers. *Ecological Engineering*, **16**: 17–29.

Kemp, J.L., Harper, D.M. and Crosa, G.A. (2002) A deeper understanding of river habitat-scale ecohydraulics: Interpreting the relationship between habitat type, depth and velocity using knowledge of sediment dynamics and macrophyte growth. *Ecohydrology and Hydrobiology*, **2**: 271–282.

Kohler, A., Sipos, V., Sonntag, E., Penska, K., Pozzi, D., Veit, U. and Björk, S. (2000) Makrophyten-Verbreitung und Standortqualität im eutrophen Björka-Kövlinge-Fluss (Skane, Südschweden). *Limnologica*, **30**: 281–298.

Luhar, M., Rominger, J. and Nepf, H. (2008) Interaction between flow, transport and vegetation spatial structure. *Environmental Fluid Mechanics*, **8**: 423–439.

Luther, H. (1949) Vorschlag zu einer ökologischen Grundeinteilung der Hydrophyten. *Acta Botanica Fennica*, **44**: 1–15.

Machata-Wenninger, C. and Janauer, G.A. (1991) The measurement of current velocities in macropyhte beds. *Aquatic Botany*, **39**: 221–230.

Madsen, T.V., Enevoldens, H.O. and Jorgensen, T.B. (1993) Effects of water velocity on photosynthesis and dark respiration in submerged stream macrophytes. *Plant, Cell and Environment*, **16**: 317–322.

Mäkirinta, U. (1978) Ein neues ökomorphologisches Lebensformen-System der aquatischen Makrophyten. *Phytocoenologia*, **4**: 446–470.

Marshall, E.J.P. and Westlake, D.F. (1990) Water velocities around water plants in chalk streams. *Folia Geobotanica and Phytotaxonomica*, **25**: 279–289.

McCune, B. and Grace, J.B. (2002) *Analysis of Ecological Communities*. MjM SoftwareDesign, Gleneden Beach, OR, USA.

McCune, B. and Mefford, M.J. (2006) *PC-ORD. Multivariate Analysis of Ecological Data. Version 5.18*. MjM Software, Gleneden Beach, OR, USA.

Newson, M.D., Harper, D.M., Padmore, C.L., Kemp, J.L. and Vogel, B. (1998) A cost-effective approach for linking habitats, flow types and species requirements. *Aquatic Conservation: Marine and Freshwater Ecosystems*, **8**: 431–446.

Ormerod, S.J. (2003) Restoration in applied ecology: editor's introduction. *Journal of Applied Ecology*, **40**: 44–50.

Pasternak, G.B., Bounrisavong, M.K. and Parikh, K.K. (2008) Backwater control on riffle-pool hydraulics, fish habitat quality, and sediment transport regime in gravel-bed rivers. *Journal of Hydrology*, **357**: 125–139.

Pitlo, R.H. (1982) Flow resistance of aquatic vegetation. In *Proceedings of the EWRS 6th Symposium on Aquatic Weeds (Novi Sad)*, Dordrecht, The Netherlands, pp. 225–234.

Pitlo, R.H. and Dawson, F.H. (1993) Flow-resistance of aquatic weeds. In Pieterse, A.H. and Murphy, K.J. (eds) *Aquatic Weeds. The Ecology and Management of Nuisance Aquatic Vegetation*, Oxford University Press, pp. 74–84.

Pott, R. and Remy, D. (2000) *Gewässer des Binnenlandes*, Eugen Ulmer, Stuttgart.

Raunkiaer, C. (1934) *The Life Forms of Plants and Statistical Plant Geography*, Clarendon Press, Oxford, UK.

Reckendorfer, W., Baranyi, C., Funk, A. and Schiemer, F. (2006) Floodplain restoration by reinforcing hydrological connectivity: expected effects on aquatic mollusc communities. *Journal of Applied Ecology*, **43**: 474–484.

Reckendorfer, W., Schmalfuss, R., Baumgartner, C., Habersack, H., Hohensinner, S., Jungwirth, M. and Schiemer, F. (2005) The Integrated River Engineering Project for the free-flowing Danube in the Austrian Alluvial Zone National Park: contradictory goals and mutual solutions. *Archiv für Hydrobiologie*, **155**: 613–630.

Rice, S.P., Little, S., Wood, P.J., Moir, H.J. and Vericat, D. (2010) The relative contributions of ecology and hydraulics to ecohydraulics. *River Research and Applications*, **26**: 363–366.

Riis, T. and Biggs, B.J.F. (2003) Hydrologic and hydraulic control of macrophyte establishment and performance in streams. *Limnology and Oceanography*, **48**: 1488–1497.

Riis, T., Sand-Jensen, K. and Larsen, S.E. (2001) Plant distribution and abundance in relation to physical conditions and location within Danish stream systems. *Hydrobiologia*, **448**: 217–228.

Rohde, S., Schütz, M., Kienast, F. and Englmaier, P. (2005) River widening: an approach to restoring riparian habitats and plant species. *River Research and Applications*, **21**: 1075–1094.

Sand-Jensen, K. (1998) Influence of submerged macrophytes on sediment composition and near-bed flow in lowland streams. *Freshwater Biology*, **39**: 663–679.

Sand-Jensen, K. and Mebus, J.R. (1996) Fine-scale patterns of water velocity within macrophyte patches in streams. *Oikos*, **76**: 169–180.

Sand-Jensen, K. and Pedersen, M.F. (1999) Velocity gradients and turbulence around macrophyte stands in streams. *Freshwater Biology*, **42**: 315–328.

Sand-Jensen, K., Pedersen, M.F. and Nielsen, S.L. (1992) Photosynthetic use of inorganic carbon among primary and secondary water plants in streams. *Freshwater Biology*, **27**: 283–293.

Sarbu, A., Janauer, G.A., Schmidt-Mumm, U., Filzmoser, P., Smarandache, D. and Pascale, G. (2011) Characterisation of the potamal Danube River and the Delta: connectivity determines indicative macrophyte assemblages. *Hydrobiologia*. doi: 10.1007/s10750-011-0705-5.

Schaffer, H. (1996) *Die Verbreitung von Characeen, aquatischen Moosen und Höheren Wasserpflanzen in der österreichischen Donau (The distribution of Charceae, aquatic mosses and higher aquatic plants in the Austrian Danube)*. Diploma thesis, Vienna.

Schiemer, F., Baumgartner, C. and Tockner, K. (1999) Restoration of floodplain rivers. The 'Danube Restoration Project'. *Regulated Rivers: Research and Management*, **15**: 231–244.

Schnauder, I. and Moggridge, H.L. (2009) Vegetation and hydraulic-morphological interactions at the individual plant, patch and channel scale. *Aquatic Sciences*, **71**: 318–330.

Schratt-Ehrendorfer, L. (1999) Geobotanisch-ökologische Untersuchungen zum Indikatorwert von Wasserpflanzen und ihren Gesellschaften in Donaualtwässern bei Wien. *Stapfia*, **64**: 23–161.

Schwartz, J.S. and Herricks, E.E. (2008) Fish use of ecohydraulic-based mesohabitat units in a low-gradient Illinois stream: Implications for stream restoration. *Aquatic Conservation: Marine and Freshwater Ecosystems*, **18**: 852–866.

Sculthorpe, C.D. (1967) *The Biology of Aquatic Vascular Plants*, Edward Arnold Publishers, London.

Thomas, E.A. and Schanz, F. (1976) Beziehungen zwischen Wasserchemismus und Primärproduktion in Fließgewässern, ein Limnologisches Problem. *Vierteljahrsschrift der Naturforschenden Gesellschaft Zürich*, **121**: 309–317.

Tockner, K. and Schiemer, F. (1997) Ecological aspects of the restoration strategy for a river-floodplain system on the Danube River in Austria. *Global Ecology and Biogeography Letters*, **6**: 321–329.

U.S. Army Corps of Engineers (2010) *Ecohydraulics: Fish Passage and Guidance*. Available at: http://el.erdc.usace.army.mil/emrrp/nfs/fishpassage.html [Accessed 28 November 2010].

van de Wolfshaar, K.E., Ruizeveld de Winter, A.C., Straatsma, M.W., van den Brink, N.G.M. and de Leeuw, J.J. (2010) Estimating spawning habitat availability in flooded areas of the river Waal, the Netherlands. *River Research and Applications*, **26**: 487–498.

Vollenweider, R.A., Talling, J.F. and Westlake, D.F. (1974) *A Manual on Methods for Measuring Primary Production in Aquatic Environments.* IBP Handbook No. 12, International Biological Programme, Blackwell Scientific Publishers, Oxford, UK.

Ward, J.V., Tockner, K. and Schiemer, F. (1999) Biodiversity of floodplain river ecosystems: ecotones and connectivity. *Regulated Rivers: Research and Management*, **15**: 125–139.

Westlake, D.F. (1973) Aquatic macrophytes in rivers: a review. *Polski Archivum Hydrobiologiae*, **20**: 31–40.

Wetzel, R.G. (2001) *Limnology*, 3rd edition, Academic Press, London, pp. 541–542.

Wiegleb, G. (1991) Die Lebens- und Wuchsformen der makrophytischen Wasserpflanzen und deren Beziehungen zur Ökologie, Verbreitung und Vergesellschaftung der Arten. *Tuexenia*, **11**: 135–147.

Willby, N.J., Abernethy, V.J. and Demars, B.O.L. (2000) Attribute-based classification of European hydrophytes and its relationship to habitat utilization. *Freshwater Biology*, **43**: 43–74.

Yen, B.C. (2002) Open channel flow resistance. *Journal of Hydraulic Engineering*, **128**: 20–39.

15 Multi-Scale Macrophyte Responses to Hydrodynamic Stress and Disturbances: Adaptive Strategies and Biodiversity Patterns

Sara Puijalon and Gudrun Bornette

Université Lyon 1, UMR 5023 Ecologie des hydrosystèmes naturels et anthropisés (Université Lyon 1; CNRS; ENTPE), 43 boulevard du 11 novembre 1918, 69622 Villeurbanne Cedex, France

15.1 Introduction

Macrophytes are aquatic angiosperms, bryophytes (mosses and liverworts), some encrusting lichens and charales. They have important cascading effects on many abiotic and biotic components of freshwater ecosystems. Macrophytes are often major primary producers in freshwater ecosystems and thus play a key role in trophic chains, providing food for invertebrates and fish (Dodds, 2002; Allan and Castillo, 2007). They also serve as substrate for periphyton and as spawning habitats and refugia for macroinvertebrates, fish, amphibians, mammals and waterfowl (Pip and Robinson, 1984; Grenouillet et al., 2000; Strayer and Malcom, 2007). In lotic ecosystems, macrophytes reduce erosion, stabilize substrate, modify stream flow and trap sediments (Sand-Jensen, 1998; Champion and Tanner, 2000; Pluntke and Kozerski, 2003). They also play important roles in biogeochemical cycles of nutrients and contaminants (e.g. through organic matter production or retention of nutrients; Engelhardt and Ritchie, 2001; Kalff, 2001). However, in some cases, particularly when proliferating, macrophytes can also interfere with human activities: they hinder boating and fishing,

enhance flood risk through increased channel resistance and are sometimes considered to decrease the aesthetic value of freshwater ecosystems (Sculthorpe, 1967; Whitton, 1975; van Nes et al., 2002). Therefore, understanding the dynamics of macrophytes, especially the extinction and colonization processes, is of primary importance for maintaining the functions and biodiversity of aquatic ecosystems.

In lotic ecosystems, such as rivers and streams, flowing water induces hydrodynamic stress on macrophytes colonizing these habitats. In standing freshwater ecosystems, macrophytes may also be exposed to episodic water movements, as floods or waves. Some effects of 'flow' (i.e. discharge or the volume of water per unit time which influences water velocity, direction and turbulence) on macrophytes have been identified for many decades, particularly with the description of biodiversity patterns according to flow conditions and the preference of macrophyte species for particular water velocities (Butcher, 1933; Gessner, 1955; Hynes, 1970; Whitton, 1975; Haslam, 1978). New research questions regarding the responses of macrophytes to flow have emerged with the recent development of studies in ecohydraulics, i.e. studies linking physical

Ecohydraulics: An Integrated Approach, First Edition. Edited by Ian Maddock, Atle Harby, Paul Kemp and Paul Wood.
© 2013 John Wiley & Sons, Ltd. Published 2013 by John Wiley & Sons, Ltd.

and ecological processes. These studies aim, for instance, to investigate the hydrodynamic or biomechanical traits of macrophytes involved in their responses to flow. These interdisciplinary approaches, involving biological, ecological and physical aspects, have revealed the complexity of the responses of macrophytes to physical influences. They embrace not only hydrodynamics and biomechanics, but also ecophysiological processes, such as the mass transfer limitation of photosynthesis. Moreover, evolutionary processes are also important in understanding the ecohydraulics of macrophytes, e.g. the phenotypic plasticity (i.e. the capacity of a genotype to express different phenotypes in different environments; Bradshaw, 1965; Sultan, 2000) observed for many species in flowing water. Finally, the processes relevant to assessing macrophyte responses to flow also depend on the scales considered, from leaf or shoot to community (Biggs, 1996; Biggs *et al.*, 2005; Nikora, 2010). This chapter reviews how hydrodynamic stress and disturbance affect macrophytes and the strategies developed by macrophytes exposed to these influences. First, the macrophyte responses will be considered at the levels of *individuals* and *patches*, outlining adaptations that minimize the risk of macrophytes encountering mechanical damage. Secondly, the responses of macrophyte *communities* to hydrodynamic disturbances and associated strategies will be reviewed. Finally, the way hydrodynamic stress and disturbance affect *biodiversity* and *succession* in lotic and flood-disturbed ecosystems will be addressed.

15.2 Individual and patch-scale response to hydrodynamic stress and disturbances

Flow exerts both direct and indirect effects on macrophytes. The main direct effect is the hydrodynamic forces it imposes on rooted macrophytes. It also exerts indirect effects, particularly through its role in the transport of oxygen, inorganic carbon, nutrients, substrate particles and propagules.

15.2.1 Response traits to hydrodynamic forces

The primary hydrodynamic force imposed on macrophytes by flow is drag. Drag is the sum of pressure and skin-frictional drag. It acts horizontally in the flow direction and tends to push macrophytes downstream (Vogel, 1994; Vogel, 2003). Drag depends on water velocity and macrophyte shape, and scales with macrophyte size, particularly surface area (Figure 15.1; Vogel, 2003). The consequences of drag on macrophytes (i.e. no breakage, breakage, uprooting or dislodgement) depend on sediment cohesive strength and macrophyte mechanical properties: breaking strength is the mechanical limit at which the stem breaks, and anchorage strength is the mechanical limit that leads to root breakage or macrophyte uprooting (Figure 15.1; Schutten and Davy, 2000; Schutten *et al.*, 2005). Macrophyte capacity to avoid mechanical

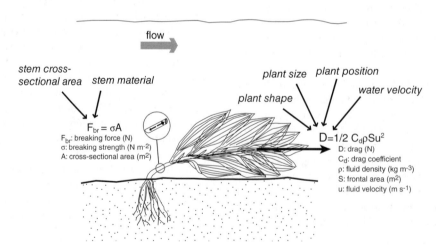

Figure 15.1 Schematic representation of a macrophyte exposed to unidirectional flow showing the force applied to the macrophyte (drag) and its resistance to breakage (breaking force). A macrophyte stem suffers mechanical failure when the stress induced by drag exceeds its tensile strength. The main factors determining drag and stem resistance to breakage are indicated in italic type. Adapted from Puijalon *et al.* (2011), by permission of John Wiley & Sons.

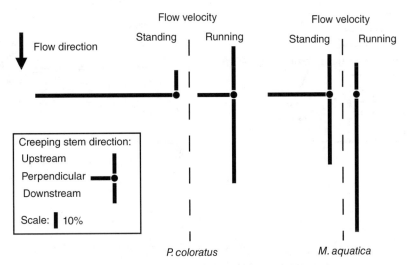

Figure 15.2 Main direction of creeping stems produced under standing versus flowing water conditions by *Potamogeton coloratus* and *Mentha aquatica*. Reproduced from Puijalon *et al.* (2008a), by permission of Oxford University Press.

damage relies, consequently, on the minimization of drag and on the maximization of breakage and anchorage strengths.

Macrophyte hydrodynamics: adaptations minimizing drag

In streams, flow is often unidirectional. Under such conditions, streamlined shapes characterized by a rounded front and an elongated rear reduce drag (Vogel, 1994; Koehl, 1996; Vogel, 2003). Streamlining is frequently observed in streams at the scale of a macrophyte patch, as a result of macrophyte development. It results from elongation and vegetative growth downstream, in the main flow direction, whereas upstream and lateral growth are reduced (Figure 15.2; Puijalon *et al.*, 2008a; Sand-Jensen and Pedersen, 2008). At the scale of individuals, streamlining is mostly due to canopy reconfiguration, i.e. changes in shape induced by forces acting on the canopy when water velocity increases (Vogel, 1984). Reconfiguration mainly results in the macrophyte bending over, being pushed parallel to flow and leaves and stems being compressed against each other into a more compact shape (Sand-Jensen, 2003). Reconfiguration can reduce drag significantly on sessile organisms (Vogel, 1984; Koehl, 1986; Vogel, 1994; Sand-Jensen, 2003; Nikora, 2010). Indeed, compaction and alignment of the canopy with the flow direction reduces the surface perpendicular to flow and lowers drag (Koehl, 1982). Moreover, bending leads to macrophyte positioning closer to the bottom where the water velocity is reduced, which also lowers drag (Koehl,

1982; Sand-Jensen, 2003). Reconfiguration may also be an important process for macrophytes that colonize still habitats episodically exposed to flowing water: it enables them to alternate a configuration maximizing photosynthesis by maximizing leaf exposition to light in standing conditions, and a streamlined configuration in flowing water (Vogel, 2003). Flexibility is the principal trait supporting reconfiguration (Schutten and Davy, 2000; Sand-Jensen, 2003; Sand-Jensen, 2008; Albayrak *et al.*, 2012) and seems, in some cases, to be an important adaptation that enables some macrophyte species to colonize fast-flowing waters (e.g. *R. fluitans*, Usherwood *et al.*, 1997).

Other morphological adaptations, such as a small size, linear strap-like or thin leaves, (e.g. *Sparganium emersum*, *Vallisneria spiralis*) are considered adaptations to flowing water, as opposed to broad, round or dissected leaves (Figure 15.3; Sculthorpe, 1967; Schutten and Davy, 2000; Sand-Jensen, 2003; Haslam, 2006; O'Hare *et al.*, 2007; Albayrak *et al.*, 2012). Flat shoots also reduce hydrodynamic forces in comparison to bushy growth forms, due to important branching or development of branches and leaves in three dimensions (Schutten and Davy, 2000; Haslam, 2006). Some species, such as *Sparganium emersum*, also form patches with open canopies that lower drag and homogenize the distribution of forces in the patch. In contrast, forces tend to be concentrated on marginal shoots in dense macrophyte patches (Sand-Jensen and Pedersen, 1999). The grouping of shoots into clumps also increases their reciprocal sheltering, resulting in reduced forces imposed on individual shoots (Haslam, 2006).

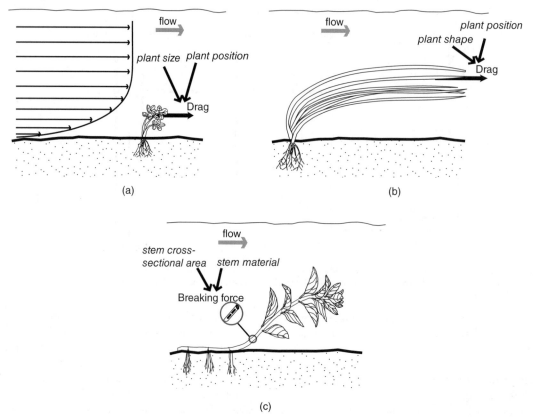

Figure 15.3 Macrophyte adaptations to flowing water: (a) dwarfing reducing drag due to scaling between drag and size and the position of the macrophyte close to the bottom; (b) linear strap-like leaves and high reconfiguration capacity, which reduce drag for a given size; and (c) high radial growth and amount of strengthening tissues which enhance stem resistance to breakage.

Macrophyte biomechanics: adaptations maximizing resistance to breakage and uprooting.

Little is known about rooting traits and mechanical properties of macrophytes (Madsen *et al.*, 2001; Schutten *et al.*, 2005; Nikora, 2010; Miler *et al.*, 2012), even if they directly determine the possible fates of drag exerted on macrophytes (Schutten *et al.*, 2005; Haslam, 2006). If stem breaking strength exceeds anchorage strength, macrophytes are more likely to suffer from uprooting than from breakage, and *vice versa* (Schutten *et al.*, 2005). These failures have different consequences for macrophyte survival, maintenance and dispersal. Indeed, shoot breakage may reduce macrophyte fitness, due to loss of biomass and meristems (Mony *et al.*, 2011), but many species are able to regenerate (Barrat-Segretain *et al.*, 1998; Barrat-Segretain and Bornette, 2000), and fragments produced by shoot breakage can disperse and be recruited (Cellot *et al.*, 1998). A weak stem, easily broken by hydrodynamic forces, could be an

adaptation protecting the below-ground organs against dislodgement and allowing both tussock regrowth and dispersal of macrophyte fragments, particularly for perennial species (Usherwood *et al.*, 1997; Schutten *et al.*, 2005; Miler *et al.*, 2012). On fine substrates easily eroded by flow, complete dislodgement occurs (Henry *et al.*, 1996), and contributes to the dispersal of whole individuals.

Some rooting systems provide high anchorage strength, such as large rooting systems consisting of horizontal stems (stolons, rhizomes) or branched roots, which may penetrate deeply into the substrate and form a tangled network that increases the area in contact with the substrate, the sediment cohesive strength and the resistance to uprooting (Sculthorpe, 1967; Schutten *et al.*, 2005; Read and Stokes, 2006). Few macrophyte species have adaptations that enable attachment to bedrock or cobbles: some bryophyte species (e.g. *Fontinalis antipyretica*) and the angiosperm family Podostemaceae attach firmly to coarse sediment and rocks through rhizoids or rhizoid-like roots (Englund, 1991; Cushing and Allan, 2001; Madsen *et al.*, 2001).

The biomechanical properties of macrophyte shoots probably differ highly from terrestrial plants or macroalgae. Indeed, many specific anatomical traits characterize macrophytes, such as the small amount of strengthening tissues, the presence of large lacunar systems and the endodermis-like structure in stems of many species (Wetzel, 1983; Raven, 1996; Rascio, 2002; Bociag *et al.*, 2009). Due to the high flexibility and consequent alignment of macrophytes in the flow direction, their shoots are mostly subjected to tensile forces induced by hydrodynamic forces (Usherwood *et al.*, 1997; Biehle *et al.*, 1998; Schutten and Davy, 2000; Schutten *et al.*, 2005). Consequently, a high tensile strength represents an important adaptation to running waters (Figure 15.3; Brewer and Parker, 1990; Usherwood *et al.*, 1997; Madsen *et al.*, 2001; Puijalon *et al.*, 2011; Miler *et al.*, 2012). The central position of strengthening tissues often observed in macrophyte stems (Whitton, 1975; Wetzel, 1983) may enhance their tensile strength (Koehl, 1982; Usherwood *et al.*, 1997; Niklas, 1999). Other structural elements usually considered to provide resistance to mechanical forces, such as silica, lignified tissues and collenchyma, may also be involved, but their role remains unclear (Eichhorn and Evert, 2006; Bociag *et al.*, 2009; Schoelynck *et al.*, 2010).

Figure 15.4 Plastic response to hydrodynamic stress (permanent exposure to current) observed for *Berula erecta*: plant dwarfing, increased allocation to below-ground parts or reduction of leaflet number.

Plastic responses to stress

Many macrophyte species do not present particular adaptations to fast water velocities, but many of them have the capacity to adapt to moderate flow stress through thigmomorphogenetic responses. Thigmomorphogenesis represents the developmental responses of sessile organisms (e.g. mosses, ferns, lichens and angiosperms) to external mechanical stimulation (Braam, 2005; Telewski, 2006). Macrophytes often present high levels of phenotypic plasticity (Santamaria, 2002), and thigmomorphogenetic responses can lead to important alterations in growth and resulting morphologies of individuals. These responses, identified a long time ago (Welch, 1935; Gessner, 1955; Sculthorpe, 1967), were frequently considered to improve resistance to current, but the adaptive consequences of these morphological changes have rarely been tested (Biehle *et al.*, 1998; Puijalon *et al.*, 2005; Puijalon *et al.*, 2008b).

Exposure to flow induces a reduction in growth for some species, leading to small or even dwarfed individuals (Figure 15.4; Whitton, 1975; Puijalon and Bornette, 2004;

Puijalon *et al.*, 2005). Such responses lead to a reduction in drag, as individuals partly escape current stress through their small size (drag scales with size) and their position close to the substrate where velocity is reduced (Vogel, 2003; Puijalon and Bornette, 2004; Puijalon *et al.*, 2005). Increased allocations to below-ground organs and to strengthening tissues observed in response to current stress probably result in enhanced resistance to breakage and uprooting (Biehle *et al.*, 1998; Fritz *et al.*, 2004; Puijalon *et al.*, 2005; Puijalon and Bornette, 2006; Bociag *et al.*, 2009). Several other responses, including reduced internode length, number of leaflets per leaf, leaf size and specific leaf area, insertion angle of leaves and dry matter content of organs, have also been observed (Gessner, 1955; Biehle *et al.*, 1998; Boeger and Poulson, 2003; Puijalon and Bornette, 2004; Puijalon and Bornette, 2006), but their adaptive value has not always been assessed. Eventually, exposure to current often alters macrophyte reproduction (both sexual reproduction and vegetative multiplication), generally reducing or even impeding sexual reproduction (decreased flower, fruit and/or seed production; Cushing and Allan, 2001; Puijalon et al., 2008b).

15.2.2 Indirect effects of flow on macrophytes

Flow also exerts indirect effects on macrophytes through its effect on nutrient supply, erosion and transport of sediment and dispersal of seeds and vegetative fragments. The most important of these is due to the modification of nutrient supply to macrophytes and, hence, metabolic processes and growth induced by water movement (Koch, 2001; Madsen et al., 2001; Biggs et al., 2005). The macrophyte surface is surrounded by a diffusive boundary layer, where transport is dominated by diffusion rather than convection. This layer extends between the surface, where the solute uptake occurs (nutrients, CO_2) and the free stream (Koch, 2001; Madsen et al., 2001; Vogel, 2003). The fluxes of molecules through this layer depend on its thickness and on the concentration and diffusion coefficients of molecules (Koch, 1994; Koch, 2001; Madsen et al., 2001). A thick boundary layer may limit the flux of CO_2 and nutrients to macrophytes and hence limit photosynthesis (Westlake, 1967; Smith and Walker, 1980; Kalff, 2001). Increasing water velocity reduces the thickness of the boundary layer, resulting in enhanced CO_2 and nutrient supply and enhanced photosynthesis (Westlake, 1967; Smith and Walker, 1980; Koch, 2001; Madsen et al., 2001). This effect of water stirring can result, for some species, in higher growth in moderate-flowing compared to standing water, despite the mechanical stress induced by flow (Whitton, 1975; Crossley et al., 2002; Puijalon et al., 2007).

Flow also influences water chemistry by mixing the nutrients in the water column, and increases the concentration of CO_2 and other dissolved gases (Cushing and Allan, 2001; Allan and Castillo, 2007). Some species are able to use only dissolved carbon dioxide but not bicarbonate as a source of inorganic carbon (Maberly and Madsen, 1998; Demars and Tremolieres, 2009), and are indirectly favoured by running waters and turbulent flow (Hynes, 1970; Cushing and Allan, 2001; Allan and Castillo, 2007).

Flow also affects macrophytes indirectly through sediment sorting (fine sediments are washed away) and particle resuspension (Rostan et al., 1987; Bornette et al., 1994b; Schwarz et al., 1996). The erosion of fine sediment results in coarse sediment, which can impede the rooting of many macrophyte species, or reduce their development because of low nutrient availability (Chambers et al., 1991;

Bornette et al., 1994a; Bornette et al., 1998; Madsen et al., 2001). Moderate siltation processes tend to favour macrophyte growth due to inputs of cohesive, nutrient-rich particles (Bornette and Large, 1995), but excessive deposition can lead to eutrophication, sediment anoxia and macrophyte burial (Bornette et al., 1998; Madsen et al., 2001; Bornette et al., 2008).

The competitive interaction between macrophytes and periphyton is also affected by flow. Periphyton (diatoms, algae) competes with macrophytes for light and nutrients (Horner et al., 1990; Koch, 2001). Slow water velocity seems to favour colonization of macrophyte shoots by epiphytic algae, whereas higher velocities reduce epiphytic development, thus promoting macrophyte growth. Finally, water movements are a major agent for dispersal of seeds and vegetative fragments of macrophytes (Boedeltje et al., 2003; Boedeltje et al., 2004; Riis and Sand-Jensen, 2006; Gurnell et al., 2008).

15.3 Community responses to temporary peaks of flow and current velocity

Spate floods, characterized by brief peaks of flow and water velocity, are disturbances, i.e. discrete and unpredictable events that cause partial or total destruction of living communities (White and Jentsch, 2001; Grime, 2001). Flood disturbances may scour substrate (Henry et al., 1996; Matthaei et al., 2003) or deposit sediment of various size classes (Shields and Abt, 1989; Schwarz et al., 1996; Olde Venterink et al., 2006). The occurrence, diversity and life-history traits of macrophytes that occur along such river floodplains depend strongly on the nature and the intensity of the flood events, as floods break or dislodge macrophytes, and select for species tolerant of such events or those able to recolonize efficiently disturbed patches (Sparks et al., 1990; Bornette et al., 1994c; Henry and Amoros, 1996; Henry et al., 1996; Bornette et al., 2008).

15.3.1 Highly disturbed communities

The resistance versus resilience of macrophyte communities to spate floods depends on the intensity and frequency of such events. When the frequency and/or intensity of flood disturbances increases, it increasingly controls plant strategies.

In heavily eroded situations, coarse substrates dominate and, consequently, few refuges occur. Community maintenance in such situations relies, consequently, on resilience, whereas resistance is low. In such situations,

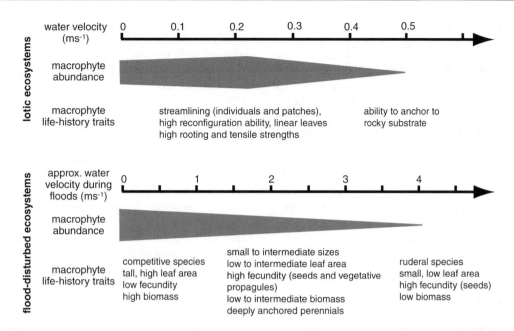

Figure 15.5 Macrophyte abundance and plant strategies according to either hydrodynamic stress or disturbances that affect freshwater ecosystems.

the opportunistic strategy is selected and the traits allowing macrophyte maintenance are usually grouped under the 'ruderal' or 'r' strategy (Figure 15.5; Southwood, 1988; Grime, 2001). Annual species regrowing from diaspores or light seeds with high dispersibility may establish (e.g. some *Potamogeton* and *Callitriche* species and *Zannichellia palustris*) and reach high cover in the newly scoured areas (Bornette and Arens, 2002; Combroux and Bornette, 2004). Charophytes are also pioneer species, which usually bloom after disturbance (Bornette and Arens, 2002), suggesting that the disturbance itself induces oospore germination. During floods, the abrasive effects of sediment movement can break cuticular dormancy (e.g. *Luronium natans* that show increasing germination if scarified, pers. obs., *Potamogeton pectinatus* or *Potamogeton pusillus*, Teltscherova and Hejny, 1973).

15.3.2 Intermediately disturbed communities

When the intensity and frequency of spates decrease, some refugia occur and disturbed patches are partly recolonized by seeds and propagules from the soil reservoir and the adjacent preserved areas (Henry *et al.*, 1996). Many macrophytes have few or no specific morphological adaptations allowing them to resist disturbances (Bor-

nette *et al.*, 1994c), and their maintenance relies mainly on their anchorage strength (deep anchorage, Henry *et al.*, 1996; Riis and Biggs, 2003) and their ability to recolonize the disturbed patch immediately after the disturbance by border effect (Figure 15.5; Barrat-Segretain, 1996).

Traits linked to vegetative and to sexual reproduction are, however, involved differently in the recolonization process, depending on patch size (Miller, 1982; Belsky, 1986). Large patches (i.e. patches that are usually generated by a high frequency and/or intensity of disturbances) are colonized mainly by seeds with high dispersal ability, whereas the edge effect tends to be low (Vandvik, 2004).

Water propagule dispersal is frequent, and increases the opportunity for propagules to reach gaps immediately after the disturbance, but it requires buoyancy of propagules (Andersson *et al.*, 2000; Boedeltje *et al.*, 2004; Riis and Sand-Jensen, 2006). For example, seeds of *Alisma plantago-aquatica*, *Carex flava*. and *Hippuris vulgaris* are able to float for > 1.5 years (Praeger, 1913). In the same way, fragments of several species that colonize disturbed habitats are able to float for several weeks (Barrat-Segretain *et al.*, 1998). Vegetative regeneration is a key function for the maintenance of species subjected to recurrent disturbance, particularly in infertile situations (Bellingham and Sparrow, 2000; Klimesova and Klimes, 2007). The role of clonal growth in species maintenance after disturbances through survival of deeply anchored roots or rhizomes, spreading from refuges or sprouting from vegetative propagules has been described (Henry *et al.*, 1996; Barsoum, 2001). This capability of regrowth may be facilitated by the production of adventitious roots that utilize nutrients in alluvial material deposited by floods, allowing for rapid rooting of fragments, as observed for terrestrial habitats (Hupp and Osterkamp, 1996). Conversely, seeds and fragments of macrophyte species that colonize undisturbed habitats tend to have low buoyancy: seeds of *Baldellia ranunculoides*, *Oenanthe fistulosa*, *Oenanthe aquatica* and *Nuphar lutea* (I. Combroux and G. Bornette, pers. obs.) or fragments of *Potamogeton coloratus* sink immediately or very soon after release (Barrat-Segretain *et al.*, 1999).

15.4 Macrophyte abundance, biodiversity and succession

15.4.1 Lotic ecosystems

In lotic ecosystems, macrophyte cover varies greatly with local conditions, but generally covers a small proportion of the streambed (Hynes, 1970; French and Chambers, 1996; Allan and Castillo, 2007). Macrophyte abundance was found to decrease (Chambers *et al.*, 1991; Grinberga, 2010) or increase (Ham *et al.*, 1981; Suren and Duncan, 1999) with increasing water velocity or to peak for intermediate velocities around 0.2–0.4 ms^{-1} (Figure 15.5; Nilsson, 1987; Riis and Biggs, 2003). The discrepancies between the patterns observed may result from the scale of the studies (patch-scaled or ecosystem-scaled studies), the description of hydraulic environment of macrophytes (water velocity or shear stress calculation) and from the nutrient level of ecosystems. Nutrients favour productivity, oxygen demand for the mineralization of organic matter, oxygen depletion and accumulation of toxic compounds around plant roots (Pezeshki, 2001). As a consequence, species tolerant to ammonium eutrophication may be favoured and plant growth may be disadvantaged until water velocity reaches values that limit anoxia and toxic compound accumulation at the water–substrate boundary. Furthermore, increasing current enhances nutrient and inorganic carbon uptake and hence photosynthesis (positive effect on macrophyte development), but also increases hydrodynamic stresses encountered by individuals (negative effect on macrophyte development; Riis and Biggs, 2003; Lacoul and Freedman, 2006; Nikora, 2010). For moderate velocities, the enhanced resource supply in macrophytes stimulates

their growth and stand development. Above a velocity threshold, sediment is eroded and hydrodynamic forces increase greatly, possibly inhibiting photosynthesis and growth and inducing tissue loss, which result in reduced stand development.

The number of species colonizing flowing water habitats is lower than species colonizing standing ones (Gessner, 1955; Hynes, 1970). Macrophyte richness and diversity are generally higher for low to moderate velocity (< 0.2–0.3 ms^{-1}), decrease for higher velocities, and are very low for fast-flowing conditions (French and Chambers, 1996; Franklin *et al.*, 2008; Grinberga, 2010; Janauer *et al.*, 2010). Macrophytes are usually not present under very fast water velocity (> 1 ms^{-1}; Whitton, 1975; Chambers *et al.*, 1991; Franklin *et al.*, 2008; Janauer *et al.*, 2010).

Macrophytes present contrasting preferences for flow conditions (Mackay *et al.*, 2003; Riis and Biggs, 2003; Haslam, 2006; Grinberga, 2010; Janauer *et al.*, 2010), which result in a general pattern of longitudinal distribution of macrophytes along a river continuum from headwater and upper reaches of streams characterized by high velocity and turbulent flow, stable, rocky substrates to lowland reaches characterized by lower velocity and finer sediments (Cushing and Allan, 2001; Allan and Castillo, 2007). The adaptations of macrophytes to velocity generally do not enable them to colonize very fast-flowing areas, i.e. current velocity above 1–1.5 ms^{-1} (Santamaria, 2002; Lacoul and Freedman, 2006; Franklin *et al.*, 2008). Such conditions often impede species colonization and maintenance due to the high hydrodynamic forces but also to the lack of suitable substrate for rooting (Madsen *et al.*, 2001; Handley and Davy, 2002; Lacoul and Freedman, 2006). Bryophytes are abundant in fast-flowing areas (> 0.6 ms^{-1}), particularly upper reaches, headwater streams and springs, because they are able to attach to rocky substrates and because some species are restricted to areas providing a good supply of carbon dioxide (Whitton, 1975; French and Chambers, 1996; Cushing and Allan, 2001; Scarlett and O'Hare, 2006; Allan and Castillo, 2007). Fast-flowing habitats and upland areas are mostly dominated by bryophytes and a few angiosperm species (e.g. some *Ranunculus* or *Myriophyllum* species; French and Chambers, 1996; Passauer *et al.*, 2002; Scarlett and O'Hare, 2006; Grinberga, 2010). Bryophyte species abundance and richness usually decrease along the upland–lowland gradient, while angiosperm richness increases (Passauer *et al.*, 2002; Scarlett and O'Hare, 2006; Grinberga, 2010). Habitats with low velocities (< 0.3 ms^{-1}) and fine substrate, often observed in mid-reaches, are characterized by richer macrophyte communities consisting of freshwater angiosperms such as *Potamogeton* species, *Sparganium emersum, Hippuris vulgaris, Elodea* sp. (Hynes, 1970; French and Chambers, 1996; Grinberga, 2010). Lower reaches, characterized by the lowest velocities (< 0.1 ms^{-1}) and fine and nutrient-rich substrates, are usually characterized by emergent and floating-leaved species (Hynes, 1970; Whitton, 1975; Allan and Castillo, 2007; Grinberga, 2010).

15.4.2 Flood-disturbed ecosystems

Species richness in aquatic habitats is also ruled by hydrodynamic disturbances, occurring both at small (local rearrangement of sediment; Englund, 1991; Suren and Duncan, 1999) and large scales (flood events).

High frequency and/or intensity of scouring floods usually leads to a decrease in, and even the absence of, macrophytes, because they are unable to recover before the next disturbance, and because of the lack of favourable habitats (Figure 15.5; Bornette and Amoros, 1991; Bornette *et al.*, 1994b; Kotschy and Rogers, 2008). Natural successional processes that occur in such ecosystems are stopped (Sparks *et al.*, 1990). According to the intermediate disturbance hypothesis (Connell, 1978; Sousa, 1984; Huston and Smith, 1987), for an intermediate frequency and/or intensity of flood, species richness reaches a maximum value and a dynamic equilibrium of macrophyte communities is sustained, maintaining a highly diverse, shifting mosaic of species (Bornette and Amoros, 1991; Roberts and Ludwig, 1991; Bornette *et al.*, 1998; Pollock *et al.*, 1998). Succession is slowed down and even stopped (Bornette *et al.*, 1994a; Bornette *et al.*, 1994b), and such disturbances allow for the maintenance of poorly competitive ruderal species (Bornette and Large, 1995; Greulich *et al.*, 2000). Low frequency and/or intensity of flood scouring is unable to impede successional processes completely, and natural accumulation of organic matter leads to the terrestrialization of aquatic ecosystems and the progressive replacement of species linked to flood scouring and groundwater supplies (e.g. *Callitriche* sp., *Groenlandia densa, Elodea canadensis,* Charophytes) by species of stable clogged habitats (e.g. *Nuphar lutea, Nymphea alba*) and finally by helophytes (e.g. *Carex pseudocyperus, Phragmites australis*; Amoros and Bornette, 1999; Amoros *et al.*, 2000).

15.5 Conclusion

Permanent or stochastic hydrodynamic forces are key factors that determine macrophyte growth, life-history traits, dispersal and biodiversity in aquatic ecosystems. Some traits involved in macrophyte resistance and adaptation to flow stress and disturbance have been identified at the individual and patch scale but many traits, such as those linked to biomechanics, anchorage and regeneration, should be considered more specifically. Studies that investigated the role of these traits either focused on a small number of species, impeding generalizations of the trends observed, or they did not encompass a sufficiently large gradient of physical constraints. Future developments in ecohydraulics should also integrate interactions between flow and other abiotic factors, particularly nutrient level, as they may modify macrophyte responses to flow stress and disturbance, both at individual and community levels.

Hydrodynamics also control the functioning of aquatic ecosystems, directly through the erosion and breakage of macrophyte communities and substrate and the leaching of nutrients, but also indirectly, through the selection of species potentially affecting the bio-geochemical fluxes in the ecosystems. This external supply of energy acts, consequently, as major auxiliary energy (*sensu* Margalef, 1974) for macrophytes, promoting matter and energy fluxes in running ecosystems (for instance, by increasing macrophyte dispersal or nutrient fluxes through the reduction of the boundary layer), regulating productivity and biodiversity, both specific and functional. Future studies in ecohydraulics should investigate how alterations of these external energy forces, for instance due to river regulation, may affect biodiversity, ecosystem functioning and/or nutrient cycling (e.g. by decreasing substrate oxidation, organic matter mineralization and macrophyte growth), and the long-term dynamics of biodiversity (e.g. through decreased diaspore fluxes) in riverine ecosystems.

Conversely, macrophyte species may act differently on the functioning of ecosystems, depending on whether they are submitted to flow stress or not. For example, depending on the conditions where they develop (flowing vs. standing), macrophytes may present contrasting degradability (due to different allocation to structural elements), effects on flow and sedimentation (due to different hydrodynamic traits) and impact on substrate oxidation and food sources for invertebrates (due to the development of an anchorage system presenting contrasting traits). All these functional consequences of macrophyte responses to flow are, as yet, still poorly understood, but may strongly contribute to the functioning and biodiversity of aquatic ecosystems. A major challenge of ecohydraulics will be to study interactions between macrophytes and these ecological, hydrological and geomorphological processes that can be modified not only by macrophytes considered as fixed structures, but also by their *responses* to flow.

References

Albayrak, I., Nikora, V., Miler, O. and O'Hare, M. (2012) Flow–plant interactions at a leaf scale: effects of leaf shape, serration, roughness and flexural rigidity. *Aquatic Sciences*, **74**: 267–286.

Allan, J.D. and Castillo, M.M. (2007) *Stream Ecology: Structure and Function of Running Waters*, Springer-Verlag, Dordrecht, The Netherlands.

Amoros, C. and Bornette, G. (1999) Antagonist and cumulative effects of connectivity: a predictive model based on aquatic vegetation in riverine wetlands. *Archiv für Hydrobiologie*, **115**(3): 311–327.

Amoros, C., Bornette, G. and Henry, C.P. (2000) Environmental auditing. A vegetation-based method for ecological diagnosis of riverine wetlands. *Environmental Management*, **25**: 211–227.

Andersson, E., Nilsson, C. and Johansson, M.E. (2000) Plant dispersal in boreal rivers and its relation to the diversity of riparian flora. *Journal of Biogeography*, **27**: 1095–1106.

Barrat-Segretain, M.H. (1996) Strategies of reproduction, dispersion, and competition in river plants: A review. *Vegetatio*, **123**: 13–37.

Barrat-Segretain, M.H. and Bornette, G. (2000) Regeneration and colonization abilities of aquatic plant fragments: effect of disturbance seasonality. *Hydrobiologia*, **421**: 31–39.

Barrat-Segretain, M.H., Bornette, G. and Hering-Vilas-Bôas, A. (1998) Comparative abilities of vegetative regeneration among aquatic plants growing in disturbed habitats. *Aquatic Botany*, **60**: 201–211.

Barrat-Segretain, M.H., Henry, C.P. and Bornette, G. (1999) Regeneration and colonization of aquatic plant fragments in relation to the distubance frequency of their habitats. *Archiv für Hydrobiologie*, **145**: 111–127.

Barsoum, N. (2001) Relative contributions of sexual and asexual regeneration strategies in *Populus nigra* and *Salix alba* during the first years of establishment on a braided gravel bed river. *Evolutionary Ecology*, **15**: 255–279.

Bellingham, P.J. and Sparrow, A.D. (2000) Resprouting as a life history strategy in woody plant communities. *Oikos*, **89**: 409–416.

Belsky, A.J. (1986) Revegetation of artificial disturbances in grasslands of the Serengeti-National-Park, Tanzania. 2. Five years of successional change. *Journal of Ecology*, **74**: 937–951.

Biehle, G., Speck, T. and Spatz, H.C. (1998) Hydrodynamics and biomechanics of the submerged water moss *Fontinalis antipyretica* – a comparison of specimens from habitats with different flow velocities. *Botanica Acta*, **111**: 42–50.

Biggs, B.J.F. (1996) Hydraulic habitat of plants in streams. *Regulated Rivers: Research and Management*, **12**: 131–144.

Biggs, B.J.F., Nikora, V.I. and Snelder, T.H. (2005) Linking scales of flow variability to lotic ecosystem structure and function. *River Research and Applications*, **21**: 283–298.

Bociag, K., Galka, A., Lazarewicz, T. and Szmeja, J. (2009) Mechanical strength of stems in aquatic macrophytes. *Acta Societatis Botanicorum Poloniae*, **78**: 181–187.

Boedeltje, G., Bakker, J.P., Bekker, R.M., Van Groenendael, J.M. and Soesbergen, M. (2003) Plant dispersal in a lowland stream in relation to occurrence and three specific life-history traits of the species in the species pool. *Journal of Ecology*, **91**: 855–866.

Boedeltje, G., Bakker, J.P., Ten Brinke, A., Van Groenendael, J.M. and Soesbergen, M. (2004) Dispersal phenology of hydrochorous plants in relation to discharge, seed release time and buoyancy of seeds: the flood pulse concept supported. *Journal of Ecology*, **92**: 786–796.

Boeger, M.R.T. and Poulson, M.E. (2003) Morphological adaptations and photosynthetic rates of amphibious *Veronica anagallis-aquatica* L. (Scrophulariaceae) under different flow regimes. *Aquatic Botany*, **75**: 123–135.

Bornette, G. and Amoros, C. (1991) Aquatic vegetation and hydrology of a braided river floodplain. *Journal of Vegetation Science*, **2**: 497–512.

Bornette, G. and Arens, M.F. (2002) Charophyte communities in cut-off river channels – The role of connectivity. *Aquatic Botany*, **73**: 149–162.

Bornette, G. and Large, A.R.G. (1995) Groundwater–surface water ecotones at the upstream part of confluences in former river channels. *Hydrobiologia*, **310**: 123–137.

Bornette, G., Amoros, C. and Chessel, D. (1994b) Effect of allogenic processes on successional rates in former river channels. *Journal of Vegetation Science*, **5**: 237–246.

Bornette, G., Amoros, C. and Lamouroux, N. (1998) Aquatic plant diversity in riverine wetlands: the role of connectivity. *Freshwater Biology*, **39**: 267–283.

Bornette, G., Amoros, C., Castella, C. and Beffy, J.L. (1994a) Succession and fluctuation in the aquatic vegetation of two former Rhône River channels. *Vegetatio*, **110**: 171–184.

Bornette, G., Henry, C., Barrat, M.H. and Amoros, C. (1994c) Theoretical habitat templets, species traits, and species richness – Aquatic macrophytes in the Upper Rhône River and its floodplain. *Freshwater Biology*, **31**: 487–505.

Bornette, G., Tabacchi, E., Hupp, C., Puijalon, S. and Rostan, J.-C. (2008) A model of plant strategies in fluvial hydrosystems. *Freshwater Biology*, **53**: 1692–1705.

Braam, J. (2005) In touch: plant responses to mechanical stimuli. *New Phytologist*, **165**: 373–389.

Bradshaw, A.D. (1965) Evolutionary significance of phenotypic plasticity in plants. *Advances in Genetics*, **13**: 115–155.

Brewer, C.A. and Parker, M. (1990) Adaptations of macrophytes to life in moving water: upslope limits and mechanical properties of stems. *Hydrobiologia*, **194**: 133–142.

Butcher, R.W. (1933) Studies on the ecology of rivers. I. On the distribution of macrophytic vegetation in the rivers of Britain. *Journal of Ecology*, **21**: 58–91.

Cellot, B., Mouillot, F. and Henry, C.P. (1998) Flood drift and propagule bank of aquatic macrophytes in a riverine wetland. *Journal of Vegetation Science*, **9**: 631–640.

Chambers, P.A., Prepas, E.E., Hamilton, H.R. and Bothwell, M.L. (1991) Current velocity and its effect on aquatic macrophytes in flowing waters. *Ecological Applications*, **1**: 249–257.

Champion, P.D. and Tanner, C.C. (2000) Seasonality of macrophytes and interaction with flow in a New Zealand lowland stream. *Hydrobiologia*, **441**: 1–12.

Combroux, I.C.S. and Bornette, G. (2004) Propagule banks and regenerative strategies of aquatic plants. *Journal of Vegetation Science*, **15**: 13–20.

Connell, J.H. (1978) Diversity in tropical rain forests and coral reefs. *Science*, **199**: 1302–1310.

Crossley, M.N., Dennison, W.C., Williams, R.R. and Wearing, A.H. (2002) The interaction of water flow and nutrients on aquatic plant growth. *Hydrobiologia*, **489**: 63–70.

Cushing, C.E. and Allan, J.D. (2001) *Streams: Their Ecology and Life*, Academic Press, San Diego, California, USA.

Demars, B.O.L. and Tremolieres, M. (2009) Aquatic macrophytes as bioindicators of carbon dioxide in groundwater fed rivers. *Science of the Total Environment*, **407**: 4752–4763.

Dodds, W.K. (2002) *Freshwater Ecology: Concepts and Environmental Applications*, Academic Press, London, UK.

Eichhorn, S.E. and Evert, R.F. (2006) *Esau's Plant Anatomy: Meristems, Cells, and Tissues of the Plant Body – Their Structure, Function, and Development*. John Wiley & Sons, Inc., Hoboken, New Jersey.

Engelhardt, K.A.M. and Ritchie, M.E. (2001) Effects of macrophyte species richness on wetland ecosystem functioning and services. *Nature*, **411**: 687–689.

Englund, G. (1991) Effects of disturbance on stream moss and invertebrate community structure. *Journal of the North American Benthological Society*, **10**: 143–153.

Franklin, P., Dunbar, M. and Whitehead, P. (2008) Flow controls on lowland river macrophytes: A review. *Science of the Total Environment*, **400**: 369–378.

French, T.D. and Chambers, P.A. (1996) Habitat partitioning in riverine macrophyte communities. *Freshwater Biology*, **36**: 509–520.

Fritz, K.M., Evans, M.A. and Feminella, J.W. (2004) Factors affecting biomass allocation in the riverine macrophyte *Justicia americana*. *Aquatic Botany*, **78**: 279–288.

Gessner, F. (1955) *Hydrobotanik, Die Physiologischen Grundlagen der Pflanzenverbreitung in Wasser. I. Energiehaushalt*. VEB Deutscher Verlag der Wissenschaften, Berlin.

Grenouillet, G., Pont, D. and Olivier, J.M. (2000) Habitat occupancy patterns of juvenile fishes in a large lowland river: interactions with macrophytes. *Archiv für Hydrobiologie*, **149**: 307–326.

Greulich, S., Bornette, G. and Amoros, C. (2000) Maintenance of a rare aquatic species through gradients of disturbance and sediment richness. *Journal of Vegetation Science*, **11**: 415–424.

Grime, J.P. (2001) *Plant Strategies, Vegetation Processes and Ecosystem Properties*, John Wiley & Sons, Ltd, Chichester, UK.

Grinberga, L. (2010) Environmental factors influencing the species diversity of macrophytes in middle-sized streams in Latvia. *Hydrobiologia*, **656**: 233–241.

Gurnell, A., Thompson, K., Goodson, J. and Moggridge, H. (2008) Propagule deposition along river margins: linking hydrology and ecology. *Journal of Ecology*, **96**: 553–565.

Ham, S.F., Wright, J.F. and Berrie, A.D. (1981) Growth and recession of aquatic macrophytes on an unshaded section of the River Lambourn, England, from 1971 to 1976. *Freshwater Biology*, **11**: 381–390.

Handley, R.J. and Davy, A.J. (2002) Seedling root establishment may limit *Najas marina* L. to sediments of low cohesive strength. *Aquatic Botany*, **73**: 129–136.

Haslam, S.M. (1978) *River Plants. The Macrophytic Vegetation of Watercourses*, Cambridge University Press, Cambridge, UK.

Haslam, S.M. (2006) *River Plants. The Macrophytic Vegetation of Watercourses*, Forrest Text, Cardigan, UK.

Henry, C.P. and Amoros, C. (1996) Are the banks a source of recolonization after disturbance: An experiment on aquatic vegetation in a former channel of the Rhone river. *Hydrobiologia*, **330**: 151–162.

Henry, C.P., Amoros, C. and Bornette, G. (1996) Species traits and recolonization processes after flood disturbances in riverine macrophytes. *Vegetatio*, **122**: 13–27.

Horner, R.R., Welch, E.B., Seeley, M.R. and Jacoby, J.M. (1990) Responses of periphyton to changes in current velocity, suspended sediment and phosphorus concentration. *Freshwater Biology*, **24**: 215–232.

Hupp, C.R. and Osterkamp, W.R. (1996) Riparian vegetation and fluvial geomorphic processes. *Geomorphology*, **14**: 277–295.

Huston, M. and Smith, T. (1987) Plant succession: life history and competition. *American Naturalist*, **130**: 168–198.

Hynes, H.B.N. (1970) *The Ecology of Running Waters*, Liverpool University Press, Liverpool, UK.

Janauer, G.A., Schmidt-Mumm, U. and Schmidt, B. (2010) Aquatic macrophytes and water current velocity in the Danube River. *Ecological Engineering*, **36**: 1138–1145.

Kalff, J. (2001) *Limnology: Inland Water Ecosystems*, Prentice Hall, Upper Saddle River, NJ, USA.

Klimesova, J. and Klimes, L. (2007) Bud banks and their role in vegetative regeneration – A literature review and proposal for simple classification and assessment. *Perspectives in Plant Ecology Evolution and Systematics*, **8**: 115–129.

Koch, E.W. (1994) Hydrodynamics, diffusion-boundary layers and photosynthesis of the seagrasses *Thalassia testudinum* and *Cymodocea nodosa*. *Marine Biology*, **118**: 767–776.

Koch, E.W. (2001) Beyond light: Physical, geological, and geochemical parameters as possible submersed aquatic vegetation habitat requirements. *Estuaries*, **24**: 1–17.

Koehl, M.A.R. (1982) The interaction of moving water and sessile organisms. *Scientific American*, **247**: 110–120.

Koehl, M.A.R. (1986) Seaweeds in moving water: Form and mechanical function. In Givnish, T.J. (ed.) *On the Economy of Plant Form and Function*, Cambridge University Press, Cambridge, UK.

Koehl, M.A.R. (1996) When does morphology matter? *Annual Review of Ecology and Systematics*, **27**: 501–542.

Kotschy, K. and Rogers, K. (2008) Reed clonal characteristics and response to disturbance in a semi-arid river. *Aquatic Botany*, **88**: 47–56.

Lacoul, P. and Freedman, B. (2006) Environmental influences on aquatic plants in freshwater ecosystems. *Environmental Reviews*, **14**: 89–136.

Maberly, S.C. and Madsen, T.V. (1998) Affinity for CO_2 in relation to the ability of freshwater macrophytes to use HCO_3^-. *Functional Ecology*, **12**: 99–106.

Mackay, S.J., Arthington, A.H., Kennard, M.J. and Pusey, B.J. (2003) Spatial variation in the distribution and abundance of submersed macrophytes in an Australian subtropical river. *Aquatic Botany*, **77**: 169–186.

Madsen, J.D., Chambers, P.A., James, W.F., Koch, E.W. and Westlake, D.F. (2001) The interaction between water movement, sediment dynamics and submersed macrophytes. *Hydrobiologia*, **444**: 71–84.

Margalef, R. (1974) *Ecologia*, Omega, Barcelona.

Matthaei, C.D., Guggelberger, C. and Huber, H. (2003) Local disturbance history affects patchiness of benthic river algae. *Freshwater Biology*, **48**: 1514–1526.

Miler, O., Albayrak, I., Nikora, V. and O'Hare, M. (2012) Biomechanical properties of aquatic plants and their effects on plant–flow interactions in streams and rivers. *Aquatic Sciences*, **74**: 31–44.

Miller, T.E. (1982) Community diversity and interactions between the size and frequency of disturbance. *American Naturalist*, **120**: 533–536.

Mony, C., Puijalon, S. and Bornette, G. (2011) Response of clonal plants to disturbances: does resprouting pattern determine ecological niche? *Folia Geobotanica*, **46**: 155–164.

Niklas, K.J. (1999) Research review. A mechanical perspective on foliage leaf form and function. *New Phytologist*, **143**: 19–31.

Nikora, V. (2010) Hydrodynamics of aquatic ecosystems: an interface between ecology, biomechanics and environmental fluid mechanics. *River Research and Applications*, **26**: 367–384.

Nilsson, C. (1987) Distribution of stream-edge vegetation along a gradient of current velocity. *Journal of Ecology*, **75**: 513–522.

O'Hare, M.T., Hutchinson, K.A. and Clarke, R.T. (2007) The drag and reconfiguration experienced by five macrophytes from a lowland river. *Aquatic Botany*, **86**: 253–259.

Olde Venterink, H., Vermaat, J.E., Pronk, M., Wiegman, F., van der Lee, G.E.M., van den Hoorn, M.W., Higler, L.W.G. and Verhoeven, J.T.A. (2006) Importance of sediment deposition and denitrification for nutrient retention in floodplain wetlands. *Applied Vegetation Science*, **9**: 163–174.

Passauer, B., Meilinger, P., Melzer, A. and Schneider, S. (2002) Does the structural quality of running waters affect the occurrence of macrophytes? *Acta Hydrochimica et Hydrobiologica*, **30**: 197–206.

Pezeshki, S.R. (2001) Wetland plant responses to soil flooding. *Environmental and Experimental Botany*, **46**: 299–312.

Pip, E. and Robinson, G.G.C. (1984) A comparison of algal periphyton composition on eleven species of submerged macrophytes. *Aquatic Ecology*, **18**: 109–118.

Pluntke, T. and Kozerski, H.-P. (2003) Particle trapping on leaves and on the bottom in simulated submerged plant stands. *Hydrobiologia*, **506–509**: 575–581.

Pollock, M.M., Naiman, R.J. and Hanley, T.A. (1998) Plant species richness in riparian wetlands – a test of biodiversity theory. *Ecology*, **79**: 94–105.

Praeger, R.L. (1913) On the buoyancy of seeds of some Britannic plants. *Proceedings of the Royal Dublin Society*, **14**: 13–62.

Puijalon, S. and Bornette, G. (2004) Morphological variation of two taxonomically distant plant species along a natural flow velocity gradient. *New Phytologist*, **163**: 651–660.

Puijalon, S. and Bornette, G. (2006) Phenotypic plasticity and mechanical stress: biomass partitioning and clonal growth of an aquatic plant species. *American Journal of Botany*, **93**: 1090–1099.

Puijalon, S., Bornette, G. and Sagnes, P. (2005) Adaptations to increasing hydraulic stress: morphology, hydrodynamics and fitness of two higher aquatic plant species. *Journal of Experimental Botany*, **56**: 777–786.

Puijalon, S., Léna, J.-P. and Bornette, G. (2007) Interactive effects of nutrient and mechanical stresses on plant morphology. *Annals of Botany*, **100**: 1297–1305.

Puijalon, S., Bouma, T.J., Van Groenendael, J. and Bornette, G. (2008a) Clonal plasticity of aquatic plant species submitted to mechanical stress: escape vs. resistance strategy. *Annals of Botany*, **102**: 989–996.

Puijalon, S., Léna, J.-P., Rivière, N., Champagne, J.-Y., Rostan, J.-C. and Bornette, G. (2008b) Phenotypic plasticity in response to mechanical stress: hydrodynamic performance and fitness of 4 aquatic plant species. *New Phytologist*, **177**: 907–917.

Puijalon, S., Bouma, T.J., Douady, C.J., van Groenendael, J., Anten, N.P.R., Martel, E. and Bornette, G. (2011) Plant resistance to mechanical stress: evidence of an avoidance–tolerance trade-off. *New Phytologist*, **191**: 1141–1149.

Rascio, N. (2002) The underwater life of secondarily aquatic plants: some problems and solutions. *Critical Reviews in Plant Sciences*, **21**: 401–427.

Raven, J.A. (1996) Into the voids: The distribution, function, development and maintenance of gas spaces in plants. *Annals of Botany*, **78**: 137–142.

Read, J. and Stokes, A. (2006) Plant biomechanics in an ecological context. *American Journal of Botany*, **93**: 1546–1565.

Riis, T. and Biggs, B.J.F. (2003) Hydrologic and hydraulic control of macrophyte establishment and performance in streams. *Limnology and Oceanography*, **48**: 1488–1497.

Riis, T. and Sand-Jensen, K. (2006) Dispersal of plant fragments in small streams. *Freshwater Biology*, **51**: 274–286.

Roberts, J. and Ludwig, J.A. (1991) Riparian vegetation along current-exposure gradients in floodplain wetlands of the River Murray, Australia. *Journal of Ecology*, **79**: 117–127.

Rostan, J.C., Amoros, C. and Juget, J. (1987) The organic content of the surficial sediment: a method for the study of ecosystems development in abandoned river channels. *Hydrobiologia*, **148**: 45–62.

Sand-Jensen, K. (1998) Influence of submerged macrophytes on sediment composition and near-bed flow in lowland streams. *Freshwater Biology*, **39**: 663–679.

Sand-Jensen, K. (2003) Drag and reconfiguration of freshwater macrophytes. *Freshwater Biology*, **48**: 271–283.

Sand-Jensen, K. (2008) Drag forces on common plant species in temperate streams: consequences of morphology, velocity and biomass. *Hydrobiologia*, **610**: 307–319.

Sand-Jensen, K. and Pedersen, O. (1999) Velocity gradients and turbulence around macrophyte stands in streams. *Freshwater Biology*, **42**: 315–328.

Santamaria, L. (2002) Why are most aquatic plants widely distributed? Dispersal, clonal growth and small-scale heterogeneity in a stressful environment. *Acta Oecologica*, **23**: 137–154.

Scarlett, P. and O'Hare, M. (2006) Community structure of instream bryophytes in English and Welsh rivers. *Hydrobiologia*, **553**: 143–152.

Schoelynck, J., Bal, K., Backx, H., Okruszko, T., Meire, P. and Struyf, E. (2010) Silica uptake in aquatic and wetland macrophytes: a strategic choice between silica, lignin and cellulose? *New Phytologist*, **186**: 385–391.

Schutten, J. and Davy, A.J. (2000) Predicting the hydraulic forces on submerged macrophytes from current velocity, biomass and morphology. *Oecologia*, **123**: 445–452.

Schutten, J., Dainty, J. and Davy, A.J. (2005) Root anchorage and its significance for submerged plants in shallow lakes. *Journal of Ecology*, **93**: 556–571.

Schwarz, W.L., Malanson, G.P. and Weirich, F.H. (1996) Effect of landscape position on the sediment chemistry of abandoned-channel wetlands. *Landscape Ecology*, **11**: 27–38.

Sculthorpe, C.D. (1967) *The Biology of Aquatic Vascular Plants*, Edward Arnold, London, UK.

Shields, J.F.D. and Abt, S.R. (1989) Sediment deposition in cutoff meander bends and implications for effective management. *Regulated Rivers: Research and Management*, **4**: 381–396.

Smith, F.A. and Walker, N.A. (1980) Photosynthesis by aquatic plants: effects of unstirred layers in relation to assimilation of CO_2 and HCO_3- and to carbon isotopic discrimination. *New Phytologist*, **86**: 245–259.

Sousa, W.P. (1984) The role of disturbance in natural communities. *Annual Review of Ecology and Systematics*, **15**: 353–391.

Southwood, T.R.E. (1988) Tactics, strategies and templets. *Oikos*, **52**: 3–18.

Sparks, R.E., Bayley, P.B., Kohler, S.L. and Osborne, L.L. (1990) Disturbance and recovery of large floodplain rivers. *Environmental Management*, **14**: 699–709.

Strayer, D.L. and Malcom, H.M. (2007) Submersed vegetation as habitat for invertebrates in the Hudson River estuary. *Estuaries and Coasts*, **30**: 253–264.

Sultan, S.E. (2000) Phenotypic plasticity for plant development, function and life-history. *Trends in Plant Science*, **5**: 537–542.

Suren, A.M. and Duncan, M. (1999) Rolling stones and mosses: effect of substrate stability on bryophyte communities in streams. *Journal of the North American Benthological Society*, **18**: 457–467.

Telewski, F.W. (2006) A unified hypothesis of mechanoperception in plants. *American Journal of Botany*, **93**: 1466–1476.

Teltscherova, L. and Hejny, S. (1973) The germination of some *Potamogeton* species from South-Bohemia in fishponds. *Folia Geobotanica Phytotaxonomica, Praha*, **8**: 231–239.

Usherwood, J.R., Ennos, A.R. and Ball, D.J. (1997) Mechanical adaptations in terrestrial and aquatic buttercups to their respective environments. *Journal of Experimental Botany*, **48**: 1469–1475.

van Nes, E.H., Scheffer, M., van den Berg, M.S. and Coops, H. (2002) Aquatic macrophytes: restore, eradicate or is there a compromise? *Aquatic Botany*, **72**: 387–403.

Vandvik, V. (2004) Gap dynamics in perennial subalpine grasslands: trends and processes change during secondary succession. *Journal of Ecology*, **92**: 86–96.

Vogel, S. (1984) Drag and flexibility in sessile organisms. *American Zoologist*, **24**: 37–44.

Vogel, S. (1994) *Life in Moving Fluids: The Physical Biology of Flow*, Princeton University Press, Princeton, New Jersey, USA.

Vogel, S. (2003) *Comparative Biomechanics: Life's Physical World*, Princeton University Press, Princeton, New Jersey, USA.

Welch, P.S. (1935) *Limnology*, McGraw-Hill Book Company Inc., New York.

Westlake, D.F. (1967) Some effects of low-velocity currents on the metabolism of aquatic macrophytes. *Journal of Experimental Botany*, **55**: 187–205.

Wetzel, R.G. (1983) *Limnology*, CBS College Publishing, Philadelphia, USA.

White, P.S. and Jentsch, A. (2001) The search for generality in studies of disturbance and ecosystem dynamics. In Esser, K., Luettge, U., Kadereit, J.W. and Beyschlag, W. (eds) *Progress in Botany*, volume **62**, Springer, Berlin, Heidelberg, pp. 399–450.

Whitton, B.A. (1975) *River Ecology. Studies in Ecology*, volume **2**, University of California Press, Berkeley and Los Angeles, California, USA.

III Management Application Case Studies

16 Application of Real-Time Management for Environmental Flow Regimes

Thomas B. Hardy[1] and Thomas A. Shaw[2]

[1] The Meadows Center for Water and Environment, Texas State University, 601 University Drive, San Marcos, Texas 78666, USA
[2] U.S. Fish and Wildlife Service, Arcata Fish and Wildlife Office, 1655 Heindon Road, Arcata, California 95521, USA[*]

16.1 Introduction

Evolution in the science of instream flows over the past three decades reflects significant advances in data acquisition technologies, telecommunications, remote sensing, computer capability, analysis methods and modelling approaches. Over this period, we have seen the emergence of 'Ecohydraulics', where the integration of hydrology, hydraulics, water quality, sediment transport, riparian processes and aquatic ecology continues to mature through efforts at the international level, such as the Ecohydraulics Committee of the International Association for Hydraulic Research and, in the United States, the Instream Flow Council (Annear et al., 2004).

River research has advanced concepts on the ecological dynamics of rivers such as the river continuum concept (Vannote et al., 1980), serial discontinuity concept (Ward and Stanford, 1983), flood pulse concept (Junk et al., 1989), riverine productivity concept (Thorp and Delong, 1994) and the 'natural flow paradigm' (Richter et al., 1996; Poff et al., 1997; Lytle and Poff, 2004). Environmental flow regime characteristics embody the natural flow regime in terms of magnitude, frequency, timing, duration, rate of change, predictability, sequencing and inter-annual vari-

ation. The natural flow regime as a 'template' for guiding flow regime recommendations is widely accepted and applied, for example in the European Union Water Framework Directive (Directive 2000/60/EC) (Dunbar et al., 1998; Acreman et al., 2006), the Instream Flow Council in the United States and Canada (Annear et al., 2004; NRC, 2005), the South African national river ecosystem protection legislative mandates (Tharme and King, 1998), the Texas Instream Flow Program (TPWD, 2008) and many others.

The implementation of these programmes relies on the application of holistic integrated multidisciplinary assessment approaches to support resource management agencies and inform the water allocation decision process. The application and integration of component physical, chemical and ecological models within multidisciplinary assessment frameworks has matured over the past few decades (Harby et al., 2004). However, recognition of the underlying uncertainty in the science that informs water allocation decisions has led to the emergence of Adaptive Management as a mechanism by which to set target flow regimes, given legal and policy constraints.

The fundamental question of how much of a change in any component(s) of the flow regime will result in what future ecological state remains unresolved and may be quantitatively irresolvable. This is not a negative critique of the accumulated science but rather a consequence of the fundamental properties of these ecosystems. They are characterized by many linkages between physical,

[*] The findings and conclusions in this article are those of the authors and do not necessarily represent the views of the U.S. Fish and Wildlife Service.

chemical and ecological components, with complex feedback loops that lead to nonlinear responses to system perturbations. This makes them inherently too complex for existing analytical techniques to fully quantify or to be adequately represented by simple reductionist approaches based on a few emergent average properties of the system. Predictability of system responses is further complicated by the inherent stochastic nature of forcing functions (e.g. climate, weather) over spatial and temporal scales such that these systems do not exhibit stable equilibrium states. The uncertainties inherent in models of these systems and the limitations of the science have, in some respect, supported the widespread adoption of adaptive management by resource managers (e.g. Peterson and Wirth, 2003; Williams *et al.*, 2007; TPWD, 2008). It is within the framework of adaptive management in which decisions are made on the basis of the status of the system and its trajectory that real-time resource management emerges as an inherent requirement for informed decision making. As illustrated below, embracing the stochastic and uncertain nature of these systems through adaptive real-time management represents a feasible strategy for managing aquatic resource systems.

In rivers such as the Klamath (Figure 16.1), where significant deviations of the natural hydrograph occur, the form and function of the river and riparian ecosystems suffer the consequences (Fischenich, 2006). Variation in flow directly influences species diversity, adaptations and the overall viability of the aquatic and riparian species and must be recognized and implemented in order to manage and restore river ecosystems successfully (Poff *et al.*, 1997; Puckridge *et al.*, 1998; Waples *et al.*, 2001; Bunn and Arthington, 2002; Beechie *et al.*, 2006). Figure 16.2 displays examples of highly altered hydrographs compared to the Williamson River, a large undammed tributary to Upper Klamath Lake (UKL) and the Grande Ronde River. The flows recorded on the Klamath below Fall Creek during WY-1961 are an example of daily peaking operations at the Copco hydropower facility. These extreme variations of the natural hydrograph persisted for the 44 years between the construction of the Copco hydropower facility in 1918 and Iron Gate Dam reregulating facility in 1962. However, despite the curtailment of the peaking effects from Copco, Iron Gate flow reregulation altered the shape of the natural hydrograph to the other extreme, with releases having long periods of 'flat-line flows' over monthly time steps.

Another sign of a river system out of ecological balance, such as the Klamath River, is the high rates of juvenile Chinook and Coho salmon mortality attributed to *Cerato-*

myxosis, a disease of trout and salmon in the Pacific Northwest caused by a myxozoan parasite *Ceratomyxa shasta* (Hendrickson *et al.*, 1989). *Ceratomyxosis* has a two-host lifecycle involving fish (salmon and trout) and an invertebrate, the freshwater polychaete *Manayunkia speciosa* (Bartholomew *et al.*, 1997; Stocking and Bartholomew, 2007). Although native to the Klamath River, these parasites and polychaete hosts have taken hold, with juvenile salmon having infection rates approaching 80% during some years (Nichols and Foott, 2006) – greatly dwarfing other river systems where *C. shasta* is also native.

16.2 Real-time management

We define real-time management to encompass resource decisions based on the status of the system and target resources at time steps that range from hours to multiyear scales. For example, seven-day weather forecasts were coupled to hourly flow and water temperatures to model future trends in hourly water temperatures in the Virgin River, which were used to trigger upstream reservoir release volumes requiring a two-day travel time to keep river temperatures below deleterious levels for several endangered fish (Nielson *et al.*, 2007). In this chapter, we explore the implications of real-time management, which spans a range of temporal scales necessary to target environmental flow regimes protective of salmon and other aquatic resource in the lower 306 km of the main stem Klamath River, CA, USA (Figure 16.1). The Klamath River was chosen because it highlights the breadth and complexity of the science, policy, legal, institutional and societal components that support the concept and show the need for real-time management.

16.3 The setting

Upper Klamath Lake (UKL) (see Figure 16.1), with a surface area of 642 km^2, is the largest freshwater body in Oregon, USA. It releases water into the Klamath River at the Link River Dam, constructed in 1921. The lake has an active storage capacity of $\sim$ 620 000 m^3. Before the establishment of the Klamath Project, the lake would enter the winter in a relatively full state, since the lake expanded over vast marshes, along with constant spring and surface inflows. During periods of freezing conditions, UKL outflows would drop, but the river and lake would eventually recover throughout the winter and spring snow

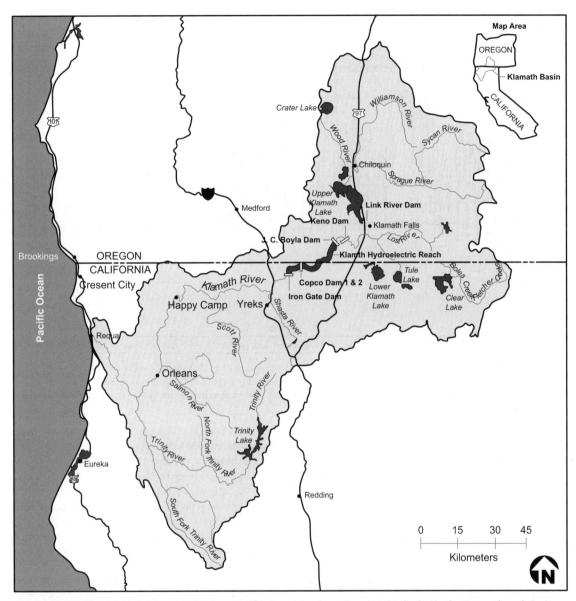

Figure 16.1 The Klamath Basin located in Northern California and Southern Oregon, USA displaying the Upper Klamath Basin, with major dams and tributaries, significant lakes and hydropower reservoirs, and Lower Klamath River basin, the anadromous river reach below Iron Gate Dam and significant tributaries. Reproduced with permission from Klamath Facilities Removal EIS/EIR Public Draft 2011.

runoff period. The runoff characteristics of the historical Klamath River hydrograph would respond to multiple within-year runoff patterns, such as a long attenuated runoff pulse with storm peaks throughout the spring runoff period, or a series of runoff pulses with varying magnitudes, or high-intensity, short-duration flows such as a rain-on-snow flood runoff event.

The U.S. Bureau of Reclamation (USBR) owns Link River Dam and the large diversion channel (A-canal) that conveys up to $33 \, \mathrm{m}^3 \, \mathrm{s}^{-1}$ to a network of distribution canals destined for the $850 \, \mathrm{km}^2$ of agricultural land and the Klamath National Wildlife Refuges complex before draining into the Klamath River at Klamath Straights. These refuges are a critical stopping point for millions of migrating birds,

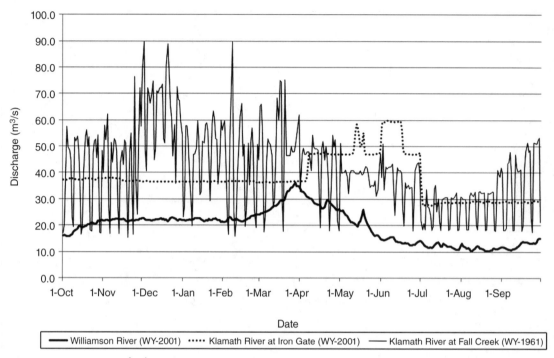

Figure 16.2 The daily flows (m^3 s^{-1}) from the Williamson River below Sprague River (WY 2001), a major tributary to Upper Klamath Lake representing a natural hydrograph in comparison to the monthly static flows below Iron Gate Dam (WY 2001), and the peaking flows experienced on the Klamath River (WY 1961) below Fall Creek prior to the construction of Iron Gate Dam between 1917 and 1962.

representing about 80% of the birds using the Pacific Flyway. The Klamath Project provides approximately 258 981 m^3 of water to 850 km^2 of irrigable croplands and the refuges. Adjacent to Link River Dam are the East-side and Westside penstocks, which send flow through two powerhouses before reentering the river below Link River Dam. The hydropower company, PacifiCorp, operates the outlet structures at Link River Dam under an agreement with the USBR and subject to Klamath Project water rights. The UKL water level fluctuates by approximately 2 m over the irrigation season. These minimum and maximum UKL elevations were established by the U.S. Fish and Wildlife Service in consultation with the USBR Klamath Project over their operations through authorities governed by the Endangered Species Act (ESA) in order to ensure protection of the federally endangered Lost River (*Deltistes luxatus*) and Shortnose (*Chasmistes brevirostris*) suckers in Klamath Lake.

Below UKL and Link River Dam, the Klamath River flows through a series of four hydropower peaking facilities, with the lowest dam, Iron Gate, located at river kilometre 306. Iron Gate Dam is a reregulating facility

constructed in 1962 to dampen the environmental effects from hydropower peaking operations upstream (Figure 16.2). Iron Gate's reservoir is relatively small, with a capacity of 72 521 m^3, of which only 4674 m^3 is available for power production. Iron Gate's hydropower hydraulic capacity is 49 m^3 s^{-1}, with flows above that capacity overtopping the spillway of the dam. Iron Gate Dam lacks any upstream fish passage facilities, effectively eliminating anadromy above this point.

Following the construction of Iron Gate Dam in 1962, the Federal Energy Regulatory Commission (FERC) established minimum Klamath River flow requirements for releases at Iron Gate. These flows met the hydraulic capacity at the dam, as well as the upper basin's water demands and Lower Klamath River main stem Fall Chinook salmon (*Oncorhynchus tschawytscha*) spawning flow requirements (37 m^3 s^{-1}). In 1997, the National Marine Fisheries Service's Protective Resource Division (NMFS) listed Coho salmon (*Oncorhynchus kisutch*) as a threatened species under the authorities of the U.S. Endangered Species Act, and, through consultation with the USBR for its Klamath Project operations, NMFS established a new

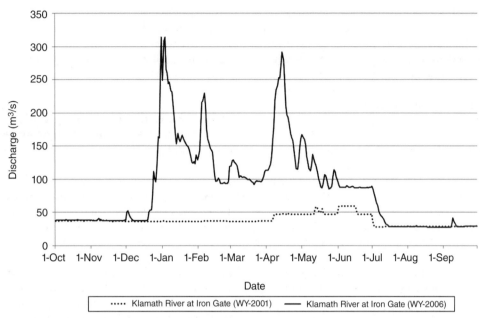

Figure 16.3 Under the current flow management regime, the Klamath River experiences both unnatural 'flat-lined flows' typical of a dry or below average water year (WY 2001) compared to a run-of-the-river experienced during an above average water year (WY 2006) followed by a severe down ramping back to the summer minimum monthly flow requirement.

set of Klamath River minimum monthly flow requirements below Iron Gate Dam. However, depending on the water year, seasonal weather pattern and prior decisions on water allocations for agricultural demands, UKL elevations may curtail the ability to provide excess flows above the minimum monthly threshold requirements below Iron Gate Dam not only for the listed Coho salmon, but also other flow-dependent native fish species and aquatic resources of the Klamath River and its tributaries. During the low flow season (July, August and September), the releases below Iron Gate Dam can represent up to 48% of total Klamath River estuary inflows.

16.4 The context and challenges with present water allocation strategies

Following the agricultural season (April through October), UKL is at its minimum water level and Klamath River flows below Iron Gate are strictly maintained at the existing minimum required monthly flows. These minimum flows become an inflexible flow target for PacifiCorp and USBR, which sustains a stable, 'flat-lined flow' management regime below Iron Gate Dam. There is little to no opportunity available for alternative flow releases

until the UKL elevation trajectory has a high likelihood of reaching the flood protection elevation requirement, established for levy protection for the adjacent private lands. PacifiCorp and USBR then increase downstream releases regardless of the hydrologic conditions of the upper basin until the threats of levy breaching subside (Figure 16.3). At that time, flows are reduced, following NMFS requirements, back to the minimum flow in order to maximize UKL storage potential.

However, the physiographic setting of the Klamath Basin, in conjunction with the stochastic nature of climate and weather, results in large variations in the runoff's timing, magnitude and duration, which often preclude the ability to meet the intended flow regime. For example, the Natural Resources Conservation Service (NRCS) April–September forecast inflow volume to UKL during an above-average water year can run off in one week, induced by a warm rain-on-snow event, with the majority of the forecast flow volume passing directly through the system as run-of-the-river. The remainder of that year then mimics a semi-drought condition in terms of UKL inflows and outflows. A different above-average water year with a long, cool spring can result in an extended runoff period, supplying ample water to the competing water allocation needs. In normal or below-normal water years, or when runoff is compressed early in the spring, flow releases below Iron Gate Dam are typically reduced and held to a minimum in order to satisfy the upstream water allocations, such as maintaining UKL elevation requirements, storage and agricultural demand.

This conflict in competing water allocations between the Upper and Lower Klamath Basins has spawned numerous lawsuits over water-management practices. Court-mediated decisions have resulted in disastrous environmental and social repercussions, such as the 2001 Klamath Project operations shutting off all irrigation supply due to ESA compliance for UKL elevations and Klamath River flows below Iron Gate Dam, resulting in protests and civil disorder. When flow releases to the lower river were severely curtailed to maintain irrigation supplies, an unprecedented Lower Klamath River adult Chinook salmon die-off of over 33 000 adults from an epizootic parasite proliferation occurred (Guillen, 2003; Figure 16.4). The resulting low number of escapes of natural Klamath

Figure 16.4 Dead adult Fall Chinook salmon line the banks of the Lower Klamath River main stem within the Yurok Tribe reservation during the mid-September 2002 fish kill of over 33 000 fish. The USFWS causative factor report attributed the die-off to a combination of factors including low flows, a large adult run concentrated in the Lower Klamath River, warm water temperatures and the outbreak of parasitic infections that led to an epizootic event. Photo by Yurok Tribe, reproduced with permission.

Basin Fall Chinook salmon had regional consequences, in the form of severe restrictions to the in-river tribal and recreational fisheries and the offshore commercial and recreational fisheries along the 1127 km zone of the California and Oregon coasts.

16.5 The issues concerning the implementation of environmental flow regimes

Implementing an environmental flow regime based on exceedance levels in the Klamath remains a subject of controversy between the state and federal resource agencies, hydropower representatives and Native American tribes. At present, the NRCS annual April 1st 70% exceedance inflow forecast for UKL is used to classify the water year into one of five types (i.e. wet, above average, average, below average, or dry). The water year then determines the bounds on the negotiated water allocation targets for the Klamath Basin agricultural season, minimum requirements for UKL elevations and minimum monthly flow requirements below Iron Gate Dam. There are differing opinions on what annual flow exceedance level should be used (e.g. 70% versus 50%) given the long lead times for the forecasts (April–September) and given that the forecasts for UKL inflow volumes are highly uncertain (Figure 16.5). Additionally, rarely do the negotiated minimum monthly flow requirements result in a managed flow regime that follows the shape of the natural hydrograph, unless the water year is wet and the river is spilling over Iron Gate Dam (Figure 16.2).

In part, this is a consequence of the physical setting of the Klamath River basin compared to other large river systems, given that the agricultural lands are at the top of the basin and adjacent to the primary storage facility of UKL. Further complications arise due to the competing needs of the endangered Lost River and Shortnose suckers within UKL, the downstream needs of multiple species and life stages of the anadromous and resident fish in the main stem Klamath River below Iron Gate Dam and the water needs of the Klamath Basin National Wildlife Refuge complex. Historically, when water deliveries to the Klamath Project were prioritized to meet agricultural demands, with limited delivery of water to the refuges and Lower Klamath River, severely overcrowded conditions for migrating birds in the refuges caused lethal outbreaks of avian botulism and cholera as well as fish kills along the Klamath River's main stem.

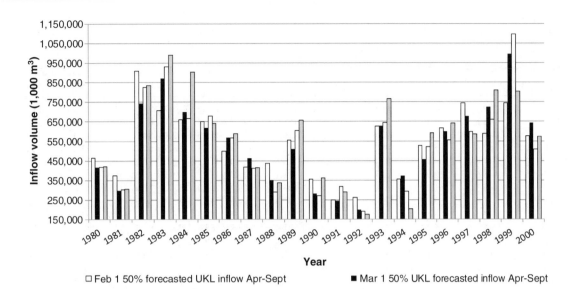

Figure 16.5 February, March and April NRCS 50% inflow forecasts for Upper Klamath Lake compared to the actual net inflows for water years 1980–2000.

16.6 Underlying science for environmental flows in the Klamath River

Hardy *et al.* (2006a) provided recommended instream flow regimes below Iron Gate Dam necessary for the protection and recovery of the aquatic resources in the main stem Klamath River. Flow recommendations incorporated consideration of over-bank flows, high-flow pulses, base flows, subsistence flows and ecological base flows. This instream flow study utilized eight intensive study sites, representative of homogeneous hydrologic and geomorphic segments, underlying physically classified mesohabitat types and the spatial and temporal distribution of species and life stages of three targeted salmonid species: Chinook, Coho and steelhead. At each study site, ranging from 1.0 to 2.3 km in length, two-dimensional hydrodynamic models at 0.25-metre resolution were integrated with habitat suitability curves to produce habitat versus discharge relationships for Chinook, Coho and steelhead life stages. The conceptual habitat models were validated against observation data independent of the data used to develop the habitat suitability curves (e.g. Hardy *et al.*, 2006a; 2006b).

Development of flow recommendations was based on averaging estimates from two approaches – hydrology and physical habitat – rooted in the natural flow paradigm

(Poff *et al.*, 1997). Both efforts employed a Periodic Autoregressive Moving Average (PARMA) model (Salas *et al.*, 2000) based on estimates of historic undepleted mean monthly natural flows for the Klamath River over a 52-year period, water years 1949–2000 (Perry *et al.*, 2005). The basic structure of the PARMA$_S(p,q)$ model is given in Equation (16.1). In the current application, a seasonal PARMA$_{12}(5,0)$ model was fitted to the data and used to generate 1000 synthetic monthly flow traces at Iron Gate Dam.

$$X_t = \sum_{j=1}^{p} \phi_t(j)X_{t-j} + \varepsilon_t - \sum_{j=1}^{q} \theta_t(j)\varepsilon_{t-j} \quad (16.1)$$

where p and q determine the order of the PARMA model in the form of lag, and S represents the seasons. X_t and X_{t-j} are the random variable values at time t and $t-j$. ε_t is the sequence of random variables with zero mean and a standard deviation of 1.0 and represents the residuals. ϕ_t and θ_t, respectively, are the autoregressive and moving average parameters.

16.6.1 Hydrology-based flow regimes

Each of the 1000 stochastic time series results for the simulated 40-year period was utilized to compute estimates of the monthly flows at each annual exceedance level. The corresponding mean and 5th and 95th percentiles of the monthly flows were then calculated and normalized

Table 16.1 Instream flow recommendations for Iron Gate Dam discharges ($m^3 s^{-1}$) by monthly exceedance levels (Hardy *et al.*, 2006a).

Percent Exceedance	October	November	December	January	February	March	April	May	June	July	August	September
5.0	49.1	69.7	95.9	113.0	126.7	126.3	135.6	108.9	90.2	62.7	44.2	44.3
10.0	48.6	68.4	92.9	108.6	121.3	123.3	129.8	105.1	86.5	60.6	43.6	43.7
15.0	48.1	67.0	90.8	107.5	119.2	121.3	125.3	102.4	84.2	58.8	42.3	42.9
20.0	47.6	65.6	88.3	104.9	119.4	117.8	119.8	98.5	80.7	56.6	39.8	42.2
25.0	47.0	64.0	85.4	103.2	115.5	113.0	115.1	96.0	78.0	54.5	38.9	41.5
30.0	46.6	62.9	83.4	99.4	111.1	111.6	111.3	91.3	75.3	51.8	37.8	40.5
35.0	46.3	61.2	81.3	96.4	103.6	109.3	104.9	88.2	71.9	49.3	37.0	39.8
40.0	46.0	59.7	79.3	91.0	97.3	104.3	98.7	83.8	69.5	46.3	35.5	38.8
45.0	44.6	58.3	76.2	85.4	91.2	101.5	91.9	79.7	66.3	42.9	34.4	37.8
50.0	44.3	56.6	72.1	79.9	85.4	95.7	85.8	75.7	63.0	37.7	33.1	37.0
55.0	43.7	54.8	67.5	74.5	79.6	89.2	79.7	71.1	58.6	35.8	31.3	36.1
60.0	43.2	53.1	63.3	68.5	72.6	82.4	73.3	67.5	56.1	34.1	29.9	35.0
65.0	42.8	51.8	59.2	62.6	66.1	74.5	68.1	61.3	52.1	32.1	28.9	33.8
70.0	42.2	50.3	55.2	57.1	60.5	66.5	64.0	58.0	46.3	30.3	28.5	32.8
75.0	41.6	48.4	51.4	51.7	55.2	58.0	57.9	53.9	41.5	28.7	27.6	31.7
80.0	41.1	47.3	46.7	45.9	50.1	52.0	54.9	47.9	37.4	26.8	26.5	30.6
85.0	40.5	45.3	43.0	41.3	45.7	44.9	49.3	40.1	32.8	25.6	25.8	29.6
90.0	40.1	43.7	39.1	35.3	42.1	39.9	43.3	34.5	30.6	23.8	25.3	28.6
95.0	39.5	42.5	35.7	32.0	40.1	36.1	37.5	33.3	29.0	22.8	24.9	27.5

to obtain the coefficients of the flow range. Multiplying the coefficients by the estimated mean monthly natural flows derived the corresponding lower and upper limits by month and exceedance level. If the existing flow in a particular month and exceedance level was lower than the lower flow limit, the recommended flow was set to the lower flow limit. If the flow was between the higher and lower limits, the existing flow was set as the recommendation. If the flow was higher than the upper limit, the flow was set to the upper limit. This approach maintains the target monthly flows at a given annual exceedance level within the expected limits of 90% of the variation in expected flows.

16.6.2 Habitat time series-based flow regimes

One hundred randomly selected stochastic flow series were used in conjunction with the habitat versus flow relationships to compute the habitat time series for each species and life stage at each study site. These habitat time series were then analyzed using the same framework as described for the hydrology-based flow regimes to derive target flows for each species and life stage at each exceedance level on a monthly basis. The final habitat time series flow recommendations were then com-

puted as the geometric mean of all the applicable species and life stages present at a site in a given month at exceedance level.

16.6.3 Flow-and-habitat-based integration for recommended flow regimes

The averages of the habitat time series and flow-based recommendations for each month and exceedance level were used as the final set of flow recommendations by study site for each Klamath River main stem reach (Table 16.1). The resulting monthly exceedance flow recommendations were proposed to be adaptive to the seasonally updated forecasts of water year exceedance levels in order to track the underlying shape of the natural hydrograph through time. Hardy *et al.* (2006a) also used two independent modelling approaches: bioenergetics (Addley, 2006) and a salmon production model (SALMOD, Williamson *et al.*, 1993) to compare recommended flows to existing operations on growth and productivity, respectively. The results of both modelling approaches demonstrated better growth rates for Chinook salmon fry and juveniles as well as higher outmigration densities of juveniles for every year in the multi-decadal simulation period when compared to historical operations.

16.6.4 Peer review of Hardy *et al.* (2006a)

Given the highly contentious nature of competing water allocation decisions within the Klamath Basin, the National Research Council convened the Committee on Hydrology, Ecology, and Fishes of the Klamath River Basin specifically to review Hardy *et al.* (2006a) and the USBR Natural Flow of the Klamath River (Perry *et al.*, 2005). The Natural Flow study provided the unimpaired flows at Iron Gate Dam used by Hardy *et al.* (2006a). The Committee's findings, *Hydrology, Ecology, and Fishes of the Klamath River Basin* (NRC, 2008), supported the work and flow recommendations in Hardy *et al.* (2006a), noting that some criticisms about the study were beyond the investigators' control. For example, the principal author had requested daily flows and was provided monthly flows by the USBR flow study. However, the Hardy *et al.* (2006a) flow recommendations have not been implemented and detrimental flow regimes continue to persist below Iron Gate Dam.

16.7 The Water Resource Integrated Modelling System for The Klamath Basin Restoration Agreement

Hardy *et al.* (2006a) also provided key technical input that supported The Klamath Basin Restoration Agreement and the Klamath Hydroelectric Settlement Agreement, which implements fixed allocations for water demands such as agriculture, refuges, municipal and natural resources. Allocations decrease in drier years, and the agreements provide for the removal of the hydropower projects and affiliated dams below Link River, fund restoration measures and salmon reintroduction efforts. The U.S. Secretary of the Interior expects to make his determination for the hydropower facility removal in 2013.

The Water Resource Integrated Modelling System (WRIMS) (formally named CALSIM) is the current USBR Klamath Project water operations model for the system and is also used to inform the water allocations negotiated within the auspices of the Klamath Basin Restoration Agreement. This water resource modelling system allows for the evaluation of operational alternatives of large, complex river basins with competing demands and priorities. WRIMS also allows for flexible operational criteria specification and a linear programming solver for efficient water allocation decisions (CDWR and USBR, 2011).

The Klamath River below Iron Gate monthly flow targets for subsequent WRIMS modelling were based on a technical advisory team's interpolation of the Hardy *et al.* (2006a) instream flow recommendation for the Klamath River between Iron Gate and the Shasta River. The team utilized the flow that provided 80% of maximum habitat for targeted salmonid species and life stages using monthly time steps. These estimated flows were then averaged with the USBR Natural Flow Study monthly exceedance values for wet, above average, average, below average and dry water year types and resulted in a temporal shifting of the annual hydrograph to mimic the pattern of the median monthly natural hydrograph and account for variation in annual water availabilities. Whenever the average flows exceeded the recommended Hardy *et al.* (2006a) 80% of maximum habitat value, they were lowered to that recommended level. The technical team also defined Upper Klamath Lake target elevations, by monthly time steps, necessary to meet Lost River and Shortnose sucker life history requirement thresholds in order to reach critical spawning and rearing habitats, as well as ensuring full UKL elevations by March 1 during all but the driest water years.

The Klamath technical team then utilized WRIMS to simulate Klamath Project operations over the 40-year historic period of record (1961–2000). The WRIMS priority structure was set to first meet the KBRA negotiated, fixed agricultural allocations as a function of each of the five defined water year types based on UKL forecast inflows. During simulations, the WRIMS priority structure then attempted to meet the secondary priority targets (National Wildlife Refuges, UKL elevations and Klamath River flows). When the WRIMS hindcast simulations did not meet the secondary target allocations, for any water year, the WRIMS simulations then attempted to balance the allocations among the secondary priority resource targets. The KBRA settlement negotiations group eventually reached agreement based on this iterative approach (designated (WRIMS-R).

16.8 The solution – real-time management

In an effort to address the flow needs of the aquatic resources below Iron Gate Dam, in the interim period prior to dam removal, we propose a framework to extend the work of Hardy *et al.* (2006a) and a structure for implementing real-time management (RTM) for flow releases below Iron Gate Dam. RTM eliminates the reliance upon a single inflow forecast to determine the one of five water year types that dictate the Klamath River minimum flow

Table 16.2 WRIMS R-32 refuge discharge at Iron Gate Dam ($m^3 s^{-1}$) outputs in percentage exceedance over targeted time steps.

Percent exceedance	October	November	December	January	February	March 1–15	March 16–31	April 1–15	April 16–30
0.0	36.8	95.6	213.4	256.1	253.9	294.0	301.2	200.0	189.8
10.0	36.8	36.8	91.6	126.9	166.9	171.2	183.2	173.2	163.1
30.0	33.7	35.9	49.8	78.7	103.7	120.1	139.8	119.2	116.3
50.0	32.0	33.1	36.8	62.2	70.8	84.6	95.8	94.7	89.7
70.0	29.9	30.7	31.5	35.8	42.2	71.9	78.5	63.3	61.4
90.0	26.4	26.7	27.4	28.0	28.2	43.0	45.8	45.5	43.6
100.0	14.7	18.0	21.4	23.8	22.9	28.7	28.4	29.6	28.5

Percent exceedance	May 1–15	May 16–31	June 1–15	June 16–30	July 1–15	July 16–31	August	September
0.0	154.6	162.4	78.1	78.1	53.2	53.2	41.9	41.6
10.0	103.4	113.7	74.9	71.2	47.8	46.3	37.1	37.1
30.0	80.1	81.6	63.8	57.1	38.3	34.5	29.4	33.9
50.0	73.1	67.8	52.2	47.4	30.3	29.8	25.3	31.1
70.0	55.1	57.6	44.3	40.5	26.2	26.1	23.3	27.9
90.0	35.2	38.8	31.3	31.0	22.6	22.2	21.6	25.5
100.0	22.5	23.2	19.0	17.4	13.7	14.0	11.7	13.5

schedule for the remainder of the water year under existing river management practices. RTM represents a transparent framework and process for assigning daily Klamath River flows below Iron Gate Dam based on the use of a hydrologic indicator stream characteristic of the natural inflow patterns of Klamath Basin above UKL. The implementation of RTM calculates the exceedance value for each day based on the real-time daily Williamson River discharge values to derive daily flows released into the Klamath River below Iron Gate Dam. Management of the river becomes a continuous function based on the hydrologic daily-derived conditions of the upper basin. We believe this concept would bring the Klamath River flows back to a regime inherent to unregulated, natural river systems. Our implementation of RTM is consistent with approaches identified in the peer review literature, supported by site-specific studies and long-term aquatic resource monitoring data (Poff et al., 1997; Annear et al., 2004; Hardy et al., 2006a; NRC, 2008).

We selected the Williamson River as the hydrologic indicator stream based on its long-term flow record, large watershed area (7950 km^2) and the fact that it is a major source of inflow above the Iron Gate Dam. The Williamson River at Chiloquin, Oregon (USGS Gauging Station 11502500; USGS, 2012), which has a real-time staff gauge covering the period 1917–2007, was utilized; however; we restricted the analysis to the 1961–2001 water

years in order to maintain consistency with the inputs and outputs of the WRIMS-R simulated outputs negotiated during the KBRA settlement process. We calculated the exceedance values for bi-weekly and monthly time steps (October, November, December, January, February, March 1–15, March 16–31, April 1–15, April 16–30, May 1–15, May 16–31, June 1–15, June 16–30, July 1–15 and July 16–31) using the Weibull distribution (Haan, 2002) (see Table 16.2) based on daily flows from the observed Williamson River gauge and the backcast WRIMS-R simulated daily flows:

$$P = 100^*[M/(n + 1)] \qquad (16.2)$$

where:

P = the probability that a given flow will be equalled or exceeded (% of time)

M = the ranked position on the listing (dimensionless)

n = the number of events for a period of record (dimensionless)

For comparison purposes, we used Hardy et al. (2006a) Klamath River instream flow recommendations based on monthly flow exceedance values (designated Hardy-RTM) (Table 16.1), the raw monthly and bi-weekly outputs of the WRIMS-R refuge simulations (designated WRIMS-RTM), the historical daily Klamath River flows below Iron

Gate Dam and the historic daily Williamson River flows below the Sprague River.

16.9 Example RTM implementation

We calculated the daily RTM-based Klamath River flow discharge requirements below Iron Gate Dam, utilizing the WRIMS-R derived discharge exceedance table and the Hardy et al. (2006a) instream flow exceedance table using the following three steps (Figure 16.6):

1 Obtain the real-time daily discharge from the hydrologic index stream gauge (Example: May 01, 2007, Williamson River below Sprague River USGS, Station: 1152500, **35.0** $m^3 s^{-1}$).

2 Calculate the corresponding exceedance value for the real-time hydrologic index stream gauge reading **Step 1** for the appropriate time step (Example: real-time discharge for the Williamson River on May 01, 2007 was **35** $m^3 s^{-1}$ and equates to a May 1–15 time step exceedance value of **63%**).

3 Calculate the discharge requirement for the target release location (Klamath River below Iron Gate Dam) by matching the hydrologic index stream's exceedance value, calculated in **Step 2**, to the target site's exceedance value for the time step to derive the corresponding discharge requirement. (Example: Williamson May 01, 2007 **Step 2** exceedance value for the time step May 1–15 was **63%** and the corresponding daily discharge requirement would be 71.2 $m^3 s^{-1}$ based on the 63% exceedance value for this time step, using Hardy-RTM exceedance values (see Table 16.1)).

In order to match the source stream and target stream exceedance and flow values within any time step, we used the following linear interpolation equation:

$$Q_{TU} = Q_{TKH} - [(E_{TKH} - E_{SK})x(Q_{TKH} - Q_{TKL})/ \atop (E_{TKH} - E_{TKL})]$$ (16.3)

where:

Q_{TU} = RTM discharge for target release unknown value

Q_{TKH} = Discharge of target at nearest known higher value

E_{TKH} = Exceedance of target at known nearest higher value

E_{SK} = Exceedance from source known value

Q_{TKL} = Discharge of target at nearest known lower value

E_{TKL} = Exceedance of target at known nearest lower value

Assumptions:

1 Uncertainty decreases as a function of operational modelled time steps to the point that daily operations equal daily RTM.

2 The use of an index stream with significant alterations from the natural hydrograph will result in RTM outputs that mimic those alterations.

3 Operational modelled outputs with fixed flows over multiple exceedances will result in similar fixed RTM outputs.

4 The maximum and minimum values from the operational modelled outputs, by time step, and hindcast over the period of record, set the RTM upper and lower bound, by time step.

In this application of RTM, when the flow from the index stream exceeded the minimum or maximum discharge value derived from the WRIMS modelled simulated values for a particular day, the corresponding minimum or maximum WRIMS discharge within that defined time period was used.

16.10 RTM performance

Our assumption was that using the WRIMS-RTM and Hardy-RTM simulations to hindcast RTM historic daily flows over the period of record would maintain an approximate balance in the total annual flow volumes. However, we expected the Hardy-RTM to utilize more volume at the Iron Gate target releases due to the intent of the instream flow recommendations prioritizing the needs of anadromous fish below Iron Gate Dam over agriculture and UKL elevations. The RTM application picks up the natural variability of inflow patterns, with releases to the Klamath River below Iron Gate fluctuating in unison with the Williamson River index stream flows (Figure 16.7). We chose water years that USBR designated as dry (1992), below average (1988), average (1986), above average (1975) and wet (1984) to compare volumes between historic Iron Gate, WRIMS-RTM and Hardy-RTM (Figure 16.8).

It is interesting that Hardy-RTM performed well, except for the driest years, with flows practically equal to historic Iron Gate releases during average and above average water years, and less during the wet years. WRIMS-RTM also performed well compared to historic Iron Gate flows, with slightly higher volumes used during the drier years, almost equal volumes during average years and less volume during the wetter water years. The difference between the RTM outputs and historic volumes is primarily due to the upper bounds that limit RTM from reaching the discharge magnitudes experienced during run-of-the-river events. Regardless, there is no management control during high-flow events, so RTM would gain

Step 1: Obtain the real-time surface-water discharge value transmitted from the chosen hydrologic index stream gauge (Example: May 01, 2007 Williamson River below Sprague River USGS station #1152500 = 35.0 m³/s(1,220cfs).

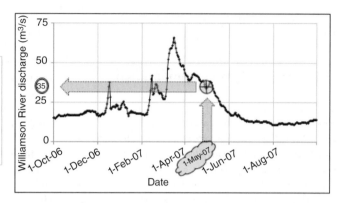

Step 2: Calculate the real-time percent exceedance by appropriate time-step using the **Step 1** discharge (example: May 01, 2007 = 35.0 m³/s) equates to a May 1-15 time-step percent exceedance value = 63%.

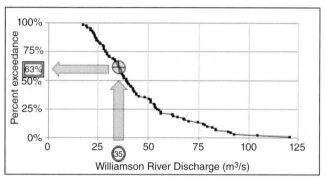

Step 3: Calculate the real-time target site discharge (cms) using the percent exceedance value calculated in **Step 2 (63%)** and the corresponding flow, percent exceedance by appropriate time-step (Example: **Hardy-RTM, May 01, 2007** discharge at iron Gate Dam equates to **71.2 m³/s**).

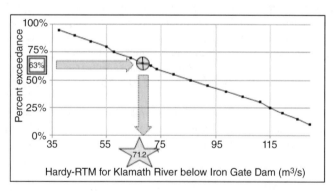

Figure 16.6 We demonstrate the sequence of steps to generate a daily Hardy-RTM Klamath River main stem release below Iron Gate Dam. Utilizing the May 01, 2007 discharge from the source stream and the corresponding exceedance linkage to the time step exceedance flow from Hardy *et al.* (2006a) to derive the May 01 release at Iron Gate Dam. This application will foster flow management ability to restore the natural flow regime and variability, the critical components of a river system's ecological integrity, back to their target river system.

implementation control, mimicking the Williamson River flows, once Klamath River flows below Iron Gate dropped to approximately 100 m³ s⁻¹.

We also plotted the historic Williamson River below Sprague River, historic Klamath River below Iron Gate Dam, WRIMS-R and the historic hindcast WRIMS-RTM and Hardy-RTM model outputs for five different water year types: dry, below average, average, above average and wet. Both the WRIMS-RTM and Hardy-RTM performed very well, with both RTM runs closely tracking the shape of the Williamson River hydrographs. We did notice that the Hardy-RTM outputs were higher in the fall and early winter months but tracked the spring natural hydrograph very well. The higher fall flows are the result of higher

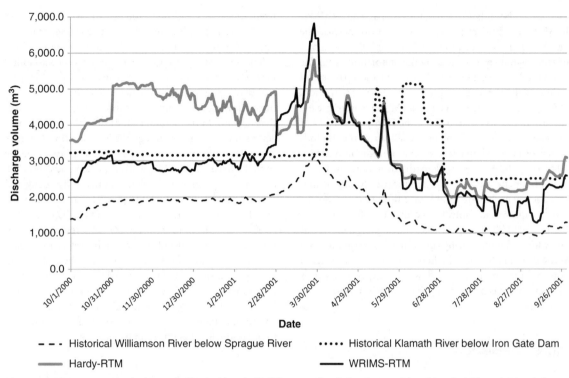

Figure 16.7 Water Year 2001 hydrographs for the historical Williamson River below Sprague River, historical Klamath River below Iron Gate Dam, Hardy-RTM and WRIMS-RTM.

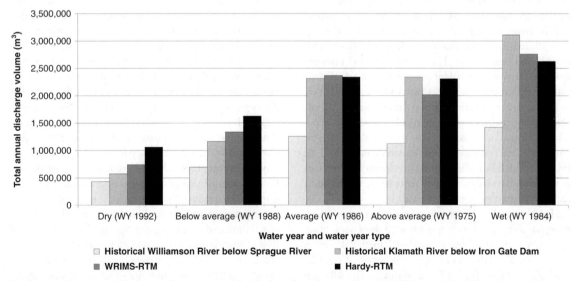

Figure 16.8 Total annual discharge volumes associated with the index tributary, historical Williamson River below Sprague River, historical Klamath River below Iron Gate Dam and the WRIMS-RTM and Hardy-RTM for the Klamath River below Iron Gate Dam, by the USBR designated water year types (dry, below average, average, above average and wet).

exceedance bounds for those time steps. The WRIMS-RTM model runs set priorities for UKL to increase the likelihood and rate of filling UKL on or before March 1 of each year, thereby increasing the opportunities for the Klamath River to experience the spring runoff patterns of inflow, as the water spills, as a run-of-the-river situation. During the wet periods, Hardy-RTM flows were greatly surpassed, due to the bounds of the Hardy *et al.* (2006a) exceedance table used to set the RTM output bounds being lower in comparison to the bounds of WRIMS-RTM outputs. However, Klamath River flows in excess of approximately 100 m^3 s^{-1} are practically run-of-the-river, so management would have little to no opportunity to capture and control these large runoff events. The WRIMS-RTM 'banking' of water during fall months did allow for additional spring runoff in comparison to both WRIMS-R and the historic flows. However, Hardy-RTM also provided more spring flow than experienced over the period of record for the Klamath River below Iron Gate Dam.

16.11 Discussion

Observed differences between the two RTM examples can be attributed to the operational balancing of demands within the WRIMS simulation that utilizes set priorities, with the model forced to meet higher priorities first, such as the fixed agricultural allocations, while balancing the losses between the secondary prioritized target flow allocations (UKL elevations, Klamath River flows, refuges and others). The Hardy *et al.* (2006a) instream flow evaluation accounted for consumptive use above Iron Gate Dam; however, they prioritized Klamath River flow needs over other consumptive uses. In addition, the Hardy *et al.* (2006a) instream recommendation process incorporated attributes of the natural flow regime, along with the annual and inter-annual variability, which influenced the final instream flow recommendations following the shape of the natural hydrograph, with recommendations being scaled over multiple inflow probabilities.

Higher diversion volumes during dry years are artifacts of agricultural demands increasing with drier conditions. This increased consumption puts the Klamath Basin water supply out of balance and historically deprived the Klamath River and UKL of sufficient volumes. The KBRA, if implemented, would help solve this ongoing water crisis by fixed allocations based upon water availability. Agriculture would receive less water during dry years, with increasing volumes during wetter years. The insurance of fixed allocations and the settlement between the Upper and Lower Klamath Basins is a positive change from past confrontations over the limited water resources.

The application of RTM clearly restores the variability in river flows below Iron Gate Dam in terms of within-year and between-year characteristics. It has the added benefit of allowing for a continuous function of flow releases that respond to the natural evolution of the runoff characteristics over time. Although the instream flow volumes and releases below Iron Gate Dam are constrained by both the fixed but variable agricultural demands and UKL elevation requirements for flood control and endangered species, they improve the flow regime characteristics compared to existing water allocation practices. The practical application of RTM to any arbitrary water resource system does require the availability of a suitable reference stream or river within the context of the system's physical setting. Our application of RTM using two different instream flow recommendations that were derived from different priorities between competing water allocation strategies (i.e. environmental versus agriculture) demonstrates the flexibility of the RTM application framework.

16.12 Conclusions

We developed a real-time management framework and implementation strategy of instream flow recommendations that are a continuous function over exceedance flow ranges and adapt to within-year changes to runoff patterns to meet anadromous species' flow needs in the main stem Klamath River. The approach relied on the natural flow paradigm to select flows that closely mimic the natural flow pattern daily, seasonally and inherently for different types of water years. We maintain that this integrated approach combining the underlying characteristic of the flow and habitat regime based on RTM meets the objectives of using the natural flow paradigm to guide the recommendation process. We believe that the RTM-derived flow regimes benefit all life stages of anadromous species during all periods of the year, including upstream migration, spawning, rearing, upstream passage and out migration, compared to existing conditions. The RTM-derived flow regime is also anticipated to reduce disease-related risk factors by increasing flow volumes and likely reducing slow backwater areas important to known disease organisms in the main stem Klamath River. The RTM supports improved benefits with increased travel times below Iron Gate Dam that move fish out of the highest thermal stress area associated with Seiad Valley.

Although RTM appears complex, we believe that effective resource management is contingent on the recognition that flow variability over different temporal scales (e.g. daily, weekly, monthly, seasonally and intra-annual) are operating differentially on physical, chemical and biological processes of the aquatic ecosystem. This requires management decisions to be responsive to the implications of actions over these same time scales. Effective management must also be reactive to the impact of stochastic processes such as floods or extreme low flows that require differential management responses given the status of the target resources. The framework and implementation of RTM to guide environmental flow regimes that support salmon restoration goals is clearly illustrated with our application to the Klamath River Basin.

We believe that past and ongoing flow management of the Klamath River continue to deviate from the natural hydrograph, disengaging the ecosystem from a functionally viable state and placing the system at a high level of ecological risk. This shift from a dynamic system of the past to the present day's hydropower peaking and long periods of reregulated steady flow endangers the aquatic fauna and flora of the river. This simplification of the flow regimes has resulted in an increase of bacterial, viral, fungal and parasitic epizootic events and the establishment of invasive species that continue to impact the aquatic resources in the Klamath River negatively and imperil the successful recovery of salmon and the overall ecological and physical processes of a functional river system.

Moving the present management of the Klamath River, and other river systems with competing interests and alterations, from the natural hydrograph to a real-time management framework, as presented in this chapter, is feasible and implementable, while meeting the multiple, competing interests for water allocations within the basin. The greatest impediment to implementation of an RTM application is not a technical issue, but the need to overcome the inherent institutional and political inertia of historical water allocation philosophies.

Acknowledgements

We would like to acknowledge the following individuals who provided conceptual or analytical support that made this work possible: Mike Cunanan, Robby VanKirk, Nicholas Hetrick, Gary Smith, Mike Belchik and Sam Williamson.

References

Acreman, M.C., Dunbar, M.J., Hannaford, J., Black, A., Bragg, O., Rowan, J. and King, J. (2006) *Development of Environmental Standards (Water Resources). Stage 3: Environmental Standards for the Water Framework Directive.* Report to the Scotland and Northern Ireland Forum for Environment Research. Centre for Ecology and Hydrology, Wallingford and University of Dundee, Dundee, UK.

Addley, R.C. (2006) *A Mechanistic Daily Net Energy Intake Approach to Modeling Habitat and Growth of Drift-feeding Salmonids (particularly the genus* Oncorhynchus*).* PhD dissertation, Department of Civil and Environmental Engineering, Utah State University, Logan, Utah.

Annear, T., Chisholm, I., Beecher, H., Locke, A., Aarrestad, P., Coomer, C., Estes, C., Hunt, J., Jacobson, R., Jobsis, G., Kauffman, J., Marshall, J., Mayes, K., Smith, G., Stalnaker, C. and Wentworth, R. (2004) *Instream Flows for Riverine Resource Stewardship,* Instream Flow Council, Cheyenne, WY.

Bartholomew, J.L., Whipple, M.J., Stevens, D.G. and Fryer, J.L. (1997) The life cycle of *Ceratomyxa shasta,* a myxosporean parasite of salmonids, requires a freshwater polychaete as an alternate host. *American Journal of Parasitology,* **83**: 859–868.

Beechie, T., Buhl, E., Ruckelshaus, M., Fullerton, A. and Holsinger, L. (2006) Hydrologic regime and the conservation of salmon life history diversity. *Biological Conservation,* **130**: 560–572.

Bunn, S.E. and Arthington, A.H. (2002) Basic principles and ecological consequences of altered flow regimes for aquatic biodiversity. *Environmental Management,* **30**(4): 492–507.

CDWR and USBR (2011) *Water Resources Integrated Modeling System.* California Department of Water Resources and U.S. Bureau of Reclamation. Technical documentation and model available at: http://modeling.water.ca.gov/hydro/model/index.html.

Dunbar, M.J., Gustard, A., Acreman, M.C. and Elliott, C.R.N. (1998) *Overseas Approaches to Setting River Flow Objectives.* R&D Technical Report W6B(96)4.

Fischenich, C.J. (2006) *Functional Objectives for Stream Restoration.* USAE Research and Development Center, Vicksburg, MS. ERDC TN-EMRRP SR-52.

Guillen, G. (2003) *Klamath River Fish Die-off September 2002, Causative Factors of Mortality.* Report Number AFWO-F-02-03.

Haan, C.T. (2002) *Statistical Methods in Hydrology,* 2nd edition, Iowa State Press.

Harby, A., Baptist, M., Dunbar, M.J. and Schmutz, S. (eds) (2004) *State-of-the-art in Data Sampling, Modelling Analysis and Applications of River Habitat Modeling.* COST Action 626 Report.

Hardy, T.B. (2008a) River restoration: Integration of sediment and riparian modeling for assessment of long-term post restoration conditions for aquatic resources. *International Symposium on the Stream Corridor Restoration,* River

Environmental Technology Institute, Seoul, Korea, November 11, 2008. English and Korean papers printed in Symposium Proceedings.

Hardy, T.B. (2008b) Applied river restoration under high flow constraints – A case study of design and implementation of fish habitat, sediment augmentation, and channel maintenance/riparian flow regimes in the Trinity River, California, USA. *4th International Workshop on River Environment – Prospects for the River Restoration Technique Harmonizing with Flood Control and River Environment*, June 1–4, 2008, Seoul, Korea. Proceedings papers printed in English and Korean.

Hardy, T.B., Addley, R.C. and Saraeva, E. (2006a) *Evaluation of Instream Flow Needs in the Lower Klamath River – Phase II Final Report*. U.S. Department of Interior, July 31, 2006.

Hardy, T.B., Shaw, T.A., Addley, R.C., Smith, G.E., Rode, M. and Belchik, M. (2006b) Validation of Chinook fry behavior-based escape cover modeling in the lower Klamath River. *International. Journal of River Basin Management*, 4(3): 169–178.

Hendrickson, G.L., Carleton, A. and Manzer, D. (1989) Geographic and seasonal distribution of the infective stage of *Ceratomyxa shasta* (Myxozoa) in Northern California. *Diseases of Aquatic Organisms*, 7: 165–169.

Junk, W.J., Bayley, P.B. and Sparks, R.E. (1989) The flood pulse concept in river floodplain systems. In Dodge, D.P. (ed.) *Proceedings of the International Large River Symposium (LARS)*, Canadian Special Publication of Fisheries and Aquatic Sciences, Ottawa, Canada, pp. 110–127.

Lytle, D.A. and Poff, N.L. (2004) Adaptation to natural flow regimes. *Trends in Ecology and Evolution*, 19: 94–100.

Nichols, K. and Foott, J.S. (2006) *FY2004 Investigational Report: Health Monitoring of Juvenile Klamath River Chinook Salmon*. U.S. Fish & Wildlife Service California–Nevada Fish Health Center, Anderson, CA.

Nielson, B., Saraeva, E. and Hardy, T.B. (2007) *Two-zone Dynamic Temperature Model Development and Calibration for the Virgin River*. Utah Water Research Laboratory, Utah State University.

NRC (2004) *Endangered and Threatened Fishes in the Klamath River Basin: Causes of Decline and Strategies for Recovery*, National Academies Press, Washington, DC.

NRC (2005) *The Science of Instream Flows – A Review of the Texas Instream Flow Program*, National Academies Press, Washington, DC.

NRC (2008) *Hydrology, Ecology, and Fishes of the Klamath River Basin*, National Academies Press, Washington, DC.

Perry, T., Leib, A., Harrison, A., Spears, M., Mull, T., Cohen, E., Rasmussen, J. and Hicks, J. (2005) *Natural flow of the Upper Klamath River – Phase I. Natural inflow to, natural losses from, and natural outfall of Upper Klamath Lake to the Link River and the Klamath River at Keno*. USDOI, Bureau of Reclamation.

Peterson, R.V. and Wirth, B.D. (2003) Glen Canyon Dam experimental flows: Employing adaptive management concepts. *Hydro Review*, 22(6): 22–27.

Poff, N. L., Allan, J.D., Bain, M.B., Karr, J.R., Prestegaard, K.L., Richter, B.D., Sparks, R.E. and Stromberg, J.C. (1997) The natural flow regime; a paradigm for river conservation and restoration. *BioScience*, 47: 769–784.

Puckridge, J.T., Sheldon, F., Walker, K.F. and Boulton, A.J. (1998) Flow variability and the ecology of large rivers. *Marine and Freshwater Research*, 49: 55–72.

Richter, B.D., Baumgartner, J.V., Powell, J. and Braun, D.P. (1996) A method for assessing hydrologic alteration within ecosystems. *Conservation Biology*, 10: 1163–1174.

Salas, J.D., Saada, N., Chung, C.H., Lane, W.L. and Frevert, D.K. (2000) *Stochastic Analysis, Modeling and Simulation (SAMS) Version 2000 User's Manual*. Water Resources, Hydrology and Environmental Research Center, Fort Collins, Colorado, USA.

Stocking, R.W. and Bartholomew, J.L. (2007) Distribution and habitat preference of *Manayunkia speciosa* (Polychaeta/Sabellidae) and infection prevalence with the parasite *Ceratomyxa shasta* (Myxozoa) in the Klamath River. *Journal of Parasitology*, 93: 78–88.

Tharme, R.E. and King, J.M. (1998) *Development of the Building Block Methodology for Instream Flow Assessments, and Supporting Research on the Effects of Different Magnitude Flows on Riverine Ecosystems*. Water Research Commission Report 576/1/98, Pretoria.

Thorp, J.H. and DeLong, M.D. (1994) The riverine productivity model: an heuristic view of carbon sources and organic processing in large river ecosystems. *Oikos*, 70: 305–308.

TPWD (2008) *Texas Instream Flow Studies: Technical Overview*. Texas Commission on Environmental Quality, Texas Parks and Wildlife Department, Texas Water Development Board. Report 369, May 2008.

USGS (2012) USGS National Water Information System: Web Interface. http://waterdata.usgs.gov/or/nwis/dv?cb_00060=on&format=rdb&begin_date=1960-10-01&end_date=2000-09-30&site_no=11502500&referred_module=sw)

Vannote, R.L., Minshall, G.W., Cummings, K.W., Sedell, J.R. and Cushing, C.E. (1980) The river continuum concept. *Canadian Journal of Fisheries and Aquatic Sciences*, 37: 130–137.

Waples, R.S., Gustafson, R.G., Weitkamp, L.A., Myers, J.M., Johnson, O.W., Busby, P.J., Hard, J.J., Bryant, G.J., Waknitz, F.W., Neely, K., Teel, D., Grant, W.S., Winans, G.A., Phelps, S., Marshall, A. and Baker, B.M. (2001) Characterizing diversity in salmon from the Pacific Northwest. *Journal of Fish Biology*, 59: 1–41.

Ward, J.V. and Stanford, J.A. (1983) The intermediate disturbance hypothesis: an explanation for biotic diversity patterns in lotic systems. In Fontaine, T.D. and Bartell, S.M. (eds) *Dynamics of Lotic Ecosystems*, Ann Arbor Sciences, Ann Arbor, Michigan, pp. 347–356.

Williams, B.K., Szaro, R.C. and Shapiro, C.D. (2007) *Adaptive Management: The U.S. Department of the Interior Technical Guide*. U.S. Department of the Interior. ISBN 1-4113-1760-2.

Williamson, S.C., Bartholow, J.M. and Stalnaker, C.B. (1993) Conceptual model for quantifying pre-smolt production from flow-dependent physical habitat and water temperature. *Regulated Rivers: Research and Management*, 8(1 and 2): 15–28.

17 Hydraulic Modelling of Floodplain Vegetation in Korea: Development and Applications

Hyoseop Woo[1] and Sung-Uk Choi[2]

[1]Korea Institute of Construction Technology, 2311 Daehwa-dong, Illsanseo-gu, Goyang-si, Gyeonggi-do, Korea
[2]Department of Civil and Environmental Engineering, Yonsei University, 134 Shinchon-dong, Seodaemun-gu, Seoul, Korea

17.1 Introduction

Rivers on the Korean Peninsula are heavily affected by the monsoon climate, which carries moist air from the Pacific Ocean to East Asia, affecting east China, Korea and Japan. Flood control during heavy rainfall periods, especially typhoons, and securing the water necessary for irrigation during dry or drought periods are, therefore, very important. The rapid progress in urbanization and industrialization in Korea since the 1960s has accelerated the degradation of natural ecosystems, including aquatic and riparian zones in rivers. This usually involved asserting control with large-scale river works such as channelization and damming. These have resulted in severe degradation of the natural ecosystems of rivers. Ecohydraulics, or ecological river engineering in a broad sense, has emerged under this situation. It is, however, at an early stage of development in Korea. Scientific knowledge concerning the process of aquatic ecosystem degeneration and methodologies for solving ecological problems in artificially altered rivers are currently under development in Korea.

Figure 17.1 shows areas of concern in ecohydraulics regarding rivers. It shows the physical interactions among the relevant factors of flow, sediment, bio and hydraulic structures. In this figure, solid lines indicate a direct influence from one factor to another, while dotted lines indicate an indirect and/or long-term effect.

As shown in this figure, the traditional hydraulics (or engineering) discipline is only concerned with the interaction between flow, sediment, river morphology and hydraulic structures (the left half of Figure 17.1), all of which are abiotic processes. Ecohydraulics (or ecological river engineering) is additionally concerned with aquatic and riparian ecosystems, and the flora and fauna living in those systems. As expressed in Figure 17.1, ecohydraulics, therefore, has several major elements such as flow resistance due to vegetation, instream flow (or environmental flow if focused on aquatic ecosystems) requirements, floodplain vegetation modelling, small dam removal and river restoration.

This chapter highlights some applications of ecohydraulic modelling to the aquatic environment, especially the river and river management systems in Korea. Among the various relations shown in Figure 17.1, this chapter highlights hydraulic modelling of vegetated flow and hydraulic analysis of floodplain vegetation establishment.

Hydraulic models for vegetated flow, which have been developed recently in Korea, are introduced after a brief state-of-the-art review of vegetated flow. Hydraulic models include bulk flow modelling, shallow water equation modelling and turbulence modelling. A case study of hydraulic analysis floodplain vegetation establishment due to dam construction in an originally sand river in Korea is then introduced after a brief explanation of the

Ecohydraulics: An Integrated Approach, First Edition. Edited by Ian Maddock, Atle Harby, Paul Kemp and Paul Wood.
© 2013 John Wiley & Sons, Ltd. Published 2013 by John Wiley & Sons, Ltd.

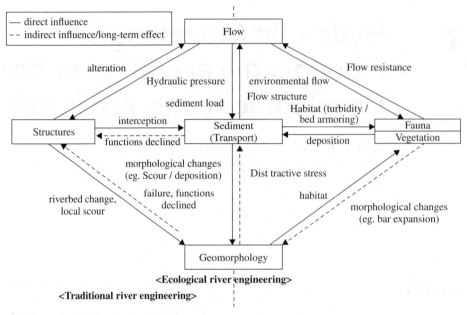

Figure 17.1 Areas of concern for traditional river engineering (or hydraulics) and ecological river engineering (or ecohydraulics) in terms of physical interactions. Modified from Tsujimoto (1999).

terms 'white river' and 'green river'. Shallow water equation modelling is used for flow analysis in the case study.

17.2 Modelling of vegetated flows

Nepf and Vivoni (2000) classified vegetated flows essentially into three types depending upon the depth ratio, defined by H/h_1 (here, H is the total flow depth and h_1 is the height of the vegetation layer). If the depth ratio is large, i.e. $H/h_1 > 5$–10, the flow is classified as terrestrial canopy flow, which is analogous to flow over a rough boundary. If the depth ratio is less than or equal to unity, then they are flows through emergent vegetation. Such flows are characterized by a reduced mean flow with suppressed turbulence. If the depth ratio has an intermediate value, the flow is then classified as a depth-limited flow with submerged vegetation (see Figure 17.2). Complications associated with these types of flows have been discussed in depth by Nepf and Ghisalberti (2008). That is, a depth-limited flow with submerged vegetation shows an inflection point in the mean velocity profile due to the absorption of momentum by the vegetation stems, which then triggers Kelvin–Helmholtz instability. The vortices at the interface dominate the mass and momentum exchanges

between the two regions, which make the bottom friction much less important than the stem drag.

Modelling vegetated flows is grouped into three approaches herein. The first approach is bulk flow modelling, which divides the water column into layers depending on flow characteristics. The flows over terrestrial canopy and flows through emergent vegetation can be modelled with a one-layer approach, since flow characteristics do not vary to any great extent over the depth. However, for flows with submerged vegetation, the mean flow and turbulence structure in the lower vegetation layer

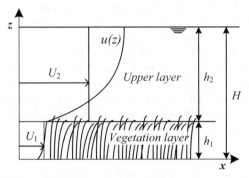

Figure 17.2 Schematic sketch of depth-limited, open-channel flows with submerged vegetation.

are too different from those in the upper layer. Thus, a one-layer approach is not appropriate for such flows. In the following section, a two-layer model for flow and suspended sediment transport with submerged vegetation is introduced. The second approach is to solve numerically one-dimensional or two-dimensional depth-averaged equations. Intrinsic features of the depth-averaged equations enable the flow over the terrestrial canopy or the flow through emergent vegetation to be modelled more easily. Experiences with hydrodynamic modelling can be utilized fully. However, the one-dimensional approach is valid only when the lateral variation of the flow can be ignored. Numerical solution of the three-dimensional equations is the third approach. This approach provides vertical profiles on the mean flow and turbulence structures. For the most part, turbulence models are required. The use of computational algorithms is complicated and the cost is rather higher than that of the other two approaches.

17.2.1 Bulk flow modelling

As stated, incorporating detailed three-dimensional flow characteristics into practical river engineering works is nearly impossible, and this necessitates the development of bulk flow models or resistance laws. In this section, flow and suspended sediment transport with submerged vegetation are given, based on bulk flow modelling.

Flow model

For depth-limited, open-channel flows with submerged vegetation, as depicted in Figure 17.2, force components such as gravity, bottom shear, stem drag and interfacial shear affect the overall flow. For uniform flows, the gravity and interfacial shear are balanced by stem drag and bottom shear in the vegetation layer, i.e.

$$\rho g h_1 S + \tau_i = F_D + \tau_b \qquad (17.1)$$

where ρ is the water density, g is the gravitational acceleration, S is the slope, τ_i is the interfacial shear stress between the vegetation and upper layers, F_D is the drag (per unit bed area) due to vegetation ($= 0.5\rho a \overline{C}_D h_1 U_1^2$) and τ_b is the bottom shear stress (here, $\overline{C}_D$ is the bulk drag coefficient averaged over h_1, a is the planting density of vegetation and U_1 is the layer-averaged velocity in the vegetation layer). Here, the vegetation density (a) is defined by the projected plant area per unit volume. A similar force balance in the upper layer leads to:

$$\rho g (H - h_1) S = \tau_i \qquad (17.2)$$

Note that the interfacial shear stress tends to reduce the mean velocity by balancing the gravity in the upper layer,

whereas it accelerates the flow in the vegetation layer. For flows with densely planted vegetation such as $\overline{C}_D a H > 0.1$, compared with bottom shear, stem drag is dominant in the vegetation layer (Nepf and Ghisalberti, 2008). Thus, if bottom shear is ignored, the velocity averaged over the vegetation layer can be obtained from Equation (17.1) with the aid of Equation (17.2). That is,

$$U_1 = \sqrt{\frac{2gHS}{a\overline{C}_D h_1}} \qquad (17.3)$$

which states that the velocity in the vegetation layer depends on the slope, depth ratio and vegetation properties. Equation (17.3) is the same as the formula derived by Huthoff et al. (2007).

Regarding the mean velocity in the upper layer, previous studies, based on laboratory experiments, revealed that the velocity profile takes a logarithmic function (Lopez and Garcia, 1998; Järvelä, 2003; Nepf and Ghisalberti, 2008). Thus, the following velocity distribution is assumed in the upper layer:

$$\frac{u_2(z)}{u_*} = \frac{1}{\kappa} \cdot \ln\left(\frac{z}{h_1}\right) + \frac{U_1}{u_*} \qquad (17.4)$$

where $u_2(z)$ is the vertical distribution of the mean velocity at the upper layer, u_* is the interfacial shear velocity atop the vegetation layer (at $z = h_1$) and κ is the von Karman constant ($= 0.41$). Note, in Equation (17.4) that u_2 at $z = h_1$ is U_1, determining the log-law constant. Similar to open-channel flows with small relative submergences (Dittrich and Koll, 1997), Yang and Choi (2009) indicated that Equation (17.4) should be modified for flows with high vegetation density, such as for $a > 5.0$ m^{-1}. Thus, Yang and Choi (2009) proposed the following relationship:

$$\frac{u_2(z)}{U_1} = \frac{u_*}{U_1} \cdot \frac{C_u}{\kappa} \cdot \ln\left(\frac{z}{h_1}\right) + 1 \qquad (17.5)$$

with $C_u = 1$ for $a < 5.0$ m^{-1} and $= 2$ for $a > 5.0$ m^{-1}. For the interfacial shear velocity in Equation (17.5), Nepf and Ghisalberti (2008) and Yang and Choi (2009) proposed the following relationship:

$$u_* = \sqrt{gh_2 S} \qquad (17.6)$$

where h_2 ($= H - h_1$) is the height of the upper layer.

Suspended sediment transport model

A model for suspended sediment transport using bulk flow modelling is introduced in this section. Under a steady equilibrium condition of suspended sediment, the vertical distribution of suspended sediment can be obtained by integrating the following equation:

$$v_s c + D_d \frac{\partial c}{\partial z} = 0 \qquad (17.7)$$

where c is the mean concentration of suspended sediment and v_s and D_d denote the particle fall velocity and eddy diffusivity of the suspended sediment, respectively. For the eddy diffusivity, Yang and Choi (2010) used the eddy viscosity concept in the upper layer and assumed a so-called 'linear bridge,' connecting the value of the eddy viscosity at h_1 to zero at the bottom, in the vegetation layer. That is, the following expressions are used for the eddy viscosity (v_t):

$$v_t = z \cdot \frac{u_* \kappa}{C_u} \cdot \left(\frac{H - z}{h_2} \right) \text{ for the upper layer} \qquad (17.8a)$$

$$v_t = z \cdot \frac{u_* \kappa}{C_u} \qquad \text{for the vegetation layer (17.8b)}$$

If the turbulent Schmidt number is assumed to be unity, then Equation (17.7) can be integrated with the help of Equations (17.8a) and (17.8b), resulting in the following expressions for the distribution of suspended sediment:

$$\frac{c}{c_b} = \left(\frac{z_b}{h_1} \right)^Z \cdot \left(\frac{H - z}{z} \cdot \frac{h_1}{h_2} \right)^{Z'} \text{ for the upper layer (17.9a)}$$

$$\frac{c}{c_b} = \left(\frac{z_b}{z} \right)^Z \qquad \text{for the vegetation layer (17.9b)}$$

where the subscript b denotes 'near-bed'. Thus, c_b is the near-bed concentration of suspended sediment, i.e. $c \, (z = z_b) = c_b$, and z_b is normally taken as the height equal to 5% of the flow depth. In Equations. (17.9a) and (17.9b), Z is the Rouse parameter ($= v_s / \kappa u_*$) and Z' is a new parameter defined by $Z \cdot h_2 / H$.

Figure 17.3 shows the vertical distribution of suspended sediment concentration. Concentration profiles predicted by Equations (17.9) are given along with the measured data of Yuuki and Okabe (2002). It can be seen that the predicted profiles are in moderately good agreement with the measured data. Specifically, the mean concentration in the upper layer predicted by the present model is higher than that obtained from the measured data. In contrast, the mean concentration by Equation (17.9b) appears to be under-predicted in the vegetation layer. This can be attributed to the linear distribution of the eddy viscosity in the vegetation layer.

17.2.2 Shallow-water equation modelling

One- and two-dimensional modelling of vegetated flows based on depth-averaged shallow water equations are

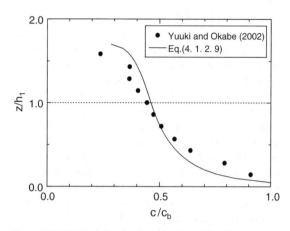

Figure 17.3 Vertical distribution of suspended sediment concentration for $Z = 0.26$.

introduced in this section. Kim et al. (2010) introduced a two-dimensional model for open-channel flows with vegetation. The governing equations of their model are the shallow-water equations with vegetation drag terms such as

$$\frac{\partial H}{\partial t} + \frac{\partial p}{\partial x} + \frac{\partial q}{\partial y} = 0 \qquad (17.10)$$

$$\frac{\partial p}{\partial t} + \frac{\partial}{\partial x} \left(\frac{p^2}{H} + \frac{gH^2}{2} \right) + \frac{\partial}{\partial y} \left(\frac{pq}{H} \right)$$

$$- \frac{\partial}{\partial x} \left(2\bar{v}_t \frac{\partial p}{\partial x} \right) - \frac{\partial}{\partial y} \left[\bar{v}_t \left(\frac{\partial p}{\partial y} + \frac{\partial q}{\partial x} \right) \right] \qquad (17.11)$$

$$+ gH \frac{\partial \eta_b}{\partial x} + \frac{1}{\rho} \tau_{bx} + \frac{1}{\rho} \frac{N_v F_{Dx}}{1 - c_v} = 0$$

$$\frac{\partial q}{\partial t} + \frac{\partial}{\partial x} \left(\frac{pq}{H} \right) + \frac{\partial}{\partial y} \left(\frac{q^2}{H} + \frac{gH^2}{2} \right)$$

$$- \frac{\partial}{\partial x} \left[\bar{v}_t \left(\frac{\partial p}{\partial y} + \frac{\partial q}{\partial x} \right) \right] - \frac{\partial}{\partial y} \left(2\bar{v}_t \frac{\partial q}{\partial x} \right) \qquad (17.12)$$

$$+ gH \frac{\partial \eta_b}{\partial y} + \frac{1}{\rho} \tau_{by} + \frac{1}{\rho} \frac{N_v F_{Dy}}{1 - c_v} = 0$$

where t denotes time, $p \, (= UH)$ and $q \, (= VH)$ denote unit discharges in the x- and y-directions, respectively (with U and V depth-averaged velocities in the respective direction), η_b denotes bed elevation from a certain datum, $\bar{v}_t$ denotes depth-averaged eddy viscosity, c_v denotes the volume ratio of vegetation and water and τ_{bi} and F_{Di} denote bed shear stress and vegetation drag in the i-direction, respectively. Integrating the parabolic distribution of the eddy viscosity from the logarithmic velocity profile leads to the following relationship for the depth-averaged eddy viscosity ($\bar{v}_t$):

$$\bar{v}_t = \frac{\kappa}{6} u_* H \qquad (17.13)$$

where, with the help of Manning's relationship, the shear velocity u_* can be estimated by

$$u_* = \frac{n\sqrt{g}}{H^{7/6}} \sqrt{p^2 + q^2} \qquad (17.14)$$

In Equations. (17.10) to (17.12), the vegetation drag per unit bed area is given by

$$N_v F_D = \frac{1}{2} \bar{C}_D \rho a \min\left(H, h_p\right) U_v^2 \qquad (17.15)$$

where N_v is the amount of vegetation in the unit bed area, h_p is the vegetation height and U_v is the apparent velocity. The relationship in Stone and Shen (2002) can be used to estimate the apparent velocity for both emergent and submerged cases. Expressions for vegetation density (a) for submerged and non-submerged stems can be found in Wu and Wang (2004).

The model was verified by comparing computed flow depth and velocity with laboratory data (Kim *et al.*, 2010). Then, the model was applied to a 7.7 km-long reach of the Han River, Korea, including the bend. The purpose of computation was to gauge the impact of planting trees on the floodplain. The discharge of 37 000 m³ s⁻¹ and the flow depth of 14.67 m, corresponding to the design flood, were used for upstream and downstream boundaries, respectively. Manning's roughness coefficient of 0.03 was used for both main channels and floodplains. It was assumed that low trees are planted at a density of 10 m⁻¹ on floodplains on both sides. The tree height and trunk diameter were assumed to be 0.5 m and 0.01 m, respectively.

Figure 17.4 shows the change in velocity after planting trees on both floodplains. The velocity distribution is normalized by the velocity without trees. It can be seen that trees on the floodplains induce the velocity to increase by up to 30% in the main channel and to decrease by up to 90% on the floodplain. Their impact on the flow depth is not serious, although not given here.

Kim *et al.* (2009) proposed a one-dimensional model for vegetated open-channel flows. Using the implicit upstream weighting method, they solved one-dimensional shallow-water equations, or Saint Venant equations, with the friction slope due to vegetation. They applied their numerical model to a laboratory experiment, a compound channel flow whose floodplains were vegetated. In fact, the experiments were carried out to

Figure 17.4 Change in depth-averaged velocity after planting trees on the floodplain.

observe the impact of floodplain trees in the Anyang Stream, Korea.

The reach is 26.0 m long, and both floodplains of a 7.5 m middle reach were vegetated. Steel cylinders with a diameter of 8 mm were planted uniformly at vegetation densities of 0.032 and 0.064 m⁻¹ on the left- and right-hand-side floodplains, respectively. For computations, a discharge of 0.6976 m³ s⁻¹ at the upstream section was imposed with a stage of 0.633 m at the downstream end. Values of $\bar{C}_D = 1.11$ and $n = 0.027$ were used.

Figure 17.5 shows the water surface profile in the longitudinal direction with and without trees. Both simulated and measured data are provided for purposes of comparison. Measured stages are the average of more than ten measurements across the width. Floodplain trees are observed to raise the upstream water level. In general, predicted profiles conform to measured data well, although they are slightly under-predicted within the vegetated zone.

17.2.3 Vertical structure modelling

In this section, a detailed modelling approach providing the vertical distributions of mean flow and turbulence structure is introduced. In particular, numerical strategies for the solution of the RANS (Reynolds Averaged Navier–Stokes) equations for vegetated flows are given.

Governing equations

These consider the vegetated open-channel flow in which cylinders of simulated vegetation are uniformly distributed. Since it is nearly impossible to take individual elements of vegetation into account in a numerical computation, even for rigid-cylinder-type vegetation, the governing equations averaged over both time and space (Raupach and Shaw, 1982) are frequently solved. Assuming that the flow is at a high Reynolds number, the equations for

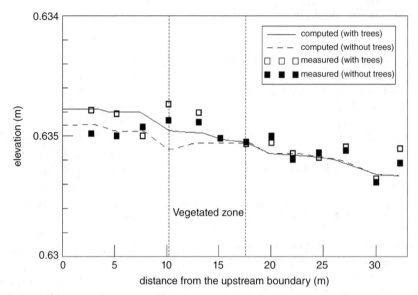

Figure 17.5 Longitudinal distribution of water surface elevation.

the conservation of mass and momentum are expressed as (in the following, angle brackets and overbars indicate the horizontal and temporal averages, respectively, and single primes denote temporal fluctuations):

$$\frac{\partial < \overline{u}_i >}{\partial x_i} = 0 \tag{17.16}$$

$$\frac{\partial < \overline{u}_i >}{\partial t} + < \overline{u}_j > \frac{\partial < \overline{u}_i >}{\partial x_j} = -\frac{1}{\rho}\frac{\partial < \overline{p} >}{\partial x_i}$$

$$+ \frac{\partial}{\partial x_j}\left(\nu\frac{\partial < \overline{u}_i >}{\partial x_j}\right) - \frac{\partial \left\langle \overline{u'_i u'_j} \right\rangle}{\partial x_j} - f_i + g_i \tag{17.17}$$

where $\langle \overline{u}_i \rangle$ is the mean velocity in the i-direction, $\overline{p}$ is the mean pressure, $\left\langle \overline{u'_i u'_j} \right\rangle$ is the Reynolds stress, g_i is the i-component of gravitational acceleration and f_i is the i-component of the drag force due to vegetation given by

$$f_i = \frac{1}{2}C_D a < \overline{u}_i > \sqrt{< \overline{u} >^2 + < \overline{v} >^2 + < \overline{w} >^2} \tag{17.18}$$

where C_D is the drag coefficient distributed vertically. For the drag coefficient, the relationship reported by Dunn (1996) is used with a value of the bulk drag coefficient $\overline{C}_D = 1.13$.

Vertical structures

Choi and Kang (2004) presented the Reynolds stress model for numerical simulations of vegetated open-

channel flows. Figure 17.6 shows the mean velocity (left) and Reynolds stress (right) of vegetated open-channel flows with a depth ratio ($= H/h_1$) of 2.85. In the figure, the profile simulated by the Reynolds stress model is compared with profiles simulated by the $k-\varepsilon$ model and algebraic stress model as well as with measured data from Lopez and Garcia (1997). A velocity defect due to vegetation is quite clearly seen below the height of vegetation. In this region, all computed profiles look identical and they agree well with measured data. However, above the height of vegetation, the velocity demonstrated by the $k-\varepsilon$ model is greater than the velocity demonstrated by the algebraic stress model and the Reynolds stress model. It is seen that the results from the Reynolds stress model match the observed data best. The Reynolds stress, which is zero at the water surface, increases up to the height of vegetation and it decreases towards the water surface. Over the entire depth, the simulated profiles generated by the Reynolds stress model and the algebraic stress model are nearly the same, and they match the measured data slightly better than the $k-\varepsilon$ model.

Compound channel flow with floodplain vegetation

Using the Reynolds stress model, Kang and Choi (2006) presented mean flow and turbulence structures in compound open-channel flows with a vegetated floodplain. The compound channel had base and total widths of

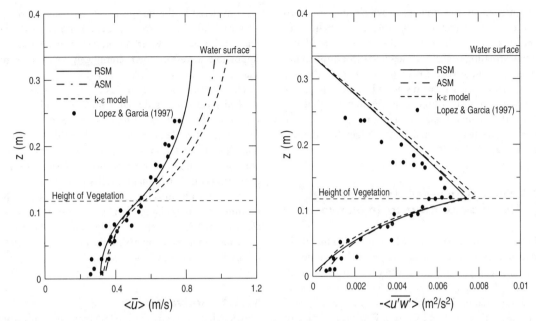

Figure 17.6 Mean flow (left) and Reynolds stress (right) of open-channel flows with submerged vegetation.

0.2 m and 0.4 m, respectively. The respective flow depths were 0.08 m and 0.04 m. Vegetation densities were varied to observe their impact from $a = 0.25$ m^{-1} to 4.0 m^{-1}.

Figure 17.7 shows the obtained contours of the streamwise mean velocity (up) and secondary current vectors (down). Note that the model clearly reproduces the velocity dip. In the figure, it is interesting to note that the mean

flow structure is considerably different from that for a plain compound channel. That is, near the junction, the bulge in contour lines towards the free surface is absent. It can also be seen that the streamwise mean velocity in the floodplain is low and the point of maximum velocity is located near the right side wall. The lateral shift observed in the figure is caused by secondary currents in

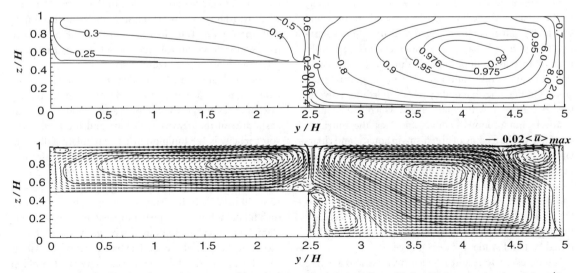

Figure 17.7 Mean flow (up) and secondary currents (down) of compound channel flow with floodplain vegetation ($a = 1.0$ m^{-1}).

the main channel, the strength of which increases with vegetation density (Naot *et al.*, 1996). The pattern of secondary currents observed in the figure shows that a clockwise-rotating vortex occupies nearly the entire main channel. This large vortex is made by merging the free surface vortex at the left-hand side and the bottom vortex of the right-hand side of the main channel. The bottom vortex at the left-hand side and free surface vortex at the right-hand side become smaller as the vegetation density on the floodplain increases.

17.3 Floodplain vegetation modelling: From white rivers to green rivers

In Korea, rivers were naturally 'white', meaning that there were very large and wide sand and/or gravel bars along the riverside, as compared to the width of the river channel, because of the high seasonal variation in flow regime. On the other hand, rivers in Europe are generally 'green', which means that the floodplains of European rivers are covered with vegetation, usually to the very edge of the river banks. In Europe, seasonal variations in river flows are much smaller compared with those in Korea (Woo and Lee, 1993).

Since river development and channelization works have accelerated since the 1960s in Korea, the flow and sediment regimes of the rivers have been significantly changed, and now they show accelerated vegetation establishment and expansion on the white sandbars of floodplains. Now, almost every river reach in Korea has 'green' riparian floodplains instead of the original 'white' ones, especially in the small and medium-sized rivers, where flow and sediment regimes can be changed easily by human activity. 'Green rivers' may pose serious engineering problems by increasing flood hazards through the increased flow resistance and fixation and heightening of riparian lands. On the other hand, they can support more diverse species compared with 'white rivers', which usually have relatively monotonous habitats and lack the nutrients required for the feeding of organisms in aquatic and riparian food chains. For example, a 'green river' can create new wetlands on once bare sandbars in the floodplain by inducing vegetation recruitment and creating backmarsh and berms in the floodplain. A good example of this is the Gudam Wetland downstream on the upper Nakdong River. The wetland has been already designated a 'Protected Wetland' under the relevant law.

Is it necessary to restore a 'green river' back to a 'white river'? The answer may be yes or no. The question should be considered in the perspective of long-term policy of national land and river management. For many of the river reaches in Korea, however, it would seem to be practically impossible to restore them completely to their original shape and composition (Woo, 2010).

The governing factors for this change are known to be the bed shear stress, sediment deposition and soil moisture content in the bars (or floodplains). Some numerical models for this phenomenon have been developed as a 'Floodplain Vegetation Model' (Benjanker *et al.*, 2007), with only the flow shear stress and succession of vegetation being considered qualitatively so far.

The case study introduced in this section describes how a 'white river' may change to a 'green river' in Korea and simulation of the recruitment and expansion of vegetation on white sandbars using hydraulic analysis. This section is based on a paper by Woo *et al.* (2011).

17.3.1 Motivation

The Nakdong River is the longest river in South Korea, with a length of 506 km. The basin area of the river is 23 394 km^2, the second largest after the Han River. It is located, as shown in Figure 17.8, in the south-east part of the Korean Peninsula and the river flows from north to south. Well-formed, narrow river valleys escort the river channels both in the main stream and tributaries from upstream to downstream. The river bed slope is low, with values of about 1/1000–1/17 000 except in the far upstream reach. The riverbed is composed mostly of sands except in the far upstream reach in the mountain area where it is composed of gravels and cobbles. In the Upper Nakdong River, two dams have been built since the 1970s, beginning with the Andong Dam in 1976 and then the Imha Dam in 1992, for flood control, water supply and hydroelectric power generation (see Figure 17.8).

Andong Dam is an 83 m-high, 612 m-long rock fill dam. It was completed in 1976 and since then no spillway discharge has been made except in a few flood events. The basin area of the dam is 1584 km^2 and the total reservoir capacity is 1284 million m^3. The regulating dam has a function of regulating the power-generating discharge from the main dam for a constant discharge to downstream. The discharge from the regulating dam is about 15 to 30 m^3 s^{-1}. Imha Dam is a 73 m-high, 515 m-long rock fill dam. It was completed in 1992 and, like Andong Dam, since then no spillway discharge has been made, except in a few flood events. The basin area of the dam is 1361 km^2 and the total reservoir capacity is 595 million m^3. The discharge from the regulating dam is about 2 to

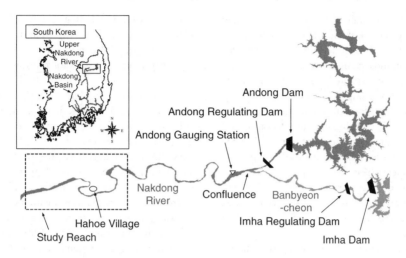

Figure 17.8 Map of the Upper Nakdong River.

10 m^3 s^{-1}. Flow duration analysis shows that, in general, the two dams reduce flood flows greater than the 12% exceedance, while they increase river flows otherwise.

Morphological changes occurred in the river when dams were constructed 25–30 km upstream – the Andong Dam on the main river course in 1976 and the Imha Dam on a major tributary in 1992. These two dams completely modified the flow and sediment regimes downstream. Since then, the river has responded morphologically to the regime changes with riverbed degradation, bed material coarsening and changes in channel width (Woo *et al.*, 2010), which are well-known phenomena caused by dams on alluvial rivers. However, a gradual change to the two large sandbars has also occurred, particularly after the second dam was constructed; there has been vegeta-

tion recruitment and establishment on the once unvegetated sandbars along the river (see Figure 17.9, right). The recruitment and establishment of this vegetation, mostly reeds and willows, has caused the 'white river' to become a 'green river'.

Many studies have been reported on the recruitment and establishment of vegetation on floodplains after the flow and sediment regimes are changed by upstream dams and/or water withdrawal. One of the first such studies was made by Williams and Wolman (1984) of USGS. They investigated ten river reaches with dams constructed upstream and found that a major cause of the vegetation recruitment on the floodplains was the decrease in bed shear stress due to flow regime change after damming. They also reported that vegetation can decrease the flood

Figure 17.9 Hahoe Village and two sandbars A and B (left: taken in the 1980s; right: taken in October 2009). Flow from lower-right to upper-left.

conveyance of the river channel. Prowse *et al.* (2002) and Graf (2006) investigated the extension of vegetative areas due to dams, which resulted in deepened and narrowed low-flow channels in most cases.

Choi *et al.* (2005) investigated a similar case at the Hwang River in Korea. Using 1D computer modelling, they found that the dam, constructed on the sandy river, has completely controlled the river discharge to an amount of about 20 m^3 s^{-1} throughout the entire period of the year, causing a dramatic decrease in the flow intensity and thus the bed shear stress on the expanded sandbars. This condition has provided for the growth of common riparian plants in Korea, such as willows and reeds, which have a good chance of being recruited and established on the sandbars in the river.

Recently, Benjankar *et al.* (2007) developed a mathematical model for the simulation of vegetation recruitment on floodplains. The hydraulic part of the model is based mainly on the distribution of the bed shear stress during floods, which has been known to directly affect the stability of the sandbar surface and thus the recruitment of vegetation.

Reviews of the existing literature on the vegetation recruitment on floodplains and sandbars show that excessive bed shear stress can cause the physical instability of germinating seeds and/or seedlings by burying or scour. Insufficient moisture, supplied either from precipitation or the capillary fringe of the groundwater beneath the floodplains, can cause the vegetation to be weathered (Amlin and Rood, 2002; 2003). Supply of fresh fine sediment may be essential to a nursery environment for some plants (Fenner *et al.*, 1985; Rood and Mahoney; 1990). Finally, extremely cold weather during the winter season can freeze plants to death (Johnson, 1994).

Here, changes in the bed shear stress due to upstream dams, as an important limiting factor for vegetation recruitment and establishment, are investigated in the study reach at each different period pre- and post-damming. Among many possible causes for vegetation recruitment on the naked sandbars, we focus mainly on bed shear stress and soil moisture content. The supply of fine sediment during the germinating season may not be crucial to vegetation recruitment on sandbars, unlike the vegetated floodplain, and the air temperature during the study reach during winter did not usually fall below $-10\,^{\circ}$C.

17.3.2 Study reach

Hahoe Village is a traditional agricultural village with a history of over six-hundred years and has been registered as a World Cultural Heritage Site by UNESCO. Included in the site are the large white sandbars denoted by A and B in Figure 17.9, considered areas of natural heritage since they served as spaces for festivals. The river, too, is considered part of the heritage site. These two sandbars, both of which are considered point bars in terms of river engineering, are thought to have maintained their shape and white colour for the last six-hundred years since the village was first founded.

As shown in Figure 17.8, the study reach of Hahoe is located downstream of the Upper Nakdong River. It is 39.1 km downstream from the confluence of the two upstream rivers, the main river with the Andong Dam and the Banbyeon-cheon stream, a major tributary with the Imha Dam. The spatial scope of this study is a reach of 15.2 km in length. The temporal scope is 47 years from 1961, before the impoundment of Andong Dam to 2008, after the impoundment of Imha Dam. For the purposes of comparison, two large sandbars, denoted by A and B, in front of Hahoe Village have been selected. It can be seen in Figure 17.9 that sandbar B has been less affected by vegetation recruitment than sandbar A, which will be investigated in detail next.

17.3.3 Hydrologic and hydraulic analyses

Soil moisture content and bed shear stress were analyzed as the possible factors causing vegetation recruitment on the previously unvegetated sandbars. First, the soil moisture content was indirectly analyzed using a comparison of the sandbar surface level and the average river-water level in May, the germinating season for reeds and willows, at some selected cross-sections. Secondly, a flow analysis was conducted in order to assess the velocity distribution in the study reach; this was followed by a sediment analysis to assess the bed shear stress distribution in the reach.

Analysis of soil moisture content – groundwater tables in the sandbars

Soil moisture content in the sandbars, especially during the germinating period for reeds and *Salix* in May, is considered an important factor for vegetation recruitment on the naked sandbars. Karrenberg *et al.* (2002) reported that the death rate of *Salix* seedlings in the first year is 77–100%, due mostly to dryness. Soil moisture content in the sandbars is affected mainly by the ground-water table in the sandbars (Mahoney and Rood, 1998).

The average river stage in the study reach in May after the completion of Imha Dam in 1992 was calculated using the water-level data in order to estimate the groundwater table in the sandbars along the river channel. The ground-water table in the sand rivers is known to be roughly equal the neighbouring river water table (Mahoney and Rood, 1998). The capillary-fringe height in the unsaturated coarse and medium sands, the common particles in the study reach, is known to be about 12.5–25 cm. The river bed topography has been changed significantly during the period from 1993 to 2007. Keeping in mind that in 1992, Imha Dam, the second dam in the reach, was completed, the topography in 1993 represents that just prior to the major change in the bed topography which occurred due to damming, while the topography in 2007 represents that after the major change occurred. Generally, the two dams caused the river beds to experience aggradation, both in the main channel and on the sandbars in the reach.

There is a difference of about 0.0–2.3 m between the sandbar surface level and the water level in sandbar A. The groundwater level in sandbar A during the germinating season in May for reeds and *Salix* approaches the sandbar surface level in the downstream direction, which implies that the environment of soil moisture in the sands is better in the downstream direction in sandbar A.

There is a difference of about 0.0–4.0 m, however, between the sandbar surface level and the water level in sandbar B. By simply comparing these results between A and B without considering other potentially critical factors for vegetation recruitment, sandbar A, especially in the downstream direction, appears to be more favourable than sandbar B in terms of soil moisture in the sandbars for the purposes of vegetation recruitment.

Flow analysis

The annual maximum daily discharges at the study reach before Andong Dam (1965–1975), between Andong and Imha Dams (1976–1990) and after the Imha Dam (1991–2009) were 1390, 970 and 470 m^3 s^{-1}, respectively. However, the annual maximum daily discharges were extremely small during the four years 1994–1997, about 115 m^3 s^{-1} on average. This implies that, during the period 1994–1997, there might have been a good chance for vegetation to be recruited and established on the naked sandbars A and B because of such a low flow intensity.

Here, shallow-water equation modelling, as described in Section 17.2.2, was applied to the study reach. That is, 'CCHE GUI' version 3.26.6 (Zhang, 2010), a two-dimensional numerical model, was used to simulate flow fields in the study reach. The rating curve measured in 2007 at Gudam Gauging Station (#622) was used for

the downstream boundary condition. Three periods were selected as

1 From 1961 to 1975 before the impoundment of Andong Dam.

2 From 1976 to 1990 before the impoundment of Imha Dam.

3 From 1991 to 2008 after the impoundment of Imha Dam.

Average annual peak flows for each period (1390, 970 and 470 m^3 s^{-1}, respectively) were used for the computation. The bed topography surveyed in 1993 was used for the simulations of the first two periods, while that surveyed in 2007 was used for the simulation of the third period.

Strong currents prevail over almost the whole reach and sandbars during floods, with a range of flow velocity of 1.0 to 1.5 m s^{-1}. On the other hand, the average annual flood discharge of 470 m^3 s^{-1} does not reach to the edges of the channel. This can act as a good opportunity for vegetation recruitment on the slightly submerged or exposed sandbars during floods. More detailed analysis on sediment is given in the following section.

Sediment analysis

The same model was used to investigate the river-bed shear stress distribution in the reach. For the bed roughness and topographic data on the study reach, the official report on the Upper Nakdong River (BRCMA, 2009) was used.

Like the flow analysis above, the same three periods and average annual peak discharges during each period were used, while the bed topographic data for 1993 and 2007 were used for the simulation of the first two periods and third period, respectively. From Equation (17.14), the river-bed shear stress τ_0 is calculated by:

$$\tau_0 = \gamma SH = \frac{\gamma n^2}{H^{1/3}} U\sqrt{U^2 + V^2} \qquad (17.19)$$

where γ is the unit weight of water. Dimensionless bed shear stress, τ_*, is calculated by:

$$\tau_* = \frac{\tau_0}{(\gamma_s - \gamma) D_s} \qquad (17.20)$$

where γ_s and D_s are the unit weights of sediment particle and particle diameter, respectively. The median diameter (D_{50}) of the sediment on the sandbars in the study reach is 1.2 mm, while $D_{10} = 0.6$ mm and $D_{90} = 2.3$ mm. According to the Shields diagram for the initiation of motion of sediment particles, large particles begin to move when the dimensionless bed shear stress τ_* is greater than 0.06. For flow and sediment particles with $\tau_* > 0.06$, the river

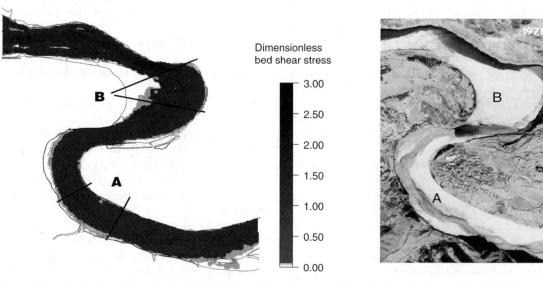

Figure 17.10 Distribution of dimensionless bed shear stress (left) and a comparable aerial photo (taken in 1971, right) before the impoundment of Andong Dam.

bed is unstable due to the motion of bed sediment particles, which results in the recruitment and establishment of vegetation hardly being possible on such river beds or sandbars.

Figure 17.10 (left) shows the distribution of τ_* for the annual average natural flood flow before Andong Dam was impounded in 1976. For comparison with the aerial photo (right), this picture focuses on the large point bars A and B in front of Hahoe Village. In this picture, the 'white'-coloured portion shows the river beds and sandbars not submerged during the floods or with very shallow water depth, where τ_* is almost 0, while the 'grey'-coloured area is where τ_* is less than 0.06. Comparing this map with the aerial photo taken in 1971 (Figure 17.10, right) or the ground photo taken in the 1980s (Figure 17.9, left), it can be seen that most of the sandbars in the reach are 'black', which means unfavourable conditions for vegetation recruitment. A small portion on the upland of sandbar B is shown in white, which means that the flow does not reach it or the flow depth is very shallow during floods. Sandbars were not formed along the grey-coloured narrow strip across the point bar A, indicating that vegetation cannot be recruited there.

Figure 17.11 shows the situation after the construction of Andong Dam and before Imha Dam. Comparing this with Figure 17.10, it can be seen that the 'white' and 'grey' portions of sandbar B were extended somewhat due to the decrease in flood discharge by Andong Dam. No

vegetation is evident in the aerial photo (right) that was taken in 1988, more than ten years after Andong Dam was built. A small 'grey'-coloured portion can be found on the upland side of sandbar B, which was known as an unfavourable area for vegetation recruitment due to the lack of soil moisture. As in the previous case, the 'grey'-coloured strip across sandbar A has no sandbars for vegetation to be recruited. The grey portion at sandbar B is also known as an unfavourable area for vegetation recruitment due to a lack of soil moisture.

Figure 17.12 shows the situation after the impoundment of Imha Dam, where the vegetative area was largely expanded on sandbar B, which was not submerged during floods. This is the result of the drastic decrease in flood discharge after the Imha Dam, i.e. from 970 m^3 s^{-1} to 470 m^3 s^{-1}. As a result, vegetation patches appear on sandbars A and B, which are close to the water, as shown in the aerial photo of Figure 17.12 (right) and, more abundantly, in the ground photo of Figure 17.9 (right), taken recently in 2009. Another significant difference between Figure17.10 and 17.12 is the vegetation shown across sandbar A. This is the outer edge of the point bar A, shown in 'white' and 'grey' in Figure 17.12 (left), which was usually submerged during floods before the impoundment of Imha Dam. Now it has numerous sandbars along the left shore and a narrow island in the channel due to the decrease in flow velocities and has become a hospitable place for vegetation recruitment, as confirmed in Figure 17.9 (right).

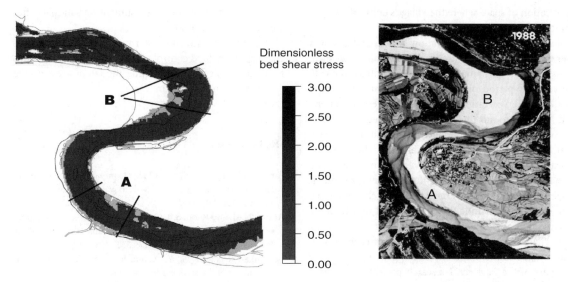

Figure 17.11 Distribution of dimensionless bed shear stress (left) and a comparable aerial photo (taken in 1988, right) after the impoundment of Andong Dam.

17.3.4 Summary

The change from a 'white river' to a 'green river' at the study reach of Hahoe Village, after dams were constructed upstream and the flow and sediment regimes were greatly changed, was especially accelerated after the second dam was constructed in 1992; this resulted in a decrease in downstream flood flows by two-thirds.

This case study revealed that the unfavourable conditions for vegetation recruitment and establishment, the unstable sandbar surface due to strong bed sediment movement, had prevailed in the two sandbars until the construction of the Imha Dam. Before the impoundment of the Imha Dam in 1992, they remained virtually 'white', without any vegetation on them, which was essential for

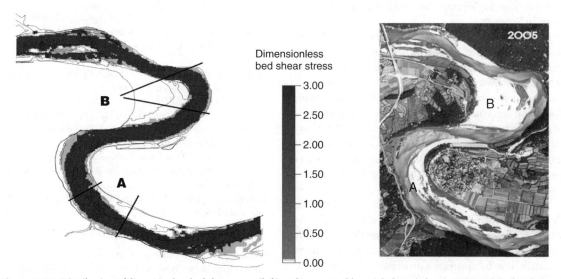

Figure 17.12 Distribution of dimensionless bed shear stress (left) and a comparable aerial photo (taken in 2005, right) after the impoundment of Imha Dam.

the creation of space where the village's cultural and traditional life was expressed.

Since 1992, however, vegetation has expanded slowly over the two sandbars, with the dimensionless bed shear stress less than 0.06 in many parts of the sandbars, and with the very dry period of 1994–1997 providing favourable conditions for vegetation recruitment and establishment.

It has been found that dimensionless bed shear stress can be a useful factor to predict the recruitment of vegetation on unvegetated sandbars in cases where the soil moisture content is not critical to vegetation recruitment.

In order to make a more rigorous model for simulating and predicting the riparian vegetation changes, however, one should include the effects of soil moisture and precipitation as well as sediment transport. In Korea, a more rigorous model for the simulation of riparian vegetation establishment and succession is being developed as a part of the *Ecoriver21* research project (for further details, please refer to the website in the references section at the end of this chapter).

17.4 Conclusions

River environments in Korea have experienced transitions, due both to dam construction and climate change. This has necessitated the scientific assessment, not based on empiricism, of vegetation within watercourses by hydraulic modelling. This chapter has introduced previous engineering attempts to model vegetated flows in rivers in Korea.

First, various modelling approaches for vegetated flows were presented. For flows with submerged vegetation, expressions for velocity and suspended sediment concentration profiles were provided using bulk flow modelling. One- and two-dimensional modelling of river flows with floodplain vegetation in Korea were introduced. Impacts of planting trees on the floodplain were assessed successfully by solving shallow-water equations. Then, detailed modelling of vegetated flows using turbulence closure was given, revealing the vertical structures of the mean flow and turbulence statistics.

Next, a case study of floodplain vegetation modelling in Korea was introduced. Hahoe Village, located downstream of the Upper Nakdong River, experienced a shift to green (vegetative) river from white (sandy) river due to regulated flows by upstream dam construction. Hydraulic modelling revealed that dimensionless shear stress rarely exceeds its critical value, indicating a higher chance of veg-

etation recruitment and establishment on sandbars after consecutive dammings.

Recently, the direction of river works in Korea has changed from flood control to river restoration. With this trend, demands for vegetated flow modelling are expected to increase continuously in Korea.

References

Amlin, N.M. and Rood, S.B. (2002) Comparative tolerances of riparian willows and cottonwoods to water table decline. *Wetlands*, **22**(2): 338–346.

Amlin, N.M. and Rood, S.B. (2003) Drought stress and recovery of riparian cottonwoods due to water table alteration along Willow Creek, Alberta. *Trees*, **17**:, 351–358.

Benjankar, R., Egger, G., Yi, X. and Jorde, K. (2007) Reservoir operations and ecosystem losses: concept and application of a dynamic floodplain vegetation model at the Kootenai River, USA. *Proceedings of the 6th International Symposium on Ecohydraulics*, Christchurch, New Zealand.

Busan Regional Construction Management and Administration (BRCMA) (2009). *Basic Plan of the Nakdong River (revised) (from Mouth of the Nakdong River to the Confluence with the Kumho River) (in Korean)*.

Choi, S.U. and Kang, H. (2004) Reynolds stress modelling of vegetated open-channel flows. *Journal of Hydraulic Research*, **42**(1): 3–11.

Choi, S.U., Yoon, B.M. and Woo, H. (2005) Effects of dam-induced flow regime change on downstream river morphology and vegetation cover in the Hwang River, Korea. *River Research and Applications*, **21**: 315–325.

Dittrich, A. and Koll, K. (1997) Velocity field and resistance of flow over rough surfaces with large and small relative submergence. *International Journal of Sediment Research*, **12**(3): 21–33.

Dunn, C.J. (1996) *Experimental Determination of Drag Coefficients in Open Channels with Simulated Vegetation*. Masters thesis, University of Illinois, Urbana-Champaign, Urbana, IL.

Fenner, P.W., Brady, W. and Patton, D.R. (1985) Effects of regulated water flows on regeneration of Fremont cottonwood. *Journal of Range Management*, **38**(2): 135–138.

Graf, W.L. (2006) Downstream hydrologic and geomorphic effects of large dams on American Rivers. *Geomorphology*, **79**(3): 336–360.

Huthoff. F., Augustijin, D.C.M. and Hulscher, S.J.M.H. (2007) Analytical solution of the depth-averaged flow velocity in case of submerged rigid cylindrical vegetation. *Water Resources Research*, **43**(6), W06413.

Järvelä, J. (2003) Influence of vegetation on flow structure in floodplains and wetlands. In Sanchez-Arcilla, A. and Bateman, A. (eds) *Proceedings of the Symposium on River, Coastal and Estuarine Morphodynamics*, Madrid, Spain, pp. 845–856.

Johnson, W.C. (1994) Woodland expansions in the Platte River, Nebraska: patterns and causes. *Ecological Monographs*, **64**: 45–84.

Kang, H. and Choi, S.U. (2006) Turbulence modelling of compound open-channel flows with and without vegetation on the floodplain using the Reynolds stress model. *Advances in Water Resources*, **29**: 1650–1664.

Karrenberg, S., Edwards, P.J. and Kollmann, J. (2002) The life history of salicaceae living in the active zone of floodplains. *Freshwater Biology*, **47**(4): 733–748.

Kim, J.S., Kim, W., Lee, H. and Kim, E.-M. (2009) One-dimensional upwind implicit scheme for estimating flow resistance by tree. *Proceedings of KWRA Annual Meeting* (in Korean).

Kim, T.B., Bae, H. and Choi, S.U. (2010) Development and application of depth-integrated 2-D numerical model for the simulation of hydraulic characteristics in vegetated open channels. *Journal of the Korean Society of Civil Engineers*, **30**(6B): 607–615 (in Korean).

Lopez, F. and Garcia, M.H. (1997) *Open-Channel Flow Through Simulated Vegetation: Turbulence Modelling and Sediment Transport*. Wetlands Research Program Technical Report WRP-CP-10, Waterway Experimental Station, Vicksburg, MS.

Lopez, F. and Garcia, M.H. (1998) Open-channel flow through simulated vegetation: Suspended sediment transport modelling. *Water Resources Research*, **34**(9): 2341–2352.

Mahoney, J.M. and Rood, S.B. (1998) Steam flow requirements for cottonwood seedling recruitment in integrative model. *Wetlands*, **18**(4): 634–645.

Naot, D., Nezu, I. and Nakagawa, H. (1996) Hydrodynamic behaviour of partly vegetated open channels. *Journal of Hydraulic Engineering*, **122**(11): 625–633.

Nepf, H. and Ghisalberti, M. (2008) Flow and transport in channels with submerged vegetation. *Acta Geophysica*, **56**(3): 753–777.

Nepf, H.M. and Vivoni, E.R. (2000) Turbulence structures in depth-limited, vegetated flow. *Journal of Geophysical Research*, **105**(C12): 28547–28557.

Prowse, T.D., Conly, F.M., Church, M. and English, M.C. (2002) A review of hydroecological results of the northern river basins study, Canada, Part 1. Peace and Slave Rivers. *River Research and Applications*, **18**(5): 429–446.

Raupach, M.R. and Shaw, R.H. (1982) Averaging procedures for flow within vegetation canopies. *Boundary Layer Meteorology*, **22**: 79–90.

Rood, S.B. and Mahoney, J.M. (1990) Collapse of riparian poplar forests downstream from dams in Western Prairies: Probable causes and prospects for mitigation. *Environmental Management*, **14**(4): 451–464.

Stone, B.M. and Shen, H.T. (2002) Hydraulic resistance of flow in channels with cylindrical roughness. *Journal of Hydraulic Engineering*, **128**(5): 500–506.

Tsujimoto, T. (1999) Fluvial processes in streams with vegetation, *Journal of Hydraulic Research*, **37**: 789–803.

Williams, G.P and Wolman, M.G. (1984) *Downstream Effects of Dams on Alluvial Channels*, USGS Professional Paper 1286, Department of the Interior, USA.

Woo, H. (2010) Trends in ecological river engineering in Korea. *Journal of Hydro-environment Research*, **4**(4): 269–278.

Woo, H. and Lee, J.-W. (1993) An analysis of flow characteristics of major rivers in Korea. *Proceedings of the 1st International Conference on Hydroscience and Engineering (ICHE)*, Washington, DC, USA, Part B, pp. 1271–1276.

Woo, H., Kim, J.-S., Cho, K.-H. and Cho, H.-J. (2011) Hydraulic and hydrologic investigations into vegetation recruitment on 'white' sandbars at the historical village of Hahoe on the Nakdong River, Korea. *Proceedings of the 34th IAHR Congress*, Brisbane, Australia, June 26 to July 1.

Woo, H., Park, M., Cho, K.-H., Cho, H. and Chung, S. (2010) Recruitment and succession of riparian vegetation in alluvial river regulated by upstream dams – focused on the Nakdong River downstream Andong and Imha Dams. *Journal of the Korean Association of Water Resources*, **43**(5): 455–470 (*with an English abstract*).

Wu, W. and Wang, S.Y. (2004) A depth-averaged two-dimensional numerical model of flow and sediment transport in open channels with vegetation. In Bennett, S.J. and Simon, A. (eds) *Riparian Vegetation and Fluvial Geomorphology*, AGU, Washington, DC.

Yang, W. and Choi, S.U. (2009) Impact of stream flexibility on mean flow and turbulence structure in open-channel flows with submerged vegetation. *Journal of Hydraulic Research*, **47**(4): 445–454.

Yang, W. and Choi, S.U. (2010) A two-layer approach for depth-limited open-channel flows with submerged vegetation. *Journal of Hydraulic Research*, **48**(4): 466–475.

Yuuki, T. and Okabe, T. (2002) Hydrodynamic mechanics of suspended load on riverbeds vegetated by woody plants. *Journal of JWRA*, **46**: 701–706 (in Japanese).

Zhang, Y. (2010) CCHE·GUI Version 3.26.6 Build 01/28/2010 (www.ncche.olemiss.edu). For further details of the *Ecoriver21* research project mentioned in this chapter, please visit: http://www.ecoriver21.re.kr/

18 A Historical Perspective on Downstream Passage at Hydroelectric Plants in Swedish Rivers

Olle Calles[1], Peter Rivinoja[2] and Larry Greenberg[1]

[1]Department of Biology, Karlstad University, S-651 88 Karlstad, Sweden
[2]Department of Wildlife, Fish and Environmental Studies, SLU (Swedish University of Agricultural Sciences), Umeå 901 83, Sweden

18.1 Introduction

Hydropower stations, with their dams and turbines, constitute barriers and sources of mortality for both upstream and downstream migrating fish. Historically, less attention has been directed towards implementing measures to assist downstream fish passage as compared to upstream fish passage (Wolf, 1946; Odeh, 1999), although the situation for downstream migrants is often more difficult than for upstream migrants and may require complex solutions. For downstream migrants, the actual route selected, which is associated with a certain risk of mortality, is thought to be influenced strongly by local flow conditions (DWA, 2005). Typically, downstream-migrating fish pass dams by swimming through turbines, spillways or some type of bypass system (Clay, 1995; Larinier, 2008). Turbine mortality depends on the type and operation of the turbine as well as on the size of the fish (Montén, 1985). Fish may also die or incur damage as they are pressed against bar racks that are placed at turbine intakes (Calles et al., 2010b). Spillways are also associated with mortality risks, either directly due to fish free-falling against concrete floors before re-entering the water (Calles and Greenberg, 2009) or indirectly due to pressure changes (Coutant and Whitney, 2000).

There are various methods for diverting fish from turbine intakes (Clay, 1995; Odeh, 1999). One such method involves the use of a 'behavioural barrier', whereby one elicits a behavioural response by the fish, using, for example, sound, light or water turbulence, so that they select a route associated with a low risk of injury or death (Welton et al., 2002). Such methods have had variable success as they depend on local conditions that permit active choice as well as on the stimulus having a strong and consistent effect. A second method of diverting fish is to construct physical barriers, using, for example, meshed grating or nets to force the fish to swim along a particular route (Odeh, 1999). Yet another way of preventing fish from entering turbines is to capture them in traps and transport them past the power plant (Muir et al., 2006).

In general, downstream passage problems have been studied mainly in Europe and North America, but not so much in the rest of the world, and have focused on anadromous species, particularly salmonids (Wolf, 1946; Larinier and Marmulla, 2004; Roscoe and Hinch, 2010). Relatively few studies have been conducted on other migratory species. Moreover, in many cases, bypass facilities for downstream migrants have been implemented without evaluating their effectiveness (Montén, 1988) or, if effectiveness has been evaluated, it has often been shown to be low (Larinier, 1998; Kemp et al., 2008; Roscoe

Ecohydraulics: An Integrated Approach, First Edition. Edited by Ian Maddock, Atle Harby, Paul Kemp and Paul Wood.
© 2013 John Wiley & Sons, Ltd. Published 2013 by John Wiley & Sons, Ltd.

and Hinch, 2010). The design of facilities for effective downstream passage must not only take into account the swimming ability and behaviour of one or more migrant species, but also the physical and hydraulic conditions at the water intake (Scruton *et al.*, 2007; Russon and Kemp, 2011). In this chapter, we provide a brief historical review of downstream bypass problems in Sweden. This is followed by a description of the Swedish situation today, with a focus on several case studies that have evaluated downstream bypass efficiency in rivers of different sizes and for several different species. We conclude this chapter by providing the reader with the needs and potential directions for future studies.

18.2 Historical review of downstream bypass problems in Sweden

18.2.1 Fish migration and 'kungsådra' (pre-1900s)

The concept of an open passageway within a river, the so-called 'kungsådra' (literally translated as 'king's vein'), is an old one, first described in a Swedish text from 2 August 1442 (Jakobsson, 1996) and in Scotland as early as the 12th century (Wolf, 1946). During the 1600s, we know that there was a general interest in preserving the kungsådra, as evinced by the recorded debates in the House of Parliament (Wolf, 1946). However, it wasn't until 1734 that it became incorporated into Swedish legislation, a law which stated that no river could be entirely blocked, and at least one-third of the river's width had to be kept open for public services such as shipping, fishing and timber floating (Montén, 1988). The term was initially applicable to all rivers with a minimum discharge > 5 m^3 s^{-1}. The king could, of course, give dispensation from this rule, but only for a minor kungsådra (i.e. a sixth of the river width), since this was considered the minimum width required for free passage of migratory species such as Atlantic salmon (*Salmo salar*), anadromous brown trout (*Salmo trutta*) and European eel (*Anguilla anguilla*). In 1906, some of the first large-scale hydroelectric projects were initiated as a result of ongoing industrialisation, and applications for dispensations from the kungsådra were submitted to the king. After many years of heated debate, a new law was formed in 1918, still stating the importance of the kungsådra, but the law also stated that this could be overruled by 'constructions of extreme importance' such as hydroelectric facilities. The new water law of 1918, in combination with a growing need for electricity, led to

the fragmentation of most Swedish rivers during the 20th century. In 1983, the concept was abandoned.

18.2.2 Fish ladders and racks with unknown function (1900–1935)

The first Swedish hydroelectric plant (HEP) was built in 1882, but the first large-scale obstacle to migration was built at Olidan in the River Göta älv in 1906–1913 (Figure 18.1), where there already existed an area of rapids considered to be a natural obstacle to upstream migration (Montén, 1988). Despite the presence of a natural barrier to migration, the appointed fisheries expert, Thorsten Ekman, recommended that several measures to facilitate eel passage should be implemented. He recommended

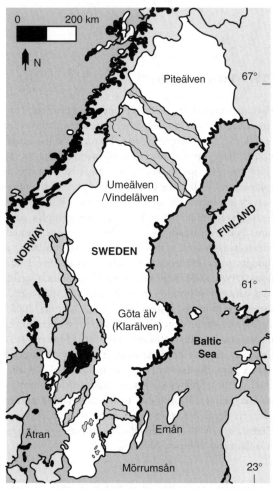

Figure 18.1 Map of Sweden showing the rivers referred to in the text and their catchment areas.

installation of a juvenile eel pass for upstream passage at the plant and 20-mm racks with a bottom gallery for downstream migration by silver eels. The resulting passage efficiencies of these measures remain unknown to this day. The complete blockage of the River Göta älv was soon followed by several similar projects in the Rivers Dalälven, Luleälven and Umeälven and, as a general rule at that time, obstacles in important rivers were equipped with technical fishways and juvenile eel passes and, in some cases, even fine-spaced racks (Montén, 1988).

In many cases, the fine-spaced racks that were installed at turbine intakes to protect fish from turbines ended up killing fish due to rack impingement, which, even today, is a problem, even though it is not widely accepted or realised (Montén, 1988; Calles and Greenberg, 2009; Calles et al., 2010b). In other words, fisheries managers have been aware of the problem of rack impingement for quite some time, as shown by the actions of the inspector of fisheries, Ossian Olofsson, who, in 1925, was in favour of placing small-spaced racks away from the turbine intake at the Norrfors HEP in the River Umeälven/Vindelälven (Figure 18.1) to prevent fish from being impinged on the racks due to water velocities exceeding the swimming capacity of the fish (Montén, 1988). The effectiveness of using fine-spaced racks for improved downstream passage of fish was rarely tested in the past; instead, work focused on what happened to fish after passing through turbines. Moreover, observations on the effects of turbines on downstream migration from the early 1900s were mostly anecdotal. One such example was the accidental passage of a wooden skiff through the Kaplan turbines at the Vargön hydroelectric facility in the River Göta älv (with a 3 m head and 8 m diameter runners), where the boat was found in the tail-race with only a small cut in the bow (Montén, 1988). Based on this observation, it was concluded that it was impossible for fish to suffer injury from turbine passage at this site, and some even suggested that fish could pass upstream through the turbines (an idea that Montén objected to).

Montén (1988) stated that very few of the rehabilitative measures implemented for passage before 1950 were effective, if they were even tested for efficiency. The first more technically advanced attempt to quantify passage, in this case upstream passage through a Swedish fishway, was through the installation, in 1935, of an automatic counter ('the electrical eye'), which monitored upstream migrating salmon in the River Umeälven and its tributary, the River Vindelälven (Montén, 1988). We have not been able to find any records of similar evaluations of downstream rehabilitative measures, and records show that there was

a general strategy shift in how to deal with passage problems, as described in the next section.

18.2.3 The compensatory stocking and turbine passage era (1935–2000)

Starting in the 1940s, the strategy to protect fish populations in regulated rivers shifted from measures enabling passage of obstacles to extensive compensatory stocking programmes, of which many are still operating today (Montén, 1988; Ackefors et al., 1991; Eriksson and Eriksson, 1993). In retrospect, this is somewhat puzzling, as a Swedish publication (Wolf, 1946) from this time stated that Scottish and Irish salmon stocks were in much better shape than Swedish stocks, which was attributed to the Scottish and Irish focus on natural reproduction rather than compensatory stocking. The observations made by Wolf (1946) describe British rehabilitation projects, mostly coordinated by the Fishery Boards, incorporating different types of fishways and even downstream guiding devices (fry-guards) with surface bypasses. Wolf's descriptions on how these fishways and bypasses were placed in relation to the main flow is in agreement with modern concepts in fish passage engineering (Clay, 1995; Larinier, 2002; Larinier and Travade, 2002; DWA, 2005).

During most of the 1900s virtually no Swedish examples exist of attempts to help downstream-migrating fish pass hydropower plants (Degerman, 2008). There are a number of studies in Sweden that have looked at downstream passage in the absence of remedial measures. Erik Montén's (1985) book about turbine mortality is a classic one in this context. Montén carried out a series of experiments, primarily on salmon and eel, where he released fish into turbines and then studied recaptured fish. In short, he found that water velocity and the orientation of the fish's body in relation to rotating turbine blades determined the likelihood that the fish would incur injury. He was able to establish relationships to determine loss rates of fish passing through turbines, based on fish length and the dimensions of the runners. Since this seminal work by Montén, scientific publications on downstream passage issues in Swedish rivers have been few, and only a handful of publications reported in the grey literature can be found (Johlander and Tielman, 1999; Degerman, 2008). Since then, others have measured losses in different types of turbines as well as refined models of turbine loss (Ferguson et al., 2008).

Research activities within the field picked up at the turn of the 21st century, when the Swedish research and development programme 'Hydropower – Environmental impact, remedial measures and costs in existing regulated waters', was initiated by Elforsk (jointly owned by Swedenergy and the Swedish National Grid Operator – Svenska Kraftnät), the Swedish Energy Agency, the National Board of Fisheries and the Swedish Environmental Protection Agency. The third phase of this programme has recently ended, and so far the programme's activities have resulted in numerous articles (e.g. Rivinoja et al., 2001; Calles and Greenberg, 2005, 2007, 2009; Rivinoja et al., 2006; Calles et al., 2007; Lundqvist et al., 2008; Greenberg and Calles, 2010), theses (e.g. Kiviloog, 2005; Rivinoja, 2005; Calles, 2006; Lindmark, 2008), and reports (e.g. Calles et al., 2003; Lundqvist et al., 2003; Malmqvist and Strasevicius, 2003; Anonymous, 2006; Kriström et al., 2010). Recent research on downstream fish passage in regulated rivers from this and other programmes is described below.

18.3 Rehabilitating downstream passage in Swedish Rivers today

18.3.1 Spill releases

Using releases of spill water through a spill gate to bypass downstream-migrating fish has often been used as a first attempt to ameliorate passage problems. In many cases, bottom-fed spill gates have been used, since this is the most common type of gate present at Swedish HEPs. The potential problem associated with the use of this kind of gate to bypass fish is that most fish species are surface-oriented and typically avoid areas with rapidly increasing water velocities, which are often created when water is released at or near the bottom. In the River Ätran (Figure 18.1) this technique was used at the Herting HEP (Figure 18.2B) when trying to increase the passage efficiency for downstream-migrating kelt and smolt of Atlantic salmon and brown trout, but when evaluated using telemetry, few fish were observed exiting through the bottom-fed spill gate used for this purpose (Karlsson, 2008; Calles et al., in press). Instead, most of the smolts passed through the intake racks and low-head Kaplan turbines, resulting in turbine-induced mortality of 13%. As the kelts could not pass through the racks, they either found other routes to pass the HEP (Figure 18.2B and 'Racks and spill gates' below) or died upstream after weeks of delay. A more successful attempt using bottom-fed spill gates at an HEP to bypass downstream-migrating salmonids was reported in a study, based on 48 Atlantic salmon smolts in the River Mörrumsån (Figure 18.1; Calles et al., 2010a; Rivinoja et al., 2012).

The practice at the HEP was to shut it down during five weeks in spring and spill all discharge through bottom-fed spill gates (Johan Tielman, 2010, personal communication).When evaluated in spring 2010–2011, the median time spent upstream of the gates for salmon smolts was < 10 minutes, and the only loss was attributed to predation in that area (four of 22 individuals). The smolts passing the gates also required less time to pass a monitoring station located downstream of the HEP than a control group released just downstream of the spill gates.

18.3.2 Fine-spaced racks without bypasses

Even though rehabilitative work on downstream passage in regulated rivers was rarely implemented during the 1900s, many HEPs are equipped with fine-spaced racks to protect fish from turbine-induced injuries. This supposedly rehabilitative measure is rather common in Swedish rivers, but few records of scientific evaluations of their function or effectiveness exist. Recent studies in the Rivers Emån, Ätran, Kävlingeån and Klarälven have focused on losses of smolts passing HEPs with fine-spaced racks, which have been installed to lower the risk of turbine-induced injuries, but without offering the smolts an alternative downstream route past the HEPs.

One such case, where downstream passage of smolts was evaluated at two hydropower dams equipped with fine-spaced racks without any bypasses, was at the upper and lower Finsjö HEPs (Figure 18.3) in the River Emån (Calles and Greenberg, 2009; Calles and Greenberg, Greenberg et al., 2012). At upper Finsjö, 72% of the trout smolts swam through spill gates, 19% through the rack and turbines and 9% did not pass the dam. This contrasts with lower Finsjö, where most of the smolts swam through the rack and the turbine (89%) and all fish managed to pass the dam. Passage success for the two dams was similar (76% at lower Finsjö and 66% at upper Finsjö) but the cause of the loss differed. At upper Finsjö, equipped with a 20-mm rack and two twin-Francis runners, turbine-induced smolt loss was about 36% (range 33–68%), whereas at lower Finsjö, equipped with a 30-mm rack and a Kaplan runner, turbine-induced mortality was about 12% (range 9–31%). We cannot distinguish possible rack-induced mortality from turbine-induced mortality, as no fish were released inside the rack at either of the HEPs. Interestingly, only 15% of the tagged smolt reached the sea, about 30 km farther downstream, suggesting there may have been delayed turbine-induced mortality (Ferguson et al., 2006).

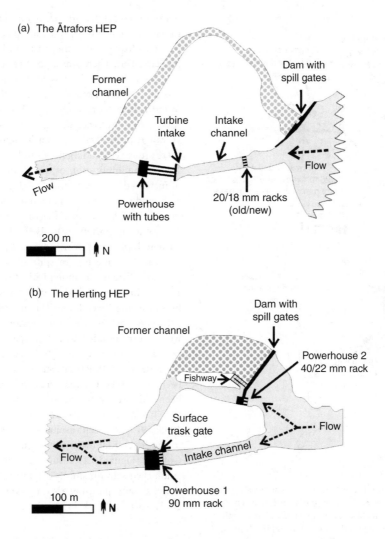

Figure 18.2 Overview of the hydroelectric plant areas at (a) Ätrafors and (b) Herting on the River Ätran. The trash racks are indicated by the black and white bars. Note that the racks were replaced in 2008 at the Ätrafors HEP (old vs. new).

Even downstream passage of kelts, which is generally not examined, was studied in the River Emån (Calles and Greenberg, 2009; Kriström *et al.*, 2010). While turbine-induced mortality of kelts was non-existent in the River Emån, due to the presence of trash racks, the kelts experienced problems passing the three dams in the river, with 48% reaching the Baltic Sea. Of the 28 days the median kelt took to reach the sea, a median of 16.8 (range: 2.0–39.6 days) of these days was spent immediately upstream of the dams, i.e. the fish were delayed. During spring, the kelts were observed moving back and forth between dams, apparently unable to find a downstream

passage route. During most of the study period, there were few passage routes available to the fish. At one of the dams, Karlshammar, the HEP was shut down and all water was released through a spill gate (4 × 6h, 100% spill), and most of the kelts that were in the vicinity of this spill gate managed to swim past the dam. In contrast, opening the spill gate at lower Finsjö, albeit not to the same extent as at Karlshammar (8h, 60% spill), did not increase passage success. The implications of fine-spaced racks for downstream-migrating fish in the River Emån were that large individuals were stopped from entering the turbines and thus could not pass the dam, whereas small

(a) Upper Finsjö HEP

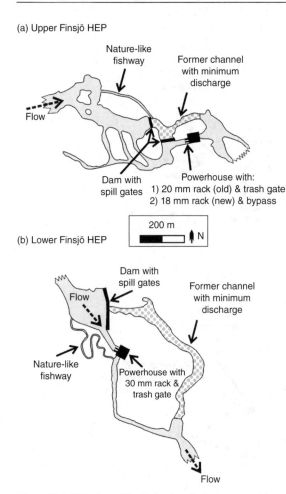

Figure 18.3 Overview of the hydroelectric plant areas at (a) upper Finsjö and (b) lower Finsjö on the River Emån. The trash racks are indicated by the black and white bars. Note that the racks were replaced in 2009 at the upper Finsjö HEP (old vs. new).

twin-Francis runners for eels released inside these 20-mm racks was high (60%), but the mortality of eels having to pass through both the racks and turbines was even higher (74%). There was a positive relationship between fish size and mortality risk and, as a result, the relative loss of silver eel biomass was higher than the relative loss of individuals. Furthermore, migratory silver eels in the River Ätran are twice as big today as compared to fifty years ago, probably due to density-dependent growth (i.e. there are fewer and larger eels today) (DeLeo and Gatto, 1996). The implication of the unfortunate combination of increased mean size and size-dependent mortality on racks and in turbines is that a much larger proportion of the eels are killed at the HEPs today than in the past.

Another important issue is the fact that many rivers have multiple dams, which means that total passage success can be quite low in spite of seemingly acceptable passage efficiency at individual HEPs (Chanseau et al., 1999). This was studied at two dams, upper and lower Finsjö, in the River Emån (Figure 18.3), where passage success at each of the dams separately was about 70%, which corresponded to a total passage success for the two dams combined of 50% (Calles, 2006; Calles and Greenberg, 2009). Passage success would, of course, have been lower if more dams were involved. This problem has been addressed in other countries, where delays have been shown to be long and passage success low (e.g. (Raymond, 1968; 1979). A recent evaluation of downstream passage in the River Klarälven, the main tributary of the River Göta älv (Figure 18.1), resulted in a 16% total survival for landlocked salmon smolts migrating past 8 HEPs, from the upstream spawning areas to Lake Vänern (Norrgård, 2011; Norrgård et al., in press).

individuals were delayed before many of them passed through the rack and turbines. All in all, the presence of fine-spaced racks was a meagre attempt to help migrating fish, as no alternative route for passage was present in the vicinity of the racks.

In some cases, the presence of fine-spaced racks actually increased total mortality of downstream-migrating fish, instead of protecting the fish as was intended. One such example was the Ätrafors HEP in the River Ätran (Figure 18.2A) where all silver eels > 680 mm were impinged and died on the 20-mm racks in the intake channel (Calles et al., 2010b). The turbine-induced mortality in the three

18.3.3 Racks and spill gates
Many measures used to divert fish are expensive, involving extensive construction costs, which can be particularly true in retrofitted solutions at existing structures. However, in some cases, alternative measures with low construction costs may be considered, which involve modifying existing structures into behavioural barriers to influence the route selected by fish. As all hydroelectric dams have spill gates, modification of these structures for fish diversion is one such alternative. In Sweden, modifications of trash deflector systems and the use of bio-acoustic fish fences have been tested as fish behavioural diverters in three rivers, the River Emån, the River Ätran and the River Mörrumsån (Figure 18.1).

In the River Emån, the use of trash deflectors and their associated spill gates was evaluated at two dams, the upper and lower Finsjö (Figure 18.3), as a way of guiding brown trout smolts away from turbine intakes (Calles and Greenberg, 2009; Greenberg et al., 2012). Originally, there existed 1-m-deep, angled trash deflectors at both power plants and these were used to lead fish to trash gates adjacent to the turbine intakes. For this to be effective, wolf traps (Wolf, 1951) were constructed at the trash gates to prevent fish from falling several metres before striking a concrete surface, leading almost certainly to death or injury (Calles, 2006). The relative discharge of water into the traps varied with total discharge through the turbines and ranged from 0 to 10% of total discharge flowing into the intake channels, depending on trap and year. The fish guidance efficiency (FGE) for the trash diverters was measured for three years (2004, 2005 and 2007) and was found to vary considerably between years, ranging from 17 to 50% at upper Finsjö and 4 to 52% at lower Finsjö. The modifications of the trash gates were not identical between years, which, in combination with different discharge conditions between years, probably explained the highly variable FGEs. One potentially important feature that differed between years was length and gap size of the trap and dewatering unit, which had a large impact on the hydraulic conditions at the gate. Hydraulic features such as high turbulence and rapid acceleration have previously been shown to negatively affect bypass attraction (Larinier, 1998).

These results were not particularly satisfying, given the variability between years, and the highest FGE achieved was never greater than 52%. Consequently, the use of overhead cover in combination with using trash deflectors was tested to determine if the response could be enhanced (Larinier, Greenberg et al., 2012). This was tested at upper Finsjö in the River Emån by mounting a tarpaulin over the intake canal, between the trash diverter and the turbine intake. The tarpaulin was constantly set up and removed approximately every two days, so that 47% of the time the tarpaulin was in place and the remaining 53% of the time it was not. Greenberg et al. (2012) found that overhead cover influenced migration path, as all smolts that exited via the trash gate did so when the entrance to the turbine intake was covered, and none did so when there was no overhead cover. Specifically, the use of a tarpaulin as overhead cover resulted in 31% of the smolts passing through the trash gate instead of the turbine. It is unclear whether the trout were reacting to a change in light associated with the overhead cover or to the physical presence of overhead structure (Kemp et al., 2005). Whatever the mechanism,

several studies have reported a reluctance by fish to enter covered or darkened structures (Glass and Wardle, 1995; Welton et al., 2002), and this behavioural trait can be used in designing diversionary structures for smolts.

At the Herting HEP on the River Ätran (Figure 18.2B), a surface gate adjacent to a 90-mm turbine intake rack was modified by building a wolf trap (Wolf, 1951) in it, following the same procedure as in the River Emån (Calles and Greenberg, 2009). In addition to the surface bypass, a siphon that discharged water from the bottom of the same 90-mm rack was constructed to bypass bottom-oriented species such as silver eels. Smolts ($N = 66$), kelts ($N = 20$) and silver eels ($N = 55$) were caught, radio-tagged and tracked whilst passing the facilities in spring (salmonids) and fall (eels) 2007. The surface gate had a fish guidance efficiency ranging from no effect for eels and trout smolts to 50% for trout kelts. The siphon did not bypass any tagged fish, and only a few untagged eels. The poor function of the measures was attributed to the failure of the rack to stop smolts and eels from entering the turbines. The kelts, on the other hand, could not pass through the rack, and instead took a long time to locate the surface bypass entrance. The conditions for downstream-migrating fish have improved at the hydroelectric plant, but the total losses remain quite high for the studied groups (smolts 10–25%; kelts 43–67%; eels 29%), and there is a need for a new rehabilitative measure that effectively allows all fish to bypass the turbines. Such a solution is currently planned and will be implemented in the near future, as described in Section 18.3.4.

In the River Mörrumsån (Figure 18.1), a bio-acoustic fish fence for diverting smolts was tested (Johlander and Tielman, 1999). A bio-acoustic fish fence is essentially a perforated tube, placed at a given depth, from which sound and compressed air are emitted and then rise up to the surface, creating a bubble 'fence' that is believed to be unpleasant for fish, to such an extent that they avoid them (Welton et al., 2002). The fence in the River Mörrumsån extended from the surface to a depth of 1.2 m (maximum water depth of 3.5 m) and was angled at about 40° to the direction of flow, guiding the fish into an opening equipped with a fish trap. Tests of the bio-acoustic fish fence showed that the FGE varied and at best reached levels around 50% for salmon smolt.

In summary, guiding structures that rely on behavioural response by the fish, i.e. that are not physical barriers to downstream passage, can be used in intake channels with velocities low enough that fish have the possibility to choose alternative routes. In many cases, these devices involve low construction costs, although the cost of the water used for the alternative route may be high. Neverthe-less, all tests of behavioural barriers conducted in Sweden have found that such measures alone attain FGEs of about 50% or less, which is well below the recommended > 80% suggested by Ferguson *et al.* (1998). From the fish point-of-view, however, a measure with 50% FGE is much better than not having an alternative route for passage.

18.3.4 Inclined racks and bypasses/traps in medium-sized rivers

The use of fine-spaced racks to stop fish from entering tur-bines has a long history in Swedish rivers, but, as outlined above, the experiences with using them have not been positive (Montén, 1988). The main reasons for their poor performance have been related to the absence of alterna-tive passageways for the fish and the lack of attention paid to the swimming capabilities of the fish in relation to the water velocities at the rack in question. The recommended approach velocity at racks to prevent fish from entering turbines is 0.50 m s^{-1} for eels and 0.25–0.30 m s^{-1} for smolts (DWA, 2005), whereas the majority of Swedish intake channels are designed for velocities considerably greater than 0.5 m s^{-1} and in some cases even higher than 1.0 m s^{-1}. In retrospect, it is not difficult to appreciate why these measures have not been very successful, resulting in fish being trapped upstream of the obstacle and/or being

impinged and killed on the racks (Calles *et al.*, 2010b; Russon *et al.*, 2010).

The key to successful HEP passage when fine-spaced racks are being used is to ensure that the pressure on the racks (the normal velocity vector) is low and at the same time there must be an alternative route past the turbines for the fish (DWA, 2005; DWA, Calles *et al.*, in press). Such a solution has been tested at two sites: one in the River Ätran (Ätrafors HEP, Figure 18.2A; Calles and Bergdahl, 2009) and one in the River Emån (Övre Finsjö HEP, Fig-ure 18.3A; Kriström *et al.*, 2010). In both cases, the risk of impingement was reduced by building a rack with an increased surface area, thereby decreasing the water veloc-ity relative to the rack even though the approach veloc-ity (V$_{\mathrm{APPROACH}}$) remained unaltered. Alternative ways of reducing the normal velocity vector (V$_{\mathrm{NORMAL}}$) and thus rack impingement could have been implemented, such as widening the intake channel or decreasing discharge into the turbines, but both of these measures were considered economically unfeasible. To fit a larger rack at the same location as the old rack, without modifying the intake, required that the rack be angled relative to the bottom, in our case at 35°. This type of rack is commonly referred to as an α-rack, whereas a rack angled in relation to the sides of the intake channel is referred to as a β-rack (Figure 18.4; DWA, 2005). The design of alternative passageways for the fish, i.e. the bypass systems, differed between the

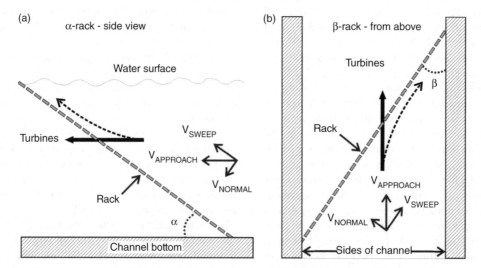

Figure 18.4 Schematic view of low-sloping (a) α-racks (side-view) and (b) β-racks (overhead view), showing the approach velocity (V$_{\mathrm{APPROACH}}$), normal velocity (V$_{\mathrm{NORMAL}}$) and the sweep velocity vectors (V$_{\mathrm{SWEEP}}$). The thick solid arrow indicates the main flow direction and the dotted arrow indicates the theoretical trajectory of the deflected fish. Modified from DWA (2005).

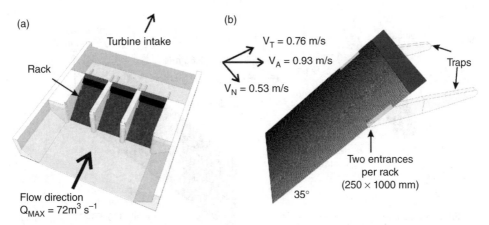

Figure 18.5 The 18-mm intake racks at Ätrafors HEP on the River Ätran, Sweden. (a) Overview of the collection facility and the three racks; (b) close-up of one of the racks and the two entrances to the fish traps. The approach (V_A), normal (V_N) and sweep (V_T) velocity vectors are shown. Each rack is 5.4 m wide and 8.4 m long.

two rivers, depending on the local conditions at the dams in each river.

In the River Ätran (mean annual discharge 50 m^3s^{-1}) at the HEP in Ätrafors, prior to rehabilitation, there were three parallel intake racks (α), with a 20-mm gap size, angled at 63° relative to the vertical, and these racks were responsible for substantial losses of silver eels due to impingement (Calles and Bergdahl, 2009; Calles *et al.*, 2010b). In 2008, the old racks were removed and replaced by three new α-racks installed at the same location (Figure 18.2A), but with an 18-mm gap size and a 35° angle relative to the bottom (Figure 18.5). The velocity vectors were changed by these modifications so that the sweeping velocity (V_{SWEEP}) became larger than the normal velocity, creating a situation opposite to the original situation with the old racks. In theory, the most important implications of this were a significantly lowered risk of impingement and a higher velocity to move fish from the rack towards the bypass entrances at the surface, but also a decreased head loss (Calles and Bergdahl, 2009; Persson and Holmberg, 2009). To provide the eels with alternative routes, exit openings, measuring 250 mm in width and 570 mm in depth (1000 mm long), were constructed on the side of each of the three racks, i.e. six exits in total (Figure 18.5A and B). The exits led to six separate fyke-nets from which the eels had to be manually removed and transported downstream for release. In summary, the new racks lowered the risk of injury for silver eels from > 70% (96% for eels > 750 mm) to < 10%, and at the same time no water was taken from hydroelectric production and the head loss was reduced from 360 mm

to 190 mm at maximum discharge and reservoir level. This was a win–win solution, with benefits both for the hydropower company (by reducing head loss and retaining all water for production) and migrating eels (substantial reduction in loss rates). The situation for the eels could, of course, be improved by replacing the traps with a bypass system, but this would reduce production of electricity, as water would be needed to transport eels back to the river.

In the River Emån (mean annual discharge 30 m^3s^{-1}) at the HEP in upper Finsjö, the turbine-induced mortality for trout smolts passing through the two small twin-Francis turbines was high, at 33–68% (Calles and Greenberg, 2009; Calles and Greenberg, 2009; Greenberg *et al.*, 2012). The total passage success, however, was strongly dependent on discharge, as many fish passed the HEP at high discharge via spill water released into the former river channel. However, in a situation where all water flows through the HEP, the passage success would be expected to be very low, as most fish would pass via the turbines. To enhance fish passage, the 20-mm α-rack was replaced by a new intake rack (α) with an 18-mm gap size and a 35° angle relative to the bottom (Figure 18.6A and B). This new set-up strongly resembled the situation in the River Ätran, with the main difference being that the bypass openings were built at the surface due to constant water levels in the intake channel. The maximum discharge through the two bypass entrances was 0.15 $m^3 s^{-1}$, and this could be reduced by a control gate in the bypass channel at the crest of the dam. The bypass system was used by numerous species, most of which were salmonids,

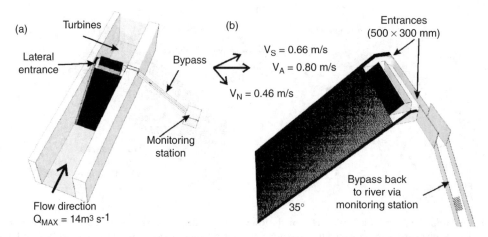

Figure 18.6   The 18-mm intake rack at upper Finsjö HEP on the River Emån, Sweden. (a) Overview of the bypass and monitoring facility; (b) close-up of the rack and the two entrances that lead the fish downstream of the dam. The approach (V_A), normal (V_N) and sweep (V_T) velocity vectors are shown. The rack is 4.8 m wide and 7.7 m long.

esocids, percids and cyprinids (Kriström *et al.*, 2010). In 2009, overall passage success for trout smolts was 84%, with 46% using the bypass system. The radio-tagged individuals that successfully passed the HEP without moving into the bypass did so by swimming through spill gates into the former channel. This bypass system needs to be tested over several years at different flow conditions, but the results so far are encouraging.

18.3.5   Guiding/skimming walls in large rivers

Two large rivers in northern Sweden, the River Umeälven/Vindelälven (mean annual discharge 480 m³s⁻¹) and the River Piteälven (mean annual discharge 180 m³s⁻¹) (Figure 18.1), are both regulated by single power stations in the lowermost reaches, about 32 and 40 km from the sea, respectively. These rivers are protected by national laws from further hydropower development and, since salmon and brown trout have vital reproduction areas located upriver of the regulated sections, fishways have been installed to allow upstream fish migration. Until now no actions have been taken to prevent downstream-migrating smolts and kelts from entering the turbines. The peak flows in these rivers occur during spring floods produced by snowmelt, often coinciding with the downstream migration of both smolts and kelts, at which time excess water is spilled from the dams. As found for most of the HEPs in northern Swedish rivers, the spill gates are bottom-oriented, while additional surface gates can be

used to divert ice, logs and debris. Despite the mandatory spill flows in the residual river sections (about 20 m³s⁻¹ in the River Umeälven/Vindelälven and 10 m³s⁻¹ in the River Piteälven), the usual spill amounts are generally too small in relation to turbine flows to effectively guide fish towards the spillways (Kiviloog, 2005; Rivinoja, 2005), which has led to a relatively high mortality of descending fish. Besides turbine-induced mortalities, the downstream migrants suffer additional losses caused by the underground turbine outlet tunnels (Montén, 1985; Rivinoja, 2005).

In 2010, surface-oriented guide walls (about 2 m deep, Figure 18.7) were installed at both of the HEP sites, with the intention of guiding the downstream-migrating fish to a modified fish ladder (River Umeälven/Vindelälven) or a spillway route (River Piteälven) (Figure 18.7). The guide walls, 100–130 m (20–26 elements) in length, consisted of a series of floating booms that were constructed in such a way that they tilted over upon contact with debris, a sort of 'self-cleaning system'.

We still don't know if these newly installed structures will be efficient enough to guide downstream migrants. However, a recent pilot study in the River Piteälven showed that there was a higher spillway guidance of salmon and brown trout smolts with the installed structure than without it (Rivinoja and Lundquist, unpublished). Whether or not these types of guidance devices will be considered generic solutions to rehabilitate downstream passages in northern European rivers remains to be evaluated.

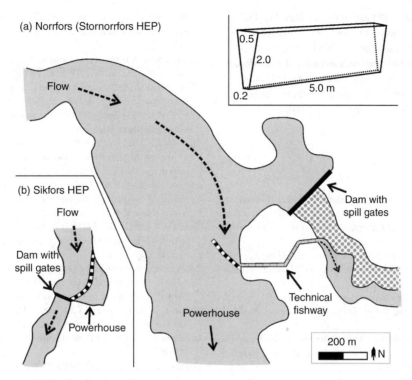

Figure 18.7 Overview of the power station areas on the River Umeälven/Vindelälven (Norrfors) and the River Piteälven (Sikfors), with the surface guide walls indicated by the black and white bars. The main flow orientations, powerhouse intakes and dams are marked by arrows. One of the guidance elements is illustrated schematically in the upper-right corner.

18.4 Concluding remarks

As human activities in Swedish rivers increased during the 18th and 19th centuries, historical documents show that the importance of intact two-way longitudinal connectivity was surprisingly well appreciated and understood (Wolf, 1946; Montén, 1985). The oldest evidence of such a 'modern approach' to river management was the use of the kungsådra concept, where at least one-third of the river's width was required to be open for public services such as shipping, fishing and timber floating. Following the implementation of hydropower at the turn of the 19th century, there was a shift from rehabilitative work aimed at maintaining connectivity and allowing natural reproduction of migrating fish species to releases of hatchery-raised fish to compensate for the damage inflicted by hydropower. Apart from the extensive work on fish turbine passage-related issues by Montén (1985), there are very few examples of rehabilitative measures that have been implemented to facilitate downstream passage for fish at hydroelectric plants, nor have there been many sci-

entific evaluations of these measures during most of the 20th century.

This situation began to change in the late 1990s when the first attempts to guide fish away from turbine intakes were initiated (Johlander and Tielman, 1999). During the last decade, there has been a shift in thinking, resembling that of the early 20th century, where rehabilitative work focuses on fish passage, both upstream and downstream, and improving conditions for natural spawning and rearing, rather than on compensatory stocking. The techniques currently being tested for improved downstream passage success at hydroelectric plants differ between rivers of different size. The most promising results in small to medium-sized rivers come from low-sloping, small-spaced racks with associated bypass systems, whereas the work in large rivers, which has focused on surface-oriented guiding walls, has only recently begun. This important rehabilitative work has gained support from recent EU legislation concerning conservation of the European eel (Council Regulation No 1100/2007/EC), and for further acceptance of such rehabilitative work, several different techniques will have to be developed and scientifically tested. Furthermore, attempts at modelling fish populations in regulated rivers in combination with cost–benefit analyses of remedial measures could provide further support to river managers when deciding how, when and where such rehabilitative work should be carried out (Lundqvist *et al.*, 2008; Kriström *et al.*, 2010).

References

Ackefors, H., Johansson, N. and Wahlberg, B. (1991) The Swedish compensatory programme for salmon in the Baltic: an action plan with biological and economic implications. *ICES Marine Science Symposia*, **192**: 109–119.

Anonymous (2006) *Hydropower – Environmental impact, remedial measures and costs in existing regulated waters*. Elforsk Rapport.

Calles, O. (2006) *Re-establishment of Connectivity for Fish Populations in Regulated Rivers*. PhD thesis, Karlstad University.

Calles, O. and Bergdahl, D. (2009) *Downstream passage of silver eels at hydroelectric facilities – before and after a remedial measure*. (In Swedish.)

Calles, O. and Greenberg, L.A. (2005) Evaluation of nature-like fishways for re-establishing connectivity in fragmented salmonid populations in the River Emån. *River Research and Applications*, **21**: 951–960.

Calles, O. and Greenberg, L.A. (2007) The use of two nature-like fishways by some fish species in the Swedish River Eman. *Ecology of Freshwater Fish*, **16**: 183–190.

Calles, O. and Greenberg, L. (2009) Connectivity is a two-way street – the need for a holistic approach to fish passage problems in regulated rivers. *River Research and Applications*, **25**: 1268–1286.

Calles, O., Nyberg, L. and Greenberg, L. (2007) Temporal and spatial variation in quality of hyporheic water in one unregulated and two regulated boreal rivers. *River Research and Applications*, **23**: 829–842.

Calles, O., Greenberg, L.A., Nyberg, L. and Löwgren, M. (2003) *Fiskvägar och flödesregimåtgärder i reglerade vatten: konsekvenser för vattendragets produktivitet och för samhällsnyttan*. (In Swedish.) Elforsk Rapport.

Calles, O., Karlsson, S., Hebrand, M. and Comoglio, C. (2012) Evaluating technical improvements for downstream migrating diadromous fish at a hydroelectric plant. *Ecological Engineering*, **48**: 30–37.

Calles, O., Kläppe, S., Rivinoja, P. and Norrgård, J. (2010a) *On the downstream migration of Atlantic salmon smolts in the River Mörrumsån 2010*. (In Swedish.) Research Report, Karlstad University.

Calles, O., Olsson, I.C., Comoglio, C., Kemp, P., Blunden, L., Schmitz, M. and Greenberg, L.A. (2010b) Size-dependent mortality of migratory silver eels at a hydropower plant, and implications for escapement to the sea. *Freshwater Biology*, **55**: 2167–2180.

Chanseau, M., Croze, O. and Larinier, M. (1999) The impact of obstacles on the Pau River (France) on the upstream migration of returning adult Atlantic salmon (*Salmo salar* L.). *Bulletin Francais de la Peche et de la Pisciculture*, **353–354**: 211–237.

Clay, C.H. (1995) *Design of Fishways and Other Fish Facilities*, Lewis Publishers, Boca Raton.

Coutant, C.C. and Whitney, R.R. (2000) Fish behavior in relation to passage through hydropower turbines: A review. *Transactions of the American Fisheries Society*, **129**: 351–380.

Degerman, E. (2008) *Ecological Restoration of Rivers*. (In Swedish.) Swedish Environmental Protection Agency and National Board of Fisheries, Stockholm.

Deleo, G.A. and Gatto, M. (1996) Trends in vital rates of the European eel: Evidence for density dependence? *Ecological Applications*, **6**: 1281–1294.

DWA (2005) *Fish Protection Technologies and Downstream Fishways. Dimensioning, Design, Effectiveness Inspection*. Hennef, German Association for Water, Wastewater and Waste (DWA).

Eriksson, T. and Eriksson, L.O. (1993) The status of wild and hatchery propagated Swedish salmon stocks after 40 years of hatchery releases in the Baltic rivers. *Fisheries Research*, **18**: 147–159.

Ferguson, J.W., Poe, T. and Carlson, T.J. (1998) Surface-oriented bypass systems for juvenile Salmonids on the Columbia River, USA. In Jungwirth, M., Schmutz, S. and Weiss, S. (eds) *Migration and Fish Bypasses*, Fishing News Books, Oxford.

Ferguson, J.W., Absolon, R.F., Carlson, T.J. and Sandford, B.P. (2006) Evidence of delayed mortality on juvenile Pacific salmon passing through turbines at Columbia River dams. *Transactions of the American Fisheries Society*, **135**: 139–150.

Ferguson, J.W., Ploskey, G.R., Leonardsson, K., Zabel, R.W. and Lundqvist, H. (2008) Combining turbine blade-strike and life cycle models to assess mitigation strategies for fish passing dams. *Canadian Journal of Fisheries and Aquatic Sciences*, **65**: 1568–1585.

Glass, C.W. and Wardle, C.S. (1995) Studies on the use of visual stimuli to control fish escape from codends. 2. The effect of a black tunnel on the reaction behavior of fish in otter trap codends. *Fisheries Research*, **23**: 165–174.

Greenberg, L. and Calles, O. (2010) Restoring ecological connectivity in rivers to improve conditions for anadromous brown trout *Salmo trutta*. In Kemp, P. (ed.) *Salmonid Fisheries: Freshwater Habitat Management*. Blackwell Publishing.

Greenberg, L.A., Calles, O., Andersson, J. and Engqvist, T. (2012) Effect of trash diverters and overhead cover on downstream migrating brown trout smolts. *Ecological Engineering*, **48**: 25–29.

Jakobsson, E. (1996) *Industrialisering av älvar: studier kring svensk vattenkraftutbyggnad 1900–1918*. PhD thesis, University of Gothenburg.

Johlander, A. and Tielman, J. (1999) *River Mörrumsån spring 1999. A study on downstream migrating salmonids at*

Hemslö upper and lower hydroelectric facilities. (In Swedish.) National Board of Fisheries and Sydkraft Hydropower AB.

Karlsson, S. (2008) *Hydro-electric Power and Downstream Migration of Atlantic Salmon and Sea Trout – Evaluation of Mortality and Movement using Telemetry in the River Ätran.* Masters thesis, Göteborgs Universitet.

Kemp, P., Gessel, M. and Williams, J. (2005) Seaward migrating subyearling chinook salmon avoid overhead cover. *Journal of Fish Biology*, **67**: 1–11.

Kemp, P.S., Gessel, M.H. and Williams, J.G. (2008) Response of downstream migrant juvenile Pacific salmonids to accelerating flow and overhead cover. *Hydrobiologia*, **609**: 205–217.

Kiviloog, J. (2005) *Three-dimensional Numerical Modelling for Studying Smolt Migration in Regulated Rivers.* Licentiate of Engineering, Chalmers University of Technology.

Kriström, B., Calles, O., Greenberg, L.A., Leonardsson, K., Paulrud, A. and Ranneby, B. (2010) *Cost–Benefit Analysis of River Regulation: The case of Emån and Ljusnan.* Scientific summary report (in Swedish with extended English summary).

Larinier, M. (1998) Upstream and downstream fish passage experience in France. In Jungwirth, M., Schmutz, S. and Weiss, S. (eds) *Migration and Fish Bypasses*, Fishing News Books, Oxford.

Larinier, M. (2002) Fishways – General considerations. *Bulletin Francais de la Peche et de la Pisciculture*, **364**: 21–27.

Larinier, M. (2008) Fish passage experience at small-scale hydro-electric power plants in France. *Hydrobiologia*, **609**: 97–108.

Larinier, M. and Marmulla, G. (2004) Fish passes: types, principles and geographical distribution – an overview. In Welcomme, R.L. and Petr, T. (eds) *Proceedings of the Second International Symposium on the Management of Large Rivers for Fisheries*, Volume II, Bangkok, Thailand, FAO Regional Office for Asia and the Pacific.

Larinier, M. and Travade, F. (2002) Downstream migration: problem and facilities. *Bulletin Francais de la Peche et de la Pisciculture*, **364**: 181–207.

Lindmark, E.M. (2008) *Flow Design for Migrating Fish.* PhD thesis, Division of Fluid Mechanics, Department of Applied Physics and Mechanical Engineering, Luleå University of Technology.

Lundqvist, H., Rivinoja, P., Leonardsson, K. and McKinnell, S. (2008) Upstream passage problems for wild Atlantic salmon (*Salmo salar*) in a flow controlled river and its effect on the population. *Hydrobiologia*, **602**: 111–127.

Lundqvist, H., Bergdahl, L., Leonardsson, K., Rivinoja, P. and Kiviloog, J. (2003) The influence of flow-change on salmon migration in a "bypass" system: observations and modelling of factors of importance for passage of power stations. (In Swedish with English summary.)

Malmqvist, B. and Strasevicius, D. (2003) *Indirekta effekter av älvregleringen i Sverige Inverkan på omgivande landekosystem genom förlusten av massutvecklande insekter från älvarnas forsar.* (In Swedish.)

Montén, E. (1985) *Fish and Turbines: Fish Injuries During Passage Through Power Station Turbines.* Stockholm, Vattenfall, Statens vattenfallsverk.

Montén, E. (1988) *Fiskodling och vattenkraft : en bok om kraftutbyggnadernas inverkan på fisket och hur man sökt kompensera skadorna genom främst fiskodling*, Vällingby, Vattenfall.

Muir, W.D., Marsh, D.M., Sandford, B.P., Smith, S.G. and Williams, J.G. (2006) Post-hydropower system delayed mortality of transported snake river stream-type Chinook salmon: Unraveling the mystery. *Transactions of the American Fisheries Society*, **135**: 1523–1534.

Norrgård, J.R. (2011) *Landlocked Atlantic Salmon* Salmo salar *L. and Trout* Salmo trutta *L. in the Regulated River Klarälven, Sweden: Implications for Conservation and Management.* Licentiate, Karlstad University.

Norrgård, J., Greenberg, L.A., Piccolo, J.J., Schmitz, M. and Bergman, E. (in press) Multiplicative loss of landlocked Atlantic salmon *Salmo salar* L. smolt during downstream migration through multiple dams. *River Research and Applications.*

Odeh, M. (1999) *Innovations in Fish Passage Technology*, American Fisheries Society, Bethesda, Maryland, USA.

Persson, F. and Holmberg, J. (2009) *Ätrafors – head loss vs. intake rack angle.* (In Swedish.) EnergoRetea Energi, Elkraft & ICT AB.

Raymond, H.L. (1968) Migration rates of yearling chinook salmon in relation to flows and impoundments in Columbia and Snake Rivers. *Transactions of the American Fisheries Society*, **97**: 356–358.

Raymond, H.L. (1979) Effects of dams and impounds on migrations of juvenile chinook salmon and steelhead from the Snake River, 1966 to 1975. *Transactions of the American Fisheries Society*, **108**: 505–529.

Rivinoja, P. (2005) *Migration Problems of Atlantic Salmon (*Salmo salar *L.) in Flow Regulated Rivers.* PhD thesis, Acta Universitatis Agriculturae Sueciae.

Rivinoja, P., Leonardsson, K. and Lundqvist, H. (2006) Migration success and migration time of gastrically radio-tagged *v.* PIT-tagged adult Atlantic salmon. *Journal of Fish Biology*, **69**: 304–311.

Rivinoja, P., McKinnell, S. and Lundqvist, H. (2001) Hindrances to upstream migration of Atlantic salmon (*Salmo salar*) in a northern Swedish river caused by a hydroelectric power-station. *Regulated Rivers: Research and Management*, **17**: 101–115.

Rivinoja, P., Christiansson, J., Norrgård, J. and Calles, O. (2012) *On the downstream migration of Atlantic salmon smolts in the River Mörrumsån 2011.* (In Swedish.) Research Report, Karlstad University.

Roscoe, D.W. and Hinch, S.G. (2010) Effectiveness monitoring of fish passage facilities: historical trends, geographic patterns and future directions. *Fish and Fisheries*, **11**: 12–33.

Russon, I.J. and Kemp, P.S. (2011) Advancing provision of multi-species fish passage: Behaviour of adult European eel (*Anguilla anguilla*) and brown trout (*Salmo trutta*) in response to accelerating flow. *Ecological Engineering*, **37**: 2018–2024.

Russon, I.J., Kemp, P.S. and Calles, O. (2010) Response of downstream migrating adult European eels (*Anguilla anguilla*) to bar racks under experimental conditions. *Ecology of Freshwater Fish*, **19**: 197–205.

Scruton, D., Pennell, C., Bourgeois, C., Goosney, R., Porter, T. and Clarke, K. (2007) Assessment of a retrofitted downstream fish bypass system for wild Atlantic salmon (*Salmo salar*) smolts and kelts at a hydroelectric facility on the Exploits River, Newfoundland, Canada. *Hydrobiologia*, **582**: 155–169.

Welton, J.S., Beaumont, W.R.C. and Clarke, R.T. (2002) The efficacy of air, sound and acoustic bubble screens in deflecting Atlantic salmon, *Salmo salar* L., smolts in the River Frome, UK. *Fisheries Management and Ecology*, **9**: 11–18.

Wolf, P. (1946) *Salmon in Sweden and England*. (In Swedish.) Swedish Salmon and Trout Association, Malmö.

Wolf, P. (1951) A trap for the capture of fish and other organisms moving downstream. *Transactions of the American Fisheries Society*, **80**: 41–51.

19 Rapid Flow Fluctuations and Impacts on Fish and the Aquatic Ecosystem

Atle Harby[1] and Markus Noack[2]

[1]SINTEF Energy Research, P.O. Box 4761 Sluppen, 7465 Trondheim, Norway
[2]Federal Institute of Hydrology, Department M3 – Groundwater, Geology, River Morphology, Mainzer Tor 1, D-56068, Koblenz, Germany

19.1 Introduction

Research on environmental impacts of river regulation and hydropower production has traditionally focused on long-term impacts of river and lake regulation in general, bypass sections, migration barriers and minimum flow. However, environmental impacts of more frequent regulation of hydropower including hydropeaking have only been studied in a limited number of cases. Increased use of renewable energy, more and more open energy markets and increased transmission capacity between power producers and customers all lead to higher demand for market-oriented regulation of hydropower (IPCC, 2011). Hydropower with reservoirs is the only renewable energy with storage possibilities, and we expect a growing interest in adjusting power production, which will lead to rapid and frequent flow fluctuations in downstream rivers. The major environmental impacts from hydropeaking are found in downstream riverine ecosystems, whereas smaller effects are identified in reservoirs.

The physical environment in natural rivers can be highly variable, with regard to hydraulic conditions, temperature, ice, visibility, cover, substrate alteration and erosion. However, anthropogenic flow alteration downstream of hydroelectric plants may lead to a harsh environment of frequent and unpredictable disturbances for freshwater organisms, in an evolutionary sense, with no natural analogue (Poff et al., 1997; Garcìa et al., 2011). Hydropeaking, or accidental stoppages in hydro operations giving rapid flow changes, strongly affects the riverine organisms through drastic alteration in habitat, leading to stranding of organisms (Hvidsten 1985; Lauters et al., 1996; Bradford, 1997; Irvine et al., 2009; Nagrodski et al., 2012). Studies on the downstream effects on fish in Norway (Harby et al., 2001; Saltveit et al., 2001; Halleraker et al., 2003; Berland et al., 2004; Flodmark et al., 2004, 2006) and by Scruton et al. (2003; 2005; 2008) and Vehanen et al. (2000; 2003) indicate that ramping rates, cover, substrate, season and light affect fish behaviour, movement and habitat use and stranding risk of young salmon and trout.

The consideration of changing morphological characteristics is a key aspect when evaluating the impact of hydropeaking on fish habitats (Clarke et al., 2008). In particular, operation over decades shifts the balance of sediment transport processes, resulting in morphological and ecological degradation downstream of hydropower plants (Sear, 1995; Osmundson et al., 2002). Typical morphological changes include the flattening of morphological features and bank and soil erosion, with subsequent alterations in the compositions of bed sediments (Moog, 1993). Furthermore, hydropeaking can cause bed armouring and colmation, with reductions in the vertical connectivity between the surface water and groundwater.

Ecohydraulics: An Integrated Approach, First Edition. Edited by Ian Maddock, Atle Harby, Paul Kemp and Paul Wood.
© 2013 John Wiley & Sons, Ltd. Published 2013 by John Wiley & Sons, Ltd.

This may lead to severe impairments to fish spawning, the benthic invertebrate community and in changes to the river's circadian rhythm (Bruno *et al.*, 2009; Meile *et al.*, 2011; Tuhtan, 2012). According to Meile (2006), there is still a gap in knowledge regarding the interaction between morphological characteristics, ecohydraulic conditons and hydropeaking. On the one hand, the high heterogeneity of instream structures may help to reduce the negative effects of hydropeaking, but, on the other hand, it can also worsen conditions. As an example, near-bank areas with shallow water depths and moderate flow velocities can provide refuge habitats for juvenile fish during high flow, but can also lead to fish stranding after the hydropeaking wave has passed (Tuhtan *et al.*, 2012).

Fish may also experience impacts from hydropeaking other than stranding. Harby *et al.* (2001) reported that juvenile salmonid fish in the River Nidelva in Trondheim would normally move to a location with more suitable habitat conditions when flow changes reduced the actual habitat quality at their location. Fish also rapidly started to use the newly wetted area after a cycle of hydropeaking, especially when water temperatures were above approximately 8 °C. At low temperatures, Harby *et al.* (2001) did not report any changes in large juvenile salmon movements during telemetry experiments in the Mandal River when fish were exposed to rapid flow changes at different magnitudes and frequencies compared to movements under stable flow conditions.

Most of the experimental studies concerning biological response to rapid flow fluctuations have been carried out for fish.

Bain (2007) summarised 43 impact studies in hydropeaked rivers. Changes in community composition through loss of taxa or shifts from sensitive to tolerant taxa were the most commonly reported impacts on invertebrates. Impacts on fish also most frequently comprised changes in species and community composition and reduced abundance. All of the studies Bain (2007) reported took place in Europe and North America, and there is a clear lack of knowledge about impacts in other regions.

Clarke *et al.* (2008) also summarised impacts of flow changes, including those of rapid changes like hydropeaking. Clarke *et al.* (2008) showed that flow fluctuations have a direct impact on many important factors for fish, such as total gas pressure, access to habitats in longitudinal and cross-sectional directions as well as into the substrate, stranding, water temperature, nutrient load and access

to food, erosion, sedimentation, sediment transport, bio-energetic impacts like growth, feeding rate, metabolism, behaviour, swimming capacity, competition and interactions among these factors.

Smokorowski *et al.* (2010) compared several biological impacts in a natural river with a regulated river with hydropeaking, the latter of which had restrictions on how fast flow could be increased and decreased. Smokorowski *et al.* (2010) concluded that composition and abundance of invertebrates, fish biomass, the condition of fish and the length of the nutrition chain were equal or greater in the regulated river compared to the natural. However, it is unclear if this is due to the restrictions in peaking or a relatively high residual flow.

Invertebrates are less able to move than fish and are at higher risk than most fish species when flow changes very quickly. A rapid increase in flow may cause catastrophic drift and changes in microhabitat which invertebrates cannot sustain. A rapid decrease in flow may cause invertebrates to strand and eventually die. Harby *et al.* (2001) showed that invertebrates needed about one month to fully recolonise an area that had been subject to wetting–drying cycles caused by hydropeaking when stable flow conditions were re-established.

Harby *et al.* (2001) investigated the impact on aquatic plants from hydropeaking in the upper part of the Mandalselva River at Smeland, observing the development of more specialised flora after repeated hydropeaking. Hydropeaking could also lead to increased plant biomass production, repeated flushing of biomass and more rapid re-growth of water vegetation.

A large research project on the environmental impacts of hydropeaking in reservoirs and lakes was carried out in Norway in the late 1990s and was summarised by Førde and Brodtkorb (2001). Only small to moderate impacts were found in the reservoir Vinjevatn, which experienced quite rapid and frequent variation. However, erosion, in particular, may increase when the drawdown zone is exposed more frequently. Hydropeaking impacts in reservoirs will not be discussed further.

The focus of this chapter is on Norwegian experiments and case studies, but we also include an example of modelling from comparable conditions in Switzerland. As most of the Norwegian hydropower system feasible for hydropeaking, discharge water directly to the sea or to salmonid rivers, most of the focus on environmental impacts is on salmonids. However, the methodology may also be highly relevant for other fish species and other organisms.

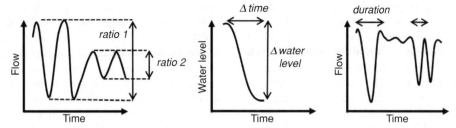

Figure 19.1 Parameters to characterise rapid flow fluctuations. The *flow ratio* (left) is the high flow divided by the low flow. The *rate of change* (middle) is the change in water level divided by the time of the change. The *duration* (right) is the time between two peaks in the hydrograph. The *frequency* of peaking is how often peaking occurs, i.e. the time between the arrows on the right diagram.

19.2 Rapid flow fluctuations

Rapid flow fluctuations can be categorised by changes in magnitude (flow ratio), the rate of change, frequency, duration and timing (Junk *et al.*, 1989). Artificial flow fluctuations should be defined as flow fluctuations induced by hydro operations, and they can simply be identified as situations where one or several of these factors exceeds natural limits. However, it is often the combination of these parameters that creates the most harmful effects.

The flow ratio is the highest flow divided by the base flow. In natural systems, the ratio may also be very high when comparing floods to low-base-flow situations, but these very seldom happen shortly after each other.

The rate of change in water level and flow in natural systems may be large when flow is increasing due to heavy rainfall events, but is very seldom so pronounced during retention of flow. Under natural conditions, a rapidly changing water level normally follows changes in weather conditions.

The duration of peaking events is often defined as the time for one 'cycle', i.e. the duration of the high-flow event until base flow is re-established. This compares to the time during which full power production or high-flow hydro operations are going on. However, for most studies of impacts, the duration between two peaks in flow or the duration of the recession in flow and the dewatering period is the most important to focus on. We recommend using the time between two peaks to describe the duration of a low-flow event rather than the duration of a high-flow event.

The frequency of peaking describes how often peaking operations are carried out. A typical frequency of peaking produces power during the daytime, when the demand

is high, and stops operations during the night. This scenario would be daily peaking. As a consequence of changing power markets and increasing shares of intermittent renewable electricity production, patterns for hydropeaking are likely to see large changes in the coming decades (IPCC, 2011).

The timing of peaking episodes is also important to consider, as they may occur at times that are favourable or unfavourable to the dynamics of the river system and the requirements of the ecosystem. For example, at the moment salmon fry emerge from the gravel, a peak flow may displace them.

Figure 19.1 illustrates the parameters we use to characterise rapid flow fluctuations and hydropeaking.

19.3 Methods to study rapid flow fluctuations and their impact

In general, methods in ecohydraulics are very well suited to studying rapid flow fluctuations and their impacts. However, rivers that are experiencing artificial flow fluctuations do not always have these fluctuations, which makes experimental sites and case study applications difficult. In many cases, there will be an overlapping effect of general river regulation impacts and hydropeaking impacts, which makes it difficult to isolate factors. Laboratory studies and model simulations may then serve as important complementary analysis.

Laboratory studies make it possible to isolate the effects of rapid flow changes, but it may sometimes be difficult to find laboratories suitable to reproduce flow ratios and rates of change in the same order of magnitude as practised in some regulated rivers. In this study, we report results from the 19.2 m-long and 3.8 m-wide 'Gurobekken' channel (described in Halleraker *et al.*, 2003) filled with river bed gravel, representing half of a river with a cross-sectional slope where the lowest edge is representing the deep mid-channel part of a river. A plastic half-pipe without gravel giving hostile habitat conditions was installed to represent the mid-channel at the lowest edge. Flow could vary between 0 and 300 l s^{-1} and water velocities could reach 50 cm s^{-1} at high flow.

It is often very difficult to reproduce natural morphological conditions (substrate, hyporheic flow, sediment dynamics, etc) in laboratories. Natural rivers or modified natural rivers may then be used for experimental studies in situ. In this study, we report findings from an experimental setup where enclosures of 75 m^2 and 63 m^2, respectively, were built in the drawdown zone of the River Nidelva (Mid-Norway) and River Dale (Western Norway). Fish leaving the area during flow reduction were trapped in a net bag (Nidelva) or a plastic half-pipe (Daleelva), both remaining under water at low flow. The enclosures were stocked with a known number of wild Atlantic salmon (*Salmo salar*) and brown trout (*Salmo trutta*). Fish were caught by electro-shocking and acclimated more than 24 hours in perforated boxes before transfer to enclosures. They were then given time to hide and establish territories within the enclosures for 6–120 hours before the stranding experiments started.

In situ case-studies of rivers affected by hydropeaking are, of course, possible to find, but it is hard to find reference conditions or comparable observations in natural rivers. It may also be important to draw comparisons with regulated rivers without rapid flow changes. As most hydropeaked sites have already been impacted by hydropeaking, it may also be difficult to distinguish between short- and long-term effects. However, case studies in rivers affected by hydropeaking have always provided more knowledge about the different impacts of fluctuating flow regimes. In this study, we have considered Norwegian rivers on a large geographical scale, ranging from southwest Norway (Suldalslågen), via eastern South Norway (Abbortjernbekken), Mid-Norway (Bævra) and Northern Norway (Alta).

The Trollheim power plant normally discharges a stable flow of 30–38 m^3 s^{-1} into the River Surna in Mid-Norway. The power station may change the production and ramp down flow to a minimum of 15 m^3 s^{-1}, normally only leading to a moderate rate of change and only a moderate flow ratio of 2–3 in Surna River. Due to accidents at the power plant, the plant was shut down with immediate effect during two occasions in 2005 and 2008, leading to flow reductions from 46 to 9 m^3 s^{-1} and from

21 to 3 m^3 s^{-1}, respectively. We used the dynamic version of the hydraulic model HEC-RAS (Anonymous, 1998) to simulate water level variations at transects, and the stationary version to model the water-covered area. These data were combined with simple surveys of relative fish abundance at different distances to shore and literature data for stranding risk and survival rates. We developed a stranding model for Atlantic salmon and brown trout for the River Surna (Halleraker *et al.*, 2005; Forseth *et al.*, 2009).

Model simulations of fish habitat with CASiMiR (see Chapter 4) have been performed on the River Saane in Switzerland in order to assess the risk of stranding of European grayling (*Thymallus thymallus*) due to hydropeaking when flow is rapidly reduced from 180 m^3 s^{-1} to 5 m^3 s^{-1} (Schneider and Noack, 2009). The multistep habitat modelling approach implemented in CASiMiR considers the flow-dependent basic habitat suitability using the variables water depth, flow velocity and substrate in combination with varying down-ramping rates and areas that become disconnected during flow reduction. The relevant input parameters and water-covered areas are generated by simulating the flow fluctuations using the 2D-hydrodynamic model Hydro_AS-2D (Nujic, 1999). Model simulations of the loss of biomass were estimated in Saane river used for hydropeaking over a period of several decades. CASiMiR simulations, based on the multistep method described in Chapter 4 of this book, delivered basic information for the assessment.

In situ experiments and experiments in nature-like laboratories, combined with model simulations, may be a good way of combining the methods above in order to achieve reliable results.

19.4 Results

19.4.1 Laboratory studies

Flodmark *et al.* (2004) investigated the level of the stress hormone cortisol in young trout in Gurobekken laboratory river subject to rapid flow changes. When fish were exposed to a rapid flow decrease without dewatering, the plasma cortisol level increased (61.3 ± 26.8 ng ml^{-1}), compared to fish exposed to constant flow (4.9 ± 3.7 ng ml^{-1}). However, after four days with daily fluctuations simulating hydropeaking, no difference in cortisol level was observed compared to fish exposed to constant flow (18.2 ± 15.8 ng ml^{-1} compared to 17.5 ± 5.6 ng ml^{-1}). Flodmark *et al.* (2004) concluded that no chronic elevated cortisol level would appear in young trout exposed to regular hydropeaking when this did not lead to dewatering of used habitats.

Experiments of stranding with individually PIT-marked young trout were also carried out in Gurobekken (Halleraker *et al.*, 2003). Fish were subject to repeated dewatering, and could escape from stranding by shifting position to the half-pipe which was always water-covered. Both stranded and escaped fish were rapidly caught and later reintroduced to the water to take part in another stranding experiment. Results show that about 40% of the fish subjected to several dewatering episodes always managed to avoid stranding and escaped to the deepest part. 55–66% of the fish stranded only once, even though they took part in several experiments with dewatering. Significantly, more fry were stranded in the first versus the second to fifth dewatering episodes (mean 22% versus 10% stranding) at rapid daytime dewatering. Stranding of larger fry older than YoY was negligible at water temperatures around 11 °C, except at rapid dewatering during daylight. Halleraker *et al.* (2003) also found a significant decrease in stranding of trout fry by reducing the dewatering speed from 60 cm h^{-1} to less than 10 cm h^{-1}.

19.4.2 Experimental studies in situ

Experiments in Nidelva in Trondheim and in Daleelva in Hordaland found similar results of stranding risk for juvenile salmonid fish (Harby *et al.*, 2001; Saltveit *et al.*, 2001). The risk of stranding is higher during daylight compared to darkness. Due to decreased mobility, the stranding risk is also higher in cold water than in warm water (water temperature above approximately 8 °C). Stranding risk is positively correlated to the presence of shelter and negatively correlated to the size of fish. Stranding risk was found to decrease when dewatering was slower than 13 cm h^{-1} measured in the vertical direction. However, if the shelter availability is high, significant stranding may also occur at slower dewatering rates.

19.4.3 Case studies

Heggenes (1988) investigated the impacts on habitat use and displacement of young trout from rapid flow changes in Abbortjernbekken in Oslo. Due to coarse substrate conditions, Heggenes (1988) did not find any impact on habitat use, and only a slight increase, most notable at night, in mobility caused by rapid flow changes.

In the early 1990s, a single stranding experiment was carried out in the Alta River (Jensen *et al.*, 1992). An average number of 1.5 stranded fish per m^2 was found

and hydropeaking was then omitted. However, several accidental shut-downs of the power plant occurred during the 1990s, and Forseth *et al.* (1996) concluded that these drops in water levels probably led to a reduced fish population.

Rapid flow changes in the bypassed section of the River Suldalslågen in south-west Norway led to low densities of young salmon (Saltveit, 1996). The effect was most pronounced close to the upstream part of the river where the flow was released. Due to an unexpected stoppage of water release, stranded and dead fish were also observed (Saltveit, 2000).

Svorka power plant in Mid-Norway discharges water directly into the Bævra River, and some hydropeaking is carried out. Low fish densities downstream of Svorka power plant are probably related to rapid changes in flow (Johnsen *et al.*, 2010). In some years, gradual and stepwise reductions in the flow have caused lower mortalities and higher fish densities.

19.4.4 Simulations and analysis: Surna River in Norway

Stationary simulations showed the difference in water-covered area before and after the flow reduction at different sections of the Surna River, showing that 194 000 m^2 and 270 000 m^2 were dewatered in 2005 and 2008, respectively. Stranding experiments in the nearby River Nidelva showed a higher stranding risk when the dewatering rate was above 13 cm h^{-1} (Saltveit *et al.*, 2001). Therefore, we distinguished between areas dewatered at a rate higher and lower than 10 cm h^{-1}, lowering the rate to 10 cm h^{-1} to include a margin. One-dimensional dynamic simulations showed that the rate of change in water level varied from 16 to 3 cm h^{-1} at different locations in Surna in 2008. Due to some dampening and the transect shape, dewatering rates were above 10 cm h^{-1} in 56% of the area (152 300 m^2) and below 10 cm h^{-1} at 117 700 m^2. In 2005, we did not do any dynamic simulations of water level changes. The dewatering ratio was estimated to vary between 62 and 11 cm h^{-1} based on recorded observations of water level changes from 30 to 6 m^3 s^{-1} during a field experiment in 2004, transect shapes and recorded average dewatering rate of 54 cm h^{-1} during the episode in 2005.

The densities of juvenile Atlantic salmon and brown trout have been recorded at nine stations of 100 m^2 close to the shore by electro-fishing since 2002 (Johnsen *et al.*, 2008). The average densities per 100 m^2 of YoY (Young-of-the-Year) for 2002–2004 were 84 for salmon and 55

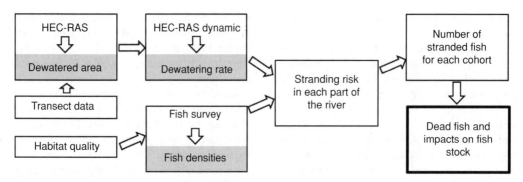

Figure 19.2 The process of estimating stranded fish and impacts on fish stock from dewatering.

for trout. For older fish, densities were 13 for salmon and 4 for trout. In 2005, we had no information about how fish densities would vary across the river and assumed that the density in the deeper parts not dewatered during the stranding episode would be 40% lower than along the shores where the electro-fishing stations were located.

The average densities per 100 m² of YoY for the 2002–2007 period were 57 for salmon and 51 for trout. For older fish, densities were 26 for salmon and 52 for trout. In 2008, we used observations of habitat quality with respect to substrate and shelter conditions to estimate the relative distribution of fish between areas close to the shore and the deeper parts not affected directly by dewatering. Combined with a mesohabitat mapping of the river, we estimated the actual densities of fish before dewatering in 2008 to be 21 YoY salmon, 19 older salmon and 9 trout per 100 m².

When combining the area dewatered, the rate of change in water level, the habitat quality of the dewatered area and estimated fish densities, stranding risk indices from Saltveit *et al.* (2001) and Harby *et al.* (2004), the number of stranded fish could be estimated (Figure 19.2). The estimated numbers of stranded fish in 2005 were 17 600 YoY salmon, 2200 older salmon, 9100 YoY trout and 700 older trout. This represents 11%, 6%, 8% and 3% of the total estimated population of YoY salmon, older salmon, YoY trout and older trout, respectively. We estimated that this would lead to a loss of 2600 salmon smolts.

The estimated numbers of stranded fish in 2008 were 14 000 YoY salmon, 3600 older salmon and between 3000 and 15 000 YoY trout. We were not able to estimate the number of stranded older trout in 2008. In 2008 we estimated that the loss of juvenile salmon would lead to a loss of about 3000 smolts, representing 3% of the estimated total smolt production in Surna downstream of the power plant.

19.4.5 Simulations and analysis: Saane River in Switzerland

The first step for analysing stranding risk in CASiMiR simulates the basic habitat suitability, which combines the conventional habitat parameters (water depth, flow velocity and dominant substrate), while in a second and third step, the effects of hydropeaking on stranding are incorporated. The second step considers the effect of high down-ramping rates on fish stranding by linking the basic habitat suitability, the down-ramping rate and a critical water depth to a stranding risk, ranging from 0 (no risk) to 1 (very high risk). The third step also considers areas that become disconnected from the main channel during flow reduction, as fish in these areas remain trapped and cannot return to the main channel. Figure 19.3 shows an example of the simulation results when flow is reduced from 45 to 36 m³ s⁻¹ within 10 minutes. Figure 19.3 (left) indicates the highest risk at the edges of the gravel bars, as, during high flow, the preferred habitats of juvenile grayling are located along the shallow inundated gravel bar. The areas disconnected from the main channel during flow reduction are located close to the right river bank and on the gravel island at the beginning of the river reach (Figure 19.3, centre). High ramping rates and hydraulically disconnected areas may not be problematic, as long as the fish are able to follow the reducing water levels. Therefore, the highest risk for fish stranding is found by superposing both stranding risk types. The magnitude of these areas and their change during the down-ramping phase is a step-function visualised in Figure 19.3 (right).

One conclusion obtained by the evaluation of the model results and collected data is that the extent of damage caused by stranding is closely related to the river morphology and even minor flow fluctuations can provoke a loss of juveniles in morphologically unfavourable sections. In flow ranges close to the base flow, in particular,

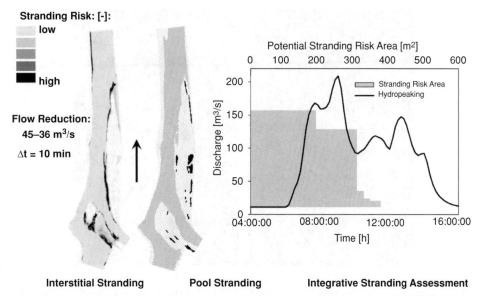

Figure 19.3 (Left) Spatial distribution of stranding risk; (centre) pool stranding – areas separated from main channel; (right) flow time series of hydropeaking event and related areas of aggregated stranding risk.

a high stranding risk is observed given the drastic change in wetted areas (Figure 19.3, right). Recommendations for the mitigation of hydropeaking impacts are: an increase in the base flow and a reduction in the peak flow as well as the limitation of down-ramping rates in predefined critical discharge ranges (hydro operations). Further, morphological restructuring to improve habitat availability would be beneficial to minimise the ecological impact due to hydropeaking (physical mitigation).

19.5 Mitigation

19.5.1 Hydro operations

The best mitigation to avoid negative impacts from hydropeaking may be to change the flow so slowly that organisms are able to adapt to the new flow situation. If possible, it may also be useful to increase or decrease the flow very slowly first, more like a warning signal to nature to show there is a flow change coming. However, changes in hydro operation are limited by the technical possibilities to start and stop machines and power plants, which may therefore make mitigation more complicated.

Based on previous studies, we give the following advice for hydro operations:
• Experiments have shown a tendency to increased stranding risk when introducing hydropeaking after a

period with stable high flows. If the power plant starts peaking operations, smooth changes in flow should be introduced at the beginning of such a period.
• Flow reduction should only be done after dark to reduce stranding risk, especially during winter (at water temperatures below 8 °C) when fish are less mobile and often hiding in the substrate during daytime.
• Fluctuations that always include a bankfull discharge have most probably no or little effect on the aquatic life.
• Decreasing flow rates less than 13 cm h^{-1} reduces the risk of stranding of salmonid juveniles, but this may still occur at locations with high water velocities, coarse substrate with an abundance of suitable interstitial spaces and side slopes less than 5%.
• Stranded fish may not always die. Humidity and water flow in the hyporheic zone may keep fish alive for several hours even if the surface seems to be dewatered. Air temperatures below zero and easy access for predators are harmful factors causing increased mortality under such conditions.
• The benthic fauna will be reduced in areas subject to frequent dewatering, and it is important to keep a certain residual flow to ensure a benthic fauna in a river subject to hydropeaking.

Clarke *et al.* (2008) suggest three important mitigation possibilities:
• Ensure that the lowest flow is high enough to maintain a sustainable ecosystem.

Figure 19.4 Restoration of the River Drau in Austria has introduced refuges and delayed the rate of change in water level during hydropeaking (right) compared to the channelised situation to the left. Credit: Regional Government of Carinthia/S. Tichy. A colour version of this image appears in the colour plate section of the book.

• The rise in flow should mimic a rapid natural flow increase, as seen in small floods.

• The timing of up- and down-ramping should be adjusted to the diurnal behaviour patterns of target species.

19.5.2 Physical mitigation

Instead of, or in addition to, mitigation by adjusting hydro operations, it is possible to build structures or modify the river channel and floodplain to reduce the negative impacts of hydropeaking. Such mitigation is referred to as physical mitigation. Artificial in-channel structures are often constructed in order to mitigate the withdrawal of water, bypassed sections and the regulation of rivers. These structures may also be important for mitigation of hydropeaking. Weirs, reflectors, submerged trees and other structures made to increase habitat diversity may also delay and reduce the magnitude of water level changes. All structures reducing the canalisation of water will also reduce the rate of change in water level and flow during peaking operations.

River restoration that includes re-introduction of side channels and the use of the whole floodplain will also provide important refuges for biota at high flows occurring during hydropeaking (Schwartz and Herricks, 2005; Ribi *et al.*, 2009). It is important that the lateral connectivity between these side channels and the main river is maintained during low flow to avoid pool trapping and

stranding. Figure 19.4 shows an example of the restoration of side channels on the River Drau in Austria.

To avoid constant wetting and drying of the river bed in hydropeaking rivers, it is possible to introduce structures which ensure that parts of the river bed are constantly either under or above water, both at high and low flows during hydropeaking. A redistribution of bed material and the introduction or removal of weirs and deflectors are useful methods to achieve this. These structures must be maintained to provide a permanent service. An example of this kind of mitigation is found in the Dordogne River in south-western France. In a reach with hydropeaking, important salmonid spawning grounds were repeatedly wetted and dried during hydropeaking (Guerri, 2008). The water level was raised by constructing a weir to ensure permanent wetting of the area (Figure 19.5). In another reach of the Dordogne River, water and potentially biota were trapped in a pool with no connection to the main stem when flow was reduced. This pool was bound to dewater gradually, and the area was changed by redistributing the bed material (Figure 19.6).

Small reservoirs or retention basins introduced at the outlet of power plants discharging water directly into downstream rivers can reduce the magnitude and rate of change in flow and water level. This kind of mitigation requires the allocation of land or river corridors and may have other impacts on the environment, and the benefits must be weighed against the costs. Figure 19.7 shows

Figure 19.5 Spawning areas in the Dordogne River (left) before and (right) after mitigation against repeated wetting and drying. After Chanseau *et al.*, 2008, reproduced with permission.

two examples of retention basins introduced (left) and planned (right) in the Alps.

Most of the structures used for physical river restoration and mitigation of hydropeaking are subject to changes due to geomorphological processes in the river. Sediment deposits and erosion may change their function, reduce their impact and it is also possible that such structures may be totally destroyed or removed. To what degree and how often this will happen is related to the dynamics of the river and to what extent the total flow and sediment regime is controllable. Even if physical mitigation leads to maintenance costs, it might be able to achieve the same mitigation effect with lower costs than by adjusting hydro operations with the potential loss of income for power production.

Relocating the outlet of peaking power plants from rivers to lakes, reservoirs and the sea will eliminate the impacts of hydropeaking in the actual river. Typical problems and potential mitigation options are summarised in Table 19.1.

19.6 Discussion and future work

The interactions between hydropeaking, hydrology, hydraulics and morphological changes and their subsequent ecological impact have not been investigated sufficiently. This is an important aspect but strongly site-specific, and more research is needed to understand processes and to gain knowledge to select proper and effective mitigation strategies.

Laboratory studies and in situ experimental studies always face the challenge of transferability of results to river conditions. The results from Flodmark *et al.* (2004) showing that daily peaking will not lead to elevated cortisol levels in young trout remain to be verified in field conditions.

Laboratory studies of stranding of fish showed large individual differences in stranding risk, as a large portion of fish managed to avoid stranding at repeated hydropeaking. This is probably due to differences in behaviour,

Figure 19.6 A reach in the Dordogne River (left) before and (right) after redistribution of bed material to prevent pool trapping. After Chanseau *et al.*, 2008, reproduced with permission.

Figure 19.7 Sketch of planned retention basin in the Hasliaare River where micro-turbines may also use the head difference between the basin and the river to produce power. From Schweizer *et al.* (2009), reproduced with permission. A colour version of this image appears in the colour plate section of the book.

as some fish seem to continue to hide in the substrate even though the water level is dropping under the surface, while others rapidly search for a new position. Hiding without moving around would normally be the most energy-efficient behaviour, but it is not a good strategy to avoid stranding. So there is a risk that the fish with the most energy-efficient behaviour in a population will die from stranding.

In situ experiments with stranding were conducted in rivers used for hydropeaking on a regular basis, which also gives natural substrate conditions, allowing fish to shelter during dewatering. Shelter availability, substrate conditions and hyporheic humidity are very important to determine stranding risk and stranding survival (Saltveit *et al.*, 2001). However, the experiments were carried out in enclosures and fish could not move around totally freely, as they were eventually stopped and guided by the fences. This means that the results must be transferred with care to other regulated rivers.

The results and conclusions from case studies in rivers experiencing hydropeaking must be evaluated in the context of what to compare them against. Should the impacts from hydropeaking be compared to natural conditions or to the traditional way of regulating hydropower?

The stranding model used in the Surna River (Figure 19.2) has many steps and sources of error. The major

Table 19.1 Environmental hydropeaking problems and potential mitigation options.

Problem (+ Positive effect ÷ Negative effect)	Slow down-ramping	Smooth up-ramping	Minimum flow close to bankfull	Retention basin	Keep peaking regular	Build refuges in channel	Creating constant river bank slopes	Allow floods
Hydromorphological impacts								
Reduced vertical connectivity (siltation)	÷	÷		÷		÷		+
Reduced lateral connectivity	+	+	+	+		÷	+	÷
Increased erosion		+		+				÷
Water quality impacts								
Water temperature changes	+	+		+		+		
Changes in turbidity	+	+					÷	
Increased biomass of water vegetation								+
Biological impacts								
Stranding of fish	+		+	+		+	+	
Stress level in fish			+		+			
Increased energy use and reduced growth in fish	+	+	+	+	+	+	+	
Reduction in reproduction areas for fish			+		+		+	+
Catastrophic drift of invertebrates and small fish		+				+		÷

sources of error are related to the quality of transect data, the calibration of both the static and the dynamic HEC-RAS model, the quality of the habitat mapping, the quality and the representativity of the fish survey data and the indices used to estimate stranding risk and mortality of stranded fish. Many other factors not included in the stranding model will also affect stranding risk and mortality, such as previous flow patterns, fish activity, intra-specific and inter-specific fish competition, water temperature and water temperature changes and other anthropogenic factors. Johnsen *et al.* (2008) have estimated the total amount of salmon smolts to 27 000 downstream of the Trollheim power plant in the Surna River. In relation to this number, the loss of 2600 (in 2005) and 3000 (in 2008) smolts is not crucial to the population, but will lead to a reduction in the total fish stock. If such accidents happen more regularly, or similar hydro operations are carried out, severe impacts on the Surna fish stocks may be the result.

The results from the Saane River indicate that, after fast down-ramping and separation of inundated areas during flow reduction, the stranding risk is strongly affected by the base flow, which should be high enough to avoid a rapid decrease in wetted areas in the low-flow phase of hydropeaking events. The proposed multistep fuzzy approach is able to locate and quantify the potential stranding risk areas and identify critical flow rates where the risk of stranding is particularly high. This provides valuable information for operational and engineering measures to minimise ecological impacts. However, the proposed approach demands detailed knowledge about both the habitat requirements of target fish species and tolerable ramping rates. The consideration of additional parameters, such as the steepness of river banks, cover characteristics and food supply by drifting macroinvertebrates in the up-ramping phase, should be performed in on-going projects and will improve the understanding of the ecological impacts of hydropeaking.

Stranding is not equal to mortality, as fish may survive several hours of dewatering under favourable conditions (Saltveit *et al.*, 2001). We did not include a correction factor for this in the simulations of the Surna and Saane Rivers, as we don't know the long-term effects on fish surviving stranding. Further research should focus on the effects of fluctuating flow and non-lethal stranding of fish in hydropeaked rivers.

In climate regions with seasonal temperature changes, river regulation often leads to changes in water temperature. If large amounts of cold water from high-altitude reservoirs are discharged into lowland downstream rivers

with initial much higher water temperatures, we will also have thermo-peaking. As many processes in aquatic ecosystems are controlled by temperature, rapid and relatively large changes in water temperature may have unknown effects. This area of research needs more focus in the future.

In Norway, the legislation and licences for hydropower production very seldom give any restrictions on the use of hydropeaking. However, many regulated rivers are operated with 'self-restrictions' to reduce the negative impacts from hydropeaking. There is still a strong need for better legislation and clear operating rules to mitigate the impacts of hydropeaking. There is also probably great potential in combining the operation of several hydropeaked rivers to avoid the most harmful operations. It is also necessary to increase the focus on using physical mitigation, often in combination with hydro operations. Until recently, environmental concerns in regulated rivers have been focusing on migration barriers and environmental and minimum flows. It is now time to include the mitigation of hydropeaking in these concerns, as we are about to witness the increased use of rapid and frequent flow regulations in rivers downstream of hydropower plants. The main knowledge gaps where we should focus our research on environmental impacts of hydropeaking in the future are:

- Impacts on the ecosystem from 'thermo-peaking'.
- Impacts on non-salmonid fish species and species other than fish (invertebrates, food webs, etc.).
- Long-term effects on fish subjected to peaking without stranding and on fish that survive stranding.
- Impact on fish at individual, population and community scales (predation, competition, age structure and composition of populations, trophic structure, biodiversity, etc.).
- Clear limits or thresholds in acceptable hydro operations (ratio, ramping rate, frequency and duration of peaking operations), specific to season and time of day.
- As in situ studies and experiments are difficult to conduct, it is important that we improve modelling tools and also focus on their validation.
- Efficiency and maintenance of physical mitigation measures.
- Guidance for improved legislation and operational rules.

Acknowledgements

The authors wish to thank Jo Halvard Halleraker, Svein Jakob Saltveit, Jo Vegar Arnekleiv, Torbjørn Forseth, Jan

Heggenes, Dave Scruton, Knut Alfredsen, Håkon Sundt, Hans-Petter Fjeldstad, Peter Borsanyi, Stein Johansen, Lars Flodmark and Teppo Vehanen for all the analysis, discussions and work we have done together on the impacts of hydropeaking during the last 15 years. The authors also wish to thank Julie Charmasson and Julian Sauterleute for providing some background material on mitigation and the definition of hydropeaking.

References

Anonymous (1998) U.S. Army Corps of Engineers, HEC-RAS, River Analysis System, User's Manual, Version 2.2, September 1998. Hydraulic Engineering Center, Davis, California, USA.

Bain, M.B. (2007) *Hydropower Operations and Environmental Conservation: St. Mary's River, Ontario and Michigan, Canada and USA.* International Lake Superior Board of Control.

Berland, G., Nickelsen, T., Heggenes, J., Økland, F., Thorstad, E.B. and Halleraker, J.H. (2004) Movements of wild Atlantic salmon parr in relation to peaking flows below a power station. *River Research and Applications,* **20**: 957–966.

Bradford, M.J. (1997) An experimental study of stranding of juvenile salmonids on gravel bars and in side channels during rapid flow decreases. *Regulated Rivers: Research and Management,* **13**: 295–401.

Bruno, M.C., Maiolini, B., Carolli, M. and Silveri, L. (2009) Impact of hydropeaking on hyporheic invertebrates in an Alpine stream (Trentino, Italy). *International Journal of Limnology,* **45**: 157–170.

Clarke, K.D., Pratt, T.C., Randall, R.G., Scruton, D.A. and Smokorowski, K.E. (2008) *Validation of the Flow Management Pathway: Effects of Altered Flow on Fish Habitat and Fishes Downstream from a Hydropower Dam.* Canadian Technical Report of Fisheries and Aquatic Sciences 2784.

Flodmark, L.E.W., Vøllestad, L.A. and Forseth, T. (2004) Performance of juvenile brown trout exposed to fluctuating water level and temperature. *Journal of Fish Biology,* **65**(2): 460–470.

Flodmark, L.E.W., Forseth, T., L'Abée-Lund, J.H. and Vøllestad, L.A. (2006) Behaviour and growth of juvenile brown trout exposed to fluctuating flow. *Ecology of Freshwater Fish,* **15**: 57–65.

Førde, E. and Brodtkorb, E. (2001) *Sluttrapport for FoU-prosjektet Effektregulering – Miljøvirkninger og konfliktreduserende tiltak.* Rapport nr 20, Statkraft Grøner.

Forseth, T., Næsje, T.F., Jensen, A.J., Saksgård, L. and Hvidsten, N.A. (1996) *Ny forbitappingsventil i Alta kraftverk: Betydning for laksebestanden.* NINA Oppdragsmelding 392.

Forseth, T., Stickler, M., Ugedal, O., Sundt, H., Bremset, G., Linnansaari, T., Hvidsten, N.A., Harby, A., Bongard, T. and Alfredsen, K. (2009) *Utfall av Trollheim Kraftverk i juli 2008. Effekter på fiskebestandene i Surna.* NINA-rapport 435.

García, A., Jorde, K., Habit, E., Caamaño, D. and Parra, O. (2011) Downstream environmental effects of dam operations: Changes in habitat quality for native fish species. *River Research and Applications,* **27**(3): 312–327.

Guerri, O. (2008) *Etude de l'impact écologique des éclusées sur la rivière Dordogne. Cahier des charges.* Etablissement Public Territorial du Bassin de la Dordogne – EPIDOR, Castelnaud-la-Chapelle, France.

Halleraker, J., Johnsen, B.O., Lund, R.A., Sundt, H., Forseth, T. and Harby, A. (2005) *Vurdering av stranding av ungfisk i Surna ved utfall av Trollheim kraftverk i august 2005.* SINTEF TR A6220.

Halleraker, J.H., Saltveit, S.J., Harby, A., Arnekleiv, J.V., Fjeldstad, H.P. and Kohler, B. (2003) Factors influencing stranding of wild juvenile brown trout (*Salmo trutta*) during rapid and frequent flow decreases in an artificial stream. *River Research and Applications,* **19**: 589–603.

Harby, A., Alfredsen, K., Arnekleiv, J.V., Flodmark, L.E.W., Halleraker, J.H., Johansen, S. and Saltveit, S.J. (2004) *Raske vannstandsendringer i elver. Virkninger på fisk, bunndyr og begroing.* SINTEF-rapport TR A5932, Trondheim, Norway.

Harby, A., Alfredsen, K., Fjeldstad, H.P., Halleraker, J.H., Arnekleiv, J.V., Borsányi, P., Flodmark, L.E.W., Saltveit, S.J., Johansen, S., Vehanen, T., Huusko, A., Clarke, K. and Scruton, D. (2001) Ecological impacts of hydropeaking in rivers. In Honningsvåg, B., Midttømme, G.H., Repp, K., Vaskinn, K. and Westeren, T. (eds) Hydropower in the New Millennium, *Proceedings of the 4th International Conference on Hydropower Development, Hydropower '01, Bergen, Norway, 20–22 June 2001.*

Heggenes, J. (1988) Effects of short-term flow fluctuations on displacement of, and habitat use by, brown trout in a small stream. *Transactions of the American Fisheries Society,* **117**: 336–344.

Hvidsten, N.A. (1985) Mortality of pre-smolt Atlantic salmon (*Salmo salar*) and brown trout (*Salmo trutta*), caused by fluctuating water levels in the regulated River Nidelva, Central Norway. *Journal of Fish Biology,* **27**: 711–718.

IPCC (2011) *Special Report on Renewable Energy Sources and Climate Change Mitigation* (eds Edenhofer, O., Pichs-Madruga, R., Sokona, Y., Seyboth, K., Matschoss, P., Kadner, S., Zwickel, T., Eickemeier, P., Hansen, G., Schlömer, S. and von Stechow, C.), Cambridge University Press, Cambridge, UK.

Irvine, R.L., Oussoren, T., Baxter, J.S. and Schmidt, D.C. (2009) The effects of flow reduction rates on fish stranding in British Columbia, Canada. *River Research and Applications,* **25**: 405–415.

Jensen, J.W., Koksvik, J.I. and Karlsen, L. (1992) Rapport fra forsøk med korttidsregulering av Altaelva. Stensil.

Johnsen, B.O., Bremset, G. and Hvidsten, N.A. (2010) *Fiskebiologiske undersøkelser i Bævra, Møre og Romsdal.* Årsrapport 2009. NINA Rapport 591.

Johnsen, B.O., Hvidsten, N.A., Bongard, T. and Bremset, G. (2008) *Ferskvannsbiologiske undersøkelser i Surna.* Årsrapport 2007. NINA Rapport 373.

Junk, W.J., Bayley, P.B. and Sparks, R.E. (1989) The flood pulse concept in river floodplain systems. *Canadian Journal of Fisheries and Aquatic Sciences*, **106**: 110–127.

Lauters, F., Lavandier, P., Lim, P., Sabaton, C. and Belaud, A. (1996) Influence of hydropeaking on invertebrates and their relationship with fish feeding habitats in a Pyrenean river. *Regulated Rivers: Research and Management*, **12**: 155–169.

Meile, T. (2006) Hydropeaking on watercourses. *Eawag News 61e*, November 2006.

Meile, T., Boillat, J.-L. and Schleiss, A.J. (2011) Hydropeaking indicators for characterization of the Upper-Rhone River in Switzerland. *Aquatic Sciences*, **73**: 171–182.

Moog, O. (1993) Quantification of daily peak hydropower effects on aquatic fauna and management to minimize environmental impacts. *Regulated Rivers: Research and Management*, **8**: 5–14.

Nagrodski, A., Raby, G.D., Hasler, C.T., Taylor, M.K. and Cooke, S.J. (2012) Fish stranding in freshwater systems: sources, consequences, and mitigation. *Journal of Environmental Management*, **103**: 133–141.

Nujic, M. (1999) Praktischer Einsatz eines hochgenauen Verfahrens für die Berechnung von tiefengemittelten Strömungen, *Mitteilungen des Instituts für Wasserwesen der Universität der Bundeswehr München*, Volume **64**.

Osmundson, D.B., Ryel, R.J., Lamarra, V.L. and Pitlick, J. (2002) Flow–sediment–biota relations: Implications for river regulation effects on native fish abundance. *Ecological Applications*, **12**: 1719–1739.

Poff, N.L., Allan, D., Bain, M., Karr, J.R., Prestegaard, K.L., Richter, B.D., Sparks, R.E. and Stromberg, J.C. (1997) The natural flow regime. A paradigm for river conservation and restoration. *BioScience*, **47**: 769–784.

Ribi, J.-M., Steffen, K., Boillat, J.-L., Peter, A. and Schleiss, A. (2009) Influence of geometry of fish shelters in river banks on their attractiveness for fishes during hydropeaking. *33rd IAHR Congress 'Water Engineering for a Sustainable Environment'*, Vancouver, Canada, 9–14 August, 2009.

Saltveit, S.J. (1996) *Skjønn Ulla-Førre. Fiskeribiologisk uttalelse. Begroing og ungfisk*. Rapport Lab. Ferskv. Økol. Innlandsfiske, Oslo, 162.

Saltveit, S.J. (2000) *Alderssammensetning, tetthet og vekst av ungfisk av laks og ørret i Suldalslågen i perioden 1976 til 1999*. Suldalslågen-Miljørapport, 7.

Saltveit, S.J., Halleraker, J.H., Arnekleiv, J.V. and Harby, A. (2001) Field experiments on stranding in juvenile Atlantic salmon (*Salmo salar*) and brown trout (*Salmo trutta*) during rapid flow decreases caused by hydropeaking. *Regulated Rivers: Research and Management*, **17**: 609–622.

Schneider, M. and Noack, M. (2009) Untersuchung der Gefährdung von Jungfischen durch Sunkereignisse mit Hilfe eines Habitatsimulationsmodells. *Wasser-Energie-Luft*, **101**: 115–120.

Schwartz, J.S. and Herricks, E.E. (2005) Fish use of stage-specific fluvial habitats as refuge patches during a flood in a low-gradient Illinois stream. *Canadian Journal of Fisheries and Aquatic Sciences*, **62**(7): 1540–1552.

Schweizer, S., Neuner, J., Ursin, M., Tscholl, H. and Meyer, M. (2009) *Trotz Ausbau der Kraftwerksleistung – deutlich geringere Pegelschwankungen in der Hasliaare.*

Scruton, D.A., Ollerhead, L.M.N., Clarke, K.D., Pennel, C., Alfredsen, K., Harby, A. and Perry, D. (2003) The behavioural response of juvenile Atlantic salmon (*Salmo salar*) and brook trout (*Salvelinus fontinalis*) to experimental hydropeaking on a Newfoundland (Canada) river. *River Research and Applications*, **19**: 577–587.

Scruton, D.A., Pennell, C.J., Robertson, M.J., Ollerhead, L.M.N., Clarke, K.D., Alfredsen, K., Harby, A. and McKinley, R.S. (2005) Seasonal response of juvenile Atlantic salmon to experimental hydropeaking power generation in Newfoundland, Canada. *North American Journal of Fisheries Management*, **25**: 964–974.

Scruton, D.A., Pennell, C.J., Ollerhead, L.M.N., Alfredsen, K., Stickler, M., Harby, A, Robertson, M.J., Clarke, K.D. and LeDrew, L.J. (2008) A synopsis of 'hydropeaking' studies on the response of juvenile Atlantic salmon to experimental flow alteration. *Hydrobiologia*, **609**: 263–275.

Sear, D.A. (1995) Morphological and sedimentological changes in a gravel-bed river following 12 years of flow regulation for hydropower. *Regulated Rivers: Research and Management*, **10**: 247–264.

Smokorowski, K.E., Metcalfe, R.A., Finucan, S.D., Jones, N., Marty, J., Power, M., Pyrce, R.S. and Steele, R. (2010) Ecosystem level assessment of environmentally based flow restrictions for maintaining ecosystem integrity: A comparison of a modified peaking versus unaltered river. *Ecohydrology*. doi: 10.1002/eco.167.

Tuhtan, J.A. (2012) *A Modeling Approach for Alpine Rivers Impacted by Hydropeaking Including the Second Law Inequality*. Dissertation No. 210, Institute for Modelling Hydraulic and Environmental Systems, University of Stuttgart, Germany.

Tuhtan, J.A., Noack, M. and Wieprecht, S. (2012) Estimating stranding risk due to hydropeaking for juvenile European grayling considering river morphology. *KSCE Journal of Civil Engineering*, **16**(2): 197–206.

Vehanen, T., Bjerke, P.L., Heggenes, J., Huusko, A. and Mäki-Petäys, A. (2000) Effect of fluctuating flow and temperature on cover type selection and behaviour by juvenile brown trout in artificial flumes. *Journal of Fish Biology*, **56**: 923–937.

Vehanen, T., Huusko, A., Yrjänä, T., Lahti, M. and Mäki-Petäys, A. (2003) Habitat preference by grayling (*Thymallus thymallus*) in an artificially modified, hydropeaking riverbed: a contribution to understand the effectiveness of habitat enhancement measures. *Journal of Applied Ichthyology*, **19**(1): 15–20.

20 Ecohydraulic Design of Riffle-Pool Relief and Morphological Unit Geometry in Support of Regulated Gravel-Bed River Rehabilitation

Gregory B. Pasternack and Rocko A. Brown

Department of Land, Air, and Water Resources, University of California, One Shields Avenue, Davis, CA 95616, USA

20.1 Introduction

Ecohydraulics is a rapidly emerging quantitative sub-discipline of river science with applications in river engineering and rehabilitation in degraded landscapes (Nestler *et al.*, 2008; Wheaton *et al.*, 2011). Ecohydraulics links ecological functions and hydrodynamic patterns at each spatio-temporal scale (Pasternack, 2011). It is often used for baseline instream flow assessment, emphasizing relations between species' physical habitat and discharge. The Instream Flow Incremental Methodology (IFIM) is a widely adopted tool for incorporating quantitative echy-draulics into flow assessment (Bovee and Milhous, 1978; Bovee, 1982; Jowett, 1997), often facilitated using PHAB-SIM software (Waddle, 2001). Normally, a statistical evaluation of flow and habitat is made, but now spatially explicit (2D) habitat modelling is practical (Ghanem *et al.*, 1996; Pasternack, 2011). Studies have compared semi-analytical, 1D and 2D ecohydraulic methods (Waddle *et al.*, 2000; Brown and Pasternack, 2009). Many rivers exhibit a spatial anisotropy of channel geometry (Merwade, 2009) capable of steering flow, thereby violating 1D-model assumptions (Brown and Pasternack, 2009). In such cases, 2D modelling and GIS-based spatial

analyses are necessary, and even for isotropic, orthogonal geometries they are preferential for evaluating mesohabitat structure (Hauer *et al.*, 2011) as well as microhabitat heterogeneity. Use of 3D models is emerging, but suffers for lack of 3D ecological relations, the high cost and complexity of 3D validation, longer 3D numerical modelling time and disproportionate GIS data volume and processing time.

In the practice of site- and reach-scale river rehabilitation, standard engineering and geomorphic methods have been highly criticized (Wissmar and Beschta, 1998; Simon *et al.*, 2007; Lave *et al.*, 2010), exacerbated by iconic failures against project goals (Kondolf and Micheli, 1995; Doyle and Harbor, 2000). Statistical ecohydraulic methods such as IFIM that depend on static channel assumptions, direct observations of channel hydraulics and static empirical parameters cannot yield predictions for alternative channel configurations, so they have limited ecohydraulic applicability as a tool for river engineering (as opposed to their strength in river assessment). Design methods that empirically mimic landforms (i.e. 'natural channel design') or hydrology (i.e. 'natural flow regime') are prescriptive and have no independent, quantitative design-testing scheme, yielding a high risk of failure when used alone.

Ecohydraulics: An Integrated Approach, First Edition. Edited by Ian Maddock, Atle Harby, Paul Kemp and Paul Wood.
© 2013 John Wiley & Sons, Ltd. Published 2013 by John Wiley & Sons, Ltd.

The philosophy underlying *ecohydraulic design for river engineering* is that channel geometry is manipulated and then mechanistically tested until it achieves a flow-dependent hydraulic regime with a palette of homogeneity and heterogeneity at different spatial scales that is suitable for the breadth of geomorphic processes and ecosystem functions characteristic of a natural river of the type undergoing diagnosis and treatment. A key aspect of this design framework is that landforms designed at multiple spatial scales are not arbitrary, but are founded on scale-dependent physical mechanisms needed for morphologic resilience, such as stage-dependent flow convergence routing (MacWilliams *et al.*, 2006; Sawyer *et al.*, 2010) and pool maintenance by turbulent vortex shedding at forcing elements (Woodsmith and Hassan, 2005; Thompson, 2006). Ecohydraulic analysis of 2D hydrodynamic models is rooted in observation, but is structurally more universal than empirical, prescriptive geomorphic methods (Brown and Pasternack, 2009; Pasternack, 2011) and can cope with synthetic channel modifications (Pasternack *et al.*, 2008; Oh *et al.*, 2010). Ecohydraulic design has been tested in different applications and found to be useful as a tool for evaluating alternatives for channel reconfiguration, gravel injection, floodplain and side channel inundation, increasing habitat complexity and spawning habitat rehabilitation (Elkins *et al.*, 2007; Manwaring *et al.*, 2009; Hoopa Valley Tribal Fisheries *et al.*, 2011).

Now that the use of spatially explicit ecohydraulics in river engineering is established, there is an opportunity to generate design principles and project guidelines through scientific testing of diverse scenarios. Pasternack *et al.* (2004) tested the value of four channel patterns for yielding high-quality Chinook salmon spawning habitat on a regulated, degraded gravel-bed river, with two of them outperforming the ad hoc project. Pasternack *et al.* (2008) used ecohydraulics to test riffle configurations, tailwater levels (imposed by the next downstream riffle that is not being altered) and discharge on physical habitat quality and morphological resilience. Instituting a backwater effect downstream of a design riffle aids both of these desired outcomes. Elkins *et al.* (2007) corroborated this in a real spawning habitat rehabilitation.

The goal of this study was to use ecohydraulic analysis of 2D model results in a numerical experiment to test the relative merits of building sequences of riffle-pool units in regulated gravel-bed rivers with different magnitudes of riffle-pool relief. Such rivers typically have bed slopes of 0.001–0.01, width-to-depth ratios of 20–100, depth-to-median-grain-size ratios of 2–60 and a Shields stress incipient motion threshold of $\sim$ 0.03–0.06. These dimensionless values express the range for which this study is relevant, except that they ignore riffle-pool relief. While metrics for evaluating (let alone designing) riffle-pool relief are limited, alternative-design morphologies were compared using the slope-detrended relief indices of amplitude (A_{RP}), the difference between the maximum riffle crest elevation and the minimum pool trough on the slope-detrended river profile (Vetter, 2011) and asymmetry (A_{RP}^*), the ratio of the absolute values of slope-detrended riffle height and pool depth about the zero-crossing line (O'Neil and Abrahams, 1984; Rayburg and Neave, 2008). A high A_{RP}^* indicates a riffle crest sticking disproportionately high above the zero-crossing line.

Riffle-pool relief is an important yet untested metric for channel design, because it could play an important role in morphologic resilience during floods (Pasternack *et al.*, 2008). There is no widely used dimensionless metric for riffle-pool relief design and there exist few pristine reference rivers free of anthropogenic influence to evaluate and mimic, especially for the design of sub-width channel features. The studies mentioned above did not systematically assess the effect of riffle-pool relief on physical habitat quality and morphological unit resilience against a range of flows. Neither did they look at a sequence of units, just an individual pool-riffle-pool unit.

The specific objectives of this study were to assess the consequences of high versus low riffle-pool relief on (1) physical habitat quality for Chinook salmon and steelhead trout in their sensitive spawning and fry life stages at the regulated discharge typical for the periods when those life stages occur and (2) sediment transport regimes during two geomorphically and ecologically significant flows, as explained below. A third objective looked beyond relief to assess how the consequences from objectives (1) and (2) vary between different riffle-pool shapes. The discharges focused on were flows associated with bed occupation during the freshwater reproductive cycle (salmon spawning in autumn, embryo incubation in late autumn and early winter, and fry development in winter) (8.5 m^3 s^{-1}) and physical-habitat rejuvenation during prescribed spring snowmelt releases, using the highest regulated release as of December 2004 (169.9 m^3 s^{-1}).

20.2 Experimental design

The approach used to assess the effect of riffle-pool relief on river rehabilitation was (a) to design six synthetic river digital elevation models (DEMs) with different riffle-pool sequence configurations for a given testbed regulated

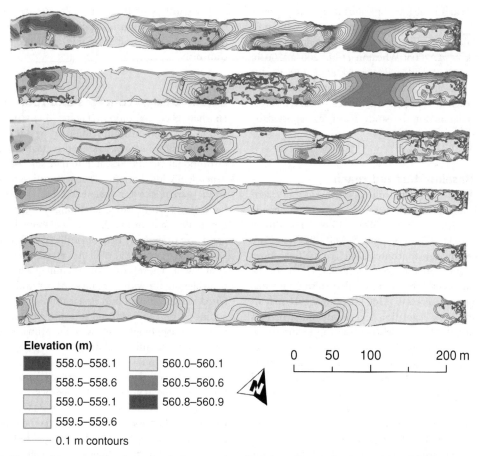

Elevation (m)

▨ 558.0–558.1	▨ 560.0–560.1
▨ 558.5–558.6	▨ 560.5–560.6
▨ 559.0–559.1	▨ 560.8–560.9
▨ 559.5–559.6	

—— 0.1 m contours

0   50   100   200 m

Figure 20.1 Topographic maps of the alternative design scenarios with 0.1-m contours. A colour version of this image appears in the colour plate section of the book.

river reach, (b) to conduct 2D modelling of the synthetic designs at two key discharges, (c) to extrapolate hydraulic predictions through physical-habitat curves and sediment transport regime equations and (d) to extract and compare performance indicators to determine designs' relative merits. Alternative designs were guided by the science about riffle-pool assemblages and the emerging knowledge of the structure, organization and function of morphological units (e.g. Padmore *et al.*, 1998; Fukushima, 2001). Previous research established performance indicators for physical habitat quality for indicator species' life stages and sediment transport regimes (Elkins *et al.*, 2007; Pasternack *et al.*, 2008; Brown and Pasternack, 2008, 2009), as explained in Sections 20.2.5–20.2.7 below.

A sequence of riffle-pool units on the gravel-bed Trinity River immediately below Lewiston Dam, California, USA (40°43′34″N, 122°47′48″W) was used as a refer-

ence assemblage (i.e. 'pre-project' or 'baseline' scenario) to prepare and validate a 2D model suitable for a straight, gravel-bed, riffle-pool stream in a confined valley (Brown and Pasternack, 2008). Then, six experimental riffle-pool sequences were fabricated using AutoCAD Land Desktop to obtain DEMs with different A_{RP} and A_{RP}^*, using the Trinity River's topography as a starting point for experimentation (Figure 20.1). These alternatives were not scientific curiosities, but actual morphologies evaluated for construction below Lewiston Dam (Pasternack, 2004a); they were consistent with fluvial landforms in straight reaches of confined gravel-bed rivers that are common in Californian rivers (e.g. Manwaring *et al.*, 2009; Sawyer *et al.*, 2009; Cramer Fish Sciences, 2010).

System response was evaluated in terms of flow pattern, fish habitat quality and sediment transport regime at two discharges within the regulated range permitted by operational rules at the time. Specific methods were previously developed and validated on the Trinity as well as on a similar shallow gravel-bed reach of the Mokelumne River (Pasternack *et al.*, 2004; Wheaton *et al.*, 2004a, 2004b; Elkins *et al.*, 2007). Unlike Pasternack *et al.*, (2008), this study only evaluated a single tailwater water surface elevation for each discharge, using the real observed stage-discharge relation for the study reach, as reported by Brown and Pasternack (2008).

20.2.1 Baseline testbed reach

Studies of hydrology, geomorphology, fisheries and reach-scale ecohydraulics of the baseline study reach already exist for the Trinity River (USFWS, 1999, 2002; Brown and Pasternack, 2008; Gaeuman, 2008, 2011) and are described briefly here for context. The Trinity River above Lewiston Dam is an 1860-km^2 basin that is part of the Klamath Mountain Province in northwestern California. Damming barred access to ~160 km of upstream spawning grounds. Flow regulation and floodplain structures limit flushing of tributary-delivered sand and geomorphic processes that maintain alluvial spawning grounds, while enabling vegetation encroachment that degrades rearing habitat. Salmonid populations have dropped sharply and are the focus of habitat rehabilitation, though systemic factors are also addressed. Since 1972, projects have included gravel augmentation, channel reconfiguration, bank vegetation removal and flow reregulation.

The 760-m Lewiston hatchery reach (LHR) that was the testbed for this study is located immediately downstream of Lewiston Dam and is the uppermost limit of salmonid spawning access. Previous work in the LHR determined that the existing artificial topography and substrates are controlled by anthropogenic and natural boundary and input controls that preclude natural geomorphic processes (Brown and Pasternack, 2008). Historically, the channel was wide with inset active alluvial gravel bars and a forested floodplain, but now it is narrow and straight. The river is confined on its south flank by a fish hatchery and on its north flank by bedrock. These forcings cause uniform high flow widths that preclude differential rates of sediment entrainment among morphological units at the 1–10 channel-width scale. The riverbed is armoured by decades of flow and sediment regulation and it has coarse artificial riffle-steps (i.e. rock weirs) buttressed by boulders, which further degrade the topography for fish habitat. These structures do not have any of the geometric, hydraulic or sedimentary attributes of riffles utilized by salmonids for spawning. The channel cannot adjust

itself regardless of flow. These factors necessitate that channel rehabilitation is dependent on physical manipulation, which is limited by non-deformable boundary controls.

20.2.2 Design concepts and tools used

Six channel configurations (Figure 20.1) were developed within the real constraints imposed by site conditions, management objectives, laws and regulations and other institutional factors. Research to date suggests that flow-dependent width variations between riffle and pool units are a prerequisite for sustainable (i.e. resilient during floods) riffle-pool units, assuming sediment supply is not limiting and in the absence of extreme bed-material grain-size differences between riffles and pools (Carling and Wood, 1994; MacWilliams *et al.*, 2006; Wheaton *et al.*, 2010; Thompson, 2011). Limited space for channel widening (or width undulation) constrained process-based design (Brown and Pasternack, 2008). Therefore, adding gravel/cobble fill and adjusting the amplitude and relief of morphologic units were the primary opportunities to enhance riffle-pool units. Although width undulation aids riffle-pool resilience (White *et al.*, 2010), rivers below most large dams are channelized, constricted and at least armoured, if not scoured to bedrock. Also, flow regulation limits inundation of high-elevation valley wall oscillations. Furthermore, constructing channel designs with gravel/cobble using front loaders involves bulk placement of the sediment mixture (Sawyer *et al.*, 2009); it is uncommon to design and install different surficial bed-material facies. Thus, evaluations of large width oscillations and differential bed texture were not considered.

The Spawning Habitat Integrated Rehabilitation Approach (SHIRA) for rehabilitating regulated gravel-bed rivers organized all phases of the project, including design development and final design selection, aided by 2D modelling and ecohydraulic analysis (Wheaton *et al.*, 2004a; 2004b). A design objective is a specific goal that is aimed for when a project plan is implemented. To achieve the objective, it is turned into a design hypothesis, which is a mechanistic inference, formulated on the basis of scientific literature and available site-specific data, and thus is assumed true as a general scientific principle (Wheaton *et al.*, 2004b). Next, specific morphological features are designed to work with the flow regime to yield the mechanism in the design hypothesis. Finally, a numerical test is formulated to determine whether the design hypothesis was appropriate for the project and the degree to which the design objective will be achieved.

Table 20.1 Riffle-pool relief metrics for each channel configuration.

Metric	Pre	Design number					
		D1	D2	D3	D4	D5	D6
Bed slope (%)	0.22	0.23	0.20	0.18	0.17	0.22	0.17
Number of units	3	3	2	4	3	3	3
Average A_{RP}[1] (m)	0.78	1.54	1.58	0.88	0.58	0.75	0.74
Average A_{RP}^*[2] (m)	0.99	2.18	2.57	2.20	3.01	2.01	0.89
Design/pre A_{RP}		1.98	2.03	1.13	0.74	0.97	0.95
Design/pre A_{RP}^*		2.21	2.61	2.24	3.06	2.04	0.91
Net fill volume (m^3)		5505	8257	7493	6116	3440	3889

[1] Pool-to-riffle amplitude
[2] Pool-to-riffle asymmetry

The six alternative designs (Figure 20.1) were created with diverse features that have many specific design hypotheses. The list of all design hypotheses is beyond the scope of this chapter, but is available on the website listed in the references section under Pasternack (2004b). A key aspect of the designs for the channel assemblages is that they span a range of different amplitudes of riffle-pool relief (Table 20.1; Figure 20.2) and different plan view patterns (Figure 20.1). For each of these design elements, a suite of concepts and objective design tools aided the creative process of conceptualizing landforms and articulating their value toward hydraulic, geomorphic and fisheries objectives in the form of design hypotheses. Riffle and bar analogues were generated by visualization from similar morphological unit scale features in unregulated rivers. Next, these analogues were scaled and overlaid on the pre-project topography. The spacing and location of riffles were determined somewhat by existing locations and more so by fixed forcing elements, but this was varied between designs. Analytical and empirical design and testing of riffle crests was performed for low-flow conditions to determine if crests had gross hydraulic properties suitable for the design objectives. The net volumetric fill (m^3) of gravel and cobble for each design was calculated by digital elevation model differencing in AutoCAD Land Desktop between the baseline topography and that for each design. Finally, 2D modelling of design surfaces was used to evaluate each design with respect to performance indicators and spatial mechanisms.

For riffle-pool relief, some of the relevant design concerns included base flow riffle and pool habitat suitability and quality, stage-dependent riffle scour potential, knickpoint migration through riffles and the resilience of riffle-pool relief. Initial riffle-crest sizing was done iteratively, aided by depth and velocity estimates using the Cipoletti weir equation and mass conservation for specified discharges, because a crest functions like a weir to cause a backwater effect (USBR, 1953; Clifford et al., 2005). Hydraulic estimates for the regulated base flow typically present during salmon spawning, embryo incubation and fry development were used to check habitat quality and gravel scour potential at that flow. Riffle-pool relief metrics (A_{RP} and A_{RP}^*) were computed for each individual pool-riffle pair in each design and averaged by design according to the method of Pasternack and Brown (2011).

For plan view morphology, design factors included the lateral distribution of mesohabitats, stage-dependent resilience of microhabitat patches, sediment routing through pools, knickpoint migration through riffle crests, the resilience of riffle-pool relief and accessibility of pools preferred for recreation fishing. Some specific design morphometrics included degree of pool constriction for flow convergence, shape of riffle exit (e.g. horseshoe shaped, straight or convex), crestline obliquity for flow divergence, partial riffle-crest notches/chutes for flow bypassing and central bars for stage-dependent microhabitat resilience.

20.2.3 Experimental riffle-pool assemblages

Due to page limits, readers are directed to Pasternack and Brown (2011) for thorough explanations of designs, with a brief overview here. Designs One, Two and Three involved high riffle-pool relief (aka 'accentuated topography') and higher bed slope, while Designs Four, Five and Six involved low riffle-pool relief (aka 'blanket fill'), lower bed slope and ~3-m widening on river right (Table 20.1). The concept for Design One was to maintain the existing pattern of features, but accentuate them by building up

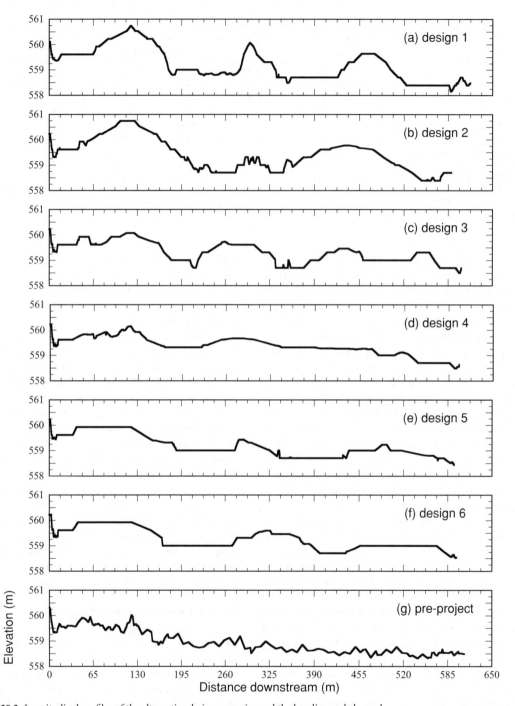

Figure 20.2 Longitudinal profiles of the alternative design scenarios and the baseline real channel.

riffles and increasing bed slope, yielding a high riffle-pool relief of 1.54 m (Figure 20.2a). The three large riffles were conceived to provide for salmon spawning, the pools for adult holding and the low-velocity bank fringe for fry habitat. Design Two had two large, broad, flat riffles with transverse orthogonal crests and accentuated pools, yielding a high riffle-pool relief of 1.58 m and a high asymmetry value of 2.57 (Figure 20.2b). The central pool was designed to be large, channel-spanning and have a lot of submerged features for juvenile rearing and adult holding. The large riffles were designed for Chinook and steelhead spawning. Using preliminary lessons learned from 'stress testing' Designs One and Two with ecohydraulic analysis, Design Three was a modification of Design One that had a significantly lower slope, longer riffles and smaller, narrower pools (Figure 20.2c). It was conjectured that the central bar and flanking pools at the end of the design would provide adult holding habitat proximal to spawning habitat and habitat heterogeneity in the face of fluctuating spawning flows. Also, the two small pools might serve as sediment traps to retain sediment placed upstream in this key spawning reach over time. Design Four had a blanket fill with three riffles; the upper two had similar shapes to those in the previous designs, with shallower intervening pools, while the third riffle was long and broken up with local crests and chutes (Figure 20.2d). Even though riffle amplitude was very low, the fact that the bed undulation occurred on top of fill created a strong asymmetry for the last pool-to-riffle unit. This design sought to increase fluvial complexity while maintaining low feature-to-feature slopes. A new feature for this design that was also used in Designs Five and Six was that the channel was widened by three metres on river right to remove encroached vegetation and provide low-velocity, shallow habitat for fry. The biggest difference from the earlier designs was the presence of a long glide that featured a pair of tightly spaced alternating bars at the end of the reach. Glides can serve for salmon spawning, especially for steelhead, so it was a worthwhile landform to evaluate. Design Five had a blanket fill with three simple riffle crests and a straight longitudinal bar (Figure 20.2e). The goal for Design Five was to have the lowest change in riffle-pool relief to limit the amount of gravel fill, while still maintaining the same slope as in the baseline topography, as well as to increase fluvial complexity and spawning habitat, while minimizing low-flow scour of that habitat. Design Six continued the evolution of the blanket-fill concept from Design Five by significantly increasing the sizes and amending the shapes of the second and third riffles, which resulted in a lower overall bed slope (Figure 20.2f). It also used two

longitudinal bars, each with a hook at the end to force flow divergence. The exit slope of the last riffle was graded convexly to cause flow divergence instead of convergence, and thus reduce scour on the riffle exit in the thalweg and shift scour energy to the rough banks where it would have little impact. Overall, the six designs use a lot of different specific elements, but share some commonality dictated by geomorphic and ecohydraulic knowledge. For comparison, the baseline real longitudinal profile is shown in Figure 20.2g. That profile shows little coherent landform organization, with just a few rock weirs peaking up above the slope-detrended median bed elevation.

20.2.4 2D numerical model

A 2D hydrodynamic model, Finite Element Surface Water Modelling System 3.1.5 (FESWMS), was used to simulate hydrodynamics for the baseline channel and the six alternative designs. FESWMS solves the vertically integrated conservation of momentum and mass equations using a finite element method to acquire depth-averaged 2D velocity components (U, V) and water depths (H) at each computational-mesh node. FESWMS simulates subcritical and supercritical flows. Froehlich (1989) described hydrodynamic equations, discretization and solution methods and other FESWMS details. This model has been frequently validated for use in shallow, regulated gravel-bed rivers (Pasternack et al., 2004; Wheaton et al., 2004b; Pasternack et al., 2006; Elkins et al., 2007; Moir and Pasternack, 2008; Manwaring et al., 2009; Sawyer et al., 2010). Brown and Pasternack (2008) developed, calibrated and validated an FESWMS model of the LDR baseline channel, which is summarized below to characterize the uncertainties that affect experimental simulations.

FESWMS was implemented using the Surface Water Modelling System v. 8.1 (Aquaveo, LLC). Computational design meshes had a typical internodal distance of 1.37 m, which was comparable to the spacing of the original topographic survey data from the reference reach (Brown and Pasternack, 2008). Based on past experience with evaluating numerical diffusion and numerical stability, the mesh resolution was high enough to avoid those problems for the finite element method. Meshes only covered the wetted channel and a few periphery dry cells, yielding slightly different final meshes for each discharge and channel configuration.

To run FESWMS in a single channel, inflowing discharge and the associated exit water surface elevation (WSE) are required. The ecologically significant discharges of 8.5 and 169.9 m^3 s^{-1} were specified at the end of the introduction section. Flow was assumed to be normal to the upstream boundary and it was distributed across the channel in proportion to the cross-sectional area of each boundary mesh element. These assumptions were validated by measuring H and U near the upstream boundary in the reference reach (Brown and Pasternack, 2008). WSEs at the downstream end of the reference reach were measured at Qs between 8.5 and 169.9 m³ s⁻¹ using a total station to obtain a stage-discharge rating curve useful for simulating any flow in this range. For the six fabricated channels, the model's downstream boundary was in a pool and corresponded with the level imposed by the next downstream riffle, which could be natural or re-engineered to any desired elevation (Elkins *et al.*, 2007; Pasternack *et al.*, 2008). Therefore, the designs upstream of the model exit had insignificant effect on the WSEs at the model's exit boundary.

The two primary model parameters in FESWMS are the eddy viscosity (E) and bed roughness (n). Pasternack and Brown (2011) explain how these were obtained. E was spatially distributed, but used a constant coefficient parameter value of 0.6. Roughness associated with resolved metre-scale bedform topography was explicitly represented in the detailed channel DEM. 2D models are highly sensitive to DEM inaccuracies (Horritt *et al.*, 2006). For unresolved roughness, a global n of 0.043 was used with all meshes (Pasternack *et al.*, 2004, 2006, 2008; Moir and Pasternack, 2008). This was not numerically calibrated; it was validated by comparing observed and predicted WSEs along the reference reach at different Qs as well as by comparing observed and predicted H and U values at cross-sections. In gravel placement, added material is well mixed and not differentiated between riffles and pools, so uniform roughness is appropriate.

In this study, FESWMS was used for exploratory experimentation using fabricated, theoretical channel configurations to improve ecohydraulic and geomorphic understanding of basic riffle-pool functioning as well as to improve the application of gravel-bed river design. Acceptance of the numerical approach requires reasonable confidence in FESWMS's predictions. Three different tests assessed model uncertainty for the LDR channel (Brown and Pasternack, 2008). First, the range of E values in model output was checked against field-based estimates and found to be similar (~ 0.02–0.1 m² s⁻¹).

Second, recognizing that in a straight confined reach, lateral and longitudinal variation in velocity magnitude in a river is highest at low discharge (Clifford and French, 1998) and that 2D-model validation performance has

been found to be insensitive to discharge (e.g. May *et al.*, 2009; Pasternack and Senter, 2011), model validation of H and U was performed at 12.9 m³ s⁻¹. Predictions evaluated in detail by Brown and Pasternack (2008) and also shown in Pasternack *et al.* (2008) yielded the typical results, with H accurately predicted and U adequately predicted. Abrupt lateral gradients in U were not predicted accurately, but at many points U was very accurately predicted.

Third, a total station was used to measure the WSE at 14 locations at 169.9 m³ s⁻¹ (vertical accuracy of <1 cm). Modelled WSE was systematically slightly higher than observed ($\sim 5\%$ of mean cross-sectional depth), but not enough to warrant iterative calibration of the n value. Overall, validation analysis showed that FESWMS is accurate enough to provide confidence that the reported spatial patterns in depth and velocity are real, but is not accurate enough to characterize poorly mapped regions with very strong lateral variation precisely, for which better mapping and 3D numerical modelling would be better. As this study used synthetic topography, map accuracy is irrelevant, while inadequate lateral velocity variation is an uncertainty.

20.2.5 Fish habitat quality

Physical habitat quality predictions were made by extrapolating 2D model depth and velocity predictions through local, independent habitat suitability curves (HSCs) for H and U from USFWS (1999). Although many local HSCs were used in the LDR rehabilitation design, this study focused on physical habitat metrics for spawning and fry-rearing life stages of anadromous Chinook salmon and steelhead trout (Figure 20.3), which are particularly sensitive to flow and topography, as expressed in 2D hydraulic patterns. Because ideal substrates would be placed, no substrate HSC was needed to compare designs.

A global habitat suitability index (GHSI) was calculated at each computational node as the geometric mean of the H and U indices (Pasternack *et al.*, 2004). To account for H and U uncertainty, GHSI values were binned with GHSI = 0 as nonhabitat, 0 < GHSI < 0.1 as very poor habitat, 0.1 < GHSI < 0.4 as low quality, 0.4 < GHSI < 0.7 as medium quality and 0.7 < GHSI < 1.0 as high quality. These broad classes reduce the impact of H and U prediction error, since they are largely insensitive to ~ 0–25% U error, unless a value is very close to a bin edge (Brown and Pasternack, 2009). Use of low-quality habitat depended on the degree of channel degradation and fish density. Elkins

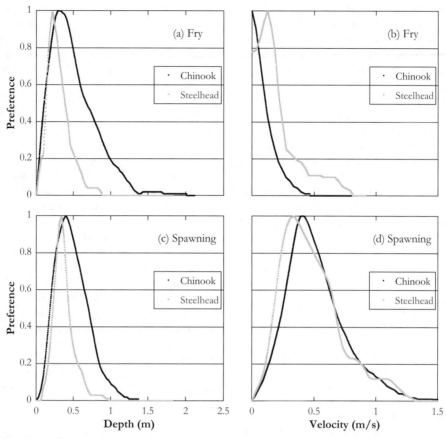

Figure 20.3 Chinook salmon and steelhead trout habitat suitability curves for sensitive fry and spawning life stages on the Trinity River. Data from USFWS (1999).

et al. (2007) reported a significant spawner preference for medium- and high-quality habitat (GHSI > 0.4) and an equally strong, statistically significant fish avoidance of nonhabitat and very poor quality habitat (GHSI < 0.1). Pasternack (2008b) reported preference for GHSI > 0.4 and avoidance for GHSI < 0.4 for three tests at a highly utilized spawning site on the lower Yuba River. As a result of these findings, the terms 'suitable habitat' and 'preferred habitat' are used to refer to all areas with GHSI > 0 and GHSI > 0.4, respectively. Although no equivalent comparison was available for fry, regional expertise suggests that fry occur in suitable habitat of any quality, avoiding just nonhabitat areas (T.R. Payne, personal communication, 2010). All performance indicators were checked for the life stages of both species, but a higher qualitative weighting was given to the habitat-suitability metric for fry and the habitat-preference metric for spawners.

20.2.6 Sediment transport regime

It is natural and unavoidable that in a regulated river with zero bedload influx, placed gravel/cobble will be entrained during floods, potentially degrading artificially contoured fluvial landforms (Merz *et al.*, 2006). The use of concepts from fluvial geomorphology (e.g. MacWilliams *et al.*, 2006; Thompson, 2006; Wilkinson *et al.*, 2008; Caamaño *et al.*, 2009; Sawyer *et al.*, 2010) in the project aimed to focus scour in appropriate locations and yield downstream deposition beyond the project area, such that the overall integrity of gravel bars remained intact, even if sub-width features adjusted. When coupled with a regular programme of a suitable quantity of gravel injection at the entrance of the reach, it ought to be feasible to sustain the constructed topography (e.g. design hypothesis tests of Wheaton *et al.*, 2004b; 2010). However, since the channel width was unavoidably constricted in the testbed reach, it was understood from the outset that flow-convergence routing was infeasible – scour would always be higher over riffles than pools in this area, tending toward diminished riffle-pool relief. Nonetheless, flow-dependent channel resilience could still vary significantly as a function of channel configuration, so that was an important basis for performance evaluation.

Pasternack *et al.* (2006) validated the suitability of FESWMS for predicting bed shear stress in shallow gravel-bed rivers, finding that the model is as good as field estimation methods most of the time; the exception being in very shallow water ($H \sim d_{90}$, size that 90% of the bed material is smaller than). In this study, Shields stress was calculated at each mesh node to evaluate the sediment transport regime and channel stability under different flow conditions:

$$\tau^* = \frac{\tau_v^b}{(\gamma_s - \gamma_w)d_{50}} \tag{20.1}$$

$$u^* = \frac{U}{5.75 \log\left(\frac{12.2H}{2d_{90}}\right)} \text{ and } \tau_v^b = \rho_w\, u^{*2} \tag{20.2}$$

where τ^* is Shields stress, τ_v^b is bed shear stress in the direction of the velocity vector, d_{50} is the median grain size, d_{90} is the size at the 90th percentile of the cumulative distribution function for bed material grain size distribution, γ_s is sediment specific weight, γ_w is the water's specific weight and ρ_w is water density. As with GHSI, τ^* was binned to account for H and U inaccuracy. Lisle *et al.* (2000) defined sediment transport regimes relative to τ^* as $\tau^* < 0.01$ represents no transport; $0.01 < \tau^* < 0.03$ represents probabilistically intermittent entrainment; $0.03 < \tau^* < 0.06$ represents the 'partial transport' domain of Wilcock *et al.* (1996); $0.06 < \tau^* < 0.1$ represents full transport of a 'carpet' of sediment 1–$2 \cdot d_{90}$ thick, and $\tau^* > 0.1$ corresponds with potentially channel-altering conditions. These threshold delineations have uncertainty but provide a reasonable basis for characterizing and comparing sediment transport conditions (Sawyer *et al.*, 2010).

20.2.7 Test analyses and outcome indicators

Tests evaluating physical habitat and sediment transport regime involved computing and comparing statistical distributions for the performance indicators for GHSI and Shields stress datasets as well as visual comparisons of the spatial patterns of GHSI and Shields stress to evaluate mechanisms. A summary table was produced in which the best-performing design(s) were identified for each GHSI indicator, and then the best overall design was evident as the one with the most occurrences as the best performer across all the metrics. An identical analysis was carried out for Shields stress. Picking the best design was a necessary part of the real-world project, while the comparison alone is more interesting to appreciate the circumstances in which each landform sequence has value for different sets of project objectives. Therefore, design response according to different performance indicators was assessed relative to differences in A_{RP} and A_{RP}^*.

For physical habitat quality, the performance indicators were the percentage distribution of GHSI bins, the habitat efficiency for habitat creation (defined below), the habitat efficiency for habitat improvement (defined below) and the spatial pattern of GHSI. Each of these indicators was used for the life stages of each species. For the baseline channel and each alternative design, the percentage of wetted area for each GHSI bin was plotted as a stacked column for the life stages of each species and the one with the highest percentages of suitable and preferred habitat (as defined in Section 20.2.5) was identified. Because each design has a different wetted area, care is required in cross-comparing this metric, as the design with the highest percentage of something is not necessarily the one with the biggest area of it. Habitat efficiency is a cost/effectiveness metric where the cost is the amount of coarse sediment needed (in m^3, but could be expressed in dollars per m^3 of gravel placed) and the effectiveness is the amount of habitat, which can be expressed in different specific metrics as well. Habitat added means the net increase in suitable habitat area (m^2) per m^3 of gravel added, while habitat improved means the net increase in preferred habitat area (m^2) per m^3 of gravel added. Designs were compared directly for their performance in habitat efficiency. Habitat efficiency is a particularly useful metric when trying to prioritize among possible projects across many different sites of widely different morphologies. Assessing the relative merits of habitat efficiency in terms of addition of suitable habitat versus improvement of habitat to the preferred state involves more professional judgement, so both metrics are worth considering as performance indicators. As justified earlier, more weight was given to habitat improvement for spawning, while more weight was given to habitat addition for fry. These choices hinge on expert judgement or stakeholder consensus – the key is that this study shows the relative merits of different landforms with different types of riffle-pool relief. Finally, visual inspection of the spatial pattern of GHSI bins relative to the DEM for each design established the link between morphology and microhabitat pattern to explain the latter.

For sediment transport regime, three performance indicators were used. First, the percentage wetted areas with $\tau^* > 0.06$ and that for $0.03 < \tau^* < 0.06$ for each scenario when discharge was $8.5\ \mathrm{m}^3\ \mathrm{s}^{-1}$ were used to compare the risk of unacceptable ecological disturbance during the period of bed occupation by redds, embryos and fry. Second, the percentage wetted areas with $\tau^* > 0.06$ and that for $0.03 < \tau^* < 0.06$ for each scenario when discharge was $169.9\ \mathrm{m}^3\ \mathrm{s}^{-1}$ were used to compare the potential of channel change for the ecological function of 'bed preparation' (Escobar and Pasternack, 2010). Finally, the stage-dependent location of peak τ^* for each scenario was inspected for evidence of flow-convergence routing of sediment through pools as a performance indicator of the resilience of riffle-pool relief under existing reservoir operations.

20.3 Results

20.3.1 Physical habitat

The statistical distribution of GHSI bins shows systematic differences in habitat benefits between designs using accentuated topography and those using blanket fills in terms of riffle-pool relief (Figure 20.4). Depending on what the habitat goals of a rehabilitation project are, clear landform preferences are evident. Designs One and Two had the most accentuated topography and they yielded the largest percentage areas of suitable and preferred Chinook fry habitat. A key result was found in comparing Chinook and steelhead fry performance in percentage area of habitat. Because the steelhead-fry velocity HSC (Figure 20.3b) has a broader range and a higher velocity of peak preference than that for Chinook fry, the blanket-fill designs yielded significantly higher percentage areas of suitable habitat compared to the accentuated-topography designs.

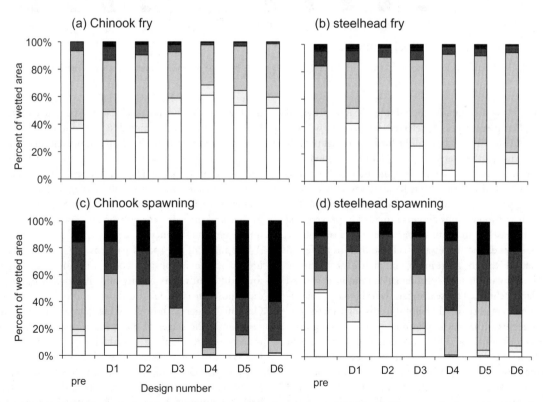

Figure 20.4 Stacked column plot comparing the percentage wetted area of each GHSI bin for life stages of each species at $8.5\ \mathrm{m}^3\ \mathrm{s}^{-1}$ among the different designs and the baseline reach. Total area varies somewhat between designs. The key for physical habitat quality is as follows: black is high, dark grey is medium, medium grey is low, and light grey is very poor quality; white is nonhabitat.

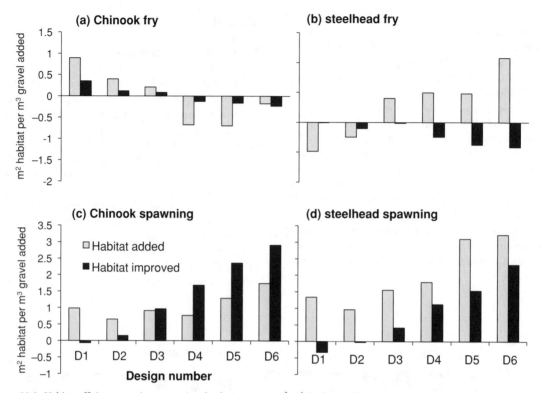

Figure 20.5 Habitat efficiency metrics comparing the designs at 8.5 m^3 s^{-1} for four different species' life stages. Higher bars are better.

However, the accentuated-topography designs provided the highest percentages of preferred steelhead fry habitat. Designs Four, Five and Six all yielded high-quality Chinook spawning habitat over > 55% of the wetted area and medium-quality habitat over another 28–39% of it. Preferred steelhead spawning habitat was also present in high percentages for these blanket-fill designs. The accentuated-topography designs provided improvements over the baseline in terms of percentage area of suitable habitat, but did not necessarily achieve that for preferred habitat, which is the more important metric for spawning.

The habitat-efficiency performance indicators yielded similar outcomes as those using the GHSI bins for spawning, but were particularly helpful in distinguishing relative benefits within each riffle-pool relief grouping (Figure 20.5). The habitat-addition and habitat-improvement efficiencies of Design One were more than double those of the next highest performer (Design Two) for Chinook fry. Also, while Design One had the least harmful effect on steelhead fry preferred-habitat loss, it caused the most loss of suitable habitat for steelhead fry. Design Six had the opposite outcome, yielding more than dou-

ble the gain in suitable steelhead fry habitat compared to the next-best-performing designs (Four and Five). For both Chinook and steelhead spawning, the blanket-fill designs performed best across habitat-efficiency metrics, and among those, Design Six was the best. Across all eight habitat-efficiency metrics for all life stages of the species (Table 20.2, efficiency columns), Design Six was the best for five and the worst for only one (Chinook fry habitat-improvement efficiency).

The maps of the spatial patterns of GHSI bins (Figure 20.6) provide insight about why the designs perform differently. For brevity and illustration purposes, only those for Chinook are shown and only for designs One and Six, since those show the sharpest contrast (the rest, for all life stages of the species are available at the website provided in the introduction to Section 20.2). For Design One, preferred spawning habitat is located in riffle entrances and, to a much smaller extent, on the periphery of channels, especially where there are lateral bars. Accentuating the topography created riffles that were too short, shallow and fast to serve as spawning habitat. Since the pools are nonhabitat for spawning, there is just

Table 20.2 Design number of the best-performing design(s) for each indicator and the best design number overall.

Life stage	GHSI bins	Habitat addition efficiency	Habitat improvement efficiency	GHSI spatial pattern	Best overall
Chinook fry	1-2	1	1	1	1
Steelhead fry	4-6	6	1	6	6
Chinook spawning	4-6	6	6	6	6
Steelhead spawning	4-6	5,6	6	6	6

too little area for spawning when topography is accentuated. An interesting nuance to that outcome is evident in the first riffle in Design Three, which had a small, high crest bounded by longer, flat shelves of riffle entrance and exit. For that riffle, the crest was only low-quality habitat, but both shelves had medium-quality habitat with high-quality habitat along both banks. That suggests some benefits to having multiple elevation tiers to a riffle. Nevertheless, the really important result was the finding that Design Six yielded nearly ubiquitous high-quality Chinook spawning habitat and widespread preferred steelhead spawning habitat. Design Six had diverse landform features at multiple spatial scales and many of them had spawning value from a hydraulic perspective.

In terms of the fry life stage, the GHSI maps show that the accentuated-topography designs yield significantly more preferred fry habitat, because they produce slackwater areas (sometimes large, slowly recirculating eddies) on the periphery behind each riffle crest. Abrupt bed-elevation increases caused flow to converge, yielding a narrower effective flow width bounded by slackwater or recirculating eddies. The higher the bed-elevation increase, the

stronger the effects. This is the concept used in whitewater park design, where bed and width constrictions focus flow to produce standing waves for kayak stunts and intervening peripheral pools for kayakers waiting in line for their turn. Numerically, the blanket-fill designs performed poorly for Chinook fry habitat, because velocities were too fast, given an exponentially decreasing velocity HSC. The blanket-fill designs in this study yielded insufficient depth and width undulations to create the sheltering observed in the accentuated-topography designs. However, steelhead fry use the widespread, moderately higher velocities found in the blanket-fill designs, so the blanket fill is beneficial for them (Figure 20.6).

20.3.2 Sediment transport regimes

The statistical distributions of Shields stress bins for $\tau^* > 0.06$ and $0.03 < \tau^* < 0.06$ indicated the relative resilience of each design across an order of magnitude of discharge (Figure 20.7). At low discharge ($8.5 \text{ m}^3 \text{ s}^{-1}$) when embryos are at risk of being scoured out, $< 5\%$ of the wetted area was in the full transport regime for each design, with the highest values associated with the

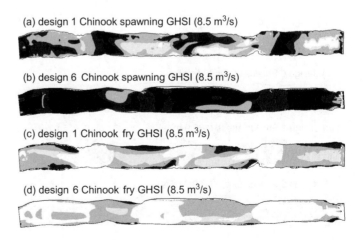

(a) design 1 Chinook spawning GHSI (8.5 m³/s)

(b) design 6 Chinook spawning GHSI (8.5 m³/s)

(c) design 1 Chinook fry GHSI (8.5 m³/s)

Figure 20.6 Examples of GHSI maps for two designs showing the spatial pattern of habitat quality for two different species' life stages at 8.5 m³ s⁻¹. The key is the same as in Figure 20.4.

(d) design 6 Chinook fry GHSI (8.5 m³/s)

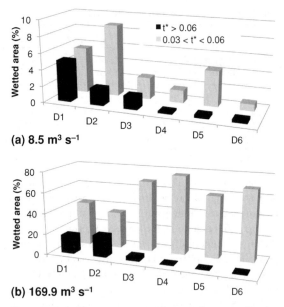

(a) 8.5 m³ s⁻¹

(b) 169.9 m³ s⁻¹

Figure 20.7 Sediment transport regime metrics comparing the designs at two discharges. Black bars show the area of full mobility and grey bars show the area of partial transport. Lower black bars are better.

designs having the highest A_{RP} (One, Two and Three). The partial-transport bin showed at-risk areas of < 9%, with the highest values associated with designs having the highest overall bed slope (One, Two and Five). The τ^* maps helped explain the difference in response of these two metrics at low flow (Figure 20.8a and b). Higher overall bed slope drives higher overall velocities, especially on riffles at within-bank flows, which is indicative of a higher area with

$\tau^* < 0.03$. Higher A_{RP} is indicative of the presence of over-steepened riffle exits, which causes an abrupt local velocity increase capable of driving knickpoint migration.

At high discharge (169.9 m³ s⁻¹), there is a marked difference in resilience between accentuated and blanket-fill topographies. Similar to low discharge, the designs with the highest A_{RP} (Designs One, Two and Three) had the largest percentage area of full bed mobility (4.4–21%). Meanwhile, the designs with the lowest slope had the highest percentage areas of partial transport (60–76%). These patterns are explained using the τ^* maps (Figure 20.8c and d) and Stewardson and McMahon's (2002) conceptual model of hydraulic variations within stream channels, as applied to the Trinity River by Brown and Pasternack (2008). According to this concept, there is a flow dependence to the transition of a river's hydraulic regime from one dominated by longitudinal velocity variation at low flow to one dominated by lateral velocity variation at high flow. As illustrated in this numerical experiment, the exact discharge required for the transition is dependent on riffle-pool relief. Specifically, for the accentuated-relief designs, the riffle crests are so high that even at 169.9 m³ s⁻¹, there is a strong longitudinal velocity variation with high velocities and shear stresses focused on riffle crests and an abrupt increase in velocity at the oversteepened riffle exits (Figure 20.8c). In contrast, this discharge is high enough for the blanket-fill channels to surpass the transition in hydraulic regime to lateral-variation dominance. As a result, these channels distribute velocity response to overall bed slope evenly along their length, avoiding focal points for scour (Figure 20.8d). Design Three is a hybrid between the two, in that its A_{RP} is just enough that the riffle crests peak up,

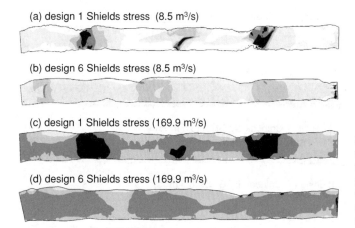

(a) design 1 Shields stress (8.5 m³/s)

(b) design 6 Shields stress (8.5 m³/s)

(c) design 1 Shields stress (169.9 m³/s)

(d) design 6 Shields stress (169.9 m³/s)

Figure 20.8 Examples of Shields stress maps at two discharges for two designs. The key is as follows: white and light grey are no transport, medium grey is probabilistically intermittent entrainment, darker medium grey is partial transport, dark grey is full mobility and black is potentially channel-altering conditions.

Table 20.3 Design number of the most resilient design(s) against scour and instability for each indicator and the best design number overall.

	Design number			
Discharge	Full bed mobility	Partial transport	τ^* spatial pattern	Best overall
Low	4	6	6	6
High	5	5	5	5

causing a longitudinal hydraulic variation at this flow and focusing scour at riffle exits somewhat.

Based on the six τ^*-bin performance indicators, all the blanket-fill designs outperform the accentuated-topography designs, but among them there is not a universal standout (Table 20.3). Design Six is the best performer at low flow, because it exhibits the lowest percentage area of partial transport, the highest percentage area of $\tau^* < 0.03$ and the best pattern of τ^* in terms of avoiding riffle-crest scour. The only problem with Design Six, which is very minor, is that the very end of the last riffle exit drops off enough to cause a sliver of partial transport, with a few spots experiencing full bed mobility. Design Four has a more gradual final riffle exit, so it avoids that problem, but it does have larger zones of partial transport in other places on riffle crests. Design Five is the best performer at high flow, because its bar configuration helps to focus flow in the centre, where peak velocity is the most even along the channel among designs and yields the strongest lateral velocity variation. Design Six also has those bars and it has just enough lateral constriction after each riffle to help induce a similar centralized focusing of velocity and scour potential over pools. However, its long, shallow riffles do experience partial transport over nearly their full width.

20.4 Discussion and conclusions

20.4.1 Lessons from numerical experimentation

Sequences of riffle-pool units designed with different riffle-pool relief attributes exhibit systematic and predictable variations in flow-dependent physical habitat patterns and geomorphic processes in a setting with a relatively uniform channel width. In general, this study found that modest riffle-pool amplitudes (<0.75 m; $A_{RP}/d_{50} \sim$

10–12, where d_{50} is median bed-material particle size) from blanket-fill designs yielded the largest habitat addition and habitat improvement efficiencies, especially for salmon spawning. The Chinook fry life stage was the exception, as high riffle-pool relief yielded large peripheral eddies in the lee of riffle crests (Figure 20.6c). Similarly, analysis of patterns of sediment transport potential found that high-relief configurations had significantly more area of full bed mobility, regardless of flow. These configurations would be expected to fall apart quickly after construction as a result of knickpoint migration through riffle crests, due to oversteepened riffle exits.

For a given riffle-pool amplitude, differences in 3D bar morphology (including A_{RP}^*, but not well explained by it) yield significant differences in habitat and channel stability. For example, Designs Five and Six had A_{RP} values of 0.75 and 0.74, respectively, but their performance indicators were significantly different. The use of longer riffle crests and curved tailouts in Design Six turned out to be important drivers of superior habitat performance. Meanwhile, the orthogonal bar forms and lower riffle exit slopes in Design Five helped avoid knickpoint migration, which is a risk in Design Six, especially at even higher flows that were not modelled in this study. Design Five had a small peak to its last riffle crest that was predicted to experience a higher velocity at the high flow, but it is still in the partial-transport regime, so it is resilient. Design Six had longer riffle-crest features to give them more resilience if knickpoint migration were to take place, but that evidently comes at the cost of increased knickpoint migration risk due to oversteepened riffle exits and the spreading of partial-transport across the full width of riffles. In evaluating these design alternatives, it was conjectured that at expected higher floods in the future, the last riffle crest in Design Five would likely transition into the full-mobility domain and not have enough crest area to absorb the erosion and avoid the 'reverse domino' riffle sequence collapse mechanism described by Pasternack et al. (2008). In contrast, it was conjectured that the evident lower-flow full mobility at the riffle exits in Design Six would simply lead to sloughing of material to naturally form a more gradual slope, because the length of full mobility and the amount of bed-material cohesion are likely insufficient to sustain a migrating knickpoint. Therefore, it was deemed more valuable to have longer riffles more resilient to the largest flood releases than to have short riffles with certain resilience to the modelled flood release. Ultimately, the lack of feasibility of significant width increases at riffle crests fundamentally constrained, instituting flow-convergence routing through this riffle-pool sequence, and that was repeatedly explained to project sponsors to ensure understanding of what Design Six would yield.

20.4.2 Merits of iterative design and construction

Ecohydraulic river rehabilitation seeks to modify river geometry to achieve specific hydraulic and sedimentary characteristics for target aquatic organisms. In doing this, multiple topographic outcomes may be possible depending on site constraints. Iterative design is the process by which a single scenario is continually modified until eco-hydraulic design criteria are satisfied. Iterative design is a key feature in the Spawning Habitat Integrated Rehabilitation Approach (Wheaton *et al.*, 2004a; Elkins *et al.*, 2007), and ought to be one in the more general practice of ecohydraulic design for river engineering. This study illustrates how iterative design can yield topographic surfaces that meet the needs of target organisms despite a wide range of potential scenarios. Without iterative design it would be difficult to determine the relationship between potential topographic manipulation and the optimal configuration of channel geometry.

Similarly, iterative construction that builds a design of a long sequence of riffles and pools over a period of years has several merits over attempting a single, massive project (Elkins *et al.*, 2007). First, institutional barriers and regulatory hurdles seem to be lower for a series of smaller projects compared to a single massive one. Stakeholders perceive small projects to be 'pilots' and are often willing to allow these with less scrutiny to see what happens, since the risks are low. Inevitable turnover in technical staff and stakeholder participants every few years often confounds large projects, whereas an incremental approach can be understood by neophytes with less effort, as they can focus on just understanding the current iteration first and learning the broader plan over time. Following the iterative approach, before people really grasp events, a significant overhaul in a river has been instituted on a transparent, scientific basis following the original plan. Second, for those who are closely monitoring and implementing the sequence of projects, iterative implementation allows for adaptive management to test design hypotheses for individual design elements and then adjust the overall design when performance indicators show that design hypotheses are not being corroborated. Finally, rivers degrade over decades, so it is sensible to rehabilitate them over a moderate duration, respecting the emerging status of the science and engineering underlying river rehabilitation as well as the uncertainty in predicting the future. No matter

how critical river rehabilitation may be to avoid systemic ecological collapse, rushing large projects is most likely to yield further ecological disturbance rather than solve outstanding problems.

20.4.3 Actual design selection

In this study, Design Six was determined to be the best alternative, but at that point additional work was performed to fine-tune the design and generate multiple flow, habitat and geomorphic predictions across a range of discharges for later evaluation. Small changes in the selected design often must be made to ensure that all available gravels and boulders are used as well as to account for new constraints that emerge when the design is vetted among all stakeholders and the public. Further thought and testing may enhance even the best of the alternatives with subtle changes. One may desire to add layers of complexity onto the basic topography using specific structures at the hydraulic-unit scale (~ 0.1–1 channel width), such as boulder clusters, riparian shade and streamwood jams. Hydraulic effects of jams are extremely difficult to simulate, but boulder clusters are feasible. Because habitat quantity and quality are stage-dependent, a more comprehensive view of the final design is obtained by modelling as wide a range of flows as the original topographic and on-going monitoring data allow. The final project design is converted into an easily followed grading plan for use by the contracted front-loader operator.

20.4.4 Outlook for ecohydraulics

Ecohydraulics is an emerging scientific subdiscipline and professional practice. Traditionally, use of heuristics in river assessments provided a deep understanding of habitat patterns and geomorphic processes, but lost favour for being opaque, non-reproducible and dependent on 'experts'. The rise of statistical analysis in ecology and geomorphology democratized river assessment and promoted greater transparency and quantification, but came at the cost of oversimplification of phenomena. Today, the combination of remote sensing, mechanistic modelling and GIS-based analysis in 'near-census' ecohydraulics (i.e. sampling at ~ 1-m resolution throughout long river segments) is poised to yield a decisive transformation in the practice of river science, in which the best features of heuristics and statistics blend to yield deep and repeatable process-based, spatially explicit predictions of river behaviour. New technologies will further enhance the capability, but the paradigm is now accessible for scientific exploration and professional practice (Pasternack, 2011).

Acknowledgements

Financial support for this work was provided by USBR Award #03FG230766 and USFWS Agreement DCN#113322G003, awarded by CALFED. We thank Michael Bounrisavong and Kaushal Parikh for their help with performing and analyzing all of the 2D model runs.

References

Bovee, K.D. (1982) A guide to stream habitat analysis using the instream flow incremental methodology. *U.S. Fish and Wildlife Service Biological Services Program*, FWS/OBS-82/26.

Bovee, K.D. and Milhous, R. (1978) Hydraulic simulation in instream flow studies: theory and techniques. *U.S. Fish and Wildlife Service Biological Services Program*, FWS/OBS-78/33.

Brown, R.A. and Pasternack, G.B. (2008) Engineered channel controls limiting spawning habitat rehabilitation success on regulated gravel-bed rivers. *Geomorphology*, **97**: 631–654.

Brown, R.A. and Pasternack, G.B. (2009) Comparison of methods for analyzing salmon habitat rehabilitation designs for regulated rivers. *River Research and Applications*, **25**: 745–772.

Caamaño, D., Goodwin, P., Buffington, J.M., Liou, J.C.P. and Daley-Laursen, S. (2009) Unifying criterion for the velocity reversal hypothesis in gravel-bed rivers. *Journal of Hydraulic Engineering*, **135**(1): 66–70.

Carling, P.A. and Wood, N. (1994) Simulation of flow over pool-riffle topography – a consideration of the velocity reversal hypothesis. *Earth Surface Processes and Landforms*, **19**(4): 319–332.

Clifford, N.J. and French, J.R. (1998) Restoration of channel physical environment in smaller, moderate gradient rivers: geomorphological bases for design criteria. In Bailey, R.G., Jose, P.V. and Sherwood, B.R. (eds) *United Kingdom Floodplains*, Westbury Academic & Scientific Publishing, Westbury, pp. 72–76.

Clifford, N.J., Soar, P.J., Harmar, O.P., Gurnell, A.M., Petts, G.E. and Emery, J.C. (2005) Assessment of hydrodynamic simulation results for eco-hydrological and eco-hydraulic applications: a spatial semivariance approach. *Hydrological Processes*, **19**: 3631–3648.

Cramer Fish Sciences (2010) *Evaluation of the 2008 and 2009 Sailor Bar gravel placements on the Lower American River, California*. Cramer Fish Sciences, Gresham, OR.

Doyle, M.W. and Harbor, J.M. (2000) Evaluation of Rosgen's streambank erosion potential assessment in northeast Oklahoma. *Journal of the American Water Resources Association*, **36**(5): 113–121.

Elkins, E.E., Pasternack, G.B. and Merz, J.E. (2007) The use of slope creation for rehabilitating incised, regulated, gravel-bed rivers. *Water Resources Research*, **43**: W05432. doi: 10.1029/2006WR005159.

Escobar, M.I. and Pasternack, G.B. (2010) A hydrogeomorphic dynamics approach to assess in-stream ecological functionality using the functional flows model, part 1 – model characteristics. *River Research and Applications*. doi: 10.1002/rra.1316.

Froehlich, D.C. (1989) *Finite Element Surface-Water Modeling System: Two-dimensional Flow in a Horizontal Plane – User's Manual*. U.S. Department of Transportation, Publication #FHWA-RD-88-177.

Fukushima, M. (2001) Salmonid habitat–geomorphology relationships in low gradient streams. *Ecology*, **82**: 1238–1246.

Fulton, A.A. (2008) *Gravel for Salmon in Bedrock Channels: Elucidating Mitigation Efficacy Through Site Characterization, 2D-Modeling, and Comparison Along the Yuba River, CA*. University of California.

Gaeuman, D. (2008) *Recommended Quantities and Gradation for Long-term Coarse Sediment Augmentation Downstream from Lewiston Dam*. Trinity River Restoration Program, Technical Monograph TM-TRRP-2008-2.

Gaeuman, D. (2011) *Water Year 2010 Implementation Monitoring Report*. Trinity River Restoration Program, Technical Monograph TM-TRRP-2011-1.

Ghanem, A., Steffler, P., Hicks, F. and Katapodis, C. (1996) Two-dimensional hydraulic simulation of physical habitat conditions in flowing streams. *Journal of Regulated Rivers: Research and Management*, **12**: 185–200.

Hanrahan, T.P. (2007) Bedform morphology of salmon spawning areas in a large gravel-bed river. *Geomorphology*, **86**: 529–536.

Hauer, C., Unfer, G., Tritthart, M., Formann, E. and Habersack, H. (2011) Variability of mesohabitat characteristics in riffle-pool reaches: testing an integrative evaluation concept (FGC) for MEM-application. *River Research and Applications*, **27**(4): 403–430.

Hoopa Valley Tribal Fisheries, McBain & Trush, Inc., Northern Hydrology and Engineering (2001) *Preliminary demonstration of the capabilities of a 2D model to assess and improve Trinity River rehabilitation designs*. McBain & Trush Technical Memorandum, Arcata, CA.

Horritt, M.S., Bates, P.D. and Mattinson, M.J. (2006) Effects of mesh resolution and topographic representation in 2D finite volume models of shallow water fluvial flow. *Journal of Hydrology*, **329**: 306–314.

Jowett, I.G. (1997) Instream flow methods: a comparison of approaches. *Regulated Rivers: Research and Management*, **13**: 115–127.

Kondolf, G.M. and Micheli, E.M. (1995) Evaluating stream restoration projects. *Environmental Management*, **19**: 1–15.

Lave, R., Doyle, M. and Robertson, M. (2010) Privatizing stream restoration in the US. *Social Studies of Science*, **40**: 677–703.

Lisle, T.E., Nelson, J.M., Pitlick, J., Madej, M.A. and Barkett, B.L. (2000) Variability of bed mobility in natural, gravel-bed channels and adjustments to sediment load at the reach and local scales. *Water Resources Research*, **36**(12): 3743–3755.

MacWilliamsJr, M.L., Wheaton, J.M., Pasternack, G.B., Street, R.L. and Kitanidis, P.K. (2006) Flow convergence and routing hypothesis for pool-riffle maintenance in alluvial rivers. *Water Resources Research*, **42**: W10427.

Manwaring, M., Cepello, S., Kennedy, S. and Pasternack, G.B. (2009) Spawning Riffle Gravel Supplementation for Anadromous Spring-Run Chinook Salmon and Steelhead. *Water-power XVI*, **27–30** July, Spokane, pp. 1–18.

May, C.L., Pryor, B.S., Lisle, T. and Lang, M. (2009) Coupling hydrodynamic modeling and empirical measures of bed mobility to predict the risk of scour and fill of salmon redds in a large regulated river. *Water Resources Research*, **45**: W05402. doi: 10.1029/2007WR006498.

McKean, J.A., Isaak, D.J. and Wright, C.W. (2008) Geomorphic controls on salmon nesting patterns described by a new, narrow-beam terrestrial-aquatic lidar. *Frontiers in Ecology and the Environment*, **6**(3): 125–130.

Merwade, V. (2009) Effect of spatial trends on interpolation of river bathymetry. *Journal of Hydrology*, **371**(1–4): 169–181.

Merz, J.E. and Ochikubo Chan, L.K. (2005) Effects of gravel augmentation on macroinvertebrate assemblages in a regulated California river. *River Research and Applications*, **21**: 61–74.

Merz, J.E., Pasternack, G.B. and Wheaton, J.M. (2006) Sediment budget for salmonid spawning habitat rehabilitation in the Mokelumne River. *Geomorphology*, **76**(1–2): 207–228.

Merz, J.E., Setka, J., Pasternack, G.B. and Wheaton, J.M. (2004) Predicting benefits of spawning habitat rehabilitation to salmonid fry production in a regulated California river. *Canadian Journal of Fisheries and Aquatic Science*, **61**: 1433–1446.

Moir, H.J. and Pasternack, G.B. (2008) Relationships between mesoscale morphological units, stream hydraulics and Chinook salmon (*Oncorhynchus tshawytscha*) spawning habitat on the Lower Yuba River, California. *Geomorphology*, **100**: 527–548.

Nestler, J.M., Goodwin, R.A., Smith, D.L. and Anderson, J.J. (2008) A mathematical and conceptual framework for ecohydraulics. In Wood, P.J., Hannah, D.M. and Sadler, J.P. (eds) *Hydroecology and Ecohydrology: Past, Present, and Future*, John Wiley & Sons, Ltd, Chichester, UK.

Oh, K., Lee, J., Rubio, C.J. and Jeong, S. (2010) Assessment of aquatic habitat effect by artificial change of streambed topography. *Water Science Technology*, **62**(12): 2872–2879.

O'Neill, M.P. and Abrahams, A.D. (1984) Objective identification of pools and riffles. *Water Resources Research*, **20**(7): 921–926. doi: 10.1029/WR020i007p00921.

Padmore, C.L., Newson, M.D. and Charlton, M.E. (1998) Instream habitat in gravel bed rivers: identification and characterisation of biotopes. In Klingeman, P.C., Beschta, R.L., Komar, P.D. and Bradley, J.B. (eds) *Gravel Bed Rivers in the Environment*, Water Resources Publications, Highlands Ranch, Colorado, pp. 345-364.

Pasternack, G.B. (2004a) http://shira.lawr.ucdavis.edu/trinity.htm

Pasternack, G.B. (2004b) http://shira.lawr.ucdavis.edu/trinity_designconcepts.htm

Pasternack, G.B. (2006) *Demonstration Project to Test a New Interdisciplinary Approach to Rehabilitating Salmon Spawning Habitat in the Central Valley*. CALFED Cooperative Agreement DCN#113322G003 Final Report, University of California at Davis, Davis, CA.

Pasternack, G.B. (2008a) Spawning habitat rehabilitation: advances in analysis tools. In Sear, D.A., DeVries, P. and Greig, S. (eds) *Salmonid spawning habitat in rivers: physical controls, biological responses, and approaches to remediation*, Symposium 65, American Fisheries Society, Bethesda, MD.

Pasternack, G.B. (2008b) *SHIRA-Based River Analysis and Field-Based Manipulative Sediment Transport Experiments to Balance Habitat and Geomorphic Goals on the Lower Yuba River*. Cooperative Ecosystems Studies Unit (CESU) 81332 6 J002 Final Report, University of California at Davis, Davis, CA.

Pasternack, G.B. (2011) 2D Modeling and Ecohydraulic Analysis, Createspace, Seattle, WA.

Pasternack, G.B. (2013) Geomorphologist's guide to participating in river rehabilitation. In Shroder, J.F. (ed) *Treatise on Geomorphology*, Vol. 9, pp. 843–860. San Diego: Academic Press.

Pasternack, G.B. and Brown, R.A. (2011) *Ecohydraulic Design of Gravel-Bed River Rehabilitation in the Lewiston Dam Reach of the Trinity River, CA*. University of California at Davis, Davis, CA.

Pasternack, G.B. and Senter, A.E. (2011) *21st Century Instream Flow Assessment Framework for Mountain Streams*. California Energy Commission, Public Interest Energy Research.

Pasternack, G.B., Bounrisavong, M.K. and Parikh, K.K. (2008) Backwater control on riffle-pool hydraulics, fish habitat quality, and sediment transport regime in gravel-bed rivers. *Journal of Hydrology*, **357**(1–2): 125–139.

Pasternack, G.B., Wang, C.L. and Merz, J. (2004) Application of a 2D hydrodynamic model to reach-scale spawning gravel replenishment on the lower Mokelumne River, California. *River Research and Applications*, **20**(2): 205–225.

Pasternack, G.B., Gilbert, A.T., Wheaton, J.M. and Buckland, E.M. (2006) Error propagation for velocity and shear stress prediction using 2D models for environmental management. *Journal of Hydrology*, **328**: 227–241.

Pate, W.D. and Rohleder, W.J. (2007) Cradle of invention. *Civil Engineering*, **77**(3): 36–43.

Payne, T.R. (2010) Personal communication. Thomas R. Payne & Associates. California, USA.

Rayburg, S. and Neave, M. (2008) Assessing morphologic complexity and diversity in river systems using three-dimensional asymmetry indices for bed elements, bedforms and bar units. *River Research and Applications*, **24**: 1343–1361.

Redd, R. and Horner, T.C. (2008) *Physical and Geochemical Characteristics of the 2008 Sailor Bar Gravel Addition*. Report to the U.S. Bureau of Reclamation, Sacramento office.

Redd, R. and Horner, T.C. (2010) *Physical and Geochemical Characteristics of the 2009 Sailor Bar Gravel Addition*. Report to the U.S. Bureau of Reclamation, Sacramento office.

Richards, K.S. (1976a) The morphology of riffle-pool sequences. *Earth Surface Processes*, **1**: 71–88. doi: 10.1002/esp.3290010108.

Richards, K.S. (1976b) Channel width and the riffle-pool sequence. *GSA Bulletin*, **87**(6): 883–890.

Sarkisian, M., Mathias, N. and Mazeika, A. (2007) Suspending the limits. *Civil Engineering*, **77**(11): 36–45.

Sawyer, A.M., Pasternack, G.B., Moir, H.J. and Fulton, A.A. (2010) Riffle-pool maintenance and flow convergence routing confirmed on a large gravel bed river. *Geomorphology*, **114**: 143–160.

Sawyer, A.M., Pasternack, G.B., Merz, J.E., Escobar, M. and Senter, A.E. (2009) Construction constraints on geomorphic-unit rehabilitation on regulated gravel-bed rivers. *River Research and Applications*, **25**: 416–437.

Senter, A.E. and Pasternack, G.B. (2010) Large wood aids spawning Chinook salmon (*Oncorhynchus tshawytscha*) in marginal habitat on a regulated river in California. *River Research and Applications*. doi: 10.1002/rra.1388.

Simon, A., Doyle, M., Kondolf, G.M., ShieldsJr., F.D., Rhoads, B. and McPhillips, M. (2007) Critical evaluation of how the Rosgen classification and associated "natural channel design" methods fail to integrate and quantify fluvial processes and channel responses. *Journal of the American Water Resources Association*, **43**(5): 1117–1131.

Stewardson, M.J. and McMahon, T.A. (2002) A stochastic model of hydraulic variations within stream channels. *Water Resources Research*, **38**(1). doi: 10.1029/2000WR000014.

Thompson, D.M. (2006) The role of vortex shedding in the scour of pools. *Advances in Water Resources*, **29**: 121–129.

Thompson, D. (2011) The velocity-reversal hypothesis revisited. *Progress in Physical Geography*, **35**(1): 123–132

USACE (2010) *Lower Yuba River Gravel Augmentation Final EA*, United States Army Corps of Engineers, Sacramento, CA.

USBR (1953) *Water Measurement Manual*, United States Bureau of Reclamation, Washington, DC.

USFWS (1999) *Trinity River Flow Evaluation*, United States Fish and Wildlife Service, Washington, DC.

USFWS (2002) *Mainstem Trinity River Chinook Salmon Spawning Survey Year 2000 and 2001*, United States Fish and Wildlife Service, Washington, DC.

Vetter, T. (2011) Riffle-pool morphometry and stage-dependent morphodynamics of a large floodplain river (Vereinigte Mulde, Sachsen-Anhalt, Germany). *Earth Surface Processes and Landforms*, **36**(12): 1647–1657.

Waddle, T.J. (2001) *PHABSIM for Windows User's Manual and Exercises*, United States Geological Survey, Fort Collins, CO.

Waddle, T.J., Steffler, P., Ghanem, A., Katopodis, C. and Locke, A. (2000) Comparison of one and two-dimensional open channel flow models for a small habitat stream. *Rivers*, **7**(3): 205–220.

Watry, C. and Merz, J.E. (2009) *Evaluation of the 2008 Sailor Bar Gravel Placement on the Lower American River, California*. Cramer Fish Sciences, Gresham, OR.

Wheaton, J.M., Pasternack, G.B. and Merz, J.E. (2004a) Spawning habitat rehabilitation – 1. Conceptual approach and methods. *International Journal of River Basin Management*, **2**(1): 3–20.

Wheaton, J.M., Pasternack, G.B. and Merz, J.E. (2004b) Spawning habitat rehabilitation – 2. Using hypothesis development and testing in design, Mokelumne River, California, U.S.A. *International Journal of River Basin Management*, **2**(1): 21–37.

Wheaton, J.M., Gibbins, C., Wainwright, J., Larsen, L. and McElroy, B. (2011) Preface: multiscale feedbacks in eco-geomorphology. *Geomorphology*, **126**(3–4): 265–268. doi: 10.1016/j.geomorph.2011.01.002.

Wheaton, J.M., Brasington, J., Darby, S., Merz, J.E., Pasternack, G.B., Sear, D.A. and Vericat, D. (2010) Linking geomorphic changes to salmonid habitat at a scale relevant to fish. *River Research and Applications*, **26**: 469–486.

White, J.Q., Pasternack, G.B. and Moir, H.J. (2010) Valley width variation influences riffle-pool location and persistence on a rapidly incising gravel-bed river. *Geomorphology*, **121**: 206–221.

Wilcock, P.R., Barta, A.F., Shea, C.C., Kondolf, G.M., Matthews, W.V.G. and Pitlick, J.C. (1996) Observations of flow and sediment entrainment on a large gravel-bed river. *Water Resources Research*, **32**: 2897–2909.

Wilkinson, S.N., Rutherfurd, I.D. and Keller, R.J. (2008) An experimental test of whether bar instability contributes to the formation, periodicity and maintenance of pool-riffle sequences. *Earth Surface Processes and Landforms*, **33**(11): 1742–1756.

Wissmar, R.C. and Beschta, R.L. (1998) Restoration and management of riparian ecosystems: a catchment perspective. *Freshwater Biology*, **40**: 571–585.

Woodsmith, R.D. and Hassan, M.A. (2005) Maintenance of an obstruction-forced pool in a gravel-bed channel: stream flow, channel morphology, and sediment transport. *Catchment Dynamics and River Processes: Mediterranean and Other Climate Regions*, **10**: 69–196.

21 Ecohydraulics for River Management: Can Mesoscale Lotic Macroinvertebrate Data Inform Macroscale Ecosystem Assessment?

Jessica M. Orlofske[1], Wendy A. Monk[2] and Donald J. Baird[2]

[1]Canadian Rivers Institute, Department of Biology, 10 Bailey Drive, P.O. Box 4400, University of New Brunswick, Fredericton, New Brunswick, E3B 5A3, Canada
[2]Environment Canada, Canadian Rivers Institute, Department of Biology, 10 Bailey Drive, P.O. Box 4400, University of New Brunswick, Fredericton, New Brunswick, E3B 5A3, Canada

21.1 Introduction

Rivers are naturally hierarchical systems, exhibiting inter-linked characteristics evident across spatial and temporal scales (Poole, 2002; Benda *et al.*, 2004; Dunbar *et al.*, 2010) and can be viewed as a set of nested filters whereby large-scale processes, such as climate, regional geology and discharge, constrain local-scale hydraulic conditions (Poff, 1997). In this way, rare, high-magnitude hydrological events can be viewed as drivers of large-scale (and thus long-term) changes in aquatic community structure, whereas low-magnitude, more frequent events control short-term, local-scale changes in community structure, often on an annual/seasonal basis. Flow management aims to incorporate biological metrics in assessments, which requires consideration of how to integrate site-level data across scales for appropriate interpretation at the management level. Sustainable river management and restoration strategies presuppose linkages between different spatio-temporal scales and relevant ecological processes, often in the absence of any direct supporting evidence. For this reason, it could be argued that river management remains a 'faith-based' discipline (*sensu* Hilborn, 2006). Robust linkages between the physical habitat templet, including hydraulic parameters and its associated biota, based on hypothesis-driven science are urgently needed.

Offering an interdisciplinary approach, research within the field of ecohydraulics is improving our understanding of the complex spatial and temporal connections between aquatic organisms and local-scale hydraulic properties of the habitat through both pure and applied research (Newson and Newson, 2000; Rice *et al.*, 2010). Yet, in a recent commentary, Lancaster and Downes (2010a; 2010b) question the value of ecohydraulic studies which rely on correlating species densities with hydrological and hydraulic variables. They highlight the need for studies to incorporate greater ecological realism: specifically information on individual organisms, biotic interactions and factors underlying population dynamics. They further suggest that this approach provides greater resolution and enhanced predictability of detected biological–ecohydraulic linkages and a framework for effective ecohydraulic research, and that the extensive datasets and considerable costs required to derive and

Ecohydraulics: An Integrated Approach, First Edition. Edited by Ian Maddock, Atle Harby, Paul Kemp and Paul Wood.
© 2013 John Wiley & Sons, Ltd. Published 2013 by John Wiley & Sons, Ltd.

maintain the sampling network may not be realistic or economically feasible. The discipline of ecohydraulics has clearly been challenged to address the immediate pressures of global threats to lotic systems without compromising scientific rigour. Lancaster and Downes (2010a; 2010b) correctly state that ecohydraulics cannot develop as a robust, predictive discipline if we ignore natural complexity, yet the value of information acquired by ecohydraulic studies for use in river management is ultimately dependent not only on appropriate study design, but also on the use of cost-effective methods, applied at appropriate scales with transferable results.

Here, we evaluate the current state of ecohydraulics science, with a specific focus on lotic macroinvertebrates. We explore the application of this knowledge for river management and identify areas of ecohydraulic research which may benefit from more innovative approaches (*sensu* Lancaster and Downes, 2010a).

21.2 Lotic macroinvertebrates in a management context

Species-rich and functionally diverse, lotic macroinvertebrates are a key component of stream and river biodiversity and significantly contribute to ecosystem function. They occupy an integral role in energy flow pathways by processing organic material and by supporting higher trophic levels as a food source. Although they exhibit diverse life-history strategies, during aquatic development they generally have limited mobility compared to larger-bodied organisms such as fish (Gore and Judy, 1981; Newson and Newson, 2000). As a consequence of this, lotic macroinvertebrate species are more ecologically diverse, each generally existing within a narrower range of habitat conditions than fish species (Gore and Judy, 1981; Newson and Newson, 2000). For these reasons, macroinvertebrates are viewed as good indicators of river habitat degradation, both in terms of water quality (e.g. organic pollution, Hilsenhoff, 1987, 1988; sedimentation, Extence *et al.*, 2013) and water quantity (e.g. flow variability, Armanini *et al.*, 2011). Thus, macroinvertebrates are routinely sampled for freshwater bioassessment and biomonitoring (Dunbar *et al.*, 2010) and are key indicators in fisheries management (Jowett and Richardson, 1990).

River biomonitoring using macroinvertebrates has always required some consideration of their response to flow, if only in terms of characterizing in-stream habitat. Recently there has been growing recognition of the importance of flow-related variables as potential predictors of lotic macroinvertebrate community structure, and an increasing awareness of threats to rivers from modification of the flow regime (Poff *et al.*, 1997). Many countries (e.g. USA, UK, Canada, Australia and South Africa) have national or regional government-led programmes for collecting macroinvertebrates to monitor the condition of river ecosystems, including ecological flows. Most programmes follow a standardized sample collection protocol including channel-habitat variables and regional geospatial information. Ideally, sites are sampled over multiple years and can be directly associated with a continuous hydrometric record from a nearby gauging station. This allows observations regarding community structure to be placed in a long-term context by monitoring natural community variability in relation to hydrological conditions over time.

Sustainable river management, in terms of biomonitoring, requires the development of rapid assessment metrics that measure the effects of hydrological alteration on aquatic community structure. This has spurred the development of the Lotic-invertebrate Index for Flow Evaluation (LIFE) in the UK (Extence *et al.*, 1999) and the Canadian Ecological Flow Index (CEFI; Armanini *et al.*, 2011). These methods use biomonitoring data and flow velocity optima of macroinvertebrate taxa to determine the influence of hydrological alteration at a site. Both LIFE and CEFI scores have demonstrated a clear response to the influence of extreme and moderate antecedent hydrological events across a range of river types (e.g. Monk *et al.*, 2006, 2008; Armanini *et al.*, 2012). In this way, such indices can be used, in the context of a reference model, to explore deviations of test sites from a regional reference state and identify ecohydrological impairment (Peters *et al.*, 2012).

Rarely measured in standard biomonitoring studies, hydraulic parameters translate hydrologic conditions to dynamic, local scale, fluid and associated substrate processes that compose the template (*sensu* Southwood, 1977) of micro- and mesoscale lotic macroinvertebrate habitats. The importance of hydraulic conditions for lotic macroinvertebrates, particularly at the mesoscale (Quinn and Hickey, 1994; Newson and Newson, 2000; Brooks *et al.*, 2005) has been well established (Nowell and Jumars, 1984; Statzner *et al.*, 1988; Hart and Finelli, 1999; Lancaster and Belyea, 2006). As a relatively new science (Rice *et al.*, 2010), ecohydraulics is only just beginning the task of quantifying the influence of such complex habitat variables, which invariably act in concert with a range of other natural and anthropogenic drivers to determine the long-term ecological status of riverine ecosystems.

21.3 Patterns in lotic macroinvertebrate response to hydraulic variables

Lotic macroinvertebrate hydraulic studies are confounded by interactions between organisms with complex lifecycles and collinear hydraulic variables resulting in complicated analyses with ambiguous interpretation (Collier, 1994). Thus, many have approached 'ecohydraulic' questions from diverse perspectives, using various methods which sometimes reveal apparently different results. We examined a range of peer-reviewed scientific publications to provide a review of common approaches, taxa and hydraulic parameters in order to elucidate patterns in the responses of lotic macroinvertebrates to hydraulic conditions (Table 21.1). The measurement or manipulation of hydraulic variables (excluding laboratory experiments) was our selection criterion, rather than focusing only on studies explicitly labelled 'ecohydraulics', as many relevant earlier works would have otherwise been excluded. Our analysis revealed that field observation is more common (85%) than field experimentation (15%) in ecohydraulic research (Table 21.1). This may reflect the difficulty associated with controlling or manipulating hydraulic conditions in the field, and has resulted in the current status, lamented by Lancaster and Downes (2010a; 2010b), where ecohydraulic knowledge is based largely on correlative rather than hypothesis-derived evidence.

The field observation studies can be further divided into community-level studies (10) which examine more than ten species or taxa, and focused studies (12) which analyze fewer than ten species or taxa (Table 21.1). Among the latter, the most frequently investigated organisms are Trichoptera (7 studies, 17 taxa), followed by Ephemeroptera (6 studies, 8 taxa), Diptera (2 studies, 2 taxa) and Hemiptera (1 study, 1 taxon). In some instances, the same taxon was included in multiple studies, as was the case for two Ephemeroptera taxa (*Rhithrogena semicolorata* and the *Deleatidium* spp. complex). Community studies ranged from those focusing on multiple species or genera of a single insect family (e.g. Trichoptera – Hydrobiosidae, Collier *et al.*, 1995) to other studies considering the entire macroinvertebrate community (e.g. Brooks *et al.*, 2005). However, many studies restricted their focus to certain taxa, including indicator groups such as the insect orders: Ephemeroptera and Trichoptera. Community studies have potentially greater overlap in taxa, but individual responses were often not reported. Field experiments followed a similar trend in terms of taxonomic breadth, with Trichoptera (Osborne and Herricks, 1987)

and Diptera (Morin *et al.*, 1986) being the focus of specific studies while a range of aquatic insect orders were the subjects of the community experimental studies (Table 21.1).

Most studies contain mixtures of hydraulic, hydrological and biotic parameters (Table 21.1). Here, we focus only on hydraulic parameters, acknowledging that this assumes other unobserved/unconsidered factors were not confounding any individual study linking hydraulic effects and community structure. This simplified approach is reasonable given the present state of our knowledge and our objective of illustrating broad patterns among ecohydraulics studies to inform future investigations. The most commonly used simple hydraulic parameters (*sensu* Statzner *et al.*, 1988) were depth, velocity and substrate characteristics (Table 21.2). Substrate characteristics are germane to a discussion of river hydraulics. Hydraulic flows and substrate possess a reciprocal relationship since both physical attributes influence and modify the other. In addition, substrate properties must be measured in order to calculate particular hydraulic parameters (e.g. shear velocity). Substrate characteristics are often substituted for hydraulic variables that are more difficult to measure in field locations and are themselves valuable explanatory variables. The derived (i.e. complex, *sensu* Statzner *et al.*, 1988) hydraulic parameters used most frequently include Froude number, shear velocity and Reynolds number (Table 21.2). Since these hydraulic parameters are derived from simple parameters, studies with derived parameters also report those corresponding simple parameters. Generally, studies combine a few simple and derived variables, although there is wide variability in study design and methods employed.

Detailed field observations typically incorporate a greater number of hydraulic variables, illustrating a common practical trade-off in ecohydraulic studies (Table 21.2). By focusing on reduced numbers/types of organisms, more effort can be dedicated to fully describing the hydraulic conditions. Field observations of the lotic macroinvertebrate community use fewer parameters in order to characterize the environment (Table 21.2). These selected parameters may be easier to measure in the field across a greater number of sampling sites or events, or they were chosen based on results of previous research.

Flow velocity and water depth are among the hydraulic variables that resulted the strongest responses from lotic macroinvertebrates. However, this trend is likely confounded by the frequency with which flow velocity and water depth are reported, and their occurrence as components of derived variables, namely Froude number, shear velocity, Reynolds number and roughness/boundary

Table 21.1 Summary of ecohydraulic studies describing the distribution patterns and association of lotic invertebrate taxa to ecohydraulic parameters arranged by type of study and year of publication.

Order	Family	Species	Location	Parameters*	Association/Response	Source
Field observations – focal taxa						
Trichoptera	Limnephilidae	*Pycnopsyche lepida*	USA	substrate particle size, velocity	Transition in substrate use from silt-organic material to larger substrate particles during development; broad velocity range	Cummins, 1964
		Pycnopsyche guttifer			Silt-organic substrate use; broad velocity range	
Trichoptera	Glossosomatidae	*Glossosoma califica*	USA	depth, velocity, rock size, rock roughness, rock slope	Species-specific microhabitat selectivity for depth, velocity, rock size and rock roughness	Teague et al, 1985
	Glossosomatidae	*Glossosoma penitum*				
	Limnephilidae	*Dicosmoecus gilvipes*				
	Uenoidae	*Neophylax rickeri*				
	Uenoidae	*Neophylax splendens*				
Trichoptera	Limnephilidae	*Oligophlebodes zelti*	Canada	depth, velocity	Weak response to velocity, lower velocities	Ogilvie and Clifford, 1986
	Uenoidae	*Neothremma alicia*			Weak response to velocity, higher velocities	
Hemiptera	Aphelocheiridae	*Aphelocheirus aestivalis*	Germany	Simple: substrate profile, water surface slope, depth, velocity, kinematic viscosity, discharge, stream width. Complex: Reynolds number, Froude number, shear velocity, shear stress, thickness of the viscous sublayer, boundary Reynolds number	Preference curves for size/age cohorts. Stronger response to complex parameters, including Froude number, thickness of the viscous sublayer, and boundary Reynolds number.	Statzner et al, 1988

Order	Family	Species	Country	Variables	Description	Reference
Ephemeroptera	Leptophlebiidae	*Deleatidium* spp. (Additional taxonomic groups)	New Zealand	depth, velocity, substrate composition (embeddedness)	Develop habitat preferences/suitability curves; depth, velocity, substrate (univariate and multivariate responses)	Jowett and Richardson, 1990
Trichoptera	Brachycentridae	*Brachycentrus occidentalis*	Canada	local velocity and flow profiles: velocities, depths, water surface slopes	Froude number	Wetmore *et al.*, 1990
Trichoptera	Glossosomatidae	*Glossosoma intermedium*			Depth	
Diptera	Simuliidae	*Simulium vittatum*				
Ephemeroptera	Leptophlebiidae	*Deleatidium*	New Zealand	velocity, depth	Froude number Velocity, depth, combination of velocity and depth; size, sex and morphotype specific responses	Collier, 1994
Ephemeroptera	Heptageniidae	*Rhithrogena semicolorata* ; *Ecdyonurus* sp. vr. *venosus*	Italy	roughness, velocity, depth, substrate composition, Froude number	Size-related habitat preferences based on roughness	Buffagni *et al.*, 1995
Ephemeroptera	Ephemeridae	*Ephemera danica*	France	depth, velocity, substrate composition	Early instars: boulders with medium velocity; later instars: sand with low velocity;	Hanquet *et al.*, 2004
	Heptageniidae	*Rhithrogena semicolorata*			Early instars: low depth and velocity; later instars: higher depth and velocity, no shift in substrate use	
Trichoptera	Hydrosphychidae	*Hydropsyche siltalai*			All instars: large substrate, high velocity; shift prior to pupation	
Trichoptera	Hydrobiosidae	*Ulmerochorema rubiconum*	Australia	velocity: surface and average (2/3 depth), substrate size index	Fast surface water velocity (instar effects)	Reich and Downes, 2004

(continued)

Table 21.1 (*Continued*)

Order	Family	Species	Location	Parameters*	Association/Response	Source
Trichoptera	Hydrobiosidae	*Ethochorema turbidum*			Slow surface water velocity (instar effects)	
Trichoptera	Hydrobiosidae	*Apsilochorema obliquum*				
Ephemeroptera	Polymitarcyidae	*Ephoron virgo*	France	shear velocity, FST hemisphere number*	Shear velocity preferences, shifts with body size	Sagnes *et al.*, 2008
Ephemeroptera	Oligoneuriidae	*Oligoneuriella rhenana*				
Ephemeroptera	Ephemerellidae	*Serratella ignita*				
Trichoptera	Hydrosphychidae	*Cheumatophsyche lepida*				
Trichoptera	Hydrosphychidae	*Hydrophsyche exocellata*				
Diptera	Blephariceridae	*Blepharicera fasciata*				
Ephemeroptera	Baetidae	*Baetis bicaudatus*	USA	velocity, depth, wetted width, discharge, bed gradient, hydraulic radius, grain size, boundary shear stress, Shields number, Reynolds number, Froude number, Darcy–Weisbach friction factor, relative submergence	Drift propensity: positively related to Reynolds number, negatively related to Shields number	Wilcox *et al.*, 2008

Field observations – invertebrate community

Order	Family	Species	Location	Parameters*	Association/Response	Source
Selected aquatic insect orders		40 taxa	Sweden	depth, velocity, substrate condition	Species-specific associations with combinations of depth, velocity and substrate conditions	Ulfstrand, 1967
Selected aquatic insect orders		18 taxa	USA	depth, velocity, substrate composition, Froude number (turbulence)	Develop habitat preferences/suitability curves; depth, velocity, substrate and turbulence (regression analysis)	Orth and Maughan, 1983

Community	Taxa	Country	Variables	Findings	Reference
Benthic community	Highlight 9 common taxa	New Zealand	Simple: mean velocity, depth, substrate size, kinematic viscosity, substrate roughness; Complex: Froude number, shear velocity, water column and boundary Reynolds number	Strongest correlation for a single variable: boundary Reynolds number; Combined variables: substrate size, depth and mean velocity; Taxa-specific responses	Quinn and Hickey, 1994
Trichoptera Hydrobiosidae	14 taxa	New Zealand	depth, velocity, substrate size categories, substrate roughness, Froude number, Reynolds number, shear velocity, boundary layer Reynolds number	Depth, velocity, Froude number	Collier et al, 1995
Selected aquatic insect orders, Amphipoda	31 taxa	Austria	substrate profile, velocity, depth, FST hemisphere number	Hydraulic stress, FST hemisphere number	Möbes-Hansen and Waringer, 1998
Benthic community	172 taxa	France	depth, velocity, Froude number, substrate roughness	Trait association with Froude number; correlations among variables	Lamouroux, et al, 2004
Benthic community	31 taxa	France	FST hemisphere number, depth, substrate particle size, bed roughness, Froude number	Strongest correlation: shear stress and Froude number; Combined variables: substrate size, depth and mean velocity; Taxa-specific responses	Mérigoux and Dolédec, 2004
Benthic community	155 species	Australia	velocity, depth, substrate roughness; roughness Reynolds number, shear velocity, Froude number	Negative associations with roughness Reynolds number, shear velocity, velocity and Froude number; highest explanatory power: roughness Reynolds number, velocity, shear velocity	Brooks et al, 2005

(continued)

Table 21.1 (*Continued*)

Taxon			Location	Parameters*	Association/Response	Source
Order	Family	Species				
Trichoptera		48 species	Slovenia	substrate, water depth, flow type	Groups of species with similar responses divided between coarse substrate, shallow water, chute flow and fine substrate in deep water	Urbanič *et al.*, 2005
Benthic community		151 taxa	Germany	FST hemisphere number	Modelling of taxa hydraulic preferences	Dolédec *et al.*, 2007
Field experiments – focal taxa						
Diptera	Simuliidae	*Prosimulium mixtum/fuscum*	Canada	velocity, depth	develop habitat preferences/suitability curves; depth, velocity (comparing different types of analysis)	Morin *et al.*, 1986
		Stegopterna mutata Simulium aureum				
Trichoptera	Hydropsychidae	*Hydropsyche bronta*	USA	velocity, turbulence (microhabitat flow pattern)	Species-specific microhabitat selectivity for velocity and turbulence; importance of pattern and turbulence of flow	Osborne and Herricks, 1987
		Hydropsyche cheilonis Hydropsyche sparna Hydropsyche betteni				
Field experiments – invertebrate community						
Selected aquatic insect orders		14 taxa	USA	substrate composition and structure, depth, velocity	Response to substrate manipulation; substrate preference, concomitant velocity preference	Minshall and Minshall, 1977
Selected aquatic insect orders		16 taxa	USA	Substrate size and composition, velocity, depth	Response to substrate manipulation; substrate preference	Rabeni and Minshall, 1977

*velocity synonymous with water velocity, current velocity; depth synonymous with water depth; FST hemisphere number (Statzner and Muller, 1989).

Table 21.2 Summary of simple and derived (complex) hydraulic parameters (*sensu* Statzner *et al.*, 1988) used in four types of lotic invertebrate ecohydraulics studies.

Hydraulic parameters	Frequency of application				
Simple	Observation, focal taxa	Observation, community	Experiment, focal taxa	Experiment, community	Total
depth	9	10	2	2	23
velocity	11	7	2	2	22
substrate composition	3	3	0	2	8
substrate particle size	4	3	0	0	7
substrate roughness	2	6	0	0	8
FST hemisphere number	1	3	0	0	4
water surface slope	2	0	0	0	2
substrate slope	2	0	0	0	2
embeddedness/relative submergence	2	0	0	0	2
kinematic viscosity	1	1	0	0	2
hydraulic radius	1	0	0	0	1
flow type	0	1	0	0	1
Complex					
Froude number	3	6	0	0	9
shear velocity	2	3	0	0	5
Reynolds number	2	2	0	0	4
boundary/roughness Reynolds number	1	3	0	0	4
shear stress	1	0	0	0	1
thickness of the viscous sublayer	1	0	0	0	1
boundary shear stress	1	0	0	0	1
Darcy–Weisbach friction factor	1	0	0	0	1
Shields number	1	0	0	0	1
turbulence	0	0	1	0	1

Reynolds number. Some studies do not describe individual taxon responses. This is particularly true of field observations of the macroinvertebrate community. In addition, the objective for several studies was the modelling of 'hydraulic' preferences, making it difficult to describe specific responses from mathematical relationships.

The variety of responses to hydraulic parameters observed in these and many other studies may be due to the diverse strategies and adaptations of lotic macroinvertebrates, which allow them to take advantage of resources available under different hydraulic conditions. River flow provides a mechanism for respiratory gas exchange and delivery of food and nutrients, aids dispersal, is a potential means for rapid escape from predators via entering the drift and defines suitable emergence and oviposition sites (Collier *et al.*, 1995; Teague *et al.*, 1985). In addition, the fluid properties of the flow may interact with the behaviour and morphology of the organism to determine their distribution in the stream (Oldmeadow *et al.*, 2010).

Therefore, it is important to consider the adaptations of the organisms themselves – in terms of their traits – with regard to the use of hydraulic habitat.

21.4 Linking ecohydraulics and lotic macroinvertebrate traits

Traits-based approaches can provide primary information for the development of alternative metrics for lotic ecosystem research and management. Such metrics could employ the functional attributes of organisms in addition to their taxonomic identity to incorporate aspects of ecological function in biomonitoring programmes. It has been argued that incorporating traits in this way could provide a more consistent means of comparing locations, yield mechanistic explanations of perceived environmental perturbations and provide earlier detection of environmental change (Culp *et al.*, 2011).

Lotic macroinvertebrate traits have been observed and quantified in order to connect functional ecological information about each taxon to environmental factors and perturbations, including hydrology (Bonada *et al.*, 2007), human impacts/land use (Dolédec *et al.*, 1999; Dolédec and Statzner, 2008; Carlisle and Hawkins, 2008), sensitivity to toxic materials (Baird and Van Den Brink, 2007; Buchwalter *et al.*, 2008), effects of restoration (Tullos *et al.*, 2009) and suites of multiple stressors (Statzner and Bêche, 2010). Horrigan and Baird (2008) identified several trait 'modalities' (the instance of the trait) that respond consistently to flow and flow-associated conditions (e.g. temperature and dissolved oxygen). However, few of the traits available in various databases (e.g. Tachet *et al.*, 2000; Vieira *et al.*, 2006) were derived to be related purposefully to hydraulic parameters. Two exceptions may be body size and body shape traits, which are subject to hydraulic properties such as Reynolds number, thickness of the boundary layer, drag, turbulence and other forces (Statzner, 1988; 2008). Some traits, for example current preference/rheophily and microhabitat preference (Vieira *et al.*, 2006), might be suitable indicators of hydraulic conditions. However these 'traits' may represent a whole organism response (e.g. a co-evolved life history strategy, Verberk *et al.*, 2008) that is being mediated by independent or linked traits (i.e. Poff *et al.*, 2006a), for example morphological traits such as armouring and shape may be correlated with rheophily (see case study below).

Hydraulic conditions change over time, as do the requirements of many lotic macroinvertebrates (Statzner, 2008; Culp *et al.*, 2011). Ecohydraulic studies demonstrate a shift in hydraulic habitat use with age or developmental stage (Collier, 1994; Collier *et al.*, 1995; Reich and Downes, 2004; Dolédec *et al.*, 2007; Sagnes *et al.*, 2008). A greater understanding of these patterns is vital to the successful management of organisms with complex lifecycles (Lancaster and Downes, 2010a), since supporting the needs of only one life stage does not guarantee the long-term success of an organism. This is particularly true for aquatic insects, which often possess multiple aquatic growth stages and terrestrial adults, some of which also have hydraulic habitat requirements for emergence and/or oviposition (Teague *et al.*, 1985; Collier *et al.*, 1995).

We have yet to fully realize the potential benefits of traits-based metrics for lotic systems management. The adoption of this approach will depend on recognizing the importance of the mesoscale and hydraulic forces, coping with multiple sources of variation, both in the system and the organisms, and clarifying how traits respond in a consistent manner to perform better than taxonomic metrics. Below we illustrate how the traits-based approach can strengthen existing metrics, focusing on the UK LIFE index (Extence *et al.*, 1999) as an example.

21.5 Trait variation among lotic macroinvertebrates in LIFE flow groups

Using national biomonitoring data paired with antecedent hydrological data from nearby gauging stations, LIFE scores have been used to quantify the response of benthic macroinvertebrate communities to both habitat modification (e.g. Dunbar *et al.*, 2010) and flow conditions (Monk *et al.*, 2006, 2008; Dunbar *et al.*, 2010) for different river types.

The association of lotic macroinvertebrate taxa to hydraulic variables is integral to LIFE. The LIFE score is based on six flow groups (based on flow velocity preferences; see Dunbar *et al.*, 2010) that can be related to perceived hydraulic habitat (microhabitat) preferences of the taxa, in terms of measured velocity and substrate composition. Taxa with lower LIFE flow group values are associated with higher velocities and large particle-size substrates (e.g. cobble); taxa with higher LIFE flow group values are associated with lower velocities and small particle size substrates (e.g. silt) (Dunbar *et al.*, 2010). The LIFE flow groups, determined thorough literature review, were used to weigh each taxon in conjunction with its abundance (log categories) for the overall calculation of the LIFE score per sample (Extence *et al.*, 1999).

To illustrate the potential association between LIFE flow groups and biological and ecological traits (*sensu* Usseglio-Polatera *et al.*, 2000), we used a trait database compiled for the UK that included 13 traits comprised of 33 modalities (Bowser, Armanini and Baird, unpublished data). We focused our analysis on Trichoptera, since they are well studied from both an ecohydraulic and ecohydrological perspective, show a variety of responses to hydraulic conditions (Table 21.1) and are often focal taxa in biomonitoring programmes. The trait database included information on 55 Trichoptera genera.

Trichoptera are a morphologically and functionally diverse order of aquatic insects (Mackay and Wiggins, 1979); however, they still possess a restricted suite of traits relative to the full complement expressed in lotic invertebrates. Therefore, some traits and some trait modalities were eliminated from this analysis because they were rare or absent (e.g. desiccation resistance, swimming

Table 21.3 Descriptions and abbreviations of larval Trichoptera traits used in LIFE flow group case study, based on those used in Horrigan and Baird (2008) and Vieira *et al.*, (2006).

Trait	Trait description	Trait code
Armouring	None (soft-bodied forms)	ArmNo
Armouring	Poor (heavily sclerotized)	ArmPoor
Armouring	Good (e.g. some cased caddisflies)	ArmGood
Attachment	Free-ranging	AtchFree
Attachment	Sessile, sedentary	AtchSessSed
Attachment	Both	AtchBoth
Adult flying strength	Weak (e.g. cannot fly into light breeze)	FlgtWeak
Adult flying strength	Strong (e.g. can fly into light breeze)	FlgtStrong
Respiration	Tegument	RespTegum
Respiration	Gills	RespGills
Rheophily	Depositional only	RheoDep
Rheophily	Depositional and erosional	RheoDepEros
Rheophily	Erosional	RheoEros
Shape	Streamlined (flat, fusiform)	ShpStreamline
Shape	Not streamlined (cylindrical, round or bluff)	ShpNonStreamline
Size at maturity	Small (< 9 mm)	SizeSmall
Size at maturity	Medium (9–16 mm)	SizeMed
Size at maturity	Large (> 16 mm)	SizeLarge
Thermal preference	Cold stenothermal or cool eurythermal	TherColdSteno
Thermal preference	Cool/warm eurythermal	TherCoolWarm
Thermal preference	Warm eurythermal	TherWarm
Voltinism	Semivoltine (< 1 generation/year)	VoltSemi
Voltinism	Univoltine (1 generation/year)	VoltUni
Voltinism	Bi- or multi-voltine (> 1 generation/year)	VoltBiMulti

ability (one taxon) and aerial respiration) or invariant (all members possessed the trait and/or modality; e.g. adult ability to exit the water) within Trichoptera. Thus, we analyzed nine observed traits, expressing 24 total modalities (Table 21.3). Due to intra- and inter-genera variation, there were several instances of multiple trait modalities within a single taxon. For analysis, each occurrence of a modality was used and we therefore employed proportions for consistency.

We observed variation in trait modalities among LIFE flow groups. Trichoptera genera in LIFE flow groups 1 and 2 showed a greater proportion of rheophily-erosional and cold/cool trait modalities (Figure 21.1). In addition, these LIFE flow groups also showed higher proportions of protective armouring and semivoltine lifecycle modalities (Figure 21.1). The reverse pattern was observed for Trichoptera genera in LIFE flow groups 3–5. Higher proportions of rheophily-depositional, warm, no or poor armouring and univoltine lifecycle modalities were shown for these taxa (Figure 21.1). However, the patterns detected in some traits were ambiguous. Traits such as respiration and attachment show less variation

across the LIFE flow groups (Figure 21.1). This is probably a consequence of focusing only on Trichoptera, since members of this insect order possess less variation in these characteristics. In addition, flight is a trait attribute of the adult form and uninformative for larvae.

Trichoptera pose a challenge for shape and size traits. Variation in these traits among LIFE flow groups was also less clear. Functionally, size and shape of case-bearing caddisflies is dependent on the shape, materials and construction of their cases. Taxonomic treatments (species accounts and keys) of case-bearing taxa typically include at least a basic description of the case and its component materials (e.g. Wallace *et al.*, 2003). The possession of a case is often not treated as a trait, but the information is incorporated into other traits (e.g. armouring) to be evaluated consistently across other non-case-bearing lotic invertebrates. Cases can also be lost or damaged during invertebrate sampling or processing, making consistent measurements difficult.

We analyzed the presence of cases in a similar fashion to traits analysis among the LIFE flow groups by first dividing these into categories/modalities, based on construction

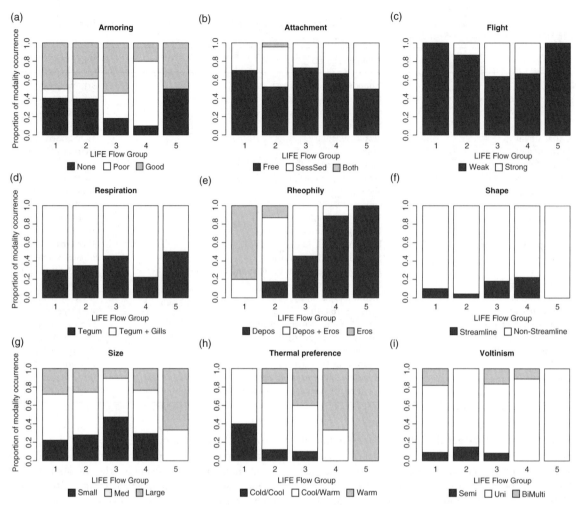

Figure 21.1 Panel of bar charts showing the proportion of each trait modality occurrence among LIFE flow groups for 55 UK Trichoptera genera. In the case that a genus could possess more than one modality for a trait, each instance was plotted. Refer to Table 21.3 for a description of traits and their abbreviations.

materials. With the exception of LIFE flow group 5, the trend was for fewer caseless taxa in higher LIFE groups (2–4) (Figure 21.2). Of the taxa possessing cases, the lower LIFE flow groups (1–3) generally had a higher proportion of stone cases, while higher LIFE flow groups (4–5) had a higher proportion of plant-material cases (Figure 21.2). LIFE flow group 5 had the most extreme pattern, with the highest proportion of caseless and plant-material cases among all LIFE flow groups, which may be due to the smaller number of taxa represented in this category.

Case construction in Trichoptera is a complex genetically-controlled, environmentally-modified behaviour which can serve multiple beneficial functions,

including enhancing respiration (Williams *et al.*, 1987) and protection from predation (Johansson, 1991). Prevailing hydrologic and hydraulic forces influence substrate characteristics and, both directly and indirectly, larval habitat selection, which may carry over into case construction (Cummins, 1964; Wetmore *et al.*, 1990). Our ad hoc analysis suggests that larval selection of case materials could reflect a shift in substrate availability and suitability for the construction of cases among flow groups. The addition of this trait information could provide further evidence of stream impairment and even aid in the diagnosis of the cause of impairment as flows and sediments are modified.

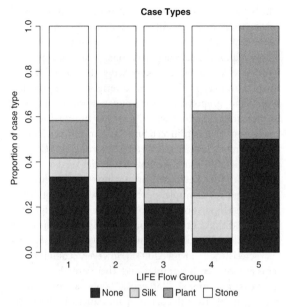

Figure 21.2 Bar chart showing the proportion of each case type occurrence among LIFE flow groups for 55 UK Trichoptera genera. In the case that a genus could possess more than one type of case, each instance was plotted. Key: None – no case present; Silk – silk case; Plant – case constructed of organic, primarily plant material; Stone – case constructed of inorganic material, primarily stones.

Not all traits reflect abiotic habitat constraints, but may be the result of dual or multiple selection pressures. Armouring, as in the Trichoptera case study above, is an example of such a trait which is influenced by the physiological requirements of the organism (respiration and thermoregulation) as well as its need for protection. Likewise, evolutionary constraints may result in 'trait syndromes' (*sensu* Poff *et al.*, 2006a) such that genetic linkages between traits may obscure individual trait responses to environmental conditions. A visual representation using non-metric multidimensional scaling (nMDS) shows trait modalities, which occur together more frequently, plotted in closer proximity than those found less frequently together (Figure 21.3). Overall, there is a lack of strong structure in the data, reflecting fewer linkages between these selected traits occurring in UK Trichoptera genera.

Since the LIFE flow groups were initially developed to reflect the organisms' biology, trends observed in the distribution of particular trait modalities among the LIFE flow groups were expected, indicating that there are common attributes of taxa that inhabit particular hydraulic habitats. This information could be used to help decipher which habitat characteristics most strongly influence lotic macroinvertebrates and to interpret the LIFE score of a stream or river site while taking into account confounding aspects of lotic macroinvertebrate biology.

Figure 21.3 nMDS plot (Jaccard's distance for binary data, stress: 23.46, vegan package in R) showing the association of trait modalities occurring among 55 UK Trichoptera genera. The plot summarizes associations among traits: modalities plotted in close proximity may occur more often together than those plotted farther apart. Refer to Table 21.3 for a description of traits and their abbreviations.

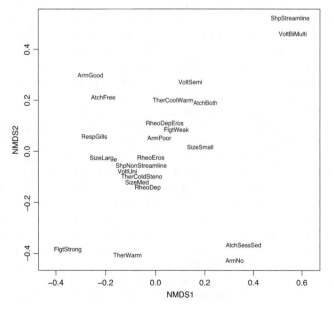

21.6 Upscaling from ecohydraulics to management

Flow variation across spatial and temporal scales from the individual patch to the watershed scale is necessary to maintain habitat diversity and sustain the biodiversity of aquatic communities within river systems. One goal of sustainable river management is to develop and apply protection strategies based on the knowledge of the ecological and habitat responses to hydraulic and hydrological processes within that system. However, the scales at which river management is applied are often quite different from the scales at which biomonitoring samples are collected (Friberg, 2010). Site scale information needs to be applicable at regional scales, allowing environmental managers to move away from extensive data collection and site-specific models (Dunbar et al., 2010).

Part of the challenge is upscaling from individual site data with finer-scale hydraulic variables based on flow velocity and individual benthic macroinvertebrate samples to the higher-level watershed scale hydrological variables based on discharge and multiple benthic macroinvertebrate samples collected across different habitats and stream orders. For example, at spatial scales greater than a habitat patch, it is often difficult to monitor habitat-specific point velocity to calculate hydraulic parameters, and the majority of flow data is collected in the form of channel discharge and summarized as hydrological indices. Wood et al. (2001) explored these linkages between scale, flow variability and macroinvertebrate community assemblages by developing models calculated at the reach scale, within-reach scale (upstream versus downstream), within-riffle scale and individual sample scale. Their results highlight the increased amount of ecological variance explained with smaller spatial scales. Newson and Newson (2000) advocate a mesoscale approach that allows an understanding of the direct linkages between ecohydraulics and ecohydrology by providing the balance between the highly detailed, resource-intensive, small-scale field studies and the often more general regional management plan. However, for many studies, a mismatch of scales may be a limitation to a direct upscaling using simple hydrological statistics (Stewardson and Gippel, 2003). For example, hydrological variables are often calculated for average changes in environmental conditions, while the biological variables may respond rapidly over much smaller spatial scales or integrate changes on short to moderate (seasonal) time scales. This mismatch of scales echoes Fausch et al. (2002),

who argued that research for management and conservation has been unsuccessful because of insufficient research centred on linking spatial and temporal scales appropriate for major management decisions, notably the need for basin-level versus process knowledge. Investigating ecohydraulic and ecohydrological patterns at the appropriate spatial and temporal scales is important for developing a greater understanding which can lead to more accurate prediction of macroinvertebrate communities (Dunbar and Acreman, 2001; Parsons et al., 2004; Biggs et al., 2005).

Historically, management strategies used highly focused approaches based on a range of different flow conditions, for example, percentile flows, wet- and dry-season baseflows, normal high flows, extreme drought and flood conditions, rates of flood rise and fall and the interannual variability using threshold limits for specific flow characteristics (Tharme, 2003). However, ecological systems are inherently complex and, therefore, do not conform to simple models. Poff et al. (2006b) raised two reoccurring problems with current management applications. First, management decisions are often needed at specific locations where detailed studies at the appropriate scale are lacking, leading to the application of sometimes inappropriate management schemes. Second, environmental regulation and management are often applied at broad geographical regional scales, which encompass substantial variation in topographic or ecological conditions. To counter these problems, we suggest that a classification of river types applied at an appropriate scale offers an approach to reduce the within-group environmental variability and, therefore, tailor the expected aquatic community to hydraulic variability for a particular river type. For example, grouping river types by hydrological regimes has provided a better understanding of the potential ecological response to environmental variability at a sample site and, consequently, to upscale and incorporate hydrological variability at the watershed scale (e.g. in Canada: Armanini et al.,2012; Monk et al., 2011; UK: Monk et al., 2006, 2008). This also highlights the clear potential for integrative ecohydrological indices to be used in an environmental flow management context, for example, the CEFI and LIFE scores, which were developed based, in part, on assumptions with hydraulic relevance. Based on biomonitoring data, CEFI was derived using average channel velocity measured at the time of sampling to calculate environmental optima associated with each taxon. Calculated at the reach scale from a single benthic macroinvertebrate sample, CEFI presents a direct link between the hydraulic habitat of the sample and the hydrological habitat of the reach when paired with discharge data

(Armanini *et al.*, 2011; 2012). For example, a site CEFI or LIFE score can be used to flag a community response at an individual site to a particular hydrological disturbance, or it could be compared to a regional score to integrate site status and assessment at the watershed scale. This level of information can support environmental flow assessment at the scales that are relevant to river managers and engineers (Dunbar *et al.*, 2010).

Effective river flow management requires adequate characterization of the ecologically relevant components of hydrological variability, which, in turn, requires rapid assessment methods (Archer and Newson, 2002), such as biological metrics (e.g. CEFI and LIFE scores) and key hydrological and hydraulic indices (e.g. Indicators of Hydrologic Alteration; Richter *et al.*, 1996). However, most rivers have been affected by some degree of anthropogenic flow alteration, for example by agricultural water withdrawals, hydropower infrastructure and operation and urbanization. These modifications can have both direct and indirect effects on the aquatic community (e.g. Peters *et al.*, 2012). Wilby *et al.* (2010) emphasize the need to understand the direct response of environmental flow management across multiple scales on the aquatic communities, particularly when planning for climate change. They suggest that more research is needed to understand the adaptation strategies across different scales and emphasize that it is critical to manage environmental flows by maintaining hydrological variability to sustain ecological integrity (e.g. Poff *et al.*, 1997; Monk *et al.*, 2008). Elucidating and integrating the mechanisms underlying observed patterns at the site (hydraulic), reach and watershed (hydrologic) scales is needed to improve understanding and prediction of system processes. This integration of hydrological, hydraulic and ecological data within management strategies still presents a major research challenge (Newson, 2002). The development of strategic national plans for paired data collection is required for sustainable management, which include co-locating ecological sample sites with gauging stations in addition to using geospatial information to develop site- and watershed-specific data.

21.7 Conclusions

Appropriate models linking mesoscale ecohydraulic observations to macroscale patterns in biota are lacking, due in part to the absence of a rigorous classification system for river habitats with global application. Upscal-

ing biological observations from the reach to watershed scale is also problematic due to a lack of biological and ecohydraulic data with appropriate spatial coverage and which can be explicitly linked to hydrological observations. Armanini *et al.* (2012) give an example for the Fraser River in British Columbia of the type of data attrition which can result when appropriate spatial matching is required: of 1600 unique site observations, 400 were classified as suitable for analysis, of these only 95 could be appropriately matched to an existing set of hydrometric observations. A further obstacle to the development of scalable models is that most studies examine hydraulic influences on community pattern without consideration of confounding variables, and thus cross-scale issues are rarely identified or commented upon. We argue that the use of correlative approaches to develop indices such as LIFE and CEFI has value, and that such indices can be rigorously tested through hypothesis-driven science, both in real, field scenarios and under controlled laboratory conditions (e.g. Armanini *et al.*, 2011). The most urgent need, however, is a sound theoretical framework to link the composition of lotic invertebrate communities to the habitat templet – going beyond physical habitat filters to incorporate or otherwise account for the influence of other natural drivers (such as competition and predation) and which allows the relative contribution of anthropogenic factors to be quantified against a background of natural variability (Peters *et al.*, 2012). In this way, the importance of hydraulic factors in influencing population and community dynamics of lotic invertebrates can be understood and managed within an appropriate spatial and temporal context. The science of ecohydraulics requires the engagement of scientists across a range of disciplines, with a strong focus on solutions to river management problems. This will ensure that key fundamental science questions can be explored and that resource managers can access monitoring and assessment tools that are reliable, robust and scientifically defensible.

References

Archer, D. and Newson, M.D. (2002) The use of indices of flow variability in assessing the hydrological and instream habitat impacts of upland afforestation and drainage. *Journal of Hydrology*, **268**: 244–258.

Armanini, D.G., Horrigan, N., Monk, W.A., Peters, D.L. and Baird, D.J. (2011) Development of a benthic macroinvertebrate flow sensitivity index for Canadian rivers. *River Research and Applications*, **27**: 723–737.

Armanini, D.G., Monk, W.A., Tenenbaum, D.E., Peters, D.L. and Baird, D.J. (2012) Influence of runoff regime type on a macroinvertebrate-based flow index in rivers of British Columbia (Canada). *Ecohydrology*, **5**: 414–423.

Baird, D.J. and Van Den Brink, P.J. (2007) Using biological traits to predict species sensitivity to toxic substances. *Ecotoxicology and Environmental Safety*, **67**: 296–301.

Benda, L., Poff, N.L., Miller, D., Dunne, T., Reeves, G., Pollock, M. and Pess, G. (2004) Network dynamics hypothesis: spatial and temporal organization of physical heterogeneity in rivers. *BioScience*, **54**: 413–427.

Biggs, B.J.F., Nikora, V.I. and Snelder, T.H. (2005) Linking scales of flow variability to lotic ecosystem structure and function. *River Research and Applications*, **21**: 283–298.

Bonada, N., Rieradevall, M. and Prat, N. (2007) Macroinvertebrate community structure and biological traits related to flow permanence in a Mediterranean river network. *Hydrobiologia*, **589**: 91–106.

Brooks, A.J., Haeusler, T., Reinfelds, I. and Williams, S. (2005) Hydraulic microhabitats and the distribution of macroinvertebrate assemblages in riffles. *Freshwater Biology*, **50**: 331–344.

Buchwalter, D.B., Cain, D.J., Martin, C.A., Xie, L., Luoma, S.N. and Garland, Jr, T. (2008) Aquatic insect ecophysiological traits reveal phylogenetically based differences in dissolved cadmium susceptibility. *Proceedings of the National Academy of Sciences*, **105**: 8321–8326.

Buffagni, A., Crosa, G. and Marchetti, R. (1995) Size-related shifts in the physical habitat of two mayfly species (Ephemeroptera). *Freshwater Biology*, **34**: 297–302.

Carlisle, D.M. and Hawkins, C.P. (2008) Land use and the structure of western US stream invertebrate assemblages: predictive models and ecological traits. *Journal of the North American Benthological Society*, **27**: 986–999.

Collier, K.J. (1994) Influence of nymphal size, sex and morphotype on microdistribution of *Deleatidium* (Ephemeroptera: Leptophlebiidae) in a New Zealand river. *Freshwater Biology*, **31**: 35–42.

Collier, K.J., Croker, G.F., Hickey, C.W., Quinn, J.M. and Smith, B.S. (1995) Effects of hydraulic conditions and larval size on the microdistribution of Hydrobiosidae (Trichoptera) in two New Zealand rivers. *New Zealand Journal of Marine and Freshwater Research*, **29**: 439–451.

Culp, J.M., Armanini, D.G., Dunbar, M.J., Orlofske, J.M., Poff, N.L., Pollard, A.I., Yates, A.G. and Hose, G.C. (2011) Incorporating traits in aquatic biomonitoring to enhance causal diagnosis and prediction. *Integrated Environmental Assessment and Management*, **7**: 187–197.

Cummins, K.W. (1964) Factors limiting the microdistribution of larvae of the caddisflies *Pycnopsyche lepida* (Hagen) and *Pycnopsyche guttifer* (Walker) in a Michigan stream (Trichoptera: Limnephilidae). *Ecological Monographs*, **34**: 271–295.

Dolédec, S. and Statzner, B. (2008) Invertebrate traits for the biomonitoring of large European rivers: An assessment of specific types of human impact. *Freshwater Biology*, **53**: 617–634.

Dolédec, S., Statzner, B. and Bournaud, M. (1999) Species traits for future biomonitoring across ecoregions: Patterns along a human-impacted river. *Freshwater Biology*, **42**: 737–758.

Dolédec, S., Lamouroux, N., Fuchs, U. and Mérigoux, S. (2007) Modelling the hydraulic preferences of benthic macroinvertebrates in small European streams. *Freshwater Biology*, **52**: 145–164.

Dunbar, M.J. and Acreman, M.C. (2001) Applied hydroecological science for the twenty-first century. *Workshop HW2: Riverine Ecological Response to Change in Hydrological Regime, Sediment Transport and Nutrient Loading* (July, 1999), IAHS, Birmingham, UK, pp. 1–18.

Dunbar, M.J., Pedersen, M.L., Cadman, D., Extence, C.A., Waddingham, J., Chadd, R.P. and Larsen, S.E. (2010). River discharge and local-scale physical habitat influence macroinvertebrate LIFE scores. *Freshwater Biology*, **55**: 226–242.

Extence, C.A., Balbi, D.M. and Chadd, R.P. (1999) River flow indexing using British benthic macroinvertebrates: A framework for setting hydroecological objectives. *Regulated Rivers: Research and Management*, **15**(6): 543–574.

Extence, C.A., Chadd, R.P., England, J., Dunbar, M.J., Wood, P.J. and Taylor, E.D. (2013) The assessment of fine sediment accumulation in rivers using macroinvertebrate community response. *River Research and Applications*, **29**: 17–55.

Fausch, K., Torgersen, C., Baxter, C. and Li, H. (2002) Landscapes to riverscapes: Bridging the gap between research and conservation of stream fishes. *BioScience*, **52**(6): 483–498.

Friberg, N. (2010) Pressure-response relationships in stream ecology: introduction and synthesis. *Freshwater Biology*, **55**: 1367–1381.

Gore, J.A. and Judy Jr, R.D. (1981) Predictive models of benthic macroinvertebrate density for use in instream flow studies and regulated flow management. *Canadian Journal of Fisheries and Aquatic Science*, **38**: 1363–1370.

Hanquet, D., Legalle, M., Garbage, S. and Céréghino, R. (2004) Ontogenetic microhabitat shifts in stream invertebrates with different biological traits. *Archiv für Hydrobiologie*, **160**: 329–346.

Hart, D.D. and Finelli, C.M. (1999) Physical–biological coupling in streams: The pervasive effects of flow on benthic organisms. *Annual Review of Ecology and Systematics*, **30**: 363–395.

Hilborn, R. (2006) Faith-based fisheries. *Fisheries*, **31**: 554–555.

Hilsenhoff, W.L. (1987) An improved biotic index of organic stream pollution. *The Great Lakes Entomologist*, **20**(1): 31–39.

Hilsenhoff, W.L. (1988) Rapid field assessment of organic pollution with a family-level biotic index. *Journal of the North American Benthological Society*, **7**(1): 65–68.

Horrigan, N. and Baird, D.J. (2008) Trait patterns of aquatic insects across gradients of flow-related factors: a multivariate analysis of Canadian national data. *Canadian Journal of Fisheries and Aquatic Science*, **65**: 670–680.

Johansson, A. (1991) Caddis larvae cases (Trichoptera; Limnephilidae) as anti-predatory devices against brown trout and sculpin. *Hydrobiologia*, **211**: 185–194.

Jowett, I. and Richardson, J. (1990) Microhabitat preferences of benthic invertebrates in a New Zealand river and the development of in-stream flow-habitat models for *Deleatidium* spp. *New Zealand Journal of Marine and Freshwater Research*, **24**: 19–30.

Lamouroux, N., Dolédec, S. and Gayraud, S. (2004) Biological traits of stream macroinvertebrate communities: effects of microhabitat, reach and basin filters. *Journal of the North American Benthological Society*, **23**(3): 449–466.

Lancaster, J. and Belyea, L.R. (2006) Defining the limits to local density: alternative views of abundance–environment relationships. *Freshwater Biology*, **51**: 783–796.

Lancaster, J. and Downes, B.J. (2010a) Linking the hydraulic world of individual organisms to ecological processes: putting ecology into ecohydraulics. *River Research and Applications*, **26**: 385–403.

Lancaster, J. and Downes, B.J. (2010b) Ecohydraulics needs to embrace ecology and sound science and to avoid mathematical artefacts. *River Research and Applications*, **26**: 921–929.

Mackay, R. and Wiggins, G.B. (1979) Ecological diversity in Trichoptera. *Annual Review of Entomology*, **24**: 185–208.

Mérigoux, S. and Dolédec, S. (2004) Hydraulic requirements of stream communities: a case study on invertebrates. *Freshwater Biology*, **49**: 600–613.

Minshall, G.W. and Minshall, J.N. (1977) Microdistribution of benthic invertebrates in a Rocky Mountain (U.S.A.) stream. *Hydrobiologia*, **55**: 231–249.

Möbes-Hansen, B. and Waringer, J.A. (1998) The influence of hydraulic stress on microdistribution patterns of zoobenthos in a sandstone brook (Weidlingbach, Lower Austria). *International Review of Hydrobiology*, **83**: 381–396.

Monk, W.A., Peters, D.L., Curry, R.A. and Baird, D.J. (2011) Quantifying trends in indicator hydroecological variables for regime-based groups of Canadian rivers. *Hydrological Processes*, **25**: 3086–3100.

Monk, W.A., Wood, P.J., Hannah, D.M. and Wilson, D.A. (2008) Macroinvertebrate community response to inter-annual and regional river flow regime dynamics. *River Research and Applications*, **24**: 988–1001.

Monk, W.A., Wood, P.J., Hannah, D.M., Wilson, D.A., Chadd, R.P. and Extence, C.A. (2006) Flow variability and macroinvertebrate community response within riverine systems. *River Research and Applications*, **22**: 595–615.

Morin, A., Harper, P-P. and Peters, R.H. (1986) Microhabitat – preference curves of blackfly larvae (Diptera: Simuliidae): A comparison of three estimation methods. *Canadian Journal of Fisheries and Aquatic Science*, **43**: 1235–1241.

Newson, M.D. (2002) Geomorphological concepts and tools for sustainable river ecosystem management. *Aquatic Conservation: Marine and Freshwater Ecosystems*, **12**: 365–379.

Newson, M.D. and Newson, C.L. (2000) Geomorphology, ecology and river channel habitat: mesoscale approaches to basin-scale challenges. *Progress in Physical Geography*, **24**: 195–217.

Nowell, A.R.M. and Jumars, P.A. (1984) Flow environments of aquatic benthos. *Annual Review of Ecology and Systematics*, **15**: 303–328.

Ogilvie, G.A. and Clifford, H.F. (1986) Life histories, production, and microdistribution of two caddisflies (Trichoptera) in a Rocky Mountain stream. *Canadian Journal of Zoology*, **64**: 2706–2716.

Oldmeadow, D.F., Lancaster, J. and Rice, S.P. (2010) Drift and settlement of stream insects in a complex hydraulic environment. *Freshwater Biology*, **55**: 1020–1035.

Orth, D.J. and Maughan, O.E. (1983) Microhabitat preferences of benthic fauna in a woodland stream. *Hydrobiologia*, **106**: 157–168.

Osborne, L.L. and Herricks, E.E. (1987) Microhabitat characteristics of *Hydropsyche* (Trichoptera: Hydropsychidae) and the importance of body size. *Journal of the North American Benthological Society*, **6**: 115–124.

Parsons, M., Thoms, M.C. and Norris, R.H. (2004) Using hierarchy to select scales of measurement in multiscale studies of stream macroinvertebrate assemblages. *Journal of the North American Benthological Society*, **23**(2): 157–170.

Peters, D.L., Baird, D.J., Monk, W.A. and Armanini, D.G. (2012) Establishing standards and assessment criteria for ecological instream flow needs for agricultural watersheds in Canada. *Journal of Environmental Quality*, **41**: 41–51.

Poff, N.L. (1997) Landscape filters and species traits: Towards mechanistic understanding and prediction in stream ecology. *Journal of the North American Benthological Society*, **16**: 391–409.

Poff, N.L., Olden, J.D., Vieira, N.K.M., Finn, D.S., Simmons, M.P. and Kondratieff, B.C. (2006a) Functional trait niches of North American lotic insects: trait-based ecological applications in light of phylogenetic relationships. *Journal of the North American Benthological Society*, **25**: 730–755.

Poff, N.L., Olden, J.D., Pepin, D.M. and Bledsoe, B.P. (2006b) Placing global stream flow variability in geographic and geomorphic contexts. *River Research and Applications*, **22**: 149–166.

Poff, N.L., Allan, J.D., Bain, M.B., Karr, J.R., Prestegaard, K.L., Richter, B.D., Sparks, R.E. and Stromberg, J.C. (1997) The natural flow regime. *BioScience*, **47**: 769–784.

Poole, G.C. (2002) Fluvial landscape ecology: addressing uniqueness within the river discontinuum. *Freshwater Biology*, **47**: 641–660.

Quinn, J.M. and Hickey, C.W. (1994) Hydraulic parameters and benthic invertebrate distributions in two gravel-bed New Zealand rivers. *Freshwater Biology*, **32**: 489–500.

Rabeni, C.F. and Minshall, G.W. (1977) Factors affecting microdistribution of stream benthic insects. *Oikos*, **29**: 33–43.

Reich, P. and Downes, B.J. (2004) Relating larval distributions to patterns of oviposition: evidence from lotic hydrobiosid caddisflies. *Freshwater Biology*, **49**: 1423–1436.

Rice, S.P., Little, S., Wood, P.J., Moir, H.J. and Vericat, D. (2010) The relative contributions of ecology and hydraulics to ecohydraulics. *River Research and Applications*, 26: 363–366.

Richter, B.D., Baumgartner, J.V., Powell, J. and Braun, D.P. (1996) A method for assessing hydrologic alteration within ecosystems. *Conservation Biology*, 10: 1163–1174.

Sagnes, P., Mérigoux, S. and Péru, N. (2008) Hydraulic habitat use with respect to body size of aquatic insect larvae: Case of six species from a French Mediterranean type stream. *Limnologica*, 38: 23–33.

Southwood, T.R.E. (1977) Habitat, the templet for ecological strategies? *Journal of Animal Ecology*, 46: 337–365.

Statzner, B. (1988) Growth and Reynolds number of lotic macroinvertebrates: A problem for adaptation of shape to drag. *Oikos*, 51: 84–87.

Statzner, B. (2008) How views about flow adaptations of benthic stream invertebrates changed over the last century. *International Review of Hydrobiology*, 93: 593–605.

Statzner, B. and Bêche, L.A. (2010) Can biological invertebrate traits resolve effects of multiple stressors on running water ecosystems? *Freshwater Biology*, 55: 80–119.

Statzner, B. and Muller, R. (1989) Standard hemispheres as indicators of flow characteristics in lotic benthos research. *Freshwater Biology*, 21: 445–459.

Statzner, B., Gore, J.A. and Resh, V.H. (1988) Hydraulic stream ecology: Observed patterns and potential applications. *Journal of the North American Benthological Society*, 7: 307–360.

Stewardson, M.J. and Gippel, C.J. (2003) Incorporating flow variability into environmental flow regimes using the flow events method. *River Research and Applications*, 19: 459–472.

Tachet, H., Richoux, P., Bournaud, M. and Usseglio-Polatera, P. (2000) *Invertebres d'Eau Douce: Systematique, Biologie, Ecologie (Freshwater Invertebrates: Taxonomy, Biology, Ecology)*, CNRS Editions, Paris.

Teague, S.A, Knight, A.W. and Teague, B.N. (1985) Stream microhabitat selectivity, resource partitioning, and niche shifts in grazing caddisfly larvae. *Hydrobiologia*, 128: 3–12.

Tharme, R.E. (2003) A global perspective on environmental flow assessment: emerging trends in the development and application of environmental flow methodologies for rivers. *River Research and Applications*, 19: 397–441.

Tullos, D.D., Penrose, D.L., Jennings, G.D. and Cope, W.G. (2009) Analysis of functional traits in reconfigured channels: implications for the bioassessment and disturbance of river restoration. *Journal of the North American Benthological Society*, 28: 80–92.

Ulfstrand, S. (1967) Microdistribution of benthic species (Ephemeroptera, Plecoptera, Trichoptera, Diptera: Simuliidae) in Lapland streams. *Oikos*, 18: 293–310.

Urbanič, G., Toman, M.J. and Krušnik, C. (2005) Microhabitat type selection of caddisfly larvae (Insecta: Trichoptera) in a shallow lowland stream. *Hydrobiologia*, 541: 1–12.

Usseglio-Polatera, P., Bournaud, M., Richoux, P. and Tachet, H. (2000) Biological and ecological traits of benthic freshwater macroinvertebrates: relationships and definition of groups with similar traits. *Freshwater Biology*, 43: 175–205.

Verberk, W.C.E.P., Siepel, H. and Esselink, H. (2008) Life-history strategies in freshwater macroinvertebrates. *Freshwater Biology*, 53: 1722–1738.

Vieira, N.K.M., Poff, N.L., Carlisle, D.M., MoultonII, S.R., Koski, M. and Kondratieff, B.C. (2006) *A Database of Lotic Invertebrate Traits for North America*. Report manuscript, Reston, Virginia.

Wallace, I.D., Wallace, B. and Philipson, G.N. (2003) *Keys to the Case-Bearing Caddis Larvae of Britain and Ireland*. Scientific Publication No. 61, Freshwater Biological Association, Ambleside, Cumbria, UK.

Wetmore, S.H., Mackay, R.J. and Newbury, R.W. (1990) Characterization of the hydraulic habitat of *Brachycentrus occidentalis*, a filter feeding caddisfly. *Journal of the North American Benthological Society*, 9: 157–169.

Wilby, R.L., Orr, H., Watts, G., Battarbee, R.W., Berry, P.M., Chadd, R., Dugdale, S.J., Dunbar, M.J., Elliott, J.A., Extence, C., Hannah, D.M., Holmes, N., Johnson, A.C., Knights, B., Milner, N.J., Ormerod, S.J., Solomon, D., Timlett, R., Whitehead, P.J. and Wood, P.J. (2010) Evidence needed to manage freshwater ecosystems in a changing climate: turning adaptation principles into practice. *The Science of the Total Environment*, 408: 4150–4164.

Wilcox, A.C., Peckarsky, B.L., Taylor, B.W. and Encalada, A.C. (2008) Hydraulic and geomorphic effects on mayfly drift in high-gradient streams at moderate discharges. *Ecohydrology*, 1: 176–186.

Williams, D.D., Tavares, A.F. and Bryant, E. (1987) Respiratory device or camouflage?: A case for the caddisfly. *Oikos*, 50(1): 42–52.

Wood, P.J., Hannah, D.M., Agnew, M.D. and Petts, G.E. (2001) Scales of hydroecological variability within a groundwater-dominated stream. *Regulated Rivers: Research and Management*, 17: 347–367.

22 Estuarine Wetland Ecohydraulics and Migratory Shorebird Habitat Restoration

José F. Rodríguez and Alice Howe

School of Engineering, The University of Newcastle, Callaghan, NSW 2308, Australia

22.1 Introduction

In the Hunter estuary of NSW, as in most of southeastern Australia, estuarine wetlands typically comprise intertidal mudflats at the seaward margin, followed farther inland first by mangrove forest and then by saltmarsh plains. Mangrove forest and saltmarsh are important for estuarine fish (Mazumder *et al.*, 2005) and provide unique habitats for many terrestrial invertebrates. In particular, saltmarsh is used by insectivorous bats as feeding habitat and by shorebirds as roosting habitat, clearly preferring it to mangrove forest (Saintilan and Rogers, 2006).

The Hunter estuary is the most important shorebird site in NSW and is recognised as a wetland of international importance under the Ramsar Convention. It is part of the East Asian–Australasian Flyway, the pathway used by around four million shorebirds that migrate to Australia for the austral summer from their breeding grounds in the Arctic Circle (Smith, 1991; Watkins, 1993). Shorebirds are drawn to the estuary for the extensive foraging habitat on the exposed intertidal mudflats of Fullerton Cove (see Figure 22.1), but they also require high-quality roosting habitat to accumulate the enormous fat reserves required to sustain them during migration (Department of the Environment and Heritage, 2005). Migratory species are particularly vulnerable to habitat degradation as they have a tendency to return to the same sites year after year, and the availability of roost habitat has been identified as a critical factor for shorebird utilisation of the estuary (Kingsford *et al.*, 1998).

Saltmarsh of the Hunter estuary, which comprises a mosaic of vegetated platforms and shallow tidal pools, provides important roosting and supplementary foraging habitat for migratory shorebirds. Shorebirds feed by probing with their beaks for benthic invertebrates in soft substrates along the shallow or exposed margins of Fullerton Cove during low tide; they then congregate in nearby open areas to roost when their feeding grounds are submerged by the tide (Geering and Winning, 1993; Watkins, 1993). These high-tide roosts consist of artificial structures, beaches, saltmarsh and shallow tidal pools (Lawler, 1996). Most shorebirds select roost sites with low predation risks, low disturbance rates and a suitable microclimate (Lawler, 1996; Luis *et al.*, 2001; Rogers *et al.*, 2006). Shorebird use of shallow water roosts is limited by their leg length, as they prefer to 'wade' rather than 'swim' (Lane, 1987). Consequently, small birds like the Red-necked Stint (*Calidris ruficollis*) prefer wet mud just above the water edge (Dann, 1987) while the large Eastern Curlew (*Numenius madagascariensis*) utilises substrates up to an inundation depth of around 0.1 m. The availability of this shallow tidal pool/intertidal flat habitat is, in general, a function of water level, topography, water manipulation, local rainfall, soil type and wind action (Skagen and Knopf, 1994). The low topographic relief of saltmarsh ensures the availability of these shallow areas

Ecohydraulics: An Integrated Approach, First Edition. Edited by Ian Maddock, Atle Harby, Paul Kemp and Paul Wood.
© 2013 John Wiley & Sons, Ltd. Published 2013 by John Wiley & Sons, Ltd.

Table 22.1 Most common migratory shorebirds in the Hunter estuary.

Common name	Scientific name	Breeding area[1]	Habitat preference in Australia[1]	Population trend 1982–2007[2]
Eastern Curlew	*Numenius madagascariensis*	Russia, NE China	Intertidal coastal mudflats, coastal lagoons, sandy spits	Decline[3]
Common Greenshank	*Tringa nebularia*	Arctic Circle, Siberia	Wide variety of inland and sheltered coastal wetlands – mudflats, saltmarshes, mangroves	Decline[3]
Marsh Sandpiper	*Tringa stagnatilis*	Eastern Europe to eastern Siberia	Coastal – permanent or ephemeral wetlands of varying degrees of salinity, commonly inland	Increase[3]
Bar-tailed Godwit	*Limosa lapponica*	Northern Russia, Scandinavia, NW Alaska	Mainly coastal, usually sheltered bays, estuaries and lagoons with large intertidal mudflats or sandflats	Decline[3]
Black-tailed Godwit	*Limosa limosa*	Iceland, North Atlantic, Europe, Russia and China	Mainly coastal, usually sheltered bays, estuaries and lagoons with large intertidal mudflats or sandflats	56% decline
Curlew Sandpiper	*Calidris ferruginea*	Arctic tundra	Intertidal mudflats of sheltered coastal areas, coastal lakes, estuaries, bays – occasionally inland wetlands	83% decline
Grey-tailed Tattler	*Heteroscelus brevipes*	Siberia	Sheltered coasts with reef or rock platforms or intertidal mudflats	Decline[3]
Lesser Sand Plover	*Charadrius mongolus*	Central and NE Asia	Usually coastal, estuaries and littoral environments – sandflats and mudflats	99% decline
Pacific Golden Plover	*Pluvialis fulva*	N Siberia, Alaska	Mainly coastal, beaches, mudflats and sandflats and other open areas such as recreational playing fields	67% decline
Terek Sandpiper	*Xenus cinereus*	Russia, eastern Europe	Intertidal coastal, – mainly saline mudflats, lagoons and sandbanks	Decline[3]

[1] from Marchant and Higgins (1993) and Higgins and Davies (1996)
[2] from Spencer (2010)
[3] trend not statistically significant due to high variability of data

for a wide range of tidal conditions, and it also provides a relatively safe environment where predators can be easily detected by birds.

As elsewhere around the world, the abundance and diversity of shorebirds and the distribution of their habitat are in decline in the Hunter estuary (Kingsford *et al.*, 1998) (Table 22.1). Evidence gathered across Australia has revealed a consistent trend of mangrove encroachment into saltmarsh (Williams and Watford, 1997; Saintilan *et al.*, 2009a, 2009b), which has been attributed to changed precipitation patterns, agricultural practices, increased sedimentation and nutrient levels, altered tidal regimes, sea level rise and land subsidence (Saintilan and Williams,

1999). Mangrove encroachment can reduce habitat diversity and overall wetland productivity, including reduced utilisation by shorebirds (Saintilan and Rogers, 2002). As a result of mangrove encroachment and a range of other threatening processes, including climate change and urbanisation, saltmarsh was declared an endangered ecological community in NSW in 2004.

In the present study, we focused on saltmarsh restoration efforts on Area E of Kooragang Island, a wetland that provides important supplementary shorebird roosting habitat for the Hunter estuary migratory shorebird population and is undergoing mangrove encroachment. The present research is aimed at establishing relationships

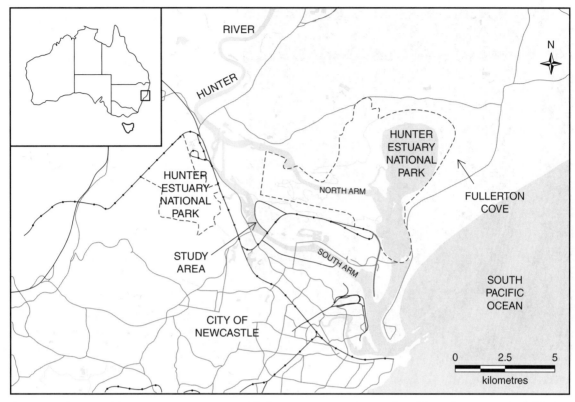

Figure 22.1 Location of the Hunter estuary, NSW, Australia (3285105200S, 15184201500E), and the study area, Area E. Modified from Howe *et al.* (2010).

between roost habitat, estuarine vegetation distribution and tidal hydrodynamics in order to guide restoration efforts. The methodology consisted of a combination of field data collection, laboratory experiments, statistical analysis and numerical hydrodynamic simulation.

22.2 Area E of Kooragang Island

The site for this research (Figures 22.1 and 22.2) is a rehabilitated wetland in the Hunter estuary, Australia, known as Area E of Kooragang Island. The site is adjacent to the Hunter Wetlands National Park and it constitutes an important shorebird roost site (Kingsford *et al.*, 1998). It consists of a 124 ha tidal sub-catchment with two defined inlets to the south arm of the Hunter River; one of which is a 0.45 m culvert (on Wader Creek) and the other an 8.5 m-wide tidal channel (on Fish Fry Creek). The area is bounded to the east by the Kooragang Island Mainline, to the north by a water supply pipeline and to the south and west by a levee on the north bank of

the Hunter River south arm. Internal flow is hydraulically complex, with a number of culverts and roads that divide the site into four major compartments: Fish Fry Creek, Wader Creek, Wader Pond and Swan Pond (Figure 22.2). Estuarine habitats include mangrove forest, saltmarsh, mudflat, tidal pools and tidal creeks. Fringing upland habitats include pasture and freshwater/brackish wetlands.

Initial rehabilitation activity involved the removal of two 0.5 m-diameter culverts on Fish Fry Creek in 1995 to increase tidal flushing and improve habitat for fisheries and shorebirds. These culverts were originally installed around 1950 to drain estuarine wetlands for agricultural activity, principally cattle grazing. The effect of culvert removal consisted primarily of a dramatic change in tidal conditions (i.e. tidal range and hydroperiod) of the area immediately upstream, but other areas of the wetland were less affected because of the presence of internal controls (culverts). Tidal channels in the Fish Fry Creek compartment were enlarged through erosion, and mangrove colonised and displaced previous marsh areas.

Figure 22.2   Area E showing inlets and areas compartmentalised by roads and culverts. A colour version of this image appears in the colour plate section of the book.

As a result, shorebird species were gradually excluded from the Fish Fry Creek compartment over the period 1997–2007.

Increased tidal flows also promoted the landward migration of saltmarsh communities, but this was limited by existing infrastructure. By 2007, there was already a 17% net loss of saltmarsh, tidal pools and other habitats used by shorebirds (Howe *et al.*, 2010).

22.3 Ecohydraulic and ecogeomorphic characterisation

Tidal wetlands evolve through dynamic interactions between physical and biological landforming processes (see review by Saco and Rodríguez, 2012). In this work the focus was on understanding the contribution of physical or abiotic properties to the distribution of estuarine wetlands, with a view to enhancing the conservation of shorebird roost habitat, particularly saltmarsh, which is endangered in NSW. The role of biotic factors, although important in defining the landward limit of estuarine wetlands and the zonation of individual species within the wetland mosaic, is secondary to the influence of tidal hydrodynam-

ics and topography in determination of broad community boundaries in these wetlands.

Transgression of mangrove into saltmarsh in Australia has been identified as a key threat to both saltmarsh and the migratory shorebird populations it supports. Saintilan and Williams (1999) suggest a range of mechanisms for mangrove transgression, including precipitation patterns, estuary agricultural practices, catchment characteristics, tidal regimes and land subsidence. They also note that, as there was no ubiquitous trend over the range of geomorphic settings analysed, the relative importance of contributing factors should be determined on an estuary-specific basis. Data presented elsewhere (Howe *et al.*, 2009; 2010) indicate that in the Hunter estuary, changes to estuary water level, resulting from eustatic sea level rise, appear to be the major factor driving landward migration of mangrove.

An extensive set of field and remote-sensing data was used for the ecohydraulic and ecogeomorphic characterisation (Howe, 2008). These include measurement of topography and soil and water properties as well as vegetation distribution of the dominant wetland vegetation species grey mangrove (*Avicennia marina*), beaded samphire (*Sarcocornia quinqueflora*) and saltwater couch

Table 22.2 Summary of field dataset collected and used in the study.

Description	Method
General landform of the study area, including spatial coordinates of critical hydraulic controls	Real-time kinematic (RTK) global positioning system (GPS)
Mapping of estuarine habitat distribution	RTK GPS and ground-truthed high-resolution aerial photography
Vegetation morphological characteristics (stem height, stem diameter, stem density and % foliage cover)	Nested vegetation quadrats
Soil surface elevation change	Surface elevation tables (SETs) and feldspar marker horizons
Soil properties (bulk density, particle size, moisture content, total organic carbon, pH and electrical conductivity)	Standard laboratory analysis on soil cores collected in the immediate vicinity of vegetation quadrats
Water properties (pH, electrical conductivity (EC), turbidity, dissolved oxygen (DO), specific gravity and temperature)	Horiba multiprobe W-22XD series field water quality analyser and gravimetric analysis of water samples
Velocity distribution at transects	2D and 3D Sontek acoustic Doppler velocimeters (ADVs)
Water surface slope at velocity transects	Sokkia SDL30 automatic level sensor and custom-made floating staves
Surface water level and temperature	Solinst MLT 3001 pressure transducers
Groundwater level	Solinst MLT 3001 pressure transducers

(*Sporobolus virginicus*). Table 22.2 summarises the experimental dataset.

22.3.1 Water level, surface elevation and vegetation dynamics

Water levels in the estuary increased during the study period, and the high-resolution soil surface elevation data from SETs (after the method of Cahoon *et al.*, 2002) were used to analyse the effects on wetland elevation from 2005 to 2008 (Howe *et al.*, 2009) (Figure 22.3). Soil surface elevation increase at mangrove and saltmarsh sites was between 2 and 2.5 mm y^{-1}, which is within the range reported for sites in southeastern Australia (Saintilan and

Rogers, 2006) and northern hemisphere studies (Cahoon *et al.*, 2002). However, the rate of local mean annual water level rise over the same period (4.70 mm y^{-1}) exceeded the increase in surface elevation, promoting a landward migration of the vegetation communities to higher grounds in order to keep up with the new levels in the estuary. A strong correlation was found between surface elevation and mean water level in the south arm of the Hunter River, which indicates that this period of high estuary water level may have driven a rapid landward transgression of mangrove into saltmarsh.

Even though Area E has been affected by recent changes in tidal conditions, other sites in the estuary show a similar correlation with water levels (Howe *et al.*, 2009).

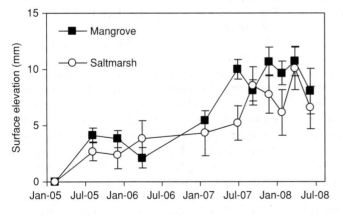
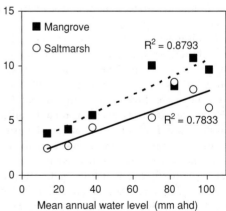

Figure 22.3 Surface elevation change and correlation with water level. Modified from Howe *et al.*, Copyright (2009), with permission from Elsevier.

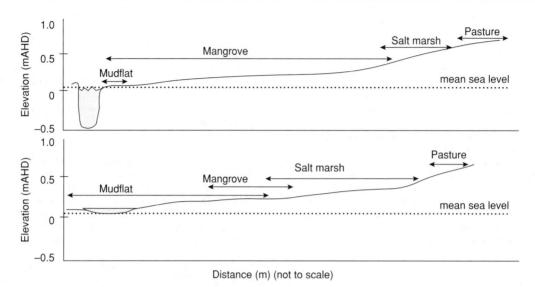

Figure 22.4 Typical hypsometric curves for (top) unattenuated and (bottom) attenuated compartments.

22.3.2 Hydraulic drivers for vegetation distribution

Results presented by Howe *et al.* (2010) identified a clear zonation of estuarine habitats in the tidal frame at all study areas. While the general zonation applied to all sites, the position in the tidal frame varied, with habitats at the attenuated compartments (Wader Pond and Swan Pond) occurring lowest, as shown by the hypsometric curves of Figure 22.4. mAHD in the figure corresponds to metres above the Australian Height Datum, approximately equal to mean sea level. In particular, hydraulic controls (culverts) lowered the position of saltmarsh in the tidal frame, which has important implications for the restoration strategy adopted in later sections of the chapter.

Multivariate statistical analysis was carried out using the software Primer v5.2.9. First, non-metric multidimensional scaling (nMDS) was applied to a range of abiotic variables in order to identify which of these variables were important drivers of estuarine habitat distribution (Howe *et al.*, 2010). Habitats could not be differentiated on the basis of soil pH, EC, bulk density, particle size, moisture content or organic carbon content. In contrast, hydraulic variables (i.e. hydroperiod, tidal range and elevation) were effective at identifying dissimilarity among a subset of sites. nMDS is an ordination tool along two axes based on distance metrics that can serve as a visual representation of dissimilarities among groups and can help identify groups without prescribing the source of

the dissimilarity. The statistical test ANOSIM was used to group data based on nMDS results to establish statistical significance and thresholds for habitat differentiation; the results are presented in Table 22.3.

22.3.3 Vegetative resistance

The main effect of vegetation on water flow consists of hydraulic resistance. Hydraulic resistance studies both in the field (Howe and Rodríguez, 2006) and in the laboratory (Howe and Rodríguez, 2007) investigated the bed and/or vegetative resistance components of mangrove pneumatophores, saltmarsh, benthic algal matting and mudflat substrates at Area E. These results are applicable to other wetlands with comparable flow and substrate characteristics.

The overall D'arcy–Weisbach resistance coefficient, f, was separated into bed f_b and vegetation f_d components and determined via the momentum equation following Nepf (1999) and Musleh and Cruise (2006). For emergent vegetation, an additional resistance term, f_s (Nepf and Vivoni, 2000; Baptist *et al.*, 2007) was considered. f_b was computed using Yen's (1991) extension of the Churchill–Barr formula to wide open channels for turbulent flow, combined with the Blasius formula for transitional flow and a laminar flow relation.

Unvegetated substrate bed resistance f_b had mean values of 0.032 for mudflat and 0.028 for benthic algal matting. These bed resistance values were consistent with those reported in the literature for unvegetated tidal

Table 22.3 Thresholds of abiotic variables Elevation, R_T (spring tidal range) and H (spring tide dimensionless hydroperiod) for each habitat type in unattenuated (Fish Fry Creek) and attenuated (Swan Pond, Wader Pond) areas of the study area in the Hunter estuary in 2004. Wader Creek presented intermediate characteristics. From Howe *et al.* (2010).

Habitat type	Unattenuated			Attenuated		
	Elevation (mAHD)	R_T (m)	H^1	Elevation (mAHD)	R_T (m)	H^1
Mangrove	< 0.40	> 0.43	< 0.32	0.34–0.42	0.29–0.30	< 0.45
Tidal pool/ Mudflat	0.3–0.45	0.38–0.53	0.09–0.17	0.22–0.45	0.27–0.5	0.9–1.0
Saltmarsh	> 0.40	< 0.43	< 0.1	> 0.42	< 0.30	< 1.0

^{1}Calculated at the mean pneumatophore mid-point for mangroves.

substrates, e.g. 0.02–1.1 reported by French and Stoddart (1992); 0.07 reported by Khatibi *et al.* (2006); and 0.03–0.25 reported by Knight (1981). Vegetative resistance was strongly dependent on Reynolds number and, to a lesser extent, on water friction slope and vegetation morphology (Figure 22.5). The range of vegetative resistance values for each substrate is shown in Table 22.4.

Vegetative resistance was at least an order of magnitude greater than bed resistance under all emergent flow conditions tested (Reynolds number range 2300 to 49 000) and under most submerged flow conditions. At high Reynolds numbers (> 80 000), the bed and vegetative resistance components for submerged vegetation were of the same order of magnitude and linear superposition of the resistance terms was warranted. Where stem density was high (> 1%) and flow conditions were emergent, the bed resistance component could be neglected.

22.3.4 Implications for wetland rehabilitation

The field studies support the findings of previous authors, e.g. Clarke and Hannon (1967), Mitsch and Gosselink (2000) and Boumans *et al.* (2002), that estuarine wetland zonation is driven by tidal flows. Mangroves are closer to the tidal source as they require frequent emersion and they outcompete saltmarsh because the length of their pneumatophores allows them to withstand larger tidal ranges. Saltmarsh occupies higher grounds due to its salt tolerance, by which it outcompetes mangroves in that zone. However, despite the well-documented connection between tidal flows and estuarine habitat distribution, there have been few previous published studies on the depth, duration and frequency of tidal inundation in mangrove and saltmarsh communities (Adam, 1990; Bockelmann *et al.*, 2002; Lewis, 2005). Uncertainty about

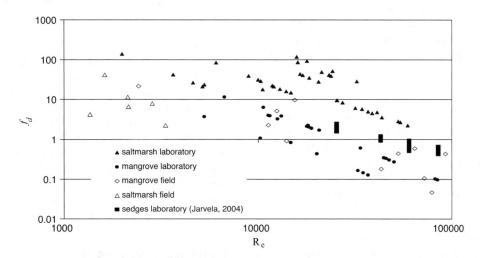

Figure 22.5 Relationship between f_d and R_e for vegetated saltmarsh and mangrove (pneumatophores) habitats compared with Järvelä's (2004) laboratory data.

Table 22.4 Vegetative resistance values for wetland substrates with various friction slopes.

	Mangrove pneumatophores		Saltmarsh	
	S_f (%)	f_d	S_f (%)	f_d
Emergent vegetation	0.014	3.3–5.2	0.093	24–163
	0.027	0.13–0.43	0.47	31–89
	0.076	3.2–11.5	2.9	41–89
	0.11	1.7–6.4	3.3	30–123
Submerged vegetation	0.014	0.17–0.80	0.073	0.48–2.1
			0.13	0.20–1.1
			0.14	0.13–0.55
			0.17	0.07–0.46

how factors such as these are related to desired management outcomes has reduced the effectiveness of wetland rehabilitation effort.

There is a general view in the ecological literature that if tidal flows are restored, vegetation, soil and animals will develop naturally (Laegdsgaard, 2006). While many successful rehabilitation projects have adopted this philosophy (Turner and Lewis, 1997; Warren *et al.*, 2002), these projects have been located in northern hemisphere wetlands where a rainfall surplus typically exists. Failure to reinstate tidal flows in these climatic regions favours development of freshwater wetlands over estuarine wetlands.

In the Hunter estuary, however, as elsewhere in southeast Australia, simply returning natural tidal flows to degraded coastal wetlands is unlikely to achieve a balance between the distribution of mangrove and saltmarsh communities. This is partly due to the extent of urbanisation of the coastline, where the built environment restricts landward migration of estuarine communities in response to increased tidal inundation, leading to a squeezing of saltmarsh between mangrove transgression and urban infrastructure.

The previous sections identified that hydraulic change, in terms of increased water levels, results in a rapid mangrove transgression into saltmarsh in the Hunter estuary. They have also quantified the hydraulic preference of saltmarsh and mangrove in terms of elevation range, tidal range and hydroperiod in different parts of the wetland (see Table 22.3) and have determined their resistance characteristics. Table 22.3 indicates that tidal range may be the most important abiotic variable for saltmarsh (can occur under different hydroperiod and elevation values), while hydroperiod is more important for mangrove (can occur under different tidal range and elevation values). This

is consistent with the findings of Montalto and Steenhuis (2004), who found a positive correlation between *Spartina alterniflora* and tidal range in New York/New Jersey tidal marshes, and Hovenden *et al.* (1995) and Lewis (2005), who all reported mangrove occurring at hydroperiods of 0.30–0.35. Results from the present study indicate that saltmarsh occurs at tidal ranges of less than 0.30 m and the mean elevation of mangrove pneumatophores occurs at hydroperiods of less than 0.30 (Table 22.3).

22.4 Modifying vegetation distribution by hydraulic manipulation

One of the primary objectives of this research was to develop management recommendations regarding the hydraulic conditions required to develop and maintain shorebird habitat at Area E. The previous section showed that this can be achieved by restricting tidal flows sufficiently to maintain a maximum tidal range of less than 0.3 m (to promote saltmarsh) and a hydroperiod of more than 0.3 (to exclude mangrove). Owing to the impracticality of testing potential management scenarios in the field, these scenarios were simulated using the proprietary hydrodynamic package, RMA-2 v7.5a developed by Resource Modelling Associates. The hydrodynamic conditions resulting from each of the scenarios modelled were then used to predict vegetation trajectories and associated shorebird roost habitat, based on the threshold criteria for each habitat type discussed above.

22.4.1 Numerical model

RMA-2 is a depth-averaged, 2D finite element model developed for implicit solution of shallow-water flow

Figure 22.6 RMA finite element mesh showing location of culverts incorporated as 1D elements (circles) and location of representative elements in each compartment (triangles) used for calibration and model output comparisons.

equations in either steady or unsteady systems. An advantage of using a finite element model such as RMA-2, as opposed to a finite difference or finite volume model, is that mesh element size may be irregular, with a high degree of detail in critical areas of the computational domain, less detail in non-critical areas and lower computational effort for a comparable level of accuracy (Finnie *et al.*, 1999).

RMA-2 is appropriate for subcritical flows with a free surface and hydrostatic pressure distribution (King, 2002), such as the flow in Area E. The model solves the depth-averaged, Reynolds-averaged Navier–Stokes equations using Boussinesq's eddy viscosity (King, 2002). Dependent variables within the model are velocity and water depth. Wetting and drying of elements is permitted. Friction coefficients are entered in either the Chézy or Manning formulations. Although both of these formulations apply strictly to uniform flow fields, Manning's equation has been widely adopted in the engineering literature for shallow water flows in tidal, river, floodplain and lake environments (Anderson and Bates, 1994; Shrestha, 1996; Crowder and Diplas, 2000; Cea *et al.*, 2006). Wind stresses may be input in one of six available formulations, but were not applied in the present study. This was because the focus of the study was on identifying inundation depths rather than suspended sediment transport. Boundary conditions may be applied as water surface ele-

vation, flow or stage–discharge relationships (King, 2002), and for the present study were applied as a water surface elevation time-series at each of the inlets to Area E.

22.4.2 Model setup

The RMA-2 model was applied to a finite element mesh of the Area E sub-catchment. Considerable effort was employed in construction of the model mesh, as it is reported to be the single most important factor in generation of accurate model predictions (Crowder and Diplas, 2000). The mesh (Figure 22.6) was developed using the associated RMAGEN graphics module from a very detailed DTM. 12 843 RTK GPS survey points were collected in the study area, at a sample density of about 100 points per hectare, with a vertical and horizontal precision of ± 20 mm and ± 10 mm, respectively. The mesh was constructed with rectangular elements where flow was concentrated in tidal channels, to ensure that flow was parallel to the element axis, and triangular elements for areas of ponded or overland flow. The initial mesh was optimised by reducing grid points where there was no significant change in slope (to improve computational efficiency) and by adding grid points where rapid changes occurred (to improve model stability). Reductions in grid size that did not contribute to a better description of the topography did not affect the results; they only slowed

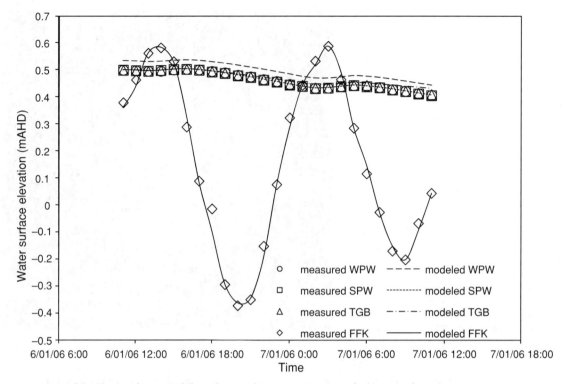

Figure 22.7 Model calibration for neap tide boundary conditions. See Figure 22.6 for location of stations.

down the computations. The optimised mesh had a total of 23 183 nodes and 10 412 elements, with typical size of rectangular channel flow elements between 5 m² and 20 m² (1 to 2 m wide and 5 to 10 m long).

The six major culverts (C4, C6, EC2, EC3, C11 and C14 in Figure 22.6) were included as 1D elements, as model convergence could not be achieved when these culverts were included as 2D elements. For simplicity, minor culverts were modelled as open channels with the same cross-sectional area. Hourly tidal water level was applied as a boundary condition at each of the inlets to Area E (Fish Fry Creek and Wader Creek). A representative 25-hour spring tide (6:00 am 1/1/6 to 6:00 am 2/2/6) and neap tide (11:00 pm 6/1/6 to 11:00 am 7/1/6) dataset was applied, based on measured data at the inlet to Fish Fry Creek. There was no rainfall during these two periods. The initial time-step for all simulations was set at 0.2 hours, with automatic time-step halving to achieve model convergence. Convergence limits were set at 0.001 m s⁻¹ for velocity in the x and y directions, and 0.0005 m for depth.

The model was calibrated by comparing neap tide model output water levels at locations in the wetland with measured water levels at the same locations. Figure 22.7 shows the selected locations including both inlets FFCk (Fish Fry Creek, elevation −0.97 mAHD) and TGB (Wader Creek, elevation 0.3 mAHD), and the wetlands WPW (Wader Pond, elevation 0.20 mAHD) and SPW (Swan Pond, elevation 0.16 mAHD). The model was then verified using spring tide conditions. Calibration and verification focused on matching water surface elevations rather than velocities, as the focus of the modelling exercise was to identify inundation depths and durations appropriate for differentiation of habitat type. The analysis presented in early sections identified tidal range, elevation and hydroperiod as key drivers of vegetation distribution in estuarine wetlands. Calibration was within the error of topographic (± 20 mm) and water level (± 5 mm) data (Figure 22.7). Verification was less precise, particularly in the attenuated compartments. This was partly because the attenuated ponds deeper in the wetland were driven by fortnightly spring–neap cycles, rather than the semi-diurnal tidal cycle used for the simulations. Also, the model did not include resistance of filamentous algae (*E. intestinalis*) observed within tidal pools in the attenuated compartments. Particularly in the attenuated Wader

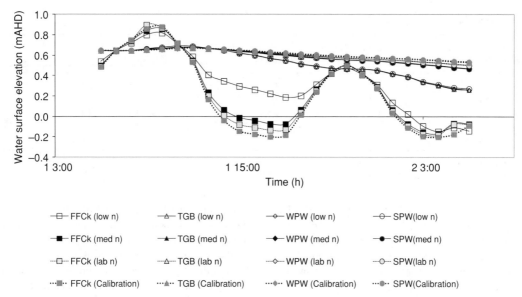

Figure 22.8 Model sensitivity to *n* values, as presented in Table 22.5.

Pond and Swan Pond compartments, the uncertainties in ground and water elevations can constitute a considerable portion of the total depth (around 10%) during low tide and model results in this area must be treated with caution. It must be pointed out, however, that rehabilitation was aimed at altering flow conditions in the unattenuated compartment, where depths were larger and calibration was more robust.

Substrate resistance type was defined using seven classes, which were generated by overlay of vegetation mapping created in MapInfo v7.5. Wetland water level was sensitive to Manning's *n*, particularly in the unattenuated Fish Fry Creek compartment, as demonstrated by sensitivity analysis (Figure 22.8). Even though calibration did not focus on water velocities, it was verified that they were generally consistent with field data, with depth-averaged velocities in the attenuated compartments of the order of 0.05 m s^{-1} and in the Fish Fry Creek channel up to 1 m s^{-1}.

Due to the RMA data requirements, the resistance coefficients *f* had to be converted to Manning's *n*. This was done using the relationship $f = 8gn^2R^{-1/3}$ with average flow depths (hydraulic radius $R \cong h$) obtained from the laboratory experiments.

The laboratory *n* value determined for saltmarsh was adopted for the model; however, calibration required an *n* value for mangrove higher than the value determined in the laboratory (Figure 22.8 and Table 22.5). A higher *n* value for mangrove was warranted, as laboratory analy-

sis included only mangrove pneumatophores at a density of 120 m^{-2}. While this density reflected the mean pneumatophore spacing recorded in the field, densities in the Fish Fry Creek compartment were up to three times higher than the mean value and pneumatophores were often covered in a dense layer of the red algae *Holothuria edulis*. In addition, mangrove stems added an extra roughness element, which was not included in the laboratory study for practical reasons. On this basis, the higher *n* value for mangrove adopted in the model is considered physically based. For the non-estuarine substrate classes not studied in the laboratory, typical Manning's *n* values were adopted.

The *n* values adopted are in agreement with values reported in the literature (Wolanski *et al.*, 1980; Knight, 1981; Kjerfve *et al.*, 1991; French and Stoddart, 1992; Mazda *et al.*, 1997; Liu *et al.*, 2003; Marani *et al.*, 2003; Khatibi *et al.*, 2006; D'Alpaos *et al.*, 2007) and do not depend on water depth. Although flow resistance in the laboratory experiments was a function of water surface slope and depth, average conditions were selected and a constant value for each substrate was assigned in the model for simplicity (see Figure 22.9 for spatial distribution of substrates).

22.4.3 Results

Comparison of vegetation distribution before and after removal of culverts at the mouth of Fish Fry Creek in 1995 indicated that lowering the bed level of the inlet

Table 22.5 Main substrates' resistance values.

Substrate type	D'Arcy–Weisbach f^1	Flow depth $(h)^2$	Lab n	Low n	Medium n	Adopted n
Saltmarsh	50	0.25	0.633	0.05	0.335	0.633
Mudflat/benthic algal matting	0.05	0.5	0.022	0.022	0.022	0.022
Mangrove	10	1.0	0.357	0.05	0.350	0.650
Pasture	—	—	—	0.015	0.015	0.025
Freshwater marsh	—	—	—	0.025	0.025	0.040

[1] determined from laboratory and field resistance data
[2] determined from field pressure transducer data

from approximately −0.4 mAHD to −0.987 mAHD led to mangrove transgression in the Fish Fry Creek compartment at the expense of the saltmarsh and shallow tidal pool habitats favoured by shorebirds. It was also noted that scour channels forming between the Fish Fry Creek and Wader Creek compartments appeared to promote mangrove encroachment into the attenuated compartments, which was predicted to substantially reduce the remaining shorebird habitat at Area E. The approximate bed elevation of these scour channels was 0.6 mAHD.

The effect of inlet configuration and scour channel development on vegetation trajectories and shorebird habitat availability was tested by identifying inunda-

tion depths under a range of hydrodynamic scenarios (Table 22.6). The simulated inlet configurations included raising the bed levels of both inlets (runs a–c) and raising the bed level of Fish Fry Creek while blocking flow through the Wader Creek inlet (runs e–f). All simulated Fish Fry Creek inlet configurations maintained the current channel width to preserve both fish passage and wetland water quality. The effect of further erosion of the channels between Fish Fry Creek and Wader Creek compartments was modelled (runs g–h) by lowering the elevation of three major scour channels between Wader Creek and Fish Fry Creek (currently at 0.6 mAHD) to 0.50 mAHD (run g) and 0.30 mAHD (run h), respectively.

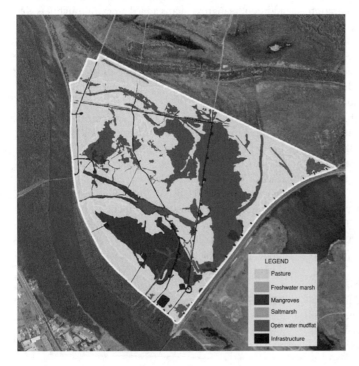

LEGEND
Pasture
Freshwater marsh
Mangroves
Saltmarsh
Open water mudflat
Infrastructure

Figure 22.9 Distribution of substrates used in the model. Resistance coefficients for each substrate are shown in Table 22.5. A colour version of this image appears in the colour plate section of the book.

Table 22.6 Model inlet configuration simulation scenarios.

Model run	Description
0	Current inlet configuration with Wader Creek inlet at −0.22 mAHD and Fish Fry Creek inlet at −1.04 mAHD
a	Elevation of both inlets at −0.22 mAHD
b	Elevation of both inlets at 0.00 mAHD
c	Elevation of both inlets at 0.20 mAHD
e	Fish Fry Creek inlet at −1.04 mAHD and no flow through Wader Creek inlet
f	Fish Fry Creek inlet at 0.20 mAHD and no flow through Wader Creek inlet
g	Current inlet configurations with channels between Wader Creek and Fish Fry Creek at 0.50 mAHD
h	Current inlet configurations with channels between Wader Creek and Fish Fry Creek at 0.30 mAHD

Compared to the existing inlet configuration, raising the bed level of the inlets (runs a–c) substantially decreased the tidal range in the Fish Fry Creek (FFCk) compartment and slightly increased water levels in Wader Pond (WPW) and Swan Pond (SPW) compartments (Figure 22.10) both for spring and neap tide scenarios. The higher water levels in WPW and SPW are due to higher water levels in FFCk that interfere with the drainage of the upper wetlands during ebb flow. The reduced tidal range at FFCk also results in lower inundation levels throughout the Fish Fry Creek compartment, as the tidal prism (i.e. the volume of water that flows in and out of the wetland) is reduced. Model results indicate that for all flow scenarios, the Wader Creek inlet contributed less than 1% of tidal exchange at Area E.

Closing the inlet at Wader Creek (run e) raised the water level in Wader Creek (TGB) and Wader Pond (WPW) by approximately 0.004 m under spring and neap tides, with no impact on Swan Pond (SPW) (Figure 22.11). Closing the Wader Creek inlet and raising Fish Fry Creek inlet to 0.20 mAHD (run f) reduced tidal inundation in Fish Fry Creek (FFCk) to 0.35 m on neap tide and 0.60 m on spring tide. It also led to a further 17-mm increase in neap and spring tide water level in Wader Pond (WPW), Swan Pond (SPW) and Wader Creek (TGB).

The effect of increased hydraulic connectivity between Wader Creek and Fish Fry Creek compartments was modelled by reducing the elevation of the scour channels between the two compartments to 0.50 mAHD (run g) and 0.30 mAHD (run h) (Figure 22.12). Model results

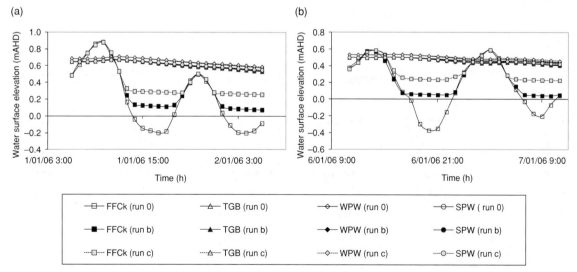

Figure 22.10 Hydrographs for existing conditions (run 0), increased inlet elevation at both inlets to 0.00 mAHD (run b) and to 0.20 mAHD (run c) for (a) spring tide and (b) neap tide.

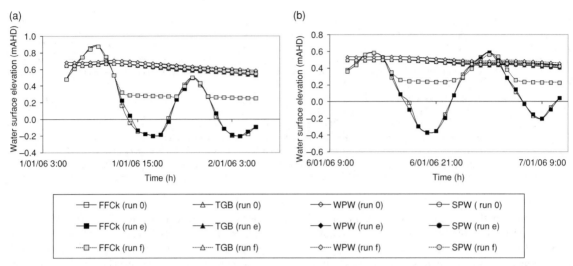

Figure 22.11 Hydrographs for existing conditions (run 0), closed Wader Creek inlet and existing Fish Fry Creek inlet elevation (−1.04 mAHD) (run e) and closed Wader Creek inlet and Fish Fry Creek inlet at −0.20 mAHD (run f) for (a) spring tide and (b) neap tide.

indicate that increasing the connectivity between these two compartments, by lowering the scour channel bed level to 0.50 m, lowered the water level in Wader Creek (TGB) and Wader Pond (WPW) by about 20 mm under neap tide, but did not affect spring tide water levels. There was no impact on Swan Pond (SPW) under either spring or neap tides. Reduction of the channel bed levels to 0.30 mAHD led to a further 10-mm fall in Wader Pond (WPW) and Wader Creek (TGB) water levels under both spring and neap tides, with no impact on Swan Pond (SPW).

22.5 Discussion

The detailed DTM used to construct the RMA-2 model mesh allowed the hydraulic complexity of Area E observed during field survey to be simulated, including the ebb-tide asymmetry driven by permanent tidal pools in the attenuated Wader Pond and Swan Pond compartments. The Fish Fry Creek compartment exhibited a semi-diurnal tidal signal, with channel flow velocities up to 1.0 m s^{-1}. Velocities

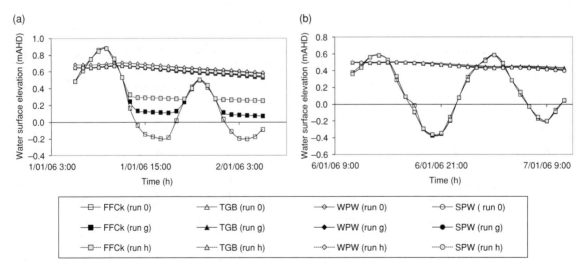

Figure 22.12 Hydrographs for existing conditions (run 0), channels at 0.50 mAHD (run g) and 0.30 mAHD (run h), with closed Wader Creek inlet and Fish Fry Creek inlet at 0.30 mAHD for (a) spring tide and (b) neap tide.

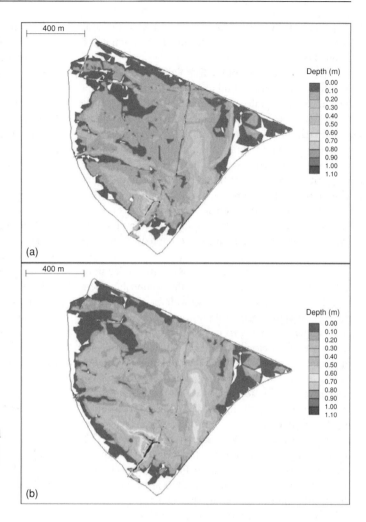

Figure 22.13 Modelled inundation depths for bed of Fish Fry Creek at 0.20 mAHD and Wader Creek closed (run f) for (a) spring tide low water and (b) spring tide high water. A colour version of this image appears in the colour plate section of the book.

in the attenuated compartments were around an order of magnitude lower than in the Fish Fry Creek compartment, and the semi-diurnal tidal signal was highly damped. The Fish Fry Creek, Wader Creek and Wader Pond compartments were highly connected by scour channels, as well as culverts, while all flow to Swan Pond was controlled by culverts.

The objective of the numerical simulation of Area E was to identify the optimum bed level elevation of the inlets at Fish Fry Creek and Wader Creek for exclusion of mangrove, and how further scour channels might influence inundation patterns throughout the study area. Ecohydraulic characterisation indicated that saltmarsh at Area E occurred where spring tidal range was less than 0.30 m, and mangrove occurred at a spring tidal range greater than 0.29 m and hydroperiod less than 0.30. These thresh-

olds provided the tools for prediction of vegetation distribution for the various spring tide inundation patterns associated with each modelled scenario. The effects of tidal range can be analysed through the spring high tide inundation depth, as this is the maximum depth of inundation, while the effects of hydroperiod can be analysed through the spring low tide inundation depth, as this is the maximum duration of inundation.

Model results indicate that raising the bed level of Fish Fry Creek to approximately 0.20 mAHD (run f in Figure 22.11) would reduce tidal range in the Fish Fry Creek compartment to less than 0.3 m (spring tide) or 0.6 m (neap tide), which was sufficient to promote saltmarsh and tidal pools over mangrove throughout the compartment, except for the Fish Fry Creek channel (Figure 22.13). Raising the bed level to this elevation would also substantially reduce ebb tide velocities in the Fish Fry Creek channel and prevent further degradation of the channel bed.

Raising the bed level of the Fish Fry Creek inlet could be achieved by construction of a broad-crested weir or a box culvert at the inlet – depending on the requirement for vehicular access – sufficiently large to allow unimpeded tidal flow above the bed level. Such an inlet configuration would maintain minimum water surface elevation in the compartment above the weir level (neglecting evaporation), without restricting high tide inundation. This configuration is predicted to maintain the existing maximum tidal inundation footprint in the compartments of Area E and would prevent regression of terrestrial vegetation into areas currently occupied by estuarine habitat.

Closing the inlet to Wader Creek is recommended since although modelling results indicate that it contributes less than 1% of current tidal exchange with the Hunter River, the current culvert bed level (−0.22 mAHD) appears to allow sufficient ebb tide water level drawdown to facilitate mangrove growth along the banks of the channel upstream of the inlet. Prior to hydraulic manipulation at Fish Fry Creek, this was one of the few places in Area E where mangrove was able to persist. Model results (runs e–f) indicate that closing the inlet at the Wader Creek location would lead to minor (< 50 mm) increases in water level in the Wader Creek and Wader Pond compartments. This increased high tide water level combined with the reduced ebb-tide drawdown is predicted to be sufficient to prevent further mangrove encroachment in this area, and lead to the eventual mortality of established mangrove.

Based on this hydraulic regime and the statistical thresholds between habitat types established for Area E, the majority of existing mangrove in the Fish Fry Creek compartment would be permanently inundated to a depth greater than 0.15 m. This inundation depth would be sufficient to prevent transpiration through pneumatophores (mean height 0.14 m) and may eventually lead to mangrove mortality. The tidal channel of Fish Fry Creek would remain deep (0.5–1.2 m) until infilled by sediment. The resulting habitat distribution in the Fish Fry Creek compartment is predicted to be similar to the pre-culvert removal distribution, except that habitats would be maintained at the current higher elevation in the tidal frame due to the increased cross-sectional area of the proposed structure relative to the former two 0.5 m-diameter culverts. Habitat would be comprised of a large tidal pool with a low tide depth range from 0–0.5 m, surrounded by a fringe of saltmarsh vegetation. Much of this pool

(approximately 6.2 ha) would be too deep for utilisation by shorebirds even under neap low tide conditions (hatched area > 0.1 m inundation in Figure 22.14). However, the remaining area of saltmarsh and shallow tidal pools (8.7 ha) is potentially suitable shorebird habitat and exceeds the area (6.0 ha) utilised by shorebirds in Fish Fry Creek compartment prior to culvert removal. Further, the proposed inlet modification is expected to prevent encroachment of mangrove into the attenuated compartments, which is predicted to occur if a 'do nothing' approach is maintained.

Raising the bed level of Fish Fry Creek (runs b and c in Figure 22.10 and runs e and f in Figure 22.11) would also prevent further scour of the channels between Wader Creek and Fish Fry Creek, which drive a stronger semi-diurnal component to water level fluctuations in Wader Pond and Wader Creek near these channels. These water level fluctuations appear to reduce soil salinity and increase soil aeration and redox potential sufficiently for successful mangrove establishment in the vicinity of the scour channels. The effect of these scour channels on wetland water levels was supported by modelling results (runs g and h in Figure 22.12), which demonstrated enhanced water level drawdown in these compartments as a result of lowering the bed elevation of the scour channels. Further degradation of these channels is expected to continue under the current inlet configuration, as the discharge capacity of the Fish Fry Creek channel exceeds the capacity of the existing twin 0.750 m-diameter culverts (C6 in Figure 22.6) that convey flow between the Fish Fry Creek and Wader Pond compartments.

22.6 Conclusions and recommendations

The extensive area of shorebird habitat available at Area E (26 ha at mean sea level and 3 ha at high water solstice spring tides), relative to other major roost sites in the estuary, is generated by hydraulic controls (culverts) which reduce the extent of tidal inundation. Hydraulic manipulation undertaken in 1995 as part of wetland rehabilitation works conducted by Kooragang Wetland Rehabilitation Project (KWRP) has led to a rapid expansion in mangrove habitat and a commensurate decline in shorebird roost habitat availability (from 31.5 ha in 1993 to 26.1 ha in 2004). If a 'do nothing' approach is maintained, it is predicted that shorebird habitat will be further reduced to approximately half of that prior to hydraulic manipulation.

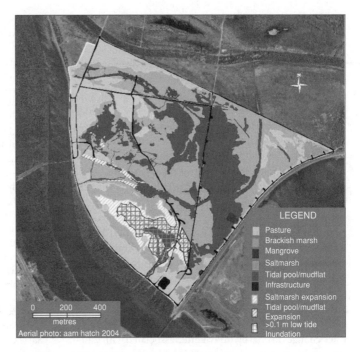

Figure 22.14 Predicted future expansion of saltmarsh and tidal pool habitats as a result of raising the inlet of Fish Fry Creek to 0.20 mAHD and closing the Wader Creek inlet. A colour version of this image appears in the colour plate section of the book.

The detailed ecohydraulic and ecogeomorphological characterisation of Area E provided some insight into the medium-term response of estuarine wetlands to the reintroduction of tidal flows. Replacement of the Fish Fry Creek culverts in 1995 with an 8.5 m-wide free span bridge increased tidal exchange with the Hunter River. Flow within Area E, however, remained highly complex, with four major compartments separated by roads to service utilities infrastructure and connected by a further 13 internal culverts.

Prior to culvert removal at Fish Fry Creek, the vegetation of Area E was dominated by terrestrial pasture species, saltmarsh and shallow tidal pools. The distribution of estuarine communities was limited by a highly attenuated tidal range and small tidal prism. After culvert removal, unattenuated tidal flow was established in the Fish Fry Creek compartment immediately upstream. This substantially increased the range, volume and velocity of tidal exchange with the Hunter River and led to a rapid increase in the area of estuarine vegetation as estuarine communities transgressed landward into terrestrial pasture. By 2004, however, much of the saltmarsh and shallow tidal pool habitats favoured by shorebirds had been invaded by mangrove. Initially, this mangrove encroachment was restricted to the Fish Fry Creek compartment, but, due to the increased conveyance capacity of the enlarged inlet, tidal channels are being eroded between

this compartment and the remainder of the wetland. Field observations indicate that erosion of these channels is increasing the tidal range in Wader Pond and Wader Creek sufficiently for further mangrove encroachment into these compartments.

Field survey data and statistical analysis indicate that hydraulic control can be used to effectively establish and maintain saltmarsh at lower elevations in the tidal frame than would occur under unattenuated flow conditions. This can be achieved by restricting tidal flows sufficiently to maintain a maximum tidal range of less than 0.3 m and a hydroperiod of more than 0.3. Hydrodynamic modelling results indicate that establishment of a broad-crested weir or box culvert at the Fish Fry Creek outlet with a bed level of approximately 0.20 mAHD would achieve this outcome. Further, modelling indicates that less than 1% of tidal flows are conveyed via Wader Creek. There is also a high level of connectivity between this creek and Fish Fry Creek. On this basis, it is proposed that the Wader Creek inlet be closed permanently to reduce ebb-tide drawdown upstream of the Wader Creek culvert which allows mangrove to survive there.

In the attenuated Wader Pond and Swan Pond compartments, modelling results indicate a limited influence of inlet conditions. Water levels are driven primarily by fortnightly and seasonal tidal cycles and by rainfall and evaporation. To investigate in more detail the relationship between vegetation distribution, tidal hydrodynamics and climatic factors on longer time scales, an ecogeomorphologic approach can be used, probably based on much simpler hydrodynamics and including other feedback mechanisms (Saco and Rodríguez, 2012).

The wetland manager, KWRP, supports ongoing wetland rehabilitation research effort at Area E and has a mandate to provide habitat for shorebirds (Svoboda, 1996). KWRP's adaptive land management framework has provided an opportunity to test the effectiveness of hydraulic manipulation to control mangrove transgression into threatened habitat (saltmarsh and associated shallow tidal pools). The alternative of manually removing mangroves, as is the current practice, is expensive, labour intensive and must be continued in perpetuity. Ground works are under way to implement hydraulic control structures based on the findings of this research.

Ongoing monitoring is required to measure the effectiveness of the proposed hydrodynamic modifications on sediment dynamics, topography, vegetation distribution and shorebird habitat availability. Monitoring should also track changes to tidal hydrodynamics associated with estuary modifications and sea level rise. Monitoring results would allow hydrodynamic model inputs and parameters, including the DTM, resistance values, hydraulic controls, climatic conditions and tidal conditions, to be revised to reflect system evolution. This updated model could then be used to inform a programme of rehabilitation works to further optimise shorebird habitat through an adaptive management approach.

References

Adam, P. (1990) *Saltmarsh Ecology*, Cambridge University Press, Cambridge.

Anderson, M.G. and Bates, P.D. (1994) Evaluating data constraints on two dimensional finite element models of floodplain flow. *Catena*, **22**: 1–15.

Baptist, M.J., Babovic, V. and Rodríguez Uthurburu, J. (2007) On inducing equations for vegetation resistance. *Journal of Hydraulic Research*, **45**(4): 435–450.

Bockelmann, A.-C., Bakker, J.P., Neuhaus, R. and Lage, J. (2002) The relation between vegetation zonation, elevation and inundation frequency in a Wadden Sea salt marsh. *Aquatic Botany*, **73**: 211–221.

Boumans, R.M.J., Burdick, D.M. and Dionne, M. (2002) Modelling habitat change in salt marshes after tidal restoration. *Restoration Ecology*, **10**: 543–555.

Cahoon, D.R., Lynch, J.C., Hensel, P., Boumans, R., Perez, B.C., Segura, B. and Day Jr, J.W. (2002) A device for high precision measurement of wetland sediment elevation: I. Recent improvements to the sedimentation–erosion table. *Journal of Sedimentary Research*, **72**: 730–733.

Cea, L., French, J.R. and Vázquez-Cendón, M.E. (2006) Numerical modelling of tidal flows in complex estuaries including turbulence: an unstructured finite volume solver and experimental validation. *International Journal for Numerical Methods in Engineering*, **67**: 1909–1932.

Clarke, L.D. and Hannon, N.J. (1967) The mangrove swamp and salt marsh communities of the Sydney district: I. Vegetation, soils and climate. *Journal of Ecology*, **55**: 753–771.

Crowder, D.W. and Diplas, P. (2000) Using two-dimensional hydrodynamic models at scales of ecological importance. *Journal of Hydrology*, **230**: 172–191.

Dann, P. (1987) The feeding behaviour and ecology of shorebirds. In Lane, B. (ed.) *Shorebirds in Australia*, Nelson Publishing, Melbourne, Australia.

Department of the Environment and Heritage (2005) Background Paper to the Wildlife Conservation Plan for Migratory Shorebirds. Australian Government, Canberra, Australia.

D'Alpaos, A., Lanzoni, S., Marani, M. and Rinaldo, A. (2007) Landscape evolution in tidal embayments: modelling the interplay of erosion, sedimentation, and vegetation dynamics. *Journal of Geophysical Research*, **112**. doi: 10.1029/2005JF000408.

Finnie, J., Donnell, B., Letter, J. and Bernard, R.S. (1999) Secondary flow correction for depth-averaged flow calculations. *Journal of Engineering Mechanics*, **125**: 848–863.

French, J.R. and Stoddart, D.R. (1992) Hydrodynamics of salt marsh creek systems: implications for marsh morphological development and material exchange. *Earth Surface Processes and Landforms*, **17**: 235–252.

Geering, D. and Winning, G. (1993) *Waders of the Hunter River Estuary: Threats to Habitat and Opportunities for Habitat Restoration*, Shortland Wetland Centre.

Higgins, P.J. and Davies, S.J.J.F. (1996) *Handbook of Australian, New Zealand and Antarctic Birds. Volume 3: Snipes to Pigeons*, Oxford University Press Australia, Melbourne.

Hovenden, M.J., Curran, M., Cole, M.A., Goulter, P.F.E., Skelton, N.J. and Allaway, W.G. (1995) Ventilation and respiration in roots of one-year-old seedlings of grey mangrove *Avicennia marina* (Forsk.) Vierh. *Hydrobiologia*, **295**: 23–29.

Howe, A. (2008) *Hydrodynamics, Geomorphology and Vegetation of Estuarine Wetlands in the Hunter, Australia: Implications for Migratory Shorebird High Tide Roost Availability*. Unpublished PhD thesis, University of Newcastle.

Howe, A. and Rodríguez, J.F. (2006) Flow resistance in saltmarsh and mangrove vegetation in an Australian coastal wetland. *ICHE 2006, Seventh International Conference on Hydro-Science and Engineering*, Philadelphia.

Howe, A. and Rodríguez, J.F. (2007) Resistance relationships for five estuarine wetland substrates. *6th International Ecohydraulics Symposium*, IAHR, Christchurch, New Zealand.

Howe, A., Rodríguez, J.F. and Saco, P.M. (2009) Vertical accretion and carbon sequestration in disturbed and undisturbed estuarine wetland soils of the Hunter estuary, southeastern Australia. *Estuarine Coastal and Shelf Science*, **84**: 75–83.

Howe, A., Rodríguez, J.F., Spencer, J., MacFarlane, G. and Saintilan, N. (2010) Response of estuarine wetlands to reinstatement of tidal flow. *Marine and Freshwater Research*, **61**: 702–713.

Järvelä, J. (2004) *Flow Resistance in Environmental Channels: Focus on Vegetation.* Unpublished PhD thesis, Helsinki University of Technology.

Khatibi, R., Williams, J.J.R. and Wormleaton, P.R. (2006) Friction parameters for flows in nearly flat tidal channels. *Journal of Hydraulic Engineering*, **126**: 741–749.

King, I. (2002) *A Short Course: Application of the RMA Suite of Models – Users' Manual for Models.* Resource Modelling Associates, Sydney.

Kingsford, R.T., Ferster Levy, R., Geering, D., Davis, S.T. and Davis, J.S.E. (1998) *Rehabilitating Estuarine Habitat on Kooragang Island for Waterbirds, including Migratory Wading Birds (May 1994 – May 1997)*, NSW National Parks and Wildlife Service.

Kjerfve, B.J., Miranda, L.E. and Wolanski, E. (1991) Modelling water circulation in an estuary and intertidal salt marsh system. *Netherlands Journal of Sea Research*, **28**: 141–147.

Knight, D.W. (1981) Some field measurements concerned with the behaviour of resistance coefficients in a tidal channel. *Estuarine, Coastal and Shelf Science*, **12**: 303–322.

Laegdsgaard, P. (2006) Ecology, disturbance and restoration of coastal saltmarsh in Australia: a review. *Wetlands Ecology and Management*, **14**: 379–399.

Lawler, W. (1996) *Guidelines for Management of Migratory Shorebird Habitat in Southern East Coast Estuaries, Australia.* Unpublished Masters thesis, University of New England.

Lewis, R.R. (2005) Ecological engineering for successful management and restoration of mangrove forests. *Ecological Engineering*, **24**: 403–418.

Liu, W.-C., Hsu, M.-H. and Wang, C.-F. (2003) Modelling of flow resistance in mangrove swamp at mouth of tidal Keelung River, Taiwan. *Journal of Waterway, Port, Coastal and Ocean Engineering*, **129**: 86–92.

Luis, A., Goss-Cutard, J.D. and Moreira, M.H. (2001) A method for assessing the quality of roosts used by waders during high tide. *Wader Study Group Bulletin 96*, pp. 71–73.

Marani, M., Belluco, E., D'Alpaos, A., Defina, A. and Lanzoni, S. (2003) On the drainage density of tidal networks. *Water Resources Research*, **39**: 1040. doi: 10.1029/2001WR001051.

Marchant, S. and Higgins, P.J. (1993) *Handbook of Australian, New Zealand and Antarctic Birds. Volume 2: Raptors to Lapwings*, Oxford University Press Australia, Melbourne.

Mazda, Y., Wolanski, E., King, B., Sase, A., Ohtsuka, D. and Magi, M. (1997) Drag force due to vegetation in mangrove swamps. *Mangroves and Salt Marshes*, **1**: 193–199.

Mazumder, D., Saintilan, N. and Williams, R.J. (2005) Temporal variations in dish catch using pop nets in mangrove and saltmarsh flats at Towra Point, NSW, Australia. *Wetlands Ecology and Management*, **13**: 457–467.

Mitsch, W.J. and Gosselink, J.G. (2000) *Wetlands*, John Wiley & Sons, Inc., New York.

Montalto, F.E. and Steenhuis, T.S. (2004) The link between hydrology and restoration of tidal marshes in the New York/New Jersey estuary. *Wetlands*, **24**: 414–425.

Musleh, F.A. and Cruise, J.F. (2006) Functional relationships of resistance in wide flood plains with rigid unsubmerged vegetation. *Journal of Hydraulic Engineering*, **132**: 163–171.

Nepf, H.M. (1999) Drag, turbulence, and diffusion in flow through emergent vegetation. *Water Resources Research*, **35**: 479–489.

Nepf, H.M. and Vivoni, E.R. (2000) Flow structure in depth-limited, vegetated flow. *Journal of Geophysical Research*, **105**(C12): 28547–28557.

Rogers, D.I., Battley, P.F., Piersma, T., Van Gils, J. and Rogers, K.G. (2006) High-tide habitat choice: insights from modelling roost selection by shorebirds around a tropical bay. *Animal Behaviour*, **72**: 563–575.

Saco, P.M. and Rodríguez, J.F. (2012) Modeling ecogeomorphic systems. In Shroder Jr, J. and Baas, A.C.W. (eds) *Treatise on Geomorphology*, volume 2, Academic Press, San Diego, California.

Saintilan, N. and Rogers, K. (2002) *The Declining Saltmarsh Resource: Coast to Coast 2002*, pp. 410–413.

Saintilan, N. and Rogers, K. (2006) Coastal wetland elevation trends in southeast Australia. Catchments to Coast. *Society of Wetland Scientists 27th International Conference*, pp. 42–54.

Saintilan, N. and Williams, R.J. (1999) Mangrove transgression into saltmarsh environments in south-east Australia. *Global Ecology and Biogeography*, **8**: 117–124.

Saintilan, N., Rogers, K. and Howe, A. (2009a) Geomorphology. In Saintilan, N. (ed.) *Australian Saltmarsh Ecology*, CSIRO Publishing.

Saintilan, N., Rogers, K. and McKee, K. (2009b) Saltmarsh–mangrove interactions in Australasia and the Americas. In Perillo, G.M.E., Wolanski, E., Cahoon, D.R. and Brinson, M.M. (eds) *Coastal Wetlands: An Integrated Ecosystems Approach*, Elsevier, pp. 855–883.

Shrestha, P.L. (1996) An integrated model suite for sediment and pollutant transport in shallow lakes. *Advances in Engineering Software*, **27**: 201–212.

Skagen, S.K. and Knopf, F.L. (1994) Migrating shorebirds and habitat dynamics at a prairie wetland complex. *Wilson Bulletin*, **106**: 91–105.

Smith, P. (1991) *The Biology and Management of Waders (Suborder Charadrii) in NSW*. Species management report 9, NSW National Parks and Wildlife Service, Australia.

Spencer, J.A. (2010) *Migratory Shorebird Ecology in the Hunter Estuary, South-eastern Australia.* PhD thesis, Australian Catholic University, Sydney.

Svoboda, P.L. (1996) *Kooragang Wetland Rehabilitation Project Management Plan.* Paterson, Hunter Catchment Management Trust.

Turner, A. and Lewis, R.R. (1997) Hydrologic restoration of coastal wetlands. *Wetlands Ecology and Management*, **4**: 65–72.

Warren, R.S., Fell, P.E., Rozsa, R., Brawley, A.H., Orsted, A.C., Olson, E.T., Swamy, V. and Niering, W.A. (2002) Salt marsh restoration in Connecticut: 20 years of science and management. *Restoration Ecology*, **10**: 497–513.

Watkins, D. (1993) *A National Plan for Shorebird Conservation in Australia*, Australasian Wader Studies Group.

Williams, R.J. and Watford, F.A. (1997) Identification of structures restricting tidal flow in New South Wales, Australia. *Wetlands Ecology and Management*, **5**: 87–97.

Wolanski, E., Jones, M. and Bunt, J.S. (1980) Hydrodynamics of a tidal creek-mangrove swamp system. *Australian Journal of Marine and Freshwater Research*, **31**: 431–450.

Yen, B.C. (1991) Hydraulic resistance in open channels. In Yen, B.C. (ed.) *Channel Flow Resistance: Centennial of Manning's Formula*, Water Resources Publications, Littleton, CO, USA.

23

Ecohydraulics at the Landscape Scale: Applying the Concept of Temporal Landscape Continuity in River Restoration Using Cyclic Floodplain Rejuvenation

Gertjan W. Geerling[1,2], Harm Duel[1], Anthonie D. Buijse[1] and Antonius J.M. Smits[2]

[1]Deltares, P.O. Box 177, 2600 MH Delft, The Netherlands
[2]DSMR, Radboud University, P.O. Box 9010, 6500 GL Nijmegen, The Netherlands

23.1 Introduction

Studies centred on ecology and hydraulics are scale dependent and interactions between the two manifest themselves at various scales. For different spatial scales, a range of definitions exists. For example, within the fluvial hydrosystems approach, scale varies between the drainage basin, a functional sector (e.g. the meandering section of a river), functional sets (e.g. groups of functional units arising from the same single process such as meander migration), functional units (e.g. point bar, pond, pool and riffle; these are also called ecotopes in Geerling et al. 2006; 2008) and mesohabitats (e.g. scour hole and submerged plant stand) (Petts and Amoros, 1996). Studies centred on ecohydraulics can focus on single species habitats by considering correlations between flow velocity or flood regime and habitat preferences at the functional unit or mesohabitat level. However, at the landscape scale (functional sector to functional set), the interaction between ecohydraulic and morphological dynamics determines habitat creation and maintenance. An understanding of these dynamics is fundamental to sustain habitat availability. This chapter considers the interaction between landscape and hydraulics in several case studies and provides an approach for landscape-scale analysis of ecohydraulic processes and patterns within riverine ecosystems.

Within riverine landscapes, the dynamics of disturbance events and the subsequent recovery are an integral part of these dynamic systems and are primarily controlled by the flow regime and sediment load (Junk et al., 1989; Amoros and Wade, 1996; Ward, 1998; Tockner et al., 2000). Fluctuating river discharge and sediment transport processes create morphologically varied landscapes and diverse habitats associated with both erosional and depositional features (Figure 23.1).

These dynamics have a significant influence on the habitat available. Disturbances create or maintain niches for individual species, which can be observed at the landscape level. For example, a shift of the river channel exposes sediment that can be colonised by pioneer woody vegetation such as Poplar (*Populus* spp.) and Willow (*Salix* spp.) (Figure 23.2). However, riverine landscape formation and the patterns observed are (partly) controlled by the presence of biotic elements such as trees and dead wood, a clear example of the influence of ecohydraulics on landscape pattern (Gurnell, 1997; Corenblit et al., 2007).

Ecohydraulics: An Integrated Approach, First Edition. Edited by Ian Maddock, Atle Harby, Paul Kemp and Paul Wood.
© 2013 John Wiley & Sons, Ltd. Published 2013 by John Wiley & Sons, Ltd.

Figure 23.1 (left) Lateral erosion and (right) sedimentation along the Allier River, France. The sediment was deposited during a single flood event in May 2003. Photos by G. W. Geerling.

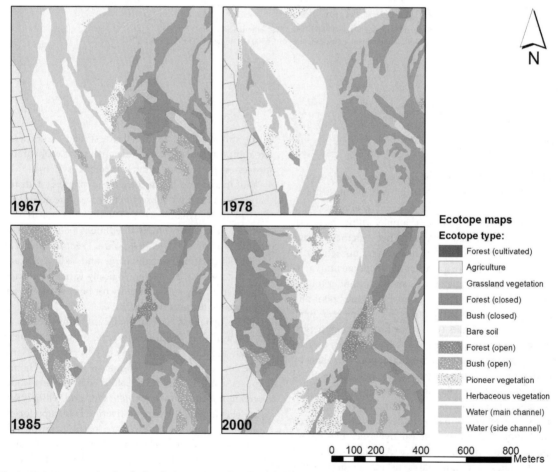

Figure 23.2 An example of ecohydraulic interactions that shape floodplain landscapes. The 1967–1978 channel migration rejuvenated older ecotopes on the right river bank and created niches for forest regeneration/succession in the abandoned channels. By 1985, these channels had been colonised by vegetation that developed into forest during the 1985–2000 period. Reproduced from Geerling (2006), with kind permission from Springer Science + Business Media. A colour version of this image appears in the colour plate section of the book.

Ecohydraulic and morphological processes shape the riverine landscape and local vegetation dynamics influence environmental conditions within habitat units at any point in time (Petts, 2000). This results in a diverse landscape with many ecotones. Riverine species have adapted to this natural diversity of habitats and it is the basis for the relatively high biodiversity recorded within natural river systems (Tockner and Ward, 1999).

Anthropogenic regulation of rivers to protect people from floods, optimise shipping logistics or to generate hydro-power influences instream hydro-morphological and hydraulic processes (see Figure 23.3 for an example of regulation taken from Geerling et al., 2008; Buijse et al. 2005;). Hydro-morphological processes can be spatially limited (e.g. embankments), spatially fixed (e.g. erosion and sedimentation in rivers with groynes) or strongly diminished (e.g. impounded rivers where sediment and water influx are largely inhibited). The net result of regulation is a disconnected and less dynamic riverine landscape, with reduced habitat diversity (Tockner and Stanford, 2002; Ward et al., 2002). At the basin scale, the net effect of regulation has been to reduce the hydrological resilience on rivers such as the Rhine. Anthropogenic modifications/regulation have resulted in unnatural fluctuations in discharge of the river, and even moderate rainfall intensity events are rapidly reflected in water levels in the Rhine (Havinga and Smits, 2000; Silva et al., 2001).

The potential ecological degradation associated with regulation is widely recognised and attempts to address it have been undertaken in many systems (Buijse et al., 2002). In The Netherlands, the large volume of flood-protection and restoration measures being undertaken provides opportunities to examine and review the results of river restoration at the scale of the individual floodplain and also at the riverine landscape scale. This has the potential to strengthen understanding of ecological diversity at larger spatial scales and, at the same time, help in the management of the system to accommodate future floods. The case studies presented in this chapter focus on the development of a scientific basis for riverine landscape restoration of regulated (i.e. heavily modified) river. The individual case studies focus on: (1) landscape dynamics in a naturally functioning river, providing a conceptual view of landscape degradation and restoration; and (2) an example of a floodplain restoration project within a regulated river. The chapter provides a synthesis of the results to underpin the development of a management strategy called 'Cyclic Floodplain Rejuvenation', which is proposed as a methodology to balance the potentially conflicting and competing demands of process-based restoration

and river regulation for flood protection and other socio-economic demands placed on contemporary rivers.

23.2 The inspiration: landscape dynamics of meandering rivers

Ecohydraulic and morphological processes and the resulting spatial and temporal patterns along a near-natural river can serve as a process example for regulated rivers. In this case study, the spatial and temporal diversity of the riverine landscape of the freely meandering Allier River (France) were analysed by Geerling et al. (2006). The Allier serves as a reference river for the Dutch/Belgian section of the Meuse (the so-called Border-Meuse) and can serve as an example for meandering rivers in general. Using a time series of aerial photographs (1954–2000), the changing landscape composition driven by erosion and deposition and the process of meander progression were analysed. The overall rejuvenation rate (pattern disturbance), the functional unit transition rate, floodplain age and the river stretch size of a steady state unit were quantified. The direct relationship between eco-hydro-morphological processes and landscape composition, resulting in the creation of niches for the colonisation of tree species and resetting of older succession stages on the opposite bank can be illustrated for the Allier River (Figure 23.2). River channel changes between 1954 and 2000 are illustrated in Figure 23.4 and indicate that the Allier floodplain rejuvenated at a rate of about 33.5 ha every five years for the 7 km-long study area (amounting to 0.97 ha km^{-1} $year^{-1}$ – river length was measured along the axis of the river) (see Geerling et al. (2006) for further details).

Land cover transitions were expressed using ecotopes as basic units, similar to the functional units identified above (Geerling et al., 2009), and transition rates from one ecotope to another were dependent on both the pattern of physical rejuvenation and ecological succession. These turnover values provide a quantitative estimate of the spatio-temporal processes identified by Tockner and Stanford (2002). As a consequence of rejuvenation, half the landscape was younger than 15 years while 24% was older than 46 years.

Some authors propose the 'steady state' or 'meta-climax concept' as the most appropriate model for a riverine ecosystem. They state that, although a system may be continually disturbed and succession reset, it is in balance at higher spatial and temporal scales (Amoros and Wade, 1996; Merritt and Cooper, 2000; Latterell et al., 2006). From a management perspective, a 'steady state unit'

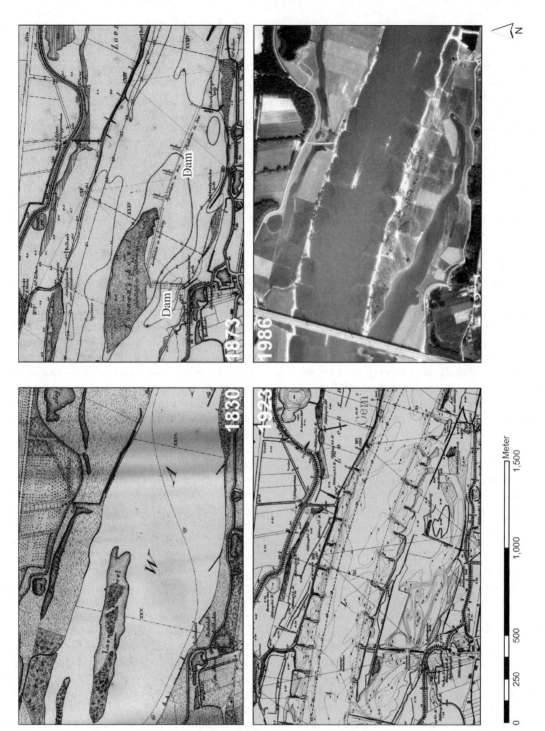

Figure 23.3 A series of maps to illustrate the changes in floodplain topography resulting from river regulation (1830 to 1986) for the River Waal, one of the Dutch Rhine branches. In 1830, the river had been regulated by dykes and groynes, although its morphology was still dynamic and allowed downstream migration of meanders, point bars and sand islands within the main channel (Middelkoop *et al.*, 2005). In 1873, regulation works commenced at the research site by connecting the island to the shore by the construction of two dams. The purpose of this regulation was to shift and subsequently confine the course of the main channel to the north. The section of the island located within the new main channel was excavated. By 1923, this process was complete and the newly constructed floodplain began to develop due to overbank sedimentation of sand; sandy levees formed alongside the river. In the 1960s, the river authorities reclaimed the floodplain and it was transformed into agricultural grassland. In 1986 it was still in agricultural use. Reprinted from Geerling *et al.*, Copyright 2008, with permission from Elsevier.

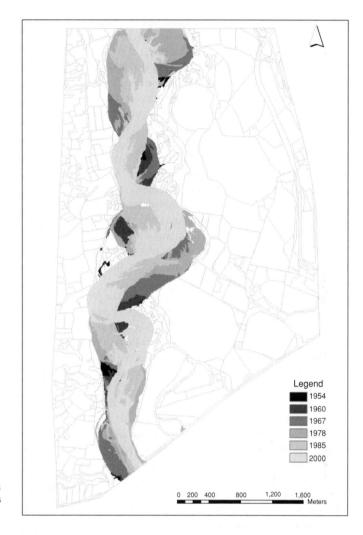

Figure 23.4 The floodplain age map for the Allier River study area, illustrating the hydro-geomorphological activity within the study area using overlays of main channel areas and exposed sediment (point bars) from 1954 to 2000. The final functional unit map for the year 2000 is presented. Reproduced from Geerling (2006), with kind permission from Springer Science + Business Media.

could be used as the basic management unit for which uniform policies, management practices, monitoring and evaluation methods could be applied. The study of the River Allier demonstrates that, within the 50-year study period, the steady state or meta-climax concept could be applicable at the scale of 1.5 to 2 meander lengths. However, this only applies when the environment within the functional sector of the river basin (i.e. one scale level higher) remains stable; long-term trends in the basin will also influence the unit size. In the River Allier, the size of the 'steady state unit' decreased from around 2 meander lengths in the 1950s to about 1.5 in 2000 (Geerling *et al.*, 2006). This decrease was caused by a change towards a more closed and structure-rich heterogeneous landscape, probably associated with a reduction in grazing intensity

following the designation of the River Allier as a nature reserve in the 1980s.

23.3 The concept: temporal continuity and discontinuity of landscapes along regulated rivers

The diversity of riverine habitats is correlated with species diversity, and lower rates of channel migration are correlated with reduced species diversity (Nilsson *et al.*, 1991; Cordes *et al.*, 1997; Hughes *et al.*, 2001; 2007). In free-flowing rivers, biodiversity varies along the course of the river and is generally high in the middle reaches, especially within meandering sections (Nilsson, 1989; Nilsson and Svedmark, 2002). Therefore, landscape diversity, or diversity in functional units, could be an indicator linked to biodiversity in formerly meandering regulated middle reaches of rivers. Data on historical dynamics and floodplain age of reference rivers, such as the Allier in France, can provide a guide or benchmark for the reinstatement of natural dynamics within regulated rivers (Middelkoop *et al.*, 2005; Greco *et al.*, 2007). In the following paragraphs we conceptualise the knowledge outlined above (Section 23.2) and the effects of regulation on rivers in terms of temporal composition of the riverine landscape, i.e. a process-based diversity index. The temporal composition of an unconstrained (unregulated) dynamic system can provide a benchmark for the restoration of a constrained system. The temporal composition of an unconstrained system is presented graphically, and the effects of regulation introduced in a step-wise manner.

Within the following section it is assumed that the variability recorded on the River Allier represents a reference for steady-state dynamics, regardless of long-term changes such as geological changes, climate change or shifts in land use. The continuous disturbance of the landscape pattern results in a mosaic of succession stages of functional units (analogous to Figure 23.2 and Figure 23.4). Based on the results for the River Allier, the distribution of patches of different ages within the active river corridor can be conceptualised (Figure 23.5a); floodplain age is defined as the time in years since the last rejuvenation event. The relative area of older succession stages depends on the floodplain boundary and, in the current example, a hypothetical flood area with a flood frequency with a historical return period (< 100 years) was selected, similar to the area of the River Allier case study.

If a river is regulated and natural hydromorphological and ecosystem dynamics are strongly inhibited, as indicated for 1923 and 1986 in Figure 23.3, rejuvenation ceases while succession continues, and thus the overall landscape age increases. The landscape composition changes dramatically in favour of species typical of less dynamic habitats (Petts and Amoros, 1996; Johnson, 1997; Hughes *et al.*, 2001). It typically takes some time for effects to become apparent in the landscape and the high biodiversity recorded in these areas is often a relict of former conditions. These may, ultimately, degrade and develop lower diversity, reflecting the shift in landscape composition (Bravard *et al.*, 1986; Tockner and Stanford, 2002). In such a setting, the distribution of patches of different ages may follow the pattern shown in Figure 23.5b.

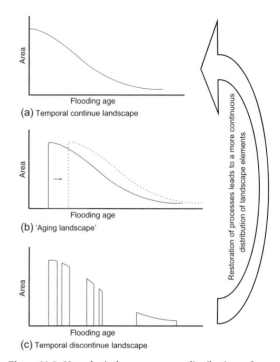

Figure 23.5 Hypothetical area versus age distributions of riverine landscapes. (a) Conceptual graph of hypothetical area versus age distribution in meandering rivers; (b) conceptual graph of hypothetical area versus age distribution of natural ecotopes in a regulated river. Existing ecotopes continue their succession, while pioneer ecotopes disappear; and (c) conceptual graph of a hypothetical area versus age distribution of natural ecotopes in a regulated river floodplain without rejuvenation and with natural ecotopes converted to agriculture or other land uses. To restore the landscape diversity, processes must be activated that rejuvenate the landscape and reinstate the continuity of succession stages (change type (c) landscapes to type (a) landscapes.

In addition to rejuvenation, natural succession will also cease if land use in the floodplains changes (partially or completely) to agriculture or other managed land uses, as is the case along most rivers in highly populated areas. Some ecotopes will be converted to pastures or arable fields, while others will remain natural or semi-natural in character. The latter will become relict ecotopes which will undergo ecological succession, such as relict disconnected side channels. Such landscapes will appear to have 'gaps' in their age distribution and will be temporally discontinuous (Figure 23.5c) and will develop into older succession stages (similar to developments in Figure 23.5b). To restore the riverine landscape interactions requires active intervention to reconnect discontinuous patches (Figure 23.5c) and to re-create a more continuous landscape (Figure 23.5a). Under ideal conditions, this will include restoring processes that disturb and rejuvenate the landscape. Note that the conceptual graphs are based on a meandering channel form as the reference process/condition. However, the pattern and rate of hydromorphological processes and the resulting landscape composition vary across rivers, and especially depend on the type of functional sector within the river basin (e.g. braiding or meandering) (Petts and Amoros, 1996; Hughes, 1997). The effort necessary to achieve, and the possibilities for, restoration will also vary with river type but are exemplified only for meandering rivers in this chapter.

23.4 Application: floodplain restoration in a heavily regulated river

The Dutch Rhine branches have been extensively regulated (as shown in Figure 23.3) and very little of this area resembles the landscape patterns and processes associated with the dynamics of a natural river (Ward *et al.*, 2001). It is often unfeasible to restore a regulated river to pristine conditions. However, it can be restored partially through the reinstatement of natural dynamic processes (Buijse *et al.*, 2002; Van der Velde *et al.*, 2006). Studies on the landscape evolution of rejuvenated restoration sites on regulated rivers are scarce. The case study on the Ewijk floodplain (Geerling *et al.*, 2008) represents an example within the constraints of a regulated setting (representing the 1986 condition of Figure 23.3). A bare ecological pioneer community situation was created by excavation of the top one to two metres of sandy soil, constituting artificial rejuvenation by removing historic sedimentation. To reinstate spontaneous processes, agricultural land use was abandoned and subsequent landscape development was not influenced by anthropogenic intervention. During the investigation, a 16-year record of land elevation measurements and six vegetation maps based on aerial photography (1986, 1989, 1992, 1997, 2000, 2005) were analysed (see Geerling *et al.*, 2008).

The study showed that reinstating pioneer stages and promoting spontaneous processes by abandoning land management rehabilitated natural levee-forming processes through local sedimentation and succession. Excavation led to colonisation and subsequent growth of softwood forest species. In the first seven years, the bare soil developed into softwood bush and the developing forest was clearly visible. Within ten years, the overall landscape diversity and patch heterogeneity increased to levels higher than before excavation (see Geerling *et al.*, 2008). The successional development of the River Waal was similar to the River Allier floodplain, with pioneer stages being replaced by herbaceous and grassland vegetation and bush by forest. Even spatial patterns of softwood forest colonisation on less dynamic shorelines of the old side channel resembled patterns along the River Allier. However, despite sedimentation and succession, the restored system did not show signs of large-scale rejuvenation during the study period, not even in response to two large flood events in 1993 and 1995.

The use of a hydraulic model suggested that flood flow velocities decreased and water surface elevations increased as sediment was deposited and vegetation established on the restored floodplain. After 16 years of landscape evolution, the flood capacity was lower than during the pre-project conditions and mean flow velocities dropped 14% below the pre-project baseline. The rate of change in flood levels and flow velocities diminished over time and flow velocity changes were strongly correlated with landscape diversity (see Geerling *et al.*, 2008). The difference in land cover, from cultivation (1986) to post-restoration (2005) is clearly visible and has had significant effects on flow patterns and velocities (Figure 23.6).

The volume of sediment deposited on the floodplain of the River Waal returned to pre-restoration levels (1982) by 2005. The rate of sedimentation was directly related to overbank flow (see Geerling *et al.*, 2008), with 40% of the total sediment being deposited in two flood events (in 1993 and 1995 and see Figure 23.7).

The floodplain topography developed in a different manner to historic conditions, when under agricultural management (i.e. the pre-restoration period). The influence of vegetation, and especially forest, on the local flow velocity and flow direction changed the patterns of sedimentation, and levee formation appeared stronger than under the previous agrarian management regime.

The Ewijk case study demonstrates that: (1) natural succession of an artificially created pioneer condition is possible and can result in a heterogeneous floodplain and a reforested landscape; (2) sedimentation patterns vary in response to changing vegetation patterns; and (3) the combined succession and sedimentation patterns strongly influence the pattern of discharge. Therefore, if applied over large areas, floodplain restoration of embanked systems may pose an increased flood risk, especially if the existing embankments have been engineered to accommodate floods over low hydraulic resistance (agrarian) floodplains, such as those that have evolved along the

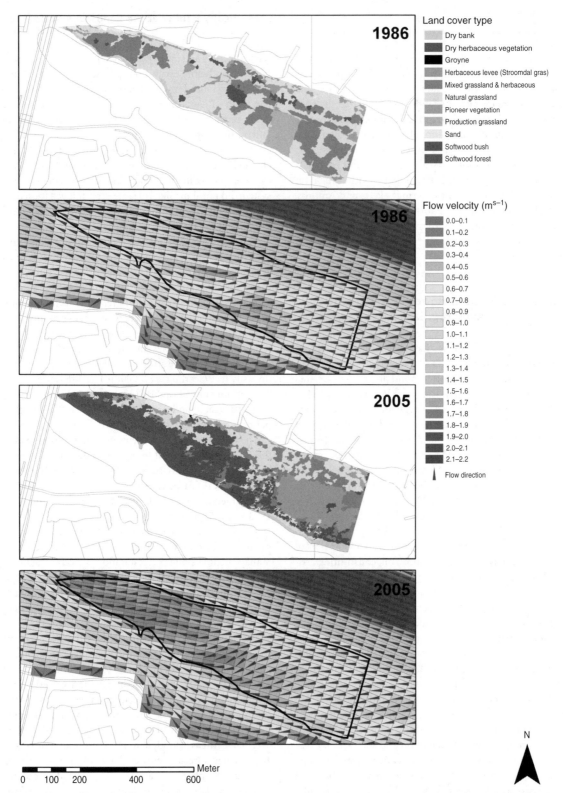

Figure 23.6 Vegetation cover, flow direction and velocity maps for 1986 and 2005 on the River Waal. Flow velocity and direction vectors were computed for levels recorded for the 1995 flood event. Flow velocities were generally reduced in the study area between surveys. Reprinted from Geerling *et al.*, Copyright 2008, with permission from Elsevier. A colour version of this image appears in the colour plate section of the book.

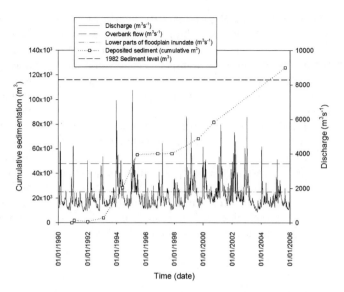

Figure 23.7 Cumulative sedimentation of the Ewijk floodplain along the Waal River (Rhine branch) 1990 to 2006 (left axis; 1990 = 0) and the discharge of the Waal River onto the floodplain 'Ewijkse Plaat' during the same period (right axis). Reference lines are given for the 1982 sediment level (i.e. before excavation/restoration), the discharge at which first inundation occurs and the discharge of first overbank flow ($3435 \text{ m}^3 \text{ s}^{-1}$). Note that 40% of the total deposition took place in the 1993/1994 and 1995 extreme flood events. Reprinted from Geerling et al., Copyright 2008, with permission from Elsevier.

Rhine over the past 500 years (Van de Ven, 2004). In these situations, recurring management is required to maintain the landscape, the intensity of which depends on the restoration targets, rates of floodplain change and the storage space between the embankments for accommodating floods. In the current case study example, management interventions have been undertaken three times since 1923, after about 40, 30 and now 20 years.

23.5 The strategy in regulated rivers: cyclic floodplain rejuvenation (CFR)

The restoration potential of a temporally fragmented landscape depends primarily on the constraints that apply to floodplain use. The general concept in restoration practice is to 'let the river do the work' and allow active rejuvenation processes to occur (Stanford et al., 1996). Generally, three degrees of restoration are recognised (FISRWG, 1998, pp. i–3):

• 'Non-intervention and undisturbed recovery: where the stream corridor is recovering rapidly and active restoration is unnecessary and even detrimental.'
• 'Partial intervention for assisted recovery: where a stream corridor is attempting to recover, but is doing so slowly or uncertainly. In such a case, action may facilitate natural processes already occurring.'
• 'Substantial intervention for managed recovery: where recovery of desired functions is beyond the repair capacity of the ecosystem and active restoration measures are needed.'

Most highly regulated systems serve multiple functions and interests and, therefore, not all ecosystem processes can realistically be restored. As a result, the third option of 'substantial intervention for managed recovery' is probably most widely applicable (Amoros et al., 1987; Geerling et al., 2008). In time, depending on how discontinuous the landscape is, a cascade of pioneer sites, restored at different times, will contribute to a temporally more heterogeneous and continuous river, with all succession stages present. This strategy is called 'cyclic floodplain rejuvenation' (CFR) and has the potential to significantly increase landscape diversity (Smits et al., 2000; Duel et al., 2001; Baptist et al., 2004).

The measures necessary should replace natural processes that are inhibited due to river regulation, but leave as much to the river in its regulated setting as required. The nature of these measures depends on the type of river system but can consist of removing historic sedimentation by excavation of floodplains, rejuvenating climax forest patches or reconnecting isolated meanders. The measures should aim to initiate new succession series to overcome and reinvigorate a landscape 'frozen in time' (Ward et al., 2002). To be most effective, CFR measures should be tailored to the existing characteristics and the ecological potential of the floodplain in question.

The potential constraints (including societal interests) can be powerful drivers for restoration in their own right. In the case of Dutch rivers, the river management authorities emphasised the need for a sustainable water management strategy based on increasing the discharge capacity whilst also emphasising the need to improve the

quality of the physical environment including riverine ecosystems and landscapes. This resulted in a new flood protection programme for Dutch rivers entitled 'Room for the river' (Ministry of Transport, Public Works and Water Management, 2000). As part of this management concept, a wide variety of measures are available, including:
• widening the main channel in combination with a reconstruction of the river banks, including removal of the groynes;
• excavating parts of the floodplain which are rarely or irregularly flooded;
• (re)constructing lateral channels and floodplain lakes;
• removing artificial levees on the floodplains;
• enlarging floodplains by relocating existing dykes.

Although the programme was not designed with the reintroduction of river dynamics in mind, the case study presented in Section 23.4 falls into the excavation category. Other measures can also contribute to 'room for processes' or rejuvenation, such as the construction of lateral side channels. The latter can have a succession from connected to semi-connected and isolated ponds.

Baptist *et al.* (2004) tested the validity of a large-scale implementation of CFR using ecological and hydraulic models. This large-scale modelling study shows that CFR can contribute to the management of hydraulic roughness and flood levels (Baptist *et al.*, 2004).

Some guiding principles in relation to the implementation of CFR within a regulated river setting with navigation and flood protection as additional functions are

Table 23.1 Guidelines for implementation of CFR.

System definition and description

To enhance and maintain the diversity of functional units (ecotopes), vegetation and habitat diversity (including different succession stages and stages of aquatic connectivity), a description of ecological reference condition or ecological potential at the scale of the river reach is essential.

Define the management area as the basic management unit for which uniform policies, management practices, monitoring and evaluation methods apply. If possible, this process should be overseen by a single agency so that restoration can be viewed from the landscape perspective.

A management area can be based on an estimated steady-state size, the minimum recommended area is a stretch of river of at least 2 meander bends.

System assessment

Develop a comprehensive monitoring programme, covering the development of morphology, vegetation and habitat diversity in relation to expected water levels during high discharges.

Map the present state of the system, incorporating successional stage (age) in maps and including relict functional units (ecotopes).

List the constraints for full restoration, such as which processes can (or cannot) be restored. Identify areas where the potential for process-based restoration is greater.

Assess whether the current regulatory measures are irreversible and/or required for other functions. If they could be removed, would this increase the potential for riverine processes to shape and restore the landscape?

Implementation

When a large-scale restoration effort consists of multiple interventions at different locations, ensure these are distributed over time to increase landscape temporal diversity. Furthermore, in the case of flood discharge limitations, this ensures that hydraulically rough climax vegetation stages stages do not accumulate over time.

The rejuvenation frequency is dependent on the development time of different morphological elements such as secondary channels – this may require multiple decades (see Figure 23.5) – but in highly restricted areas also dependson the total floodplain conveyance capacity.

Do not underestimate the impact of morphological developments on water levels, especially when the required time for ecological development is greater than 30 years.

Floodplain lowering over large areas (> 200 ha) will reduce water level considerably and, as a result, create many opportunities for ecological succession from pioneer stages. Unfortunately, this will be at the expense of existing ecological value and a trade-off has to be made using maps of succession stage.

When changes in hydraulic roughness of floodplains are not acceptable because of flood safety reasons, and the impact of CFR measures on lowering the water levels is limited (in many cases there are hydraulic bottlenecks present within the river reach), opportunities for increasing forest cover on the floodplains will be restricted.

outlined in Table 23.1. In other systems, the guidelines could be modified.

Many river restoration projects have been undertaken without prior knowledge of the reference condition, but these have greatly contributed to our present understanding of riverine ecosystems (Buijse *et al.*, 2002). CFR requires an interdisciplinary approach, where river managers, civil engineers, ecologists and other stakeholders work together. CFR appears a perfect example of an adaptive management strategy, managing and evaluating in regular cycles while increasing system knowledge step by step (Walters and Holling, 1990).

23.6 General conclusions

Ecohydraulic and morphological processes in rivers shape the riverine landscape. In river systems where these processes are capable of forming and rejuvenating the floodplains, this leads to a mosaic of succession stages (temporal diversity) and an ecologically diverse landscape. It can be regarded as a temporally continuous landscape, in which young ecotopes are formed and older patches are rejuvenated or succession is reset. River regulation hampers the processes that act on the landscape and disrupts the balance of processes. This results in a temporally discontinuous and aging landscape that lacks pioneer stages and may be subject to significant biodiversity loss over time.

Based on the case studies presented in this chapter, it can be concluded that the concept of temporal continuity and the cyclic floodplain rejuvenation strategy are potentially valuable tools to help implement river restoration and flood protection for western European lowland rivers such as the River Waal. However, CFR is a process that needs to be planned carefully in space (i.e. taking the steady state into consideration) and in time (i.e. the periodicity depends on factors such as rates of sedimentation, flood risk management objectives and restoration targets). As more data are required on the appropriate spatial and temporal dimensions of cyclic floodplain rejuvenation, the implementation of adaptive management strategies is advised in the future to ensure that management actions implemented can be used to help inform the management of other sites where restoration measures are proposed. Furthermore, the response and monitoring of flora and fauna (or biodiversity) to various anthropogenic restoration methods must be considered an essential part of this process.

References

Amoros, C. and Wade, P.M. (1996) Ecological successions. In Petts, G.E. and Amoros, C. (eds) *Fluvial Hydrosystems*, Chapman & Hall, London, pp. 211–241.

Amoros, C., Rostan, J.-C., Pautou, G. and Bravard, J.-P. (1987) The reversible process concept applied to the environmental management of large river systems. *Environmental Management*, **11**: 607–617.

Baptist, M., Penning, W.E., Duel, H., Smits, A.J.M., Geerling, G.W., van der Lee, G.E.M. and van Alphen, J.S.L. (2004) Assessment of the effects of cyclic floodplain rejuvenation on flood levels and biodiversity along the Rhine River. *River Research*, **20**: 285–297.

Bravard, J.-P., Amoros, C. and Pautou, G. (1986) Impact of civil engineering works on the successions of communities in a fluvial system: a methodological and predictive approach applied to a section of the Upper Rhône River, France. *Oikos*, **47**(1): 92–111.

Buijse, A.D., Klijn, F., Leuven, R.S.E.W., Middelkoop, H., Schiemer, F., Thorp, J.H. and Wolfert, H.P. (2005) Rehabilitation of large rivers: references, achievements and integration into river management. *Archiv für Hydrobiologie Supplement*, **155**(1–4): 1–4.

Buijse, A., Coops, H., Staras, M., Jans, L.H., van Geest, G.J., Grift, R.E., Ibelings, B.W., Oosterberg, W. and Roozen, C.J.M. (2002) Restoration strategies for river floodplains along large lowland rivers in Europe. *Freshwater Biology*, **47**(4): 889–907.

Cordes, L., Hughes, F.M.R. and Getty, M. (1997) Factors affecting the regeneration and distribution of riparian woodlands along a northern prairie river: the Red Deer River, Alberta, Canada. *Journal of Biogeography*, **24**: 675–695.

Corenblit, D., Tabacchi, E., Steiger, J. and Gurnell, A.M. (2007) Reciprocal interactions and adjustments between fluvial landforms and vegetation dynamics in river corridors: A review of complementary approaches. *Earth Science Reviews*, **84**(1–2): 56–86.

Duel, H., Baptist, M.J. and Penning, W.E. (2001) *Cyclic Floodplain Rejuvenation. A New Strategy Based on Floodplain Measures for Both Flood Risk Management and Enhancement of the Biodiversity of the River Rhine.* Publication 14-2001, Netherlands Centre for River Studies, Delft.

FISRWG (1998) *Stream Corridor Restoration: Principles, Processes, and Practices.* Interagency Stream Restoration Working Group (FISRWG).

Geerling, G.W., Vreeken-Buijs, M.J., Jesse, P. and Ragas, A.M.J. (2009) Mapping river floodplain ecotopes by segmentation of spectral (CASI) and structural (LiDAR) remote sensing data. *River Research*, **813**: 795–813.

Geerling, G.W., Kater, E., van den Brink, C., Baptist, M.J., Ragas, A.M.J. and Smits, A.J.M. (2008) Nature rehabilitation by floodplain excavation: The hydraulic effect of 16 years of sedimentation and vegetation succession along the Waal River, NL. *Geomorphology*, **99**(1–4): 317–328.

Geerling, G.W., Ragas, A.M.J., Leuven, R.S.E.W., van den Berg, J.H., Breedveld, M., Liefhebber, D. and Smits, A.J.M. (2006) Succession and rejuvenation in floodplains along the river Allier (France). *Hydrobiologia*, **565**(1): 71–86.

Greco, S.E., Fremier, A.K., Larsen, E.W. and Plant, R.E. (2007) A tool for tracking floodplain age land surface patterns on a large meandering river with applications for ecological planning and restoration design. *Landscape and Urban Planning*, **81**: 354–373.

Gurnell, A. (1997) The hydrological and geomorphological significance of forested floodplains. *Global Ecology and Biogeography Letters*, **6**: 219–229.

Havinga, H. and Smits, A. (2000) River management along the Rhine: a retrospective view. In Smits, A.J.M., Nienhuis, P. and Leuven, R. (eds) *New Approaches to River Management*, Backhuys Publishers, Leiden, pp. 15–32.

Hughes, F.M.R. (1997) Floodplain biogeomorphology. *Progress in Physical Geography*, **21**(4): 501–529.

Hughes, F.M.R., Adams, W.M., Muller, E., Nilsson, C., Richards, K.S., Barsoum, N., Decamps, H., Foussadier, R., Girel, J., Guilloy, H., Hayes, A., Johansson, M., Lambs, L., Pautou, G., Peiry, J.L., Perrow, M., Vautier, F. and Winfield, M. (2001) The importance of different scale processes for the restoration of floodplain woodlands. *Regulated Rivers*, **17**: 325–345.

Hupp, C.R. and Rinaldi, M. (2007) Riparian vegetation patterns in relation to fluvial landforms and channel evolution along selected rivers of Tuscany (central Italy). *Annals of the Association of American Geographers*, **97**: 12–30.

Johnson, W.C. (1997) Equilibrium response of riparian vegetation to flow regulation in the Platte River, Nebraska. *Regulated Rivers: Research and Management*, **13**: 403–415.

Junk, W., Bayley, P. and Sparks, R. (1989) The flood pulse concept in river-floodplain systems. In Dodge, D. (ed.) *Proceedings of the International Large River Symposium*. Canadian Special Publication of Fisheries and Aquatic Sciences, 106, pp. 110–127.

Latterell, J.J., Bechtold, J.S., Naiman, R.J., O'Keefe, T.C. and van Pelt, R. (2006) Dynamic patch mosaics and channel movement in an unconfined river valley of the Olympic Mountains. *Freshwater Biology*, **51**: 523–544.

Merritt, D.M. and Cooper, D.J. (2000) Riparian vegetation and channel change in response to river regulation: a comparative study of regulated and unregulated streams in the Green River Basin, USA. *Regulated Rivers: Research and Management*, **16**: 543–564.

Middelkoop, H., Schoor, M.M., Wolfert, H.P., Maas, G.J. and Stouthamer, E. (2005) Targets for ecological rehabilitation of the lower Rhine and Meuse based on a historic geomorphologic reference. *Archiv für Hydrobiologie Supplement*, **155**(1–4): 63–88.

Ministry of Transport, Public Works and Water Management (2000) Nota Ruimte voor de rivier (in Dutch). Ministry of Transport, Public Works and Water Management, Directorate-General Rijkswaterstaat, Den Haag.

Nilsson, C. (1989) Patterns of plant species richness along riverbanks. *Ecology*, **70**(1): 77–84.

Nilsson, C. and Svedmark, M. (2002) Basic principles and ecological consequences of changing water regimes: riparian plant communities. *Environmental Management*, **30**: 468–480.

Nilsson, C., Grelsson, G., Dynesius, M., Johansson, M. and Sperens, U. (1991) Small rivers behave like large rivers: effects of post-glacial history on plant species richness along riverbanks. *Journal of Biogeography*, **18**: 533–541.

Petts, G.E. (2000) A perspective on the abiotic processes sustaining the ecological integrity of running waters. *Hydrobiologia*, **422**(1961): 15–27.

Petts, G.E. and Amoros, C. (1996) *Fluvial Hydrosystems*, Chapman & Hall, London.

Silva, W., Klijn, F. and Dijkman, J. (2001) *Room for the Rhine Branches in the Netherlands: What the Research has Taught Us*. Ministry of Transport, Public Works and Water Management, Directorate-General for Public Works and Water Management, Institute for Inland Water Management and Waste Water Treatment/RIZA.

Smits, A.J.M., Havinga, H. and Marteijn, E.C.L. (2000) New concepts in river and water management in the Rhine river basin: how to live with the unexpected. In Smits, A.J.M., Nienhuis, P.H. and Leuven, R.S.E.W. (eds) *New Approaches to River Management*, Backhuys Publishers, Leiden, pp. 267–286.

Stanford, J.A., Ward, J., Liss, W.J., Frissell, C.A., Williams, R.N., Lichatowich, J.A. and Coutant, C.C. (1996) A general protocol for restoration of regulated rivers. *Regulated Rivers: Research and Management*, **12**(4–5): 391–413.

Tockner, K. and Stanford, J.A. (2002) Riverine flood plains: present state and future trends. *Environmental Conservation*, **29**(3): 308–330.

Tockner, K. and Ward, J.V. (1999) Biodiversity along riparian corridors. *Archiv für Hydrobiologie Supplement on Large Rivers*, **11**(3): 293–310.

Tockner, K., Malard, F. and Ward, J.V. (2000) An extension of the flood pulse concept. *Hydrological Processes*, **14**: 2861–2883.

Van de Ven, G.P. (2004) *Man-made Lowlands. History of Water Management and Land Reclamation in The Netherlands*, Stichting Matrijs, Utrecht.

Van der Velde, G., Leuven, R.S.E.W., Ragas, A.M.J. and Smits, A.J.M. (2006) Living rivers: trends and challenges in science and management. *Hydrobiologia*, **565**: 359–367.

Walters, C. and Holling, C. (1990) Large scale management experiments and learning by doing. *Ecology*, **71**(6): 2060–2068.

Ward, J.V. (1998) Riverine landscapes: biodiversity patterns, disturbance regimes, and aquatic conservation. *Biological Conservation*, **83**: 269–278.

Ward, J.V., Malard, F. and Tockner, K. (2002) Landscape ecology: a framework for integrating pattern and process in river corridors. *Landscape Ecology*, **17**: 35–45.

Ward, J.V., Tockner, K., Uehlinger, U. and Malard, F. (2001) Understanding natural patterns and processes in river corridors as the basis for effective river restoration. *Regulated Rivers: Research and Management*, **17**(4–5), 311–323.

24 Embodying Interactions Between Riparian Vegetation and Fluvial Hydraulic Processes Within a Dynamic Floodplain Model: Concepts and Applications

Gregory Egger[1], Emilio Politti[1], Virginia Garófano-Gómez[2], Bernadette Blamauer[3], Teresa Ferreira[4], Rui Rivaes[4], Rohan Benjankar[5] and Helmut Habersack[3]

[1]Environmental Consulting Ltd, Bahnhofstrasse 39, 9020 Klagenfurt, Austria
[2]Institut d'Investigació per a la Gestió Integrada de Zones Costaneres (IGIC), Universitat Politècnica de València, C/Paranimf, 1, 46730 Grau de Gandia (València), Spain
[3]Christian Doppler Laboratory for Advanced Methods in River Monitoring, Modelling and Engineering, Institute of Water Management, Hydrology and Hydraulic Engineering, University of Natural Resources and Life Sciences Vienna, Muthgasse 107, 1190 Vienna, Austria
[4]Forest Research Centre, Instituto Superior de Agronomia, Technical University of Lisbon, Tapada da Ajuda 1349-017, Lisbon, Portugal
[5]Center for Ecohydraulics Research, University of Idaho, 322 E. Front Street, Boise, ID 83702, USA

24.1 Introduction

Riparian ecosystems are complex and dynamic environments where living organisms experience disturbances distributed unevenly over space and time (Naiman *et al.*, 1993; Naiman and Décamps, 1997). Interactions between riparian vegetation and fluvial hydraulic processes are regulated by the flow regime, which is one of the primary drivers shaping morphological patterns and ecological integrity (Poff *et al.*, 1997). One of the outcomes of these complex interactions is discrete vegetation units, which represent snapshots in time, characterized by stands of different age and structural features.

Over recent decades, significant efforts have been directed towards the implementation of riparian vegetation simulation models, the complexity of which reflects the development of both information technology and scientific research within the domain of Ecohydraulics. Early models can be traced to the late 70s (Franz and Bazzaz, 1977; Phipps, 1979) followed by more sophisticated models in subsequent decades (Pearlstine *et al.*, 1985; Poiani and Johnson, 1993; Auble *et al.*, 1994; Glenz *et al.*, 2008). All of the models were based on functional relationships between riparian vegetation and physical habitat, and demonstrated that some characteristics of these ecosystems could be partially simulated. However,

Ecohydraulics: An Integrated Approach, First Edition. Edited by Ian Maddock, Atle Harby, Paul Kemp and Paul Wood.
© 2013 John Wiley & Sons, Ltd. Published 2013 by John Wiley & Sons, Ltd.

the majority of models were developed for specific boundary conditions and were not easily transferable to other geographical regions (Merritt *et al.*, 2010). Further limitations included the coarse spatial resolution and lack of validation with ground-truth data.

As fluvial hydraulic processes and riparian dynamics have become increasingly well studied and quantified in a wide range of situations, the potential to make generalizations regarding ecohydraulics has increased. A major challenge for riparian ecosystem modellers is the development and implementation of models with broad geographical coverage of physical habitat factors, which characterize flood-pulse drivers and which can account for vegetation establishment, succession or resetting over a range of river conditions and climatic regions. Such an holistic model, 'CASiMiR Vegetation' (Computer Aided Simulation Model for In-stream flow and Riparia; www.casimir-software.de), was proposed by Benjankar *et al.* (2011). This model is potentially capable of representing the spatio-temporal landscape variability typically observed across natural riparian ecosystems. The number of potential applications for a dynamic floodplain model is great, ranging from fundamental river research to river management, encompassing biodiversity conservation, river restoration, guidelines for reservoir operation and adaptation to climate change (Haney and Power, 1996; Wieringa and Morton, 1996; Downs and Kondolf, 2002; Perona *et al.*, 2009).

In this chapter, causal interactions between physical habitat and floodplain vegetation will be examined. The complex network of hierarchical relationships will be used to illustrate our present understanding of riparian ecosystem functioning. The relationships between riparian vegetation and fluvial hydraulic processes are considered and used to build a conceptual dynamic floodplain vegetation model able to replicate riparian habitat characteristics spatially and temporally for different river conditions and eco-regions. The model, an upgrade of the original CASiMiR Vegetation model (Benjankar *et al.*, 2011), has been calibrated and tested on three continents and for different climate settings, and has been found to be a valuable guiding tool for river management and conservation purposes.

24.2 Physical habitat and its effects on floodplain vegetation

Riparian zones occur from the edge of water bodies to upland communities situated on hill-slopes and terraces

(Naiman *et al.*, 2005). Four zones can be defined with respect to their transverse position on the river corridor and the degree to which they are influenced by river dynamics (Goodwin *et al.*, 1997; Ward *et al.*, 2002):

• The **aquatic zone** is confined within the wetted perimeter of the channel and is inundated under mean water conditions. The dominant vegetation is composed of herbaceous layers, which can be displaced easily. This zone is the most prone to disturbances resulting from hydraulic forces of the flowing stream.

• The **bank zone** is the area immediately adjacent to the aquatic zone. It is subject to a large degree of disturbance as a result of the proximity to the active channel and relatively high groundwater levels. Due to the regular disturbance of this zone, occurring at bankfull discharge, vegetation exhibits short lifecycles and vegetation stands are typically young in age. The dominant landscape features are sand and gravel bars, and vegetation types include woody pioneers (*Salix* spp., *Populus* spp.), bands of herbaceous vegetation (forbs) and reeds (*Phalaris* spp., *Phragmites* spp., *Epilobium* spp., *Scirpus* spp., etc.) and low shrubs (*Tamarix* spp., *Salix* spp.) (Karrenberg *et al.*, 2002).

• The **floodplain zone** is affected by both small and large floods. The floodplain is situated beyond the bank zone at greater distances from the active channel. Under natural conditions, its outer margin conventionally corresponds to the extent of a 100-year flood. Vegetation in this zone tends to be more robust and well developed. The increased soil depth and nutrient pools tend to encourage the development of long-living and flood-resistant softwood forest species (*Populus* spp., *Salix* spp., *Alnus* spp.).

• The associated riverine **wetland zone** is located within the floodplain. In most instances, such habitats originate in former side/cut-off channels of the river. The wetlands may be periodically connected to the main channel and are characterized by hydric soils. Groundwater and overbank flooding are more important in these areas than in the rest of the floodplain. Characteristic habitats can include deep and shallow open-water habitats and marshes (Naiman *et al.*, 2005).

Additional information and the salient characteristics of three of these zones are summarized in Table 24.1.

Riparian ecosystems are diverse but share common characteristics – being dependent on the flow regime of the river. The flow regime is characterized by hydrological variability, which controls physical habitat and, hence, riparian characteristics, such as groundwater level, soil moisture, nutrient supply and disturbance regime. All of these habitat characteristics influence the establishment, growth and mortality of floodplain vegetation. Low flows

Table 24.1 Ecological characteristics of the different riverine zones: (i) bank zone, (ii) floodplain zone and (iii) riverine wetland zone.

Riverine zone	Bank	Floodplain	Riverine wetland
Substrate	Aggradation of gravel and sand	Sedimentation of sand and silt	Sedimentation of silt and clay
Location and habitats formed	Islands and bars within the active channel	Fluvial terraces on alluvial plains	Ponds, flood troughs and depressions, which are principally formed by the river on the floodplain
Morphodynamics and disturbance regime	High – intense deposition and erosion processes over short time scales (several times a year to several years)	Moderate – sedimentation during floods with long return periods (many years to decades)	Low – low level and constant sedimentation. Most deposition occurs during floods with additional biogenic sedimentation at other times
Soil moisture	Wet to dry; highly variable	Wet to moderately dry; moderate variability	Wet to completely inundated; more or less constant moisture
Key ecological factors	(1) Mechanical disturbances due to flooding; (2) periods of extreme drought; (3) high flow pulses	(1) Competition for light and water; (2) physiological constraints due to flood inundation; (3) good growing conditions for vegetation	(1) Long flood inundation; (2) low soil aeration due to waterlogged ground

have a double effect on riparian biocoenosis: first, they enable recruitment of floodplain species to habitats that under mean water conditions would be submerged and therefore too wet or hydraulically active to allow vegetation establishment (Richter and Thomas, 2007); second, low water conditions and receding groundwater levels dry the soil, leading, ultimately, to desiccation and the exclusion of vegetation.

The mean flow/discharge represents the primary hydraulic control on floodplain vegetation. Water supply determines the growth conditions for vegetation and the mean water level during the summer. Below mean level, only pioneer taxa are likely to be recorded, since they are physiologically well adapted to permanent submersion and high mechanical disturbance. Above mean water level, reeds and emergent flora are more common (Ellenberg, 2009).

Bankfull discharge shapes the dimension and geometry of natural alluvial channels. Short-term discharge variability does not necessarily cause channel morphology to adjust during each event, because the channel responds to a range of competent discharges (Richards, 1982). The recurrence interval of bankfull discharge typically varies between 1.5 and 2 years for natural rivers (Leopold, 1994; 1997). This relatively high-frequency event contributes the largest volume of transported material (sediment) when compared with larger and less-frequent events, and

generates the dominant channel-forming/changing processes.

High-flow pulses prevent riparian vegetation from encroaching into the channel and determine streambed and bank substrate (Richter and Thomas, 2007). In addition, they control the establishment of seedlings on the bank zone by creating nursery sites formed by sediment deposition or by the disruption of pre-existing vegetation cover ('score disturbance') (Mahoney and Rood, 1998; Polzin and Rood, 2006; Braatne et al., 2007).

Floods connect the river to the surrounding floodplain through cyclic expansion and contraction. Floods recharge the floodplain water table and provide seedlings with access to moist soils, create new sites for plant recruitment, flush organic material such as woody debris, deposit nutrients onto the floodplain and disperse seeds of riparian plants (Richter and Thomas, 2007). Flooding is one of the most important factors accounting for the remarkably high biodiversity/species richness associated with riparian zones (Ward et al., 2002). The effect of large floods (>10-year events) on riverine ecosystems is determined primarily by the degree of mechanical disturbance. In contrast, small floods (i.e. 2–10 year events) are characterized by more subtle effects, yet they are potentially of greater physiological significance to riparian vegetation in determining the establishment, growth and survival of vegetation assemblages (Richter and Thomas, 2007).

The physiological effect of a flood event on floodplain vegetation is inversely proportional to the tolerance of individual taxa and reflects the capacity of taxa to survive anoxic conditions.

Water depth influences the magnitude of flood damage and is dependent on whether inundation is complete or partial, and whether soil saturation is temporary or permanent (Glenz *et al.*, 2006). Few plant species are capable of withstanding extended periods of soil saturation. While some plants can tolerate root anoxia, they may not be able to survive submersion of all leaves. Thus, tolerance of complete submersion is much lower than that for partial submersion. The response of vegetation to submersion can be quantified in terms of inhibition of seed germination, decline in vegetative and reproductive growth, promotion of early senescence and even mortality. However, the most significant symptom recorded in most trees is a decline in shoot growth. Flood stress responses result from a combination of flood factors, including flood level and relative water depth, flood duration, flood timing and time since the last flood event. Some biotic factors, however, can mitigate flood impacts, including morphological adaptations such as the development of lenticels, adventitious roots and aerenchyma tissues, and physiological adaptations such as the regulation of anoxic metabolism.

Flooding is potentially more harmful if it occurs during the growing season than if it occurs during the dormant season. Irrespective of its duration, flooding has less or no effect during the dormant season because of a reduced demand for oxygen by roots and microorganisms. Due to the capacity of cold water to hold more dissolved oxygen, it is less damaging to vegetation than warmer water. Similarly, rapidly flowing water is chemically less harmful than stagnant water, as it has a higher dissolved oxygen concentration due to mixing throughout the water column. However, if flooding is recurrent, damage or mortality may occur due to reduced oxygen supply, increasing the effects of the trees' biochemical response.

The influence of groundwater on riparian vegetation largely reflects the supply of moisture, which ultimately makes vegetation less dependent on precipitation (Stromberg *et al.*, 1996). In semi-arid regions, groundwater levels are vital in sustaining floodplain vegetation and biodiversity during the extended dry season. Groundwater level, like river stage, rises and falls (Naiman *et al.*, 2005) and, as a consequence, soil aeration is promoted, creating favourable growing conditions. The dynamic nature of groundwater is a feature of most riparian habitats and distinguishes them from other groundwater-dependent ecosystems such as bogs and fens. The soil moisture balance is linked to groundwater level, topography and pedological properties (such as soil water storage capacity, soil texture and thickness) and also to climatic parameters such as evaporation and precipitation.

Morphodynamic processes triggered by floods, such as sediment deposition/aggradation and fluvial erosion, result in physical disturbances. These disturbances determine species diversity, population recruitment and survival and the age structure of stands (Denslow, 1980; Plachter, 1998; Edwards *et al.*, 1999). The extent and intensity of flood disturbance can be measured by its destructive impact on the vegetation. Three degrees of damage can be distinguished: (1) partial disruption of the herbaceous layer; (2) partial disruption of the tree and shrub layer due to sedimentation and/or shearing of plant fronds and branches; and (3) complete destruction of all plant layers including the erosion of roots (Canham and Marks, 1985; Pickett and White, 1985).

24.3 Succession phases and their environmental context

Riverscapes can be viewed as complex networks of patches, resulting from small-scale gradients in height above the bed and local substrate characteristics, and related gradients of water/moisture availability to the root system (Naiman *et al.*, 2005). Organizing riparian ecosystems into recognizable units, integrating arrays of natural vegetation, soil characteristics and fluvial geomorphology, and the vegetation community response to different disturbances, allows comparison between different rivers (Forman, 1995; Kovalchik and Clausnitzer, 2004; Naiman *et al.*, 2005); assuming that similar morpho-edaphic conditions yield similar vegetation structure. A modelling approach based on recognition of discrete riparian vegetation patches, as shaped by fluvial hydraulic processes, is potentially transferable to other geographical regions and climates. The relationships between fluvial hydraulic processes and vegetation structure could then be used to build a widely applicable scheme of succession-regression pathways rather than undertaking species-specific comparisons. Succession of taxa may differ between regions or case studies, although it is essentially driven by the same processes.

In general terms, three different succession stages can be recognized (colonization, transition and mature stages) and several phases can be defined within each stage, representing temporally consecutive levels of vegetation development. The differences between stages reside in the dominant development stage, habitat characteristics and the flow disturbance regime they have experienced.

Primary succession starts when seedlings begin to colonize bare and fresh sediments or when the germination of seeds is observed. The first phase of the 'Colonization Stage', known as the 'Initial Phase', consists of bare soil where few plants are present, and in some instances seedlings only become established for a short time period before the next disturbance. Groundwater is typically near the surface and there is a high frequency of disturbance (see also Table 24.1). The second phase of the 'Colonization Stage' is the 'Pioneer Phase' and is characterized by relatively sparse vegetation. It consists primarily of species with a ruderal or stress-tolerant strategy which are adapted to frequent disturbance and strong hydrological variability (wetting and drying). In headwater reaches, low fertility substrates support limited vegetation, seed bank and organic matter (Karrenberg et al., 2002). In contrast, in lowland rivers, seed banks and nutrient levels do not represent a limiting factor and colonizing processes occur rapidly (Naiman et al., 2005). Annual and bi-annual plant species have a competitive advantage immediately after disturbance events, since they grow rapidly and complete their lifecycles before other plants are able to replace them (Oliver and Larson, 1996).

Disturbance clearings provide suitable habitat conditions for seedlings or vegetative growth and the establishment of woody species adapted to frequent disturbance, usually Salicaceae in Europe and North America (e.g. Auble and Scott, 1998; Karrenberg et al., 2002). Large pieces of wood (tree trunks, roots and branches) deposited during the 'Initial Phase' allow the rapid re-establishment of vegetative biomass in the early stages of post-flood riparian forest renewal (Naiman et al., 2008), as they are able to trap seeds and fine sediments (Corenblit et al., 2009; Francis et al., 2009). During the 'Pioneer Phase', biomass production, standing biomass and the average age of vegetation patches is low, whereas species diversity is high. Local habitat characteristics largely determine vegetation patterns, and biotic interactions (e.g. competition) are of minor importance during this phase.

The 'Transition Stage' (also referred to as the 'Consolidation Stage'; Naiman et al., 2005) takes place during the years following vegetation colonization. Vegetation cover, biomass production and standing crop increase, while the average vegetation stand age increases and species diversity remains high. Several phases can be distinguished within this stage. The 'Herb Phase' is characterized by short-lived herbaceous species with a ruderal or competitive strategy, but in many cases it may comprise monospecific vegetation of sedges and/or reeds. Second, the 'Shrub Phase' can develop directly from the 'Pioneer Phase' or follow the 'Herb Phase'. The 'Shrub Phase' comprises woody and long-lived species with stress-tolerant and competitive development strategies. The 'Early Successional Woodland Phase' (also called the 'Stem Exclusion Phase'; Naiman et al., 2005) occurs when trees are able to replace the shrubs as the dominant life form in the patch. The recruitment of dominant trees typically occurs on sand and gravel bars (during the 'Initial Phase') and includes pioneer species like willows and cottonwoods ('softwood forest'), such as Populus nigra and Salix alba in Europe (Barsoum, 2002) and Salix sitchensis, Alnus rubra and Populus trichocarpa in the Pacific Northwest (Polzin and Rood, 2000; Van Pelt et al., 2006). These species are fast growing and saplings do not develop in shady locations (except when sprouting). The standing biomass during this phase is greater than that recorded in the previous stage, and the plant community is more persistent and complex, increasing species biodiversity.

The 'Transition Stage' finishes with the onset of the 'Established Forest Phase', also known as the 'Understorey Re-initiation Phase' (Oliver and Larson, 1996; Naiman et al., 2005). In most instances, the species present during this phase are recruited to the understorey in the Early Successional Woodland Phase. They typically have slow growth rates and many may die before the onset of the next succession phase. When a portion of the old softwood forest is dying, a gap is created in the canopy and shade-tolerant ('hardwood forest') understorey species may become established. Typical species recorded during this phase are Fraxinus spp., Quercus spp., Ulmus spp. (middle Europe) and Picea spp. (Pacific Northwest). These patches are more stable, less prone to disturbance and usually have multiple canopy layers.

Despite low biomass production, the standing biomass is large during the 'Mature Stage'. The old growth stage is characterized by high species diversity, though it may be lower than that recorded during the 'Transition Stage'. The 'Mature Mixed Forest Phase' is typical of this stage, including a combination of woody and long-lived riparian and terrestrial species. Individual over-storey trees may die as the forest stand ages, opening up canopy space which leads rapidly to regeneration of the understorey. This is an autogenic process whereby trees regenerate and grow without the influence of external disturbances. Live old trees and dead standing trees occur together with a diverse understorey and several layers of vegetation. The 'Climax Stage' should comprise terrestrial upland forest,

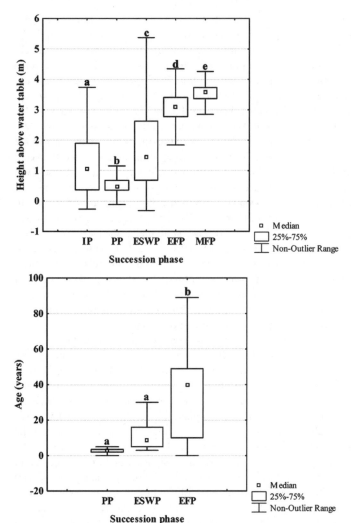

Figure 24.1 Height above water table and age differentiation of succession phases measured in riparian patches of the Odelouca River (southern Portugal) (letters represent significantly different groups, ANOVA, $p < 0.001$). IP – Initial Phase; PP – Pioneer Phase; ESWP – Early Successional Woodland Phase; EFP – Established Forest Phase; MFP – Mature Forest Phase. Distance to water table increases from younger to older phases while, conversely, the disturbance associated with the water table decreases (Rivaes *et al.*, in press).

which reflects the dominant vegetation in the area beyond the influence of the flood regime of the river.

An example of the temporal duration of different succession phases and their relationship with mean water level is displayed in Figure 24.1. The spatial arrangement of succession phases produces a mosaic on the riparian landscape and is typical of natural rivers (Figure 24.2). The local climate, valley shape and hydro-morphological setting determine the relative proportion of the different succession phases and, in many cases, some only occupy a small proportion of the floodplain or may be absent. Furthermore, the phases are shaped by flow regime variability and are embedded within the succession process, the disturbance regime and local morpho-edaphic conditions. For each river, vegetation patches can be classified,

described and allocated to a phase according to the prevailing abiotic characteristics. Vegetation patches can then be organized in succession series, which reflect current hydroecological and hydraulic processes.

In general, three vegetation succession series have been described for riverine environments (Kovalchik and Clausnitzer, 2004): woodland, reed and wetland series. Succession series may change spatially and temporally along the river valley, reflecting changes in soil and fluvial characteristics associated with erosion or deposition processes resulting from hydraulic forces/flood events. The woodland riparian series is the most common riparian sequence and succeeds from 'Pioneer' to 'Mature Mixed Forest Phase'; it also has a trajectory to the regional woodland climax, though the latter is seldom achieved. The reed

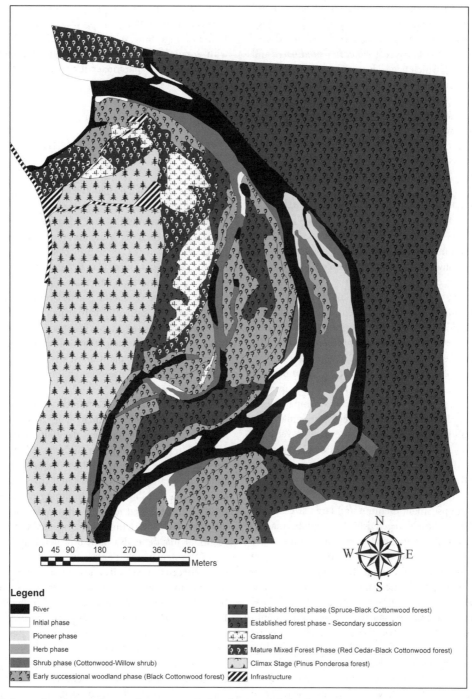

Figure 24.2 Succession phases and vegetation types on the largely natural Flathead River (Montana, USA). Patches close to the river channel, where disturbances are frequent, are dominated by phases belonging to the Colonization Stage (e.g. Initial and Pioneer Phases) or by the younger phases of the Transitional Stage (e.g. Shrub Phase). In the patches lying far from the channel, the most mature phases of the Transitional Stage (e.g. Established Forest Phase) are more abundant, together with the Mature Mixed Forest Phase and the Climax Stage.

series can occur on permanently waterlogged habitats and may be stable over many years, displaying high resilience to flood events. This succession series is usually dominated by competitive herbs (e.g. *Phalaris arundinacea* and *Phragmites* spp.), which have dense herbaceous layers and strong rhizomatous systems, preventing the germination of woody species. Finally, the wetland series is recorded in some riverine wetlands located in areas inundated or saturated by surface or groundwater at sufficient frequency and duration to support vegetation typically adapted to saturated soil conditions.

24.4 Response of floodplain vegetation to fluvial processes

Vegetation has major impacts on water conveyance and flood levels (Kouwen and Fathi-Moghadam, 2000) by dissipating kinetic energy (Tabacchi *et al.*, 2000). Vegetation also causes channel narrowing by obstructing flow (Osterkamp and Hupp, 2010), which results in increasing flow velocities and the concentration of flows around vegetated areas as well as slower flow velocities within the vegetated patches (Rominger *et al.*, 2010).

The effects of vegetation on in-stream hydraulics are principally associated with energy loss, which may be expressed as vegetation roughness and is comparable to bed roughness. Bed roughness is often approximated by empirical values like Manning's n, the Darcy–Weisbach friction factor f or Chezy's resistance factor C (Wu *et al.*, 1999). The determination of vegetation roughness is required to derive changes in hydraulic parameters such as flow velocity. In this context, vegetation can be classified into three groups: small, medium and large-sized forms (DVWK, 1991). 'Small' vegetation (e.g. aquatic macrophytes or grasses) is very small compared to the water depth, and its roughness can be approximated as surface roughness. The height of 'medium' vegetation forms is typically within the same range as the average water depth, whereas 'large' vegetation forms are characterized by a plant height greater than the average water depth (DVWK, 1991). The latter two classes are characterized by a mixture of form and surface roughness and dependent on vegetation properties. To determine vegetation roughness for medium and large plant forms, the drag force equation (Equation (24.1)), which describes the resisting force of a disturbing element within the flow, can be used.

$$F_\mathrm{D} = \frac{1}{2}\rho C_\mathrm{d} A_\mathrm{p} U^2 \qquad (24.1)$$

where F_D is the drag force, ρ is the water density, C_d the drag coefficient, A_p the projected vegetation area and U the average flow velocity. By resolving this equation with gravitational force, the flow velocity, based on the Chezy roughness parameter for vegetation C_veg, can be determined (Equation (24.2)):

$$U = \sqrt{\frac{2g}{C_\mathrm{d} A_\mathrm{p} \sigma_{v,n}}} \sqrt{RS} = C_\mathrm{veg}\sqrt{RS} \qquad (24.2)$$

$$C_\mathrm{veg} = \sqrt{\frac{2g}{C_\mathrm{d} A_\mathrm{p} \sigma_{v,n}}} \qquad (24.3)$$

where g is the gravitational acceleration, $\sigma_{v,n}$ the lateral spacing of plants, R the hydraulic radius and S the energy slope. In Equation (24.2) and Equation (24.3), the vegetation roughness is dependent on the following variables: the drag coefficient C_d, the projected vegetation area A_p and the lateral spacing of plants $\sigma_{v,n}$. With increasing values of C_d, A_p and $\sigma_{v,n}$, which are highly variable within plant species and vegetation patches, roughness increases. For parameterization of vegetation patches, the initial projected vegetation area A_p, can be determined in the field by measurements of the height and width of the plants. The determination of the plant density is also important, for which some authors use photographs (Schneider, 2010; Blamauer *et al.*, 2011).

Flume studies conducted to measure single plant drag forces (Oplatka, 1998; Freeman *et al.*, 2000; Wilson *et al.*, 2010) have revealed that stem flexibility and the density of leaves are dominant factors influencing the drag force and, hence, the roughness. Plant flexibility has a major influence on the projected vegetated cross-sectional area by allowing bending and streamlining of branches and leaves when exposed to higher flow velocities. For rigid vegetation (e.g. tree trunks), the projected vegetated cross-sectional area depends on the flow depth and the diameter of the stem/trunk. For flexible vegetation, a relationship between flow velocity and compaction can be applied which reduces the projected vegetated cross-sectional area (Oplatka, 1998; Kouwen and Fathi-Moghadam, 2000; Järvelä, 2004). This relationship can be used as a correction factor for streamlining behaviour based on flow velocity (Schneider, 2010).

Van de Wiel (2003) used the bending angle of vegetation to determine projected vegetated cross-sectional area. This approach was also used by Blamauer *et al.* (2011), who combined bending with width contraction for area calculation within an iterative bending model. The iterative process was required as the streamlining and bending, and thus the contraction of the vegetated cross-sectional area, depends on flow velocity, which, in turn, is dependent on roughness. This iterative model was implemented in a two-dimensional hydrodynamic-numerical model to calculate vegetation roughness based on plant properties (Blamauer *et al.*, 2011).

The lateral spacing of plants $\sigma_{v,n}$ is also an important ecohydraulic factor and can be measured directly in the field by determining the abundance of plants per unit area. The definition of the drag coefficient C_d is, however, more problematical. Freeman *et al.* (2000) identified that increasing flow velocities, water depths, plant densities, plant flexibilities and plant spacing decreased the drag coefficient and that the Reynolds number is also important. DVWK (1991) proposed that values of C_d for plant populations ranged from 0.6 to 2.4, with values from 1.2 to 1.55 being applicable for willows (Van de Wiel, 2003; Järvelä, 2004). Wunder *et al.* (2011) found that leafy willows exhibited a lower drag coefficient, from 0.35 to 0.85, than leafless willows, which ranged from 0.5 to 1.0 for higher flow velocities and from 1.2 to 2.0 for lower velocities under experimental conditions. This may reflect the higher streamlining capacity of leaves subject to higher flow, and thus a more hydro-dynamically efficient shape.

The presence of vegetation modifies roughness and represents an obstacle to flow, which causes a deceleration of longitudinal flow velocities within vegetation patches and flow acceleration in the open channel (Zong and Nepf, 2011). The deceleration in a longitudinal direction is balanced by lateral acceleration which causes a diversion of the flow away from the patch. Between the different flow velocities, shear-layer vortices develop (Zong and Nepf, 2011). Aquatic vegetation, therefore, significantly influences the vertical velocity profile. In non-vegetated channels, a logarithmic velocity distribution curve develops, whereas in vegetated patches the shape of the velocity profile depends on the flow condition, plant properties (emergent or submerged), vegetation density and the vertical distribution of the plants (Schneider, 2011). Under submerged conditions, Stephan and Gutknecht (2002) observed the development of a logarithmic velocity profile above vegetation.

The impact of vegetation on water elevation during a flood depends on flow conditions, vegetation density and the alignment of the plants within the channel (Habersack *et al.*, 2008; Schneider, 2010). During flume experiments, Schneider (2010) found that higher vegetation densities resulted in greater backwater effects due to the reduction of flow velocity and cross-sectional area. For submerged plants, or for low vegetation densities, the alignment of plants was almost insignificant as a determinant of water

level, but for emergent vegetation or high vegetation densities, a nonlinear array resulted in greater water depth than for a linear alignment in the direction of flow.

Plants clearly influence the hydraulic environment and, thus, erosion and sedimentation processes (Simon *et al.*, 2004) by acting as protection/cover for benthic sediments (Tabacchi *et al.*, 2000) and decreasing re-suspension of fine material via a reduction of shear stress within the patch (Nepf *et al.*, 2010). Vegetation also enhances deposition and trapping of fine sediments and nutrients (Nepf *et al.*, 2010; Osterkamp and Hupp, 2010), leading to increased bed elevation and changes in soil moisture properties on exposed in-stream bars (Tabacchi *et al.*, 2000).

24.5 Linking fluvial processes and vegetation: the disturbance regime approach as the backbone for the dynamic model

The interaction between natural disturbances and successional processes leads to the formation of a mosaic of vegetation patches on the riverscape (Richter and Richter, 2000; Ward *et al.*, 2002) and the allocation of each one of these patches to a phase within the succession series. Disturbances are defined as '*relatively discrete events that disrupt the structure of an ecosystem, community, or population and change resource availability or the physical environment*' (White and Pickett, 1985, p. 7). Important disturbance descriptors include the event magnitude, duration and recurrence interval. For floods, disturbance descriptors also include hydraulic parameters such as critical bed shear stress (Parker, 1996), specific stream power (Komar and Carling, 1991), sediment transport capacity (Yang and Schenggan, 1991) and bank stability (Rinaldi and Darby, 2005).

The intensity of flood disturbances can be measured by their destructive impact on vegetation derived from hydraulic descriptors. Disturbance intensity also reflects mortality due to physiological damage resulting from either drought or submersion. Other causes of mortality should also be acknowledged, including other disturbances such as disease, fire and wind. The effects of disturbances on plants/vegetation depend on their properties (such as size and life form) and lifecycle characteristics. Disturbance severity and frequency are functions of stand resistance and recovery time (Huston, 1994). Recovery time is also related to the position of the community on the succession series.

Applying the disturbance regime approach to the vegetation of riparian ecosystems, succession and resetting can be simplified into three major process types, described using the quadrangular relationship between the disturbance intensity (I), vegetation resistance to disturbance (R) and the recovery time ($TREC$) vis-à-vis the disturbance interval ($TDIS$) (Egger *et al.*, 2007; Formann, 2009):

• A metastable process ('low-metastable systems' of Forman and Godron, 1986), where the disturbance intensity (I) and resistance (R) are in equilibrium ($I = R$). The disturbance interval ($TDIS$) is shorter than the recovery time of the maturation stage ($TREC$-mat), so the mature stage will never be reached ($TDIS \ll TREC$-mat). Such stages/systems are called metastable because, although oscillatory and gap dynamics are observed, no real successional development takes place. Constant interruption of the succession sequence causes the biocoenosis to remain at the same early stage, which is adapted to the disturbance (inherent disturbance; Böhmer, 1999). Metastable processes are typical of the colonization stage and the young transitional phases.

• The oscillation process, where disturbance intensity (I) and resistance (R) are not in equilibrium ($I > R$). Moreover, disturbance intervals are shorter than the recovery time of the mature stage ($TDIS < TREC$-mat). The sites experience vegetation succession, although they never reach maturity due to the high frequency of disturbance. Disturbance severity is high, recovery time is high and the system is always in an unstable state (succession and retrogression; cyclic or acyclic).

• The acyclic process, where disturbance intervals are longer than the recovery time. Succession continues until the mature stage is reached. Stable and non-stable phases, less adapted to the disturbance regime, alternate within the system. This is typical of the floodplain zone rarely affected by river dynamics.

However, the temporal dynamics of living ecosystems and their processes cannot be wholly replicated by computer models. Indicator values for different elements of ecosystem processes and pathways must be an input to models with simulated timings and frequencies replicating the real world, either for vegetation development or for flood disturbance. An attempt to build such a process-based dynamic floodplain model, 'CASiMiR Vegetation', was undertaken by Benjankar *et al.* (2011).

The CASiMiR model assumes that vegetation development depends on the functional relationship between hydrology, physical processes and vegetation communities. The main challenge in developing CASiMiR Vegetation (henceforth CASiMiR) was the dynamic coupling of biotic factors and disturbances to replicate the disturbance regime approach. Biotic factors are efficiently represented by the succession time series concept and vegetation stages/phases. Physical processes and disturbances can be represented by quantified indicator values to represent a gradient of disturbance magnitude: spring mean water level, shear stress (as indicator of morphodynamic disturbance) and flood duration (as indicator for physiological stress). CASiMiR simulates the interaction between vegetation succession drivers and disturbance regime with respect to metastable, oscillation and acyclic processes.

A primary feature of CASiMiR is the spatio-temporal simulation of vegetation succession or resetting that can also be visualized. The model is divided into modules: recruitment, morphodynamic disturbance and flood duration (Figure 24.3). CASiMiR is referred to as dynamic because it takes different inputs for each simulated year, and the outputs of each model run form an input to the next iteration.

The rules upon which CASiMiR is based can be described briefly as follows. The recruitment sub-model implements the 'Recruitment Box Model' concept (Mahoney and Rood, 1998), and has been extended to non-woody species. Spring mean water level influences the recruitment and scour disturbance applied to seedlings. Morphodynamic disturbance, when strong enough, causes resetting of the vegetation stands to an earlier phase. Flood duration causes physiological stress to the root system and, depending on the duration of inundation, it may lead to total or partial resetting of the vegetation community. If none of these disturbances is strong enough to influence vegetation succession, the system may develop towards a more mature stage, until the climax is reached (Figure 24.4). Furthermore, during each model run it verifies whether the topography of the study area has changed from the previous iteration (simulated year).

The CASiMiR model is primarily rule-based, although it also makes use of parameters based on expert knowledge. Driving forces are applied in an automatic fashion: the effect of each driving force is computed in sequence and, as a result, synergistic effects are not considered. In addition, the model adopts a Boolean approach and relies on hard (parameterized) thresholds. For hydraulic modelling inputs, such as the creation of shear stress maps, an external hydraulic model must be used. Calibration is achieved by comparing simulated results and field measurements. CASiMiR Vegetation is implemented in C# programming language based on Microsoft. Net (dot net) framework 3.5 technology.

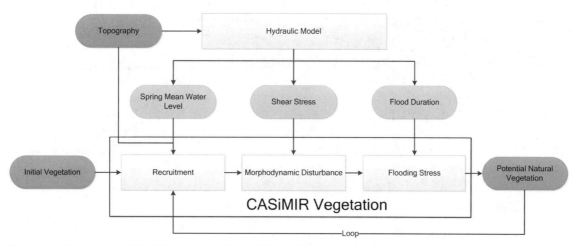

Figure 24.3 Flow diagram of the CASiMiR Vegetation model. Light blocks (e.g. Recruitment) represent CASiMiR sub-models, while grey blocks (e.g. Shear Stress) represent inputs and outputs of the model. The order of sub-model blocks is arranged according to CASiMiR execution order.

24.6 Model applications

Simulation models allow prediction of system changes with respect to variables driving ecosystem variability. Models can be used as a 'virtual laboratory' (Perona *et al.*, 2009) to run experiments otherwise impossible to perform within acceptable timeframes and economic constraints. Models, therefore, allow the testing of hypotheses and the visualization of potential decision outcomes. Within the specific domain of Ecohydraulics, river

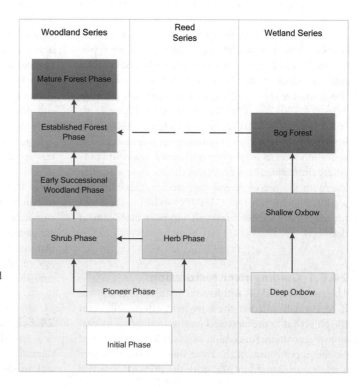

Figure 24.4 Succession series and phases of progression. Rectangular blocks represent the different succession phases of the three series (woodland, reed and wetland). Woodland and reed series both recruit in the bank zone and therefore share the Initial and Pioneer Phases. When terminal phases of the reed and wetland series (namely the Herb Phase and Bog Forest) reach the end of their lifecycle, they converge to the woodland series, which is normally the climax series of the floodplain.

Figure 24.5 Location of the river systems where the model CASiMiR Vegetation has been tested and calibrated.

management decisions can include morphological or hydrological modifications performed either for river conservation or river management purposes. In both cases, such modifications affect riparian biology and river health.

The disturbance regime approach means that CASiMiR is applicable to most climatic regions and river types. So far, CASiMiR has been tested and calibrated on river systems located in the Austrian Alps, Pacific Northwest (Benjankar *et al.*, 2011), the Iberian peninsula (García-Arias *et al.*, 2012) and the Korean peninsula (Egger *et al.*, 2012) (Figure 24.5). Its application includes guidance for implementation of river restoration measures, reservoir operation rules to support environmental flows and providing simulations under different climate change scenarios. With the use of several case studies, the following section will illustrate how CASiMiR can be used as a tool for supporting scientific research and river management decisions.

24.6.1 Guiding river restoration

The role of CASiMiR within river restoration initiatives is to evaluate whether the long-term effects will meet the objectives to promote and reactivate riparian succession/regeneration. An example of its application is located on the upper course of the Drau River (Austria), where a reach was restored in 2002 with the primary objective

of improving the bank zone and sustaining the habitat of *Myricaria germanica*, an endangered indicator plant of natural Alpine river systems (Egger *et al.*, 2007). Two alternative channel geometry scenarios were proposed, within a simulated timeframe of 20 years. In the first scenario – 'small in-stream bars' – channel morphology improvements could be achieved by creating a narrow channel with deep water. In contrast, within the second scenario – 'large in-stream bars' – channel morphology changes could be characterized by shallow water depth and broad channel width. After model calibration, the two scenarios were projected over time (Figure 24.6).

In the 'small in-stream bars' scenario, the bank zone was subjected to a degree of disturbance too high to facilitate vegetation establishment. Conversely, in the 'large in-stream bars' scenario, the low degree of disturbance allowed extensive vegetation colonization which encroached onto the bank zone and limited *Myricaria germanica* regeneration. At the end of both model simulations, neither of the scenarios resulted in a morphology suitable for the long-term sustainability of habitat required by *Myricaria germanica*.

24.6.2 Climate change effects on floodplain vegetation

Climatic variations include not only changes to the thermal regime but also modification of the volume and

Small in-stream bars   Large in-stream bars

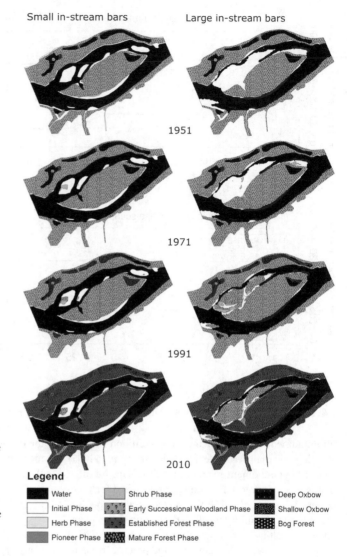

1951

1971

1991

2010

Figure 24.6 Model output selection for the two scenarios (one map every 20 simulated years). On the left, a selection of simulated maps for the 'small in-stream bars' scenario and on the right, a selection from the 'large in-stream bars' scenario. Maps from the 'small in-stream bars' scenario show how the bank zone remains largely bare over time, while in the 'large in-stream bars' scenario, the bank zone is steadily colonized by vegetation.

Legend

Water	Shrub Phase	Deep Oxbow
Initial Phase	Early Successional Woodland Phase	Shallow Oxbow
Herb Phase	Established Forest Phase	Bog Forest
Pioneer Phase	Mature Forest Phase	

timing of precipitation, which, in turn, influence a wide range of hydrological factors (Kundzewicz, 2008). These factors influence and drive the establishment, development and resetting of riparian vegetation. To evaluate the long-term effects of climate change on Alpine riparian vegetation on the River Drau (Austria), CASiMiR was used to simulate the spatial and quantitative distribution of vegetation over a 31-year time period. Hydrodynamics were simulated using the two-dimensional numerical flow model RSim-2D (Tritthart, 2005). Climate change impacts were evaluated based on two sub-scenarios of the IPCC storyline A1B (IPCC, 2001). The first scenario (S1) consisted of a 25% increase in annual peak discharge, while the second scenario (S2) consisted of a 25% reduction in annual peak discharge.

Simulations indicated that the climate change scenario suggesting an increase in peak discharge (S1), led to the resetting of successional processes (Figure 24.7); the opposite phenomenon was observed for the climate change scenario with a reduction in peak discharge (S2), i.e. the active channel area was reduced as a consequence of extensive vegetation colonization. As a result, climate change effects on riparian vegetation would result in significant modifications to the proportion of succession phases under both scenarios, and hence a loss of ecological functionality.

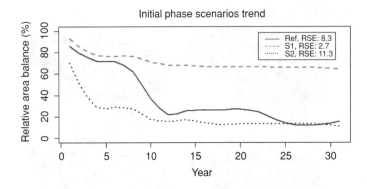

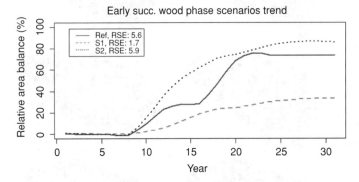

Figure 24.7 Lines represent the relative area trend of the Initial and Early Successional Woodland Phases in the reference period (Ref) and two climate change scenarios (S1: peak discharge increase and S2: peak discharge decrease), RSE: residual standard error. Trend lines were obtained applying the Loess interpolation technique to the relative area of the different succession phases. The technique employed is a smooth interpolation method based on locally weighted polynomial regression (Cleveland *et al.*, 1992).

24.6.3 Flow requirements and dam operations

Impoundments are one of the most common impacts affecting riparian ecosystems (Nilsson *et al.*, 2005) and pre-impoundment reference/baseline data are often unavailable or cover only a limited time period for vegetation (Braatne *et al.*, 2008; Burke *et al.*, 2009). The timing and quantity of water released can mitigate negative impacts resulting from reservoir operations on riparian ecosystems. However, for economic and practical reasons, the definition of environmentally sound operating rules requires decision support tools able to present results of alternative flow regime operating rules. For assessing the impacts of impoundment operating rules, CASiMiR simulations provide an opportunity to replicate natural reference conditions and the outcome of alternative flow regimes provided by varying the operating rules.

In the Operational Loss Assessment Kootenai project (Idaho, USA), CASiMiR was applied to simulate natural reference conditions and succession dynamics before and after the construction of the Libby Dam in 1973 (Benjankar, 2009). The assessment of the impact of the dam was derived from the difference between succession phases relative to the area simulated in historic and contempo-

rary scenarios. Statistical differences were tested by means of a two-tailed *t*-test (Dytham, 1999) (Table 24.2). In the historic scenario, the Initial and Pioneer Phases were more prevalent and so were Shrub and Early Successional Woodland Phases. Mature Succession Phases were more

Table 24.2 Comparison of the simulated normalized mean area with standard deviation, under different scenarios, for the reed and woodland series. 1: Initial Phase (IP), 2: Pioneer Phase (PP), 3: Shrub Phase (SP) 4: Early Successional Woodland Phase (ESWP), 5: Mature Phase (MF). St. Dev.: standard deviation. Area balances between scenarios differ significantly (*t*-test, $\alpha = 0.05$).

Historic scenario (1911–1938)		
IP[1] and PP[2]	SP[3] and ESWP[4]	MF[5]
Mean 0.15	0.073	0.585
St. Dev. 0.024	0.010	0.012

Contemporary scenario (1973–2007)		
IP[1] and PP[2]	SP[3] and ESWP[4]	MF[5]
Mean 0.029	0.062	0.553
St. Dev. 0.013	0.009	0.018

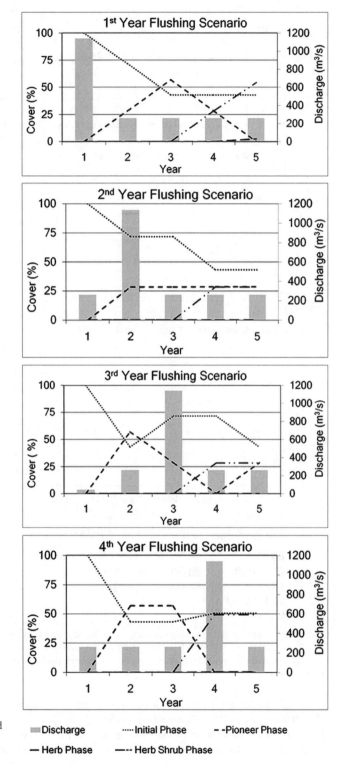

Figure 24.8 Area of vegetation for different flushing flow scenarios. Each chart represents a different scenario: grey bars represent the peak discharge applied in the simulated scenario, while the lines represent the relative area of the different succession phases.

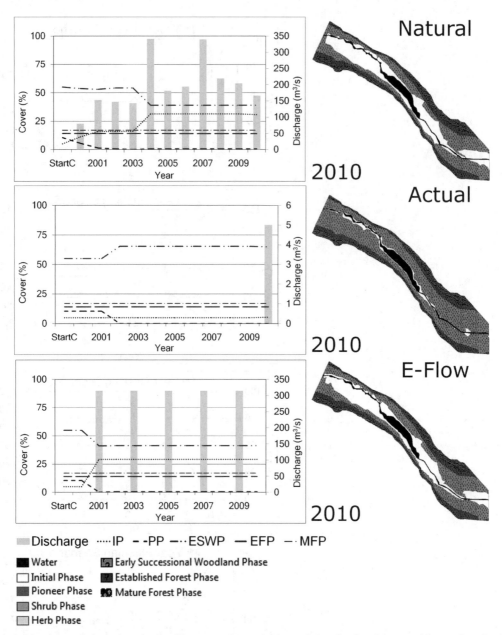

Figure 24.9 Riparian patch dynamics according (from top to bottom) to natural, actual and environmental flow regimes for the Sado River (Portugal). In the charts, the lines represent riparian succession phases' relative area balance for each simulated year, while the grey bars represent the maximum discharge applied to simulate the year. Beside the charts, the maps portray the last year simulated in the three scenarios.

common in the historic than the contemporary scenario (Merritt and Cooper, 2000). The main reason for this was that a significant proportion of the Mature Succession Phases were recycled into Initial and Pioneer Phases

during the pre-dam scenario period (1939–1972) and the vegetation had not reached the Mature Succession Phases (Benjankar et al., 2012). The Mature Succession Phases gradually increased during the contemporary scenario

(0.553 in Table 24.2) from the post-dam scenario (0.526) (Benjankar *et al.*, 2012), to levels typical of the pattern in other rivers impacted by impoundments (Merritt and Cooper, 2000).

In South Korea, CASiMiR was adapted as part of Ecoriver21, the national river restoration programme, and used to analyze changes in riparian vegetation due to dam operation. The study site was the Nakdong River, where the bank zone was historically composed of sand (Egger *et al.*, 2012). The goal of the simulation was to identify the most appropriate timing for flushing flows to restore former habitat conditions, to maximize the area of bare ground and minimize the extent of the vegetation cover. Scenarios differed in relation to the year in which the simulated flushing flow was applied (Figure 24.8). Simulations indicated that it would be more appropriate to undertake the flushing flows every third year because it would result in the increment of the Pioneer Phase. In addition, it would allow full exploitation of the vegetation recovery time while potential economic and bed degradation issues could be avoided for longer compared with a management strategy based on flushing flows during the second year. The key management consideration arising from the results was the ten-fold increase in vegetation morphodynamic resistance when succession from the 'Pioneer' to 'Herb Phase' occurred.

To underline the role of environmental flows as a riparian vegetation management tool on regulated rivers, CASiMiR was applied to a reach of the Sado River (Portugal). The study reach is located downstream from the Monte da Rocha Dam, where the current no-flow policy has allowed riparian vegetation to succeed towards the Mature stage, with significant vegetation encroachment within the historic channel. The ability of different reservoir flows in preventing riparian vegetation encroachment into the channel as well as their capabilities to maintain riparian habitats were tested using the simulation of three different flow regimes. First, the natural flow regime with no human impacts; second, the current situation and dam operation; and third, an environmental flow regime defined by Alves *et al.* (2003).

When the different flow regimes were compared in terms of their influence in shaping riparian vegetation (Figure 24.9), it was noticeable that the natural flow regime caused the highest disturbance of riparian vegetation patches, with the resetting of bank zone patches and maintenance of the area of river channel unoccupied by vegetation. The simulated natural flow regime was the only flow regime allowing some degree of pioneer vegetation establishment. The same trend was observed

using the simulated environmental flow regime, while the actual flow regime promoted vegetation succession within the channel (Figure 24.9). For the environmental flow method, a bi-annual release of a discharge with a two-year return period was recommended to flush/erode encroaching vegetation and to maintain channel characteristics. Nevertheless, the advised flushing flow appeared to be significantly higher than the average discharge and could lead to bed degradation. It was therefore decided to use the model results as a starting point for further discussions for optimization of flushing flows.

24.7 Conclusion

Riparian vegetation and fluvial hydraulic processes act as both active and passive factors involved in riparian landscape creation, maintenance and regeneration (Corenblit *et al.*, 2007). Fluvial hydraulic processes are the regulatory agents maintaining the diversity of riparian corridor features, including their creation, temporal persistence and turnover (Church, 2002). Conversely, vegetation modifies flow patterns and the substrate characteristics of the river channel and corridor, influencing hydraulic and geomorphological processes (Tsujimoto, 1999; Gurnell *et al.*, 2002). The continual interaction of such dynamic drivers accounts for the spatial and temporal variability typically observed in riparian landscapes (Gordon *et al.*, 2000; Fierke and Kauffman, 2005; Lyon and Gross, 2005; Willms *et al.*, 2006). Such variability is at best summarized by the shifting mosaic concept, which postulates a temporally constant but spatially shifting quantity of riverscape features (Tockner *et al.*, 2000; Ward *et al.*, 2002). Under the riparian study and analysis operational perspective, the shifting mosaic concept implies long-term observations, which unfortunately are seldom possible. Such observations aim to quantify spatial variations in vegetation distribution over time as an indicator of the fluvial processes underpinning the shifting mosaic paradigm. CASiMiR attempts to provide solutions to such problems and its results allow variations on a yearly basis to be observed and different spatial resolutions to be examined. Simulations also allow the formulation of 'what if' scenarios and replication of altered and natural conditions, with the latter providing comparisons with the effects of the former.

The dynamic behaviour of riparian ecosystems makes it difficult to simulate the inherently complex aspects of their functioning. For modelling, broad knowledge and an array of data are required and attention must be given to observable system elements. Based on this premise, CASiMiR Vegetation uses succession phases as basic model elements and consistent indicators of the underlying fluvial hydraulic processes. The simulated development of riverine vegetation can be conceptualized as a function of a disturbance regime where vegetation growth is counterbalanced by the recycling/resetting ability of fluvial hydraulic processes, acting as regulatory drivers (Bendix and Hupp, 2000). Despite the apparent simplicity of such a statement, nature is far more complex than any conceptualization so far affordable by computer models. As a consequence, the results generated by computer models must be evaluated and considered in relation to the goals of the process and must be based upon process replication rather than reproduction of observed patterns in the landscape.

The greatest challenge modellers face is to capture and replicate widely observed ecosystem parameters with a sufficient level of detail. The development philosophy of CASiMiR Vegetation is to attain this goal by introducing generalizations and simplifications (e.g. succession phases) consistent with observations from heterogeneous riparian ecosystems which are realistic and relevant across different river types and climates. The generalizations and simplifications upon which CASiMiR has been built include disturbance forces and their cumulative effect on simulated vegetation growth and patch evolution through the use of locally relevant physical and biological inputs. Ultimately, the advancements in science and computing technology, together with increasing social awareness regarding ecohydraulics, hydromorphology and ecohydrological issues associated with lotic ecosystem management, provide the motivation to develop tools to support the sustainable management and conservation of riparian floodplain ecosystems.

Acknowledgements

The authors would like to acknowledge:
- The financial support by the Austrian Federal Ministry of Economy, Family and Youth, the National Foundation for Research, Technology and Development and via Donau.
- The Korea Institute of Construction Technology (KICT) and its vice president, Dr Hyoseop Woo, for offering the opportunity to test CASiMiR in South Korea.
- The Bonnerville Power Administration, Norm Merz and Scott Soults from The Kootenai Tribe of Idaho (USA) for their role in the research development.
- The Carinthian government for authorizing the field work at the Drau River and the Austrian climate and energy fund for providing funding through the ACRP programme (project no. A963615).
- The Austrian Lebensministerium (Project 100403) and the Austrian Forschungsförderungsgesellschaft (Project 8296330/9103 SCK/KUG) for the granted funds.
- The sponsor of the RIPFLOW project (Era-Net IWRM Funding Initiative; Spanish MEC CGL2008-03076-E/BTE) and SCARCE project (CONSOLIDER-Ingenio, CSD2009-00065).
- Patricia Rodríguez-González and António Albuquerque for their support with field work and data analysis, and António N. Pinheiro for supporting hydrological analysis and hydraulic modelling of the Portuguese case studies.
- Rui Rivaes benefited from a PhD grant sponsored by Universidade Técnica de Lisboa (UTL).

References

Alves, M.H., Bernardo, J.M., Matias, P. and Martins, J.P. (2003) *Caudais ecologicos em Portugal*. Portugal Instituto da Água, Lisbon.

Auble, G.T. and Scott, M.L. (1998) Fluvial disturbance patches and cottonwood recruitment along the upper Missouri River, Montana. *Wetlands*, **18**: 546–556.

Auble, G.T., Friedman, J.M. and Scott, M.L. (1994) Relating riparian vegetation to present and future stream flows. *Ecological Applications*, **4**(3): 544–554.

Barsoum, N. (2002) Relative contributions of sexual and asexual regeneration strategies in *Populus nigra* and *Salix alba* during the first years of establishment on a braided gravel bed river. *Evolutionary Ecology*, **15**: 255–279.

Bendix, J. and Hupp, C.R. (2000) Hydrological and geomorphological impacts on riparian plant communities. *Hydrological Processes*, **14**(16–17): 2977–2990.

Benjankar, R. (2009) *Quantification of Reservoir Operation-Based Losses to Floodplain Physical Processes and Impact on the Floodplain Vegetation at Kootenai River, USA*. PhD dissertation, University of Idaho, Idaho, USA.

Benjankar, R., Egger, G., Jorde, K., Goodwin, P. and Glenn, N.F. (2011) Dynamic floodplain vegetation model development for the Kootenai River, USA. *Journal of Environmental Management*, **92**: 3058–3070.

Benjankar, R., Jorde, K., Yager, E., Egger, G., Goodwin, P. and Glenn, N.F. (2012) The impact of river modification and dam operation on floodplain vegetation succession trends in the Kootenai River, USA, *Ecological Engineering*, **46**: 88–97.

Blamauer, B., Klösch, M., Tritthart, M. and Habersack, M. (2011) Refining parameterization of bar vegetation roughness based on in situ-measurements of vegetation bending during flood events. *Engineers Australia: Proceedings of the 34th IAHR World Congress*, Brisbane, 26 June–1st July, 2011, pp. 3388–3395.

Böhmer, H.-J. (1999) Vegetationsdynamik im Hochgebirge unter dem Einfluss natürlicher Störungen. *Dissertationes Botanicae*, **311**: 180.

Braatne, J.H., Jamieson, R., Gill, K.M. and Rood, S.B. (2007) Instream flows and the decline of riparian cottonwoods along the Yakima River, Washington, USA. *River Research and Applications*, **23**(3): 247–267.

Braatne, J.H., Rood, S.B., Goater, L.A. and Blair, C.L. (2008) Analyzing the impacts of dams on riparian ecosystems: A review of research strategies and their relevance to the Snake River through Hells Canyon. *Environmental Management*, **41**: 267–281.

Burke, M., Jorde, K. and Buffington, J.M. (2009) Application of a hierarchical framework for assessing environmental impacts of dam operation: changes in stream flow, bed mobility and recruitment of riparian trees in a western North American river. *Journal of Environmental Management*, **90**(3): S224–S236.

Canham, C.D. and Marks, P.L. (1985) The response of woody plants to disturbance: patterns of establishment and growth. In Pickett, S.T.A. and White, P.S. (eds) *The Ecology of Natural Disturbance and Patch Dynamics*, Academic Press, Orlando, pp. 197–216.

Church, M. (2002) Geomorphic thresholds in riverine landscapes. *Freshwater Biology*, **47**(4): 541–557.

Clements, F.E. (1916) *Plant Succession: An Analysis of the Development of Vegetation*. Publication N. 242, Carnegie Institute, Washington, DC.

Cleveland, W.S., Grosse, E. and Shyu, W.M. (1992) Local regression models. In Chambers, J.M. and Hastie, T.J. (eds) *Statistical Models in S*, Wadsworth & Brooks/Cole.

Corenblit, D., Steiger, J., Gurnell, A.M. and Naiman, R.J. (2009) Plants intertwine fluvial landform dynamics with ecological succession and natural selection: a niche construction perspective for riparian systems. *Global Ecology and Biogeography*, **18**: 507–520.

Corenblit, D., Tabacchi, E., Steiger, J. and Gurnell, A.M. (2007) Reciprocal interactions and adjustments between fluvial landforms and vegetation dynamics in river corridors: A review of complementary approaches. *Earth-Science Reviews*, **84**(1–2): 56–86.

Denslow, J.S. (1980) Patterns of plant species diversity during succession under different disturbance regimes. *Oecologia*, **46**: 18–21.

Downs, P.W. and Kondolf, G.M. (2002) Post-project appraisals in adaptive management of river channel restoration. *Environmental Management*, **29**(4): 477–496.

DVWK (1991) *Hydraulische Berechnungen von Fließgewässern*. DVWK Merkblätter 220, Parey Verlag, Berlin.

Dytham, C. (1999) *Choosing and Using Statistics: A Biologist's Guide*, Wiley-Blackwell, pp. 92–96.

Edwards, P.J., Kollmann, J., Gurnell, A.M., Petts, G.E., Tockner, K. and Ward, J.V. (1999) A conceptual model of vegetation dynamics on gravel bars of a large Alpine river. *Wetlands Ecology and Management*, **7**: 141–153.

Egger, G., Aigner, S. and Angermann, K. (2007) Vegetationsdynamik einer alpinen Wildflusslandschaft und Auswirkungen von Renaturierungsmaßnahmen auf das Störungsregime, dargestellt am Beispiel des Tiroler Lechs. *Verein zum Schutz der Bergwelt* e.V., Jahrbuch Verein zum Schutz der Bergwelt (München), **72**: 5–54.

Egger, G., Politti, E., Woo, H., Cho, K.H., Park, M., Cho, H. and Benjankar, R. (2012) Dynamic vegetation model as a tool for ecological impact assessments of dam operation. *Journal of Hydro-environment Research*, **6**: 151–161.

Ellenberg, H. (2009) *Vegetation Ecology of Central Europe*, Cambridge University Press, Cambridge, UK.

Fierke, M.K. and Kauffman, J.B. (2005) Structural dynamics of riparian forests along a black cottonwood successional gradient. *Forest Ecology and Management*, **215**: 149–162.

Forman, R.T.T. (1995) *Land Mosaics: The Ecology of Landscapes and Regions*, Cambridge University Press, Cambridge, UK.

Forman, R. and Godron, M. (1986) *Landscape Ecology*, John Wiley & Sons, Inc.

Formann, E. (2009) *Evaluation of Riverine Processes and their Interrelationships in River Restoration Based on Reduced Complexity Modelling and the Dynamic Disturbance Regime Approach*. PhD dissertation, University of Natural Resources and Life Sciences, Vienna, Austria.

Francis, R.A., Gurnell, A.M., Petts, G. E. and Edwards, P.J. (2009) Riparian tree establishment on gravel bars: Interactions between plant growth strategy and the physical environment. In Smith, G.H.S., Best, J.L., Bristow, C.S. and Petts, G.E. (eds) *Braided Rivers: Process, Deposits, Ecology and Management*, Blackwell Publishing Ltd, Oxford, UK, pp. 361–380.

Franz, H. and Bazzaz, F.A. (1977) Simulation of vegetation response to modified hydrologic regimes: A probabilistic model based on niche differentiation in a floodplain forest. *Ecology*, **58**: 176–183.

Freeman, G.E., Rahmeyer, W.H. and Copeland, R.R. (2000) *Determination of Resistance Due to Shrubs and Woody Vegetation*. Technical Report, U.S. Army Corps of Engineers, Washington.

García-Arias, A., Francés, F., Ferreira, M.T., Egger, G., Martínez-Capel, F., Garófano-Gómez, V., Andrés-Doménech, I., Politti, E., Rivaes, R. and Rodríguez-González, P. Implementing a dynamic riparian vegetation model in three European river systems. *Ecohydrology*, in press. doi:10.1002/eco.1331.

Glenz, C., Iorgulescu, I., Kienast, F. and Schlaepfer, R. (2008) Modelling the impact of flooding stress on the growth performance of woody species using fuzzy logic. *Ecological Modelling*, **218**(1–2): 18–28.

Glenz, C., Schlaepfer, R., Iorgulescu, I. and Kienast, F. (2006) Flooding tolerance of Central European tree and shrub species. *Forest Ecology and Management*, **235**(1–3): 1–13.

Goodwin, C.N., Hawkins, C.P. and Kershner, J.L. (1997) Riparian restoration in the Western United States: Overview and perspective. *Restoration Ecology*, **5**(4): 4–14.

Gordon, C., Cooper, C., Senior, C.A., Banks, H., Gregory, J.M., Johns, T.C., Mitchell, J.F.B. and Wood, R.A. (2000) The simulation of SST, sea ice extents and ocean heat transports in a version of the Hadley Centre coupled model without flux adjustments. *Climate Dynamics*, **16**(2–3): 147–168.

Gurnell, A.M., Piegay, H., Swanson, F.J. and Gregory, S.V. (2002) Large wood and fluvial processes. *Freshwater Biology*, **47**(4): 601–619.

Habersack, H., Hofbauer, S. and Hauer, C. (2008) Vegetation impacts on flood flows – evaluation of flow resistance based on a hydraulic scale model and numerical hydrodynamic modelling. In Altinakar, M.S., Kokpinar, M.A., Darama, Y. Yegen, E.B. and Harmancioglu, N. (eds) *River Flow 2008*, pp. 425–433.

Haney, A. and Power, R.L. (1996) Adaptive management for sound ecosystem management. *Environmental Management*, **20**(6): 879–886.

Huston, M.A. (1994) *Biological Diversity – The Coexistence of Species on Changing Landscapes*, Cambridge University Press, Cambridge, UK.

IPCC (2001) *Climate Change 2001: The Scientific Basis*, Cambridge University Press, Cambridge, UK.

Järvelä, J. (2004) Determination of flow resistance caused by non-submerged woody vegetation. *International Journal of River Basin Management*, **2**(1): 61–70.

Johnson, W.C. (1992) Dams and riparian forests: case study from the Upper Missouri River. *Rivers*, **3**: 229–242.

Karrenberg, S., Edwards, P.J. and Kollmann, J. (2002) The life history of Salicaceae living in the active zone of floodplains. *Freshwater Biology*, **47**(4): 733–748.

Komar, P.D. and Carling, P.A. (1991) Grain sorting in gravel-bed streams and the choice of particle sizes for flow competence evaluations. *Sedimentology*, **35**: 681–695.

Kouwen, N. and Fathi-Moghadam, M. (2000) Friction factors for coniferous trees along rivers. *Journal of Hydraulic Engineering*, **126**(10): 732–740.

Kovalchik, B.L. and Clausnitzer, R.R. (2004) *Classification and Management of Aquatic, Riparian, and Wetland Sites on the National Forests of Eastern Washington: Series description*. U.S. Department of Agriculture, Forest Service, Portland, U.S.

Kundzewicz, Z.W. (2008) Climate change impacts on the hydrological cycle. *Ecohydrology and Hydrobiology*, **8**(2–4): 195–203.

Leopold, L.B. (1994) *A View of the River*, Harvard University Press, Cambridge, MA.

Leopold, L.B. (1997) *Water, Rivers, and Creeks*, University Science Books, Sausalito, CA.

Lyon, J. and Gross, N. (2005) Patterns of plant diversity and plant environmental relationships across three riparian corridors. *Forest Ecology and Management*, **204**(2–3): 267–278.

Mahoney, J.M. and Rood, S.B. (1998) Stream flow requirements for cottonwood seedling recruitment – An integrative model. *Wetlands*, **18**(4): 634–645.

Merritt, D.M. and Cooper, D.J. (2000) Riparian vegetation and channel change response to river regulation: A comparative study of regulated and unregulated streams in the Green River Basin, USA. *Regulated Rivers: Research and Management*, **16**: 543–564.

Merritt, D.M., Scott, M.L., Poff, N.L., Auble, G.T. and Lytle, D.A. (2010) Theory, methods and tools for determining environmental flows for riparian vegetation: Riparian vegetation – flow response guilds. *Freshwater Biology*, **55**: 206–225.

Naiman, R.J. and Décamps, H. (1997) The ecology of interfaces: Riparian zones. *Annual Review of Ecology and Systematics*, **28**: 621–658.

Naiman, R.J., Décamps, H. and McClain, M.E. (2005) *Riparia – Ecology, Conservation and Management of Streamside Communities*, Elsevier Academic Press.

Naiman, R.J., Décamps, H. and Pollock, M. (1993) The role of riparian corridors in maintaining regional biodiversity. *Ecological Applications*, **3**:209-212.

Naiman, R.J., Latterell, J.J., Pettit, N.E. and Olden, J.D. (2008) Flow variability and the biophysical vitality of river systems. *Comptes Rendus Geoscience*, **340**: 629–643.

Nepf, H., Zong, L. and Rominger, J. (2010) Flow, deposition, and erosion near finite patches of vegetation. In Dittrich, A., Koll, K., Aberle, J. and Geisenhainer, P. (eds) *River Flow 2010*, pp. 33–40.

Nilsson, C., Reidy, C.A., Dynesius, M. and Revenga, C. (2005) Fragmentation and flow regulation of the world's large river systems. *Science*, **308**: 405–407.

Oliver, C.D. and Larson, B.C. (1996) *Forest Stand Dynamics*, updated edition, John Wiley & Sons, Inc., New York.

Oplatka, M. (1998) Stabilität von Weidenverbauungen an Flussufern. *Mitteilungen*, **156**, ETH Zürich.

Osterkamp, W.R. and Hupp, C.R. (2010) Fluvial processes and vegetation – Glimpses of the past, the present, and perhaps the future. *Geomorphology*, **116**: 274–285.

Parker, G. (1996) Gravel-bed channel stability. In Nakato, T. and Ettema, R. (eds) *Issues and Directions in Hydraulics*, A A Balkema, Rotterdam, The Netherlands, pp. 115–133.

Pearlstine, L., McKellar, H. and Kitchens, W. (1985) Modelling the impacts of a river diversion on bottomland forest communities in the Santee River floodplain, South Carolina. *Ecological Modelling*, **29**(1–4): 283–302.

Perona, P., Camporeale, C., Perucca, E., Savina, M., Molnar, P., Burlando, P. and Ridolfi, L. (2009) Modelling river and riparian vegetation interactions and related importance for sustainable ecosystem management. *Aquatic Sciences*, **71**: 266–278.

Phipps, R.L. (1979) Simulation of wetland forest dynamics. *Ecological Modelling*, **7**: 257–288.

Pickett, S.T.A. and White, P.S. (1985) Patch dynamics: A synthesis. In Pickett, S.T.A. and White, P.S. (eds) *The Ecology of Natural Disturbance and Patch Dynamics*, Academic Press, Orlando, pp. 371–384.

Plachter, H. (1998) Die Auen alpiner Wildflüsse als Modelle störungsgeprägter ökologischer Systeme. In Finck, P., Klein, M., Riecken, U. and Schröder, E. (eds) *Schutz und Förderung dynamischer Prozesse in der Landschaft. Schriftenreihe für Landschaftspflege und Naturschutz 56*, Bundesamt für Naturschutz, Bonn-Bad Godesberg, Bonn, pp. 21–66.

Poff, N.L., Allan, J.D., Bain, M.B., Karr, J.R., Prestegaard, K.L., Richter, B.D., Sparks, R.E. and Stromberg, J.C. (1997) The natural flow regime: A paradigm for river conservation and management. *BioScience*, **47**: 769–784.

Poiani, K.A. and Johnson, W.C. (1993) A spatial simulation model of hydrology and vegetation dynamics in semi-permanent prairie wetlands. *Ecological Applications*, **3**: 279–293.

Polzin, M.L. and Rood, S.B. (2000) Effects of damming and flow stabilization on riparian processes and black cottonwoods along the Kootenay River. *Rivers*, **7**(3): 221–232.

Polzin, M.L. and Rood, S.B. (2006) Effective disturbance: seedling safe sites and patch recruitment of riparian cottonwoods after a major flood of a mountain river. *Wetlands*, **26**(4): 965–980.

Richards, K. (1982) *Rivers – Form and Process in Alluvial Channels*, Methuen, New York.

Richter, B.D. and Richter, H.E. (2000) Prescribing flood regimes to sustain riparian ecosystems along meandering rivers. *Conservation Biology*, **14**: 1467–1478.

Richter, B.D. and Thomas, G.A. (2007) Restoring environmental flows by modifying dam operations. *Ecology And Society*, **12**(1), online.

Rinaldi, M. and Darby, S.E. (2005) Advances in modelling river bank erosion processes. *6th Gravel Bed River Workshop*, Austria.

Rominger, J.T., Lightbody, A.F. and Nepf, H.M. (2010) Effects of added vegetation on sand bar stability and stream hydrodynamics. *Journal of Hydraulic Engineering*, **136**: 994–1002.

Rivaes, R. Rodríguez-González, P.M. Albuquerque, A. Pinheiro, A. Egger, G. and Ferreira, M.T. (2012) Riparian vegetation responses to altered flow regimes driven by climate change in Mediterranean rivers. *Ecohydrology*. In press. doi:10.1002/eco.1287.

Schneider, S. (2010) *Widerstandsverhalten von holziger Auenvegetation*. PhD dissertation, Karlsruher Institute of Technology (KIT), Karlsruhe, Germany.

Schneider, S. (2011) *Hocwassersichere Entwicklung und Unterhaltung von Fließgewässern im urbanen Bereich Maßnahmen und ihre hydraulischen Wirkungen*. LUBW Landesanstalt für Umwelt, Messungen und Naturschutz Baden-Wüttemberg, 76231 Karlsruhe, Germany.

Simon, A., Bennett, S.J. and Neary, V.S. (2004) Riparian vegetation and fluvial geomorphology: Problems and opportunities. In Bennett, S.J. and Simon, A. (eds) *Riparian Vegetation and Fluvial Geomorphology*, American Geophysical Union, Washington, DC, pp. 1–10.

Stephan, U. and Gutknecht, D. (2002) Hydraulic resistance of submerged flexible vegetation. *Journal of Hydrology*, **269**: 27–43.

Stromberg, J.C, Tiller, R. and Richter, B. (1996) Effects of groundwater decline on riparian vegetation of semiarid regions: the San Pedro, Arizona. *Ecological Applications*, **6**(1): 113–131.

Tabacchi, E., Lambs, L., Guilloy, H., Planty-Tabacchi, A.-M., Muller, E. and Décamps, H. (2000) Impacts of riparian vegetation on hydrological processes. *Hydrological Processes*, **14**(16–17): 2959–2976.

Tockner, K., Malard, F. and Ward, J.V. (2000) An extension of the flood pulse concept. *Hydrological Processes*, **14**: 2861–2883.

Tritthart, M. (2005) *Three-dimensional Numerical Modeling of Turbulent River Flow Using Polyhedral Finite Volumes*. Wiener Mitteilungen, Technical University Vienna, Vienna.

Tsujimoto, T. (1999) Fluvial processes in streams with vegetation. *Journal of Hydraulic Research*, **37**(6): 789–803.

Van de Wiel, M. (2003) *Numerical Modelling of Channel Adjustment in Alluvial Meandering Rivers with Riparian Vegetation*. PhD dissertation, Department of Geography, University of Southampton, UK.

Van Pelt, R., O'Keefe, T.C., Latterell, J.J. and Naiman, R.J. (2006) Riparian forest stand development along the Queets River in Olympic National Park, Washington. *Ecological Monographs*, **76**: 277–298.

Ward, J.V., Malard, F. and Tockner, K. (2002) Landscape ecology: A framework for integrating pattern and process in river corridors. *Landscape Ecology*, **17**: 35–45.

White, P.S. and Pickett, S.T.A. (1985) Natural disturbance and patch dynamics: An introduction. In Pickett, S.T.A. and White, P.S. (eds) *The Ecology of Natural Disturbance and Patch Dynamics*, Academic Press, Orlando, pp. 3–13.

Wieringa, M.J. and Morton, A.G. (1996) Hydropower, adaptive management, and biodiversity. *Environmental Management*, **20**(6): 831–840.

Willms, C.R., Pearce, D.W. and Rood, S.B. (2006) Growth of riparian cottonwoods: A developmental pattern and the influence of geomorphic context. *Trees*, **20**: 210–218.

Wilson, C.A.M.E., Xavier, P., Schoneboom, T., Aberle, J., Rauch, H.-P., Lammeranner, W., Weissteiner, C. and Thomas, H. (2010) The hydrodynamic drag of full scale trees. In Dittrich, A., Koll, K., Aberle, J. and Geisenhainer, P. (eds) *River Flow 2010*, pp. 453–460.

Wu, F.-C., Shen, H.W. and Chou, Y.-J. (1999) Variation of roughness coefficients for unsubmerged and submerged vegetation. *Journal of Hydraulic Engineering*, **125**(9): 934–942.

Wunder, S., Lehmann, B. and Nestmann, F. (2011) Determination of the drag coefficients of emergent and just submerged willows. *International Journal of River Basin Management*, **9**(3–4): 231–236.

Yang, C. and Schenggan, W. (1991) Comparison of selected bed-material load formulas. *Journal of Hydraulic Engineering*, **117**(8): 973–989.

Zong, L. and Nepf, H. (2011) Spatial distribution of deposition within a patch of vegetation. *Water Resources Research*, **47**: 1–12.

IV Conclusion

25 Research Needs, Challenges and the Future of Ecohydraulics Research

Ian Maddock[1], Atle Harby[2], Paul Kemp[3] and Paul Wood[4]

[1] Institute of Science and the Environment, University of Worcester, Henwick Grove, Worcester, WR2 6AJ, UK
[2] SINTEF Energy Research, P.O. Box 4761 Sluppen, 7465 Trondheim, Norway
[3] International Centre for Ecohydraulics Research, University of Southampton, Highfield, Southampton, SO17 1BJ, UK
[4] Department of Geography, Loughborough University, Leicestershire, LE11 3TU, UK

25.1 Introduction

Research developments within the field of ecohydraulics have been driven by a range of factors, including the need for fundamental understanding of the role of hydrological regimes and geomorphology in determining the hydraulic properties of flow and its influence on biota and ecology, the need to assess, manage and mitigate the impacts of river regulation from dams, water abstraction and inter-basin transfers through the designation of environmental flows, the design and evaluation of river restoration schemes and the ability to facilitate the effective passage of migratory species throughout catchments (Dyson *et al.*, 2003; Darby and Sear, 2008; Kingsford, 2011; Kemp, 2012; Shenton *et al.*, 2012).

The chapters and detailed case studies contained within this book provide an overview of the current status and recent developments in the interdisciplinary field of eco-hydraulics. Research presented has highlighted how technological developments have increased scientific understanding and the ability to characterise and analyse hydraulic information, with specific reference to the management, maintenance and restoration of lotic and riparian systems and the ecological communities they support. The hydraulic characteristics of specific riverine sites, reaches or catchments are determined by the interaction between geomorphology and the flow regime, and the

importance of this interplay at a range of spatial scales is at the heart of ecohydraulics research (see Chapter 7).

The focus of a majority of the chapters in the book has been on lotic systems, reflecting the primary influence that the hydraulic properties of flow exert on fluvial habitats (e.g. through erosion, scour, sediment transport, deposition and siltation processes) and the organisms and communities inhabiting them (e.g. through species-specific tolerance limits to flow forces and associated biophysical and chemical processes). However, hydraulic forces and processes are not unique to rivers and parallel developments and advances are taking place within tidal estuarine (see Chapter 22) and marine environments (Folkard, 2005), and also in connection with the hydrodynamics of lentic water bodies (Ambrosetti *et al.*, 2012). Major opportunities exist to enhance ecohydraulic understanding in these environments in the future.

Recent developments in ecohydraulics research and understanding have been characterised by a shift in emphasis from standard hydraulic variables (e.g. water depth and mean column velocity) to more complex parameters to assess higher order turbulent flow properties (see Chapter 2) and advances in hydraulic modelling (see Chapter 3) through the use of two- and three-dimensional numerical models and fuzzy logic. The chapters within this volume clearly illustrate the development of different types of ecohydraulic numerical models at

Ecohydraulics: An Integrated Approach, First Edition. Edited by Ian Maddock, Atle Harby, Paul Kemp and Paul Wood.
© 2013 John Wiley & Sons, Ltd. Published 2013 by John Wiley & Sons, Ltd.

varying spatial and temporal resolutions. For example, Chapter 20 illustrates the use of a two-dimensional model to assess the efficacy of various designs of pool-riffle scale units on physical habitat availability for selected target species as part of a river restoration project. Chapters 4 and 5 outline the use of one- and two-dimensional hydraulic models that utilise fuzzy logic at multiple scales (reach to basin) to simulate habitat availability, fragmentation and connectivity, while Chapter 19 examines the stranding risk of fish. The MesoHABSIM model described in Chapter 6 can be applied at multiple hierarchies of scale (point/micro to river landscape) to assess discharge–habitat relationships, habitat time series and compare the impact of various flow and operational scenarios on physical habitat availability. These represent just a selection of examples of the use and application of ecohydraulic modelling approaches used for river management and assessment. Undoubtedly, further advances in ecohydraulic modelling are likely to continue at a rapid pace.

Understanding the interactions between the hydraulic environment and the ecology of riverine ecosystems remains at the core of ecohydraulics, and significant advances in primary knowledge of these relationships have been made for floral (Chapters 13 and 14) and faunal groups (e.g. Chapters 9, 11 and 12). The chapters within this volume also clearly demonstrate how this growing knowledge can be used to help manage, monitor and mitigate the impacts of water resource developments for a wide range of biota including periphyton (Chapter 13), macrophytes (Chapters 14 and 15), macroinvertebrates (Chapter 21), fish (Chapters 8, 10, 16, 18 and 19), amphibians (Chapter 11), birds (Chapter 22) and riparian and floodplain vegetation (Chapters 17, 23 and 24). This rapidly evolving field of research is being driven by a number of legislative drivers at both the national and international levels (Acreman and Ferguson, 2010), the growing recognition of the need for sustainable management strategies for water resources, the protection of habitats, ecological communities and individual species, and the need to actively restore degraded systems (Kingsford, 2011).

25.2 Research needs and future challenges

Demand for the application of ecohydraulics research is being driven by river regulators seeking to maintain and enhance aquatic communities and manage river ecosystems to meet environmental standards and legislative drivers like the EU Water Framework Directive, the South African National Water Act (1998) and the Australian National Water Initiative (2004). However, ecohydraulics is a relatively young interdisciplinary research arena that needs to address a number of challenges to demonstrate its true potential and allow it to develop in order to provide answers to the complex management and operation issues for freshwater resources under increasing pressure in a changing climate (see Chapter 1).

Four key challenges need to be overcome if ecohydraulics is to be more widely accepted as both a pure and applied science with critical impact within scientific (physical and biological), engineering, water resource management and industrial communities:

• we need to develop a better understanding of the relationship between the turbulent properties of flowing water and its influence on individual organisms and ecological communities;
• we need to be able to effectively integrate hydraulically realistic information with ecological data for an interdisciplinary and applied end user community;
• we need to improve the transferability of data, variables and results in ecohydraulics across temporal and spatial scales;
• we need to bridge the gap between scientific research and the development and implementation of management recommendations (e.g. for environmental flows or river restoration).

The interrelated nature of these themes is illustrated below.

25.2.1 Measuring and modelling turbulent flow properties and their influence on ecological communities

Technological advances in the collection and management of complex data and modelling potential have provided an impetus for research in ecohydraulics. This can be clearly illustrated with reference to the changes in hydraulic data collection techniques that have occurred in recent decades. During the 1980s, hydraulic data collection for ecohydraulic research typically utilised equipment and sampling techniques that had been designed decades earlier for spot gauging of river discharge. Impeller or cup-type current meters were used and recorded mean column velocity at a single point (standardised to 0.6 depth from the surface down), normally averaged over a time period with a corresponding water depth reading. The 1990s saw the introduction of electromagnetic current meters, still measuring velocity in one dimension (streamwise) and time averaged at a point but with the advantage of having no moving parts; hence, they could assess flow velocities within macrophyte beds and worked at low flow velocities. Since the start of the 21st century, advances in acoustic technology have led to the use of acoustic Doppler velocimeters (ADVs) and acoustic Doppler current profilers (ADCPs). An ADCP can provide a high-resolution snapshot of the spatial distribution of velocities (vertical and lateral) for a cross-section within the space of a few minutes without the need to enter the stream. Modern ADVs can provide velocity measurements in three dimensions (streamwise, vertical, lateral) with 200 measurements per second in each dimension (200 Hz) and for a vertical column of water several cm deep near the probe sensor at mm increments within the measurement column. An ADV, recording at 100 Hz over a 3 cm water column at 1 mm increments for 30 seconds in three dimensions creates 270 000 data records. Of course, this enhanced data collection ability generates huge amounts of raw data that relies heavily on the parallel developments in electronic data storage capacity. The result is that these data can be utilised to calculate higher order turbulent flow properties, such as turbulence intensity and turbulent kinetic energy (see Chapter 2), and to calibrate and validate two- and three-dimensional numerical hydraulic models. When these data and modelling efforts are coupled with ecological survey data, it is possible to investigate the role of turbulence on biotic communities, as shown in Chapters 8 and 17.

In parallel to advances in hydraulic data collection, recording of biological/ecological data has witnessed similar developments. For example, in the 1970s, it was impossible to use telemetry to survey fish movements for small or juvenile fish due to the size and weight of transmitters and tags used to mark fish. The subsequent advances in telemetry to study fish movements and distribution have been remarkable. The development of RFID (Radio Frequency Identification) technology with passive integrated tags (PITs) during the 1980s (Prentice et al., 1987) means that, today, we can monitor fish smaller than 6 cm in size (Linnansaari et al., 2007; 2009). Current acoustic telemetry can provide three-dimensional fine-scale trajectories of fish movement, while Didson (Dual-frequency identification sonar) 'acoustic cameras' survey fish and fish movements even in turbid waters and at night when conventional cameras don't work (Crossman et al., 2011).

Developments in ecohydraulic research are intrinsically linked to these technological advances in field and laboratory flume measurement of flow properties, animal telemetry, computer processing and data storage capabilities, and computer modelling. However, because these instruments are developing rapidly and are relatively recent innovations, their use to examine the influence of these higher order hydraulic properties on ecological communities or individual species is still arguably in its infancy. There remains a gap between our knowledge of the characterisation of the hydraulic environment and our understanding of how these influence habitats, biotic communities and individual species.

25.2.2 Interdisciplinary integration and co-operation

In most fields of research, there has been a continuous development from single-disciplinary studies towards multi- and inter-disciplinary studies (Porter and Rafols, 2009). In the case of ecohydraulics, there were historic parallel developments within physical sciences (hydrology, hydraulics, water resource management, engineering and fluvial geomorphology) and biological sciences (ecology and biology) and 'ecohydraulic' research as we recognise it today clearly pre-dates the use of the term. Historically, research was undertaken within clearly defined disciplinary boundaries with distinct conceptual frameworks, terminologies and literatures. However, the contemporary use of the term 'ecohydraulics' has been most widely utilised by physical scientists thus far and, given this background, it may not be surprising that some biological scientists have questioned its 'ecological' credentials due to a perceived lack of biological realism and use of simplified biological proxies in many modelling exercises (Lancaster and Downes, 2010; Shenton et al., 2012).

Ecohydraulics has evolved in response to the need to solve river management problems (e.g. to determine environmental flows, design fish passage systems and implement river restoration) and to respond to national and international legislation. As new disciplines evolve at the boundary of well-established ones, there is a natural propensity for researchers to demand definitions of the exact scope and limits of the new area of study. This can be detrimental to the progress of the new discipline, as researchers debate the semantics of disciplinary boundaries. Engaging with a more flexible approach that has dynamic pervious boundaries to facilitate interdisciplinary research which can evolve over time will ensure ecohydraulic research can respond to the changing nature of the problems it is addressing. It also needs to bridge the gap between traditional disciplines, each with their own conceptual understanding, modes of operation and scientific terminology and literature. Enabling effective communication between such disciplines is not a simple

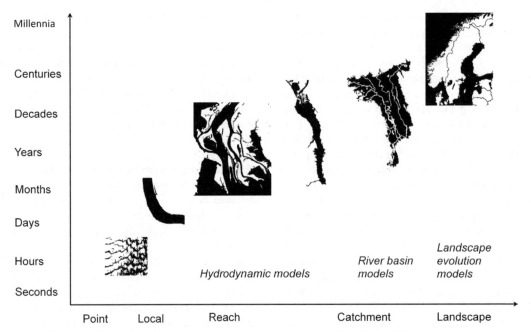

Figure 25.1 Different spatial (*x*-axis) and temporal (*y*-axis) scales in river science. After Zinke (2011).

task, yet a very important one if there is to be successful research at this interdisciplinary boundary.

The chapters included in this volume reflect the breadth of research currently being undertaken from a range of disciplinary backgrounds, many involving interdisciplinary collaboration. Some chapters are centred primarily on hydraulic properties or modelling the conveyance of water and changes to habitat over varying spatial scales (e.g. Chapters 3 and 20) whilst others examine the response to individual species or communities to hydraulic parameters across a range of spatial and temporal scales (e.g. Chapters 11, 14 and 21). Future developments within the field will undoubtedly require greater interdisciplinary collaboration and co-operation between engineers, hydrologists, ecologists, social, economic and political scientists to address and provide solutions to management which are centred on habitats, communities and species-specific research questions. This will ensure that the outputs are rigorous, widely applicable and understood, and are of direct relevance and use to the scientific, social science and end-user communities.

25.2.3 Transferability across spatial and temporal scales

Ecohydraulic studies don't frequently consider multiple spatial and temporal scales, although results are often used to highlight relationships or impacts at other sites

over time. It remains a major challenge to collect sufficient data over relevant spatial (site, reach or catchment) and temporal scales that are of direct interest to researchers and managers. Improvements in data collection methods to address larger spatial scales through the use of manned and unmanned aircraft and drones with remote sensing techniques equipped with optical instruments, LiDAR or video cameras have already been applied in ecohydraulic (McKean *et al.*, 2009; Carbonneau *et al.*, 2012, Carbonneau and Piégay, 2012; Flener *et al.*, 2012). As remote sensing techniques are used in a wide range of applications outside river science, we are likely to witness rapid and significant developments within this area, facilitating the collection of physical parameters over large areas, which again will help to transfer and compare data and results across spatial scales.

It is also challenging to apply detailed numerical modelling tools with a fine spatial resolution to large sections of rivers. Even with modern computational techniques, this requires a lot of resources and remains a future challenge. The use of different types of numerical models interlinked to cover varying spatial scales is a useful way to overcome this challenge. Figure 25.1 illustrates the different spatial scales covered by ecohydraulics.

Data collection, modelling and analysis strategies must not only cover different spatial scales, but also different temporal scales. Many physical processes vary with season and between years, but also at smaller time scales such as day–night variations, hourly or even minutes and seconds for example when considering the influence of turbulent eddies. For instance, flow may vary from day to day and water temperature may vary from early morning to midday in most natural rivers. In addition to variations in physical factors, temporal variations are also dominant and highly important for most species and ecosystems. These variations in requirements and behaviour patterns are related to lifecycle or life history and other biological factors, but are also a direct response to changes in physical conditions. This linkage is another example of the strong need for interdisciplinary collaboration within ecohydraulics. Natural variability may be high in biological responses to temporal variations in physical conditions, and the establishment of long time series with sufficient temporal resolution is essential if we want to draw general conclusions from the ecohydraulic interplay between physical and biological drivers and responses.

25.2.4 Ecohydraulics and management

From the initiation of research that has subsequently been recognised as laying the foundations of ecohydraulics (e.g. Gore, 1978; Bovee, 1982; Statzner *et al.*, 1988), it has always been an applied science with real-world management applications and implications. However, the balance between fundamental understanding and the advances in science and technology enabling us to undertake our research is not always easily or readily transferable to the management arena. A number of challenges remain regarding:
• the ecohydraulic properties we measure;
• the complexity, volume and relevance to end users of the data collected;
• the transferability of models and flume investigations into management solutions.

First, it is therefore vitally important that future studies explore and quantify which hydraulic properties are ecologically significant for individuals, populations, species and communities and to bridge that gap between hydraulic characterisation and its relevance for aquatic ecology. Whilst many of the parameters we currently use are considered 'ecologically relevant', how they influence ecology is frequently unknown. Second, as new technological innovations have developed in recent years, there has been a growth in high-resolution ecohydraulic data. Many of the investigations involved the deployment of considerable resources, in terms of both finances and person-hours, and the outcomes of this research are increasingly

complex and difficult to present to non-specialists. River managers often require recommendations and solutions to applied problems to be relatively straightforward to implement, inexpensive to undertake on very strict budget constraints and, in many cases, the results need to be applied across many sites, rivers or wide geographic areas. In many instances, the gap between academic research and the needs of river managers is widening and the challenge of converting the results of this research into useful management recommendations has to be addressed. By involving river managers within the decision-making process during the research phase of projects, and not just when management recommendations are being finalised towards the end, ecohydraulics can move from an interdisciplinary to a 'transdisciplinary' science.

Finally, it is important that the results of ecohydraulic models and laboratory investigations are validated in the field wherever possible to demonstrate their applicability and ability to characterise the complexity of the real world. Understanding the fundamental associations and relationships between hydraulic forces and floral and faunal communities, species and their habitats, modelling them and using this information to provide management recommendations continue to be important and challenging goals for river scientists, managers and other end users. If those engaged with ecohydraulics research are to make a significant contribution to the sustainable development of riverine ecosystems, then focusing of efforts towards reaching these goals is paramount. Only then can the results of this research be translated into meaningful management recommendations.

References

Acreman, M. and Ferguson, A.J.D. (2010) Environmental flows and the European Water Framework Directive. *Freshwater Biology*, **55**: 32–48.
Ambrosetti, W., Barbanti, L., Rolla, A., Castellano, L. and Sala, N. (2012) Hydraulic paths and estimation of the real residence time of water in Lago Maggiore (N. Italy): application of massless markers transported in 3D motion fields. *Journal of Limnology*, **71**: 23–33.
Bovee, K. (1982) *A Guide to Stream Habitat Analysis Using the Instream Flow Incremental Methodology*. Instream Flow Information Paper 12, USDI Fish and Wildlife Service, FWS/OBS-82126.
Carbonneau, P. and Piégay, H. (2012) *Fluvial Remote Sensing for Science and Management*, Wiley-Blackwell.
Carbonneau, P., Fonstad, M.A., Marcus, W.A. and Dugdale, S.J. (2012) Making riverscapes real. *Geomorphology*, **137**: 74–86.

Crossman, J.A., Martel, G., Johnson, P.N. and Bray, K. (2011) The use of Dual-frequency IDentification SONar (DIDSON) to document white sturgeon activity in the Columbia River, Canada. *Journal of Applied Ichthyology*, **27**: 53–57.

Darby, S. and Sear, D. (eds) (2008) *River Restoration: Managing the Uncertainty in Restoring Physical Habitat*, John Wiley & Sons, Ltd, Chichester, UK.

Dyson, M., Bergkamp, G. and Scanlon, J. (eds) (2003) *Flow: The Essentials of Environmental Flows*, IUCN, Gland, Switzerland and Cambridge, UK.

Flener, C., Lotsari, E., Alho, P. and Kayhko, J. (2012) Comparison of empirical and theoretical remote sensing based bathymetry models in river environments. *River Research and Applications*, **28**: 118–133.

Folkard, A.M. (2005) Hydrodynamic of model *Posidonia oceanica* in shallow water. *Limnology and Oceanography*, **50**: 1592–1600.

Gore, J.A. (1978) Technique for predicting in-stream flow requirements for benthic macroinvertebrates. *Freshwater Biology*, **8**: 141–151.

Kemp, P. (2012) Bridging the gap between fish behaviour, performance and hydrodynamics: an ecohydraulics approach to fish passage research. *River Research and Applications*, **28**: 403–406.

Kingsford, R.T. (2011) Conservation management of rivers and wetlands under climate change – a synthesis. *Marine and Freshwater Research*, **62**: 217–222.

Lancaster, J. and Downes, B.J. (2010) Linking the hydraulic world of individual organisms to ecological processes: putting ecology into ecohydraulics. *River Research and Applications*, **26**: 385–403.

Linnansaari, T., Roussel, J.M., Cunjak, R.A. and Halleraker, J.H. (2007) Efficacy and accuracy of portable PIT-antennae when locating fish in ice-covered streams. *Hydrobiologia*, **582**: 281–287.

Linnansaari, T., Alfredsen, K., Stickler, M., Arnekleiv, J.V., Harby, A. and Cunjak, R.A. (2009) Does ice matter? Site fidelity and movements by Atlantic salmon (*Salmo salar* L.) parr during winter in a substrate enhanced river reach. *River Research and Applications*, **25**: 773–787.

McKean, J., Nagel, D., Tonina, D., Bailey, P., Wright, C.W., Bohn, C. and Nayegandhi, A. (2009) Remote sensing of channels and riparian zones with a narrow-beam aquatic-terrestrial LIDAR. *Remote Sensing*, **4**: 1065–1096.

Porter, A.L. and Rafols, I. (2009) Is science becoming more inter-disciplinary? Measuring and mapping six research fields over time. *Scientometrics*, **81**: 719–745.

Prentice, E.F., Flagg, T.A. and McCutcheon, S. (1987) *A Study to Determine the Biological Feasibility of a New Fish Tagging System, 1986–87*. Department of Energy, Bonneville Power Administration Division of Fish and Wildlife.

Shenton, W., Bond, N.R., Yen, J.D.L. and MacNally, R. (2012) Putting the "Ecology" into environmental flows: ecological dynamics and demographic modelling. *Environmental Management*, **50**: 1–10.

Statzner, B., Gore, J.A. and Resh, V.H. (1988) Hydraulic stream ecology – observed patterns and potential applications. *Journal of the North American Benthological Society*, **7**: 307–360.

Zinke, P. (2011) *Modelling of Flow and Levee Depositions in a Freshwater Delta with Natural Vegetation*. Unpublished doctoral thesis at NTNU.

Index

Ecohydraulics: An Integrated Approach, First Edition. Edited by Ian Maddock, Atle Harby, Paul Kemp and Paul Wood.
© 2013 John Wiley & Sons, Ltd. Published 2013 by John Wiley & Sons, Ltd.